Student's Solutions Manual

Jeffery A. Cole
Anoka-Ramsey Community College

Intermediate Algebra

Ninth Edition

Margaret L. Lial
American River College

John Hornsby
University of New Orleans

Terry McGinnis

Addison-Wesley
is an imprint of

Reproduced by Pearson Addison-Wesley from electronic files supplied by the author.

Publishing as Pearson Addison-Wesley, 75 Arlington Street, Boston, MA 02116.

ISBN-13: 978-0-321-57629-3
ISBN-10: 0-321-57629-2

2 3 4 5 6 CRS 12 11 10 09

Addison-Wesley
is an imprint of

www.pearsonhighered.com

Preface

This *Student's Solutions Manual* contains solutions to selected exercises in the text *Intermediate Algebra, Ninth Edition* by Margaret L. Lial, John Hornsby, and Terry McGinnis. It contains solutions to all margin exercises, the odd-numbered exercises in each section, all Relating Concepts exercises, as well as solutions to all the exercises in the review sections, the chapter tests, and the cumulative review sections.

This manual is a text supplement and should be read along *with* the text. You should read all exercise solutions in this manual because many concept explanations are given and then used in subsequent solutions. All concepts necessary to solve a particular problem are not reviewed for every exercise. If you are having difficulty with a previously covered concept, refer back to the section where it was covered for more complete help.

A significant number of today's students are involved in various outside activities, and find it difficult, if not impossible, to attend all class sessions; this manual should help meet the needs of these students. In addition, it is my hope that this manual's solutions will enhance the understanding of all readers of the material and provide insights to solving other exercises.

I appreciate feedback concerning errors, solution correctness or style, and manual style. Any comments may be sent directly to me at the address below, at jeff.cole@anokaramsey.edu, or in care of the publisher, Pearson Addison-Wesley.

I would like to thank Ken Grace, of Anoka-Ramsey Community College, and Jeannine Grace, for typesetting the manuscript and providing assistance with many features of the manual; Mary Johnson and Marv Riedesel, formerly of Inver Hills Community College, for their careful accuracy checking and valuable suggestions; Jim McLaughlin, for his help with the entire art package; and the authors and Maureen O'Connor and Courtney Slade, of Pearson Addison-Wesley, for entrusting me with this project.

Jeffery A. Cole
Anoka-Ramsey Community College
11200 Mississippi Blvd. NW
Coon Rapids, MN 55433

Preface

Table of Contents

CHAPTER 1 REVIEW OF THE REAL NUMBER SYSTEM

1.1 Basic Concepts

1.1 Margin Exercises

1. $\{0, 10, \frac{3}{10}, 52, 98.6\}$

(a) The natural numbers are $\{1, 2, 3, 4, 5, 6, \dots\}$. In the given set, the elements 10 and 52 are natural numbers.

(b) The whole numbers are $\{0, 1, 2, 3, 4, 5, \dots\}$. In the given set, the elements 0, 10, and 52 are whole numbers.

2. **(a)** $\{x \mid x \text{ is a whole number less than } 5\}$

The whole numbers are $\{0, 1, 2, 3, 4, 5, 6, \dots\}$. The whole numbers less than 5 would be the set $\{0, 1, 2, 3, 4\}$.

(b) $\{y \mid y \text{ is a whole number greater than } 12\}$

$\{13, 14, 15, \dots\}$ is the set of whole numbers greater than 12.

3. **(a)** $\{0, 1, 2, 3, 4, 5\}$

In set-builder notation, one answer is $\{x \mid x \text{ is a whole number less than } 6\}$.

(b) $\{7, 14, 21, 28, \dots\}$

In set-builder notation, one answer is $\{x \mid x \text{ is a multiple of 7 greater than } 0\}$.

4. **(a)** $\{-4, -2, 0, 2, 4, 6\}$

These are the even integers between and including -4 and 6. Place dots for $-4, -2, 0, 2, 4$, and 6 on a number line.

−4 −2 0 2 4 6

(b) $\{-1, 0, \frac{2}{3}, 2.5\}$

Place dots at -1 and 0 on a number line. To graph $\frac{2}{3}$, divide the interval between 0 and 1 into three parts. $\frac{2}{3}$ is located two-thirds of the distance from 0 to 1. To graph 2.5, divide the interval between 2 and 3 into two equal parts. 2.5 is located halfway between 2 and 3.

−2 −1 0 1 2 3

(c) $\{5, \frac{16}{3}, 6, \frac{13}{2}, 7, \frac{29}{4}\}$

Place dots at 5, 6, and 7 on a number line. To locate the fractions, first change each to a mixed number and place dots at $\frac{16}{3} = 5\frac{1}{3}$, $\frac{13}{2} = 6\frac{1}{2}$, and $\frac{29}{4} = 7\frac{1}{4}$.

4 5 6 7 8

5. **(a)** -6 is a *rational* number because

$$-6 = -\frac{6}{1}.$$

-6 is also a *real* number.

(b) 12 is a *whole* number, a *rational* number $(12 = \frac{12}{1})$, and a *real* number.

(c) $0.\overline{3}$, a repeating decimal, is a *rational* number and a *real* number.

(d) $-\sqrt{15}$ is an *irrational* number and a *real* number.

(e) π, a nonrepeating, nonterminating decimal, is an *irrational* number and a *real* number.

(f) $\frac{22}{7}$ is a *rational* number and a *real* number.

(g) 3.14, a terminating decimal, is a *rational* number and a *real* number.

6. **(a)** The statement "All whole numbers are integers" is *true*. As shown in Figure 5 in the textbook, the set of integers includes the set of whole numbers.

(b) The statement "Some integers are whole numbers" is *true* since 0 and the positive integers are whole numbers. The negative integers are not whole numbers.

(c) The statement "Every real number is irrational" is *false*. Some real numbers such as $\frac{4}{9}$, $0.\overline{3}$, and $\sqrt{16}$ are rational. Others such as $\sqrt{2}$ and π are irrational.

7.

	Number	Additive Inverse
(a)	9	-9
(b)	-12	12
(c)	$-\frac{6}{5}$	$\frac{6}{5}$
(d)	0	0
(e)	1.5	-1.5

8. **(a)** $|6| = 6$

(b) $|-3| = -(-3) = 3$

(c) $-\left|\frac{1}{4}\right|$

Evaluate the absolute value first. Then find the additive inverse.

$$-\left|\frac{1}{4}\right| = -\left(\frac{1}{4}\right) = -\frac{1}{4}$$

(d) $-|-2| = -(2) = -2$

(e) $-|-7.25| = -(7.25) = -7.25$

(f) $|-6| + |-3|$

Evaluate each absolute value first, and then add.

$$|-6| + |-3| = 6 + 3 = 9$$

(g) $|-9| - |-4| = 9 - 4 = 5$

(h) $-|9 - 4| = -|5| = -5$

9. Software publishers: $|5.3| = 5.3$

Fabric mills: $|-5.9| = 5.9$

Since $5.9 > 5.3$, fabric mills will show the greater change.

10. **(a)** $3 < 7$ since 3 is to the left of 7 on a number line.

(b) $9 > 2$ since 9 is to the right of 2.

(c) $-4 > -8$ since -4 is to the right of -8.

(d) $-2 < -1$ since -2 is to the left of -1.

(e) $0 > -3.5$ since 0 is to the right of -3.5.

(f) $\frac{5}{8} < \frac{3}{4}$ since $\frac{5}{8}$ $(= 0.625)$ is to the left of $\frac{3}{4}$ $(= 0.75)$.

(g) $-0.3 > -0.5$ since -0.3 is to the right of -0.5.

11. **(a)** $-2 \le -3$ is *false*.

Since -2 is to the right of -3 on a number line, it must be greater.

(b) $0.5 \le 0.5$ is *true*.

The statement is "0.5 is less than *or* equal to 0.5." Since 0.5 is equal to 0.5, the statement is true.

(c) $-9 \ge -1$ reads "-9 is greater than or equal to -1." This is *false*, since -9 is to the left of -1 on a number line.

(d) $5 \cdot 8 \le 7 \cdot 7$

Since $5 \cdot 8 = 40$, $7 \cdot 7 = 49$, and $40 \le 49$, we must have $5 \cdot 8 \le 7 \cdot 7$. The statement is *true*.

(e) $3(4) > (2)6$

Since $3(4) = 12$ and $2(6) = 12$, the statement becomes $12 > 12$, which is *false*.

1.1 Section Exercises

1. $\{x \mid x$ is a natural number less than $6\}$
The set of natural numbers is $\{1, 2, 3, \dots\}$, so the set of natural numbers less than 6 is $\{1, 2, 3, 4, 5\}$.

3. $\{z \mid z$ is an integer greater than $4\}$
The set of integers is $\{\dots, -3, -2, -1, 0, 1, 2, 3, \dots\}$, so the set of integers greater than 4 is $\{5, 6, 7, 8, \dots\}$.

5. $\{a \mid a$ is an even integer greater than $8\}$
The set of even integers is $\{\dots, -2, 0, 2, 4, \dots\}$, so the set of even integers greater than 8 is $\{10, 12, 14, 16, \dots\}$.

7. $\{x \mid x$ is an irrational number that is also rational$\}$
Irrational numbers cannot also be rational numbers. The set of irrational numbers that are also rational is $\emptyset$.

9. $\{p \mid p$ is a number whose absolute value is $4\}$
This is the set of numbers that lie a distance of 4 units from 0 on the number line. Thus, the set of numbers whose absolute value is 4 is $\{-4, 4\}$.

11. $\{2, 4, 6, 8\}$ can be described by $\{x \mid x$ is an even natural number less than or equal to $8\}$.

13. $\{4, 8, 12, 16, \dots\}$ can be described by $\{x \mid x$ is a multiple of 4 greater than $0\}$.

15. Graph $\{-3, -1, 0, 4, 6\}$.
Place dots for $-3, -1, 0, 4$, and 6 on a number line.

-4 -2 0 2 4 6 8

17. Graph $\{-\frac{2}{3}, 0, \frac{4}{5}, \frac{12}{5}, \frac{9}{2}, 4.8\}$.

Place dots for $-\frac{2}{3} = -0.\overline{6}$, 0, $\frac{4}{5} = 0.8$, $\frac{12}{5} = 2.4$, $\frac{9}{2} = 4.5$, and 4.8 on a number line.

$-\frac{2}{3}$ $\frac{4}{5}$ $\frac{12}{5}$ $\frac{9}{2}$ 4.8

-1 0 1 2 3 4 5

19. $\{-8, -\sqrt{5}, -0.6, 0, \frac{3}{4}, \sqrt{3}, \pi, 5, \frac{13}{2}, 17, \frac{40}{2}\}$

(a) The elements $5, 17$, and $\frac{40}{2}$ (or 20) are natural numbers.

(b) The elements $0, 5, 17$, and $\frac{40}{2}$ are whole numbers.

(c) The elements $-8, 0, 5, 17$, and $\frac{40}{2}$ are integers.

(d) The elements $-8, -0.6, 0, \frac{3}{4}, 5, \frac{13}{2}, 17$, and $\frac{40}{2}$ are rational numbers.

(e) The elements $-\sqrt{5}$, $\sqrt{3}$, and π are irrational numbers.

(f) All the elements are real numbers.

21. The statement "Every rational number is an integer" is *false*. Some are integers, but others, like $\frac{3}{4}$, are not.

23. The statement "Every irrational number is an integer" is *false*. Irrational numbers have decimal representations that neither terminate nor repeat, so no irrational numbers are integers.

25. The statement "Every natural number is a whole number" is *true*. The whole numbers consist of the natural numbers and zero.

27. The statement "Some rational numbers are whole numbers is *true*. Every whole number is rational.

29. The statement "The absolute value of any number is the same as the absolute value of its additive inverse" is *true*. The distance on a number line from 0 to a number is the same as the distance from 0 to the additive inverse of that number.

31. **(a)** The additive inverse of 6 is -6.

(b) $6 > 0$, so $|6| = 6$.

33. **(a)** The additive inverse of -12 is $-(-12) = 12$.

(b) $-12 < 0$, so $|-12| = -(-12) = 12$.

35. **(a)** The additive inverse of $\frac{6}{5}$ is $-\frac{6}{5}$.

(b) $\frac{6}{5} > 0$, so $\left|\frac{6}{5}\right| = \frac{6}{5}$.

37. $|-8|$
Use the definition of absolute value.

$$-8 < 0, \text{ so } |-8| = -(-8) = 8.$$

39. $\frac{3}{2} > 0$, so $\left|\frac{3}{2}\right| = \frac{3}{2}$.

41. $-|5| = -(5) = -5$

43. $-|-2| = -[-(-2)] = -(2) = -2$

45. $-|4.5| = -(4.5) = -4.5$

47. $|-2| + |3| = 2 + 3 = 5$

49. $|-9| - |-3| = 9 - 3 = 6$

51.
$$\begin{aligned}|-1| + |-2| - |-3| &= 1 + 2 - 3\\ &= 3 - 3\\ &= 0\end{aligned}$$

53. **(a)** The greatest absolute value is $|-4.1| = 4.1$.

Therefore, Louisiana had the greatest change in population. The population decreased 4.1%.

(b) The smallest absolute value is $|0.6| = 0.6$.

Therefore, West Virginia had the smallest change in population. The population increased 0.6%.

55. Compare the depths of the bodies of water. The deepest, that is, the body of water whose depth has the greatest absolute value, is the Pacific Ocean ($|-12{,}925| = 12{,}925$) followed by the Indian Ocean, the Caribbean Sea, the South China Sea, and the Gulf of California.

57. True; the absolute value of the depth of the Pacific Ocean is

$$|-12{,}925| = 12{,}925$$

which is greater than the absolute value of the depth of the Indian Ocean,

$$|-12{,}598| = 12{,}598.$$

59. True; since -6 is to the left of -2 on a number line, -6 is less than -2.

61. False; since -4 is to the left of -3 on a number line, -4 is *less* than -3, not greater.

63. True; since 3 is to the right of -2 on a number line, 3 is greater than -2.

65. True; since $-3 = -3$, -3 is greater than *or equal to* -3.

67. "7 is greater than y" can be written as $7 > y$.

69. "5 is greater than or equal to 5" can be written as $5 \geq 5$.

71. "$3t - 4$ is less than or equal to 10" can be written as $3t - 4 \leq 10$.

73. "$5x + 3$ is not equal to 0" can be written as $5x + 3 \neq 0$.

75. $-6 \stackrel{?}{<} 7 + 3$
$-6 < 10$ *True*

The last statement is true since -6 is to the left of 10 on a number line.

77. $2 \cdot 5 \stackrel{?}{\geq} 4 + 6$
$10 \geq 10$ *True*

79. $-|-3| \stackrel{?}{\geq} -3$
$-3 \geq -3$ *True*

81. $-8 \stackrel{?}{>} -|-6|$
$-8 > -6$ *False*

83. The 2005 egg production in IA was 12,978 million and in CA it was 5082 million. Thus, the egg production in IA was greater than in CA in 2005.

85. For which states is the bar shorter (and number smaller) for 2006 than 2005?

$$\text{FL: } 2940 < 2985$$

Thus, 2006 egg production was less than 2005 production for Florida.

87. 2005: OH $= x = 7506$,
2006: OH $= y = 7507$
Since $7506 < 7507$, $x < y$ is true.

1.2 Operations on Real Numbers

1.2 Margin Exercises

1. Add the absolute values of the numbers. The answer is negative since two negative numbers are being added.

(a) $-2+(-7)=-(|-2|+|-7|)$
$=-(2+7)=-9$

(b) $-15+(-6)=-(15+6)=-21$

(c) $-1.1+(-1.2)=-(1.1+1.2)=-2.3$

(d) $-\frac{3}{4}+\left(-\frac{1}{2}\right)=-\left(\frac{3}{4}+\frac{1}{2}\right)$
$=-\left(\frac{3}{4}+\frac{2}{4}\right)$
$=-\frac{5}{4}$

2. **(a)** $12+(-1)$

Subtract the lesser absolute value from the greater absolute value $(12-1)$, and take the sign of the number with the greater absolute value.

$$12+(-1)=12-1=11$$

(b) $3+(-7)$

Subtract the lesser absolute value from the greater absolute value $(7-3)$, and take the sign of the greater.

$$3+(-7)=-(7-3)=-4$$

(c) $-17+5=-(17-5)=-12$

(d) $-1.5+3.2=1.7$

(e) $-\frac{3}{4}+\frac{1}{2}=-\frac{3}{4}+\frac{2}{4}=-\frac{1}{4}$

3. **(a)** $9-12$

Change the sign of the second number and add.

$$9-12=9+(-12)=-3$$

(b) $-7-2=-7+(-2)=-9$

(c) $-8-(-2)$

Change the sign of the -2, then add.

$$-8-(-2)=-8+2=-6$$

(d) $12-(-5)=12+5=17$

(e) $-6.3-(-11.5)=-6.3+11.5=5.2$

(f) $\frac{3}{4}-\left(-\frac{2}{3}\right)$
$=\frac{3}{4}+\frac{2}{3}$ *Add the opposite.*
$=\frac{9}{12}+\frac{8}{12}$ *LCD = 12*
$=\frac{17}{12}$, or $1\frac{5}{12}$ *Add numerators.*

4. **(a)** $-6+9-2$

Add and subtract in order from left to right.

$$-6+9-2=(-6+9)-2$$
$$=3-2$$
$$=1$$

(b) $12-(-4)+8=(12+4)+8$
$=16+8$
$=24$

(c) $-6-(-2)-8-1=-6+2-8-1$
$=-4+(-8)-1$
$=-12-1$
$=-13$

(d) $-3-[(-7)+15]+6=-3-8+6$
$=-11+6$
$=-5$

5. **(a)** $-7(-5)=7\cdot 5=35$

The numbers have the same sign, so the product is positive.

(b) $-0.9(-15)=13.5$

(c) $-\frac{4}{7}\left(-\frac{14}{3}\right)=\frac{4\cdot 2\cdot 7}{7\cdot 3}=\frac{4\cdot 2}{3}=\frac{8}{3}$

(d) $7(-2)=-14$

The numbers have different signs, so the product is negative.

(e) $-0.8(0.006)=-0.0048$

(f) $\frac{5}{8}(-16)=-\frac{5\cdot 2\cdot 8}{8}=-10$

(g) $-\frac{2}{3}(12)=-\frac{2\cdot 4\cdot 3}{3}=-8$

6.

	Number	Reciprocal
(a)	15	$\frac{1}{15}$
(b)	-7	$\frac{1}{-7} = -\frac{1}{7}$
(c)	$\frac{8}{9}$	$\frac{9}{8}$
(d)	$-\frac{1}{3}$	$-\frac{3}{1} = -3$
(e)	$0.125 = \frac{1}{8}$	$\frac{8}{1} = 8$

7. **(a)** $\frac{9}{0}$ is undefined.

(b) $\frac{0}{9} = 0$

(c) $\frac{-9}{0}$ is undefined.

(d) $\frac{0}{-9} = 0$

8. **(a)** $\frac{-16}{4} = -16\left(\frac{1}{4}\right) = -4$

The numbers have different signs, so the quotient is negative.

(b) $\frac{8}{-2} = 8\left(\frac{1}{-2}\right) = -4$

(c) $\frac{-15}{-3} = -15\left(-\frac{1}{3}\right) = 5$

(d) $\frac{\frac{3}{8}}{-\frac{11}{16}} = \frac{3}{8} \div \left(-\frac{11}{16}\right)$

$= \frac{3}{8} \cdot \left(-\frac{16}{11}\right)$ *Multiply by the reciprocal.*

$= -\frac{3 \cdot 2 \cdot 8}{8 \cdot 11} = -\frac{6}{11}$

(e) $-\frac{3}{4} \div \frac{7}{16}$

$= -\frac{3}{4} \cdot \frac{16}{7}$ *Multiply by the reciprocal.*

$= -\frac{3 \cdot 4 \cdot 4}{4 \cdot 7}$ *Factor; multiply numerators and multiply denominators.*

$= -\frac{12}{7}$ *Lowest terms*

9. **A.** $\frac{3}{5}$ **B.** $\frac{3}{-5} = -\frac{3}{5} = \frac{-3}{5}$

C. $-\frac{3}{5} = \frac{-3}{5}$ **D.** $\frac{-3}{-5} = \frac{3}{5}$

B and **C** are equal to $\frac{-3}{5}$.

1.2 Section Exercises

1. The sum of a positive number and a negative number is 0 if the numbers are additive inverses. For example, $4 + (-4) = 0$.

3. The sum of two negative numbers is a negative number. For example, $-7 + (-21) = -28$.

5. The sum of a positive number and a negative number is positive if the positive number has greater absolute value. For example, $15 + (-2) = 13$.

7. The difference between two negative numbers is negative if the number with lesser absolute value is subtracted from the one with greater absolute value. For example, $-15 - (-3) = -12$.

9. The product of two numbers with different signs is negative. For example, $-5(15) = -75$.

11. $13 + (-4) = 13 - 4 = 9$

13. $-6 + (-13) = -(6 + 13) = -19$

15. $-\frac{7}{3} + \frac{3}{4} = -\frac{28}{12} + \frac{9}{12} = -\frac{19}{12}$

17. The difference between 2.3 and 0.45 is 1.85. The number with the greater absolute value, -2.3, is negative, so the answer is negative. Thus, $-2.3 + 0.45 = -1.85$.

19. $-6 - 5 = -6 + (-5) = -(6 + 5) = -11$

21. $8 - (-13) = 8 + 13 = 21$

23. $-16 - (-3) = -16 + 3 = -13$

25. $-12.31 - (-2.13) = -12.31 + 2.13$
$= -(12.31 - 2.13)$
$= -10.18$

27. $\frac{9}{10} - \left(-\frac{4}{3}\right) = \frac{9}{10} + \frac{4}{3} = \frac{27}{30} + \frac{40}{30} = \frac{67}{30}$

29. $|-8 - 6| = |-14| = -(-14) = 14$

31. $-|-4 + 9| = -|5| = -5$

33. $-2 - |-4| = -2 - 4 = -2 + (-4) = -6$

35. $-7 + 5 - 9 = (-7 + 5) - 9 = -2 - 9 = -11$

37. $6 - (-2) + 8 = 6 + 2 + 8 = 8 + 8 = 16$

39. $-9 - 4 - (-3) + 6 = (-9 - 4) + 3 + 6$
$= -13 + 3 + 6$
$= -10 + 6$
$= -4$

41. $-0.382 + 4 - (-0.6) = 3.618 + 0.6$
$= 4.218$

43. $-\frac{3}{4}-\left(\frac{1}{2}-\frac{3}{8}\right)=-\frac{6}{8}-\left(\frac{4}{8}-\frac{3}{8}\right)$
$=-\frac{6}{8}-\frac{1}{8}$
$=-\frac{7}{8}$

45. $-4-[(-4-6)+12]-13$
$=-4-(-10+12)-13$
$=-4-2-13$
$=-4+(-2)+(-13)$
$=-6+(-13)$
$=-19$

47. $|-11|-|-5|-|7|+|-2|$
$=11-5-7+2$
$=6-7+2$
$=-1+2=1$

49. The product of two numbers with *different* signs is *negative*, so

$$5(-7)=-35.$$

51. The product of two numbers with the *same* sign is *positive*, so

$$-8(-5)=40.$$

53. The product of two numbers with the *same* sign is *positive*, so

$$-10\left(-\frac{1}{5}\right)=2.$$

55. The product of two numbers with *different* signs is *negative*, so

$$\frac{3}{4}(-16)=-\frac{3}{4}\cdot 4\cdot 4=-12.$$

57. The product of two numbers with the *same* sign is *positive*, so

$$-\frac{5}{2}\left(-\frac{12}{25}\right)=\frac{5\cdot 2\cdot 6}{2\cdot 5\cdot 5}=\frac{6}{5}.$$

59. The product of two numbers with the *same* sign is *positive*, so

$$-\frac{3}{8}\left(-\frac{24}{9}\right)=\frac{3\cdot 3\cdot 8}{8\cdot 9}=1.$$

61. $-2.4(-2.45)=5.88$

63. $3.4(-3.14)=-10.676$

65. The reciprocal of $6=\frac{6}{1}$ is $\frac{1}{6}$.

67. The reciprocal of $-7=-\frac{7}{1}$ is $-\frac{1}{7}$. Remember that a number and its reciprocal always have the same sign.

69. The reciprocal of $-\frac{2}{3}$ is $-\frac{3}{2}$.

71. The reciprocal of $\frac{1}{5}$ is $\frac{5}{1}=5$.

73. The reciprocal of $0.02=\frac{2}{100}=\frac{1}{50}$ is $\frac{50}{1}=50$.

75. The reciprocal of $-0.001=-\frac{1}{1000}$ is -1000.

77. The quotient of two nonzero real numbers with *different* signs is *negative*, so

$$\frac{-14}{2}=-14\cdot\frac{1}{2}=-2\cdot 7\cdot\frac{1}{2}=-7.$$

79. The quotient of two nonzero real numbers with the *same* sign is *positive*, so

$$\frac{-24}{-4}=24\cdot\frac{1}{4}=4\cdot 6\cdot\frac{1}{4}=6.$$

81. The quotient of two nonzero real numbers with *different* signs is *negative*, so

$$\frac{100}{-25}=-100\cdot\frac{1}{25}=-4\cdot 25\cdot\frac{1}{25}=-4.$$

83. $\frac{0}{-8}=0\cdot\left(-\frac{1}{8}\right)=0$

85. Division by 0 is undefined, so $\frac{5}{0}$ is *undefined.*

87. The quotient of two nonzero real numbers with the *same* sign is *positive*, so

$$-\frac{10}{17}\div\left(-\frac{12}{5}\right)=\frac{10}{17}\cdot\frac{5}{12}=\frac{2\cdot 5\cdot 5}{17\cdot 2\cdot 6}=\frac{25}{102}.$$

89. $\frac{\frac{12}{13}}{-\frac{4}{3}}=\frac{12}{13}\div\left(-\frac{4}{3}\right)$

$=\frac{12}{13}\left(-\frac{3}{4}\right)$ *Multiply by the reciprocal.*

$=-\frac{3\cdot 4\cdot 3}{13\cdot 4}$ *Factor; mult. numerators and mult. denominators.*

$=-\frac{9}{13}$ *Lowest terms*

91. $-\frac{27.72}{13.2}=-2.1$

93. $\frac{-100}{-0.01}=\frac{100}{\frac{1}{100}}=100\div\frac{1}{100}$
$=100\cdot 100=10{,}000$

95. $\frac{1}{6}-\left(-\frac{7}{9}\right)=\frac{1}{6}+\frac{7}{9}=\frac{3}{18}+\frac{14}{18}=\frac{17}{18}$

97. $-\frac{1}{9}+\frac{7}{12}=-\frac{4}{36}+\frac{21}{36}=\frac{17}{36}$

99. $-\frac{3}{8}-\frac{5}{12}=-\frac{9}{24}-\frac{10}{24}=-\frac{19}{24}$

101. $-\frac{7}{30}+\frac{2}{45}-\frac{3}{10}=-\frac{21}{90}+\frac{4}{90}-\frac{27}{90}=-\frac{44}{90}=-\frac{22}{45}$

103. $\frac{8}{25}\left(-\frac{5}{12}\right)=-\frac{2\cdot 4\cdot 5}{5\cdot 5\cdot 3\cdot 4}=-\frac{2}{5\cdot 3}=-\frac{2}{15}$

105. $\frac{5}{6}\left(-\frac{9}{10}\right)\left(-\frac{4}{5}\right)=\frac{5\cdot 3\cdot 3\cdot 2\cdot 2}{2\cdot 3\cdot 2\cdot 5\cdot 5}=\frac{3}{5}$

107. $\frac{7}{6} \div \left(-\frac{9}{10}\right) = \frac{7}{6} \cdot \left(-\frac{10}{9}\right) = -\frac{7 \cdot 2 \cdot 5}{2 \cdot 3 \cdot 9}$

$= -\frac{7 \cdot 5}{3 \cdot 9} = -\frac{35}{27}, \text{ or } -1\frac{8}{27}$

109. $\frac{-\frac{8}{9}}{2} = -\frac{8}{9} \div \frac{2}{1} = -\frac{8}{9} \cdot \frac{1}{2} = -\frac{2 \cdot 4}{9 \cdot 2} = -\frac{4}{9}$

111. $-8.6 - 3.751 = -(8.6 + 3.751) = -12.351$

113. $(-4.2)(1.4)(2.7) = (-5.88)(2.7) = -15.876$

115. $-24.84 \div 6 = -4.14$

117. $-2496 \div (-0.52) = 4800$

119. $-14.23 + 9.81 + 74.63 - 18.715$
$= -4.42 + 74.63 - 18.715$
$= 70.21 - 18.715$
$= 51.495$

121. To find the difference between these two temperatures, subtract the low temperature from the high temperature.

$90° - (-22°) = 90° + 22° = 112°\text{F}$

123. $15.79 - (-9.10) = 15.79 + 9.10 = 24.89$

The difference is 24.89%.

125. $48.35 - 35.99 - 20.00 - 28.50 + 66.27$
$= 12.36 - 20.00 - 28.50 + 66.27$
$= -7.64 - 28.50 + 66.27$
$= -36.14 + 66.27$
$= 30.13$

His balance is \$30.13.

127. $-382.45 + 25.10 + 34.50 - 45.00 - 98.17$
$= -466.02$

His balance is −\$466.02.

(a) To pay off the balance, his payment should be \$466.02.

(b) $-466.02 + 300 - 24.66 = -190.68$

His balance is −\$190.68.

129. (a) $-142 - 225 - 185 + 77 = -475$

The total loss for 2003–2006 was \$475 thousand.

(b) $77 - (-185) = 77 + 185 = 262$

The difference between the profit or loss from 2005 to 2006 was \$262 thousand.

(c) $-225 - (-142) = -225 + 142 = -83$

The difference between the profit or loss from 2003 to 2004 was −\$83 thousand.

131.

Year	Difference (in billions)
2000	\$538 − \$409 = \$129
2010	\$916 − \$710 = \$206
2020	\$1479 − \$1405 = \$74
2030	\$2041 − \$2542 = −\$501

1.3 Exponents, Roots, and Order of Operations

1.3 Margin Exercises

1. (a) $\underbrace{3 \cdot 3 \cdot 3 \cdot 3 \cdot 3}_{\text{5 factors of 3}} = 3^5$

(b) $\frac{2}{7} \cdot \frac{2}{7} \cdot \frac{2}{7} \cdot \frac{2}{7} = \left(\frac{2}{7}\right)^4$

(c) $(-10)(-10)(-10) = (-10)^3$

(d) $(0.5)(0.5) = (0.5)^2$

(e) $y \cdot y \cdot y \cdot y \cdot y \cdot y \cdot y \cdot y = y^8$

2. (a) $5^3 = 5 \cdot 5 \cdot 5$
$= 25 \cdot 5 = 125$

(b) $3^4 = 3 \cdot 3 \cdot 3 \cdot 3$
$= 9 \cdot 3 \cdot 3$
$= 27 \cdot 3 = 81$

(c) $(-4)^5 = (-4)(-4)(-4)(-4)(-4)$
$= 16(-4)(-4)(-4)$
$= -64(-4)(-4)$
$= 256(-4)$
$= -1024$

(d) $(-3)^4 = (-3)(-3)(-3)(-3) = 81$

(e) $(0.75)^2 = (0.75)(0.75)$
$= 0.5625$

(f) $\left(\frac{2}{5}\right)^4 = \frac{2}{5} \cdot \frac{2}{5} \cdot \frac{2}{5} \cdot \frac{2}{5} = \frac{16}{625}$

3.

		Exponent	Base	Value
(a)	7^3	3	7	343
(b)	$(-5)^4$	4	−5	625
(c)	-5^4	4	5	−625
(d)	$-(0.9)^5$	5	0.9	−0.59049

4. (a) $\sqrt{9} = 3$ since 3 is positive and $3^2 = 9$.

(b) $\sqrt{49} = 7$ since 7 is positive and $7^2 = 49$.

(c) $-\sqrt{81} = -9$ since the negative sign is outside the radical sign.

(d) $\sqrt{\frac{121}{81}} = \frac{11}{9}$ since $(\frac{11}{9})^2 = \frac{121}{81}$.

(e) $\sqrt{0.25} = 0.5$ since $(0.5)^2 = 0.25$.

(f) $\sqrt{-9}$ is not a real number.

(g) $-\sqrt{-169}$ is not a real number.

5. **(a)** $5 \cdot 9 + 2 \cdot 4 = 45 + 8$ *Multiply.*
$= 53$ *Add.*

(b) $4 - 12 \div 4 \cdot 2 = 4 - 3 \cdot 2$ *Divide.*
$= 4 - 6$ *Multiply.*
$= -2$ *Subtract.*

6. **(a)** $(4+2) - 3^2 - (8-3)$

$= 6 - 3^2 - 5$ *Add and subtract inside parentheses.*

$= 6 - 9 - 5$ *Evaluate the power.*

$= -3 - 5$ *Subtract from left to right.*

$= -8$

(b) $6 + \frac{2}{3}(-9) - \frac{5}{8} \cdot 16$

$= 6 + (-6) - 10$ *Do the multiplications first.*

$= 0 - 10$ *Add.*

$= -10$ *Subtract.*

7. **(a)** $\dfrac{10 - 6 + 2\sqrt{9}}{11 \cdot 2 - 3(2)^2}$

$= \dfrac{10 - 6 + 2(3)}{11 \cdot 2 - 3(4)}$ *Evaluate powers and roots.*

$= \dfrac{10 - 6 + 6}{22 - 12}$ *Multiply.*

$= \dfrac{10}{10}$ *Add and subtract.*

$= 1$ *Reduce.*

(b) $\dfrac{-4(8) + 6(3)}{3\sqrt{49} - \frac{1}{2}(42)} = \dfrac{-4(8) + 6(3)}{3(7) - \frac{1}{2}(42)}$ *Evaluate the root.*

$= \dfrac{-32 + 18}{21 - 21}$ *Multiply.*

$= \dfrac{-14}{0}$, which is undefined.

8. Substitute; $w = 4$, $x = -12$, $y = 64$, and $z = -3$.

(a) $5x - 2w = 5(-12) - 2(4)$
$= -60 - 8$
$= -68$

(b) $-6(x - \sqrt{y}) = -6(-12 - \sqrt{64})$
$= -6(-12 - 8)$
$= -6(-20)$
$= 120$

(c) $\dfrac{5x - 3 \cdot \sqrt{y}}{x - 1} = \dfrac{5(-12) - 3\sqrt{64}}{-12 - 1}$

$= \dfrac{5(-12) - 3(8)}{-12 - 1}$

$= \dfrac{-60 - 24}{-12 - 1}$

$= \dfrac{-84}{-13}$

$= \dfrac{84}{13}$

(d) $w^2 + 2z^3 = 4^2 + 2(-3)^3$
$= 16 + 2(-27)$
$= 16 + (-54)$
$= -38$

9. $1.909x - 3791$

2002: $1.909(2002) - 3791 \approx 30.8$
2008: $1.909(2008) - 3791 \approx 42.3$

In 2002, Americans spent about \$30.8 billion on their pets. In 2008, Americans spent about \$42.3 billion on their pets.

1.3 Section Exercises

1. $-4^6 = (-4)^6$ is *false.*
$-4^6 = -(4 \cdot 4 \cdot 4 \cdot 4 \cdot 4 \cdot 4) = -4096$,
whereas
$(-4)^6 = (-4)(-4)(-4)(-4)(-4)(-4)$
$= 4096.$

3. The statement "$\sqrt{16}$ is a positive number" is *true.* The symbol $\sqrt{}$ always gives a positive square root when the radicand is positive.

5. The statement "$(-2)^7$ is a negative number" is *true.* $(-2)^7$ has an odd number of negative factors, so the product is negative.

7. The statement "The product of 8 positive factors and 8 negative factors is positive" is *true.* The product of 8 positive factors is positive and the product of 8 negative factors is positive, so the product of these two products is positive.

9. The statement "In the exponential -3^5, -3 is the base" is *false.* The base is 3, not -3. If the problem were written $(-3)^5$, then -3 would be the base.

11. **(a)** $8^2 = 64$

(b) $-8^2 = -(8 \cdot 8) = -64$

(c) $(-8)^2 = (-8)(-8) = 64$

(d) $-(-8)^2 = -(64) = -64$

13. $8 \cdot 8 \cdot 8 = 8^3$

15. $\frac{1}{2} \cdot \frac{1}{2} = \left(\frac{1}{2}\right)^2$

17. $(-4)(-4)(-4)(-4) = (-4)^4$

19. $z \cdot z \cdot z \cdot z \cdot z \cdot z \cdot z = z^7$

21. $4^2 = 4 \cdot 4 = 16$

23. $0.28^3 = (0.28)(0.28)(0.28) = 0.021952$

25. $\left(\frac{1}{5}\right)^3 = \frac{1}{5} \cdot \frac{1}{5} \cdot \frac{1}{5} = \frac{1}{125}$

27. $\left(\frac{4}{5}\right)^4 = \left(\frac{4}{5}\right)\left(\frac{4}{5}\right)\left(\frac{4}{5}\right)\left(\frac{4}{5}\right)$
$= \frac{4 \cdot 4 \cdot 4 \cdot 4}{5 \cdot 5 \cdot 5 \cdot 5} = \frac{256}{625}$

29. $(-5)^3 = (-5)(-5)(-5)$
$= 25(-5) = -125$

31. $(-2)^8 = (-2)(-2)(-2)(-2)(-2)(-2)(-2)(-2)$
$= 256$

33. $-3^6 = -(3 \cdot 3 \cdot 3 \cdot 3 \cdot 3 \cdot 3) = -729$

35. $-8^4 = -(8 \cdot 8 \cdot 8 \cdot 8) = -4096$

37. In $(-4.1)^7$, the exponent is 7; the base is -4.1.

39. In -4.1^7, the exponent is 7; the base is 4.1.

41. $\sqrt{81} = 9$ since 9 is positive and $9^2 = 81$.

43. $\sqrt{169} = 13$ since 13 is positive and $13^2 = 169$.

45. $-\sqrt{400} = -(\sqrt{400}) = -(20) = -20$

47. $\sqrt{\frac{100}{121}} = \frac{10}{11}$ since $\frac{10}{11}$ is positive and $\left(\frac{10}{11}\right)^2 = \frac{100}{121}$.

49. $-\sqrt{0.49} = -\sqrt{(0.7)^2} = -(0.7) = -0.7$

51. There is no real number whose square is negative, so $\sqrt{-36}$ is *not a real number*.

53. **(a)** $\sqrt{144} = 12$; choice **B**

(b) $\sqrt{-144}$ is not a real number; choice **C**

(c) $-\sqrt{144} = -12$; choice **A**

55. If a is a positive number, then $-a$ is a negative number. Therefore, $\sqrt{-a}$ is not a real number, so $-\sqrt{-a}$ is *not a real number*.

57. $12 + 3 \cdot 4 = 12 + 12$ *Multiply.*
$= 24$ *Add.*

59. $2[-5 - (-7)] = 2(-5 + 7)$
$= 2(2) = 4$

61. $-12(-\frac{3}{4}) - (-5) = 9 - (-5)$
$= 9 + 5 = 14$

63. $6 \cdot 3 - 12 \div 4 = 18 - 12 \div 4$ *Multiply.*
$= 18 - 3$ *Divide.*
$= 15$ *Subtract.*

65. $10 + 30 \div 2 \cdot 3 = 10 + 15 \cdot 3$ *Divide.*
$= 10 + 45$ *Multiply.*
$= 55$ *Add.*

67. $-3(5)^2 - (-2)(-8)$
$= -3(25) - (-2)(-8)$ *Evaluate power.*
$= -75 - 16$ *Multiply.*
$= -91$ *Subtract.*

69. $5 - 7 \cdot 3 - (-2)^3$
$= 5 - 7 \cdot 3 - (-8)$ *Evaluate power.*
$= 5 - 21 + 8$ *Multiply; change subtraction to addition.*
$= -16 + 8$ *Subtract.*
$= -8$ *Add.*

71. $-7(\sqrt{36}) - (-2)(-3)$
$= -7(6) - (-2)(-3)$ *Evaluate root.*
$= -42 - 6$ *Multiply.*
$= -48$ *Subtract.*

73. $-14(-\frac{2}{7}) \div (2 \cdot 6 - 10)$
$= -14(-\frac{2}{7}) \div (12 - 10)$ *Work inside parentheses.*
$= -14(-\frac{2}{7}) \div 2$
$= 4 \div 2$ *Multiply.*
$= 2$ *Divide.*

75. $6|4 - 5| - 24 \div 3$
$= 6|-1| - 24 \div 3$ *Simplify within absolute value bars.*
$= 6(1) - 24 \div 3$ *Take absolute value.*
$= 6 - 8$ *Multiply and divide.*
$= -2$ *Subtract.*

77. $|-6 - 5|(-8) + 3^2$
$= |-11|(-8) + 3^2$ *Simplify within absolute value bars.*
$= 11(-8) + 3^2$ *Take absolute value.*
$= 11(-8) + 9$ *Evaluate power.*
$= -88 + 9$ *Multiply.*
$= -79$ *Add.*

79. $\dfrac{(-5+\sqrt{4})(-2^2)}{-5-1}$

$= \dfrac{(-5+2)(-4)}{-6}$ *Evaluate root and power; subtract.*

$= \dfrac{(-3)(-4)}{-6}$ *Work inside parentheses.*

$= \dfrac{12}{-6}$ *Multiply.*

$= -2$ *Divide.*

81. $\dfrac{2(-5)+(-3)(-2)}{-8+3^2-1}$

$= \dfrac{-10+6}{-8+9-1}$ *Evaluate power; multiply.*

$= \dfrac{-4}{1-1}$ *Add.*

$= \dfrac{-4}{0}$ *Subtract.*

Since division by 0 is undefined, the given expression is *undefined.*

In Exercises 83–90, $a=-3$, $b=64$, and $c=6$.

83. $3a+\sqrt{b} = 3(-3)+\sqrt{64}$
$= 3(-3)+8$
$= -9+8$
$= -1$

85. $\sqrt{b}+c-a = \sqrt{64}+6-(-3)$
$= 8+6+3$
$= 14+3 = 17$

87. $4a^3+2c = 4(-3)^3+2(6)$
$= 4(-27)+12$
$= -108+12 = -96$

89. $\dfrac{2c+a^3}{4b+6a} = \dfrac{2(6)+(-3)^3}{4(64)+6(-3)}$

$= \dfrac{12+(-27)}{256+(-18)}$

$= \dfrac{-15}{238} = -\dfrac{15}{238}$

In Exercises 91–94, $w=4$, $x=-\frac{3}{4}$, $y=\frac{1}{2}$, and $z=1.25$.

91. $wy-8x = 4(\frac{1}{2})-8(-\frac{3}{4})$
$= 2+6$
$= 8$

93. $xy+y^4 = -\frac{3}{4}(\frac{1}{2})+(\frac{1}{2})^4$
$= -\frac{3}{8}+\frac{1}{16}$
$= -\frac{6}{16}+\frac{1}{16} = -\frac{5}{16}$

95. $(v\times 0.5485-4850)\div 1000\times 31.44$
$= (150{,}000\times 0.5485-4850)\div 1000\times 31.44$
$= (82{,}275-4850)\div 1000\times 31.44$
$= 77{,}425\div 1000\times 31.44$
$= 77.425\times 31.44$
$= 2434.242 \approx 2434$

The owner would pay $2434 in property taxes.

97. $(v\times 0.5485-4850)\div 1000\times 31.44$
$= (250{,}000\times 0.5485-4850)\div 1000\times 31.44$
$= (137{,}125-4850)\div 1000\times 31.44$
$= 132{,}275\div 1000\times 31.44$
$= 132.275\times 31.44$
$= 4158.726 \approx 4159$

The owner would pay $4159 in property taxes.

99. number of oz $\times$ % alcohol $\times$ 0.075 $\div$ body weight in lb $-$ hr of drinking $\times$ 0.015

$= 36\times 4.0\times 0.075\div 135-3\times 0.015$
$= 144\times 0.075\div 135-3\times 0.015$
$= 10.8\div 135-3\times 0.015$
$= 0.08-3\times 0.015$
$= 0.08-0.045$
$= 0.035$

101. Decreased weight will result in higher BACs.

BAC for the man
$= 48\times 3.2\times 0.075\div(190-25)-2\times 0.015$
≈ 0.040

BAC for the woman
$= 36\times 4.0\times 0.075\div(135-25)-3\times 0.015$
≈ 0.053

103. (a)

Year	Average Price (in dollars)
1977	$0.1399(1977)-274.4\approx 2.18$
1987	3.58
1997	$0.1399(1997)-274.4\approx 4.98$
2007	$0.1399(2007)-274.4\approx 6.38$

(b) The average price of a theater ticket in the United States almost tripled from 1997 to 2007.

1.4 Properties of Real Numbers

1.4 Margin Exercises

1. **(a)** Let $a=8$, $b=m$, and $c=n$ in the statement of the distributive property. Then,

$$8(m+n) = 8m+8n.$$

(b) $-4(p-5) = -4[p+(-5)]$
$= -4(p)+(-4)(-5)$
$= -4p+20$

(c) Use the second form of the distributive property.

$$3k + 6k = (3 + 6)k = 9k$$

(d) Use the second form of the distributive property.

$$\begin{aligned} -6m + 2m &= (-6 + 2)m \\ &= -4m \end{aligned}$$

(e) $2r + 3s$

Since there is no common number or variable here, the distributive property cannot be used.

(f) $5(4p - 2q + r) = 5(4p) + 5(-2q) + 5(r)$
$= 20p - 10q + 5r$

2. (a) $14 \cdot 5 + 14 \cdot 85 = 14(5 + 85)$
$= 14(90)$
$= 1260$

(b) $78 \cdot 33 + 22 \cdot 33 = (78 + 22)33$
$= 100(33)$
$= 3300$

3. (a) The number that must be added to 4 to get 0 is -4.

(b) The number that must be added to -7.1 to get 0 is 7.1.

(c) The sum of -9 and 9 is 0.

(d) The number that must be multiplied by 5 to get 1 is the reciprocal of 5, which is $\frac{1}{5}$.

(e) The number that must be multiplied by $-\frac{3}{4}$ to get 1 is the reciprocal of $-\frac{3}{4}$, which is $-\frac{4}{3}$.

(f) The product of 7 and $\frac{1}{7}$ is 1.

4. (a) $p - 3p = 1p - 3p$ *Identity property*
$= (1 - 3)p$ *Distributive property*
$= -2p$

(b) $r + r + r$
$= 1r + 1r + 1r$ *Identity property*
$= (1 + 1 + 1)r$ *Distributive property*
$= 3r$

(c) $-(3 + 4p)$
$= -1(3 + 4p)$ *Identity property*
$= -1(3) + (-1)(4p)$ *Distributive property*
$= -3 + (-4p)$
$= -3 - 4p$

(d) $-(k - 2)$
$= -1(k - 2)$ *Identity property*
$= -1[k + (-2)]$
$= -1(k) + (-1)(-2)$ *Distributive property*
$= -k + 2$

5. (a) $-3w + 7 - 8w - 2$
$= -3w - 8w + 7 - 2$ *Commutative and associative properties*
$= -11w + 5$ *Combine like terms.*

(b) $12b - 9 + 4b - 7b + 1$
(See Example 4 for more detailed steps.)
$= 12b + 4b - 7b - 9 + 1$
$= (12 + 4 - 7)b - 9 + 1$
$= 9b - 8$

6. (a) $4x - 7x - 10x + 5x$
$= (4 - 7 - 10 + 5)x$ *Distributive property*
$= -8x$

(b) $9 - 2(a - 3) + 4 - a$
$= 9 - 2a + 6 + 4 - a$ *Distributive property*
$= -2a - a + 9 + 6 + 4$ *Commutative and associative properties*
$= -3a + 19,$ *Combine like terms.*
or $19 - 3a$

(c) $10 - 3(6 + 2t)$
$= 10 - 18 - 6t$ *Distributive property*
$= -8 - 6t$ *Subtract.*

(d) $7x - (4x - 2)$
$= 7x - 4x + 2$ *Distributive property*
$= 3x + 2$ *Combine like terms.*

(e) $(4m)(2n)$
$= [(4m)(2)]n$ *Order of operations*
$= [4(m \cdot 2)]n$ *Associative property*
$= [4(2m)]n$ *Commutative property*
$= [(4 \cdot 2)m]n$ *Associative property*
$= (8m)n$ *Multiply.*
$= 8(mn)$ *Associative property*
$= 8mn$

7. **(a)** Using the multiplication property of zero, $197 \cdot 0 = 0$.

(b) Using the multiplication property of zero, $0(-\frac{8}{9}) = 0$.

(c) $0 \cdot \_\_\_\_\_ = 0$

Any real number will be a solution, since zero times any number will equal zero.

1.4 Section Exercises

1. The identity element for addition is 0 since, for any real number a, $a + 0 = 0 + a = a$. Choice **B** is correct.

3. The additive inverse of a is $-a$ since, for any real number a, $a + (-a) = 0$ and $-a + a = 0$. Choice **A** is correct.

5. The multiplication property of 0 states that the <u>product</u> of 0 and any real number is <u>0</u>.

7. The associative property is used to change the <u>grouping</u> of three terms or factors.

9. When simplifying an expression, only <u>like</u> terms can be combined.

11. Using the distributive property,

$$2(m + p) = 2m + 2p.$$

13. $$\begin{aligned} -5(2d - f) &= -5[2d + (-f)] \\ &= -5(2d) + (-5)(-f) \\ &= -10d + 5f \end{aligned}$$

15. Using the second form of the distributive property,

$$\begin{aligned} 5k + 3k &= (5 + 3)k \\ &= 8k. \end{aligned}$$

17. $$\begin{aligned} 7r - 9r &= (7 - 9)r \quad \textit{Distributive property} \\ &= -2r \end{aligned}$$

19. $$\begin{aligned} a + 7a &= 1a + 7a \quad \textit{Identity property} \\ &= (1 + 7)a \quad \textit{Distributive property} \\ &= 8a \end{aligned}$$

21. $-8z + 4w$
Since there is no common variable factor here, we cannot use the distributive property to simplify the expression.

23. $$\begin{aligned} -(4b - c) &= -1(4b - c) \quad \textit{Identity property} \\ &= -4b + c \quad \textit{Distributive property} \end{aligned}$$

25. $$\begin{aligned} 96 \cdot 19 + 4 \cdot 19 &= (96 + 4)19 \\ &= (100)19 \\ &= 1900 \end{aligned}$$

27. $$\begin{aligned} 58 \cdot \frac{3}{2} - 8 \cdot \frac{3}{2} &= (58 - 8)\frac{3}{2} \\ &= (50)\frac{3}{2} = \frac{50}{1} \cdot \frac{3}{2} \\ &= \tfrac{150}{2} = 75 \end{aligned}$$

29. $$\begin{aligned} 4.31(69) + 4.31(31) &= 4.31(69 + 31) \\ &= 4.31(100) \\ &= 431 \end{aligned}$$

31. $$\begin{aligned} &-12y + 4y + 3 + 2y \\ &= -12y + 4y + 2y + 3 \\ &= (-12 + 4 + 2)y + 3 \\ &= -6y + 3 \end{aligned}$$

33. $$\begin{aligned} &-6p + 11p - 4p + 6 + 5 \\ &= (-6 + 11 - 4)p + (6 + 5) \\ &= p + 11 \end{aligned}$$

35. $$\begin{aligned} &3(k + 2) - 5k + 6 + 3 \\ &= 3k + 6 - 5k + 6 + 3 \\ &= 3k - 5k + 6 + 6 + 3 \\ &= (3 - 5)k + 6 + 6 + 3 \\ &= -2k + 15 \end{aligned}$$

37. $$\begin{aligned} &-2(m + 1) + 3(m - 4) \\ &= -2(m) + (-2)(1) + 3(m) + (3)(-4) \\ &= -2m - 2 + 3m - 12 \\ &= -2m + 3m - 2 - 12 \\ &= (-2 + 3)m + [-2 + (-12)] \\ &= m - 14 \end{aligned}$$

39. $$\begin{aligned} &0.25(8 + 4p) - 0.5(6 + 2p) \\ &= 0.25(8) + 0.25(4p) \\ &\quad + (-0.5)(6) + (-0.5)(2p) \\ &= 2 + p - 3 - p \\ &= p - p + 2 - 3 \\ &= (1 - 1)p + 2 - 3 \\ &= 0p - 1 \\ &= -1 \end{aligned}$$

41. $$\begin{aligned} &-(2p + 5) + 3(2p + 4) - 2p \\ &= -2p - 5 + 6p + 12 - 2p \\ &= (-2 + 6 - 2)p + (-5 + 12) \\ &= 2p + 7 \end{aligned}$$

43. $$\begin{aligned} &2 + 3(2z - 5) - 3(4z + 6) - 8 \\ &= 2 + 6z - 15 - 12z - 18 - 8 \\ &= 6z - 12z + 2 - 15 - 18 - 8 \\ &= (6 - 12)z - 39 \\ &= -6z - 39 \end{aligned}$$

45. $$\begin{aligned} 5x + 8x &= (5 + 8)x \quad \textit{Distributive property} \\ &= 13x \end{aligned}$$

47. $$\begin{aligned} 5(9r) &= (5 \cdot 9)r \quad \textit{Associative property} \\ &= 45r \end{aligned}$$

49. $5x + 9y = 9y + 5x$ *Commutative property*

51. $1 \cdot 7 = 7$ *Identity property*

53. $8(-4 + x)$
$= 8(-4) + 8x$ *Distributive property*
$= -32 + 8x$

55. Answers will vary. Commutative: one example is washing your face and brushing your teeth. The activities can be carried out in either order.

57. The terms have been grouped using the *associative property* of addition.

58. The terms have been regrouped using the *associative property* of addition.

59. The order of the terms inside the parentheses has been changed using the *commutative property* of addition.

60. The terms have been regrouped using the *associative property* of addition.

61. The second form of the *distributive property* was used to rewrite the sum as a product.

62. The numbers in parentheses have been added to simplify the expression.

Chapter 1 Review Exercises

1. $\{-4, -1, 2, \frac{9}{4}, 4\}$

Place dots for $-4, -1, 2, \frac{9}{4} = 2.25$, and 4 on a number line.

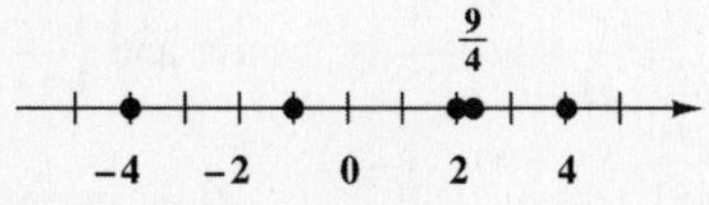

2. $\{-5, -\frac{11}{4}, -0.5, 0, 3, \frac{13}{3}\}$

Place dots for $-5, -\frac{11}{4} = -2.75, -0.5, 0, 3$, and $\frac{13}{3} = 4.\overline{3}$ on a number line.

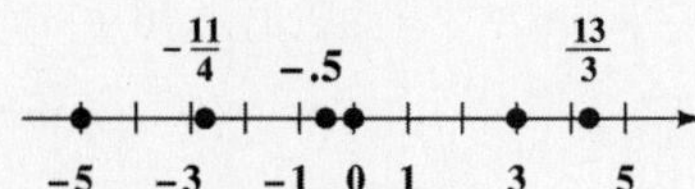

3. $|-16| = -(-16) = 16$

4. $|23| = 23$

5. $-|-4| = -[-(-4)] = -(4) = -4$

6. $|-8| - |-3| = -(-8) - [-(-3)]$
$= 8 - [3]$
$= 5$

In Exercises 7–10,

$$S = \left\{-9, -\frac{4}{3}, -\sqrt{4}, -0.25, 0, 0.\overline{35}, \frac{5}{3}, \sqrt{7}, \sqrt{-9}, \frac{12}{3}\right\}.$$

7. The elements 0 and $\frac{12}{3}$ (or 4) are whole numbers.

8. The elements -9, $-\sqrt{4}$ (or -2), 0, and $\frac{12}{3}$ (or 4) are integers.

9. The elements $-9, -\frac{4}{3}, -\sqrt{4}$ (or -2), $-0.25, 0, 0.\overline{35}, \frac{5}{3}$, and $\frac{12}{3}$ (or 4) are rational numbers. (Remember that terminating and repeating decimals are rational numbers.)

10. All the elements in the set are real numbers except $\sqrt{-9}$.

11. $\{x \mid x$ is a natural number between 3 and 9$\}$
The natural numbers between 3 and 9 are $4, 5, 6, 7,$ and 8. Therefore, the set is $\{4, 5, 6, 7, 8\}$.

12. $\{y \mid y$ is a whole number less than 4$\}$
The whole numbers less than 4 are $0, 1, 2,$ and 3. Therefore, the set is $\{0, 1, 2, 3\}$.

13. $4 \cdot 2 \leq |12 - 4|$
$8 \leq |8|$
$8 \leq 8$ *True*

14. $2 + |-2| > 4$
$2 + 2 > 4$
$4 > 4$ *False*

15. $4(3 + 7) > -|40|$
$4(10) > -40$
$40 > -40$ *True*

16. The longest bar represents the greatest change. Chrysler had the greatest change of 13.7%.

17. The shortest bar represents the least change. Honda had the least change of -2.5%.

18. Ford: $|5.2| = 5.2$

General Motors: $|-5.6| = 5.6$

The absolute value of the percent change for Ford was *less* than the absolute value of the percent change for General Motors, so the statement is *false*.

19. Toyota $\overset{?}{>}$ 4(Mazda)
$13.4 \overset{?}{>} 4(3.0)$
$13.4 > 12.0$ *True*

20. $$-\frac{5}{8} - \left(-\frac{7}{3}\right) = -\frac{5}{8} + \frac{7}{3}$$
$$= -\frac{15}{24} + \frac{56}{24} = \frac{41}{24}$$

21. $-\frac{4}{5}-\left(-\frac{3}{10}\right)=-\frac{4}{5}+\frac{3}{10}$
$=-\frac{8}{10}+\frac{3}{10}$
$=-\frac{5}{10}=-\frac{1}{2}$

22. $-5+(-11)+20-7$
$=-16+20-7$
$=4-7$
$=-3$

23. $-9.42+1.83-7.6-1.9$
$=-7.59-7.6-1.9$
$=-15.19-1.9$
$=-17.09$

24. $-15+(-13)+(-11)$
$=-28+(-11)$
$=-39$

25. $-1-3-(-10)+(-7)$
$=-4+10+(-7)$
$=6+(-7)$
$=-1$

26. $\frac{3}{4}-\left(\frac{1}{2}-\frac{9}{10}\right)=\frac{3}{4}-\left(\frac{5}{10}-\frac{9}{10}\right)$
$=\frac{3}{4}-\left(-\frac{4}{10}\right)$
$=\frac{3}{4}+\frac{4}{10}$
$=\frac{15}{20}+\frac{8}{20}=\frac{23}{20}$

27. $-\frac{2}{3}-\left(\frac{1}{6}-\frac{5}{9}\right)=-\frac{2}{3}-\left(\frac{3}{18}-\frac{10}{18}\right)$
$=-\frac{2}{3}-\left(-\frac{7}{18}\right)$
$=-\frac{2}{3}+\frac{7}{18}$
$=-\frac{12}{18}+\frac{7}{18}$
$=-\frac{5}{18}$

28. $-|-12|-|-9|+(-4)-|10|$
$=-(12)-9+(-4)-10$
$=-21+(-4)-10$
$=-25-10$
$=-35$

29. $11{,}049-(-282)=11{,}049+282$
$=11{,}331$

The difference is 11,331 feet.

30. $2(-5)(-3)(-3)=(-10)(-3)(-3)$
$=(30)(-3)$
$=-90$

31. $-\frac{3}{7}\left(-\frac{14}{9}\right)=\frac{3}{7}\cdot\frac{2\cdot 7}{3\cdot 3}=\frac{2}{3}$

32. $-4.6(2.48)=-11.408$

33. $\frac{75}{-5}=15\cdot 5\left(\frac{1}{-5}\right)=-15$

34. $\frac{-2.3754}{-0.74}=3.21$

35. $\frac{5}{7-7}=\frac{5}{0}$, which is *undefined.*
$\frac{7-7}{5}=\frac{0}{5}=0$

36. $10^4=10\cdot 10\cdot 10\cdot 10=10{,}000$

37. $\left(\frac{3}{7}\right)^3=\frac{3}{7}\cdot\frac{3}{7}\cdot\frac{3}{7}=\frac{27}{343}$

38. $(-5)^3=(-5)(-5)(-5)=-125$

39. $-5^3=-(5\cdot 5\cdot 5)=-125$

40. $(1.7)^2=(1.7)(1.7)=2.89$

41. $\sqrt{400}=20$, because 20 is positive and $20^2=400$.

42. $-\sqrt{196}=-(\sqrt{196})=-(14)=-14$

43. $\sqrt{\frac{64}{121}}=\frac{8}{11}$ since $\frac{8}{11}$ is positive and $\left(\frac{8}{11}\right)^2=\frac{64}{121}$.

44. $-\sqrt{0.81}=-\sqrt{(0.9)^2}=-(0.9)=-0.9$

45. Since there is no real number whose square is -64, $\sqrt{-64}$ is *not a real number.*

46. $-14\left(\frac{3}{7}\right)+6\div 3=-2(3)+(6\div 3)$
$=-6+2=-4$

47. $-\frac{2}{3}[5(-2)+8-4^3]$
$=-\frac{2}{3}[5(-2)+8-64]$ *Evaluate the power.*
$=-\frac{2}{3}[-10+8-64]$
$=-\frac{2}{3}[-66]=\frac{2}{3}(22\cdot 3)=44$

48. $\dfrac{-5(3^2)+9(\sqrt{4})-5}{6-5(-2)}$
$=\dfrac{-5(9)+9(2)-5}{6+10}$
$=\dfrac{-45+18-5}{16}$
$=\dfrac{-32}{16}=-2$

In Exercises 49–51, let $k=-4$, $m=2$, and $n=16$.

49. $4k-7m=4(-4)-7(2)$
$=-16-14$
$=-30$

50. $-3\sqrt{n}+m+5k$
$=-3(\sqrt{16})+2+5(-4)$
$=-3(4)+2-20$
$=-12+2-20$
$=-30$

51. $\dfrac{4m^3-3n}{7k^2-10}=\dfrac{4(2)^3-3(16)}{7(-4)^2-10}$
$=\dfrac{4(8)-3(16)}{7(16)-10}$
$=\dfrac{32-48}{112-10}$
$=\dfrac{-16}{102}=-\dfrac{8}{51}$

52. **(a)** 6 ft 1 in. = $(6\cdot 12+1)$ in. = 73 in.

704 × (weight in pounds) ÷ (height in inches)^2
$=704\times 205\div 73^2$
$=704\times 205\div 5329$
$=144{,}320\div 5329$
≈ 27

Carlos Beltran's BMI is 27.

(b) Answers will vary.

53. $2q+19q$
$=(2+19)q$ *Distributive property*
$=21q$

54. $13z-17z$
$=(13-17)z$ *Distributive property*
$=-4z$

55. $-m+6m$
$=-1m+6m$ *Identity property*
$=(-1+6)m$ *Distributive property*
$=5m$

56. $5p-p$
$=5p+(-1)p$ *Identity property*
$=[5+(-1)]p$ *Distributive property*
$=4p$

57. $-2(k+3)$
$=-2(k)+(-2)(3)$ *Distributive property*
$=-2k-6$

58. $6(r+3)$
$=6(r)+6(3)$ *Distributive property*
$=6r+18$

59. $9(2m+3n)$
$=9(2m)+9(3n)$ *Distributive property*
$=18m+27n$

60. $-(-p+6q)-(2p-3q)$
$=-1(-p+6q)+(-1)(2p-3q)$
$=-1(-p)+(-1)(6q)$
$\quad+(-1)(2p)+(-1)(-3q)$
$=p-6q-2p+3q$
$=p-2p-6q+3q$
$=-p-3q$

61. $-3y+6-5+4y$
$=-3y+4y+6-5$
$=y+1$

62. $2a+3-a-1-a-2$
$=2a-a-a+3-1-2$
$=0$

63. $-3(4m-2)+2(3m-1)-4(3m+1)$
$=-12m+6+6m-2-12m-4$
$=-12m+6m-12m+6-2-4$
$=-18m$

64. $2x+3x=(2+3)x$ *Distributive property*
$=5x$

65. $-4\cdot 1=-4$ *Identity property*

66. $2(4x)=(2\cdot 4)x$ *Associative property*
$=8x$

67. $-3+13=13+(-3)$ *Commutative property*
$=10$

68. $-3+3=0$ *Inverse property*

69. $5(x+z)=5x+5z$ *Distributive property*

70. $0+7=7$ *Identity property*

71. $8\cdot\dfrac{1}{8}=1$ *Inverse property*

72. $3a+5a+6a$
$=(3+5+6)a$ *Distributive property*
$=14a$

73. $\frac{9}{28} \cdot 0 = 0$ *Multiplication property of 0*

74. **[1.2]** For 2003 (in millions of dollars):

$|169{,}924 - 221{,}595| = |-51{,}671| = 51{,}671$

The balance of trade (exports minus imports) is *negative* since the amount of money spent on imports is greater than the amount of money received for exports.

75. **[1.2]** For 2004 (in millions of dollars):

$|189{,}880 - 256{,}360| = |-66{,}480| = 66{,}480$

The balance of trade (exports minus imports) is *negative* since the amount of money spent on imports is greater than the amount of money received for exports.

76. **[1.2]** For 2005 (in millions of dollars):

$|211{,}349 - 287{,}870| = |-76{,}521| = 76{,}521$

The balance of trade (exports minus imports) is *negative* since the amount of money spent on imports is greater than the amount of money received for exports.

77. **[1.3]** $\left(-\frac{4}{5}\right)^4 = \left(-\frac{4}{5}\right)\left(-\frac{4}{5}\right)\left(-\frac{4}{5}\right)\left(-\frac{4}{5}\right)$
$= \frac{256}{625}$

78. **[1.2]** $-\frac{5}{8}(-40) = -\frac{5}{8} \cdot \frac{-40}{1}$
$= 5 \cdot 5 = 25$

79. **[1.3]** $-25\left(-\frac{4}{5}\right) + 3^3 - 32 \div \sqrt{4}$
$= -25(-\frac{4}{5}) + 27 - 32 \div 2$
$= 20 + 27 - 16$
$= 31$

80. **[1.2]** $-8 + |-14| + |-3| = -8 + 14 + 3$
$= 9$

81. **[1.3]** $\frac{6 \cdot \sqrt{4} - 3 \cdot \sqrt{16}}{-2 \cdot 5 + 7(-3) - 10}$
$= \frac{6 \cdot 2 - 3 \cdot 4}{-2 \cdot 5 + 7(-3) - 10}$
$= \frac{12 - 12}{-10 - 21 - 10}$
$= \frac{0}{-41} = 0$

82. **[1.3]** $-\sqrt{25} = -(5) = -5$

83. **[1.2]** $-\frac{10}{21} \div -\frac{5}{14} = -\frac{10}{21} \cdot -\frac{14}{5}$
$= \frac{2 \cdot 5}{3 \cdot 7} \cdot \frac{2 \cdot 7}{5}$
$= \frac{2 \cdot 2}{3} = \frac{4}{3}$

84. **[1.2]** $0.8 - 4.9 - 3.2 + 1.14$
$= -4.1 - 3.2 + 1.14$
$= -7.3 + 1.14$
$= -6.16$

85. **[1.3]** $-3^2 = -(3 \cdot 3) = -9$

86. **[1.2]** $\frac{-38}{-19} = 38\left(\frac{1}{19}\right) = 2 \cdot 19\left(\frac{1}{19}\right) = 2$

87. **[1.4]** $-2(k - 1) + 3k - k$
$= -2k + 2 + 3k - k$
$= -2k + 3k - k + 2$
$= (-2 + 3 - 1)k + 2$
$= 0k + 2 = 2$

88. **[1.3]** Since there is no real number whose square is -100, $-\sqrt{-100}$ is *not a real number*.

89. **[1.3]**
$-m(3k^2 + 5m)$
$= -2[3(-4)^2 + 5(2)]$ *Let k = –4, m = 2.*
$= -2[3(16) + 5(2)]$
$= -2[48 + 10]$
$= -2[58]$
$= -116$

90. **[1.3]** In order to evaluate $(3 + 2)^2$, you should work within the parentheses first.

Chapter 1 Test

1. $\{-3, 0.75, \frac{5}{3}, 5, 6.3\}$

Place dots at $-3, 0.75, \frac{5}{3} = 1.\overline{6}, 5$, and 6.3.

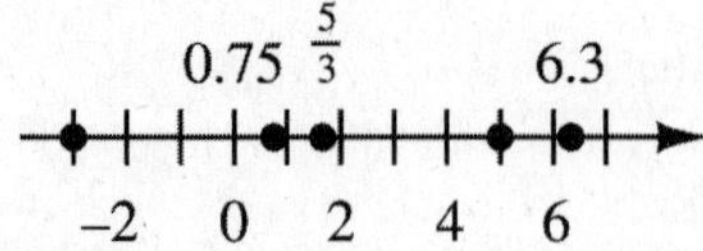

In Exercises 2–5,

$$A = \left\{-\sqrt{6}, -1, -0.5, 0, 3, \sqrt{25}, 7.5, \tfrac{24}{2}, \sqrt{-4}\right\}.$$

2. The elements 0, 3, $\sqrt{25}$ (or 5), and $\frac{24}{2}$ (or 12) are whole numbers.

3. The elements -1, 0, 3, $\sqrt{25}$ (or 5), and $\frac{24}{2}$ (or 12) are integers.

4. The elements -1, -0.5, 0, 3, $\sqrt{25}$ (or 5), 7.5, and $\frac{24}{2}$ (or 12) are rational numbers.

5. All the elements in the set are real numbers except $\sqrt{-4}$.

6. $-6+14+(-11)-(-3)$
$=8+(-11)+3$
$=-3+3=0$

7. $10-4\cdot 3+6(-4)$
$=10-12+(-24)$
$=-2+(-24)=-26$

8. $7-4^2+2(6)+(-4)^2$
$=7-16+12+16$
$=19$

9. $\dfrac{10-24+(-6)}{\sqrt{16}(-5)}$
$=\dfrac{-14+(-6)}{4(-5)}$
$=\dfrac{-20}{-20}=1$

10. $\dfrac{-2[3-(-1-2)+2]}{\sqrt{9}(-3)-(-2)}$
$=\dfrac{-2[3-(-3)+2]}{3(-3)-(-2)}$
$=\dfrac{-2[8]}{-9-(-2)}$
$=\dfrac{-16}{-7}=\dfrac{16}{7}$

11. $\dfrac{8\cdot 4-3^2\cdot 5-2(-1)}{-3\cdot 2^3+1}$
$=\dfrac{8\cdot 4-9\cdot 5-2(-1)}{-3\cdot 8+1}$
$=\dfrac{32-45+2}{-24+1}$
$=\dfrac{-11}{-23}=\dfrac{11}{23}$

12. $17{,}400-(-32{,}995)$
$=17{,}400+32{,}995$
$=50{,}395$

The difference between the height of Mt. Foraker and the depth of the Philippine Trench is 50,395 feet.

13. $14{,}110-(-23{,}376)$
$=14{,}110+23{,}376$
$=37{,}486$

The difference between the height of Pikes Peak and the depth of the Java Trench is 37,486 feet.

14. $-24{,}721-(-23{,}376)$
$=-24{,}721+23{,}376$
$=-1345$

The Cayman Trench is 1345 feet deeper than the Java Trench.

15. $\sqrt{196}=14$, because 14 is positive and $14^2=196$.

16. $-\sqrt{225}=-(\sqrt{225})=-(15)=-15$

17. Since there is no real number whose square is -16, $\sqrt{-16}$ is *not a real number.*

18. **(a)** If a is positive, then $\sqrt{a}$ will represent a positive number.

(b) If a is negative, then $\sqrt{a}$ will not represent a real number.

(c) If a is 0, then $\sqrt{a}$ will be 0.

In Exercises 19–20, let $k=-3$, $m=-3$, and $r=25$.

19. $\sqrt{r}+2k-m$
$=\sqrt{25}+2(-3)-(-3)$
$=5+2(-3)-(-3)$
$=5-6-(-3)$
$=5-6+3$
$=-1+3$
$=2$

20. $\dfrac{8k+2m^2}{r-2}$
$=\dfrac{8(-3)+2(-3)^2}{25-2}$
$=\dfrac{8(-3)+2(9)}{23}$
$=\dfrac{-24+18}{23}$
$=\dfrac{-6}{23}$, or $-\dfrac{6}{23}$

21. $-3(2k-4)+4(3k-5)-2+4k$
$=-3(2k)+(-3)(-4)+4(3k)$
$\quad+4(-5)-2+4k$
$=-6k+12+12k-20-2+4k$
$=-6k+12k+4k+12-20-2$
$=10k-10$

22. When simplifying

$$(3r+8)-(-4r+6),$$

the subtraction sign in front of $(-4r+6)$ changes the sign of the terms $-4r$ and 6.

$$\begin{aligned}&(3r+8)-(-4r+6)\\&=3r+8-(-4r)-6\\&=3r+8+4r-6\\&=3r+4r+8-6\\&=7r+2\end{aligned}$$

23. $6+(-6)=0$

The answer is **B**, *Inverse property*.
The sum of 6 and its inverse, -6, equals zero.

24. $4+5=5+4$

The answer is **E**, *Commutative property*.
The order of the terms is reversed.

25. $-2+(3+6)=(-2+3)+6$

The answer is **D**, *Associative property*.
The order of the terms is the same, but the grouping has changed.

26. $5x+15x=(5+15)x$

The answer is **A**, *Distributive property*.
This is the second form of the distributive property.

27. $13\cdot 0=0$

The answer is **F**, *Multiplication property of* 0.
Multiplication by 0 always equals 0.

28. $-9+0=-9$

The answer is **C**, *Identity property*.
The addition of 0 to any number does not change the number.

29. $4\cdot 1=4$

The answer is **C**, *Identity property*.
Multiplication of any number by 1 does not change the number.

30. $(a+b)+c=(b+a)+c$

The answer is **E**, *Commutative property*.
The order of the terms a and b is reversed.

CHAPTER 2 LINEAR EQUATIONS AND APPLICATIONS

2.1 Linear Equations in One Variable

2.1 Margin Exercises

1. **(a)** $9x = 10$ is an *equation* because it contains an equals sign.

   (b) $9x + 10$ is an *expression* because it does not contain an equals sign.

   (c) $3 + 5x - 8x + 9$ is an *expression* because it does not contain an equals sign.

   (d) $3 + 5x = -8x + 9$ is an *equation* because it contains an equals sign.

2. To decide if a given number is a solution, substitute that number for the variable in the equation to see if the resulting statement is true or false.

   (a) $3k = 15$; 5

   The number 5 is a solution since $3 \cdot 5 = 15$ and $15 = 15$ is true.

   (b) $r + 5 = 4$; 1

   The number 1 is not a solution since $1 + 5 = 6$ and $6 = 4$ is false.

   (c) $-8m = 12$; $\frac{3}{2}$

   The number $\frac{3}{2}$ is not a solution since $-8(\frac{3}{2}) = -12$ and $-12 = 12$ is false.

3. **(a)**

$3p + 2p + 1 = -24$	*Original equation*
$5p + 1 = -24$	*Combine terms.*
$5p + 1 - 1 = -24 - 1$	*Subtract 1.*
$5p = -25$	*Combine terms.*
$\frac{5p}{5} = \frac{-25}{5}$	*Divide by 5.*
$p = -5$	*Proposed solution*

Check by substituting -5 for p in the *original* equation.

$3p + 2p + 1 = -24$	*Original equation*
$3(-5) + 2(-5) + 1 \stackrel{?}{=} -24$	*Let p = −5.*
$-15 - 10 + 1 \stackrel{?}{=} -24$	
$-24 = -24$	*True*

The true statement indicates that $\{-5\}$ is the solution set.

(b)

$3p = 2p + 4p + 5$	*Original equation*
$3p = 6p + 5$	*Combine terms.*
$3p - 6p = 6p + 5 - 6p$	*Subtract 6p.*
$-3p = 5$	*Combine terms.*
$\frac{-3p}{-3} = \frac{5}{-3}$	*Divide by −3.*
$p = -\frac{5}{3}$	*Proposed solution*

Check by substituting $-\frac{5}{3}$ for p in the *original* equation.

$3p = 2p + 4p + 5$	*Original equation*
$3(-\frac{5}{3}) \stackrel{?}{=} 2(-\frac{5}{3}) + 4(-\frac{5}{3}) + 5$	*Let p = −5/3.*
$-5 \stackrel{?}{=} -\frac{10}{3} - \frac{20}{3} + 5$	
$-5 \stackrel{?}{=} -\frac{30}{3} + 5$	
$-5 \stackrel{?}{=} -10 + 5$	
$-5 = -5$	*True*

Solution set: $\{-\frac{5}{3}\}$

(c)

$4x + 8x = 17x - 9 - 1$	*Original equation*
$12x = 17x - 10$	*Combine terms.*
$12x - 17x = 17x - 10 - 17x$	*Subtract 17x.*
$-5x = -10$	*Combine terms.*
$\frac{-5x}{-5} = \frac{-10}{-5}$	*Divide by −5.*
$x = 2$	*Proposed solution*

Check by substituting 2 for x in the *original* equation.

$4x + 8x = 17x - 9 - 1$	*Original equation*
$4(2) + 8(2) \stackrel{?}{=} 17(2) - 9 - 1$	*Let x = 2.*
$8 + 16 \stackrel{?}{=} 34 - 9 - 1$	
$24 = 24$	*True*

Solution set: $\{2\}$

(d) $-7+3t-9t=12t-5$

$-7-6t=12t-5$ *Combine terms.*

$-7-6t+6t+5=12t-5+6t+5$ *Add 6t; add 5.*

$-2=18t$

$\frac{-2}{18}=\frac{18t}{18}$ *Divide by 18.*

$-\frac{1}{9}=t$ *Proposed solution*

We will use the following notation to indicate the value of each side of the original equation after we have substituted the proposed solution and simplified.

Check $t=-\frac{1}{9}$: $-\frac{19}{3}=-\frac{19}{3}$ *True*

Solution set: $\{-\frac{1}{9}\}$

4. **(a)** $5p+4(3-2p)=2+p-10$

$5p+12-8p=2+p-10$ *Distributive property*

$12-3p=p-8$ *Combine terms.*

$12-3p+3p+8=p-8+3p+8$ *Add 3p; add 8.*

$20=4p$ *Combine terms.*

$\frac{20}{4}=\frac{4p}{4}$ *Divide by 4.*

$5=p$ *Proposed solution*

Check $p=5$: $-3=-3$ *True*

Solution set: $\{5\}$

(b) $3(z-2)+5z=2$

$3z-6+5z=2$ *Distributive property*

$8z-6=2$ *Combine terms.*

$8z-6+6=2+6$ *Add 6.*

$8z=8$ *Combine terms.*

$\frac{8z}{8}=\frac{8}{8}$ *Divide by 8.*

$z=1$ *Proposed solution*

Check $z=1$: $2=2$ *True*

Solution set: $\{1\}$

(c) $-2+3(x+4)=8x$

$-2+3x+12=8x$ *Distributive property*

$3x+10=8x$ *Combine terms.*

$3x+10-3x=8x-3x$ *Subtract 3x.*

$10=5x$ *Combine terms.*

$\frac{10}{5}=\frac{5x}{5}$ *Divide by 5.*

$2=x$ *Proposed solution*

Check $x=2$: $16=16$ *True*

Solution set: $\{2\}$

(d) $6-(4+m)=8m-2(3m+5)$

$6-4-m=8m-6m-10$ *Distributive property*

$2-m=2m-10$ *Combine terms.*

$2-m+m+10=2m-10+m+10$ *Add m; add 10.*

$12=3m$ *Combine terms.*

$\frac{12}{3}=\frac{3m}{3}$ *Divide by 3.*

$4=m$ *Proposed solution*

Check $m=4$: $-2=-2$ *True*

Solution set: $\{4\}$

5. **(a)** $\frac{2p}{7}-\frac{p}{2}=-3$

Multiply each side by the LCD, 14.

$14\left(\frac{2p}{7}-\frac{p}{2}\right)=14(-3)$

$14\left(\frac{2p}{7}\right)-14\left(\frac{p}{2}\right)=14(-3)$ *Distributive property*

$4p-7p=-42$ *Multiply.*

$-3p=-42$ *Combine terms.*

$\frac{-3p}{-3}=\frac{-42}{-3}$ *Divide by −3.*

$p=14$ *Proposed solution*

Check $p=14$: $-3=-3$ *True*

Solution set: $\{14\}$

(b) $\frac{k+1}{2}+\frac{k+3}{4}=\frac{1}{2}$

Multiply each side by the LCD, 4, and use the distributive property.

$$4\left(\frac{k+1}{2}\right)+4\left(\frac{k+3}{4}\right)=4\left(\frac{1}{2}\right)$$
$$2(k+1)+1(k+3)=2$$
$$2k+2+k+3=2$$
$$3k+5=2$$
$$3k=-3 \quad \textit{Subtract 5.}$$
$$k=-1 \quad \textit{Divide by 3.}$$

Check $k=-1$: $\frac{1}{2}=\frac{1}{2}$ *True*

Solution set: $\{-1\}$

6. $0.04x+0.06(20-x)=0.05(50)$

Multiply each side by 100, and use the distributive property.

$$4x+6(20-x)=5(50)$$
$$4x+120-6x=250$$
$$-2x+120=250$$
$$-2x=130 \quad \textit{Subtract 120.}$$
$$x=-65 \quad \textit{Divide by } -2.$$

Check $x=-65$: $2.5=2.5$ *True*

Solution set: $\{-65\}$

7. $0.10(x-6)+0.05x=0.06(50)$

$$0.10x-0.6+0.05x=3 \quad \textit{Dist. prop.}$$
$$0.15x-0.6=3 \quad \textit{Combine.}$$
$$0.15x=3.6 \quad \textit{Add 0.6.}$$
$$x=24 \quad \textit{Div. by 0.15.}$$

Check $x=24$: $1.8+1.2=3$ *True*

Solution set: $\{24\}$

8. **(a)** $5(x+2)-2(x+1)=3x+1$

$$5x+10-2x-2=3x+1$$
$$3x+8=3x+1$$
$$3x+8-3x=3x+1-3x \quad \textit{Subtract 3x.}$$
$$8=1 \quad \textit{False}$$

Since the result, $8=1$, is *false*, the equation has no solution and is called a *contradiction.*

Solution set: $\emptyset$

(b) $\frac{x+1}{3}+\frac{2x}{3}=x+\frac{1}{3}$

Multiply each side by the LCD, 3, and use the distributive property.

$$3\left(\frac{x+1}{3}\right)+3\left(\frac{2x}{3}\right)=3\left(x+\frac{1}{3}\right)$$
$$x+1+2x=3x+1$$
$$3x+1=3x+1$$

This is an *identity.* Any real number will make the equation true.

Solution set: {all real numbers}

(c) $5(3x+1)=x+5$

$$15x+5=x+5$$
$$14x+5=5 \quad \textit{Subtract x.}$$
$$14x=0 \quad \textit{Subtract 5.}$$
$$x=0 \quad \textit{Divide by 14.}$$

This is a *conditional equation.*

Check $x=0$: $5=5$ *True*

Solution set: $\{0\}$

2.1 Section Exercises

1. **A.** $3x+x-2=0$ can be written as $4x-2=0$, so it is linear.

C. $9x-4=9$ is in linear form.

3. $3(x+4)=5x$ *Original equation*

$$3(6+4)\stackrel{?}{=}5\cdot 6 \quad \textit{Let } x=6.$$
$$3(10)\stackrel{?}{=}30 \quad \textit{Add.}$$
$$30=30 \quad \textit{True}$$

Since a true statement is obtained, 6 is a solution.

5. Suppose your last name is Lincoln. Then $x=7$ and both sides are evaluated as -48. The equation is an identity, so any number is a solution.

7. **(a)** $5x=10$ is an *equation* because it contains an equals sign.

(b) $5x+10$ is an *expression* because it does not contain an equals sign.

(c) $5x+6(x-3)=12x+6$ is an *equation* because it contains an equals sign.

(d) $5x+6(x-3)-(12x+6)$ is an *expression* because it does not contain an equals sign.

9. The student made a sign error when the distributive property was applied. The left side of the second line should be $8x-4x+6$. This gives us $4x+6=3x+7$ and then $x=1$. Thus, the correct solution is 1.

In the following exercises, we do not show the checks of the solutions. To be sure that your solutions are correct, check them by substituting into the original equations.

11. $9x + 10 = 1$

$9x + 10 - 10 = 1 - 10$ *Subtract 10.*

$9x = -9$

$\frac{9x}{9} = \frac{-9}{9}$ *Divide by 9.*

$x = -1$

Solution set: $\{-1\}$

13. $5x + 2 = 3x - 6$

$5x + 2 - 3x = 3x - 6 - 3x$ *Subtract 3x.*

$2x + 2 = -6$

$2x + 2 - 2 = -6 - 2$ *Subtract 2.*

$2x = -8$

$\frac{2x}{2} = \frac{-8}{2}$ *Divide by 2.*

$x = -4$

Solution set: $\{-4\}$

15. $7x - 5x + 15 = x + 8$

$2x + 15 = x + 8$ *Combine terms.*

$2x = x - 7$ *Subtract 15.*

$x = -7$ *Subtract x.*

Solution set: $\{-7\}$

17. $12w + 15w - 9 + 5 = -3w + 5 - 9$

$27w - 4 = -3w - 4$ *Combine terms.*

$30w - 4 = -4$ *Add 3w.*

$30w = 0$ *Add 4.*

$w = 0$ *Divide by 30.*

Solution set: $\{0\}$

19. $3(2t - 4) = 20 - 2t$

$6t - 12 = 20 - 2t$ *Distributive property*

$8t - 12 = 20$ *Add 2t.*

$8t = 32$ *Add 12.*

$t = 4$ *Divide by 8.*

Solution set: $\{4\}$

21. $-5(x + 1) + 3x + 2 = 6x + 4$

$-5x - 5 + 3x + 2 = 6x + 4$ *Distributive property*

$-2x - 3 = 6x + 4$ *Combine terms.*

$-3 = 8x + 4$ *Add 2x.*

$-7 = 8x$ *Subtract 4.*

$-\frac{7}{8} = x$ *Divide by 8.*

Solution set: $\{-\frac{7}{8}\}$

23. $2(x + 3) = -4(x + 1)$

$2x + 6 = -4x - 4$ *Remove parentheses.*

$6x + 6 = -4$ *Add 4x.*

$6x = -10$ *Subtract 6.*

$x = \frac{-10}{6} = -\frac{5}{3}$ *Divide by 6.*

Solution set: $\{-\frac{5}{3}\}$

25. $3(2w + 1) - 2(w - 2) = 5$

$6w + 3 - 2w + 4 = 5$ *Remove parentheses.*

$4w + 7 = 5$ *Combine terms.*

$4w = -2$ *Subtract 7.*

$w = \frac{-2}{4}$ *Divide by 4.*

$w = -\frac{1}{2}$

Solution set: $\{-\frac{1}{2}\}$

27. $2x + 3(x - 4) = 2(x - 3)$

$2x + 3x - 12 = 2x - 6$

$5x - 12 = 2x - 6$

$3x = 6$

$x = \frac{6}{3} = 2$

Solution set: $\{2\}$

29. $6p - 4(3 - 2p) = 5(p - 4) - 10$

$6p - 12 + 8p = 5p - 20 - 10$

$14p - 12 = 5p - 30$

$9p = -18$

$p = -2$

Solution set: $\{-2\}$

31. $2[w - (2w + 4) + 3] = 2(w + 1)$

$2[w - 2w - 4 + 3] = 2(w + 1)$

$2[-w - 1] = 2(w + 1)$

$-w - 1 = w + 1$ *Divide by 2.*

$-1 = 2w + 1$ *Add w.*

$-2 = 2w$ *Subtract 1.*

$-1 = w$ *Divide by 2.*

Solution set: $\{-1\}$

33. $-[2z - (5z + 2)] = 2 + (2z + 7)$

$-[2z - 5z - 2] = 2 + 2z + 7$

$-[-3z - 2] = 2 + 2z + 7$

$3z + 2 = 2z + 9$

$z = 7$

Solution set: $\{7\}$

35.
$$\begin{aligned} -3m+6-5(m-1) &= -5m-(2m-4)+5 \\ -3m+6-5m+5 &= -5m-2m+4+5 \\ -8m+11 &= -7m+9 \\ -m+11 &= 9 \\ -m &= -2 \\ m &= 2 \end{aligned}$$

Solution set: $\{2\}$

37.
$$\begin{aligned} -3(x+2)+4(3x-8) &= 2(4x+7)+2(3x-6) \\ -3x-6+12x-32 &= 8x+14+6x-12 \\ 9x-38 &= 14x+2 \\ -38 &= 5x+2 \\ -40 &= 5x \\ -8 &= x \end{aligned}$$

Solution set: $\{-8\}$

39. The denominators of the fractions are 4, 3, 6, and 1. The LCD is $(4)(3)(6)(1) = 12$, since it is the smallest number into which each denominator can divide without a remainder.

41. **(a)** We need to make the coefficient of the first term on the left an integer. Since $0.05 = \frac{5}{100}$, we multiply by 10^2 or 100. This will also take care of the second term.

(b) We need to make 0.006, 0.007, and 0.009 integers. These numbers can be written as $\frac{6}{1000}$, $\frac{7}{1000}$, and $\frac{9}{1000}$. Multiplying by 10^3 or 1000 will eliminate the decimal points (the denominators) so that all the coefficients are integers.

43.
$$\frac{m}{2}+\frac{m}{3} = 10$$

Multiply each side by the LCD, 6.

$$\begin{aligned} 6\left(\frac{m}{2}+\frac{m}{3}\right) &= 6(10) \\ 6\left(\frac{m}{2}\right)+6\left(\frac{m}{3}\right) &= 60 && \textit{Distributive property} \\ 3m+2m &= 60 \\ 5m &= 60 && \textit{Add.} \\ m &= 12 && \textit{Divide by 5.} \end{aligned}$$

Check $m = 12$: $6+4 = 10$ *True*

Solution set: $\{12\}$

45.
$$\frac{3}{4}x+\frac{5}{2}x = 13$$

Multiply each side by the LCD, 4.

$$\begin{aligned} 4\left(\frac{3}{4}x+\frac{5}{2}x\right) &= 4(13) \\ 4\left(\frac{3}{4}x\right)+4\left(\frac{5}{2}x\right) &= 4(13) && \textit{Distributive property} \\ 3x+10x &= 52 \\ 13x &= 52 && \textit{Combine terms.} \\ x &= 4 && \textit{Divide by 13.} \end{aligned}$$

Check $x = 4$: $13 = 13$ *True*

Solution set: $\{4\}$

47.
$$\frac{1}{5}x-2 = \frac{2}{3}x-\frac{2}{5}x$$

Multiply each side by the LCD, 15, and use the distributive property.

$$\begin{aligned} 15\left(\tfrac{1}{5}x\right)-15(2) &= 15\left(\tfrac{2}{3}x\right)-15\left(\tfrac{2}{5}x\right) \\ 3x-30 &= 10x-6x \\ 3x-30 &= 4x \\ -30 &= x && \textit{Subtract 3x.} \end{aligned}$$

Check $x = -30$: $-8 = -8$ *True*

Solution set: $\{-30\}$

49.
$$\frac{x-8}{5}+\frac{8}{5} = -\frac{x}{3}$$

Multiply each side by the LCD, 15, and use the distributive property.

$$\begin{aligned} 15\left(\frac{x-8}{5}\right)+15\left(\frac{8}{5}\right) &= 15\left(-\frac{x}{3}\right) \\ 3(x-8)+3(8) &= -5x \\ 3x-24+24 &= -5x \\ 3x &= -5x \\ 8x &= 0 && \textit{Add 5x.} \\ x &= 0 && \textit{Divide by 8.} \end{aligned}$$

Check $x = 0$: $0 = 0$ *True*

Solution set: $\{0\}$

51.
$$\frac{3x-1}{4}+\frac{x+3}{6} = 3$$

Multiply each side by the LCD, 12.

$$\begin{aligned} 12\left(\frac{3x-1}{4}+\frac{x+3}{6}\right) &= 12(3) \\ 3(3x-1)+2(x+3) &= 36 \\ 9x-3+2x+6 &= 36 \\ 11x+3 &= 36 \\ 11x &= 33 \\ x &= 3 \end{aligned}$$

Check $x = 3$: $2+1 = 3$ *True*

Solution set: $\{3\}$

53.
$$\frac{4t+1}{3} = \frac{t+5}{6}+\frac{t-3}{6}$$

Multiply each side by the LCD, 6.

$$\begin{aligned} 6\left(\frac{4t+1}{3}\right) &= 6\left(\frac{t+5}{6}+\frac{t-3}{6}\right) \\ 2(4t+1) &= (t+5)+(t-3) \\ 8t+2 &= 2t+2 \\ 6t &= 0 \\ t &= 0 \end{aligned}$$

Check $t = 0$: $\frac{1}{3} = \frac{5}{6}-\frac{3}{6}$ *True*

Solution set: $\{0\}$

55. $0.05x + 0.12(x + 5000) = 940$

Multiply each side by 100.

$$\begin{aligned} 5x + 12(x + 5000) &= 100(940) \\ 5x + 12x + 60{,}000 &= 94{,}000 \\ 17x &= 34{,}000 \\ x &= 2000 \end{aligned}$$

Check $x = 2000$: $100 + 840 = 940$ *True*

Solution set: $\{2000\}$

57. $0.02(50) + 0.08r = 0.04(50 + r)$

Multiply each side by 100.

$$\begin{aligned} 2(50) + 8r &= 4(50 + r) \\ 100 + 8r &= 200 + 4r \\ 4r &= 100 \\ r &= 25 \end{aligned}$$

Check $r = 25$: $1 + 2 = 3$ *True*

Solution set: $\{25\}$

59. $0.05x + 0.10(200 - x) = 0.45x$

Multiply each side by 100.

$$\begin{aligned} 5x + 10(200 - x) &= 45x \\ 5x + 2000 - 10x &= 45x \\ 2000 - 5x &= 45x \\ 2000 &= 50x \\ 40 &= x \end{aligned}$$

Check $x = 40$: $2 + 16 = 18$ *True*

Solution set: $\{40\}$

61. $0.006(x + 2) = 0.007x + 0.009$

Multiply each side by 1000.

$$\begin{aligned} 6(x + 2) &= 7x + 9 \\ 6x + 12 &= 7x + 9 \\ 3 &= x \end{aligned}$$

Check $x = 3$: $0.03 = 0.021 + 0.009$ *True*

Solution set: $\{3\}$

63. A conditional equation is true only for certain value(s), an identity has infinitely many solutions, and a contradiction has no solutions.

65. **(a)** $7 = 7$ is true and the original equation has solution set {all real numbers}, choice **B**.

(b) $x = 0$ indicates the original equation has solution set $\{0\}$, choice **A**.

(c) $7 = 0$ is false and the original equation has solution set $\emptyset$, choice **C**.

67.
$$\begin{aligned} -x + 4x - 9 &= 3(x - 4) - 5 \\ 3x - 9 &= 3x - 12 - 5 \\ 3x - 9 &= 3x - 17 \\ -9 &= -17 \quad \textit{False} \end{aligned}$$

The equation is a *contradiction*.

Solution set: $\emptyset$

69.
$$\begin{aligned} -11x + 4(x - 3) + 6x &= 4x - 12 \\ -11x + 4x - 12 + 6x &= 4x - 12 \\ -x - 12 &= 4x - 12 \\ 0 &= 5x \\ 0 &= x \end{aligned}$$

This is a *conditional* equation.

Solution set: $\{0\}$

71.
$$\begin{aligned} -2(t + 3) - t - 4 &= -3(t + 4) + 2 \\ -2t - 6 - t - 4 &= -3t - 12 + 2 \\ -3t - 10 &= -3t - 10 \end{aligned}$$

The equation is an *identity*.

Solution set: {all real numbers}

73.
$$\begin{aligned} 7[2 - (3 + 4x)] - 2x &= -9 + 2(1 - 15x) \\ 7[2 - 3 - 4x] - 2x &= -9 + 2 - 30x \\ 7[-1 - 4x] - 2x &= -7 - 30x \\ -7 - 28x - 2x &= -7 - 30x \\ -7 - 30x &= -7 - 30x \end{aligned}$$

The equation is an *identity*.

Solution set: {all real numbers}

2.2 Formulas and Percent

2.2 Margin Exercises

1. **(a)** To solve $I = prt$ for p, treat p as the only variable.

$$\begin{aligned} I &= prt \\ I &= p(rt) && \textit{Associative property} \\ \frac{I}{rt} &= \frac{p(rt)}{rt} && \textit{Divide by rt.} \\ \frac{I}{rt} &= p, \text{ or } p = \frac{I}{rt} \end{aligned}$$

(b) To solve $I = prt$ for r, treat r as the only variable.

$$\begin{aligned} I &= prt \\ \frac{I}{pt} &= \frac{r(pt)}{pt} && \textit{Divide by pt.} \\ \frac{I}{pt} &= r, \text{ or } r = \frac{I}{pt} \end{aligned}$$

2. **(a)** Solve $P = a + b + c$ for a.

$$P - (b + c) = a + (b + c) - (b + c) \quad \textit{Subtract } (b + c).$$

$$P - b - c = a, \text{ or } a = P - b - c$$

(b) Solve $V = \frac{1}{3}\pi r^2 h$ for h.

$$3V = \pi r^2 h \quad \textit{Multiply by 3.}$$

$$\frac{3V}{\pi r^2} = h, \text{ or } h = \frac{3V}{\pi r^2} \quad \textit{Divide by } \pi r^2.$$

3. Solve $M = \frac{1}{3}(a + b + c)$ for b.

$$3M = a + b + c \quad \textit{Multiply by 3.}$$

$$3M - a - c = b \quad \textit{Subtract a \& c.}$$

4. **(a)** Solve each equation for y.

$$2x + 7y = 5$$

$$2x + 7y - 2x = 5 - 2x \quad \textit{Subtract } 2x.$$

$$7y = 5 - 2x$$

$$\frac{7y}{7} = \frac{5 - 2x}{7} \quad \textit{Divide by 7.}$$

$$y = \frac{5 - 2x}{7}$$

(b) $5x - 6y = 12$

$$-6y = 12 - 5x \quad \textit{Subtract } 5x.$$

$$y = \frac{12 - 5x}{-6}, \quad \textit{Divide by } -6.$$

$$\text{or } y = \frac{5x - 12}{6}$$

5. **(a)** Use the formula for the area of a triangle. Solve for h.

$$A = \frac{1}{2}bh$$

$$2A = bh \quad \textit{Multiply by 2.}$$

$$\frac{2A}{b} = h, \text{ or } h = \frac{2A}{b} \quad \textit{Divide by } b.$$

Now substitute $A = 36$ and $b = 12$.

$$h = \frac{2(36)}{12} = 6$$

The height is 6 in.

(b) Use $d = rt$. Solve for r.

$$\frac{d}{t} = \frac{rt}{t} \quad \textit{Divide by } t.$$

$$\frac{d}{t} = r \text{ or } r = \frac{d}{t}$$

Now substitute $d = 500$ and $t = 20$.

$$r = \frac{500}{20} = 25$$

The rate is 25 mph.

(c) Use $d = rt$. Solve for t.

$$\frac{d}{r} = \frac{rt}{r} \quad \textit{Divide by } r.$$

$$\frac{d}{r} = t \text{ or } t = \frac{d}{r}$$

Now substitute $d = 500$ and $r = 157.085$.

$$t = \frac{500}{157.085} \approx 3.183$$

His time was about 3.183 hr.

6. **(a)** The given amount of mixture is 20 oz. The part that is oil is 1 oz. Thus, the percent of oil is

$$\frac{\text{amount}}{\text{base}} = \frac{1}{20} = 0.05 = 5\%.$$

(b) Let x represent the amount of commission earned.

$$\frac{x}{22{,}000} = 0.06 \quad \frac{\textit{amount } a}{\textit{base } b} = \textit{percent}$$

$$x = 0.06(22{,}000) \quad \textit{Multiply by 22,000.}$$

$$x = 1320$$

The salesman earns \$1320.

7. Let x represent the amount spent on pet supplies/medicine.

$$\frac{x}{41.2} = 0.238 \quad \textit{23.8\% = 0.238}$$

$$x = 0.238(41.2) \quad \textit{Multiply by 41.2.}$$

$$x = 9.8056$$

Therefore, about \$9.8 billion was spent on pet supplies/medicine.

8. **(a)** Let $x =$ the percent decrease (as a decimal).

$$\text{percent decrease} = \frac{\text{amount of decrease}}{\text{base}}$$

$$x = \frac{80 - 56}{80}$$

$$x = \frac{24}{80}$$

$$x = 0.3$$

The percent markdown was 30%.

(b) Let $x =$ the percent increase (as a decimal).

$$\text{percent increase} = \frac{\text{amount of increase}}{\text{base}}$$

$$x = \frac{689 - 650}{650}$$

$$x = \frac{39}{650}$$

$$x = 0.06$$

The percent increase was 6%.

2.2 Section Exercises

1. **(a)**
$$\frac{7x+8}{3} = 12$$
$$3\left(\frac{7x+8}{3}\right) = 3(12)$$
$$7x+8 = 36$$

(b)
$$\frac{ax+k}{c} = t\,(c \neq 0)$$
$$c\left(\frac{ax+k}{c}\right) = tc$$
$$ax+k = tc$$

2. **(a)**
$$7x+8 = 36$$
$$7x+8-8 = 36-8$$

(b)
$$ax+k = tc$$
$$ax+k-k = tc-k$$

3. **(a)** $7x = 28$ **(b)** $ax = tc - k$

4. **(a)** $\frac{7x}{7} = \frac{28}{7}$, $x = 4$ **(b)** $\frac{ax}{a} = \frac{tc-k}{a}$, $x = \frac{tc-k}{a}$

5. The restriction $a \neq 0$ must be applied. If $a = 0$, the denominator becomes 0 and division by 0 is undefined.

6. To solve an equation for a particular variable, such as solving the second equation for x, go through the same steps as you would in solving for x in the first equation. Treat all other variables as constants.

7. Solve $A = LW$ for W.
$$\frac{A}{L} = \frac{LW}{L} \quad \textit{Divide by L.}$$
$$\frac{A}{L} = W, \text{ or } W = \frac{A}{L}$$

9. Solve $P = 2L + 2W$ for L.
$$P - 2W = 2L \quad \textit{Subtract 2W.}$$
$$\frac{P-2W}{2} = \frac{2L}{2} \quad \textit{Divide by 2.}$$
$$\frac{P-2W}{2} = L, \text{ or } L = \frac{P}{2} - W$$

11. **(a)** Solve for $V = LWH$ for W.
$$\frac{V}{LH} = \frac{LWH}{LH}$$
$$\frac{V}{LH} = W, \text{ or } W = \frac{V}{LH}$$

(b) Solve for $V = LWH$ for H.
$$\frac{V}{LW} = \frac{LWH}{LW}$$
$$\frac{V}{LW} = H, \text{ or } H = \frac{V}{LW}$$

13. Solve $C = 2\pi r$ for r.
$$\frac{C}{2\pi} = \frac{2\pi r}{2\pi} \quad \textit{Divide by } 2\pi.$$
$$\frac{C}{2\pi} = r$$

15. **(a)** Solve $A = \frac{1}{2}h(b+B)$ for h.
$$2A = h(b+B) \quad \textit{Multiply by 2.}$$
$$\frac{2A}{b+B} = h \quad \textit{Divide by } b+B.$$

(b) Solve $A = \frac{1}{2}h(b+B)$ for B.
$$2A = h(b+B) \quad \textit{Multiply by 2.}$$
$$\frac{2A}{h} = b+B \quad \textit{Divide by h.}$$
$$\frac{2A}{h} - b = B \quad \textit{Subtract b.}$$

OR Solve $A = \frac{1}{2}h(b+B)$ for B.
$$2A = hb + hB \quad \textit{Multiply by 2.}$$
$$2A - hb = hB \quad \textit{Subtract hb.}$$
$$\frac{2A-hb}{h} = B \quad \textit{Divide by h.}$$

17. Solve $F = \frac{9}{5}C + 32$ for C.
$$F - 32 = \tfrac{9}{5}C \quad \textit{Subtract 32.}$$
$$\tfrac{5}{9}(F-32) = \tfrac{5}{9}(\tfrac{9}{5}C) \quad \textit{Multiply by } \tfrac{5}{9}.$$
$$\tfrac{5}{9}(F-32) = C$$

19.
$$4x + 9y = 11$$
$$4x + 9y - 4x = 11 - 4x \quad \textit{Subtract 4x.}$$
$$9y = 11 - 4x$$
$$\frac{9y}{9} = \frac{11-4x}{9} \quad \textit{Divide by 9.}$$
$$y = \frac{11-4x}{9}$$

21.
$$-3x + 2y = 5$$
$$2y = 5 + 3x \quad \textit{Add 3x.}$$
$$y = \frac{5+3x}{2} \quad \textit{Divide by 2.}$$

23.
$$6x - 5y = 7$$
$$-5y = 7 - 6x \quad \textit{Subtract 6x.}$$
$$y = \frac{7-6x}{-5}, \quad \textit{Divide by } -5.$$
$$\text{or } y = \frac{6x-7}{5}$$

24. Solve $k = dF - DF$ for F.
$$k = F(d-D)$$
Distributive property in reverse
$$\frac{k}{d-D} = F, \text{ or } F = \frac{k}{d-D}$$

25. Solve $Mv = mv - Vm$ for m.

$$Mv = m(v - V) \quad \textit{Distributive property in reverse}$$

$$\frac{Mv}{v - V} = m, \text{ or } m = \frac{Mv}{v - V}$$

26. Solve $A = 2HW + 2LW + 2LH$ for W.

$$A - 2LH = 2HW + 2LW \quad \textit{Get the W-terms on one side.}$$

$$A - 2LH = W(2H + 2L) \quad \textit{Distributive property in reverse}$$

$$\frac{A - 2LH}{2H + 2L} = W, \text{ or } W = \frac{A - 2LH}{2H + 2L}$$

27. Solve $d = rt$ for t.

$$t = \frac{d}{r}$$

To find t, substitute $d = 500$ and $r = 152.672$.

$$t = \frac{500}{152.672} \approx 3.275$$

His time was about 3.275 hours.

29. Use the formula $F = \frac{9}{5}C + 32$.

$$\begin{aligned} F &= \tfrac{9}{5}(45) + 32 \quad \textit{Let } C = 45. \\ &= 81 + 32 \\ &= 113 \end{aligned}$$

The corresponding temperature is 113°F.

31. Solve $P = 4s$ for s.

$$s = \frac{P}{4}$$

To find s, substitute 920 for P.

$$s = \frac{920}{4} = 230$$

The length of each side is 230 m.

33. Use the formula $C = 2\pi r$.

$$370\pi = 2\pi r \quad \textit{Let } C = 370\pi.$$

$$\frac{370\pi}{2\pi} = \frac{2\pi r}{2\pi} \quad \textit{Divide by } 2\pi.$$

$$185 = r$$

So the radius of the circle is 185 inches and the diameter is twice that length, that is, 370 inches.

35. Use $V = LWH$.
Let $V = 187$, $L = 11$, and $W = 8.5$.

$$187 = 11(8.5)H$$

$$187 = 93.5H$$

$$2 = H \quad \textit{Divide by 93.5.}$$

The ream is 2 inches thick.

37. The mixture is 36 oz and that part which is alcohol is 9 oz. Thus, the percent of alcohol is

$$\tfrac{9}{36} = \tfrac{1}{4} = \tfrac{25}{100} = 25\%.$$

The percent of water is

$$100\% - 25\% = 75\%.$$

39. Find what percent \$6900 is of \$230,000.

$$\tfrac{6900}{230{,}000} = 0.03 = 3\%$$

The agent received a 3% rate of commission.

In Exercises 41–44, use the rule of 78:

$$u = f \cdot \frac{k(k+1)}{n(n+1)}$$

41. Substitute 700 for f, 4 for k, and 36 for n.

$$\begin{aligned} u &= 700 \cdot \frac{4(4+1)}{36(36+1)} \\ &= 700 \cdot \frac{4(5)}{36(37)} \approx 10.51 \end{aligned}$$

The unearned interest is \$10.51.

43. Substitute 380.50 for f, 8 for k, and 24 for n.

$$\begin{aligned} u &= (380.50) \cdot \frac{8(8+1)}{24(24+1)} \\ &= (380.50) \cdot \frac{8(9)}{24(25)} \approx 45.66 \end{aligned}$$

The unearned interest is \$45.66.

45. **(a)** Detroit:

$$\text{Pct.} = \frac{W}{W+L} = \frac{88}{88+74} = \frac{88}{162} \approx .543$$

(b) Minnesota:

$$\text{Pct.} = \frac{W}{W+L} = \frac{79}{79+83} = \frac{79}{162} \approx .488$$

(c) Chicago:

$$\text{Pct.} = \frac{W}{W+L} = \frac{72}{72+90} = \frac{72}{162} \approx .444$$

(d) Kansas City:

$$\text{Pct.} = \frac{W}{W+L} = \frac{69}{69+93} = \frac{69}{162} \approx .426$$

47. $\dfrac{57.9 \text{ million}}{111.4 \text{ million}} \approx 0.52$

In 2006, about 52% of the U.S. households that owned at least one TV set owned at least 3 TV sets.

49. $0.34(242{,}070) = 82{,}303.80$

To the nearest dollar, \$82,304 will be spent to provide housing.

51. $\dfrac{\$41{,}000}{\$242{,}070} \approx 0.1694$

So the food cost is about 17%, which agrees with the percent shown in the graph.

53. Let $x =$ the percent increase (as a decimal).

$$\begin{aligned} \text{percent increase} &= \frac{\text{amount of increase}}{\text{base}} \\ x &= \frac{11.34 - 10.50}{10.50} \\ x &= \frac{0.84}{10.50} \\ x &= 0.08 \end{aligned}$$

The percent increase was 8%.

55. Let $x =$ the percent decrease (as a decimal).

$$\begin{aligned} \text{percent decrease} &= \frac{\text{amount of decrease}}{\text{base}} \\ x &= \frac{134{,}953 - 129{,}798}{134{,}953} \\ x &= \frac{5155}{134{,}953} \\ x &= 0.038 \end{aligned}$$

The percent decrease was 3.8%.

57. $$\begin{aligned} \text{percent decrease} &= \frac{\text{amount of decrease}}{\text{base}} \\ &= \frac{18.98 - 9.97}{18.98} \\ &= \frac{9.01}{18.98} = 0.475 \end{aligned}$$

The percent discount was 47.5%.

2.3 Applications of Linear Equations

2.3 Margin Exercises

1. **(a)** "9 added to a number" translates as

$$9 + x, \text{ or } x + 9.$$

(b) "The difference between 7 and a number" translates as

$$7 - x.$$

Note: $x - 7$ is the difference between a number and 7.

(c) "Four times a number" translates as

$$4 \cdot x \text{ or } 4x.$$

(d) "The quotient of 7 and a nonzero number" translates as

$$\frac{7}{x} \quad (x \neq 0).$$

2. **(a)**

The sum of a number and 6	is	28.
↓	↓	↓
$x + 6$	$=$	28

An equation is $x + 6 = 28$.

(b)

If twice a number	decreased	by 3,	is	the result 17.
↓	↓	↓	↓	↓
$2x$	$-$	3	$=$	17

An equation is $2x - 3 = 17$.

(c)

The product of a number and 7	is	twice the number	plus	12.
↓	↓	↓	↓	↓
$7x$	$=$	$2x$	$+$	12

An equation is $7x = 2x + 12$.

(d)

The quotient of a number and 6,	added to	twice the number,	is	7.
↓	↓	↓	↓	↓
$\frac{x}{6}$	$+$	$2x$	$=$	7

An equation is $\frac{x}{6} + 2x = 7$.

3. **(a)** $5x - 3(x + 2) = 7$ is an *equation* because it has an equals sign.

(b) $5x - 3(x + 2)$ is an *expression* because there is no equals sign.

4. *Step 2*
The length and perimeter are given in terms of the width W. The length L is 5 cm more than the width, so

$$L = W + 5.$$

The perimeter P is 5 times the width, so

$$P = 5W.$$

Step 3
Use the formula for perimeter of a rectangle.

$$\begin{aligned} P &= 2L + 2W \\ 5W &= 2(W + 5) + 2W \quad P = 5W;\ L = W + 5 \end{aligned}$$

Step 4
Solve the equation.

$$\begin{aligned} 5W &= 2W + 10 + 2W && \textit{Distributive property} \\ 5W &= 4W + 10 && \textit{Combine terms.} \\ W &= 10 && \textit{Subtract } 4W. \end{aligned}$$

Step 5
The width is 10 and the length is

$$L = W + 5 = 10 + 5 = 15.$$

The rectangle is 10 cm by 15 cm.

Step 6
15 is 5 more than 10 and $P = 2(10) + 2(15) = 50$ is five times 10, as required.

5. *Step 2*
Let $x =$ the number of RBIs for Rodriguez.
Then $x - 19 =$ the number of RBIs for Holliday.

Step 3
The sum of their RBIs is 293, so an equation is

$$x + (x - 19) = 293.$$

Step 4
Solve the equation.

$$\begin{aligned} 2x - 19 &= 293 \\ 2x &= 312 && \textit{Add 19.} \\ x &= 156 && \textit{Divide by 2.} \end{aligned}$$

Step 5
Rodriguez had 156 RBIs and Holliday had $156 - 19 = 137$ RBIs.

Step 6
137 is 19 less than 156, and the sum of 137 and 156 is 293.

6. **(a)** Let x be the store's cost, which is increased by 25% of x, or $0.25x$. Then an equation is

$$\begin{aligned} x + 0.25x &= 2375. \\ 1x + 0.25x &= 2375 && \textit{Identity property} \\ 1.25x &= 2375 && \textit{Combine terms} \\ x &= 1900 && \textit{Divide by 1.25.} \end{aligned}$$

The store's cost was \$1900.

(b) Let x be the amount she earned before deductions. Then 10% of x, or $0.10x$, is the amount of her deductions. An equation is

$$\begin{aligned} x - 0.10x &= 162. \\ 1x - 0.10x &= 162 && \textit{Identity property} \\ 0.90x &= 162 && \textit{Combine terms} \\ x &= 180 && \textit{Divide by 0.90.} \end{aligned}$$

She earned \$180 before deductions were made.

7. **(a)** Let $x =$ the amount invested at 5%.
Then $72{,}000 - x =$ the amount invested at 3%.

Use $I = prt$ with $t = 1$.
Make a table to organize the information.

Principal	Rate (as a Decimal)	Interest
x	0.05	$0.05x$
$72{,}000 - x$	0.03	$0.03(72{,}000 - x)$
72,000	← Totals →	3160

The last column gives the equation.

$$\begin{aligned} 0.05x + 0.03(72{,}000 - x) &= 3160 \\ 0.05x + 2160 - 0.03x &= 3160 && \textit{Distributive property} \\ 0.02x + 2160 &= 3160 && \textit{Combine terms.} \\ 0.02x &= 1000 && \textit{Subtract 2160.} \\ x &= 50{,}000 && \textit{Divide by 0.02.} \end{aligned}$$

The woman invested \$50,000 at 5% and \$72,000 – \$50,000 = \$22,000 at 3%.

Check 5% of \$50,000 is \$2500 and 3% of \$22,000 is \$660. The sum is \$3160, as required.

(b) Let $x =$ the amount invested at 5%.
Then $34{,}000 - x =$ the amount invested at 4%.

Use $I = prt$ with $t = 1$.
Make a table to organize the information.

Principal	Rate (as a Decimal)	Interest
x	0.05	$0.05x$
$34{,}000 - x$	0.04	$0.04(34{,}000 - x)$
34,000	← Totals →	1545

The last column gives the equation.

$$\begin{aligned} 0.05x + 0.04(34{,}000 - x) &= 1545 \\ 0.05x + 1360 - 0.04x &= 1545 && \textit{Distributive property} \\ 0.01x + 1360 &= 1545 && \textit{Combine terms.} \\ 0.01x &= 185 && \textit{Subtract 1360.} \\ x &= 18{,}500 && \textit{Divide by 0.01.} \end{aligned}$$

The man invested \$18,500 at 5% and \$34,000 – \$18,500 = \$15,500 at 4%.

Check 5% of \$18,500 is \$925 and 4% of \$15,500 is \$620. The sum is \$1545, as required.

8. **(a)** Let $x =$ the number of liters of the 10% solution. Then $x + 60 =$ the number of liters of the 15% solution.

Make a table to organize the information.

Number of Liters	Percent (as a Decimal)	Liters of Pure Solution
x	$10\% = 0.10$	$0.10x$
60	$25\% = 0.25$	$0.25(60)$
$x + 60$	$15\% = 0.15$	$0.15(x + 60)$

The last column gives the equation.

$$0.10x + 0.25(60) = 0.15(x + 60)$$

$$0.10x + 15 = 0.15x + 9 \quad \textit{Distributive property}$$

$$15 = 0.05x + 9 \quad \textit{Subtract 0.10x.}$$

$$6 = 0.05x \quad \textit{Subtract 9.}$$

$$120 = x \quad \textit{Divide by 0.05.}$$

120 L of 10% solution should be used.

Check 10% of 120 L is 12 L and 25% of 60 L is 15 L. The sum is $12 + 15 = 27$ L, which is the same as 15% of 180 L, as required.

(b) Let $x =$ the amount of \$8 per lb candy. Then $x + 100 =$ the amount of \$7 per lb candy.

Make a table to organize the information.

Number of Pounds	Price per Pound	Value
x	\$8	$8x$
100	\$4	400
$x + 100$	\$7	$7(x + 100)$

The last column gives the equation.

$$8x + 400 = 7(x + 100)$$

$$8x + 400 = 7x + 700 \quad \textit{Distributive property}$$

$$x + 400 = 700 \quad \textit{Subtract 7x.}$$

$$x = 300 \quad \textit{Subtract 400.}$$

300 lb of candy worth \$8 per lb should be used.

Check 300 lb of candy worth \$8 per lb is worth \$2400. 100 lb of candy worth \$4 per lb is worth \$400. The sum is $2400 + 400 = \$2800$, which is the same as 400 lb of candy worth \$7 per lb, as required.

9. **(a)** Let $x =$ the number of liters of pure acid.

Number of Liters	Percent (as a Decimal)	Liters of Pure Acid
x	$100\% = 1$	x
6	$30\% = 0.30$	$0.30(6)$
$x + 6$	$50\% = 0.50$	$0.50(x + 6)$

The last column gives the equation.

$$x + 0.30(6) = 0.50(x + 6)$$

$$1x + 1.8 = 0.5x + 3$$

$$0.5x + 1.8 = 3 \quad \textit{Subtract 0.5x.}$$

$$0.5x = 1.2 \quad \textit{Subtract 1.8.}$$

$$x = 2.4 \quad \textit{Divide by 0.5.}$$

2.4 L of pure acid are needed.

Check 100% of 2.4 L is 2.4 L and 30% of 6 L is 1.8 L. The sum is $2.4 + 1.8 = 4.2$ L, which is the same as 50% of $2.4 + 6 = 8.4$ L, as required.

(b) Let $x =$ the number of liters of water.

Number of Liters	Percent (as a Decimal)	Liters of Pure Antifreeze
x	$0\% = 0$	0
20	$50\% = 0.50$	$0.50(20)$
$x + 20$	$40\% = 0.40$	$0.40(x + 20)$

The last column gives the equation.

$$0 + 0.50(20) = 0.40(x + 20)$$

$$10 = 0.4x + 8$$

$$2 = 0.4x \quad \textit{Subtract 8.}$$

$$5 = x \quad \textit{Divide by 0.4.}$$

5 L of water are needed.

Check 50% of 20 L is 10 L as is 40% of $20 + 5 = 25$ L, as required.

2.3 Section Exercises

1. **(a)** 12 more than a number $\underline{x + 12}$

(b) 12 is more than a number. $\underline{12 > x}$

3. **(a)** 4 less than a number $\underline{x - 4}$

(b) 4 is less than a number. $\underline{4 < x}$

5. 20% can be written as $0.20 = 0.2 = \frac{20}{100} = \frac{2}{10} = \frac{1}{5}$, so "20% of a number" can be written as $0.20x$, $0.2x$, or $\frac{x}{5}$. We see that "20% of a number" cannot be written as $20x$, choice **D**.

7. Twice a number, increased by 18 $\underline{2x + 18}$

9. 15 decreased by four times a number $\underline{15 - 4x}$

11. The product of 10 and 6 less than a number $\underline{10(x-6)}$

13. The quotient of five times a number and 9 $\underline{\frac{5x}{9}}$

15. The sentence "the sum of a number and 6 is -31" can be translated as

$$x + 6 = -31.$$
$$x = -37 \quad \textit{Subtract 6.}$$

The number is -37.

17. The sentence "if the product of a number and -4 is subtracted from the number, the result is 9 more than the number" can be translated as

$$\begin{aligned} x-(-4x) &= x+9. \\ x+4x &= x+9 \\ 4x &= 9 \\ x &= \tfrac{9}{4} \end{aligned}$$

The number is $\frac{9}{4}$.

19. The sentence "when $\frac{2}{3}$ of a number is subtracted from 12, the result is 10" can be translated as

$$\begin{aligned} 12 - \tfrac{2}{3}x &= 10. \\ 36 - 2x &= 30 && \textit{Multiply by 3.} \\ -2x &= -6 && \textit{Subtract 36.} \\ x &= 3 && \textit{Divide by } -2. \end{aligned}$$

The number is 3.

21. $5(x+3) - 8(2x-6)$ is an *expression* because there is no equals sign.

23. $5(x+3) - 8(2x-6) = 12$ has an equals sign, so this represents an *equation.*

25. $\dfrac{t}{2} - \dfrac{t+5}{6} - 8$ is an *expression* because there is no equals sign.

27. *Step 1*
We are asked to find the number of patents each university secured.

Step 2
Let $x =$ the number of patents MIT secured. Then $x - 38 =$ the number of patents Stanford secured.

Step 3
A total of 230 patents were secured, so

$$\underline{x} + \underline{x - 38} = 230.$$

Step 4
$$\begin{aligned} 2x - 38 &= 230 \\ 2x &= 268 \\ x &= \underline{134} \end{aligned}$$

Step 5
MIT secured 134 patents and Stanford secured $134 - 38 =$ 96 patents.

Step 6
The number of Stanford patents was 38 fewer than the number of MIT patents and the total number of patents was $134 + \underline{96} = \underline{230}$.

29. *Step 2*
Let $W =$ the width of the base. Then $2W - 65$ is the length of the base.

Step 3
The perimeter of the base is 860 feet. Using $P = 2L + 2W$ gives us

$$2(2W - 65) + 2W = 860.$$

Step 4
$$\begin{aligned} 4W - 130 + 2W &= 860 \\ 6W - 130 &= 860 \\ 6W &= 990 \\ W &= \tfrac{990}{6} = 165 \end{aligned}$$

Step 5
The width of the base is 165 feet and the length of the base is $2(165) - 65 = 265$ feet.

Step 6
$2L + 2W = 2(265) + 2(165) = 530 + 330 = 860$, which is the perimeter of the base, and the length, 265 ft, is 65 ft less than twice the base, 330 ft.

31. *Step 2*
Let $x =$ the length of the middle side. Then the shortest side is $x - 75$ and the longest side is $x + 375$.

Step 3
The perimeter of the Bermuda Triangle is 3075 miles. Using $P = a + b + c$ gives us

$$x + (x - 75) + (x + 375) = 3075.$$

Step 4
$$\begin{aligned} 3x + 300 &= 3075 \\ 3x &= 2775 && \textit{Subtract 300.} \\ x &= 925 && \textit{Divide by 3.} \end{aligned}$$

Step 5
The length of the middle side is 925 miles. The length of the shortest side is $x - 75 = 925 - 75 = 850$ miles. The length of the longest side is $x + 375 = 925 + 375 = 1300$ miles.

Step 6
$925 + 850 + 1300 = 3075$ miles (the correct perimeter), the shortest side measures 75 miles less than the middle side, and the longest side measures 375 miles more than the middle side, so the answer checks.

33. *Step 2*
Let $x =$ the height of the Eiffel Tower.
Then $x - 804 =$ the height of the Leaning Tower of Pisa.

Step 3
Together these heights are 1164 ft, so

$$x + (x - 804) = 1164.$$

Step 4

$$\begin{aligned} 2x - 804 &= 1164 \\ 2x &= 1968 \\ x &= 984 \end{aligned}$$

Step 5
The height of the Eiffel Tower is 984 feet and the height of the Leaning Tower of Pisa is $984 - 804 =$ 180 feet.

Step 6
180 feet is 804 feet shorter than 984 feet and the sum of 180 feet and 984 feet is 1164 feet.

35. *Step 2*
Let $x =$ the Yankees' payroll (in millions).
Then $x - 70.4 =$ the Tigers' payroll (in millions).

Step 3
The two payrolls totaled \$347.8 million, so

$$x + (x - 70.4) = 347.8$$

Step 4

$$\begin{aligned} 2x - 70.4 &= 347.8 \\ 2x &= 418.2 \\ x &= 209.1 \end{aligned}$$

Step 5
In 2008, the Yankees' payroll was \$209.1 million and the Tigers' payroll was $209.1 - 70.4 =$ \$138.7 million.

Step 6
\$138.7 million is \$70.4 million less than \$209.1 million and the sum of \$138.7 million and \$209.1 million is \$347.8 million.

37. Let $x =$ the 2004 cost. Then

$$\begin{aligned} x + 3.1\%(x) &= 36.78. \\ x + 3.1(0.01)(x) &= 36.78 \\ 1x + 0.031x &= 36.78 \\ 1.031x &= 36.78 \\ x &= \tfrac{36.78}{1.031} \approx 35.67 \end{aligned}$$

The 2004 cost was \$35.67.

39. Let $x =$ the 2007 population.

The 2007 population was 106.6% of the 2000 population.

$$\begin{aligned} x &= (106.6\%)(237{,}230) \\ &= 1.066(237{,}230) \\ &= 252{,}887.18 \end{aligned}$$

The 2007 population was about 252,887.

41. Let $x =$ the amount of the receipts excluding tax. Since the sales tax is 9% of x, the total amount is

$$\begin{aligned} x + 0.09x &= 2725 \\ 1x + 0.09x &= 2725 \\ 1.09x &= 2725 \\ x &= \frac{2725}{1.09} = 2500 \end{aligned}$$

Thus, the tax was $0.09(2500) =$ \$225.

43. Let $x =$ the amount invested at 3%. Then $12{,}000 - x =$ the amount invested at 4%.
Complete the table. Use $I = prt$ with $t = 1$.

Principal	Rate (as a Decimal)	Interest
x	0.03	$0.03x$
$12{,}000 - x$	0.04	$0.04(12{,}000 - x)$
12,000	← Totals →	440

The last column gives the equation.

Interest at 3%	+	interest at 4%	=	total interest.
$0.03x$	+	$0.04(12{,}000 - x)$	=	440

$$\begin{aligned} 3x + 4(12{,}000 - x) &= 44{,}000 \quad \textit{Multiply by 100.} \\ 3x + 48{,}000 - 4x &= 44{,}000 \\ -x &= -4000 \\ x &= 4000 \end{aligned}$$

He should invest \$4000 at 3% and $12{,}000 - 4000 =$ \$8000 at 4%.

Check \$4000 @ 3% = \$120 and \$8000 @ 4% = \$320; \$120 + \$320 = \$440.

45. Let $x =$ the amount invested at 4.5%. Then $2x - 1000 =$ the amount invested at 3%.
Use $I = prt$ with $t = 1$. Make a table.

Principal	Rate (as a Decimal)	Interest
x	0.045	$0.045x$
$2x - 1000$	0.03	$0.03(2x - 1000)$
	Total →	1020

The last column gives the equation.

Interest at 4.5%	+	interest at 3%	=	total interest.
$0.045x$	+	$0.03(2x - 1000)$	=	1020

$$\begin{aligned} 45x + 30(2x - 1000) &= 1{,}020{,}000 \quad \textit{Multiply by 1000.} \\ 45x + 60x - 30{,}000 &= 1{,}020{,}000 \\ 105x &= 1{,}050{,}000 \\ x &= \frac{1{,}050{,}000}{105} = 10{,}000 \end{aligned}$$

She invested \$10,000 at 4.5% and $2x - 1000 = 2(10{,}000) - 1000 =$ \$19,000 at 3%.

Check \$19,000 is \$1000 less than two times \$10,000. \$10,000 @ 4.5% = \$450 and \$19,000 @ 3% = \$570; \$450 + \$570 = \$1020.

47. Let $x =$ the amount of additional money to be invested at 4%.
Use $I = prt$ with $t = 1$. Make a table.
Use the fact that the total return on the two investments is 6%.

Principal	Rate (as a decimal)	Interest
27,000	0.07	0.07(27,000)
x	0.04	$0.04x$
$27{,}000 + x$	0.06	$0.06(27{,}000 + x)$

The last column gives the equation.

$$\begin{array}{ccccc} \text{Interest at 7\%} & + & \text{interest at 4\%} & = & \text{interest at 6\%}. \\ 0.07(27{,}000) & + & 0.04x & = & 0.06(27{,}000 + x) \end{array}$$

$$\begin{aligned} 7(27{,}000) + 4x &= 6(27{,}000 + x) && \textit{Multiply by 100.} \\ 189{,}000 + 4x &= 162{,}000 + 6x \\ 27{,}000 &= 2x \\ 13{,}500 &= x \end{aligned}$$

They should invest \$13,500 at 4%.

Check \$27,000 @ 7% = \$1890 and \$13,500 @ 4% = \$540; \$1890 + \$540 = \$2430, which is the same as (\$27,000 + \$13,500) @ 6%.

49. Let $x =$ the number of liters of 10% acid solution needed. Make a table.

Liters of Solution	Percent (as a Decimal)	Liters of Pure Acid
10	0.04	$0.04(10) = 0.4$
x	0.10	$0.10x$
$x + 10$	0.06	$0.06(x + 10)$

Write the equation from the last column in the table.

$$\begin{array}{ccccc} \text{Acid in 4\%} & + & \text{acid in 10\%} & = & \text{acid in 6\%}. \\ 0.4 & + & 0.10x & = & 0.06(x + 10) \end{array}$$

$$\begin{aligned} 0.4 + 0.10x &= 0.06x + 0.6 && \textit{Distributive property} \\ 0.04x &= 0.2 && \textit{Subtract 0.06x and 0.4.} \\ x &= 5 && \textit{Divide by 0.04.} \end{aligned}$$

Five liters of the 10% solution are needed.

Check 4% of 10 is 0.4 and 10% of 5 is 0.5; $0.4 + 0.5 = 0.9$, which is the same as 6% of $(10 + 5)$.

51. Let $x =$ the number of liters of the 20% alcohol solution. Make a table.

Liters of Solution	Percent (as a Decimal)	Liters of Pure Alcohol
12	0.12	$0.12(12) = 1.44$
x	0.20	$0.20x$
$x + 12$	0.14	$0.14(x + 12)$

Write the equation from the last column in the table.

$$\begin{array}{ccccc} \text{Alcohol in 12\%} & + & \text{alcohol in 20\%} & = & \text{alcohol in 14\%}. \\ 1.44 & + & 0.20x & = & 0.14(x + 12) \end{array}$$

$$\begin{aligned} 144 + 20x &= 14(x + 12) && \textit{Multiply by 100.} \\ 144 + 20x &= 14x + 168 && \textit{Distributive property} \\ 6x &= 24 && \textit{Subtract 14x and 144.} \\ x &= 4 && \textit{Divide by 6.} \end{aligned}$$

4L of 20% alcohol solution are needed.

Check 12% of 12 is 1.44 and 20% of 4 is 0.8; $1.44 + 0.8 = 2.24$, which is the same as 14% of $(12 + 4)$.

53. Let $x =$ the amount of pure dye used (pure dye is 100% dye). Make a table.

Gallons of Solution	Percent (as a Decimal)	Gallons of Pure Dye
x	1	$1x = x$
4	0.25	$0.25(4) = 1$
$x + 4$	0.40	$0.40(x + 4)$

Write the equation from the last column in the table.

$$\begin{aligned} x + 1 &= 0.4(x + 4) \\ x + 1 &= 0.4x + 1.6 && \textit{Distributive property} \\ 0.6x &= 0.6 && \textit{Subtract 0.4x and 1.} \\ x &= 1 && \textit{Divide by 0.6.} \end{aligned}$$

One gallon of pure (100%) dye is needed.

Check 100% of 1 is 1 and 25% of 4 is 1; $1 + 1 = 2$, which is the same as 40% of $(1 + 4)$.

55. Let $x =$ the amount of \$6 per lb nuts. Make a table.

Cost per lb	Pounds of Nuts	Total Cost
\$2	50	$2(50) = 100$
\$6	x	$6x$
\$5	$x + 50$	$5(x + 50)$

The total value of the \$2 per lb nuts and the \$6 per lb nuts must equal the value of the \$5 per lb nuts.

continued

$$\begin{aligned} 100 + 6x &= 5(x + 50) \\ 100 + 6x &= 5x + 250 \\ x &= 150 \end{aligned}$$

She should use 150 lb of $6 nuts.

Check 50 pounds of the $2 per lb nuts are worth $100 and 150 pounds of the $6 per lb nuts are worth $900; $100 + $900 = $1000, which is the same as $(50 + 150)$ pounds worth $5 per lb.

57. We cannot expect the final mixture to be worth more than each of the ingredients. Answers will vary.

59. **(a)** Let $x =$ the amount invested at 5%.
$800 - x =$ the amount invested at 10%.

(b) Let $y =$ the amount of 5% acid used.
$800 - y =$ the amount of 10% acid used.

60. Organize the information in a table.

(a)

Principal	Percent (as a Decimal)	Interest
x	0.05	$0.05x$
$800 - x$	0.10	$0.10(800 - x)$
800	0.0875	$0.0875(800)$

The amount of interest earned at 5% and 10% is found in the last column of the table, $0.05x$ and $0.10(800 - x)$.

(b)

Liters of Solution	Percent (as a Decimal)	Liters of Pure Acid
y	0.05	$0.05y$
$800 - y$	0.10	$0.10(800 - y)$
800	0.0875	$0.0875(800)$

The amount of pure acid in the 5% and 10% mixtures is found in the last column of the table, $0.05y$ and $0.10(800 - y)$.

61. Refer to the tables for Exercise 60. In each case, the last column gives the equation.

(a) $0.05x + 0.10(800 - x) = 0.0875(800)$

(b) $0.05y + 0.10(800 - y) = 0.0875(800)$

62. In both cases, multiply by 10,000 to clear the decimals.

(a)
$$\begin{aligned} 0.05x + 0.10(800 - x) &= 0.0875(800) \\ 500x + 1000(800 - x) &= 875(800) \\ 500x + 800{,}000 - 1000x &= 700{,}000 \\ -500x &= -100{,}000 \\ x &= 200 \end{aligned}$$

Jack invested $200 at 5% and
$800 - x = 800 - 200 = \$600$ at 10%.

(b)
$$\begin{aligned} 0.05y + 0.10(800 - y) &= 0.0875(800) \\ 500y + 1000(800 - y) &= 875(800) \\ 500y + 800{,}000 - 1000y &= 700{,}000 \\ -500y &= -100{,}000 \\ y &= 200 \end{aligned}$$

Jill used 200 L of 5% acid solution and $800 - y = 800 - 200 = 600$ L of 10% acid solution.

63. The processes used to solve Problems A and B were virtually the same. Aside from the variables chosen, the problem information was organized in similar tables and the equations solved were the same. The amounts of money in Problem A correspond to the amounts of solution in Problem B.

2.4 Further Applications of Linear Equations

2.4 Margin Exercises

1. Let $x =$ the number of dimes.
Then $26 - x =$ the number of half-dollars.

	Number of Coins	Denom-ination	Value
Dimes	x	$0.10	$0.10x$
Halves	$26 - x$	$0.50	$0.50(26 - x)$
	26	← Totals →	8.60

Multiply the number of coins by the denominations, and add the results to get 8.60.

$$\begin{aligned} 0.10x + 0.50(26 - x) &= 8.60 \\ 1x + 5(26 - x) &= 86 \quad \textit{Multiply by 10.} \\ 1x + 130 - 5x &= 86 \\ -4x &= -44 \\ x &= 11 \end{aligned}$$

The cashier has 11 dimes and $26 - 11 = 15$ half-dollars.

Check The number of coins is $11 + 15 = 26$ and the value of the coins is $0.10(11) + $0.50(15) = $8.60, as required.

2. Let $x =$ the amount of time needed for the cars to be 420 mi apart.

Make a table. Use the formula $d = rt$, that is, find each distance by multiplying rate by time.

	Rate	Time	Distance
Northbound Car	60	x	$60x$
Southbound Car	45	x	$45x$
Total →			420

The total distance traveled is the sum of the distances traveled by each car, since they are traveling in opposite directions. This total is 420 mi.

$$
\begin{aligned}
60x + 45x &= 420 \\
105x &= 420 \\
x &= \tfrac{420}{105} = 4
\end{aligned}
$$

The cars will be 420 mi apart in 4 hr.

Check The northbound car travels $60(4) = 240$ miles and the southbound car travels $45(4) = 180$ miles for a total of 420 miles, as required.

3. Let $x =$ the time it takes Clay to catch up to Elayn. Then $x + \frac{1}{2} =$ Elayn's time.

Make a table. Use the formula $d = rt$, that is, find each distance by multiplying rate by time.

	Rate	Time	Distance
Elayn	3	$x + \frac{1}{2}$	$3(x + \frac{1}{2})$
Clay	5	x	$5x$

The distance traveled by Elayn is equal to the distance traveled by Clay.

$$
\begin{aligned}
3(x + \tfrac{1}{2}) &= 5x \\
3x + \tfrac{3}{2} &= 5x \\
6x + 3 &= 10x \quad \textit{Multiply by 2.} \\
3 &= 4x \\
\tfrac{3}{4} &= x
\end{aligned}
$$

It takes Clay $\frac{3}{4}$ hr or 45 min to catch up to Elayn.

Check Elayn travels $3(\frac{3}{4} + \frac{1}{2}) = \frac{15}{4}$ miles and Clay also travels $5(\frac{3}{4}) = \frac{15}{4}$ miles, as required.

4. Let $x =$ the measure of the second angle. Then $x + 15 =$ the measure of the first angle, and $2x + 25 =$ the measure of the third angle.

The sum of the three measures must equal 180°.

$$
\begin{aligned}
x + (x + 15) + (2x + 25) &= 180 \\
4x + 40 &= 180 \\
4x &= 140 \\
x &= 35
\end{aligned}
$$

The angles measure 35°, $35 + 15 = 50°$, and $2(35) + 25 = 95°$.

Check $35° + 50° + 95° = 180°$, as required.

2.4 Section Exercises

1. The total amount is

$$
\begin{aligned}
38(0.05) + 26(0.10) &= 1.90 + 2.60 \\
&= \$4.50.
\end{aligned}
$$

3. Use $d = rt$, or $r = \dfrac{d}{t}$. Substitute 1320 for d and 24 for t.

$$r = \frac{1320}{24} = 55$$

His rate was 55 mph.

5. Let $x =$ the number of pennies. Then x is also the number of dimes, and $44 - 2x$ is the number of quarters.

Number of Coins	Denomination	Value
x	0.01	$0.01x$
x	0.10	$0.10x$
$44 - 2x$	0.25	$0.25(44 - 2x)$
44	← Totals →	4.37

The sum of the values must equal the total value.

$$
\begin{aligned}
0.01x + 0.10x + 0.25(44 - 2x) &= 4.37 \\
x + 10x + 25(44 - 2x) &= 437 \\
&\quad \textit{Multiply by 100.} \\
x + 10x + 1100 - 50x &= 437 \\
-39x + 1100 &= 437 \\
-39x &= -663 \\
x &= 17
\end{aligned}
$$

There are 17 pennies, 17 dimes, and $44 - 2(17) = 10$ quarters.

Check The number of coins is $17 + 17 + 10 = 44$ and the value of the coins is $\$0.01(17) + \$0.10(17) + \$0.25(10) = \$4.37$, as required.

7. Let $x =$ the number of loonies. Then $37 - x$ is the number of toonies.

Number of Coins	Denomination	Value
x	1	$1x$
$37 - x$	2	$2(37 - x)$
37	← Totals →	51

The sum of the values must equal the total value.

$$
\begin{aligned}
1x + 2(37 - x) &= 51 \\
x + 74 - 2x &= 51 \\
-x + 74 &= 51 \\
23 &= x
\end{aligned}
$$

She has 23 loonies and $37 - 23 = 14$ toonies.

Check The total number of coins is 37 and the value of the coins is $\$1(23) + \$2(14) = \$51$, as required.

9. Let $x =$ the number of \$10 coins.
Then $53 - x$ is the number of \$20 coins.

Number of Coins	Denomination	Value
x	10	$10x$
$53 - x$	20	$20(53 - x)$
53	← Totals →	780

The sum of the values must equal the total value.

$$\begin{aligned} 10x + 20(53 - x) &= 780 \\ 10x + 1060 - 20x &= 780 \\ -10x &= -280 \\ x &= 28 \end{aligned}$$

He has 28 \$10 coins and $53 - 28 = 25$ \$20 coins.

Check The number of coins is $28 + 25 = 53$ and the value of the coins is
$\$10(28) + \$20(25) = \$780$, as required.

11. Let $x =$ the number of adult tickets sold. Then $2010 - x =$ the number of children and senior tickets sold.

Cost of Ticket	Number Sold	Amount Collected
\$14	x	$14x$
\$11	$2010 - x$	$11(2010 - x)$
Totals	2010	\$24,726

Write the equation from the last column of the table.

$$\begin{aligned} 14x + 11(2010 - x) &= 24{,}726 \\ 14x + 22{,}110 - 11x &= 24{,}726 \\ 3x &= 2616 \\ x &= 872 \end{aligned}$$

There were 872 adult tickets sold and $2010 - 872 = 1138$ children and senior tickets sold.

Check The amount collected was
$\$14(872) + \$11(1138)$
$= \$12{,}208 + \$12{,}518 = \$24{,}726$, as required.

13. $d = rt$, so

$$r = \frac{d}{t} = \frac{100}{12.37} \approx 8.08$$

Her rate was about 8.08 m/sec.

15. $d = rt$, so

$$r = \frac{d}{t} = \frac{400}{47.63} \approx 8.40$$

His rate was about 8.40 m/sec.

17. Let $t =$ the time until they are 110 mi apart. Make a table. Use the formula $d = rt$, that is, find each distance by multiplying rate by time.

	Rate	Time	Distance
First Steamer	22	t	$22t$
Second Steamer	22	t	$22t$
Total →			110

The total distance traveled is the sum of the distances traveled by each steamer, since they are traveling in opposite directions. This total is 110 mi.

$$\begin{aligned} 22t + 22t &= 110 \\ 44t &= 110 \\ t &= \tfrac{110}{44} = \tfrac{5}{2}, \text{ or } 2\tfrac{1}{2} \end{aligned}$$

It will take them $2\frac{1}{2}$ hr.

Check Each steamer traveled $22(2.5) = 55$ miles for a total of $2(55) = 110$ miles, as required.

19. Let $t =$ Mulder's time.
Then $t - \frac{1}{2} =$ Scully's time.

	Rate	Time	Distance
Mulder	65	t	$65t$
Scully	68	$t - \frac{1}{2}$	$68(t - \frac{1}{2})$

The distances are equal.

$$\begin{aligned} 65t &= 68(t - \tfrac{1}{2}) \\ 65t &= 68t - 34 \\ -3t &= -34 \\ t &= \tfrac{34}{3}, \text{ or } 11\tfrac{1}{3} \end{aligned}$$

Mulder's time will be $11\frac{1}{3}$ hr. Since he left at 8:30 A.M., $11\frac{1}{3}$ hr or 11 hr 20 min later is 7:50 P.M.

Check Mulder's distance was $65(\frac{34}{3}) = 736\frac{2}{3}$ miles. Scully's distance was $68(\frac{34}{3} - \frac{1}{2}) = 68(\frac{65}{6}) = 736\frac{2}{3}$, as required.

21. Let $x =$ her average speed on Sunday.
Then $x + 5 =$ her average speed on Saturday.

	Rate	Time	Distance
Saturday	$x + 5$	3.6	$3.6(x + 5)$
Sunday	x	4	$4x$

The distances are equal.

$$\begin{aligned} 3.6(x + 5) &= 4x \\ 3.6x + 18 &= 4x \\ 18 &= 0.4x \qquad \textit{Subtract 3.6x.} \\ x &= \tfrac{18}{0.4} = 45 \end{aligned}$$

Her average speed on Sunday was 45 mph.

Check On Sunday, 4 hours @ 45 mph = 180 miles. On Saturday, 3.6 hours @ 50 mph = 180 miles. The distances are equal.

23. Let $x =$ Anne's time.
Then $x + \frac{1}{2} =$ Johnny's time.

	Rate	Time	Distance
Anne	60	x	$60x$
Johnny	50	$x + \frac{1}{2}$	$50(x + \frac{1}{2})$

The total distance is 80.

$$\begin{aligned} 60x + 50(x + \tfrac{1}{2}) &= 80 \\ 60x + 50x + 25 &= 80 \\ 110x &= 55 \\ x &= \tfrac{55}{110} = \tfrac{1}{2} \end{aligned}$$

They will meet $\frac{1}{2}$ hr after Anne leaves.

Check Anne travels $60(\frac{1}{2}) = 30$ miles. Johnny travels $50(\frac{1}{2} + \frac{1}{2}) = 50$ miles. The sum of the distances is 80 miles, as required.

25. The sum of the measures of the three angles of a triangle is 180°.

$$\begin{aligned} (x - 30) + (2x - 120) + (\tfrac{1}{2}x + 15) &= 180 \\ \tfrac{7}{2}x - 135 &= 180 \\ 7x - 270 &= 360 \end{aligned}$$

Multiply by 2.

$$\begin{aligned} 7x &= 630 \\ x &= 90 \end{aligned}$$

With $x = 90$, the three angle measures become

$$\begin{aligned} (90 - 30)^\circ &= 60^\circ, \\ (2 \cdot 90 - 120)^\circ &= 60^\circ, \\ \text{and} \quad (\tfrac{1}{2} \cdot 90 + 15)^\circ &= 60^\circ. \end{aligned}$$

Check $60^\circ + 60^\circ + 60^\circ = 180^\circ$, as required.

27. The sum of the measures of the three angles of a triangle is 180°.

$$\begin{aligned} (3x + 7) + (9x - 4) + (4x + 1) &= 180 \\ 16x + 4 &= 180 \\ 16x &= 176 \\ x &= 11 \end{aligned}$$

With $x = 11$, the three angle measures become

$$\begin{aligned} (3 \cdot 11 + 7)^\circ &= 40^\circ, \\ (9 \cdot 11 - 4)^\circ &= 95^\circ, \\ \text{and} \quad (4 \cdot 11 + 1)^\circ &= 45^\circ. \end{aligned}$$

Check $40^\circ + 95^\circ + 45^\circ = 180^\circ$, as required.

29. The sum of the measures of the angles of a triangle is 180°.

$$\begin{aligned} x + 2x + 60 &= 180 \\ 3x + 60 &= 180 \\ 3x &= 120 \\ x &= 40 \end{aligned}$$

The measures of the unknown angles are 40° and $2x = 80^\circ$.

30. The sum of the measures of the marked angles, $60^\circ + y^\circ$, must equal 180°. Thus, the measure of the unknown angle is 120°.

31. The sum of the measures of the unknown angles in Exercise 29 is $40^\circ + 80^\circ = 120^\circ$. This is equal to the measure of the angle in Exercise 30.

32. The sum of the measures of angles ① and ② is equal to the measure of angle ③.

33. Vertical angles have equal measure.

$$\begin{aligned} 8x + 2 &= 7x + 17 \\ x &= 15 \end{aligned}$$

$8 \cdot 15 + 2 = 122$ and $7 \cdot 15 + 17 = 122$.

The angles are both 122°.

35. The sum of the two angles is 90°.

$$\begin{aligned} (5x - 1) + 2x &= 90 \\ 7x - 1 &= 90 \\ 7x &= 91 \\ x &= 13 \end{aligned}$$

The angles are $(5 \cdot 13 - 1)^\circ = 64^\circ$ and $(2 \cdot 13)^\circ = 26^\circ$.

37. Let $x =$ the first consecutive integer. Then $x + 1$ will be the second consecutive integer, and $x + 2$ will be the third consecutive integer.

The sum of the first and twice the second is 22 more than twice the third.

$$\begin{aligned} x + 2(x + 1) &= 2(x + 2) + 22 \\ x + 2x + 2 &= 2x + 4 + 22 \\ 3x + 2 &= 2x + 26 \\ x &= 24 \end{aligned}$$

Since $x = 24$, $x + 1 = 25$, and $x + 2 = 26$. The three consecutive integers are 24, 25, and 26.

39. Let $x=$ the current age. Then $x+1$ will be the age next year. The sum of these ages will be 95 years.

$$\begin{aligned} x+(x+1)&=95\\ 2x+1&=95\\ 2x&=94\\ x&=47 \end{aligned}$$

If my current age is 47, in 10 years I will be

$$47+10=57 \text{ years old.}$$

Summary Exercises on Solving Applied Problems

1. Let $x=$ the width of the rectangle. Then $x+3$ is the length of the rectangle.

If the length were decreased by 2 inches and the width were increased by 1 inch, the perimeter would be 24 inches. Use the formula $P=2L+2W$, and substitute 24 for P, $(x+3)-2$ or $x+1$ for L, and $x+1$ for W.

$$\begin{aligned} P&=2L+2W\\ 24&=2(x+1)+2(x+1)\\ 24&=2x+2+2x+2\\ 24&=4x+4\\ 20&=4x\\ 5&=x \end{aligned}$$

The width of the rectangle is 5 inches, and the length is $5+3=8$ inches.

3. Let $x=$ the regular price of the item. The sale price after a 37% (or 0.37) discount was \$35.87, so an equation is

$$\begin{aligned} x-0.37x&=35.87.\\ 0.63x&=35.87\\ x&\approx 56.94 \end{aligned}$$

To the nearest cent, the regular price was \$56.94.

5. Let $x=$ the amount invested at 4%. Then $2x$ is the amount invested at 5%. Use $I=prt$ with $t=1$ yr. Make a table.

Principal	Rate (as a Decimal)	Interest
x	0.04	$0.04x$
$2x$	0.05	$0.05(2x)=0.10x$
	Total →	77

The last column gives the equation.

$$\begin{array}{ccccc} \text{Interest at 4\%} & + & \text{interest at 5\%} & = & \text{total interest.}\\ 0.04x & + & 0.10x & = & 77 \end{array}$$

$$\begin{aligned} 4x+10x&=7700 \quad \textit{Multiply by 100.}\\ 14x&=7700\\ x&=550 \end{aligned}$$

\$550 is invested at 4% and $2(\$550)=\$1100$ is invested at 5%.

Check \$550 @ 4% = \$22 and \$1100 @ 5% = \$55; \$22 + \$55 = \$77

7. Let $x=$ the number of points he scored in 2005–2006. Then $x-402=$ the number of points he scored in 2006–2007. The total number of points he scored was 5262.

$$\begin{aligned} x+(x-402)&=5262\\ 2x-402&=5262\\ 2x&=5664\\ x&=2832 \end{aligned}$$

He scored in 2832 points in 2005–2006 and $2832-402=2430$ points 2006–2007.

9. Let $x=$ the side length of the square cut out of each corner. Then the width is $12-2x$ and the length is $16-2x$. We want the length to be 5 cm less than twice the width.

$$\begin{aligned} \text{length}&=2(\text{width})-5\\ 16-2x&=2(12-2x)-5\\ 16-2x&=24-4x-5\\ 16-2x&=19-4x\\ 2x&=3\\ x&=\tfrac{3}{2}, \text{ or } 1\tfrac{1}{2} \end{aligned}$$

The square should be $1\frac{1}{2}$ cm on each side.

Check The width is $12-2(\frac{3}{2})=9$ and the length is $16-2(\frac{3}{2})=13$. Two times the width is $2(9)=18$, which is 5 more than the length, 13.

11. Let $x=$ the number of liters of the 5% drug solution.

Liters of Solution	Percent (as a decimal)	Liters of Pure Drug
20	0.10	$20(0.10)=2$
x	0.05	$0.05x$
$20+x$	0.08	$0.08(20+x)$

$$\begin{array}{ccccc} \text{Drug in 10\%} & + & \text{drug in 5\%} & = & \text{drug in 8\%.}\\ 2 & + & 0.05x & = & 0.08(20+x) \end{array}$$

$$\begin{aligned} 200+5x&=8(20+x) \quad \textit{Multiply by 100.}\\ 200+5x&=160+8x\\ 40&=3x\\ x&=\tfrac{40}{3}, \text{ or } 13\tfrac{1}{3} \end{aligned}$$

The pharmacist should add $13\frac{1}{3}$ L.

Check 10% of 20 is 2 and 5% of $\frac{40}{3}$ is $\frac{2}{3}$; $2+\frac{2}{3}=\frac{8}{3}$, which is the same as 8% of $(20+\frac{40}{3})$.

13. Let $x =$ the number of \$5 bills. Then $126 - x$ is the number of \$10 bills.

Number of Bills	Denomination	Value
x	5	$5x$
$126 - x$	10	$10(126 - x)$
126	← Totals →	840

The sum of the values must equal the total value.

$$\begin{aligned} 5x + 10(126 - x) &= 840 \\ 5x + 1260 - 10x &= 840 \\ -5x &= -420 \\ x &= 84 \end{aligned}$$

There are 84 \$5 bills and $126 - 84 = 42$ \$10 bills.

Check The value of the bills is \$5(84) + \$10(42) = \$840, as required.

15. Let $x =$ the least integer. Then $x + 1$ is the middle integer and $x + 2$ is the greatest integer.

"The sum of the least and greatest of three consecutive integers is 45 more than the middle integer" translates to

$$\begin{aligned} x + (x + 2) &= 45 + (x + 1). \\ 2x + 2 &= x + 46 \\ x &= 44 \end{aligned}$$

The three consecutive integers are 44, 45, and 46.

Check The sum of the least and greatest integers is $44 + 46 = 90$, which is the same as 45 more than the middle integer.

17. The sum of the measures of the three angles of a triangle is 180°.

$$\begin{aligned} x + (6x - 50) + (x - 10) &= 180 \\ 8x - 60 &= 180 \\ 8x &= 240 \\ x &= 30 \end{aligned}$$

With $x = 30$, the three angle measures become

$$(6 \cdot 30 - 50)^\circ = 130^\circ, \quad (30 - 10)^\circ = 20^\circ, \text{ and } 30^\circ.$$

Chapter 2 Review Exercises

1.
$$\begin{aligned} -(8 + 3x) + 5 &= 2x + 6 \\ -8 - 3x + 5 &= 2x + 6 \\ -3x - 3 &= 2x + 6 \\ -5x &= 9 \\ x &= -\tfrac{9}{5} \end{aligned}$$

Solution set: $\{-\frac{9}{5}\}$

2.
$$\begin{aligned} -(r + 5) - (2 + 7r) + 8r &= 3r - 8 \\ -r - 5 - 2 - 7r + 8r &= 3r - 8 \\ -7 &= 3r - 8 \\ 1 &= 3r \\ \tfrac{1}{3} &= r \end{aligned}$$

Solution set: $\{\frac{1}{3}\}$

3.
$$\frac{m - 2}{4} + \frac{m + 2}{2} = 8$$

Multiply each side by the LCD, 4.

$$\begin{aligned} 4\left(\frac{m - 2}{4} + \frac{m + 2}{2}\right) &= 4(8) \\ (m - 2) + 2(m + 2) &= 32 \\ m - 2 + 2m + 4 &= 32 \\ 3m + 2 &= 32 \\ 3m &= 30 \\ m &= 10 \end{aligned}$$

Solution set: $\{10\}$

4.
$$\begin{aligned} \frac{2q + 1}{3} - \frac{q - 1}{4} &= 0 \\ 4(2q + 1) - 3(q - 1) &= 0 \quad \textit{Multiply by 12.} \\ 8q + 4 - 3q + 3 &= 0 \\ 5q + 7 &= 0 \\ 5q &= -7 \\ q &= -\tfrac{7}{5} \end{aligned}$$

Solution set: $\{-\frac{7}{5}\}$

5.
$$\begin{aligned} 5(2x - 3) &= 6(x - 1) + 4x \\ 10x - 15 &= 6x - 6 + 4x \\ 10x - 15 &= 10x - 6 \\ -15 &= -6 \quad \textit{False} \end{aligned}$$

This is a false statement, so the equation is a *contradiction.*

Solution set: $\emptyset$

6.
$$\begin{aligned} -3x + 2(4x + 5) &= 10 \\ -3x + 8x + 10 &= 10 \\ 5x + 10 &= 10 \\ 5x &= 0 \\ x &= 0 \quad \textit{Divide by 5.} \end{aligned}$$

Solution set: $\{0\}$

7.
$$\begin{aligned} \frac{1}{2}x - \frac{3}{8}x &= \frac{1}{4}x + 2 \\ 4x - 3x &= 2x + 16 \quad \textit{Multiply by 8.} \\ -x &= 16 \quad \textit{4 - 3 - 2 = -1} \\ x &= -16 \end{aligned}$$

Solution set: $\{-16\}$

8. $0.05x + 0.03(1200 - x) = 42$

Multiply by 100 to clear all decimals.

$$\begin{aligned} 5x + 3(1200 - x) &= 4200 \\ 5x + 3600 - 3x &= 4200 \\ 2x + 3600 &= 4200 \\ 2x &= 600 \\ x &= 300 \end{aligned}$$

Solution set: $\{300\}$

9. Solve each equation.

A. $x - 7 = 7$

$x = 14$ *Add 7.*

Solution set: $\{14\}$

B. $9x = 10x$

$0 = x$ *Subtract 9x.*

Solution set: $\{0\}$

C. $x + 4 = -4$

$x = -8$ *Subtract 4.*

Solution set: $\{-8\}$

D. $8x - 8 = 8$

$8x = 16$ *Add 8.*

$x = 2$ *Divide by 8.*

Solution set: $\{2\}$

Equation **B** has $\{0\}$ as its solution set.

10. Solve $-2x + 5 = 7$.

Begin by subtracting 5 from each side. Then divide each side by -2.

11.

$$\begin{aligned} 7r - 3(2r - 5) + 5 + 3r &= 4r + 20 \\ 7r - 6r + 15 + 5 + 3r &= 4r + 20 \\ 4r + 20 &= 4r + 20 \\ 20 &= 20 \quad \textit{True} \end{aligned}$$

This equation is an *identity*.

Solution set: $\{\text{all real numbers}\}$

12.

$$\begin{aligned} 8p - 4p - (p - 7) + 9p + 13 &= 12p \\ 8p - 4p - p + 7 + 9p + 13 &= 12p \\ 12p + 20 &= 12p \\ 20 &= 0 \quad \textit{False} \end{aligned}$$

This equation is a *contradiction*.

Solution set: $\emptyset$

13.

$$\begin{aligned} -2r + 6(r - 1) + 3r - (4 - r) &= -(r + 5) - 5 \\ -2r + 6r - 6 + 3r - 4 + r &= -r - 5 - 5 \\ 8r - 10 &= -r - 10 \\ 9r &= 0 \\ r &= 0 \end{aligned}$$

This equation is a *conditional* equation.

Solution set: $\{0\}$

14. Solve $V = LWH$ for L.

$$\frac{V}{WH} = \frac{LWH}{WH} \qquad \textit{Divide by WH.}$$

$$\frac{V}{WH} = L, \text{ or } L = \frac{V}{WH}$$

15. Solve $A = \frac{1}{2}h(b + B)$ for b.

$$\begin{aligned} 2A &= h(b + B) && \textit{Multiply by 2.} \\ \frac{2A}{h} &= b + B && \textit{Divide by h.} \\ \frac{2A}{h} - B &= b && \textit{Subtract B.} \end{aligned}$$

OR Solve $A = \frac{1}{2}h(b + B)$ for b.

$$\begin{aligned} 2A &= hb + hB && \textit{Multiply by 2.} \\ 2A - hB &= hb && \textit{Subtract hB.} \\ \frac{2A - hB}{h} &= b && \textit{Divide by h.} \end{aligned}$$

16. Solve $4x + 7y = 9$ for y.

$$\begin{aligned} 7y &= 9 - 4x && \textit{Subtract 4x.} \\ y &= \frac{9 - 4x}{7} && \textit{Divide by 7.} \end{aligned}$$

17. Use the formula $V = LWH$ and substitute 180 for V, 9 for L, and 4 for W.

$$\begin{aligned} 180 &= 9(4)H \\ 180 &= 36H \\ 5 &= H \end{aligned}$$

The height is 5 feet.

18.

$$\begin{aligned} \text{percent increase} &= \frac{\text{amount of increase}}{\text{base}} \\ &= \frac{17.5\text{ M} - 15.3\text{ M}}{15.3\text{ M}} \\ &= \frac{2.2\text{ M}}{15.3\text{ M}} \approx 0.144 \end{aligned}$$

The percent increase was 14.4%.

19. Use the formula $I = prt$, and solve for r.

$$\frac{I}{pt} = \frac{prt}{pt}$$
$$\frac{I}{pt} = r$$

Substitute 30,000 for p, 6600 for I, and 4 for t.

$$r = \frac{6600}{30{,}000(4)} = \frac{6600}{120{,}000} = 0.055$$

The rate is 5.5%.

20. Use the formula $C = \frac{5}{9}(F - 32)$ and substitute 77 for F.

$$C = \tfrac{5}{9}(77 - 32)$$
$$= \tfrac{5}{9}(45) = 25$$

The Celsius temperature is 25°.

21. **(a)** The amount of money spent on Social Security in 2005 was about

$$0.21(\$2500 \text{ billion}) = \$525 \text{ billion.}$$

(b) The amount of money spent on education and social services in 2005 was about

$$0.039(\$2500 \text{ billion}) = \$97.5 \text{ billion}$$

22.
$$C = 2\pi r$$
$$200\pi = 2\pi r \quad \textit{Substitute } 200\pi \textit{ for C.}$$
$$\frac{200\pi}{2\pi} = \frac{2\pi r}{2\pi} \quad \textit{Divide by } 2\pi.$$
$$100 = r$$

The radius is 100 mm.

23. "One-fifth of a number, subtracted from 14" is written

$$14 - \tfrac{1}{5}x.$$

24. "The product of 6 and a number, divided by 3 more than the number" is written

$$\frac{6x}{x+3}.$$

25. Let $x =$ the width of the rectangle.
Then $2x - 3 =$ the length of the rectangle.

Use the formula $P = 2L + 2W$ with $P = 42$.

$$42 = 2(2x - 3) + 2x$$
$$42 = 4x - 6 + 2x$$
$$48 = 6x$$
$$8 = x$$

The width is 8 meters and the length is $2(8) - 3 = 13$ meters.

26. Let $x =$ the length of each equal side. Then $2x - 15 =$ the length of the third side.

Use the formula $P = a + b + c$ with $P = 53$.

$$53 = x + x + (2x - 15)$$
$$53 = 4x - 15$$
$$68 = 4x$$
$$17 = x$$

The lengths of the three sides are 17 inches, 17 inches, and $2(17) - 15 = 19$ inches.

27. Let $x =$ the number of kilograms of peanut clusters. Then $3x$ is the number of kilograms of chocolate creams. The clerk has a total of 48 kg.

$$x + 3x = 48$$
$$4x = 48$$
$$x = 12$$

The clerk has 12 kilograms of peanut clusters.

28. Let $x =$ the number of liters of the 20% solution. Make a table.

Liters of Solution	Percent (as a decimal)	Liters of Pure Chemical
x	0.20	$0.20x$
15	0.50	$0.50(15) = 7.5$
$x + 15$	0.30	$0.30(x + 15)$

The last column gives the equation.

$$0.20x + 7.5 = 0.30(x + 15)$$
$$0.20x + 7.5 = 0.30x + 4.5$$
$$3 = 0.10x$$
$$30 = x$$

30 L of the 20% solution should be used.

29. Let $x =$ the number of liters of water.

Liters of Solution	Percent (as a decimal)	Liters of Pure Acid
30	0.40	$0.40(30) = 12$
x	0	$0(x) = 0$
$30 + x$	0.30	$0.30(30 + x)$

The last column gives the equation.

$$12 + 0 = 0.30(30 + x)$$
$$12 = 9 + 0.3x$$
$$3 = 0.3x$$
$$10 = x$$

10 L of water should be added.

30. Let $x =$ the amount invested at 6%. Then $x - 4000 =$ the amount invested at 4%.

Principal	Rate (as a decimal)	Interest
x	0.06	$0.06x$
$x - 4000$	0.04	$0.04(x - 4000)$
	Total →	$840

The last column gives the equation.

$$\begin{aligned} 0.06x + 0.04(x - 4000) &= 840 \\ 6x + 4(x - 4000) &= 84{,}000 \quad \textit{Multiply by 100.} \\ 6x + 4x - 16{,}000 &= 84{,}000 \\ 10x &= 100{,}000 \\ x &= 10{,}000 \end{aligned}$$

Anna should invest $10,000 at 6% and $10,000 – $4000 = $6000 at 4%.

31. Use the formula $d = rt$ or $r = \frac{d}{t}$. Here, d is about 400 mi and t is about 8 hr. Since $\frac{400}{8} = 50$, the best estimate is choice **A**.

32. Use the formula $d = rt$.

(a) Here, $r = 53$ mph and $t = 10$ hr.

$$d = 53(10) = 530$$

The distance is 530 miles.

(b) Here, $r = 164$ mph and $t = 2$ hr.

$$d = 164(2) = 328$$

The distance is 328 miles.

33. Let $x =$ the time it takes for the trains to be 297 mi apart.

Make a table. Use the formula $d = rt$.

	Rate	Time	Distance
Passenger Train	60	x	$60x$
Freight Train	75	x	$75x$
Total →			297

The total distance traveled is the sum of the distances traveled by each train.

$$\begin{aligned} 60x + 75x &= 297 \\ 135x &= 297 \\ x &= 2.2 \end{aligned}$$

It will take the trains 2.2 hours before they are 297 miles apart.

Check $2.2(60) + 2.2(75) = 297$

34. Let $x =$ the speed of the faster car and $x - 15 =$ the speed of the slower car. Make a table. Use the formula $d = rt$.

	Rate	Time	Distance
Faster Car	x	2	$2x$
Slower Car	$x - 15$	2	$2(x - 15)$
Total →			230

The total distance traveled is the sum of the distances traveled by each car.

$$\begin{aligned} 2x + 2(x - 15) &= 230 \\ 2x + 2x - 30 &= 230 \\ 4x &= 260 \\ x &= 65 \end{aligned}$$

The faster car travels at 65 km per hr, while the slower car travels at 65 – 15 = 50 km per hr.

Check $2(65) + 2(50) = 230$

35. Let $x =$ amount of time spent averaging 45 miles per hour. Then $4 - x =$ amount of time at 50 mph.

	Rate	Time	Distance
First Part	45	x	$45x$
Second Part	50	$4 - x$	$50(4 - x)$
Total →			195

From the last column:

$$\begin{aligned} 45x + 50(4 - x) &= 195 \\ 45x + 200 - 50x &= 195 \\ -5x &= -5 \\ x &= 1 \end{aligned}$$

The automobile averaged 45 mph for 1 hour.

Check 45 mph for 1 hour = 45 miles and 50 mph for 3 hours = 150 miles; 45 + 150 = 195.

36. Let $x =$ the average speed for the first hour. Then $x - 7 =$ the average speed for the second hour. Using $d = rt$, the distance traveled for the first hour is $x(1)$ miles, for the second hour is $(x - 7)(1)$ miles, and for the whole trip, 85 miles.

$$\begin{aligned} x + (x - 7) &= 85 \\ 2x - 7 &= 85 \\ 2x &= 92 \\ x &= 46 \end{aligned}$$

The average speed for the first hour was 46 mph.

Check 46 mph for 1 hour = 46 miles and 46 – 7 = 39 mph for 1 hour = 39 miles; 46 + 39 = 85.

37. **[2.1]**
$$\begin{aligned} (7 - 2k) + 3(5 - 3k) &= k + 8 \\ 7 - 2k + 15 - 9k &= k + 8 \\ -11k + 22 &= k + 8 \\ -12k + 22 &= 8 \\ -12k &= -14 \\ k &= \tfrac{-14}{-12} = \tfrac{7}{6} \end{aligned}$$

Solution set: $\{\frac{7}{6}\}$

38. **[2.1]** $\frac{4x+2}{4}+\frac{3x-1}{8}=\frac{x+6}{16}$

Clear fractions by multiplying by the LCD, 16.

$$\begin{aligned}4(4x+2)+2(3x-1)&=x+6\\16x+8+6x-2&=x+6\\22x+6&=x+6\\21x&=0\\x&=0\end{aligned}$$

Solution set: $\{0\}$

39. **[2.1]**
$$\begin{aligned}-5(6p+4)-2p&=-32p+14\\-30p-20-2p&=-32p+14\\-32p-20&=-32p+14\\-20&=14 \quad \textit{False}\end{aligned}$$

The equation is a *contradiction.*

Solution set: $\emptyset$

40. **[2.1]**
$$\begin{aligned}0.08x+0.04(x+200)&=188\\8x+4(x+200)&=18{,}800 \quad \textit{Multiply by 100.}\\8x+4x+800&=18{,}800\\12x+800&=18{,}800\\12x&=18{,}000\\x&=1500\end{aligned}$$

Solution set: $\{1500\}$

41. **[2.1]**
$$\begin{aligned}5(2r-3)+7(2-r)&=3(r+2)-7\\10r-15+14-7r&=3r+6-7\\3r-1&=3r-1\\3r&=3r\\0&=0 \quad \textit{True}\end{aligned}$$

Solution set: {all real numbers}

42. **[2.2]** $Ax+By=C$ for x
$$\begin{aligned}Ax&=C-By \quad \textit{Subtract By.}\\x&=\frac{C-By}{A} \quad \textit{Divide by A.}\end{aligned}$$

43. **[2.3]** Let $x=$ the length of each side of the original square;
$x+4=$ the length of each side of the enlarged square.

The original perimeter is $4x$. The perimeter of the enlarged square is $4(x+4)$. The perimeter of the enlarged square is 8 in. less than twice the perimeter of the original square.

$$\begin{aligned}4(x+4)&=2(4x)-8\\4x+16&=8x-8\\16&=4x-8\\24&=4x\\6&=x\end{aligned}$$

The length of a side of the original square is 6 in.

44. **[2.4]** Let $x=$ the time traveled by eastbound car. Then $x-1=$ the time traveled by westbound car.

	Rate	Time	Distance
Eastbound Car	40	x	$40x$
Westbound Car	60	$x-1$	$60(x-1)$

Their total distance is 240 mi.

$$\begin{aligned}40x+60(x-1)&=240\\40x+60x-60&=240\\100x-60&=240\\100x&=300\\x&=3\end{aligned}$$

The eastbound car traveled for 3 hr and the westbound car traveled for $3-1=2$ hr.

45. **[2.3]** *Step 2*
Let $x=$ the number of visits to the Golden Gate National Recreation Area (in millions).
Then $x+5.46=$ the number of visits to the Blue Ridge Parkway (in millions).

Step 3
The total number of visits was 32.44 million, so

$$x+(x+5.46)=32.44$$

Step 4
$$\begin{aligned}2x+5.46&=32.44\\2x&=26.98\\x&=13.49\end{aligned}$$

Step 5
In 2006, there were 13.49 million visits to the Golden Gate National Recreation Area and $13.49+5.46=18.95$ million visits to the Blue Ridge Parkway.

Step 6
18.95 million is 5.46 million more than 13.49 million and the sum of 13.49 million and 18.95 million is 32.44 million.

46. **[2.3]** Let $x=$ the amount invested at 3%.
Then $x+600=$ the amount invested at 5%.

Principal	Rate (as a Decimal)	Interest
x	0.03	$0.03x$
$x+600$	0.05	$0.05(x+600)$

The total interest is \$126.

$$\begin{aligned}0.03x+0.05(x+600)&=126\\0.03x+0.05x+30&=126\\0.08x+30&=126\\0.08x&=96\\x&=1200\end{aligned}$$

\$1200 was invested at 3% and $1200+600=$ \$1800 was invested at 5%.

Check 5% of \$1800 is \$90 and 3% of \$1200 is \$36. The sum is \$126, as required.

Chapter 2 Test

1.
$$\begin{aligned} 3(2x-2)-4(x+6) &= 4x+8 \\ 6x-6-4x-24 &= 4x+8 \\ 2x-30 &= 4x+8 \\ -2x-30 &= 8 \\ -2x &= 38 \\ x &= -19 \end{aligned}$$

Check $x=-19$: $-120+52=-68$ *True*

Solution set: $\{-19\}$

2.
$$\begin{aligned} 0.08x+0.06(x+9) &= 1.24 \\ 8x+6(x+9) &= 124 \end{aligned}$$

Multiply each side by 100 to eliminate the decimals.

$$\begin{aligned} 8x+6x+54 &= 124 \\ 14x+54 &= 124 \\ 14x &= 70 \\ x &= 5 \end{aligned}$$

Check $x=5$: $0.40+0.84=1.24$ *True*

Solution set: $\{5\}$

3.
$$\frac{x+6}{10}+\frac{x-4}{15}=1$$

Multiply each side by the LCD, 30.

$$\begin{aligned} 3(x+6)+2(x-4) &= 30 \\ 3x+18+2x-8 &= 30 \\ 5x+10 &= 30 \\ 5x &= 20 \\ x &= 4 \end{aligned}$$

Check $x=4$: $1+0=1$ *True*

Solution set: $\{4\}$

4. **(a)**
$$\begin{aligned} 3x-(2-x)+4x+2 &= 8x+3 \\ 3x-2+x+4x+2 &= 8x+3 \\ 8x &= 8x+3 \\ 0 &= 3 \quad \textit{False} \end{aligned}$$

The false statement indicates that the equation is a *contradiction.*

Solution set: $\emptyset$

(b)
$$\frac{x}{3}+7=\frac{5x}{6}-2-\frac{x}{2}+9$$

Multiply each side by the LCD, 6.

$$\begin{aligned} 2x+42 &= 5x-12-3x+54 \\ 2x+42 &= 2x+42 \\ 0 &= 0 \quad \textit{True} \end{aligned}$$

The true statement indicates that the equation is an *identity.*

Solution set: {all real numbers}

(c)
$$\begin{aligned} -4(2x-6) &= 5x+24-7x \\ -8x+24 &= -2x+24 \\ 24 &= 6x+24 \\ 0 &= 6x \\ 0 &= x \end{aligned}$$

This is a *conditional equation.*

Check $x=0$: $24=0+24-0$ *True*

Solution set: $\{0\}$

5. Solve $S=-16t^2+vt$ for v.

$$\begin{aligned} S+16t^2 &= vt && \textit{Add } 16t^2. \\ \frac{S+16t^2}{t} &= v, && \textit{Divide by } t. \end{aligned}$$

or $v=\dfrac{S+16t^2}{t}$

6. Solve $-3x+2y=6$ for y.

$$\begin{aligned} 2y &= 6+3x && \textit{Add } 3x. \\ y &= \frac{6+3x}{2} && \textit{Divide by 2.} \end{aligned}$$

7. Solve $d=rt$ for t and substitute 500 for d and 149.335 for r.

$$t=\frac{d}{r}=\frac{500}{149.335}\approx 3.348$$

Harvik's time was about 3.348 hr.

8. Use $I=Prt$ and substitute \$1733.75 for I, \$36,500 for P, and 1 for t.

$$\begin{aligned} 1733.75 &= 36{,}500r(1) \\ r &= \frac{1733.75}{36{,}500}=0.0475 \end{aligned}$$

The rate of interest is 4.75%.

9. $\dfrac{27{,}318}{36{,}826}\approx 0.742$

About 74.2% were classified as post offices.

10. Let $x=$ the amount invested at 3%.
Then $32{,}000-x=$ the amount invested at 5%.

Principal	Rate (as a Decimal)	Interest
x	0.03	$0.03x$
$32{,}000-x$	0.05	$0.05(32{,}000-x)$
\$32,000	← Totals →	\$1320

We can write an equation from the last column.

$$
\begin{aligned}
0.03x + 0.05(32{,}000 - x) &= 1320 \\
3x + 5(32{,}000 - x) &= 132{,}000 \quad \textit{Multiply each side by 100.} \\
3x + 160{,}000 - 5x &= 132{,}000 \\
-2x &= -28{,}000 \\
x &= 14{,}000
\end{aligned}
$$

He invested \$14,000 at 3% and \$32,000 – \$14,000 = \$18,000 at 5%.

11. Let x = the speed of the faster car. Then $x - 15$ = the speed of the slower car.

Make a table. Use the formula $d = rt$.

	Rate	Time	Distance
Slower Car	$x - 15$	6	$6(x - 15)$
Faster Car	x	6	$6x$
Total →			630

The total distance traveled is the sum of the distances traveled by each car.

$$
\begin{aligned}
6(x - 15) + 6x &= 630 \\
6x - 90 + 6x &= 630 \\
12x &= 720 \\
x &= 60
\end{aligned}
$$

The faster car traveled at 60 mph, while the slower car traveled at $60 - 15 = 45$ mph.

12. The sum of the three angle measures is 180°.

$$
\begin{aligned}
(2x + 20) + x + x &= 180 \\
4x + 20 &= 180 \\
4x &= 160 \\
x &= 40
\end{aligned}
$$

The three angle measures are 40°, 40°, and $(2 \cdot 40 + 20)° = 100°$.

13.
$$
\begin{aligned}
A &= \frac{24f}{b(p+1)} \\
A &= \frac{24(200)}{1920(24+1)} \quad \textit{Let } f = 200,\ b = 1920, \textit{ and } p = 24. \\
&= \frac{4800}{48{,}000} \\
&= 0.1
\end{aligned}
$$

The approximate annual interest rate is 10%.

14.
$$
\begin{aligned}
A &= \frac{24f}{b(p+1)} \\
A &= \frac{24(740)}{3600(36+1)} \quad \textit{Let } f = 740,\ b = 3600, \textit{ and } p = 36. \\
&= \frac{17{,}760}{133{,}200} \\
&\approx 0.1333
\end{aligned}
$$

The approximate annual interest rate is 13.33%.

15. 21% of $5000 = 0.21(5000) = 1050$

We would expect 1050 white-collar workers in a group of 5000 stockholders.

Cumulative Review Exercises (Chapters 1–2)

Exercises 1–6 refer to set A.
Let $A = \{-8, -\frac{2}{3}, -\sqrt{6}, 0, \frac{4}{5}, 9, \sqrt{36}\}$.

Note that $\sqrt{36} = 6$.

1. The elements 9 and 6 are natural numbers.

2. The elements 0, 9, and 6 are whole numbers.

3. The elements -8, 0, 9, and 6 are integers.

4. The elements $-8, -\frac{2}{3}, 0, \frac{4}{5}, 9,$ and 6 are rational numbers.

5. The element $-\sqrt{6}$ is an irrational number.

6. All the elements in set A are real numbers.

7.
$$
\begin{aligned}
-\frac{4}{3} - \left(-\frac{2}{7}\right) &= -\frac{4}{3} + \frac{2}{7} \\
&= -\frac{28}{21} + \frac{6}{21} \\
&= -\frac{22}{21}
\end{aligned}
$$

8.
$$
\begin{aligned}
|-4.2| + |5.6| - |-1.9| &= 4.2 + 5.6 - 1.9 \\
&= 9.8 - 1.9 \\
&= 7.9
\end{aligned}
$$

9. $(-2)^4 + (-2)^3 = 16 + (-8) = 8$

10.
$$
\begin{aligned}
\sqrt{25} - \frac{\sqrt{100}}{2} &= 5 - \frac{10}{2} \\
&= 5 - 5 \\
&= 0
\end{aligned}
$$

11. $(-3)^5 = (-3)(-3)(-3)(-3)(-3) = -243$

12. $\left(\frac{6}{7}\right)^3 = \frac{6}{7} \cdot \frac{6}{7} \cdot \frac{6}{7} = \frac{216}{343}$

13. $4^6 = 4 \cdot 4 \cdot 4 \cdot 4 \cdot 4 \cdot 4 = 4096$

14. $-4^6 = -(4 \cdot 4 \cdot 4 \cdot 4 \cdot 4 \cdot 4) = -4096$

15. $-\sqrt{49} = -(6) = -6$, which *is* a real number.

$\sqrt{-49}$ *is not* a real number.

16. $\frac{4-4}{4+4} = \frac{0}{8} = 0$

$\frac{4+4}{4-4} = \frac{8}{0}$, which is *undefined.*

For Exercises 17–20, let $a = 2$, $b = -3$, and $c = 4$.

17. $$\begin{aligned} -3a + 2b - c &= -3(2) + 2(-3) - 4 \\ &= -6 - 6 - 4 \\ &= -16 \end{aligned}$$

18. $$\begin{aligned} -2b^2 - c^2 &= -2(-3)^2 - 4^2 \\ &= -2(9) - 16 \\ &= -18 - 16 \\ &= -34 \end{aligned}$$

19. $$\begin{aligned} -8(a^2 + b^3) &= -8[2^2 + (-3)^3] \\ &= -8[4 + (-27)] \\ &= -8(-23) \\ &= 184 \end{aligned}$$

20. $$\begin{aligned} \frac{3a^3 - b}{4 + 3c} &= \frac{3(2)^3 - (-3)}{4 + 3(4)} \\ &= \frac{3(8) - (-3)}{4 + 3(4)} \\ &= \frac{24 + 3}{4 + 12} \\ &= \frac{27}{16} \end{aligned}$$

21. $$\begin{aligned} &-7r + 5 - 13r + 12 \\ &= -7r - 13r + 5 + 12 \\ &= (-7 - 13)r + (5 + 12) \\ &= -20r + 17 \end{aligned}$$

22. $$\begin{aligned} &-(3k + 8) - 2(4k - 7) + 3(8k + 12) \\ &= -3k - 8 - 8k + 14 + 24k + 36 \\ &= -3k - 8k + 24k - 8 + 14 + 36 \\ &= 13k + 42 \end{aligned}$$

23. $(a + b) + 8 = 8 + (a + b)$

The order of the terms $(a + b)$ and 8 have been reversed. This is an illustration of the *commutative property*.

24. $5x + 13x = (5 + 13)x$

The common variable, x, has been removed from each term. This is an illustration of the *distributive property*.

25. $-13 + 13 = 0$

The sum of a number and its opposite is equal to 0. This is an illustration of the *inverse property*.

26. $$\begin{aligned} -4x + 7(2x + 3) &= 7x + 36 \\ -4x + 14x + 21 &= 7x + 36 \\ 10x + 21 &= 7x + 36 \\ 3x &= 15 \\ x &= 5 \end{aligned}$$

Solution set: $\{5\}$

27. $$\begin{aligned} -\frac{3}{5}x + \frac{2}{3}x &= 2 \\ 3(-3x) + 5(2x) &= 15(2) \quad \textit{Multiply by 15.} \\ -9x + 10x &= 30 \\ x &= 30 \end{aligned}$$

Solution set: $\{30\}$

28. $$\begin{aligned} 0.06x + 0.03(100 + x) &= 4.35 \\ 6x + 3(100 + x) &= 435 \quad \textit{Multiply by 100.} \\ 6x + 300 + 3x &= 435 \\ 9x &= 135 \\ x &= 15 \end{aligned}$$

Solution set: $\{15\}$

29. Solve $P = a + b + c$ for c.

$$P - (a + b) = a + b + c - (a + b)$$

Subtract (a + b).

$$P - a - b = c$$

30. $$\begin{aligned} 4(2x - 6) + 3(x - 2) &= 11x + 1 \\ 8x - 24 + 3x - 6 &= 11x + 1 \\ 11x - 30 &= 11x + 1 \\ -30 &= 1 \quad \textit{False} \end{aligned}$$

Solution set: $\emptyset$

31. $$\begin{aligned} \frac{2}{3}x + \frac{5}{8}x &= \frac{31}{24}x \\ 8(2x) + 3(5x) &= 31x \quad \textit{Multiply by the LCD, 24.} \\ 16x + 15x &= 31x \\ 31x &= 31x \quad \textit{True} \end{aligned}$$

Solution set: {all real numbers}

32. Let x = the amount of pure alcohol that should be added.

Liters of Solution	Percent (as a Decimal)	Liters of Pure Alcohol
x	1.00	$1.00x$
7	0.10	0.10(7)
$x + 7$	0.30	$0.30(x + 7)$

The last column gives the equation.

$$\begin{aligned} 1.00x + 0.10(7) &= 0.30(x + 7) \\ 10x + 1(7) &= 3(x + 7) \quad \textit{Multiply by 10.} \\ 10x + 7 &= 3x + 21 \\ 7x &= 14 \\ x &= 2 \end{aligned}$$

2 L of pure alcohol should be added to the solution.

33. Let $x =$ the number of nickels. Then $x - 4 =$ the number of quarters. The collection contains 29 coins, so the number of pennies is

$$29 - x - (x - 4) = 33 - 2x.$$

	Number of Coins	Denomination	Value
Pennies	$33 - 2x$	0.01	$0.01(33 - 2x)$
Nickels	x	0.05	$0.05x$
Quarters	$x - 4$	0.25	$0.25(x - 4)$
	29		$2.69

From the last column:

$$\begin{aligned} 0.01(33 - 2x) + 0.05x + 0.25(x - 4) &= 2.69 \\ 1(33 - 2x) + 5x + 25(x - 4) &= 269 \\ &\textit{Multiply by 100.} \\ 33 - 2x + 5x + 25x - 100 &= 269 \\ 28x - 67 &= 269 \\ 28x &= 336 \\ x &= 12 \end{aligned}$$

There are $33 - 2(12) = 9$ pennies, 12 nickels, and $12 - 4 = 8$ quarters.

34. Let $x =$ the amount invested at 5%. Then $x + 2000 =$ the amount invested at 6%.

Principal	Rate (as a Decimal)	Interest
x	0.05	$0.05x$
$x + 2000$	0.06	$0.06(x + 2000)$

The total interest is $670.

$$\begin{aligned} 0.05x + 0.06(x + 2000) &= 670 \\ 0.05x + 0.06x + 120 &= 670 \\ 0.11x + 120 &= 670 \\ 0.11x &= 550 \\ x = \frac{550}{0.11} &= 5000 \end{aligned}$$

$5000 was invested at 5% and $5000 + 2000 =$ $7000 was invested at 6%.

35. Let $x =$ the time for Jack to be $\frac{1}{4}$ mile ahead of Jill.

	Rate	Time	Distance
Jack	7	x	$7x$
Jill	5	x	$5x$

Jack's distance is $\frac{1}{4}$ mile more than Jill's distance.

$$\begin{aligned} 7x &= 5x + \tfrac{1}{4} \\ 2x &= \tfrac{1}{4} \\ x &= \tfrac{1}{8} \end{aligned}$$

Jack will be $\frac{1}{4}$ mile ahead of Jill in $\frac{1}{8}$ hr.

36. Clark's rule:

$$\frac{\text{Weight of child in pounds}}{150} \times \begin{matrix}\text{adult}\\\text{dose}\end{matrix} = \begin{matrix}\text{child's}\\\text{dose}\end{matrix}$$

If the child weighs 55 lb and the adult dosage is 120 mg, then

$$\frac{55}{150} \times 120 = 44.$$

The child's dosage is 44 mg.

37. 5 feet, 8 inches $= 5(12) + 8 = 68$ inches

$$\begin{aligned} \text{BMI} &= \frac{704 \times (\text{weight in pounds})}{(\text{height in inches})^2} \\ &= \frac{704 \times 160}{68^2} = \frac{112{,}640}{4624} \approx 24.4 \end{aligned}$$

His BMI is about 24.4.

38. **(a)** 1975: 1756
2005: 1452

$1756 - 1452 = 304$

The number decreased by 304 newspapers.

(b) $\frac{304}{1756} \approx 0.173$ or 17.3%.

The number decreased by approximately 17.3%.

CHAPTER 3 LINEAR INEQUALITIES AND ABSOLUTE VALUE

3.1 Linear Inequalities in One Variable

3.1 Margin Exercises

1. **(a)** $x < -1$ is written in interval notation as $(-\infty, -1)$. The parenthesis next to -1 means that -1 is not in the interval and is not part of the graph.

   −3 −1 0 1

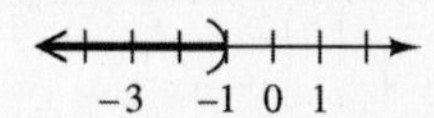

   (b) $x \geq -3$ is written in interval notation as $[-3, \infty)$. The bracket next to -3 means that -3 is in the interval and is part of the graph.

   −3 −1 0

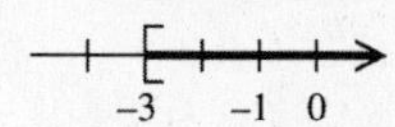

   (c) $-4 \leq x < 2$ is written in interval notation as $[-4, 2)$. Here, -4 is in the interval and 2 is not.

   −4 0 2 4

   (d) $0 < x < 3.5$ is written in interval notation as $(0, 3.5)$. Here, 0 and 3.5 are not in the interval.

   0 1 2 3.5

2. **(a)**
$$x - 3 < -9$$
$$x - 3 + 3 < -9 + 3 \quad \textit{Add 3.}$$
$$x < -6$$

Check: Substitute -6 for x in the equation $x - 3 = -9$.

$$x - 3 = -9$$
$$-6 - 3 \stackrel{?}{=} -9 \quad \textit{Let } x = -6.$$
$$-9 = -9 \quad \textit{True}$$

This shows that -6 is the boundary point. Now test a number on each side of -6 to verify that numbers *less than* -6 make the *inequality* true. We choose -8 and 0.

$$x - 3 < -9$$

Let $x = -8$.	*Let* $x = 0$.
$-8 - 3 \stackrel{?}{<} -9$	$0 - 3 \stackrel{?}{<} -9$
$-11 < -9$ *True*	$-3 < -9$ *False*
-8 is in the solution set.	0 is not in the solution set.

The check confirms that $(-\infty, -6)$ is the solution set.

−6 −2 0 2

(b)
$$p + 6 < 8$$
$$p + 6 - 6 < 8 - 6 \quad \textit{Subtract 6.}$$
$$p < 2$$

Check: Substitute 2 for p in the equation $p + 6 = 8$.

$$p + 6 = 8$$
$$2 + 6 \stackrel{?}{=} 8 \quad \textit{Let } p = 2.$$
$$8 = 8 \quad \textit{True}$$

This shows that 2 is the boundary point. Now test a number on each side of 2 to verify that numbers *less than* 2 make the *inequality* true. We choose 0 and 3.

$$p + 6 < 8$$

$0 + 6 \stackrel{?}{<} 8$ *Let* $p = 0$.	$3 + 6 \stackrel{?}{<} 8$ *Let* $p = 3$.
$6 < 8$ *True*	$9 < 8$ *False*
0 is in the solution set.	3 is not in the solution set.

The check confirms that $(-\infty, 2)$ is the solution set.

−3 −2 −1 0 1 2 3

3.
$$2k - 5 \geq 1 + k$$
$$2k - k \geq 5 + 1 \quad \textit{Subtract k; add 5.}$$
$$k \geq 6$$

Check: Substitute 6 for k in $2k - 5 = 1 + k$.

$$2k - 5 = 1 + k$$
$$2(6) - 5 \stackrel{?}{=} 1 + 6 \quad \textit{Let } k = 6.$$
$$12 - 5 \stackrel{?}{=} 7$$
$$7 = 7 \quad \textit{True}$$

So 6 satisfies the equality part of $\geq$. Choose 0 and 7 as test points.

$$2k - 5 \geq 1 + k$$

Let $k = 0$.	*Let* $k = 7$.
$2(0) - 5 \stackrel{?}{\geq} 1 + 0$	$2(7) - 5 \stackrel{?}{\geq} 1 + 7$
$-5 \geq 1$ *False*	$14 - 5 \stackrel{?}{\geq} 1 + 7$
	$9 \geq 8$ *True*
0 is not in the solution set.	7 is in the solution set.

The check confirms that $[6, \infty)$ is the solution set.

−2 0 2 4 6

4. This exercise is designed to reinforce the idea that multiplying each side of an inequality by a negative number reverses the direction of the inequality symbol.

(a) $7 < 8$

$7(-5) > 8(-5)$ *Multiply by −5; reverse inequality.*

$-35 > -40$ *True*

(b) $-1 > -4$

$-1(-5) < -4(-5)$ *Multiply by −5; reverse inequality.*

$5 < 20$ *True*

5. **(a)** $2x < -10$

$\frac{2x}{2} < \frac{-10}{2}$ *Divide by 2 > 0; do not reverse the inequality.*

$x < -5$

Check that the solution set is the interval $(-\infty, -5)$.

−5 −3 −1 0

(b) $-7k \geq 8$

$\frac{-7k}{-7} \leq \frac{8}{-7}$ *Divide by −7 < 0; reverse the inequality.*

$k \leq -\frac{8}{7}$

Check that the solution set is the interval $(-\infty, -\frac{8}{7}]$.

$-\frac{8}{7}$

−3 −2 −1 0 1 2

(c) $-9m < -81$

$\frac{-9m}{-9} > \frac{-81}{-9}$ *Divide by −9 < 0; reverse the inequality.*

$m > 9$

Check that the solution set is the interval $(9, \infty)$.

−3 0 3 6 9

6. **(a)** $x + 4(2x - 1) \geq x + 2$

$x + 8x - 4 \geq x + 2$ *Distributive property*

$9x - 4 \geq x + 2$ *Combine terms.*

$8x - 4 \geq 2$ *Subtract x.*

$8x \geq 6$ *Add 4.*

$x \geq \frac{3}{4}$ *Divide by 8.*

Check that the solution set is the interval $[\frac{3}{4}, \infty)$.

$\frac{3}{4}$

−1 0 1 2 3

(b) $m - 2(m - 4) \leq 3m$

$m - 2m + 8 \leq 3m$ *Dist. property*

$-m + 8 \leq 3m$ *Combine terms.*

$-4m + 8 \leq 0$ *Subtract 3m.*

$-4m \leq -8$ *Subtract 8.*

$m \geq 2$ *Divide by −4 < 0; reverse inequality.*

Check that the solution set is the interval $[2, \infty)$.

−1 0 1 2 3 4

7. **(a)** $5 - 3(m - 1) \leq 2(m + 3) + 1$

$5 - 3m + 3 \leq 2m + 6 + 1$ *Dist. property*

$8 - 3m \leq 2m + 7$

$8 - 5m \leq 7$ *Subtract 2m.*

$-5m \leq -1$ *Subtract 8.*

$\frac{-5m}{-5} \geq \frac{-1}{-5}$ *Divide by −5; reverse the inequality.*

$m \geq \frac{1}{5}$

Check that the solution set is the interval $[\frac{1}{5}, \infty)$.

$\frac{1}{5}$

0 1 2 3 4 5

(b) $\frac{1}{4}(m + 3) + 2 \leq \frac{3}{4}(m + 8)$

$4\left[\frac{1}{4}(m + 3) + 2\right] \leq 4\left[\frac{3}{4}(m + 8)\right]$

Multiply by 4.

$(m + 3) + 8 \leq 3(m + 8)$

$m + 11 \leq 3m + 24$

$-2m + 11 \leq 24$ *Subtract 3m.*

$-2m \leq 13$ *Subtract 11.*

$m \geq -\frac{13}{2}$ *Divide by −2; reverse symbol.*

Check that the solution set is the interval $[-\frac{13}{2}, \infty)$.

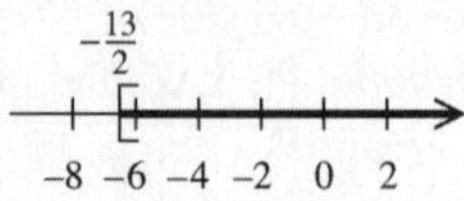

8. **(a)** $1 > x > -1$ is written $-1 < x < 1$.

(b) $16 \geq p \geq 11$ is written $11 \leq p \leq 16$.

(c) $-2 > t \geq -8$ is written $-8 \leq t < -2$.

9. **(a)** $-3 \leq x - 1 \leq 7$

$-2 \leq x \leq 8$ *Add 1 to each part.*

Check that the solution set is the interval $[-2, 8]$.

−2 0 2 4 6 8

(b) $5 < 3x - 4 < 9$

$9 < 3x < 13$ *Add 4 to each part.*

$\frac{9}{3} < \frac{3x}{3} < \frac{13}{3}$ *Divide each part by 3.*

$3 < x < \frac{13}{3}$

Check that the solution set is the interval $(3, \frac{13}{3})$.

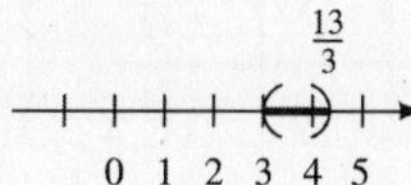

10. *Step 2*
Let $h =$ the number of hours she can rent the leaf blower.

Step 3
She must pay \$10, plus \$7.50h, to rent the leaf blower for h hours, and this amount must be *no more than* \$40.

Cost of renting	is no more than	40 dollars.
↓	↓	↓
$10 + 7.50h$	$\leq$	40

Step 4 $7.50h \leq 30$ *Subtract 10.*

$h \leq 4$ *Divide by 7.50.*

Step 5
She can use the leaf blower for a maximum of 4 hr. (She may use it for less time, as indicated by the inequality $h \leq 4$.)

Step 6
If Dona uses the leaf blower for 4 hr, she will spend $10 + 7.50(4) = 40$, the maximum amount.

11. Let $x =$ the score Alex must make on the fourth test.

To find the average of the four scores, add the scores and divide by 4. This average must be at least 90, that is, greater than or equal to 90.

$$\frac{92 + 90 + 84 + x}{4} \geq 90$$

$$\frac{266 + x}{4} \geq 90$$

$266 + x \geq 360$ *Multiply by 4.*

$x \geq 94$ *Subtract 266.*

Alex must score at least 94 on the fourth test.

Check: $\frac{92 + 90 + 84 + 94}{4} = \frac{360}{4} = 90$

A score of 94 or more will give an average of at least 90, as required.

3.1 Section Exercises

1. $x \leq 3$

In interval notation, this inequality is written $(-\infty, 3]$. The bracket indicates that 3 is included. The answer is choice **D**.

3. $x < 3$

In interval notation, this inequality is written $(-\infty, 3)$. The parenthesis indicates that 3 is not included. The graph of this inequality is shown in choice **B**.

5. $-3 \leq x \leq 3$

In interval notation, this inequality is written $[-3, 3]$. The brackets indicates that -3 and 3 are included. The answer is choice **F**.

7. **(a)** The wind speed s of a Category 4 hurricane is between 131 mph and 155 mph [inclusive], which can be described by the three-part inequality $131 \leq s \leq 155$.

(b) The wind speed s of a Category 5 hurricane is greater than 155 mph, which can be described by the inequality $s > 155$.

(c) The storm surge x of a Category 3 hurricane is between 9 ft and 12 ft [inclusive], which can be described by the three-part inequality $9 \leq x \leq 12$.

(d) The storm surge x of a Category 5 hurricane is greater than 18 ft, which can be described by the inequality $x > 18$.

9. Since $4 > 0$, the student should not have reversed the direction of the inequality symbol when dividing by 4. We reverse the inequality symbol only when multiplying or dividing by a *negative* number. The solution set is $[-16, \infty)$.

For Exercises 11–38 and 45–56, the check is shown for Exercises 13 and 23, and left for the student for the other exercises.

11. $x - 4 \leq 3$

$x \leq 7$ *Add 4.*

Check that the solution set is the interval $(-\infty, 7]$.

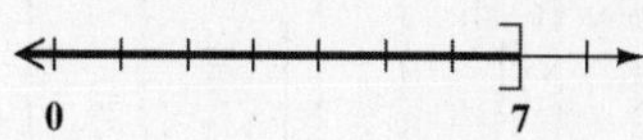

13. $4x + 1 \geq 21$

$4x \geq 20$ *Subtract 1.*

$\frac{4x}{4} \geq \frac{20}{4}$ *Divide by 4.*

$x \geq 5$

Check: Substitute 5 for x in the *equation* $4x + 1 = 21$.

$4(5) + 1 \stackrel{?}{=} 21$

$21 = 21$ *True*

So 5 satisfies the equality part of $\geq$. Choose 0 and 6 as test points.

$4x + 1 \geq 21$

Let x = 0.	*Let x = 6.*
$4(0) + 1 \stackrel{?}{\geq} 21$	$4(6) + 1 \stackrel{?}{\geq} 21$
$1 \geq 21$ *False*	$25 \geq 21$ *True*
0 is not in the solution set.	6 is in the solution set.

The check confirms that $[5, \infty)$ is the solution set.

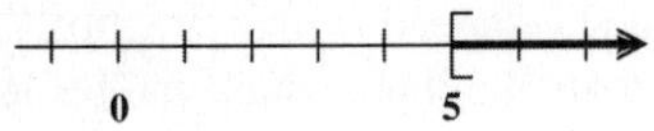

15. $5x > -25$

$x > -5$ *Divide by 5.*

Check that the solution set is the interval $(-5, \infty)$.

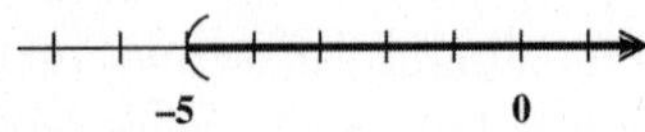

17. $-4x < 16$

Divide each side by -4 and reverse the inequality symbol.

$\frac{-4x}{-4} > \frac{16}{-4}$

$x > -4$

Check that the solution set is the interval $(-4, \infty)$.

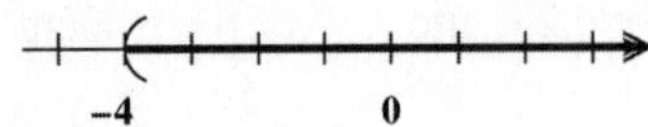

19. $-\frac{3}{4}r \geq 30$

Multiply each side by $-\frac{4}{3}$ and reverse the inequality symbol.

$-\frac{4}{3}(-\frac{3}{4}r) \leq -\frac{4}{3}(30)$

$r \leq -40$

Check that the solution set is the interval $(-\infty, -40]$.

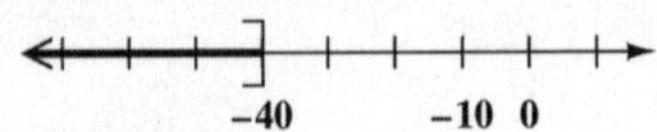

21. $-1.3m \geq -5.2$

Divide each side by -1.3 and reverse the inequality symbol.

$\frac{-1.3m}{-1.3} \leq \frac{-5.2}{-1.3}$

$m \leq 4$

Check that the solution set is the interval $(-\infty, 4]$.

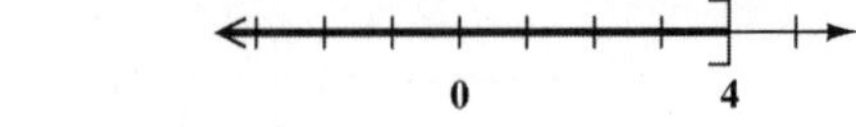

23. $\frac{3k-1}{4} > 5$

$4\left(\frac{3k-1}{4}\right) > 4(5)$ *Multiply by 4.*

$3k - 1 > 20$

$3k > 21$ *Add 1.*

$k > 7$ *Divide by 3.*

Check: Let $k = 7$ in the *equation* $\frac{3k-1}{4} = 5$.

$\frac{3(7)-1}{4} \stackrel{?}{=} 5$

$\frac{20}{4} \stackrel{?}{=} 5$

$5 = 5$ *True*

This shows that 7 is the boundary point. Now test a number on each side of 7. We choose 0 and 10.

$\frac{3k-1}{4} > 5$

Let k = 0.	*Let k = 10.*
$\frac{3(0)-1}{4} \stackrel{?}{>} 5$	$\frac{3(10)-1}{4} \stackrel{?}{>} 5$
$\frac{-1}{4} > 5$ *False*	$\frac{29}{4} > 5$ *True*
0 is not in the solution set.	10 is in the solution set.

The check confirms that $(7, \infty)$ is the solution set.

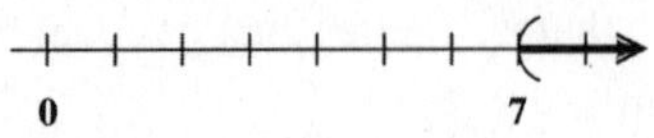

25. $\frac{2k-5}{-4} > 5$

Multiply each side by -4 and reverse the inequality symbol.

$-4\left(\frac{2k-5}{-4}\right) < -4(5)$

$2k - 5 < -20$

$2k < -15$ *Add 5.*

$k < -\frac{15}{2}$ *Divide by 2.*

Check that the solution set is the interval $(-\infty, -\frac{15}{2})$.

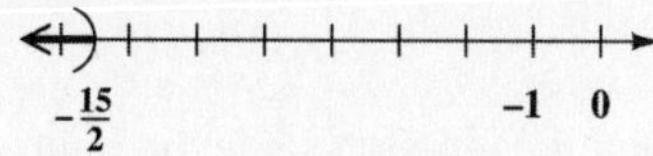

27. $3k + 1 < -20$

$3k < -21$ *Subtract 1.*

$k < -7$ *Divide by 3.*

Check that the solution set is the interval $(-\infty, -7)$.

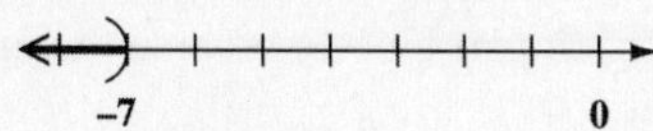

29. $x + 4(2x - 1) \geq x$

$x + 8x - 4 \geq x$

$9x - 4 \geq x$

$9x \geq x + 4$ *Add 4.*

$8x \geq 4$ *Subtract x.*

$x \geq \frac{1}{2}$ *Divide by 8.*

Check that the solution set is the interval $[\frac{1}{2}, \infty)$.

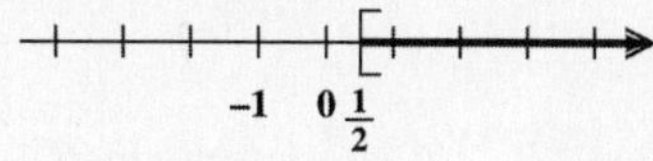

31. $-(4 + r) + 2 - 3r < -14$

$-4 - r + 2 - 3r < -14$ *Distributive property*

$-4r - 2 < -14$ *Combine terms.*

$-4r < -12$ *Add 2.*

Divide each side by -4 and reverse the inequality symbol.

$r > 3$

Check that the solution set is the interval $(3, \infty)$.

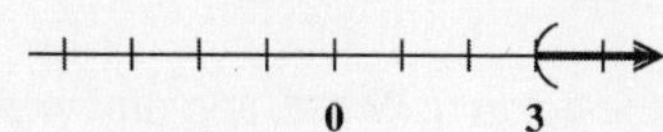

33. $-3(z - 6) > 2z - 2$

$-3z + 18 > 2z - 2$ *Distributive property*

$-5z > -20$ *Subtract 2z; subtract 18.*

Divide each side by -5 and reverse the inequality symbol.

$z < 4$

Check that the solution set is the interval $(-\infty, 4)$.

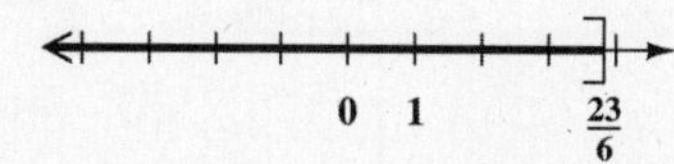

35. $\frac{2}{3}(3k - 1) \geq \frac{3}{2}(2k - 3)$

Multiply each side by 6 to clear the fractions.

$6 \cdot \frac{2}{3}(3k - 1) \geq 6 \cdot \frac{3}{2}(2k - 3)$

$4(3k - 1) \geq 9(2k - 3)$

$12k - 4 \geq 18k - 27$ *Distributive prop.*

$-6k \geq -23$ *Subtract 18k; add 4.*

Divide each side by -6 and reverse the inequality symbol.

$k \leq \frac{23}{6}$

Check that the solution set is the interval $(-\infty, \frac{23}{6}]$.

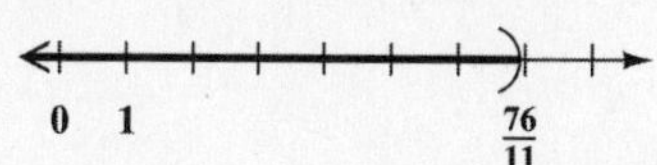

37. $-\frac{1}{4}(p + 6) + \frac{3}{2}(2p - 5) < 10$

Multiply each term by 4 to clear the fractions.

$-1(p + 6) + 6(2p - 5) < 40$

$-p - 6 + 12p - 30 < 40$

$11p - 36 < 40$

$11p < 76$

$p < \frac{76}{11}$

Check that the solution set is the interval $(-\infty, \frac{76}{11})$.

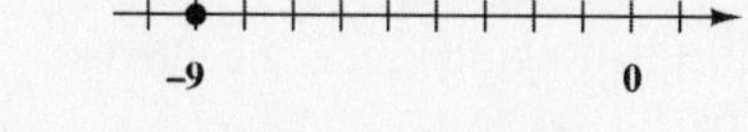

39. $5(x + 3) - 2(x - 4) = 2(x + 7)$

$5x + 15 - 2x + 8 = 2x + 14$

$3x + 23 = 2x + 14$

$x = -9$

Solution set: $\{-9\}$

The graph is the point -9 on a number line.

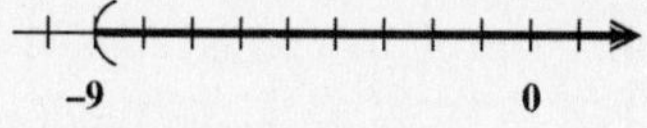

40. $5(x + 3) - 2(x - 4) > 2(x + 7)$

$5x + 15 - 2x + 8 > 2x + 14$

$3x + 23 > 2x + 14$

$x > -9$

Check that the solution set is the interval $(-9, \infty)$. The graph extends from -9 to the right on a number line; -9 is not included in the graph.

41. $5(x+3)-2(x-4)<2(x+7)$
$5x+15-2x+8<2x+14$
$3x+23<2x+14$
$x<-9$

Check that the solution set is the interval $(-\infty,-9)$. The graph extends from -9 to the left on a number line; -9 is not included in the graph.

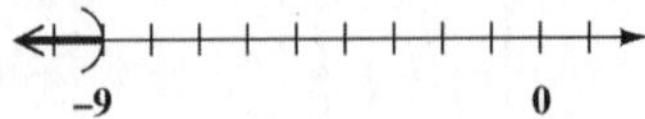

42. If we graph all the solution sets from Exercises 39–41; that is, $\{-9\}$, $(-9,\infty)$, and $(-\infty,-9)$, on the same number line, we will have graphed the set of all real numbers.

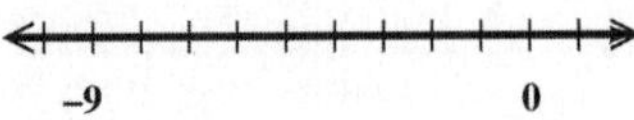

43. The solution set of the given equation is the point -3 on a number line. The solution set of the first inequality extends from -3 to the right (toward ∞) on the same number line. Based on Exercises 39–41, the solution set of the second inequality should then extend from -3 to the left (toward $-\infty$) on the number line. Complete the statement with $\underline{(-\infty,-3)}$.

45. $-4<x-5<6$

Add 5 to each part of the inequality to isolate the variable x.

$-4+5<x-5+5<6+5$
$1<x<11$

Check that the solution set is the interval $(1,11)$.

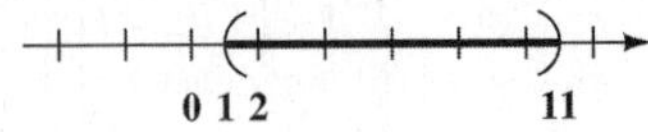

47. $-9\le k+5\le 15$

Subtract 5 from each part.

$-9-5\le k+5-5\le 15-5$
$-14\le k\le 10$

Check that the solution set is the interval $[-14,10]$.

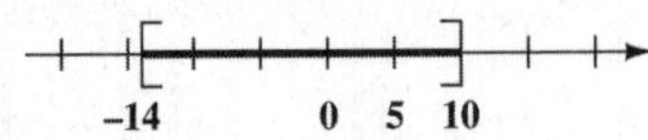

49. $-6\le 2(z+2)\le 16$
$-6\le 2z+4\le 16$ *Distributive property*
$-10\le 2z\le 12$ *Subtract 4.*
$-5\le z\le 6$ *Divide by 2.*

Check that the solution set is the interval $[-5,6]$.

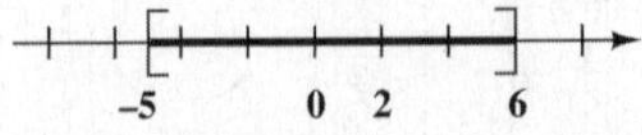

51. $-16<3t+2<-10$
$-18<3t<-12$ *Subtract 2.*
$-6<t<-4$ *Divide by 3.*

Check that the solution set is the interval $(-6,-4)$.

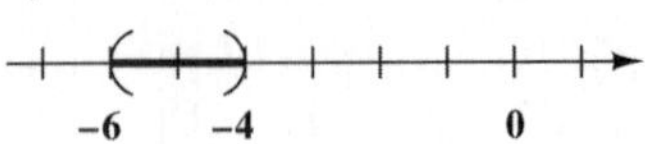

53. $4<-9x+5\le 8$
$-1<-9x\le 3$ *Subtract 5.*

Divide each part by -9; reverse the inequality symbols.

$\frac{1}{9}>x\ge -\frac{1}{3}$

The last inequality may be written as

$$-\frac{1}{3}\le x<\frac{1}{9}.$$

Check that the solution set is the interval $[-\frac{1}{3},\frac{1}{9})$.

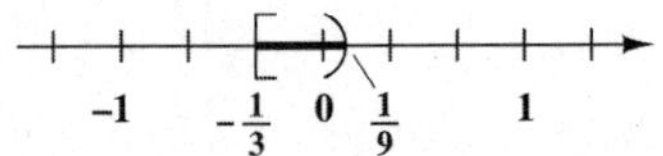

55. $-1\le \dfrac{2x-5}{6}\le 5$
$-6\le 2x-5\le 30$ *Multiply by 6.*
$-1\le 2x\le 35$ *Add 5.*
$-\frac{1}{2}\le x\le \frac{35}{2}$ *Divide by 2.*

Check that the solution set is the interval $[-\frac{1}{2},\frac{35}{2}]$.

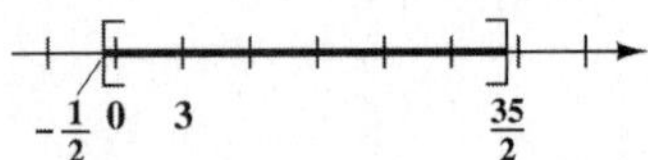

57. Draw a horizontal line at the 90°F mark. It intersects the *upper* boundary of the forecasted highs in two places. The temperature is expected to be at least 90°F from about 2:30 P.M. to 6:00 P.M.

59. Locate the dot at 5:39 P.M. and draw a vertical line through it. The line will intersect the lower and upper boundaries of the forecasted highs at about 84°F and 91°F, so the range of predicted temperatures is 84°F–91°F.

61. Let x = the number of months. The cost of Plan A is $54.99x$ and the cost of Plan B is $49.99x+129$. To determine the number of months that would be needed to make Plan B less expensive, solve the following inequality.

Plan B (cost) < Plan A (cost)
$49.99x+129<54.99x$
$129<5x$ *Subtract 49.99x.*
$5x>129$ *Equivalent*
$x>\frac{129}{5}$ $[=25.8]$ *Divide by 5.*

It will take 26 months for Plan B to be the better deal.

63. Let $x =$ her score on the third test. Her average must be at least 84 (≥ 84). To find the average of three numbers, add them and divide by 3.

$$\frac{90+82+x}{3} \geq 84$$
$$\frac{172+x}{3} \geq 84 \quad \textit{Add.}$$
$$172 + x \geq 252 \quad \textit{Multiply by 3.}$$
$$x \geq 80 \quad \textit{Subtract 172.}$$

She must score at least 80 on her third test.

65. Cost $C = 20x + 100$; Revenue $R = 24x$
The business will show a profit only when $R > C$. Substitute the given expressions for R and C.

$$R > C$$
$$24x > 20x + 100$$
$$4x > 100$$
$$x > 25$$

The company will show a profit upon selling 26 DVDs.

67. $\text{BMI} = \dfrac{704 \times (\text{weight in pounds})}{(\text{height in inches})^2}$

(a) Let the height equal 72.

$$19 \leq \text{BMI} \leq 25$$
$$19 \leq \frac{704w}{72^2} \leq 25$$
$$19(72^2) \leq 704w \leq 25(72^2)$$
$$\frac{19(72^2)}{704} \leq w \leq \frac{25(72^2)}{704}$$
$$(\approx 139.91) \leq w \leq (\approx 184.09)$$

According to the BMI formula, the healthy weight range (rounded to the nearest pound) for a person who is 72 inches tall is 140 to 184 pounds.

(b) Answers will vary.

69. Six times a number is between -12 and 12.

$$-12 < 6x < 12$$
$$-2 < x < 2 \quad \textit{Divide by 6.}$$

This is the set of all numbers between -2 and 2, or, $(-2, 2)$.

71. When 1 is added to twice a number, the result is greater than or equal to 7.

$$2x + 1 \geq 7$$
$$2x \geq 6 \quad \textit{Subtract 1.}$$
$$x \geq 3 \quad \textit{Divide by 2.}$$

This is the set of all numbers greater than or equal to 3, or, $[3, \infty)$.

73. One third of a number is added to 6, giving a result of at least 3.

$$6 + \tfrac{1}{3}x \geq 3$$
$$\tfrac{1}{3}x \geq -3 \quad \textit{Subtract 6.}$$
$$x \geq -9 \quad \textit{Multiply by 3.}$$

This is the set of all numbers greater than or equal to -9, or, $[-9, \infty)$.

3.2 Set Operations and Compound Inequalities

3.2 Margin Exercises

1. **(a)** Let $A = \{3, 4, 5, 6\}$ and $B = \{5, 6, 7\}$.

The set $A \cap B$, the intersection of A and B, contains those elements that belong to both A *and* B; that is, the numbers 5 and 6. Therefore,

$$A \cap B = \{5, 6\}.$$

(b) Let $N = \{s, d, c, g\}$ and $O = \{i, m, h, g\}$.

The set $N \cap O$ contains those elements that belong to both N *and* O; that is, symptom g. Therefore,

$$N \cap O = \{g\}.$$

2. **(a)** $x < 10$ and $x > 2$
Graph each inequality.

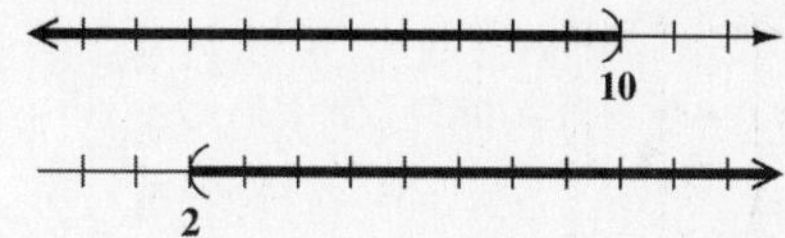

The word *and* means to take the values that satisfy both inequalities; that is, the numbers between 2 and 10, not including 2 and 10.

Solution set: $(2, 10)$

0 2 4 6 8 10

(b) $x + 3 \leq 1$ and $x - 4 \geq -12$
Solve each inequality.

$$x + 3 \leq 1 \quad \text{and} \quad x - 4 \geq -12$$
$$x + 3 - 3 \leq 1 - 3 \quad \text{and} \quad x - 4 + 4 \geq -12 + 4$$
$$x \leq -2 \quad \text{and} \quad x \geq -8$$

−2

−8

The values that satisfy both inequalities are the numbers between -8 and -2, including -8 and -2.

Solution set: $[-8, -2]$

−8 −4 −2 0

3.

$$\begin{aligned} 2x &\geq x-1 \quad &&\text{and} \quad & 3x &\geq 3+2x \\ 2x-x &\geq x-1-x \quad &&\text{and} \quad & 3x-2x &\geq 3+2x-2x \\ x &\geq -1 \quad &&\text{and} \quad & x &\geq 3 \end{aligned}$$

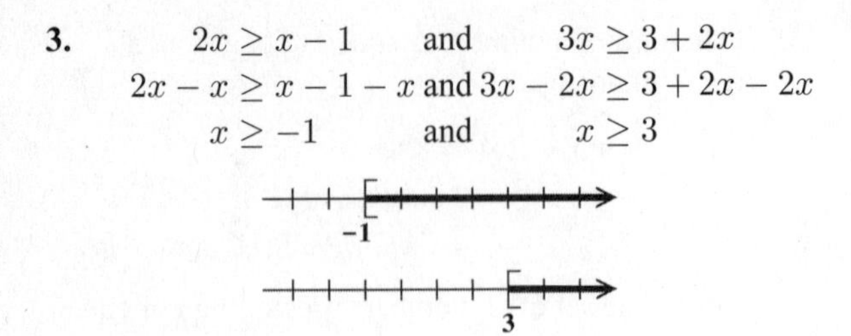

The overlap of the two graphs consists of the numbers that are greater than or equal to -1 and are also greater than or equal to 3; that is, the numbers greater than or equal to 3.

Solution set: $[3, \infty)$

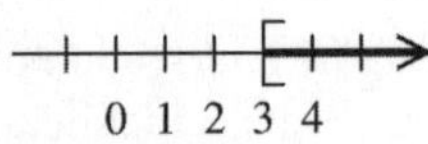

4. **(a)** $x < 5$ and $x > 5$

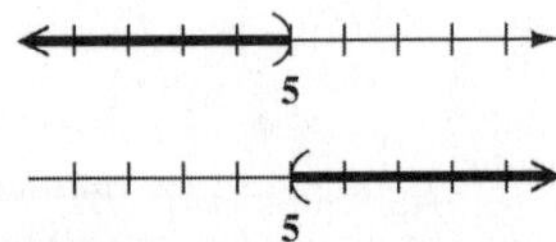

The numbers that satisfy both inequalities must be less than 5 *and* greater than 5. There are no such numbers.

Solution set: $\emptyset$

(b) $x + 2 > 3$ and $2x + 1 < -3$

$$\begin{aligned} & & 2x &< -4 \\ x &> 1 \quad \text{and} & x &< -2 \end{aligned}$$

The numbers that satisfy both inequalities must be greater than 1 *and* less than -2. There are no such numbers.

Solution set: $\emptyset$

5. **(a)** Let $A = \{3, 4, 5, 6\}$ and $B = \{5, 6, 7\}$.

The set $A \cup B$, the union of A and B, consists of all elements in either A *or* B (or both). Start by listing the elements of set A: $3, 4, 5, 6$. Then list any additional elements from set B. In this case, the elements 5 and 6 are already listed, so the only additional element is 7. Therefore,

$$A \cup B = \{3, 4, 5, 6, 7\}.$$

(b) The set $N \cup O$ consists of all elements in either N or O (or both).

$$\begin{aligned} N \cup O &= \{s, d, c, g\} \cup \{i, m, h, g\} \\ &= \{s, d, c, g, i, m, h\} \end{aligned}$$

6. **(a)** $x + 2 > 3$ or $2x + 1 < -3$

$$\begin{aligned} & & 2x &< -4 \\ x &> 1 \quad \text{or} & x &< -2 \end{aligned}$$

The graph of the solution set consists of all numbers greater than 1 *or* less than -2.

Solution set: $(-\infty, -2) \cup (1, \infty)$

(b) $x - 1 > 2$ or $3x + 5 < 2x + 6$

$$x > 3 \quad \text{or} \quad x < 1$$

The graph of the solution set consists of all numbers greater than 3 *or* less than 1.

Solution set: $(-\infty, 1) \cup (3, \infty)$

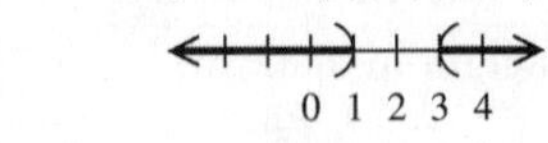

7. **(a)** $2x + 1 \leq 9$ or $2x + 3 \leq 5$

$$\begin{aligned} 2x &\leq 8 \quad \text{or} & 2x &\leq 2 \\ x &\leq 4 \quad \text{or} & x &\leq 1 \end{aligned}$$

The solution set is all numbers less than or equal to 4 *or* less than or equal to 1. Note that this is simply the first inequality, $x \leq 4$.

Solution set: $(-\infty, 4]$

(b) $3x - 4 > 2$ or $-2x + 5 < 3$

$$\begin{aligned} 3x &> 6 & -2x &< -2 \\ x &> 2 \quad \text{or} & x &> 1 \end{aligned}$$

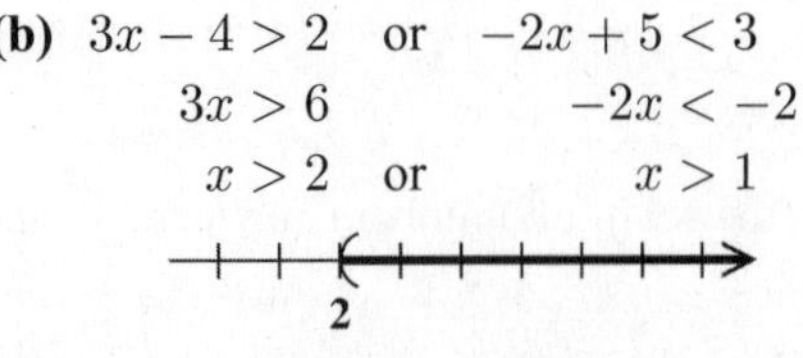

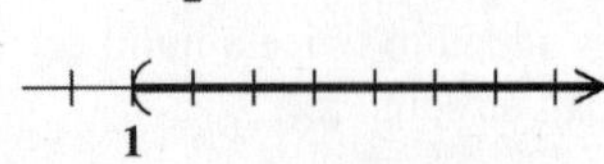

The graph of the solution set consists of all numbers greater than 2 *or* greater than 1.

Solution set: $(1, \infty)$

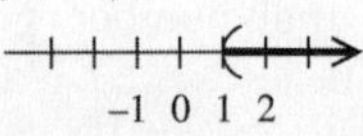

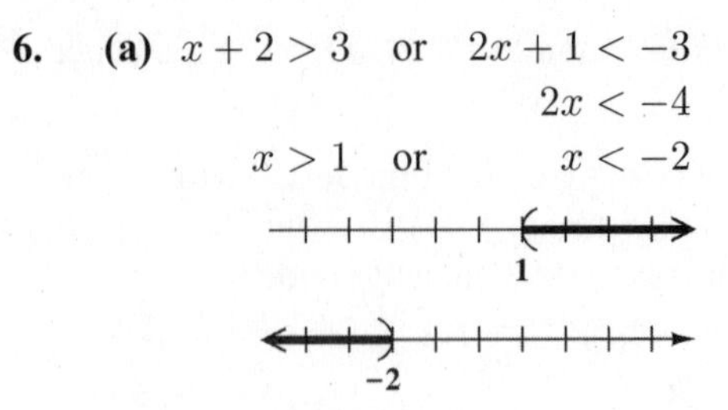

8. $3x - 2 \le 13$ or $x + 5 \ge 7$
$3x \le 15$
$x \le 5$ or $x \ge 2$

The solution set is all numbers that are either less than or equal to 5 *or* greater than or equal to 2. All real numbers are included.

Solution set: $(-\infty, \infty)$

9. **(a)** All films had admissions greater than 130,000,000, but no films had a gross income of less than \$800,000,000. Thus, there are no elements (films) that satisfy each set, so the required set is the empty set, symbolized by $\emptyset$.

(b) Since all the films had admissions greater than 130,000,000 and we have an *or* statement, the second condition doesn't have an effect on the solution set. The required set is the set of all films; that is, {*Gone with the Wind, Star Wars, The Sound of Music, E.T., The Ten Commandments*}.

3.2 Section Exercises

1. **(i)** $2x + 1 = 3$
$2x = 2$
$x = 1$ **Solution set: {1}**

(ii) $2x + 1 > 3$
$2x > 2$
$x > 1$ **Solution set: $(1, \infty)$**

(iii) $2x + 1 < 3$
$2x < 2$
$x < 1$ **Solution set: $(-\infty, 1)$**

The union of the solution sets $\{1\}$, $(1, \infty)$, and $(-\infty, 1)$ is the set of all real numbers; that is, $(-\infty, \infty)$. The statement is *true.* (See Section 3.1, Exercises 39–43, for a discussion of this concept.)

3. The union of $(-\infty, 6)$ and $(6, \infty)$ is $(-\infty, 6) \cup (6, \infty)$, which is all numbers except 6. This statement is *false*.

In Exercises 5–12, let $A = \{1, 2, 3, 4, 5, 6\}$, $B = \{1, 3, 5\}$, $C = \{1, 6\}$, and $D = \{4\}$.

5. The intersection of sets A and D is the set of all elements in both set A and D. Therefore,

$$A \cap D = \{4\} \text{ or set } D.$$

7. The intersection of set B and the set of no elements (empty set), $B \cap \emptyset$, is the set of no elements or $\emptyset$.

9. The union of sets A and B is the set of all elements that are in either set A or set B or both sets A and B. Since all numbers in set B are also in set A, the set $A \cup B$ will be the same as set A.

$$A \cup B = \{1, 2, 3, 4, 5, 6\} \text{ or set } A$$

11. The set $B \cup C$ is made up of all numbers that are either in set B *or* in set C (or both).

$$B \cup C = \{1, 3, 5, 6\}$$

13. The first graph represents the set $(-\infty, 5)$. The second graph represents the set $(0, \infty)$. The intersection includes the elements common to both sets, that is, $(0, 5)$.

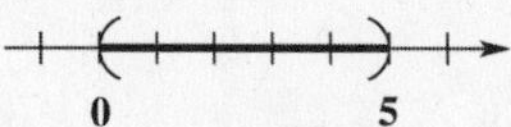

15. The first graph represents the set $[1, \infty)$. The second graph represents the set $[4, \infty)$. The intersection includes the elements common to both sets, that is $[4, \infty)$.

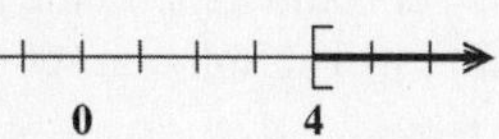

17. The first graph represents the set $(-\infty, 1]$. The second graph represents the set $(-\infty, 3]$. The union includes all elements in either set, or in both, that is, $(-\infty, 3]$.

19. Answers will vary. One example is: The intersection of two streets is the region common to *both* streets.

21. $x < 2$ and $x > -3$

The graph of the solution set will be all numbers that are both less than 2 and greater than -3. The solution set is $(-3, 2)$.

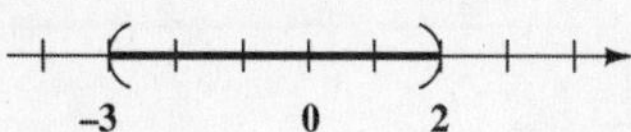

23. $x \le 2$ and $x \le 5$

The graph of the solution set will be all numbers that are both less than or equal to 2 and less than or equal to 5. The overlap is the numbers less than or equal to 2. The solution set is $(-\infty, 2]$.

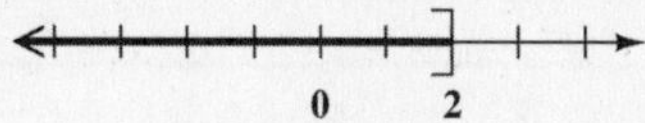

25. $x \le 3$ and $x \ge 6$

The graph of the solution set will be all numbers that are both less than or equal to 3 and greater than or equal to 6. There are no such numbers. The solution set is $\emptyset$.

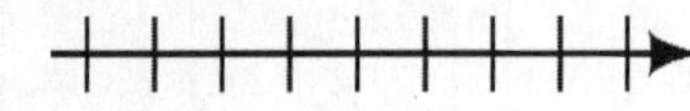

27. $x - 3 \le 6$ and $x + 2 \ge 7$
$x \le 9$ and $x \ge 5$

The graph of the solution set is all numbers that are both less than or equal to 9 and greater than or equal to 5. This is the intersection. The elements common to both sets are the numbers between 5 and 9, including the endpoints. The solution set is $[5, 9]$.

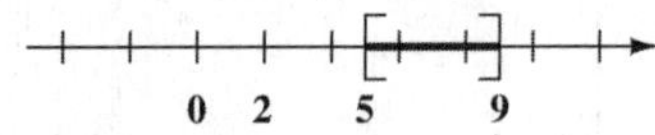

29. $3x - 4 \le 8$ and $4x - 1 \le 15$
$3x \le 12$ and $4x \le 16$
$x \le 4$ and $x \le 4$

Since both inequalities are identical, the graph of the solution set is the same as the graph of one of the inequalities. The solution set is $(-\infty, 4]$.

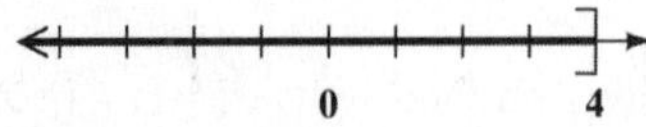

31. $x \le 1$ or $x \le 8$

The graph of the solution set will be all numbers that are either less than or equal to 1 or less than or equal to 8. The solution set is $(-\infty, 8]$.

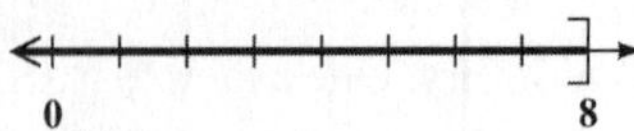

33. $x \ge -2$ or $x \ge 5$

The graph of the solution set will be all numbers that are either greater than or equal to -2 or greater than or equal to 5. The solution set is $[-2, \infty)$.

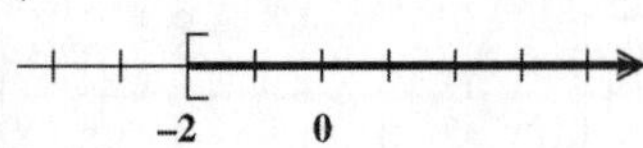

35. $x + 3 \ge 1$ or $x - 8 \le -4$
$x \ge -2$ or $x \le 4$

The graph of the solution set will be all numbers that are either greater than or equal to -2 or less than or equal to 4. This is the set of all real numbers. The solution set is $(-\infty, \infty)$.

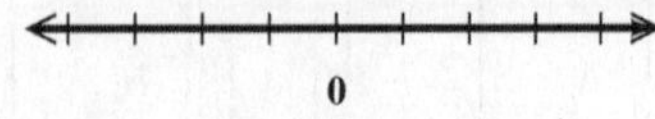

37. $x + 2 > 7$ or $1 - x > 6$
$-5 > x$
$x > 5$ or $x < -5$

The graph of the solution set is all numbers either greater than 5 or less than -5. This is the union. The solution set is $(-\infty, -5) \cup (5, \infty)$.

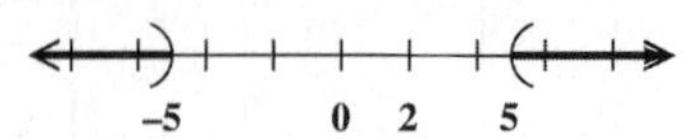

39. $x + 1 > 3$ or $-4x + 1 \ge 5$
$-4x \ge 4$
$x > 2$ or $x \le -1$

The graph of the solution set is all numbers either less than -1 or greater than 2. This is the union. The solution set is $(-\infty, -1] \cup (2, \infty)$.

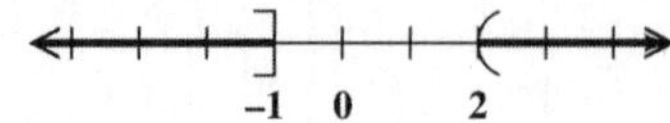

41. $4x - 8 > 0$ or $4x - 1 < 7$
$4x > 8$ or $4x < 8$
$x > 2$ or $x < 2$

The graph of the solution set is all numbers either greater than 2 or less than 2. This is all real numbers except 2. The solution set is $(-\infty, 2) \cup (2, \infty)$.

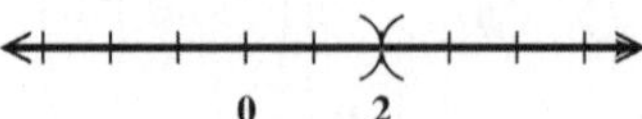

43. $(-\infty, -1] \cap [-4, \infty)$

The intersection is the set of numbers less than or equal to -1 and greater than or equal to -4. The numbers common to both original sets are between, and including, -4 and -1. The simplest interval form is $[-4, -1]$.

45. $(-\infty, -6] \cap [-9, \infty)$

The intersection is the set of numbers less than or equal to -6 and greater than or equal to -9. The numbers common to both original sets are between, and including, -9 and -6. The simplest interval form is $[-9, -6]$.

47. $(-\infty, 3) \cup (-\infty, -2)$

The union is the set of numbers that are either less than 3 or less than -2, or both. This is all numbers less than 3. The simplest interval form is $(-\infty, 3)$.

49. $[3, 6] \cup (4, 9)$

The union is the set of numbers between, and including, 3 and 6, or between, but not including, 4 and 9. This is the set of numbers greater than or equal to 3 and less than 9. The simplest interval form is $[3, 9)$.

51. $x < -1$ and $x > -5$

The word "and" means to take the intersection of both sets. $x < -1$ and $x > -5$ is true only when

$$-5 < x < -1.$$

The graph of the solution set is all numbers greater than -5 *and* less than -1. This is all numbers between -5 and -1, not including -5 or -1. The solution set is $(-5, -1)$.

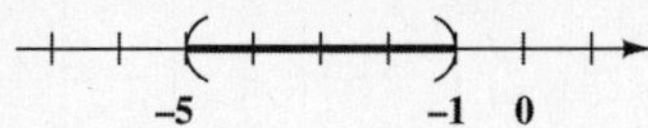

53. $x < 4$ or $x < -2$

The word "or" means to take the union of both sets. The graph of the solution set is all numbers that are either less than 4 *or* less than -2, or both. This is all numbers less than 4. The solution set is $(-\infty, 4)$.

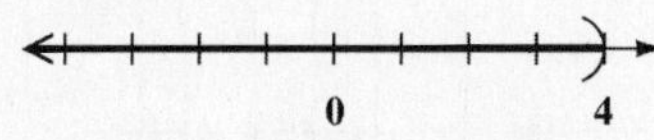

55. $x + 1 \geq 5$ and $x - 2 \leq 10$
$x \geq 4$ and $x \leq 12$

The word "and" means to take the intersection of both sets. The graph of the solution set is all numbers that are both greater than or equal to 4 *and* less than or equal to 12. This is all numbers between, and including, 4 and 12. The solution set is $[4, 12]$.

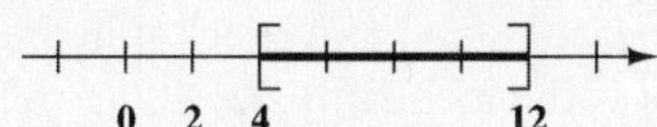

57. $-3x \leq -6$ or $-3x \geq 0$
$x \geq 2$ or $x \leq 0$

The word "or" means to take the union of both sets. The graph of the solution set is all numbers that are either greater than or equal to 2 *or* less than or equal to 0. The solution set is $(-\infty, 0] \cup [2, \infty)$.

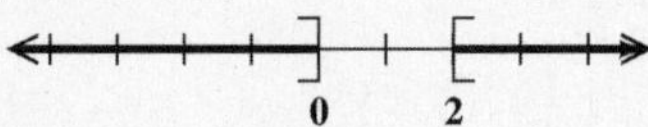

For Exercises 59–62, find the area and perimeter of each of the given yards.

For Luigi's, Mario's, and Than's yards, use the formulas $A = LW$ and $P = 2L + 2W$.

Luigi's yard
$A = 50(30) = 1500 \text{ ft}^2$
$P = 2(50) + 2(30) = 160 \text{ ft}$

Mario's yard
$A = 40(35) = 1400 \text{ ft}^2$
$P = 2(40) + 2(35) = 150 \text{ ft}$

Than's yard
$A = 60(50) = 3000 \text{ ft}^2$
$P = 2(60) + 2(50) = 220 \text{ ft}$

For Joe's yard, use the formulas $A = \frac{1}{2}bh$ and $P = a + b + c$.

Joe's yard
$A = \frac{1}{2}(40)(30) = 600 \text{ ft}^2$
$P = 30 + 40 + 50 = 120 \text{ ft}$

To be fenced, a yard must have a perimeter $P \leq 150$ ft. To be sodded, a yard must have an area $A \leq 1400 \text{ ft}^2$.

59. Find "the yard can be fenced *and* the yard can be sodded."

A yard that can be fenced has $P \leq 150$. Mario and Joe qualify.

A yard that can be sodded has $A \leq 1400$. Again, Mario and Joe qualify.

Find the intersection. Mario's and Joe's yards are common to both sets, so Mario and Joe can have their yards both fenced and sodded.

60. Find "the yard can be fenced *and* the yard cannot be sodded."

A yard that can be fenced has $P \leq 150$. Mario and Joe qualify.

A yard that cannot be sodded has $A > 1400$. Luigi and Than qualify.

Find the intersection. There are no yards common to both sets, so none of them qualify.

61. Find "the yard cannot be fenced *and* the yard can be sodded."

A yard that cannot be fenced has $P > 150$. Luigi and Than qualify.

A yard that can be sodded has $A \leq 1400$. Mario and Joe qualify.

Find the intersection. There are no yards common to both sets, so none of the qualify.

62. Find "the yard cannot be fenced *and* the yard cannot be sodded."

A yard that cannot be fenced has $P > 150$. Luigi and Than qualify.

A yard that cannot be sodded has $A > 1400$. Again, Luigi and Than qualify.

Find the intersection. Luigi's and Than's yards are common to both sets, so Luigi and Than qualify.

63. Find "the yard can be fenced *or* the yard can be sodded." From Exercise 59, Maria's and Joe's yards qualify for both conditions, so the union is Maria and Joe.

64. Find "the yard cannot be fenced *or* the yard can be sodded." From Exercise 61, Luigi's and Than's yards cannot be fenced, and from Exercise 59, Maria's and Joe's yards can be sodded. The union includes all of them.

65. The set of expenses that are less than \$2500 for public schools *and* are greater than \$5000 for private schools is {Tuition and fees}.

67. The set of expenses that are less than \$2300 for public schools *or* are greater than \$10,000 for private schools is {Tuition and fees, Dormitory charges}.

3.3 Absolute Value Equations and Inequalities

3.3 Margin Exercises

1. **(a)** $|x| = 3$

The solution set consists of the numbers that are 3 units from 0; that is, the numbers 3 and -3.

Solution set: $\{-3, 3\}$

-3 -2 -1 0 1 2 3

(b) $|x| > 3$

The solution set consists of the numbers that are more than 3 units from 0. This is all numbers greater than 3 *or* all numbers less than -3.

Solution set: $(-\infty, -3) \cup (3, \infty)$

-3 -2 -1 0 1 2 3

(c) $|x| < 3$

The solution set consists of the numbers that are less than 3 units from 0. This is all numbers between -3 and 3.

Solution set: $(-3, 3)$

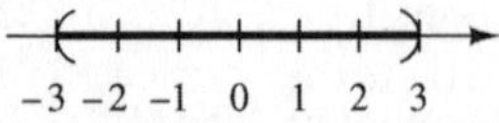

2. **(a)** $|x + 2| = 3$

$$x + 2 = 3 \quad \text{or} \quad x + 2 = -3$$
$$x = 1 \quad \text{or} \quad x = -5$$

Check $x = 1$: $|3| = 3$ *True*

Check $x = -5$: $|-3| = 3$ *True*

Solution set: $\{-5, 1\}$

-5 -4 -3 -2 -1 0 1

(b) $|3x - 4| = 11$

$$3x - 4 = 11 \quad \text{or} \quad 3x - 4 = -11$$
$$3x = 15 \quad \text{or} \quad 3x = -7$$
$$x = 5 \quad \text{or} \quad x = -\tfrac{7}{3}$$

Check $x = 5$: $|11| = 11$ *True*

Check $x = -\frac{7}{3}$: $|-11| = 11$ *True*

Solution set: $\{-\frac{7}{3}, 5\}$

$-\frac{7}{3}$ 0 2 4 5

3. **(a)** $|x + 2| > 3$

$$x + 2 > 3 \quad \text{or} \quad x + 2 < -3$$
$$x > 1 \quad \text{or} \quad x < -5$$

We'll assume that we have found the correct boundary points, -5 and 1. These points partition the real number line into 3 intervals. We'll pick a point in each region and check it in the original absolute value inequality, $|x + 2| > 3$.

Check $x = -6$: $|-4| > 3$ *True*
Check $x = 0$: $|2| > 3$ *False*
Check $x = 2$: $|4| > 3$ *True*

Solution set: $(-\infty, -5) \cup (1, \infty)$

-5 -4 -3 -2 -1 0 1

(b) $|3x - 4| \geq 11$

$$3x - 4 \geq 11 \quad \text{or} \quad 3x - 4 \leq -11$$
$$3x \geq 15 \quad \text{or} \quad 3x \leq -7$$
$$x \geq 5 \quad \text{or} \quad x \leq -\tfrac{7}{3}$$

Check $x = -3, 0$, and 6 in $|3x - 4| \geq 11$.

Check $x = -3$: $|-13| \geq 11$ *True*
Check $x = 0$: $|-4| \geq 11$ *False*
Check $x = 6$: $|14| \geq 11$ *True*

Solution set: $(-\infty, -\frac{7}{3}] \cup [5, \infty)$

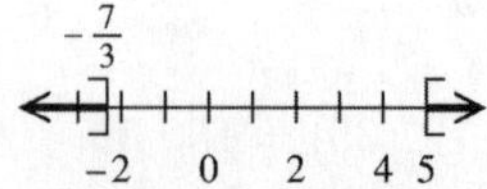

4. **(a)** $|x+2|<3$

$$-3<x+2<3$$
$$-5<x<1$$

Check $x=-6$, 0, and 2 in $|x+2|<3$.

Check $x=-6$: $|-4|<3$ *False*
Check $x=0$: $|2|<3$ *True*
Check $x=2$: $|4|<3$ *False*

Solution set: $(-5,1)$

−5 −4 −3 −2 −1 0 1

(b) $|3x-4|\le 11$

$$-11\le 3x-4\le 11$$
$$-7\le 3x\le 15$$
$$-\tfrac{7}{3}\le x\le 5$$

Check $x=-3$, 0, and 6 in $|3x-4|\le 11$.

Check $x=-3$: $|-13|\le 11$ *False*
Check $x=0$: $|-4|\le 11$ *True*
Check $x=6$: $|14|\le 11$ *False*

Solution set: $[-\tfrac{7}{3},5]$

$-\frac{7}{3}$ 0 2 4 5

5. $|5x+2|-9=-7$
We first *isolate* the absolute value expression, that is, rewrite the equation so that the absolute value expression is alone on one side of the equals sign.

$$|5x+2|=2$$

$5x+2=2$ or $5x+2=-2$
$5x=0$ or $5x=-4$
$x=0$ or $x=-\frac{4}{5}$

Check $x=0$: $|2|-9=-7$ *True*
Check $x=-\frac{4}{5}$: $|-2|-9=-7$ *True*

Solution set: $\{-\frac{4}{5},0\}$

6. **(a)** $|x+2|-3>2$
$|x+2|>5$ *Isolate.*

$x+2>5$ or $x+2<-5$
$x>3$ or $x<-7$

Solution set: $(-\infty,-7)\cup(3,\infty)$

−7 −4 −2 0 3

(b) $|3x+2|+4\le 15$
$|3x+2|\le 11$ *Isolate.*

$$-11\le 3x+2\le 11$$
$$-13\le 3x\le 9$$
$$-\tfrac{13}{3}\le x\le 3$$

Solution set: $[-\frac{13}{3},3]$

$-\frac{13}{3}$ −2 0 2 3

7. **(a)** $|k-1|=|5k+7|$

$k-1=5k+7$ or $k-1=-(5k+7)$
$-8=4k$ or $k-1=-5k-7$
$-2=k$ or $6k=-6$
$k=-1$

Check $k=-2$: $|-3|=|-3|$ *True*
Check $k=-1$: $|-2|=|2|$ *True*

Solution set: $\{-2,-1\}$

(b) $|4r-1|=|3r+5|$

$4r-1=3r+5$ or $4r-1=-(3r+5)$
$r=6$ or $4r-1=-3r-5$
$7r=-4$
$r=-\frac{4}{7}$

Check $r=6$: $|23|=|23|$ *True*
Check $r=-\frac{4}{7}$: $|-\frac{23}{7}|=|\frac{23}{7}|$ *True*

Solution set: $\{-\frac{4}{7},6\}$

8. **(a)** $|6x+7|=-5$

Since the absolute value of an expression can never be negative, there are no solutions for this equation.

Solution set: $\emptyset$

(b) $|\frac{1}{4}x-3|=0$
The expression $\frac{1}{4}x-3$ will equal 0 *only* if

$\frac{1}{4}x-3=0.$
$x-12=0$ *Multiply by 4.*
$x=12$

Solution set: $\{12\}$

9. **(a)** $|x|>-1$

The absolute value of a number is always greater than or equal to 0. Therefore, the inequality is true for all real numbers.

Solution set: $(-\infty,\infty)$

(b) $|x| < -5$

There is no number whose absolute value is less than a negative number, so this inequality has no solution.

Solution set: $\emptyset$

(c) $|x+2| \le 0$

The value of $|x+2|$ will never be less than 0. However, $|x+2|$ will equal 0 when $x=-2$.

Solution set: $\{-2\}$

(d) $|t-10| - 2 \le -3$

$$|t-10| \le -1$$

There is no number whose absolute value is less than a negative number, so this inequality has no solution.

Solution set: $\emptyset$

3.3 Section Exercises

1. $|x| = 5$ has two solutions, $x=5$ or $x=-5$. The graph is Choice **E**.

$|x| < 5$ is written $-5 < x < 5$. Notice that -5 and 5 are not included. The graph is Choice **C**, which uses parentheses.

$|x| > 5$ is written $x < -5$ or $x > 5$. The graph is Choice **D**, which uses parentheses.

$|x| \le 5$ is written $-5 \le x \le 5$. This time -5 and 5 are included. The graph is Choice **B**, which uses brackets.

$|x| \ge 5$ is written $x \le -5$ or $x \ge 5$. The graph is Choice **A**, which uses brackets.

3. **(a)** $|ax+b| = k,\ k=0$
This means the distance from $ax+b$ to 0 is 0, so $ax+b=0$, which has one solution.

(b) $|ax+b| = k,\ k>0$
This means the distance from $ax+b$ to 0 is a positive number k, so $ax+b=k$ or $ax+b=-k$. There are two solutions.

(c) $|ax+b| = k,\ k<0$
This means the distance from $ax+b$ to 0 is a negative number k, which is impossible because distance is always non-negative. There are no solutions.

5. $|x| = 12$

$x = 12$ or $x = -12$

Solution set: $\{-12, 12\}$

7. $|4x| = 20$

$4x = 20$ or $4x = -20$

$x = 5$ or $x = -5$

Solution set: $\{-5, 5\}$

9. $|x-3| = 9$

$x-3 = 9$ or $x-3=-9$

$x = 12$ or $x = -6$

Solution set: $\{-6, 12\}$

11. $|2x+1| = 9$

$2x+1 = 9$ or $2x+1 = -9$

$2x = 8$ or $2x = -10$

$x = 4$ or $x = -5$

Solution set: $\{-5, 4\}$

13. $|4r-5| = 17$

$4r-5 = 17$ or $4r-5=-17$

$4r = 22$ or $4r = -12$

$r = \frac{22}{4} = \frac{11}{2}$ or $r = -3$

Solution set: $\{-3, \frac{11}{2}\}$

15. $|2x+5| = 14$

$2x+5 = 14$ or $2x+5 = -14$

$2x = 9$ or $2x = -19$

$x = \frac{9}{2}$ or $x = -\frac{19}{2}$

Solution set: $\{-\frac{19}{2}, \frac{9}{2}\}$

17. $\left|\frac{1}{2}x + 3\right| = 2$

$\frac{1}{2}x + 3 = 2$ or $\frac{1}{2}x+3 = -2$

$\frac{1}{2}x = -1$ or $\frac{1}{2}x = -5$

$x = -2$ or $x = -10$

Solution set: $\{-10, -2\}$

19. $\left|1 - \frac{3}{4}k\right| = 7$

$1 - \frac{3}{4}k = 7$ or $1 - \frac{3}{4}k = -7$

Multiply all sides by 4.

$4 - 3k = 28$ or $4-3k = -28$

$-3k = 24$ or $-3k = -32$

$k = -8$ or $k = \frac{32}{3}$

Solution set: $\{-8, \frac{32}{3}\}$

21. $|x| > 3$

$x > 3$ or $x < -3$

Solution set: $(-\infty, -3) \cup (3, \infty)$

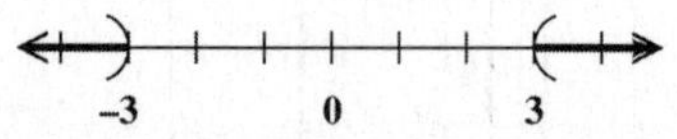

23. $|k| \geq 4$

$k \geq 4$ or $k \leq -4$

Solution set: $(-\infty, -4] \cup [4, \infty)$

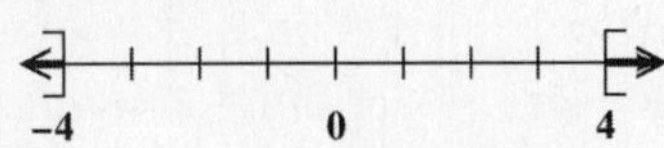

25. $|t + 2| > 8$

$t + 2 > 8$ or $t + 2 < -8$

$t > 6$ or $t < -10$

Solution set: $(-\infty, -10) \cup (6, \infty)$

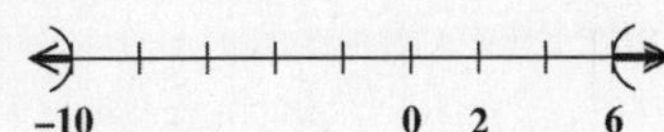

27. $|3x - 1| \geq 8$

$3x - 1 \geq 8$ or $3x - 1 \leq -8$

$3x \geq 9$ or $3x \leq -7$

$x \geq 3$ or $x \leq -\frac{7}{3}$

Solution set: $(-\infty, -\frac{7}{3}] \cup [3, \infty)$

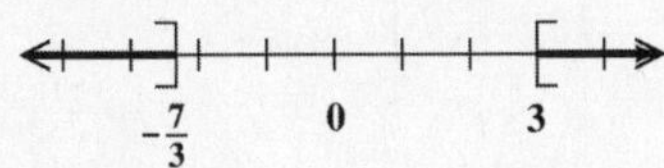

29. $|3 - x| > 5$

$3 - x > 5$ or $3 - x < -5$

$-x > 2$ or $-x < -8$

Multiply by -1,
and reverse the inequality symbols.

$x < -2$ or $x > 8$

Solution set: $(-\infty, -2) \cup (8, \infty)$

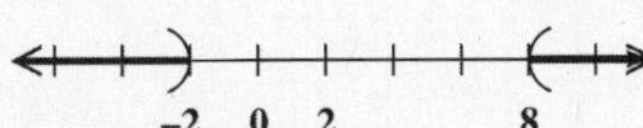

31. **(a)** $|2x + 1| < 9$

The graph of the solution set will be all numbers between -5 and 4, since the absolute value is less than 9.

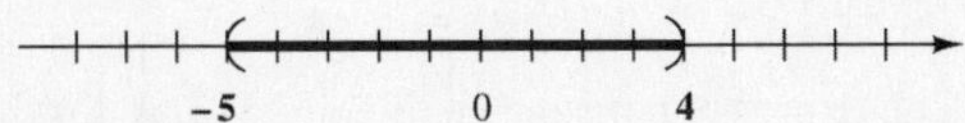

(b) $|2x + 1| > 9$

The graph of the solution set will be all numbers less than -5 or greater than 4, since the absolute value is greater than 9.

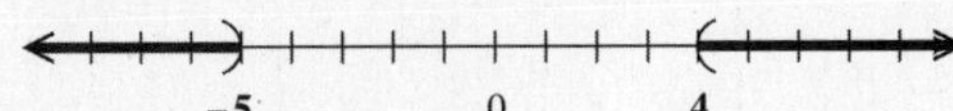

33. $|x| \leq 3$

$-3 \leq x \leq 3$

Solution set: $[-3, 3]$

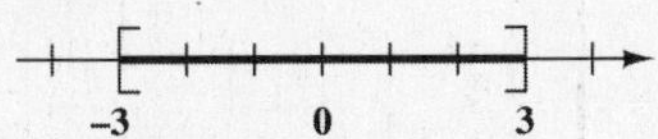

35. $|k| < 4$

$-4 < k < 4$

Solution set: $(-4, 4)$

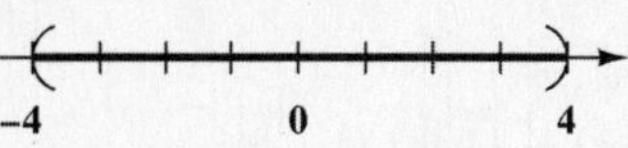

37. $|t + 2| \leq 8$

$-8 \leq t + 2 \leq 8$

$-10 \leq t \leq 6$

Solution set: $[-10, 6]$

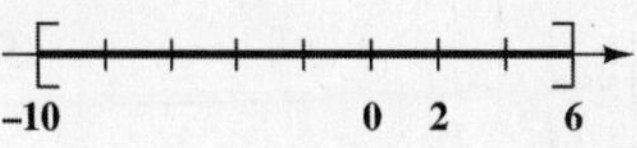

39. $|3x - 1| < 8$

$-8 < 3x - 1 < 8$

$-7 < 3x < 9$ *Add 1.*

$-\frac{7}{3} < x < 3$ *Divide by 3.*

Solution set: $(-\frac{7}{3}, 3)$

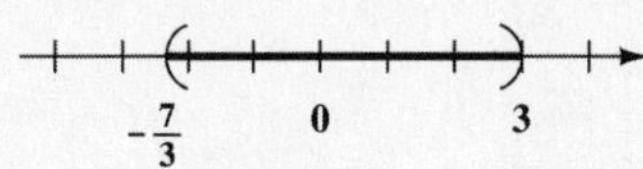

41. $|3 - x| \leq 5$

$-5 \leq 3 - x \leq 5$

$-8 \leq -x \leq 2$

Multiply each part by -1, and reverse the inequality symbols.

$8 \geq x \geq -2$ or $-2 \leq x \leq 8$

Solution set: $[-2, 8]$

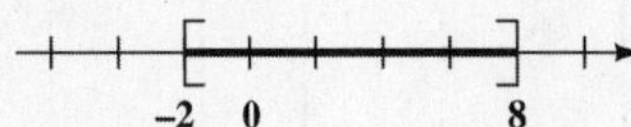

43. $|-4 + k| > 6$

$-4 + k > 6$ or $-4 + k < -6$

$k > 10$ or $k < -2$

Solution set: $(-\infty, -2) \cup (10, \infty)$

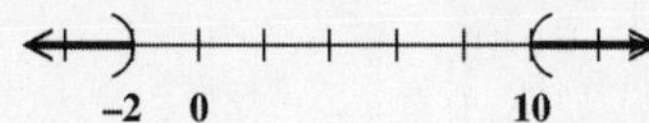

45. $|7 + 2z| = 5$

$7 + 2z = 5$ or $7 + 2z = -5$

$2z = -2$ or $2z = -12$

$z = -1$ or $z = -6$

Solution set: $\{-6, -1\}$

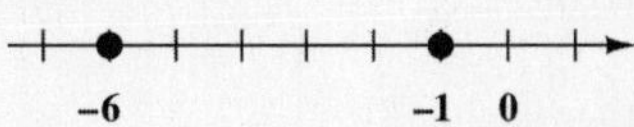

47. $|3r-1| \le 11$
$-11 \le 3r-1 \le 11$
$-10 \le 3r \le 12$
$-\frac{10}{3} \le r \le 4$

Solution set: $[-\frac{10}{3}, 4]$

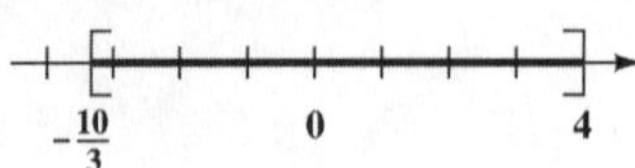

49. $|-3x-8| \le 4$
$-4 \le -3x-8 \le 4$
$4 \le -3x \le 12$
Divide each side by -3 and reverse the inequality symbols.
$-\frac{4}{3} \ge x \ge -4$ or $-4 \le x \le -\frac{4}{3}$

Solution set: $[-4, -\frac{4}{3}]$

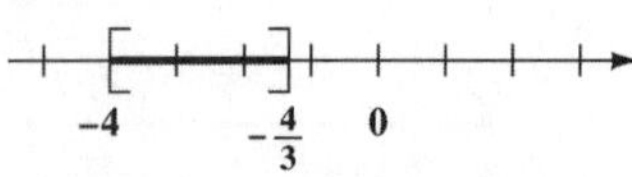

51. $|x|-1=4$
$|x|=5$

$x=5$ or $x=-5$

Solution set: $\{-5,5\}$

53. $|x+4|+1=2$
$|x+4|=1$
$x+4=1$ or $x+4=-1$
$x=-3$ or $x=-5$

Solution set: $\{-5,-3\}$

55. $|2x+1|+3>8$
$|2x+1|>5$
$2x+1>5$ or $2x+1<-5$
$2x>4$ or $2x<-6$
$x>2$ or $x<-3$

Solution set: $(-\infty,-3) \cup (2,\infty)$

57. $|x+5|-6 \le -1$
$|x+5| \le 5$
$-5 \le x+5 \le 5$
$-10 \le x \le 0$

Solution set: $[-10,0]$

59. $|3x+1|=|2x+4|$
$3x+1=2x+4$ or $3x+1=-(2x+4)$
$3x+1=-2x-4$
$5x=-5$
$x=3$ or $x=-1$

Solution set: $\{-1,3\}$

61. $\left|m-\dfrac{1}{2}\right| = \left|\dfrac{1}{2}m-2\right|$
$m-\frac{1}{2}=\frac{1}{2}m-2$ or $m-\frac{1}{2}=-(\frac{1}{2}m-2)$
Multiply by 2. $m-\frac{1}{2}=-\frac{1}{2}m+2$
$2m-1=m-4$ or $2m-1=-m+4$
$3m=5$
$m=-3$ or $m=\frac{5}{3}$

Solution set: $\{-3,\frac{5}{3}\}$

63. $|6x|=|9x+1|$
$6x=9x+1$ or $6x=-(9x+1)$
$-3x=1$ or $6x=-9x-1$
$15x=-1$
$x=-\frac{1}{3}$ or $x=-\frac{1}{15}$

Solution set: $\{-\frac{1}{3},-\frac{1}{15}\}$

65. $|2p-6|=|2p+11|$
$2p-6=2p+11$ or $2p-6=-(2p+11)$
$-6=11$ *False* or $2p-6=-2p-11$
$4p=-5$
No solution or $p=-\frac{5}{4}$

Solution set: $\{-\frac{5}{4}\}$

67. $|x| \ge -10$

The absolute value of a number is always greater than or equal to 0. Therefore, the inequality is true for all real numbers.

The solution set is $(-\infty,\infty)$.

69. $|12t-3|=-8$

Since the absolute value of an expression can never be negative, there are no solutions for this equation.
Solution set: $\emptyset$

71. $|4x+1|=0$

The expression $4x+1$ will equal 0 *only* for the solution of the equation

$4x+1=0.$
$4x=-1$
$x=\frac{-1}{4}$ or $-\frac{1}{4}$

Solution set: $\{-\frac{1}{4}\}$

73. $|2q-1|<-6$

There is no number whose absolute value is less than a negative number, so this inequality has no solution.

Solution set: $\emptyset$

75. $|x+5| > -9$

Since the absolute value of an expression is always nonnegative (positive or zero), the inequality is true for any real number x.

Solution set: $(-\infty, \infty)$

77. $|7x+3| \le 0$

The absolute value of an expression is always nonnegative (positive or zero), so this inequality is true only when

$$\begin{aligned} 7x+3 &= 0 \\ 7x &= -3 \\ x &= -\tfrac{3}{7}. \end{aligned}$$

Solution set: $\{-\frac{3}{7}\}$

79. $|5x-2| \ge 0$

The absolute value of an expression is always nonnegative, so the inequality is true for every possible value of x.

Solution set: $(-\infty, \infty)$

81. $|10z+7| > 0$

Since an absolute value of an expression is always nonnegative, there is only one possible value of z that makes this statement false. Solving the equation $10z+7=0$ will give that value of z.

$$\begin{aligned} 10z+7 &= 0 \\ 10z &= -7 \\ z &= -\tfrac{7}{10} \end{aligned}$$

The solution set of the inequality is

$$(-\infty, -\tfrac{7}{10}) \cup (-\tfrac{7}{10}, \infty).$$

83. $|x-2|+3 \ge 2$

$|x-2| \ge -1$

Since the absolute value of an expression is always nonnegative (positive or zero), the inequality is true for any real number x.

The solution set is $(-\infty, \infty)$.

85. Let x represent the calcium intake for a specific female. For x to be within 100 mg of 1000 mg, we must have

$$|x-1000| \le 100.$$

$$\begin{aligned} -100 \le x &- 1000 \le 100 \\ 900 \le \quad &x \quad \le 1100 \end{aligned}$$

87. Add the given heights with a calculator to get 4756. There are 10 numbers, so divide the sum by 10.

$$\frac{4756}{10} = 475.6$$

The average height is 475.6 ft.

88. $|x-k| < 50$

Substitute 475.6 for k and solve the inequality.

$$|x-475.6| < 50$$

$$\begin{aligned} -50 < x &- 475.6 < 50 \\ 425.6 < \quad &x \quad < 525.6 \end{aligned}$$

The buildings with heights between 425.6 ft and 525.6 ft are the 1201 Walnut, City Hall, Fidelity Bank and Trust Building, Kansas City Power and Light, and the Hyatt Regency Crown Center.

89. $|x-k| < 75$

Substitute 475.6 for k and solve the inequality.

$$|x-475.6| < 75$$

$$\begin{aligned} -75 < x &- 475.6 < 75 \\ 400.6 < \quad &x \quad < 550.6 \end{aligned}$$

The buildings with heights between 400.6 ft and 550.6 ft are City Center Square, Commerce Tower, Federal Office Building, 1201 Walnut, City Hall, Fidelity Bank and Trust Building, Kansas City Power and Light, and the Hyatt Regency Crown Center.

90. **(a)** This would be the opposite of the inequality in Exercise 89, that is,

$$|x-475.6| \ge 75.$$

(b) $|x-475.6| \ge 75$

$$\begin{aligned} x-475.6 &\ge 75 \quad \text{or} \quad & x-475.6 &\le -75 \\ x &\ge 550.6 \quad \text{or} \quad & x &\le 400.6 \end{aligned}$$

(c) The buildings that are not within 75 ft of the average have height less than or equal to 400.6 or greater than or equal to 550.6. They are Town Pavillion and One Kansas City Place.

(d) The answer makes sense because it includes all the buildings *not* listed earlier which had heights within 75 ft of the average.

Summary Exercises on Solving Linear and Absolute Value Equations and Inequalities

1. $4z+1=49$

$$\begin{aligned} 4z+1 &= 49 \\ 4z &= 48 \\ z &= 12 \end{aligned}$$

Solution set: $\{12\}$

3. $6q - 9 = 12 + 3q$
$3q = 21$
$q = 7$

Solution set: $\{7\}$

5. $|a + 3| = -4$

Since the absolute value of an expression is always nonnegative, there is no number that makes this statement true. Therefore, the solution set is $\emptyset$.

7. $8r + 2 \geq 5r$
$3r \geq -2$
$r \geq -\frac{2}{3}$

Solution set: $[-\frac{2}{3}, \infty)$

9. $2q - 1 = -7$
$2q = -6$
$q = -3$

Solution set: $\{-3\}$

11. $6z - 5 \leq 3z + 10$
$3z \leq 15$
$z \leq 5$

Solution set: $(-\infty, 5]$

13. $9x - 3(x + 1) = 8x - 7$
$9x - 3x - 3 = 8x - 7$
$6x - 3 = 8x - 7$
$4 = 2x$
$2 = x$

Solution set: $\{2\}$

15. $9x - 5 \geq 9x + 3$
$-5 \geq 3$ *False*

This is a false statement, so the inequality is a contradiction.

Solution set: $\emptyset$

17. $|q| < 5.5$
$-5.5 < q < 5.5$

Solution set: $(-5.5, 5.5)$

19. $\frac{2}{3}x + 8 = \frac{1}{4}x$
$8x + 96 = 3x$ *Multiply by 12.*
$5x = -96$
$x = -\frac{96}{5}$

Solution set: $\{-\frac{96}{5}\}$

21. $\frac{1}{4}p < -6$
$4(\frac{1}{4}p) < 4(-6)$
$p < -24$

Solution set: $(-\infty, -24)$

23. $\frac{3}{5}q - \frac{1}{10} = 2$
$6q - 1 = 20$ *Multiply by 10.*
$6q = 21$
$q = \frac{21}{6} = \frac{7}{2}$

Solution set: $\{\frac{7}{2}\}$

25. $r + 9 + 7r = 4(3 + 2r) - 3$
$8r + 9 = 12 + 8r - 3$
$8r + 9 = 8r + 9$
$0 = 0$ *True*

The last statement is true for any real number r.

Solution set: $(-\infty, \infty)$

27. $|2p - 3| > 11$
$2p - 3 > 11$ or $2p - 3 < -11$
$2p > 14$ or $2p < -8$
$p > 7$ or $p < -4$

Solution set: $(-\infty, -4) \cup (7, \infty)$

29. $|5a + 1| \leq 0$

The expression $|5a + 1|$ is never less than 0 since an absolute value expression must be nonnegative. However, $|5a + 1| = 0$ if

$5a + 1 = 0$
$5a = -1$
$a = \frac{-1}{5} = -\frac{1}{5}$

Solution set: $\{-\frac{1}{5}\}$

31. $-2 \leq 3x - 1 \leq 8$
$-1 \leq 3x \leq 9$
$-\frac{1}{3} \leq x \leq 3$

Solution set: $[-\frac{1}{3}, 3]$

33. $|7z - 1| = |5z + 3|$
$7z - 1 = 5z + 3$ or $7z - 1 = -(5z + 3)$
$2z = 4$ or $7z - 1 = -5z - 3$
$12z = -2$
$z = 2$ or $z = \frac{-2}{12} = -\frac{1}{6}$

Solution set: $\{-\frac{1}{6}, 2\}$

35. $|1-3x| \geq 4$

$1-3x \geq 4$ or $1-3x \leq -4$

$-3x \geq 3$ or $-3x \leq -5$

$x \leq -1$ or $x \geq \frac{5}{3}$

Solution set: $(-\infty, -1] \cup [\frac{5}{3}, \infty)$

37. $-(m+4)+2 = 3m+8$

$-m-4+2 = 3m+8$

$-m-2 = 3m+8$

$-10 = 4m$

$m = \frac{-10}{4} = -\frac{5}{2}$

Solution set: $\{-\frac{5}{2}\}$

39. $-6 \leq \frac{3}{2} - x \leq 6$

$-12 \leq 3-2x \leq 12$

$-15 \leq -2x \leq 9$

$\frac{15}{2} \geq x \geq -\frac{9}{2}$ or $-\frac{9}{2} \leq x \leq \frac{15}{2}$

Solution set: $[-\frac{9}{2}, \frac{15}{2}]$

41. $|x-1| \geq -6$

The absolute value of an expression is always nonnegative, so the inequality is true for any real number x.

Solution set: $(-\infty, \infty)$

43. $8q-(1-q) = 3(1+3q)-4$

$8q-1+q = 3+9q-4$

$9q-1 = 9q-1$ *True*

This is an identity.

Solution set: $(-\infty, \infty)$

45. $|r-5| = |r+9|$

$r-5 = r+9$ or $r-5 = -(r+9)$

$-5 = 9$ *False* $\quad r-5 = -r-9$

$2r = -4$

No solution or $r = -2$

Solution set: $\{-2\}$

47. $2x+1 > 5$ or $3x+4 < 1$

$2x > 4$ or $3x < -3$

$x > 2$ or $x < -1$

Solution set: $(-\infty, -1) \cup (2, \infty)$

Chapter 3 Review Exercises

1. $-\frac{2}{3}x < 6$

$-2x < 18$ *Multiply by 3.*

Divide by -2; reverse the inequality symbol.

$x > -9$

Solution set: $(-9, \infty)$

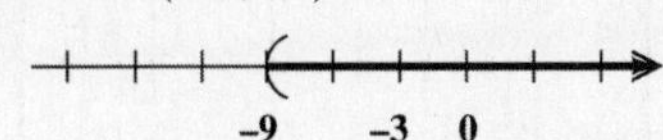

2. $-5x-4 \geq 11$

$-5x \geq 15$

Divide by -5; reverse the inequality symbol.

$x \leq -3$

Solution set: $(-\infty, -3]$

3. $\frac{6a+3}{-4} < -3$

Multiply by -4; reverse the inequality symbol.

$6a+3 > 12$

$6a > 9$

$a > \frac{9}{6} = \frac{3}{2}$

Solution set: $(\frac{3}{2}, \infty)$

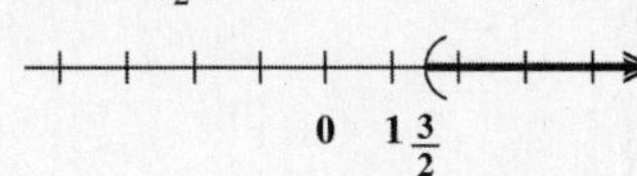

4. $\frac{9x+5}{-3} > 3$

Multiply by -3; reverse the inequality symbol.

$9x+5 < -9$

$9x < -14$

$x < -\frac{14}{9}$

Solution set: $(-\infty, -\frac{14}{9})$

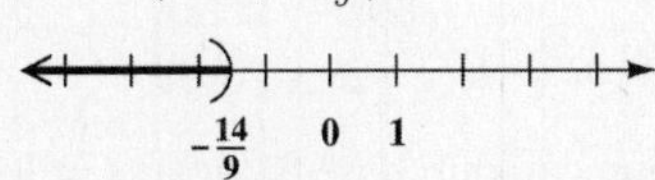

5. $5-(6-4t) \geq 2t-7$

$5-6+4t \geq 2t-7$

$4t-1 \geq 2t-7$

$2t \geq -6$

$t \geq -3$

Solution set: $[-3, \infty)$

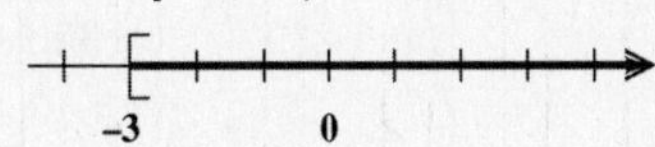

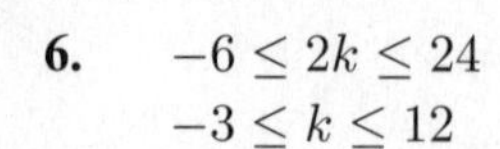

6. $-6 \leq 2k \leq 24$
$-3 \leq k \leq 12$

Solution set: $[-3, 12]$

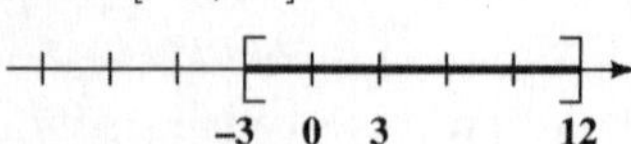

7. $8 \leq 3x - 1 < 14$
$9 \leq 3x < 15$
$3 \leq x < 5$

Solution set: $[3, 5)$

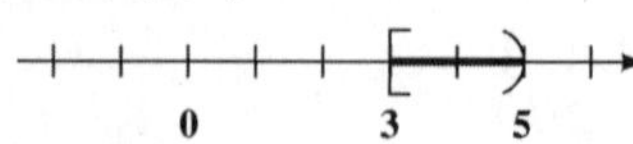

8. $-4 < 3 - 2z < 9$
$-7 < -2z < 6$
Divide all parts by -2; reverse the inequality symbols.
$\frac{7}{2} > z > -3$ or $-3 < z < \frac{7}{2}$

Solution set: $(-3, \frac{7}{2})$

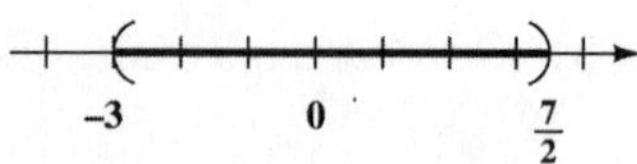

9. Let $x =$ the length of the playground. The width is 22 m, so the perimeter is $2x + 2 \cdot 22$. The perimeter must be no greater than 120 m, so

$$\begin{aligned} 2x + 44 &\leq 120 \\ 2x &\leq 76 \\ x &\leq 38 \end{aligned}$$

The length must be 38 m or less.

10. Let $x =$ the number of tickets that can be purchased. The total cost of the tickets is $\$89x$. Including the \$50 discount and staying within the available \$2000, we have

$$\begin{aligned} 89x - 50 &\leq 2000. \\ 89x &\leq 2050 \\ x &\lesssim 23.03 \end{aligned}$$

The group can purchase 23 tickets or less.

11. Let $x =$ the student's score on the fifth test. The average of the five test scores must be at least 70. The inequality is

$$\begin{aligned} \frac{75 + 79 + 64 + 71 + x}{5} &\geq 70. \\ 75 + 79 + 64 + 71 + x &\geq 350 \\ 289 + x &\geq 350 \\ x &\geq 61 \end{aligned}$$

The student will pass algebra if any score greater than or equal to 61 on the fifth test is achieved.

12. The result, $-8 < -13$, is a false statement. There are no real numbers that make this inequality true. The solution set is $\emptyset$.

For Exercises 13–16, let $A = \{\text{a, b, c, d}\}$, $B = \{\text{a, c, e, f}\}$, and $C = \{\text{a, e, f, g}\}$.

13. $A \cap B = \{\text{a, b, c, d}\} \cap \{\text{a, c, e, f}\}$
$= \{\text{a, c}\}$

14. $A \cap C = \{\text{a, b, c, d}\} \cap \{\text{a, e, f, g}\}$
$= \{\text{a}\}$

15. $B \cup C = \{\text{a, c, e, f}\} \cup \{\text{a, e, f, g}\}$
$= \{\text{a, c, e, f, g}\}$

16. $A \cup C = \{\text{a, b, c, d}\} \cup \{\text{a, e, f, g}\}$
$= \{\text{a, b, c, d, e, f, g}\}$

17. $x > 4$ and $x < 7$

The graph of the solution set will be all numbers which are both greater than 4 and less than 7. The overlap is the numbers between 4 and 7, not including the endpoints.

Solution set: $(4, 7)$

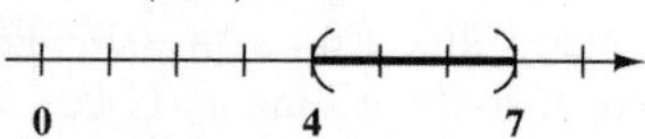

18. $x + 4 > 12$ and $x - 2 < 12$
$x > 8$ and $x < 14$

The graph of the solution set will be all numbers between 8 and 14, not including the endpoints.

Solution set: $(8, 14)$

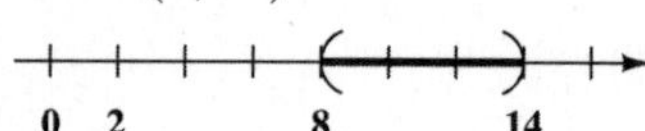

19. $x > 5$ or $x \leq -3$

The graph of the solution set will be all numbers that are either greater than 5 or less than or equal to -3.

Solution set: $(-\infty, -3] \cup (5, \infty)$

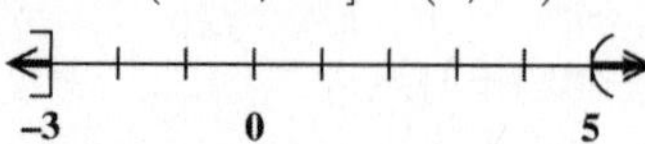

20. $x \geq -2$ or $x < 2$

The graph of the solution set will be all numbers that are either greater than or equal to -2 or less than 2. All real numbers satisfy these criteria.

Solution set: $(-\infty, \infty)$

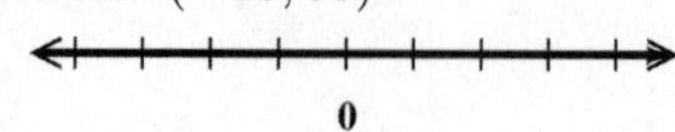

21. $x - 4 > 6$ and $x + 3 \leq 10$
$x > 10$ and $x \leq 7$

The graph of the solution set will be all numbers that are both greater than 10 and less than or equal to 7. There are no real numbers satisfying these criteria.

Solution set: $\emptyset$

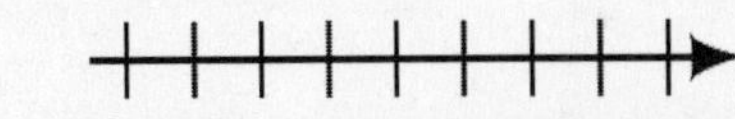

22. $-5x + 1 \geq 11$ or $3x + 5 \geq 26$
$-5x \geq 10$ or $3x \geq 21$
$x \leq -2$ or $x \geq 7$

The graph of the solution set will be all numbers that are either less than or equal to -2 or greater than or equal to 7.

Solution set: $(-\infty, -2] \cup [7, \infty)$

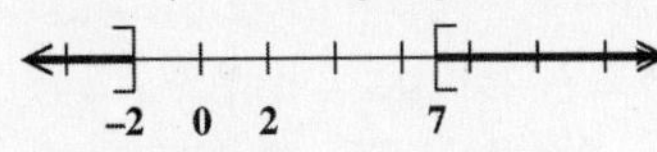

23. $(-3, \infty) \cap (-\infty, 4)$

$(-3, \infty)$ includes all real numbers greater than -3.
$(-\infty, 4)$ includes all real numbers less than 4.
Find the intersection. The numbers common to both sets are greater than -3 and less than 4.

$$-3 < x < 4$$

Solution set: $(-3, 4)$

24. $(-\infty, 6) \cap (-\infty, 2)$

$(-\infty, 6)$ includes all real numbers less than 6.
$(-\infty, 2)$ includes all real numbers less than 2.
Find the intersection. The numbers common to both sets are less than 2.

Solution set: $(-\infty, 2)$

25. $(4, \infty) \cup (9, \infty)$

$(4, \infty)$ includes all real numbers greater than 4.
$(9, \infty)$ includes all real numbers greater than 9.
Find the union. The numbers in the first set, the second set, or in both sets are all the real numbers that are greater than 4.

Solution set: $(4, \infty)$

26. $(1, 2) \cup (1, \infty)$

$(1, 2)$ includes the real numbers between 1 and 2, not including 1 and 2.
$(1, \infty)$ includes all real numbers greater than 1.
Find the union. The numbers in the first set, the second set, or in both sets are all real numbers greater than 1.

Solution set: $(1, \infty)$

27. **(a)** The set of states with less than 1 million female workers is {Maine, Oregon, Utah}. The set of states with more than 1 million male workers is {Illinois, North Carolina, Oregon, Wisconsin}. Oregon is the only state in both sets, so the set of states with less than 1 million female workers *and* more than 1 million male workers is {Oregon}.

(b) The set of states with less than 1 million female workers *or* more than 2 million male workers is {Illinois, Maine, North Carolina, Oregon, Utah}.

(c) It is easy to see that the sum of the female and male workers for each state doesn't exceed 7 million, so the set of states with a total of more than 7 million civilian workers is {}, or $\emptyset$.

28. $|x| = 7$
$x = 7$ or $x = -7$

Solution set: $\{-7, 7\}$

29. $|x + 2| = 9$

$x + 2 = 9$ or $x + 2 = -9$
$x = 7$ or $x = -11$

Solution set: $\{-11, 7\}$

30. $|3k - 7| = 8$
$3k - 7 = 8$ or $3k - 7 = -8$
$3k = 15$ or $3k = -1$
$k = 5$ or $k = -\frac{1}{3}$

Solution set: $\{-\frac{1}{3}, 5\}$

31. $|z - 4| = -12$

Since the absolute value of an expression can never be negative, there are no solutions for this equation.

Solution set: $\emptyset$

32. $|2k - 7| + 4 = 11$
$|2k - 7| = 7$
$2k - 7 = 7$ or $2k - 7 = -7$
$2k = 14$ or $2k = 0$
$k = 7$ or $k = 0$

Solution set: $\{0, 7\}$

33. $|4a + 2| - 7 = -3$
$|4a + 2| = 4$
$4a + 2 = 4$ or $4a + 2 = -4$
$4a = 2$ or $4a = -6$
$a = \frac{2}{4}$ or $a = -\frac{6}{4}$
$a = \frac{1}{2}$ or $a = -\frac{3}{2}$

Solution set: $\{-\frac{3}{2}, \frac{1}{2}\}$

34. $|3p+1| = |p+2|$

$3p+1 = p+2$ or $3p+1 = -(p+2)$

$2p = 1$ or $3p+1 = -p-2$

$4p = -3$

$p = \frac{1}{2}$ or $p = -\frac{3}{4}$

Solution set: $\{-\frac{3}{4}, \frac{1}{2}\}$

35. $|2m-1| = |2m+3|$

$2m-1 = 2m+3$ or $2m-1 = -(2m+3)$

$-1 = 3$ *False* or $2m-1 = -2m-3$

$4m = -2$

No solution or $m = -\frac{2}{4} = -\frac{1}{2}$

Solution set: $\{-\frac{1}{2}\}$

36. $|x| < 12$

$-12 < x < 12$

Solution set: $(-12, 12)$

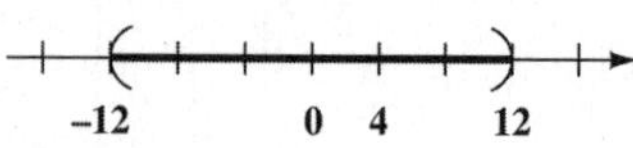

37. $|-x+6| \le 7$

$-7 \le -x+6 \le 7$

$-13 \le -x \le 1$

Multiply by -1; reverse the inequality symbols.

$13 \ge x \ge -1$ or $-1 \le x \le 13$

Solution set: $[-1, 13]$

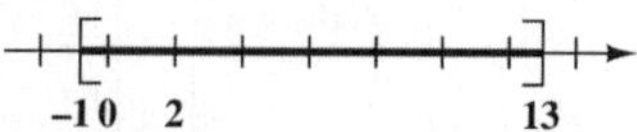

38. $|2p+5| \le 1$

$-1 \le 2p+5 \le 1$

$-6 \le 2p \le -4$

$-3 \le p \le -2$

Solution set: $[-3, -2]$

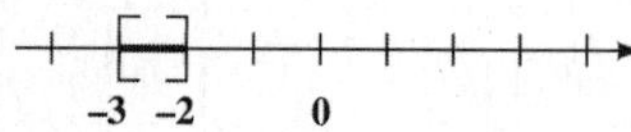

39. $|x+1| \ge -3$

Since the absolute value of an expression is always nonnegative (positive or zero), the inequality is *true* for any real number x.

Solution set: $(-\infty, \infty)$

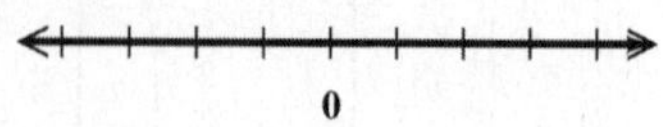

40. $|5r-1| > 9$

$5r-1 > 9$ or $5r-1 < -9$

$5r > 10$ or $5r < -8$

$r > 2$ or $r < -\frac{8}{5}$

Solution set: $(-\infty, -\frac{8}{5}) \cup (2, \infty)$

$-\frac{8}{5}$ 0 2

41. $|3x+6| \ge 0$

The absolute value of an expression is always nonnegative, so the inequality is true for any real number x.

Solution set: $(-\infty, \infty)$

0

42. **[3.1]** $(7-2x) + 3(5-3x) \ge x+8$

$7 - 2x + 15 - 9x \ge x + 8$

$-11x + 22 \ge x + 8$

$-12x \ge -14$

$x \le \frac{-14}{-12}$

$x \le \frac{7}{6}$

Solution set: $(-\infty, \frac{7}{6}]$

43. **[3.2]** $x < 5$ and $x \ge -4$

The real numbers that are common to both sets are the numbers greater than or equal to -4 and less than 5.

$-4 \le x < 5$

Solution set: $[-4, 5)$

44. **[3.1]** $\dfrac{3}{4}(a-2) - \dfrac{1}{3}(5-2a) < -2$

$9(a-2) - 4(5-2a) < -24$

Multiply by 12.

$9a - 18 - 20 + 8a < -24$

$17a - 38 < -24$

$17a < 14$

$a < \frac{14}{17}$

Solution set: $(-\infty, \frac{14}{17})$

45. **[3.1]** Let $x =$ the employee's earnings during the fifth month. The average of the five months must be at least \$1000.

$$\frac{900+1200+1040+760+x}{5} \geq 1000$$
$$900+1200+1040+760+x \geq 5000$$
$$3900+x \geq 5000$$
$$x \geq 1100$$

Any amount greater than or equal to \$1100 will qualify the employee for the pension plan.

46. **[3.1]** $-5r \geq -10$

$r \leq 2$ *Divide by -5; reverse inequality.*

Solution set: $(-\infty, 2]$

47. **[3.3]** $|7x-2| > 9$

$7x-2 > 9$ or $7x-2 < -9$
$7x > 11$ or $7x < -7$
$x > \frac{11}{7}$ or $x < -1$

Solution set: $(-\infty, -1) \cup (\frac{11}{7}, \infty)$

48. **[3.3]** $|2x-10| = 20$

$2x-10 = 20$ or $2x-10 = -20$
$2x = 30$ or $2x = -10$
$x = 15$ or $x = -5$

Solution set: $\{-5, 15\}$

49. **[3.3]** $|m+3| \leq 13$

$-13 \leq m+3 \leq 13$
$-16 \leq m \leq 10$

Solution set: $[-16, 10]$

50. **[3.2]** $x \geq -2$ or $x < 4$

The solution set includes all numbers either greater than or equal to -2 or all numbers less than 4. This is the union and is the set of all real numbers. The solution set is $(-\infty, \infty)$.

51. **[3.3]** $|m-1| = |2m+3|$

$m-1 = 2m+3$ or $m-1 = -(2m+3)$
$m-1 = -2m-3$
$3m = -2$
$-4 = m$ or $m = -\frac{2}{3}$

Solution set: $\{-4, -\frac{2}{3}\}$

52. **[3.2]** $x > 6$ and $x < 8$

The graph of the solution set is all numbers both greater than 6 *and* less than 8. This is the intersection. The elements common to both sets are the numbers between 6 and 8, not including the endpoints. The solution set is $(6, 8)$.

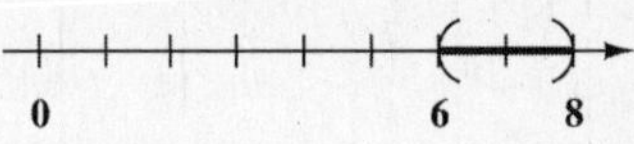

53. **[3.2]** $-5x+1 \geq 6$ or $3x+5 \geq 26$
$-5x \geq 5$ or $3x \geq 21$
$x \leq -1$ or $x \geq 7$

The graph of the solution set is all numbers either less than or equal to -1 *or* greater than or equal to 7. This is the union. The solution set is $(-\infty, -1] \cup [7, \infty)$.

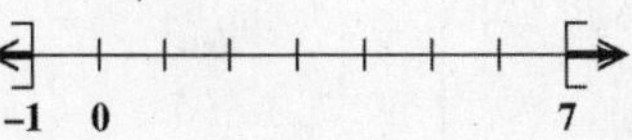

54. **[3.3]** **(a)** $|5x+3| < k$

If $k < 0$, then $|5x+3|$ would be less than a negative number. Since the absolute value of an expression is always nonnegative (positive or zero), the solution set is $\emptyset$.

(b) $|5x+3| > k$

If $k < 0$, then $|5x+3|$ would be greater than a negative number. Since the absolute value of an expression is always nonnegative (positive or zero), the solution set is the set of all real numbers, $(-\infty, \infty)$.

(c) $|5x+3| = k$

If $k < 0$, then $|5x+3|$ would be equal to a negative number. Since the absolute value of an expression is always nonnegative (positive or zero), the solution set is $\emptyset$.

Chapter 3 Test

1. When multiplying or dividing each side of an inequality by a negative number, we must reverse the direction of the inequality symbol.

2. $4-6(x+3) \leq -2-3(x+6)+3x$
$4-6x-18 \leq -2-3x-18+3x$
$-6x-14 \leq -20$
$-6x \leq -6$

Divide each side by -6 and reverse the direction of the inequality symbol.

$x \geq 1$

Solution set: $[1, \infty)$

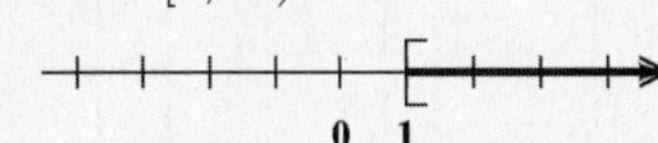

3. $-\frac{4}{7}x > -16$

$-4x > -112$ *Multiply by 7.*

Divide each side by -4 and reverse the direction of the inequality symbol.

$x < 28$

Solution set: $(-\infty, 28)$

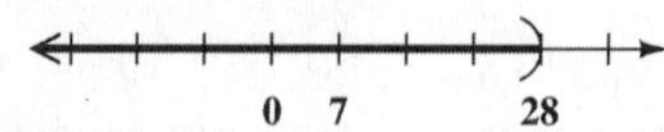

4. $-6 \le \frac{4}{3}x - 2 \le 2$

$-18 \le 4x - 6 \le 6$ *Multiply by 3.*

$-12 \le 4x \le 12$ *Add 6.*

$-3 \le x \le 3$ *Divide by 4.*

Solution set: $[-3, 3]$

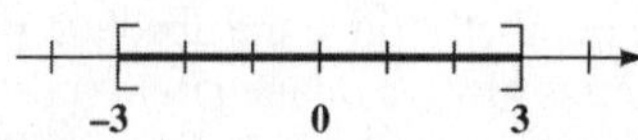

5. Since all the choices have -3 for the coefficient of x, we'll multiply each side of the given inequality by -3, remembering to reverse the direction of the inequality symbol.

$$x < -3$$
$$-3(x) > -3(-3)$$
$$-3x > 9$$

The answer is choice **C.**

6. Let $x =$ the percentage of the U.S. population that was foreign born.

(a) From the graph, $x \ge 7$ in 1990, 2000, and 2006.

(b) From the graph, $x < 6$ in 1960 and 1970.

(c) From the graph, $10 < x < 12$ in 2000.

7. Let $x =$ the score on the fourth test.

$$\frac{83 + 76 + 79 + x}{4} \ge 80$$
$$\frac{238 + x}{4} \ge 80$$
$$238 + x \ge 320$$
$$x \ge 82$$

A score of 82 or more will guarantee him a B.

8. $C = 50x + 5000$; $R = 60x$

To determine the values of x such that R is at least equal to C, we will solve the inequality $R \ge C$.

$$R \ge C$$
$$60x \ge 50x + 5000$$
$$10x \ge 5000$$
$$x \ge 500$$

R is at least equal to C when x is in the interval $[500, \infty)$.

9. **(a)** $A \cap B = \{1, 2, 5, 7\} \cap \{1, 5, 9, 12\}$
$= \{1, 5\}$

(b) $A \cup B = \{1, 2, 5, 7\} \cup \{1, 5, 9, 12\}$
$= \{1, 2, 5, 7, 9, 12\}$

10. $x \le 2$ and $x \ge 2$

The only number common to *both* inequalities is 2.

Solution set: $\{2\}$

11. $3k \ge 6$ and $k - 4 < 5$

$k \ge 2$ and $k < 9$

The solution set is all numbers both greater than or equal to 2 *and* less than 9. This is an intersection. The numbers common to both sets are between 2 and 9, including 2 but not 9. The solution set is $[2, 9)$.

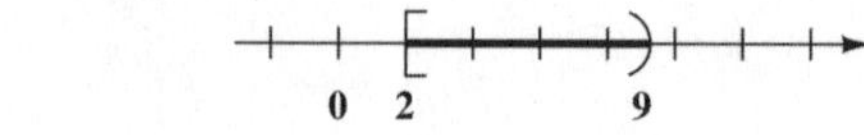

12. $-4x \le -24$ or $4x - 2 < 10$

$4x < 12$

$x \ge 6$ or $x < 3$

The solution set is all numbers less than 3 *or* greater than or equal to 6. This is a union. The solution set is $(-\infty, 3) \cup [6, \infty)$.

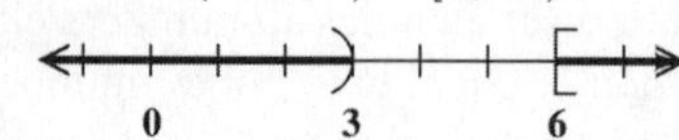

13. $|4x + 3| \le 7$

$-7 \le 4x + 3 \le 7$

$-10 \le 4x \le 4$

$-\frac{10}{4} \le x \le 1$ or $-\frac{5}{2} \le x \le 1$

Solution set: $[-\frac{5}{2}, 1]$

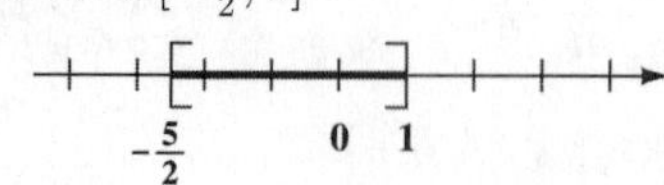

14. $|5-6x|>12$

$$5-6x>12 \quad \text{or} \quad 5-6x<-12$$
$$-6x>7 \quad \text{or} \quad -6x<-17$$
$$x<-\tfrac{7}{6} \quad \text{or} \quad x>\tfrac{17}{6}$$

Solution set: $(-\infty,-\frac{7}{6})\cup(\frac{17}{6},\infty)$

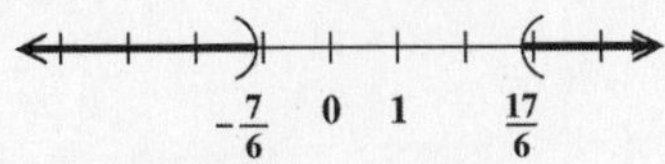

15. $|-3x+4|-4<-1$

$|-3x+4|<3$ *Isolate.*

$$-3<-3x+4<3$$
$$-7<-3x<-1$$
$$\tfrac{7}{3}>x>\tfrac{1}{3}, \text{ or } \tfrac{1}{3}<x<\tfrac{7}{3}$$

Solution set: $(\frac{1}{3},\frac{7}{3})$

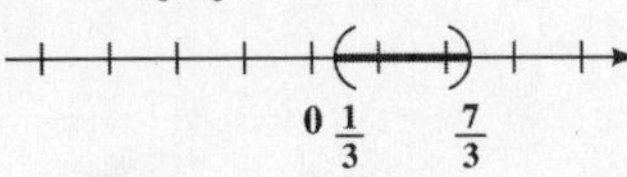

16. $|7-x|\le -1$

Since the absolute value of an expression is always nonnegative (positive or zero), the inequality is *false* for any real number x.

Solution set: $\emptyset$

17. $|3k-2|+1=8$

$|3k-2|=7$ *Isolate.*

$$3k-2=7 \quad \text{or} \quad 3k-2=-7$$
$$3k=9 \quad \text{or} \quad 3k=-5$$
$$k=3 \quad \text{or} \quad k=-\tfrac{5}{3}$$

Solution set: $\{-\frac{5}{3},3\}$

18. $|3-5x|=|2x+8|$

$$3-5x=2x+8 \quad \text{or} \quad 3-5x=-(2x+8)$$
$$-7x=5 \quad \text{or} \quad 3-5x=-2x-8$$
$$-3x=-11$$
$$x=-\tfrac{5}{7} \quad \text{or} \quad x=\tfrac{11}{3}$$

Solution set: $\{-\frac{5}{7},\frac{11}{3}\}$

19. $|4x+3|+5=4$

$|4x+3|=-1$

Since the absolute value of an expression can never be negative, there are no solutions for this equation.

Solution set: $\emptyset$

20. **(i)** $|8x-5|<k$

If $k<0$, then $|8x-5|$ would be less than a negative number. Since the absolute value of an expression is always nonnegative (positive or zero), the solution set is $\emptyset$.

(ii) $|8x-5|>k$

If $k<0$, then $|8x-5|$ would be greater than a negative number. Since the absolute value of an expression is always nonnegative (positive or zero), the solution set is the set of all real numbers, $(-\infty,\infty)$.

(iii) $|8x-5|=k$

If $k<0$, then $|8x-5|$ would be equal to a negative number. Since the absolute value of an expression is always nonnegative (positive or zero), the solution set is $\emptyset$.

Cumulative Review Exercises (Chapters 1–3)

1. $$\frac{108}{144}=\frac{3\cdot 36}{4\cdot 36}=\frac{3}{4}$$

2. $$\frac{8(7)-5(6+2)}{3\cdot5+1}\overset{?}{\ge}1$$
$$\frac{8(7)-5(8)}{3\cdot5+1}\overset{?}{\ge}1$$
$$\frac{56-40}{15+1}\overset{?}{\ge}1$$
$$\frac{16}{16}\overset{?}{\ge}1$$
$$1\ge1 \quad \textit{True}$$

The inequality is *true*.

3. $$\frac{5}{6}+\frac{1}{4}-\frac{7}{15}$$
$$=\frac{10\cdot5}{10\cdot6}+\frac{15\cdot1}{15\cdot4}-\frac{4\cdot7}{4\cdot15} \quad \textit{LCD = 60}$$
$$=\frac{50}{60}+\frac{15}{60}-\frac{28}{60}$$
$$=\frac{37}{60}$$

4. $$\frac{9}{8}\cdot\frac{16}{3}\div\frac{5}{8}=\left(\frac{9}{8}\cdot\frac{16}{3}\right)\cdot\frac{8}{5}$$
$$=\frac{3}{1}\cdot\frac{16}{1}\cdot\frac{1}{5}$$
$$=\frac{3\cdot16}{5}$$
$$=\frac{48}{5}$$

5. $$9-(-4)+(-2)=9+4+(-2)$$
$$=13+(-2)$$
$$=11$$

6. $$\frac{-4(9)(-2)}{-3^2}=\frac{-36(-2)}{-(9)}=\frac{72}{-9}=-8$$

7. $|-7-1|(-4)+(-4)$
$= |-8|(-4)+(-4)$
$= 8(-4)+(-4)$
$= -32+(-4)$
$= -36$

8. $(-5)^3 = (-5)(-5)(-5) = -125$

9. $\left(\frac{3}{2}\right)^4 = \left(\frac{3}{2}\right)\left(\frac{3}{2}\right)\left(\frac{3}{2}\right)\left(\frac{3}{2}\right) = \frac{81}{16}$

10. Let $x = 2$, $y = -3$, and $z = 4$.

$$\begin{aligned} -2y + 4(x - 3z) &= -2(-3) + 4(2 - 3\cdot 4) \\ &= 6 + 4(2 - 12) \\ &= 6 + 4(-10) \\ &= 6 - 40 \\ &= -34 \end{aligned}$$

11. Let $x = 2$, $y = -3$, and $z = 4$.

$$\begin{aligned} \frac{3x^2 - y^2}{4z} &= \frac{3(2)^2 - (-3)^2}{4(4)} \\ &= \frac{3(4) - 9}{16} \\ &= \frac{12 - 9}{16} \\ &= \frac{3}{16} \end{aligned}$$

12. $7(k+m) = 7k + 7m$ illustrates the *distributive property* because 7 was distributed to each element within the parentheses.

13. $3 + (5+2) = 3 + (2+5)$ illustrates the *commutative property* because the order of addition within the parentheses was changed.

14. $-4(k+2) + 3(2k-1)$
$= -4k - 8 + 6k - 3$ *Distributive property*
$= -4k + 6k - 8 - 3$
$= 2k - 11$

15.
$$\begin{aligned} 4 - 5(a+2) &= 3(a+1) - 1 \\ 4 - 5a - 10 &= 3a + 3 - 1 \\ -5a - 6 &= 3a + 2 \\ -8a - 6 &= 2 \\ -8a &= 8 \\ a &= -1 \end{aligned}$$

Check $a = -1$: $4 - 5 = 0 - 1$ *True*

Solution set: $\{-1\}$

16.
$$\begin{aligned} \frac{2}{3}x + \frac{3}{4}x &= -17 \\ 12\left(\tfrac{2}{3}x + \tfrac{3}{4}x\right) &= 12(-17) \quad \textit{Multiply by LCD, 12.} \\ 4(2x) + 3(3x) &= 12(-17) \\ 8x + 9x &= -204 \\ 17x &= -204 \\ x &= -12 \end{aligned}$$

Check $x = -12$: $-8 - 9 = -17$ *True*

Solution set: $\{-12\}$

17.
$$\begin{aligned} \frac{2x+3}{5} &= \frac{x-4}{2} \\ 10\left(\frac{2x+3}{5}\right) &= 10\left(\frac{x-4}{2}\right) \quad \textit{Multiply by LCD, 10.} \\ 2(2x+3) &= 5(x-4) \\ 4x + 6 &= 5x - 20 \\ -x + 6 &= -20 \\ -x &= -26 \\ x &= 26 \end{aligned}$$

Check $x = 26$: $\frac{55}{5} = \frac{22}{2}$ *True*

Solution set: $\{26\}$

18. $|3m - 5| = |m + 2|$

$3m - 5 = m + 2$ or $3m - 5 = -(m+2)$
$2m - 5 = 2$ or $3m - 5 = -m - 2$
$2m = 7$ or $4m - 5 = -2$
$m = \frac{7}{2}$ or $4m = 3$
$m = \frac{3}{4}$

Check $m = \frac{7}{2}$: $\left|\frac{11}{2}\right| = \left|\frac{11}{2}\right|$ *True*

Check $m = \frac{3}{4}$: $\left|-\frac{11}{4}\right| = \left|\frac{11}{4}\right|$ *True*

Solution set: $\{\frac{3}{4}, \frac{7}{2}\}$

19. Solve $3x + 4y = 24$ for y.

$4y = 24 - 3x$ *Subtract 3x.*

$y = \frac{24 - 3x}{4}$ *Divide by 4.*

20. Solve $A = P(1 + ni)$ for n.

$A = P + Pni$ *Distributive property*

$A - P = Pni$ *Subtract P.*

$\frac{A - P}{Pi} = n$, or $n = \frac{A - P}{iP}$ *Divide by Pi.*

21. $3-2(x+7) \le -x+3$

$3-2x-14 \le -x+3$

$-2x-11 \le -x+3$

$-x \le 14$

Multiply each side by -1 and reverse the inequality symbol.

$x \ge -14$

Solution set: $[-14, \infty)$

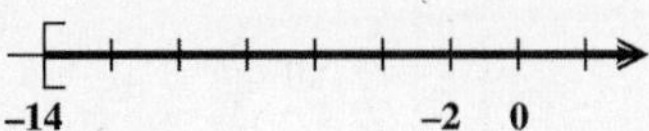

22. $-4 < 5-3x \le 0$

$-9 < -3x \le -5$

Divide each part by -3 and reverse the inequality symbol.

$3 > x \ge \frac{5}{3}$ or $\frac{5}{3} \le x < 3$

Solution set: $[\frac{5}{3}, 3)$

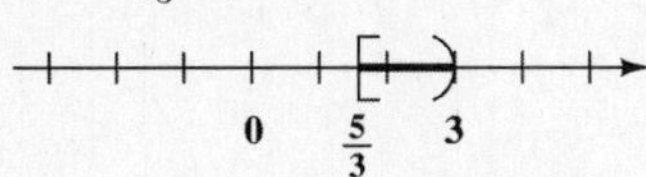

23. $2x+1 > 5$ or $2-x > 2$

$2x > 4$ or $-x > 0$

$x > 2$ or $x < 0$

Solution set: $(-\infty, 0) \cup (2, \infty)$

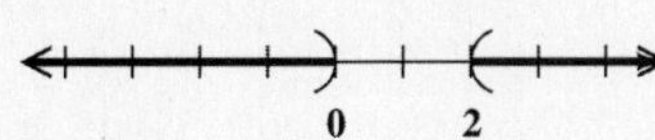

24. $|-7k+3| \ge 4$

$-7k+3 \ge 4$ or $-7k+3 \le -4$

$-7k \ge 1$ $\quad$ $-7k \le -7$

$k \le -\frac{1}{7}$ or $k \ge 1$

Solution set: $(-\infty, -\frac{1}{7}] \cup [1, \infty)$

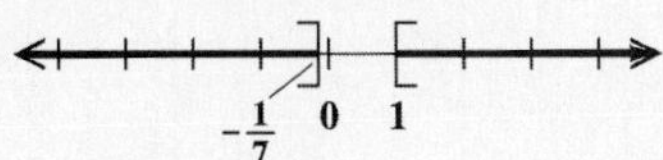

25. Let $x =$ the amount invested at 7% and at 10%.

Principal	Rate (as a Decimal)	Interest
x	0.07	$0.07x$
x	0.10	$0.10x$

The total amount invested is $2x$, and the total interest is

$$0.1(2x) - 150 = 0.2x - 150.$$

$$\begin{array}{ccccc} \text{Interest at 7\%} & + & \text{interest at 10\%} & = & \text{total interest.} \\ 0.07x & + & 0.10x & = & 0.2x - 150 \\ & & 0.17x & = & 0.20x - 150 \\ & & 150 & = & 0.03x \\ & & 5000 & = & x \end{array}$$

He invested \$5000 at each rate.

26. Let $x =$ the number of grams of food C. Then $2x$ is the number of grams of food A. There are 5 grams of food B and the total is at most 24 grams.

$$x + 2x + 5 \le 24$$
$$3x \le 19$$
$$x \le \tfrac{19}{3}, \text{ or } 6\tfrac{1}{3}$$

He may use at most $6\frac{1}{3}$ grams of food C.

27. Let $x =$ the grade the student must make on the third test. To find the average of the three tests, add them and divide by 3. This average must be at least 80.

$$\frac{88+78+x}{3} \ge 80$$
$$\frac{166+x}{3} \ge 80$$
$$166 + x \ge 240$$
$$x \ge 74$$

He must score 74 or greater on his third test.

28. Let x = the speed of the slower car.

	Rate	Time	Distance
Slower Car	x	4	$4x$
Faster Car	$x+20$	4	$4(x+20)$

The sum of their distances is 400.

$$\begin{aligned} 4x + 4(x+20) &= 400 \\ 4x + 4x + 80 &= 400 \\ 8x + 80 &= 400 \\ 8x &= 320 \\ x &= 40 \end{aligned}$$

Since $x = 40$, $x + 20 = 60$.

The cars' speeds are 40 mph and 60 mph.

29. **(a)** 1995: 1533; 2005: 1452
Difference: $1533 - 1452 = 81$

The number of daily newspapers decreased by 81 between 1995 and 2005.

(b) $\frac{81}{1533} \approx 0.053$ or 5.3%

The number decreased by approximately 5.3%.

30. Let x = the length of the middle-sized piece. Then $3x$ = the length of the longest piece, and $x - 5$ = the length of the shortest piece.

The total length is 40 cm.

$$\begin{aligned} x + 3x + (x-5) &= 40 \\ 5x - 5 &= 40 \\ 5x &= 45 \\ x &= 9 \end{aligned}$$

Since $x = 9$, $3x = 27$, and $x - 5 = 4$.

The three pieces should be cut with lengths 4 cm, 9 cm, and 27 cm.

CHAPTER 4 GRAPHS, LINEAR EQUATIONS, AND FUNCTIONS

4.1 The Rectangular Coordinate System

4.1 Margin Exercises

1. **(a)** To plot $(-4, 2)$, move four units from the origin to the left along the x-axis, and then two units up parallel to the y-axis. The point is located in quadrant II.

(b) To plot $(3, -2)$, move three units from the origin to the right along the x-axis, and then two units down parallel to the y-axis. The point is located in quadrant IV.

(c) To plot $(-5, -6)$, move five units from the origin to the left along the x-axis, and then six units down parallel to the y-axis. The point is located in quadrant III.

(d) To plot $(4, 6)$, move four units from the origin to the right along the x-axis, and then six units up parallel to the y-axis. The point is located in quadrant I.

(e) To plot $(-3, 0)$, move three units from the origin to the left along the x-axis. Do not move up or down since the y-coordinate is 0. The point belongs to *no quadrant.*

(f) To plot $(0, -5)$, do not move along the x-axis since the x-coordinate is 0. Move five units from the origin down the y-axis. The point belongs to *no quadrant.*

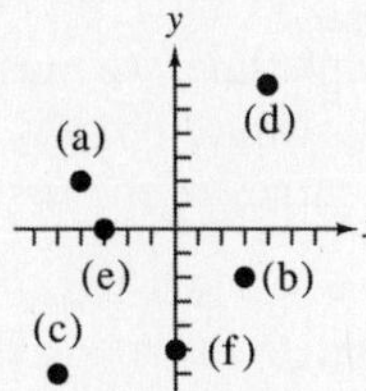

2. **(a)** $3x - 4y = 12$

To complete the ordered pairs, substitute the given value of x or y in the equation.
For $(0, \_\_\_\_)$, let $x = 0$.

$$\begin{aligned} 3x - 4y &= 12 \\ 3(0) - 4y &= 12 \\ -4y &= 12 \\ y &= -3 \end{aligned}$$

The ordered pair is $(0, -3)$.

For $(\_\_\_\_, 0)$ let $y = 0$.

$$\begin{aligned} 3x - 4y &= 12 \\ 3x - 4(0) &= 12 \\ 3x &= 12 \\ x &= 4 \end{aligned}$$

The ordered pair is $(4, 0)$.

For $(\_\_\_\_, -2)$, let $y = -2$.

$$\begin{aligned} 3x - 4y &= 12 \\ 3x - 4(-2) &= 12 \\ 3x + 8 &= 12 \\ 3x &= 4 \\ x &= \tfrac{4}{3} \end{aligned}$$

The ordered pair is $(\frac{4}{3}, -2)$.

For $(-4, \_\_\_\_)$, let $x = -4$.

$$\begin{aligned} 3x - 4y &= 12 \\ 3(-4) - 4y &= 12 \\ -12 - 4y &= 12 \\ -4y &= 24 \\ y &= -6 \end{aligned}$$

The ordered pair is $(-4, -6)$.

(b) The completed table follows.

x	y
0	-3
4	0
$\frac{4}{3}$	-2
-4	-6

3. Plot the intercepts, $(4, 0)$ and $(0, -3)$ from Margin Exercise 2 and draw the line through them.

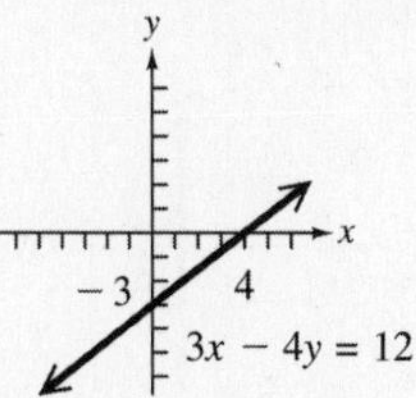

4. $2x - y = 4$
To find the x-intercept, let $y = 0$.

$$\begin{aligned} 2x - y &= 4 \\ 2x - 0 &= 4 \\ 2x &= 4 \\ x &= 2 \end{aligned}$$

The x-intercept is $(2, 0)$.

continued

To find the y-intercept, let $x = 0$.

$$\begin{aligned} 2x - y &= 4 \\ 2(0) - y &= 4 \\ -y &= 4 \\ y &= -4 \end{aligned}$$

The y-intercept is $(0, -4)$.

Plot the intercepts, and draw the line through them.

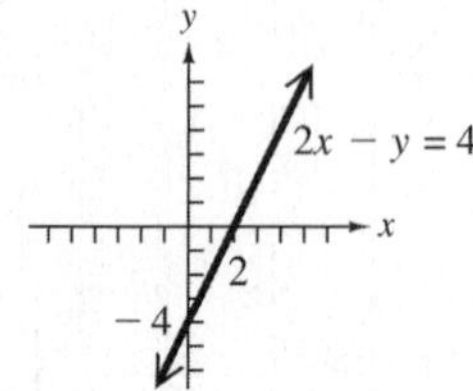

5. $y + 4 = 0$

In standard form, the equation is $0x + y = -4$. Every value of x leads to $y = -4$, so the y-intercept is $(0, -4)$. There is no x-intercept. The graph is the horizontal line through $(0, -4)$.

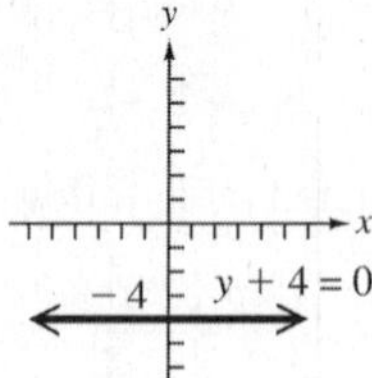

6. $x = 2$

In standard form, the equation is $x + 0y = 2$. Every value of y leads to $x = 2$, so the x-intercept is $(2, 0)$. There is no y-intercept. The graph is the vertical line through $(2, 0)$.

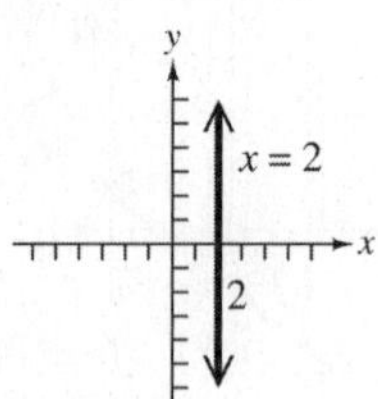

7. $3x - y = 0$

To find the x-intercept, let $y = 0$.

$$\begin{aligned} 3x - 0 &= 0 \\ 3x &= 0 \\ x &= 0 \end{aligned}$$

Since the x-intercept is $(0, 0)$, the y-intercept is also $(0, 0)$.

Find another point. Let $x = 1$.

$$\begin{aligned} 3(1) - y &= 0 \\ 3 - y &= 0 \\ y &= 3 \end{aligned}$$

This gives the ordered pair $(1, 3)$.

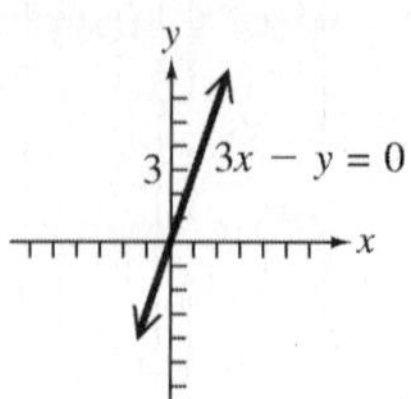

8. By the midpoint formula, the midpoint of the segment with endpoints $(-5, 8)$ and $(2, 4)$ is

$$\left(\frac{-5+2}{2}, \frac{8+4}{2}\right) = \left(\frac{-3}{2}, \frac{12}{2}\right) = (-1.5, 6).$$

4.1 Section Exercises

1. **(a)** x represents the year; y represents the revenue in billions of dollars.

(b) The dot above the year 2006 appears to be at about 2400, so the revenue in 2006 was about \$2400 billion.

(c) The ordered pair is $(2006, 2400)$.

(d) In the context of this graph, the ordered pair $(2004, 1880)$ means that in the year 2004, federal tax revenues were about \$1880 billion.

3. The point with coordinates $(0, 0)$ is called the <u>origin</u> of a rectangular coordinate system.

5. To find the x-intercept of a line, we let <u>y</u> equal 0 and solve for <u>x</u>.

7. To graph a straight line, we must find a minimum of <u>two</u> points. A third point is sometimes found to check the accuracy of the first two points.

9. Every point on the x-axis has y-coordinate equal to 0, so the equation of the x-axis is <u>$y = 0$</u>.

11. **(a)** The point $(1, 6)$ is located in quadrant I, since the x- and y-coordinates are both positive.

(b) The point $(-4, -2)$ is located in quadrant III, since the x- and y-coordinates are both negative.

(c) The point $(-3, 6)$ is located in quadrant II, since the x-coordinate is negative and the y-coordinate is positive.

(d) The point $(7, -5)$ is located in quadrant IV, since the x-coordinate is positive and the y-coordinate is negative.

(e) The point $(-3, 0)$ is located on the x-axis, so it does not belong to any quadrant.

13. **(a)** If $xy > 0$, then both x and y have the same sign.
(x, y) is in quadrant I if x and y are positive.
(x, y) is in quadrant III if x and y are negative.

(b) If $xy < 0$, then x and y have different signs.
(x, y) is in quadrant II if $x < 0$ and $y > 0$.
(x, y) is in quadrant IV if $x > 0$ and $y < 0$.

(c) If $\frac{x}{y} < 0$, then x and y have different signs.
(x, y) is in either quadrant II or IV. (See part (b).)

(d) If $\frac{x}{y} > 0$, then x and y have the same sign.
(x, y) is in either quadrant I or III. (See part (a).)

For Exercises 15–24, see the rectangular coordinate system after Exercise 23.

15. To plot $(2, 3)$, go two units from zero to the right along the x-axis, and then go three units up parallel to the y-axis.

17. To plot $(-3, -2)$, go three units from zero to the left along the x-axis, and then go two units down parallel to the y-axis.

19. To plot $(0, 5)$, do not move along the x-axis at all since the x-coordinate is 0. Move five units up along the y-axis.

21. To plot $(-2, 4)$, go two units from zero to the left along the x-axis, and then go four units up parallel to the y-axis.

23. To plot $(-2, 0)$, go two units to the left along the x-axis. Do not move up or down since the y-coordinate is 0.

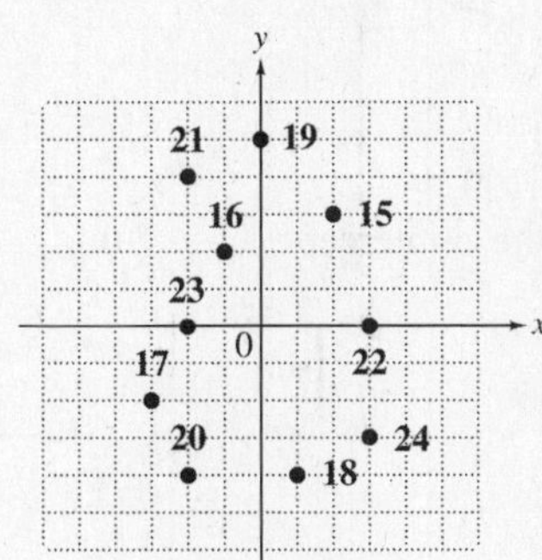

25. $x - y = 3$

For $x = 0$: $x - y = 3$
$0 - y = 3$
$y = -3$ $(0, -3)$

For $y = 0$: $x - y = 3$
$x - 0 = 3$
$x = 3$ $(3, 0)$

For $x = 5$: $x - y = 3$
$5 - y = 3$
$-y = -2$
$y = 2$ $(5, 2)$

For $x = 2$: $x - y = 3$
$2 - y = 3$
$-y = 1$
$y = -1$ $(2, -1)$

Plot the ordered pairs and draw the line through them.

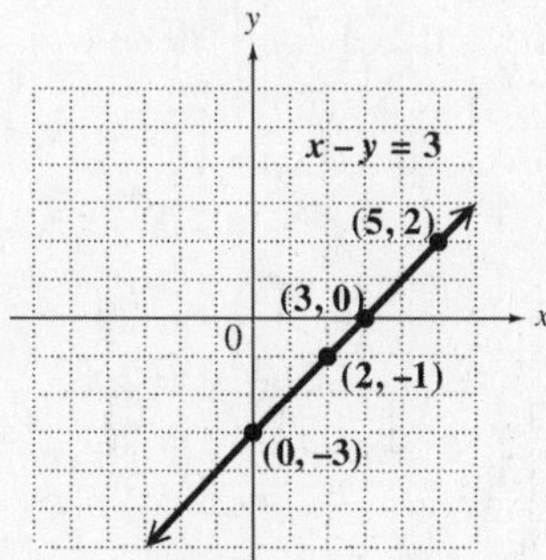

27. $x + 2y = 5$

For $x = 0$: $x + 2y = 5$
$0 + 2y = 5$
$2y = 5$
$y = \frac{5}{2}$ $(0, \frac{5}{2})$

For $y = 0$: $x + 2y = 5$
$x + 2(0) = 5$
$x + 0 = 5$
$x = 5$ $(5, 0)$

For $x = 2$: $x + 2y = 5$
$2 + 2y = 5$
$2y = 3$
$y = \frac{3}{2}$ $(2, \frac{3}{2})$

For $y = 2$: $x + 2y = 5$
$x + 2(2) = 5$
$x + 4 = 5$
$x = 1$ $(1, 2)$

Plot the ordered pairs and draw the line through them.

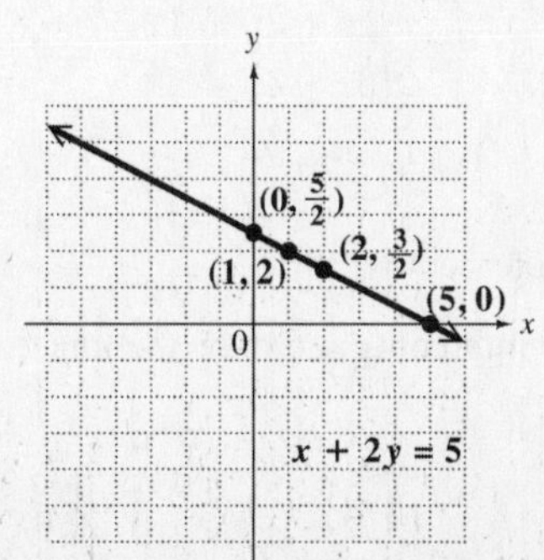

29. $4x - 5y = 20$

For $x = 0$:
$$\begin{aligned} 4x - 5y &= 20 \\ 4(0) - 5y &= 20 \\ -5y &= 20 \\ y &= -4 \quad (0, -4) \end{aligned}$$

For $y = 0$:
$$\begin{aligned} 4x - 5y &= 20 \\ 4x - 5(0) &= 20 \\ 4x &= 20 \\ x &= 5 \quad (5, 0) \end{aligned}$$

For $x = 2$:
$$\begin{aligned} 4x - 5y &= 20 \\ 4(2) - 5y &= 20 \\ 8 - 5y &= 20 \\ -5y &= 12 \\ y &= -\tfrac{12}{5} \quad (2, -\tfrac{12}{5}) \end{aligned}$$

For $y = -3$:
$$\begin{aligned} 4x - 5y &= 20 \\ 4x - 5(-3) &= 20 \\ 4x + 15 &= 20 \\ 4x &= 5 \\ x &= \tfrac{5}{4} \quad (\tfrac{5}{4}, -3) \end{aligned}$$

Plot the ordered pairs and draw the line through them.

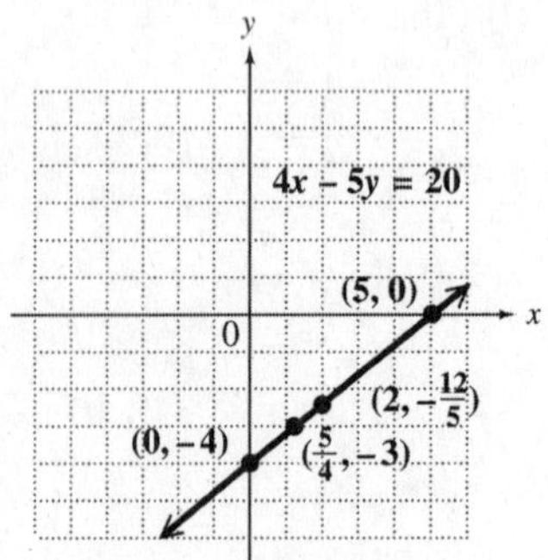

31. $2x + 3y = 12$
To find the x-intercept, let $y = 0$.
$$\begin{aligned} 2x + 3y &= 12 \\ 2x + 3(0) &= 12 \\ 2x &= 12 \\ x &= 6 \end{aligned}$$

The x-intercept is $(6, 0)$.

To find the y-intercept, let $x = 0$.
$$\begin{aligned} 2x + 3y &= 12 \\ 2(0) + 3y &= 12 \\ 3y &= 12 \\ y &= 4 \end{aligned}$$

The y-intercept is $(0, 4)$.

Plot the intercepts and draw the line through them.

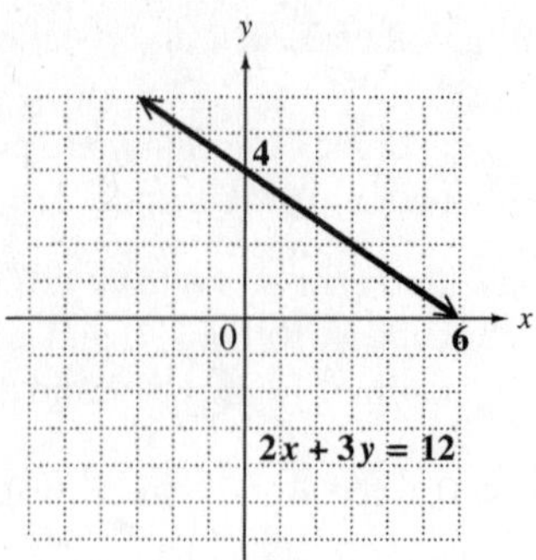

33. $x - 3y = 6$
To find the x-intercept, let $y = 0$.
$$\begin{aligned} x - 3y &= 6 \\ x - 3(0) &= 6 \\ x - 0 &= 6 \\ x &= 6 \end{aligned}$$

The x-intercept is $(6, 0)$.

To find the y-intercept, let $x = 0$.
$$\begin{aligned} x - 3y &= 6 \\ 0 - 3y &= 6 \\ -3y &= 6 \\ y &= -2 \end{aligned}$$

The y-intercept is $(0, -2)$.

Plot the intercepts and draw the line through them.

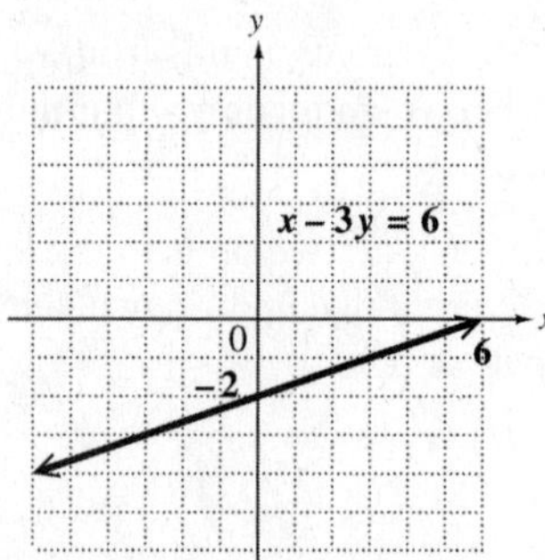

35. $3x - 7y = 9$
To find the x-intercept, let $y = 0$.
$$\begin{aligned} 3x - 7y &= 9 \\ 3x - 7(0) &= 9 \\ 3x &= 9 \\ x &= 3 \end{aligned}$$

The x-intercept is $(3, 0)$.

To find the y-intercept, let $x = 0$.
$$\begin{aligned} 3x - 7y &= 9 \\ 3(0) - 7y &= 9 \\ -7y &= 9 \\ y &= -\tfrac{9}{7} \end{aligned}$$

The y-intercept is $(0, -\tfrac{9}{7})$.

Plot the intercepts and draw the line through them.

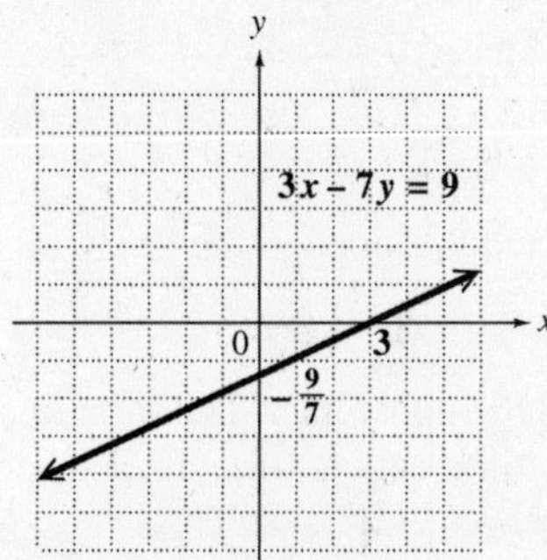

37. $y = 5$
This is a horizontal line. Every point has y-coordinate 5, so no point has y-coordinate 0. There is no x-intercept.
Since every point of the line has y-coordinate 5, the y-intercept is $(0, 5)$. Draw the horizontal line through $(0, 5)$.

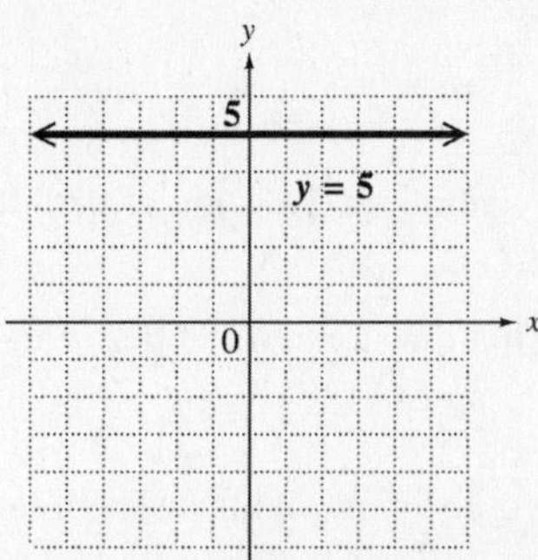

39. $x = 5$
This is a vertical line. Every point has x-coordinate 5, so the x-intercept is $(5, 0)$.
Since every point of the line has x-coordinate 5, no point has x-coordinate 0. There is no y-intercept.
Draw the vertical line through $(5, 0)$.

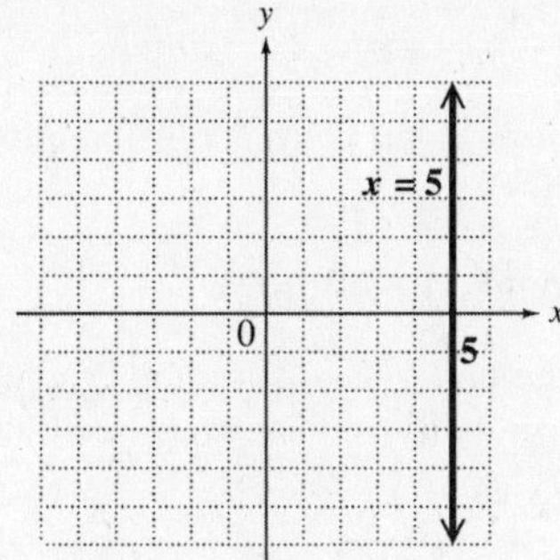

41. $x + 5y = 0$
To find the x-intercept, let $y = 0$.

$$\begin{aligned} x + 5y &= 0 \\ x + 5(0) &= 0 \\ x &= 0 \end{aligned}$$

The x-intercept is $(0, 0)$, and since $x = 0$, this is also the y-intercept.

Since the intercepts are the same, another point is needed to graph the line. Choose any number for y, say $y = -1$, and solve the equation for x.

$$\begin{aligned} x + 5y &= 0 \\ x + 5(-1) &= 0 \\ x &= 5 \end{aligned}$$

This gives the ordered pair $(5, -1)$. Plot $(5, -1)$ and $(0, 0)$, and draw the line through them.

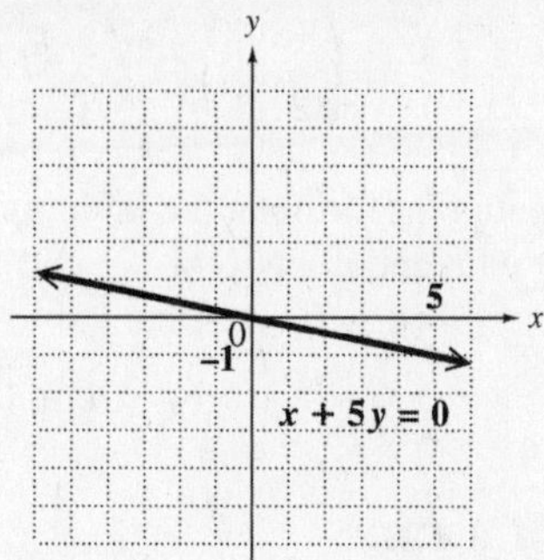

43. $2x = 3y$
If $x = 0$, then $y = 0$, so the x- and y-intercepts are $(0, 0)$. To get another point, let $x = 3$.

$$\begin{aligned} 2(3) &= 3y \\ 2 &= y \end{aligned}$$

Plot $(3, 2)$ and $(0, 0)$, and draw the line through them.

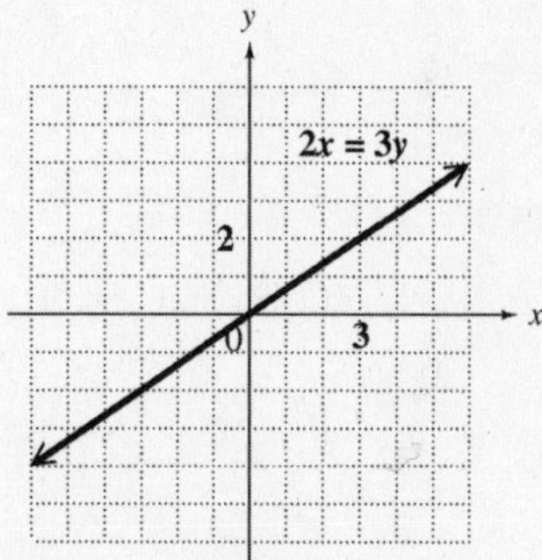

45. By the midpoint formula, the midpoint of the segment with endpoints $(-8, 4)$ and $(-2, -6)$ is

$$\left(\frac{-8 + (-2)}{2}, \frac{4 + (-6)}{2}\right) = \left(\frac{-10}{2}, \frac{-2}{2}\right) = (-5, -1).$$

47. By the midpoint formula, the midpoint of the segment with endpoints $(3, -6)$ and $(6, 3)$ is

$$\left(\frac{3 + 6}{2}, \frac{-6 + 3}{2}\right) = \left(\frac{9}{2}, \frac{-3}{2}\right) = \left(\frac{9}{2}, -\frac{3}{2}\right).$$

49. By the midpoint formula, the midpoint of the segment with endpoints $(-9, 3)$ and $(9, 8)$ is

$$\left(\frac{-9 + 9}{2}, \frac{3 + 8}{2}\right) = \left(\frac{0}{2}, \frac{11}{2}\right) = \left(0, \frac{11}{2}\right).$$

51. By the midpoint formula, the midpoint of the segment with endpoints $(2.5, 3.1)$ and $(1.7, -1.3)$ is

$$\left(\frac{2.5+1.7}{2}, \frac{3.1+(-1.3)}{2}\right) = \left(\frac{4.2}{2}, \frac{1.8}{2}\right) = (2.1, 0.9).$$

53. By the midpoint formula, the midpoint of the segment with endpoints $\left(\frac{1}{2}, \frac{1}{3}\right)$ and $\left(\frac{3}{2}, \frac{5}{3}\right)$ is

$$\left(\frac{\frac{1}{2}+\frac{3}{2}}{2}, \frac{\frac{1}{3}+\frac{5}{3}}{2}\right) = \left(\frac{\frac{4}{2}}{2}, \frac{\frac{6}{3}}{2}\right) = \left(\frac{2}{2}, \frac{2}{2}\right) = (1, 1).$$

55. By the midpoint formula, the midpoint of the segment with endpoints $\left(-\frac{1}{3}, \frac{2}{7}\right)$ and $\left(-\frac{1}{2}, \frac{1}{14}\right)$ is

$$\left(\frac{-\frac{1}{3}+(-\frac{1}{2})}{2}, \frac{\frac{2}{7}+\frac{1}{14}}{2}\right) = \left(\frac{-\frac{5}{6}}{2}, \frac{\frac{5}{14}}{2}\right) = \left(-\frac{5}{12}, \frac{5}{28}\right).$$

4.2 Slope of a Line

4.2 Margin Exercises

1. **(a)** rise = vertical change = 2 ft

(b) run = horizontal change = 10 ft

(c) $\text{slope} = \frac{\text{rise}}{\text{run}} = \frac{2 \text{ ft}}{10 \text{ ft}} = \frac{2}{10}$, or $\frac{1}{5}$

2. **(a)** Let $(x_1, y_1) = (-2, 7)$ and $(x_2, y_2) = (4, -3)$ Then

$$m = \frac{y_2 - y_1}{x_2 - x_1} = \frac{-3-7}{4-(-2)} = \frac{-10}{6} = -\frac{5}{3}.$$

The slope is $-\frac{5}{3}$.

(b) Let $(x_1, y_1) = (1, 2)$ and $(x_2, y_2) = (8, 5)$ Then

$$m = \frac{y_2 - y_1}{x_2 - x_1} = \frac{5-2}{8-1} = \frac{3}{7}.$$

The slope is $\frac{3}{7}$.

(c) Let $(x_1, y_1) = (8, -4)$ and $(x_2, y_2) = (3, -2)$ Then

$$m = \frac{y_2 - y_1}{x_2 - x_1} = \frac{-2-(-4)}{3-8} = \frac{2}{-5} = -\frac{2}{5}.$$

The slope is $-\frac{2}{5}$.

3. **(a)** To find the slope of the line with equation

$$2x + y = 6,$$

first find the intercepts. Replace y with 0 to find that the x-intercept is $(3, 0)$; replace x with 0 to find that the y-intercept is $(0, 6)$. The slope is then

$$m = \frac{6-0}{0-3} = \frac{6}{-3} = -2.$$

The slope is -2.

(b) To find the slope of the line with equation

$$3x - 4y = 12,$$

first find the intercepts. The x-intercept is $(4, 0)$, and the y-intercept is $(0, -3)$. The slope is then

$$m = \frac{-3-0}{0-4} = \frac{-3}{-4} = \frac{3}{4}.$$

The slope is $\frac{3}{4}$.

4. **(a)** To find the slope of the line with equation

$$x = -6,$$

select two different points on the line, such as $(-6, 0)$ and $(-6, 3)$, and use the slope formula.

$$m = \frac{3-0}{-6-(-6)} = \frac{3}{0}$$

Since division by zero is undefined, the slope is *undefined.*

(b) To find the slope of the line with equation

$$y + 5 = 0,$$

select two different points on the line, such as $(0, -5)$ and $(2, -5)$, and use the slope formula.

$$m = \frac{-5-(-5)}{2-0} = \frac{0}{2} = 0$$

The slope is 0.

5. Solve the equation for y.

$$\begin{aligned} 2x - 5y &= 8 \\ -5y &= -2x + 8 && \textit{Subtract } 2x. \\ y &= \tfrac{2}{5}x - \tfrac{8}{5} && \textit{Divide by } -5. \end{aligned}$$

The slope is given by the coefficient of x, so the slope is $\frac{2}{5}$.

6. **(a)** Through $(1, -3)$; $m = -\frac{3}{4}$

Locate the point $(1, -3)$ on the graph. Use the slope formula to find a second point on the line (write $-\frac{3}{4}$ as $\frac{-3}{4}$).

$$m = \frac{\text{change in } y}{\text{change in } x} = \frac{-3}{4}$$

From $(1, -3)$, move *down* 3 units and then 4 units to the *right* to $(5, -6)$. Draw the line through the two points. (Note that we could also think of the slope as $\frac{3}{-4}$, and then move *up* 3 units and *left* 4 units.)

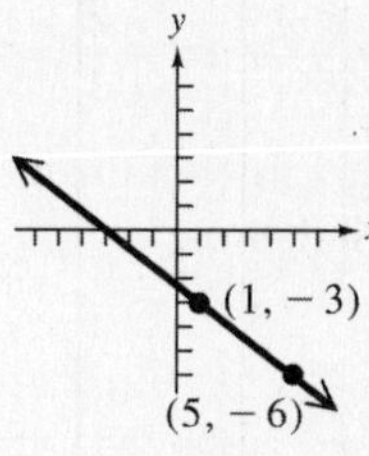

(b) Through $(-1,-4)$; $m=2$

Locate the point $(-1,-4)$ on the graph. Use the slope formula to find a second point on the line (write 2 as $\frac{2}{1}$).

$$m=\frac{\text{change in } y}{\text{change in } x}=\frac{2}{1}$$

From $(-1,-4)$, move *up* 2 units and then 1 unit to the *right* to $(0,-2)$. Draw the line through the two points.

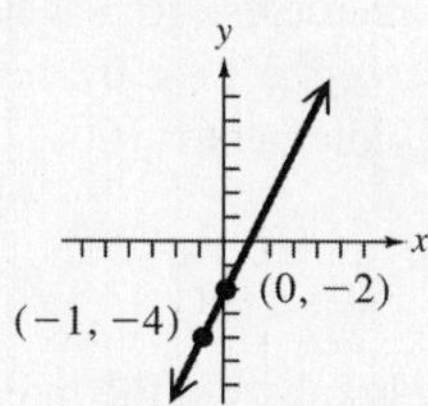

7. **(a)** Find the slope of each line.
The line through $(-1,2)$ and $(3,5)$ has slope

$$m_1=\frac{5-2}{3-(-1)}=\frac{3}{4}.$$

The line through $(4,7)$ and $(8,10)$ has slope

$$m_2=\frac{10-7}{8-4}=\frac{3}{4}.$$

The slopes are the same, so the lines are *parallel.*

(b) Find the slope of each line.
The line through $(5,-9)$ and $(3,7)$ has slope

$$m_1=\frac{7-(-9)}{3-5}=\frac{16}{-2}=-8.$$

The line through $(0,2)$ and $(8,3)$ has slope

$$m_2=\frac{3-2}{8-0}=\frac{1}{8}.$$

The product of the slopes is

$$(-8)\left(\frac{1}{8}\right)=-1,$$

so the lines are *perpendicular.*

(c) Solve each equation for y.

$$\begin{aligned}2x-y&=4\\ -y&=-2x+4\\ y&=2x-4\\ (\text{so } m_1&=2)\end{aligned}\qquad\qquad \begin{aligned}2x+y&=6\\ y&=-2x+6\\ (\text{so } m_2&=-2)\end{aligned}$$

Since $m_1\neq m_2$, the lines are not parallel. Since $m_1m_2=2(-2)=-4$, the lines are not perpendicular either. Therefore, the answer is *neither.*

(d) Solve each equation for y.

$$\begin{aligned}3x+5y&=6\\ 5y&=-3x+6\\ y&=-\tfrac{3}{5}x+\tfrac{6}{5}\\ (\text{so } m_1&=-\tfrac{3}{5})\end{aligned}\qquad\qquad \begin{aligned}5x-3y&=2\\ -3y&=-5x+2\\ y&=\tfrac{5}{3}x-\tfrac{2}{3}\\ (\text{so } m_2&=\tfrac{5}{3})\end{aligned}$$

Since $m_1m_2=(-\frac{3}{5})(\frac{5}{3})=-1$, the lines are *perpendicular.*

8. **(a)** $(x_1,y_1)=(2000,690)$ and $(x_2,y_2)=(2003,886)$.

$$\begin{aligned}\text{average rate of change}&=\frac{y_2-y_1}{x_2-x_1}\\ &=\frac{886-690}{2003-2000}\\ &=\frac{196}{3}=65\tfrac{1}{3}\end{aligned}$$

The average rate of change is $65.\overline{3}$.

(b) It is greater than 58, which is the average rate of change for 2000–2005 found in Example 9.

9. $(x_1,y_1)=(2000,942.5)$ and $(x_2,y_2)=(2006,614.9)$.

$$\begin{aligned}\text{average rate of change}&=\frac{y_2-y_1}{x_2-x_1}\\ &=\frac{614.9-942.5}{2006-2000}\\ &=\frac{-327.6}{6}=-54.6\end{aligned}$$

Thus, the average rate of change from 2000 to 2006 was -54.6 million CDs per year.

4.2 Section Exercises

1.
$$\begin{aligned}\text{slope}&=\frac{\text{change in vertical position}}{\text{change in horizontal position}}\\ &=\frac{30\text{ feet}}{100\text{ feet}}\end{aligned}$$

Choices **A**, 0.3, **B**, $\frac{3}{10}$, and **D**, $\frac{30}{100}$ are all correct.

3. To get to B from A, we must go up 2 units and move right 1 unit. Thus,

$$\text{slope of } AB=\frac{\text{rise}}{\text{run}}=\frac{2}{1}=2.$$

5. slope of $CD = \frac{\text{rise}}{\text{run}} = \frac{-7}{0}$, which is undefined.

7. $m = \frac{6-2}{5-3} = \frac{4}{2} = 2$

9. $m = \frac{-4-(-4)}{-3-(-5)} = \frac{-4+4}{-3+5} = \frac{0}{2} = 0$

11. $m = \frac{-6-0}{-3-(-3)} = \frac{-6}{-3+3} = \frac{-6}{0}$, which is *undefined.*

13. Let $(x_1, y_1) = (-2, -3)$ and $(x_2, y_2) = (-1, 5)$. Then

$$m = \frac{y_2 - y_1}{x_2 - x_1} = \frac{5-(-3)}{-1-(-2)} = \frac{8}{1} = 8.$$

The slope is 8.

15. Let $(x_1, y_1) = (-4, 1)$ and $(x_2, y_2) = (2, 6)$. Then

$$m = \frac{y_2 - y_1}{x_2 - x_1} = \frac{6-1}{2-(-4)} = \frac{5}{6}.$$

The slope is $\frac{5}{6}$.

17. Let $(x_1, y_1) = (2, 4)$ and $(x_2, y_2) = (-4, 4)$. Then

$$m = \frac{y_2 - y_1}{x_2 - x_1} = \frac{4-4}{-4-2} = \frac{0}{-6} = 0.$$

The slope is 0.

19. The points shown on the line are $(-3, 3)$ and $(-1, -2)$. The slope is

$$m = \frac{-2-3}{-1-(-3)} = \frac{-5}{2} = -\frac{5}{2}.$$

21. The points shown on the line are $(3, 3)$ and $(3, -3)$. The slope is

$$m = \frac{-3-3}{3-3} = \frac{-6}{0}, \text{ which is undefined.}$$

23. "The line has positive slope" means that the line goes up from left to right. This is line *B*.

25. "The line has slope 0" means that there is no vertical change; that is, the line is horizontal. This is line *A*.

27. To find the slope of the line with equation

$$x + 2y = 4,$$

first find the intercepts. Replace y with 0 to find that the x-intercept is $(4, 0)$; replace x with 0 to find that the y-intercept is $(0, 2)$. The slope is then

$$m = \frac{2-0}{0-4} = -\frac{2}{4} = -\frac{1}{2}.$$

To sketch the graph, plot the intercepts and draw the line through them.

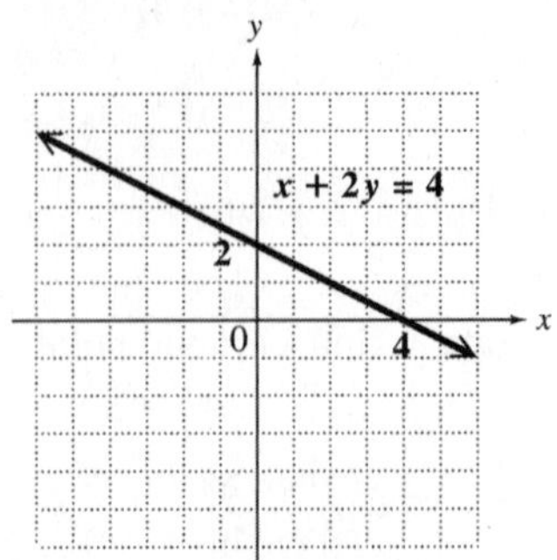

29. To find the slope of the line with equation

$$-x + y = 4,$$

first find the intercepts. Replace y with 0 to find that the x-intercept is $(-4, 0)$; replace x with 0 to find that the y-intercept is $(0, 4)$. The slope is then

$$m = \frac{4-0}{0-(-4)} = \frac{4}{4} = 1.$$

To sketch the graph, plot the intercepts and draw the line through them.

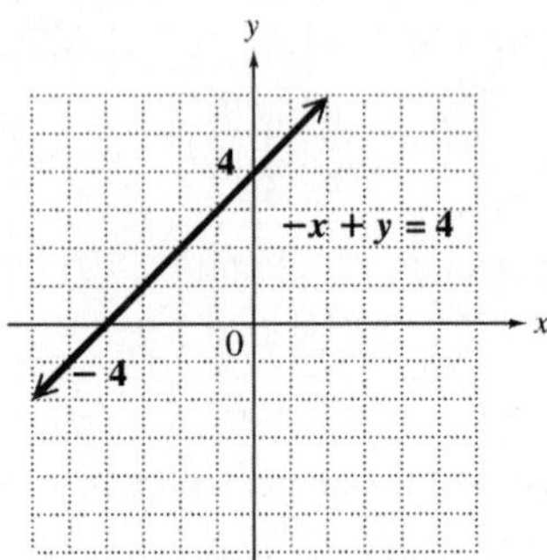

31. To find the slope of the line with equation

$$6x + 5y = 30,$$

first find the intercepts. Replace y with 0 to find that the x-intercept is $(5, 0)$; replace x with 0 to find that the y-intercept is $(0, 6)$. The slope is then

$$m = \frac{6-0}{0-5} = -\frac{6}{5}.$$

To sketch the graph, plot the intercepts and draw the line through them.

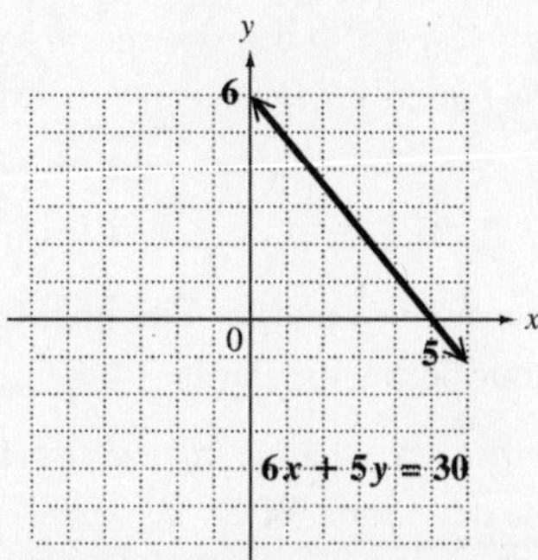

33. The graph of $x + 2 = 0$, or $x = -2$, is a vertical line with x-intercept $(-2, 0)$. The slope of a vertical line is undefined because the denominator equals zero in the slope formula.

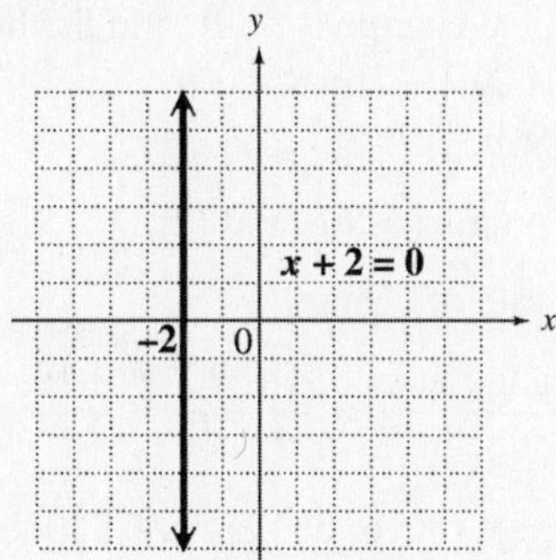

35. In the equation

$$y = 4x,$$

replace x with 0 and then x with 1 to get the ordered pairs $(0, 0)$ and $(1, 4)$, respectively. (There are other possibilities for ordered pairs.) The slope is then

$$m = \frac{4 - 0}{1 - 0} = \frac{4}{1} = 4.$$

To sketch the graph, plot the two points and draw the line through them.

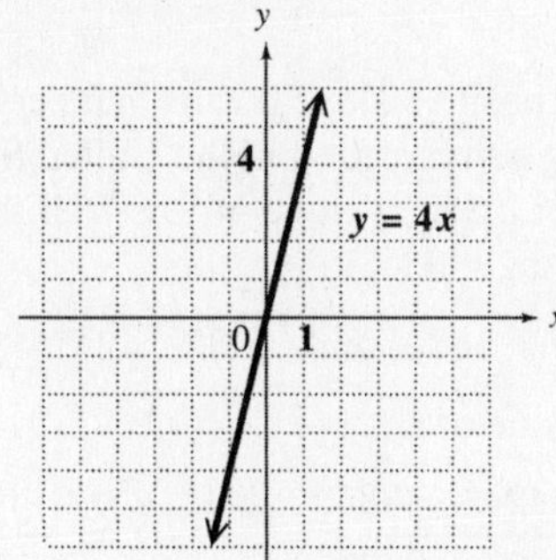

37. $y - 3 = 0 \quad (y = 3)$
The graph of $y = 3$ is the horizontal line with y-intercept $(0, 3)$. The slope of a horizontal line is 0.

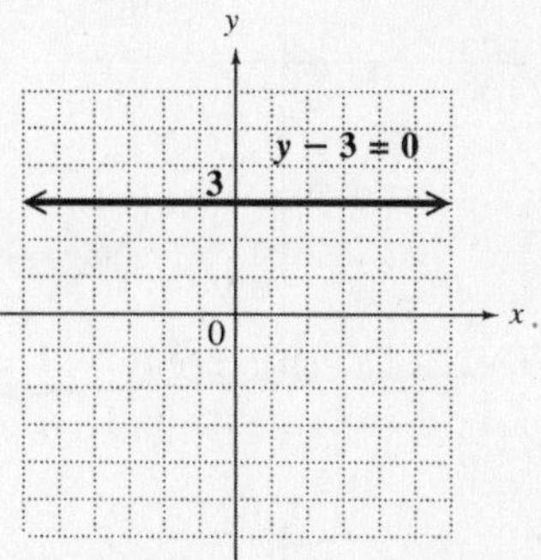

39. To graph the line through $(-4, 2)$ with slope $m = \frac{1}{2}$, locate $(-4, 2)$ on the graph. To find a second point, use the definition of slope.

$$m = \frac{\text{change in } y}{\text{change in } x} = \frac{1}{2}$$

From $(-4, 2)$, go up 1 unit. Then go 2 units to the right to get to $(-2, 3)$. Draw the line through $(-4, 2)$ and $(-2, 3)$.

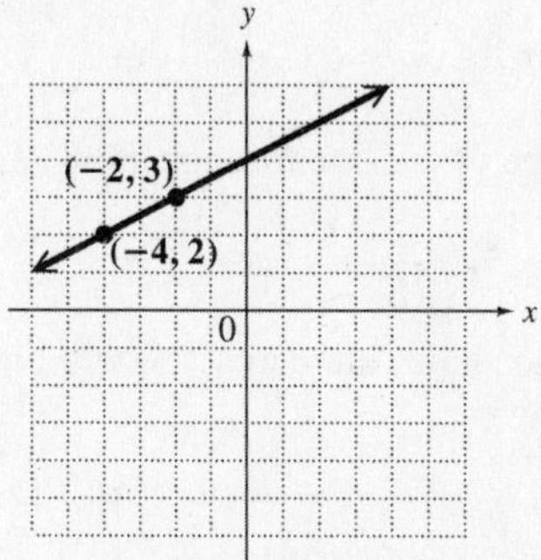

41. To graph the line through $(0, -2)$ with slope $m = -\frac{2}{3}$, locate the point $(0, -2)$ on the graph. To find a second point on the line, use the definition of slope, writing $-\frac{2}{3}$ as $\frac{-2}{3}$.

$$m = \frac{\text{change in } y}{\text{change in } x} = \frac{-2}{3}$$

From $(0, -2)$, move 2 units down and then 3 units to the right to get to $(3, -4)$. Draw the line through $(3, -4)$ and $(0, -2)$. (Note that the slope could also be written as $\frac{2}{-3}$. In this case, move 2 units up and 3 units to the left to get another point on the same line.)

continued

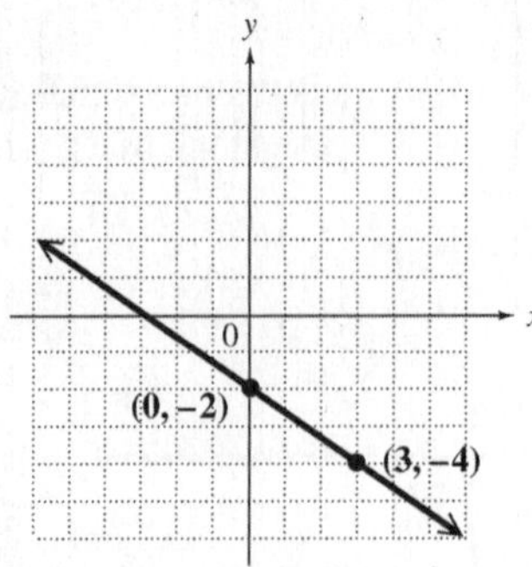

43. Locate $(-1,-2)$. Then use $m = 3 = \frac{3}{1}$ to go 3 units up and 1 unit right to $(0,1)$.

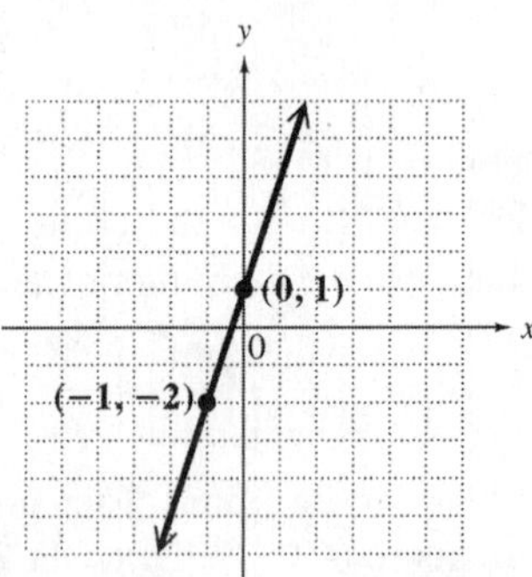

45. The slope of the line through $(4,6)$ and $(-8,7)$ is

$$m = \frac{7-6}{-8-4} = \frac{1}{-12} = -\frac{1}{12}.$$

The slope of the line through $(7,4)$ and $(-5,5)$ is

$$m = \frac{5-4}{-5-7} = \frac{1}{-12} = -\frac{1}{12}.$$

Since the slopes are equal, the two lines are *parallel.*

47. $2x + 5y = -7$ and $5x - 2y = 1$
Solve the equations for y.

$$5y = -2x - 7 \qquad -2y = -5x + 1$$
$$y = -\tfrac{2}{5}x - \tfrac{7}{5} \qquad y = \tfrac{5}{2}x - \tfrac{1}{2}$$

The slopes are the coefficients of x, $-\frac{2}{5}$ and $\frac{5}{2}$. These are negative reciprocals of one another, so the lines are *perpendicular.*

49. $2x + y = 6$ and $x - y = 4$
Solve the equations for y.

$$y = -2x + 6 \qquad -y = -x + 4$$
$$y = x - 4$$

The slopes are -2 and 1. The lines are *neither* parallel nor perpendicular.

51. $3x = y$ and $2y - 6x = 5$
The slope of the first line is the coefficient of x, namely 3. Solve the second equation for y.

$$2y = 6x + 5$$
$$y = 3x + \tfrac{5}{2}$$

The slope of the second line is also 3, so the lines are *parallel.*

53. $2x + 5y = -8$ and $6 + 2x = 5y$
Solve the equations for y.

$$5y = -2x - 8 \qquad 5y = 2x + 6$$
$$y = -\tfrac{2}{5}x - \tfrac{8}{5} \qquad y = \tfrac{2}{5}x + \tfrac{6}{5}$$

The slopes are $-\frac{2}{5}$ and $\frac{2}{5}$. The lines are *neither* parallel nor perpendicular.

55. $4x - 3y = 8$ and $4y + 3x = 12$
Solve the equations for y.

$$-3y = -4x + 8 \qquad 4y = -3x + 12$$
$$y = \tfrac{4}{3}x - \tfrac{8}{3} \qquad y = -\tfrac{3}{4}x + 3$$

The slopes, $\frac{4}{3}$ and $-\frac{3}{4}$, are negative reciprocals of one another, so the lines are *perpendicular.*

57. The vertical change is 63 ft, and the horizontal change is $250 - 160 = 90$ ft.
The slope is $\frac{63}{90} = \frac{7}{10}$.

59. Use the points $(0,20)$ and $(4,4)$.

average rate of change

$$= \frac{\text{change in } y}{\text{change in } x} = \frac{4-20}{4-0} = \frac{-16}{4} = -4$$

The average rate of change is $-\$4000$ per year, that is, the value of the machine is decreasing \$4000 each year during these years.

61. We can see that there is no change in the percent of pay raise. Thus, the average rate of change is 0% per year, that is, the percent of pay raise is not changing—it is 3% each year during these years.

63. **(a)** Use $(2000, 268.3)$ and $(2006, 381.0)$.

average rate of change

$$= \frac{381.0 - 268.3}{2006 - 2000} = \frac{112.7}{6} \approx 18.78$$

The average rate of change is about \$18.78 billion per year.

(b) The positive slope means that personal spending on recreation in the United States *increased* by an average of \$18.78 billion each year from 2000 to 2006.

65. Use $(2003, 5836)$ and $(2006, 1424)$.

average rate of change

$$= \frac{1424 - 5836}{2006 - 2003} = \frac{-4412}{3} \approx -1470.67$$

The average rate of change in sales is about $-\$1470.67$ million per year, that is, the sales of analog TVs in the United States decreased by an average of \$1470.67 million each year from 2003 to 2006.

66. For $A(3,1)$ and $B(6,2)$, the slope of $\overline{AB}$ is

$$m = \frac{2-1}{6-3} = \frac{1}{3}.$$

67. For $B(6,2)$ and $C(9,3)$, the slope of $\overline{BC}$ is

$$m = \frac{3-2}{9-6} = \frac{1}{3}.$$

68. For $A(3,1)$ and $C(9,3)$, the slope of $\overline{AC}$ is

$$m = \frac{3-1}{9-3} = \frac{2}{6} = \frac{1}{3}.$$

69. The slope of $\overline{AB}$ = slope of $\overline{BC}$
= slope of $\overline{AC}$
$= \frac{1}{3}.$

70. For $A(1,-2)$ and $B(3,-1)$, the slope of $\overline{AB}$ is

$$m = \frac{-1-(-2)}{3-1} = \frac{1}{2}.$$

For $B(3,-1)$ and $C(5,0)$, the slope of $\overline{BC}$ is

$$m = \frac{0-(-1)}{5-3} = \frac{1}{2}.$$

For $A(1,-2)$ and $C(5,0)$, the slope of $\overline{AC}$ is

$$m = \frac{0-(-2)}{5-1} = \frac{2}{4} = \frac{1}{2}.$$

Since the three slopes are the same, the three points are collinear.

71. For $A(0,6)$ and $B(4,-5)$, the slope of $\overline{AB}$ is

$$m = \frac{-5-6}{4-0} = \frac{-11}{4} = -\frac{11}{4}.$$

For $B(4,-5)$ and $C(-2,12)$, the slope of $\overline{BC}$ is

$$m = \frac{12-(-5)}{-2-4} = \frac{17}{-6} = -\frac{17}{6}.$$

Since these two slopes are not the same, the three points are not collinear.

4.3 Linear Equations in Two Variables

4.3 Margin Exercises

1. **(a)** Slope 2; y-intercept $(0,-3)$
Here $m=2$ and $b=-3$. Substitute these values into the slope-intercept form.

$$\begin{aligned} y &= mx+b \\ y &= 2x+(-3) \\ y &= 2x-3 \end{aligned}$$

(b) Slope $-\frac{2}{3}$; y-intercept $(0,0)$

Here $m=-\frac{2}{3}$ and $b=0$.

$$\begin{aligned} y &= mx+b \\ y &= -\tfrac{2}{3}x+0 \\ y &= -\tfrac{2}{3}x \end{aligned}$$

2. **(a)** $y = 2x+3$

Here $m=2$ and $b=3$. Plot the y-intercept $(0,3)$. The slope 2 can be interpreted as

$$m = \frac{\text{rise}}{\text{run}} = \frac{\text{change in } y}{\text{change in } x} = \frac{2}{1}.$$

From $(0,3)$, move *up* 2 units and to the *right* 1 unit, and plot a second point at $(1,5)$. Draw a line through the two points.

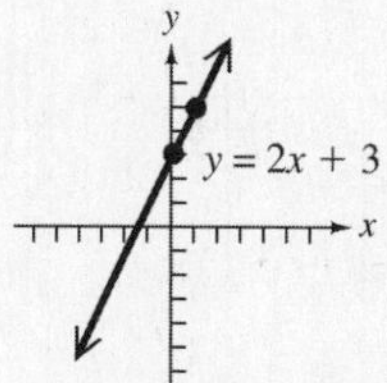

(b) $3x+4y=8$
Solve the equation for y.

$$\begin{aligned} 4y &= -3x+8 \\ y &= -\tfrac{3}{4}x+2 \end{aligned}$$

Plot the y-intercept $(0,2)$. The slope can be interpreted as either $\frac{-3}{4}$ or $\frac{3}{-4}$. Using $\frac{-3}{4}$, move from $(0,2)$ *down* 3 units and to the *right* 4 units to locate the point $(4,-1)$. Draw a line through the two points.

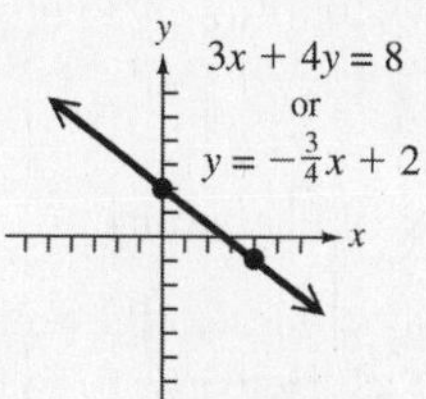

3. **(a)** Through $(-2,7)$ with $m=3$
Use the point-slope form with $(x_1,y_1)=(-2,7)$ and $m=3$.

$$\begin{aligned} y-y_1 &= m(x-x_1) \\ y-7 &= 3[x-(-2)] \\ y-7 &= 3(x+2) \\ y-7 &= 3x+6 \\ y &= 3x+13 \end{aligned}$$

In standard form, $Ax+By=C$,

$$3x-y=-13.$$

(b) Through $(1, 3)$ with $m = -\frac{5}{4}$

$$\begin{aligned} y - y_1 &= m(x - x_1) \\ y - 3 &= -\tfrac{5}{4}(x - 1) \\ y - 3 &= -\tfrac{5}{4}x + \tfrac{5}{4} \\ y &= -\tfrac{5}{4}x + \tfrac{5}{4} + \tfrac{12}{4} \\ y &= -\tfrac{5}{4}x + \tfrac{17}{4} \end{aligned}$$

4. First find the slope of the line. Then use the slope and one of the points in the point-slope form.

(a) Through $(-1, 2)$ and $(5, 7)$

$$m = \frac{7 - 2}{5 - (-1)} = \frac{5}{6}$$

Let $(x_1, y_1) = (5, 7)$.

$$\begin{aligned} y - y_1 &= m(x - x_1) \\ y - 7 &= \tfrac{5}{6}(x - 5) \\ 6y - 42 &= 5x - 25 && \textit{Multiply by 6.} \\ 5x - 6y &= -17 && \textit{Standard form} \end{aligned}$$

(b) Through $(-2, 6)$ and $(1, 4)$

$$m = \frac{4 - 6}{1 - (-2)} = \frac{-2}{3} = -\frac{2}{3}$$

Let $(x_1, y_1) = (1, 4)$.

$$\begin{aligned} y - y_1 &= m(x - x_1) \\ y - 4 &= -\tfrac{2}{3}(x - 1) \\ 3y - 12 &= -2x + 2 && \textit{Multiply by 3.} \\ 2x + 3y &= 14 && \textit{Standard form} \end{aligned}$$

5. **(a)** Through $(8, -2)$; $m = 0$

A line with slope $m = 0$ is horizontal. The horizontal line through the point (c, d) has equation $y = d$. Here $d = -2$, so the equation is $y = -2$.

(b) The vertical line through $(3, 5)$

The vertical line through the point (c, d) has equation $x = c$. Here $c = 3$, so the equation is $x = 3$.

6. **(a)** Through $(-8, 3)$; parallel to the line $2x - 3y = 10$

Find the slope of the given line.

$$\begin{aligned} 2x - 3y &= 10 \\ -3y &= -2x + 10 \\ y &= \tfrac{2}{3}x - \tfrac{10}{3} \end{aligned}$$

The slope is $\frac{2}{3}$, so a line parallel to it also has slope $\frac{2}{3}$. Use $m = \frac{2}{3}$ and $(x_1, y_1) = (-8, 3)$ in the point-slope form.

$$\begin{aligned} y - y_1 &= m(x - x_1) \\ y - 3 &= \tfrac{2}{3}[x - (-8)] \\ y - 3 &= \tfrac{2}{3}(x + 8) \\ y - 3 &= \tfrac{2}{3}x + \tfrac{16}{3} \\ y &= \tfrac{2}{3}x + \tfrac{16}{3} + \tfrac{9}{3} \\ y &= \tfrac{2}{3}x + \tfrac{25}{3} \end{aligned}$$

(b) Through $(-8, 3)$; perpendicular to $2x - 3y = 10$

The slope of $2x - 3y = 10$ is $\frac{2}{3}$. The negative reciprocal of $\frac{2}{3}$ is $-\frac{3}{2}$, so the slope of the line through $(-8, 3)$ is $-\frac{3}{2}$.

$$\begin{aligned} y - y_1 &= m(x - x_1) \\ y - 3 &= -\tfrac{3}{2}[x - (-8)] \\ y - 3 &= -\tfrac{3}{2}(x + 8) \\ y - 3 &= -\tfrac{3}{2}x - 12 \\ y &= -\tfrac{3}{2}x - 9 \end{aligned}$$

7. **(a)** An equation is $y = 0.1x$.

(b) Since the price you pay is \$0.10 per minute plus a flat rate of \$0.20, an equation for x minutes is

$$y = 0.1x + 0.2.$$

(c) The ordered pair $(15, 1.7)$ indicates that the price of a 15-minute call is \$1.70.

8. **(a)** $1970 - 1950 = 20$ and $2000 - 1950 = 50$, so use $(20, 52.3)$ and $(50, 84.1)$.

$$m = \frac{84.1 - 52.3}{50 - 20} = \frac{31.8}{30} = 1.06$$

$$\begin{aligned} y - y_1 &= m(x - x_1) \\ y - 52.3 &= 1.06(x - 20) \\ y - 52.3 &= 1.06x - 21.2 \\ y &= 1.06x + 31.1 \end{aligned}$$

(b) For 1995, $x = 1995 - 1950 = 45$.

$$\begin{aligned} y &= 1.06x + 31.1 \\ y &= 1.06(45) + 31.1 && \textit{Let } x = 45. \\ y &= 47.7 + 31.1 \\ y &= 78.8 \end{aligned}$$

About 78.8% of the United States population 25 yr or older were at least high school graduates in 1995.

4.3 Section Exercises

1. Choice **A**, $3x - 2y = 5$, is in the form $Ax + By = C$ with $A \geq 0$ and integers A, B, and C having no common factor (except 1).

3. Choice **A**, $y = 6x + 2$, is in the form $y = mx + b$.

5. $y = -3x + 10$
$3x + y = 10$ *Standard form*

7. $y = 2x + 3$
This line is in slope-intercept form with slope $m = 2$ and y-intercept $(0, b) = (0, 3)$. The only graph with positive slope and with a positive y-coordinate of its y-intercept is **A**.

9. $y = -2x - 3$
This line is in slope-intercept form with slope $m = -2$ and y-intercept $(0, b) = (0, -3)$. The only graph with negative slope and with a negative y-coordinate of its y-intercept is **C**.

11. $y = 2x$
This line has slope $m = 2$ and y-intercept $(0, b) = (0, 0)$. The only graph with positive slope and with y-intercept $(0, 0)$ is **H**.

13. $y = 3$
This line is a horizontal line with y-intercept $(0, 3)$. The only graph that has these characteristics is **B**.

15. $m = 5; b = 15$
Substitute these values in the slope-intercept form.

$$y = mx + b$$
$$y = 5x + 15$$

17. $m = -\frac{2}{3}$; y-intercept $(0, \frac{4}{5})$

Here $m = -\frac{2}{3}$ and $b = \frac{4}{5}$. Substitute these values into the slope-intercept form.

$$y = mx + b$$
$$y = -\frac{2}{3}x + \frac{4}{5}$$

19. Slope $\frac{2}{5}$; y-intercept $(0, 5)$
Here $m = \frac{2}{5}$ and $b = 5$. Substitute these values in the slope-intercept form.

$$y = mx + b$$
$$y = \frac{2}{5}x + 5$$

21. $-x + y = 2$
(a) Solve for y to get the equation in slope-intercept form.

$$-x + y = 2$$
$$y = x + 2$$

(b) The slope is the coefficient of x, which is 1.

(c) The y-intercept is the point $(0, b)$, or $(0, 2)$.

(d)

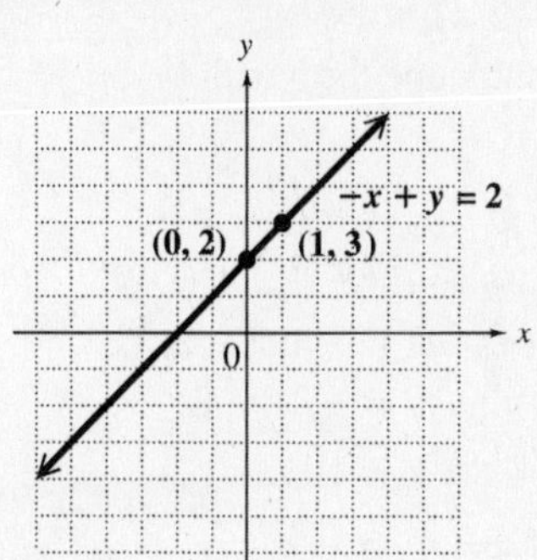

23. $4x - 5y = 20$
(a) Solve for y to get the equation in slope-intercept form.

$$4x - 5y = 20$$
$$-5y = -4x + 20$$
$$y = \frac{4}{5}x - 4$$

(b) The slope is the coefficient of x, which is $\frac{4}{5}$.

(c) The y-intercept is the point $(0, b)$, or $(0, -4)$.

(d)

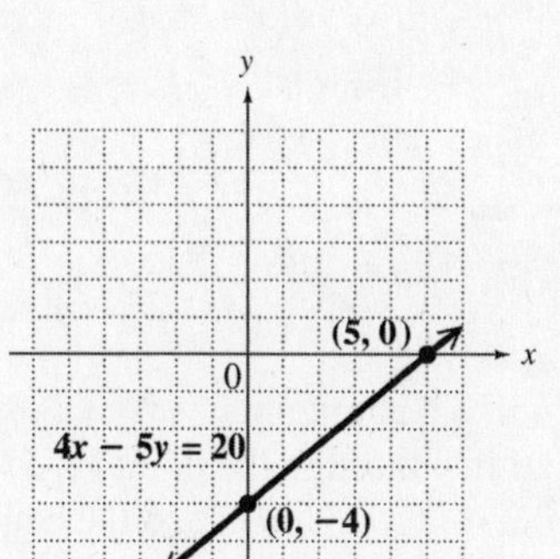

25. $x + 2y = -4$
(a) Solve for y to get the equation in slope-intercept form.

$$x + 2y = -4$$
$$2y = -x - 4$$
$$y = -\frac{1}{2}x - 2$$

(b) The slope is the coefficient of x, which is $-\frac{1}{2}$.

(c) The y-intercept is the point $(0, b)$, or $(0, -2)$.

(d)

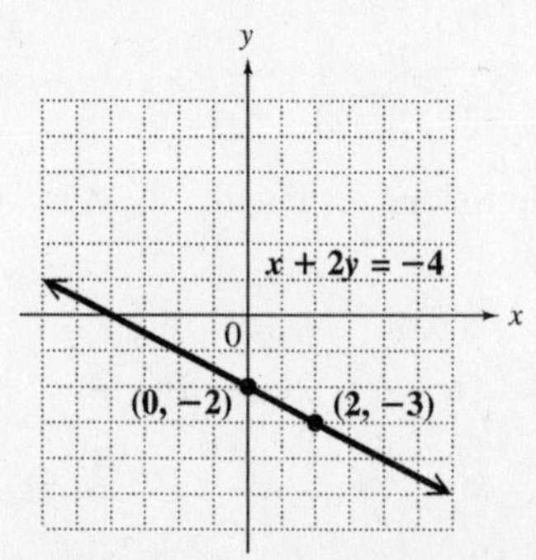

27. Through $(-2, 4)$; $m = -\frac{3}{4}$
Use the point-slope form with $(x_1, y_1) = (-2, 4)$ and $m = -\frac{3}{4}$.

$$\begin{aligned} y - y_1 &= m(x - x_1) \\ y - 4 &= -\tfrac{3}{4}[x - (-2)] \\ 4(y - 4) &= -3(x + 2) \\ 4y - 16 &= -3x - 6 \\ 3x + 4y &= 10 \end{aligned}$$

29. Through $(5, 8)$; $m = -2$
Use the point-slope form with $(x_1, y_1) = (5, 8)$ and $m = -2$.

$$\begin{aligned} y - y_1 &= m(x - x_1) \\ y - 8 &= -2(x - 5) \\ y - 8 &= -2x + 10 \\ 2x + y &= 18 \end{aligned}$$

31. Through $(-5, 4)$; $m = \frac{1}{2}$
Use the point-slope form with $(x_1, y_1) = (-5, 4)$ and $m = \frac{1}{2}$.

$$\begin{aligned} y - y_1 &= m(x - x_1) \\ y - 4 &= \tfrac{1}{2}[x - (-5)] \\ 2(y - 4) &= 1(x + 5) \\ 2y - 8 &= x + 5 \\ -x + 2y &= 13 \\ x - 2y &= -13 \end{aligned}$$

33. Through $(-4, 12)$; horizontal
A horizontal line through the point (c, d) has equation $y = d$. Here $d = 12$, so the equation is $y = 12$.

35. Through $(9, 10)$; undefined slope
A vertical line has undefined slope and equation $x = c$. Since the x-value in $(9, 10)$ is 9, the equation is $x = 9$.

37. Through $(0.5, 0.2)$; horizontal
A horizontal line through the point (c, d) has equation $y = d$. Since the y-value in $(0.5, 0.2)$ is 0.2, the equation of this line is $y = 0.2$.

39. $(3, 4)$ and $(5, 8)$
Find the slope.

$$m = \frac{8 - 4}{5 - 3} = \frac{4}{2} = 2$$

Use the point-slope form with $(x_1, y_1) = (3, 4)$ and $m = 2$.

$$\begin{aligned} y - y_1 &= m(x - x_1) \\ y - 4 &= 2(x - 3) \\ y - 4 &= 2x - 6 \\ -2x + y &= -2 \\ 2x - y &= 2 \end{aligned}$$

41. $(6, 1)$ and $(-2, 5)$
Find the slope.

$$m = \frac{5 - 1}{-2 - 6} = \frac{4}{-8} = -\frac{1}{2}$$

Use the point-slope form with $(x_1, y_1) = (6, 1)$ and $m = -\frac{1}{2}$.

$$\begin{aligned} y - y_1 &= m(x - x_1) \\ y - 1 &= -\tfrac{1}{2}(x - 6) \\ 2(y - 1) &= -1(x - 6) \\ 2y - 2 &= -x + 6 \\ x + 2y &= 8 \end{aligned}$$

43. $\left(-\frac{2}{5}, \frac{2}{5}\right)$ and $\left(\frac{4}{3}, \frac{2}{3}\right)$
Find the slope.

$$m = \frac{\frac{2}{3} - \frac{2}{5}}{\frac{4}{3} - \left(-\frac{2}{5}\right)} = \frac{\frac{10 - 6}{15}}{\frac{20 + 6}{15}} = \frac{\frac{4}{15}}{\frac{26}{15}} = \frac{4}{26} = \frac{2}{13}$$

Use the point-slope form with $(x_1, y_1) = (-\frac{2}{5}, \frac{2}{5})$ and $m = \frac{2}{13}$.

$$\begin{aligned} y - \tfrac{2}{5} &= \tfrac{2}{13}\left[x - \left(-\tfrac{2}{5}\right)\right] \\ 13\left(y - \tfrac{2}{5}\right) &= 2\left(x + \tfrac{2}{5}\right) \\ 13y - \tfrac{26}{5} &= 2x + \tfrac{4}{5} \\ -2x + 13y &= \tfrac{30}{5} \\ 2x - 13y &= -6 \end{aligned}$$

45. $(2, 5)$ and $(1, 5)$
Find the slope.

$$m = \frac{5 - 5}{1 - 2} = \frac{0}{-1} = 0$$

A line with slope 0 is horizontal. A horizontal line through the point (x, k) has equation $y = k$, where k is the common y-value, so the equation is $y = 5$.

47. $(7, 6)$ and $(7, -8)$
Find the slope.

$$m = \frac{-8 - 6}{7 - 7} = \frac{-14}{0} \quad \textit{Undefined}$$

A line with undefined slope is a vertical line. The equation of a vertical line is $x = k$, where k is the common x-value. So the equation is $x = 7$.

49. Through $(7, 2)$; parallel to $3x - y = 8$
Find the slope of the line with equation

$$\begin{aligned} 3x - y &= 8. \\ -y &= -3x + 8 \\ y &= 3x - 8 \end{aligned}$$

The slope is 3, so a line parallel to it also has slope 3. Use $m = 3$ and $(x_1, y_1) = (7, 2)$ in the point-slope form.

$$\begin{aligned} y - y_1 &= m(x - x_1) \\ y - 2 &= 3(x - 7) \\ y - 2 &= 3x - 21 \\ y &= 3x - 19 \end{aligned}$$

51. Through $(-2, -2)$; parallel to $-x + 2y = 10$
Find the slope of the line with equation

$$\begin{aligned} -x + 2y &= 10. \\ 2y &= x + 10 \\ y &= \tfrac{1}{2}x + 5 \end{aligned}$$

The slope is $\frac{1}{2}$, so a line parallel to it also has slope $\frac{1}{2}$. Use $m = \frac{1}{2}$ and $(x_1, y_1) = (-2, -2)$ in the point-slope form.

$$\begin{aligned} y - y_1 &= m(x - x_1) \\ y - (-2) &= \tfrac{1}{2}[x - (-2)] \\ y + 2 &= \tfrac{1}{2}(x + 2) \\ y + 2 &= \tfrac{1}{2}x + 1 \\ y &= \tfrac{1}{2}x - 1 \end{aligned}$$

53. Through $(8, 5)$; perpendicular to $2x - y = 7$
Find the slope of the line with equation

$$\begin{aligned} 2x - y &= 7. \\ -y &= -2x + 7 \\ y &= 2x - 7 \end{aligned}$$

The slope of the line is 2. Therefore, the slope of the line perpendicular to it is $-\frac{1}{2}$ since $2(-\frac{1}{2}) = -1$. Use $m = -\frac{1}{2}$ and $(x_1, y_1) = (8, 5)$ in the point-slope form.

$$\begin{aligned} y - y_1 &= m(x - x_1) \\ y - 5 &= -\tfrac{1}{2}(x - 8) \\ y - 5 &= -\tfrac{1}{2}x + 4 \\ y &= -\tfrac{1}{2}x + 9 \end{aligned}$$

55. Through $(-2, 7)$; perpendicular to $x = 9$
$x = 9$ is a vertical line so a line perpendicular to it will be a horizontal line. It goes through $(-2, 7)$ so its equation is

$$y = 7.$$

57. Distance = (rate)(time), so

$$y = 45x.$$

x	$y = 45x$	Ordered Pair
0	$45(0) = 0$	$(0, 0)$
5	$45(5) = 225$	$(5, 225)$
10	$45(10) = 450$	$(10, 450)$

59. Total cost = (cost/gal)(number of gallons), so

$$y = 5.00x.$$

x	$y = 5.00x$	Ordered Pair
0	$5.00(0) = 0$	$(0, 0)$
5	$5.00(5) = 25.00$	$(5, 25.00)$
10	$5.00(10) = 50.00$	$(10, 50.00)$

61. **(a)** The fixed cost is \$99, so that is the value of b. The variable cost is \$41, so

$$y = mx + b = 41x + 99.$$

(b) If $x = 5$, $y = 41(5) + 99 = 304$. The ordered pair is $(5, 304)$. The cost of a 5-month membership is \$304.

(c) If $x = 12$, $y = 41(12) + 99 = 591$. The cost of the first year's membership is \$591.

63. **(a)** The fixed cost is \$36, so that is the value of b. The variable cost is \$60, so

$$y = mx + b = 60x + 36.$$

(b) If $x = 5$, $y = 60(5) + 36 = 336$. The ordered pair is $(5, 336)$. The cost of the plan for 5 months is \$336.

(c) For a 2-year contract, $x = 24$, so $y = 60(24) + 36 = 1476$. The cost of the plan for 2 years is \$1476.

65. **(a)** The fixed cost is \$50, so that is the value of b. The variable cost is \$0.20, so

$$y = mx + b = 0.20x + 50.$$

(b) If $x = 5$, $y = 0.20(5) + 50 = 51$. The ordered pair is $(5, 51)$. The charge for driving 5 miles is \$51.

(c)
$$\begin{aligned} 84.60 &= 0.20x + 50 \qquad \textit{Let } y = 84.60. \\ 34.60 &= 0.20x \\ x &= \frac{34.60}{0.20} = 173 \end{aligned}$$

The car was driven 173 miles.

67. **(a)** Use $(0, 3921)$ and $(3, 7805)$.

$$m = \frac{7805 - 3921}{3 - 0} = \frac{3884}{3} \approx 1294.7$$

The equation is $y = 1294.7x + 3921$. The slope tells us that the sales of digital cameras in the United States increased by \$1294.7 million per year from 2003 to 2006.

(b) The year 2007 corresponds to $x = 4$, so $y = 1294.7(4) + 3921 = \$9099.8$ million.

69. **(a)** Use $(3, 99{,}059)$ and $(7, 95{,}898)$.

$$m = \frac{95{,}898 - 99{,}059}{7 - 3} = \frac{-3161}{4} = -790.25$$

Now use the point-slope form.

$$\begin{aligned} y - 99{,}059 &= -790.25(x - 3) \\ y - 99{,}059 &= -790.25x + 2370.75 \\ y &\approx -790.25x + 101{,}430 \end{aligned}$$

(b) The year 2005 corresponds to $x = 5$, so $y = -790.25(5) + 101{,}430 \approx 97{,}479$. This value is slightly less than the actual value of 98,071.

71. When $C = 0°$, $F = \underline{32}°$, and when $C = 100°$, $F = \underline{212}°$. These are the freezing and boiling temperatures for water.

72. **(a)** The two points of the form (C, F) would be $(0, 32)$ and $(100, 212)$.

(b) $m = \frac{212 - 32}{100 - 0} = \frac{180}{100} = \frac{9}{5}$

73. Let $m = \frac{9}{5}$ and $(x_1, y_1) = (0, 32)$.

$$\begin{aligned} y - y_1 &= m(x - x_1) \\ F - 32 &= \tfrac{9}{5}(C - 0) \\ F - 32 &= \tfrac{9}{5}C \\ F &= \tfrac{9}{5}C + 32 \end{aligned}$$

74.

$$\begin{aligned} F &= \tfrac{9}{5}C + 32 \\ F - 32 &= \tfrac{9}{5}C \\ \tfrac{5}{9}(F - 32) &= C \end{aligned}$$

75. $F \approx 2C + 30$

$F \approx 2(15) + 30$ *Let C = 15.*

$= 30 + 30 = 60°$

76. $F = \frac{9}{5}C + 32$

$F = \frac{9}{5}(15) + 32$ *Let C = 15.*

$= 27 + 32$

$= 59°$

$60° - 59° = 1°$

This exact answer is one less than the approximation found in Exercise 75. They differ by 1°.

77. $F \approx 2C + 30$

$F \approx 2(30) + 30$ *Let C = 30.*

$= 60 + 30$

$= 90°$

$F = \frac{9}{5}C + 32$

$F = \frac{9}{5}(30) + 32$ *Let C = 30.*

$= 54 + 32$

$= 86°$

$90° - 86° = 4°$

The exact answer is four less than the approximation. They differ by 4°.

78. Since $\frac{9}{5}$ is a little less than 2, and 32 is a little more than 30, $\frac{9}{5}C + 32 \approx 2C + 30$.

Summary Exercises on Slopes and Equations of Lines

1. The slope of the line through $(3, -3)$ and $(8, -6)$ is

$$m = \frac{-6 - (-3)}{8 - 3} = \frac{-3}{5} = -\frac{3}{5}.$$

3. $3x - 7y = 21$

Solve the equation for y.

$$\begin{aligned} -7y &= -3x + 21 \\ y &= \tfrac{3}{7}x - 3 \end{aligned}$$

The slope is $\frac{3}{7}$.

5. Through $(4, -2)$ with slope -3

(a) Use the point-slope form with $(x_1, y_1) = (4, -2)$ and $m = -3$.

$$\begin{aligned} y - y_1 &= m(x - x_1) \\ y - (-2) &= -3(x - 4) \\ y + 2 &= -3x + 12 \\ y &= -3x + 10 \end{aligned}$$

(b)
$$\begin{aligned} y &= -3x + 10 \\ 3x + y &= 10 \end{aligned}$$

7. Through $(-2, 6)$ and $(4, 1)$

(a) The slope is

$$m = \frac{1 - 6}{4 - (-2)} = \frac{-5}{6} = -\frac{5}{6}.$$

Use the point-slope form.

$$y - y_1 = m(x - x_1)$$
$$y - 6 = -\tfrac{5}{6}[x - (-2)]$$
$$y - 6 = -\tfrac{5}{6}x - \tfrac{5}{3}$$
$$y = -\tfrac{5}{6}x - \tfrac{5}{3} + \tfrac{18}{3}$$
$$y = -\tfrac{5}{6}x + \tfrac{13}{3}$$

(b) $y = -\tfrac{5}{6}x + \tfrac{13}{3}$

$6y = -5x + 26$ *Multiply by 6.*

$5x + 6y = 26$

9. Through $(-2, 5)$; parallel to $3x - y = 4$

(a) Find the slope of $3x - y = 4$.

$$-y = -3x + 4$$
$$y = 3x - 4$$

The slope is 3, so a line parallel to it also has slope 3. Use $m = 3$ and $(x_1, y_1) = (-2, 5)$ in the point-slope form.

$$y - y_1 = m(x - x_1)$$
$$y - 5 = 3[x - (-2)]$$
$$y - 5 = 3(x + 2)$$
$$y - 5 = 3x + 6$$
$$y = 3x + 11$$

(b) $y = 3x + 11$

$-3x + y = 11$

$3x - y = -11$

11. Through $(5, -8)$; parallel to $y = 4$

(a) $y = 4$ is a horizontal line, so a line parallel to it must be horizontal and have an equation of the form $y = k$. It goes through $(5, -8)$, so its equation is

$$y = -8.$$

(b) $y = -8$ is already in standard form.

13. Through $(-4, 2)$; parallel to the line through $(3, 9)$ and $(6, 11)$

(a) The slope of the line through $(3, 9)$ and $(6, 11)$ is

$$m = \frac{11 - 9}{6 - 3} = \frac{2}{3}.$$

Use the point-slope form with $(x_1, y_1) = (-4, 2)$ and $m = \frac{2}{3}$ (since the slope of the desired line must equal the slope of the given line).

$$y - y_1 = m(x - x_1)$$
$$y - 2 = \tfrac{2}{3}[x - (-4)]$$
$$y - 2 = \tfrac{2}{3}(x + 4)$$
$$y - 2 = \tfrac{2}{3}x + \tfrac{8}{3}$$
$$y = \tfrac{2}{3}x + \tfrac{8}{3} + \tfrac{6}{3}$$
$$y = \tfrac{2}{3}x + \tfrac{14}{3}$$

(b) $y = \tfrac{2}{3}x + \tfrac{14}{3}$

$3y = 2x + 14$

$-2x + 3y = 14$

$2x - 3y = -14$

15. Through $(-4, 12)$ and the midpoint of the segment with endpoints $(5, 8)$ and $(-3, 2)$

(a) The midpoint of the segment with endpoints $(5, 8)$ and $(-3, 2)$ is

$$\left(\frac{5 + (-3)}{2}, \frac{8 + 2}{2}\right) = \left(\frac{2}{2}, \frac{10}{2}\right) = (1, 5).$$

The slope of the line through $(-4, 12)$ and $(1, 5)$ is

$$m = \frac{5 - 12}{1 - (-4)} = \frac{-7}{5} = -\frac{7}{5}.$$

Use the point-slope form with $(x_1, y_1) = (-4, 12)$ and $m = -\frac{7}{5}$.

$$y - y_1 = m(x - x_1)$$
$$y - 12 = -\tfrac{7}{5}[x - (-4)]$$
$$y - 12 = -\tfrac{7}{5}(x + 4)$$
$$y - 12 = -\tfrac{7}{5}x - \tfrac{28}{5}$$
$$y = -\tfrac{7}{5}x - \tfrac{28}{5} + \tfrac{60}{5}$$
$$y = -\tfrac{7}{5}x + \tfrac{32}{5}$$

(b) $y = -\tfrac{7}{5}x + \tfrac{32}{5}$

$5y = -7x + 32$

$7x + 5y = 32$

17. Through $(2, -1)$; parallel to $y = \frac{1}{5}x + \frac{7}{4}$

(a) The slope of the desired line is the same as the slope of the given line, so use the point-slope form with $(x_1, y_1) = (2, -1)$ and $m = \frac{1}{5}$.

$$y - y_1 = m(x - x_1)$$
$$y - (-1) = \tfrac{1}{5}(x - 2)$$
$$y + 1 = \tfrac{1}{5}x - \tfrac{2}{5}$$
$$y = \tfrac{1}{5}x - \tfrac{2}{5} - \tfrac{5}{5}$$
$$y = \tfrac{1}{5}x - \tfrac{7}{5}$$

(b) $y = \tfrac{1}{5}x - \tfrac{7}{5}$

$5y = x - 7$ *Multiply by 5.*

$-x + 5y = -7$

$x - 5y = 7$

19. Through $(0.3, 1.5)$ and $(0.4, 1.7)$

(a) $m = \frac{1.7 - 1.5}{0.4 - 0.3} = \frac{0.2}{0.1} = 2$

Use the point-slope form with $(x_1, y_1) = (0.3, 1.5)$ and $m = 2$.

$$\begin{aligned} y - y_1 &= m(x - x_1) \\ y - 1.5 &= 2(x - 0.3) \\ y - 1.5 &= 2x - 0.6 \\ y &= 2x + 0.9 \end{aligned}$$

(b)

$$\begin{aligned} y &= 2x + 0.9 \\ 10y &= 20x + 9 \quad \textit{Multiply by 10.} \\ -20x + 10y &= 9 \\ 20x - 10y &= -9 \end{aligned}$$

21. Slope -0.5, $b = -2$

The slope-intercept form of a line, $y = mx + b$, becomes $y = -0.5x - 2$, or $y = -\frac{1}{2}x - 2$, which is choice **B**.

23. Passes through $(4, -2)$ and $(0, 0)$

$$m = \frac{0 - (-2)}{0 - 4} = \frac{2}{-4} = -\frac{1}{2}$$

Using $m = -\frac{1}{2}$ and a y-intercept of $(0, 0)$, we get $y = -\frac{1}{2}x + 0$, which is choice **A**.

25. $m = \frac{1}{2}$, passes through the origin
Use the point-slope form with $(x_1, y_1) = (0, 0)$ and $m = \frac{1}{2}$.

$$\begin{aligned} y - y_1 &= m(x - x_1) \\ y - 0 &= \tfrac{1}{2}(x - 0) \\ y &= \tfrac{1}{2}x \quad \text{or} \quad 2y = x \end{aligned}$$

This is choice **E**.

4.4 Linear Inequalities in Two Variables

4.4 Margin Exercises

1. **(a)** $x + y \le 4$

Step 1
Graph the line, $x + y = 4$, which has intercepts $(4, 0)$ and $(0, 4)$, as a solid line since the inequality involves " $\le$ ".

Step 2
Test $(0, 0)$.

$$\begin{aligned} x + y &\le 4 \\ 0 + 0 &\overset{?}{\le} 4 \\ 0 &\le 4 \quad \textit{True} \end{aligned}$$

Step 3
Since the result is true, shade the region that contains $(0, 0)$.

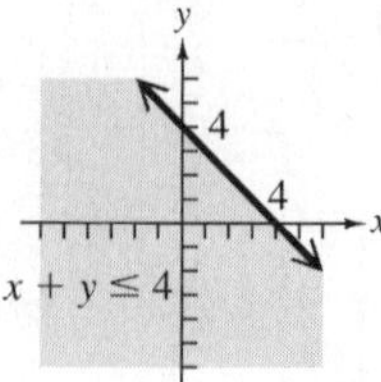

(b) $3x + y \ge 6$

Step 1
Graph the line $3x + y = 6$ as a solid line through $(2, 0)$ and $(0, 6)$.

Step 2
Test $(0, 0)$.

$$\begin{aligned} 3x + y &\ge 6 \\ 3(0) + 0 &\overset{?}{\ge} 6 \\ 0 &\ge 6 \quad \textit{False} \end{aligned}$$

Step 3
Since the result is false, shade the region that does not contain $(0, 0)$.

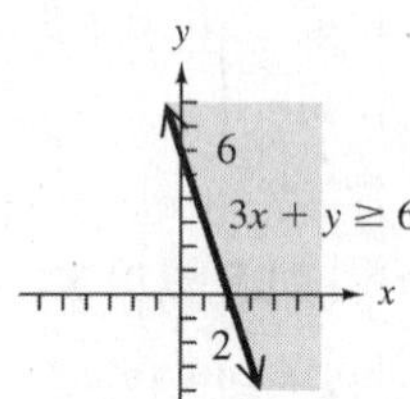

2. **(a)** $x + y > 0$

Solve the inequality for y.

$$y > -x \quad \textit{Subtract x.}$$

Graph the boundary line, $y = -x$ [which has slope -1 and y-intercept $(0, 0)$], as a dashed line because the inequality symbol is $>$. Since the inequality is solved for y and the inequality symbol is $>$, we shade the half-plane above the boundary line.

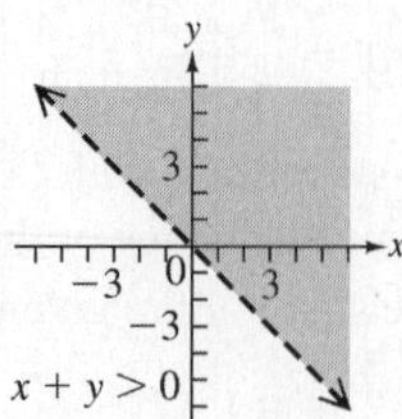

(b) $3x - 2y > 0$

Solve the inequality for y.

$$-2y > -3x \quad \textit{Subtract } 3x.$$
$$y < \frac{3}{2}x \quad \textit{Divide by } -2; \textit{ change } > \textit{ to } <.$$

Graph the boundary line, $y = \frac{3}{2}x$ [which has slope $\frac{3}{2}$ and y-intercept $(0, 0)$], as a dashed line because the inequality symbol is $<$. Since the inequality is solved for y and the inequality symbol is $<$, we shade the half-plane below the boundary line.

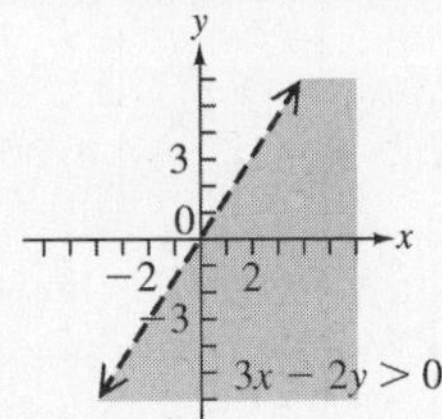

3. $x - y \le 4$ and $x \ge -2$
Graph $x - y = 4$, which has intercepts $(4, 0)$ and $(0, -4)$, as a solid line since the inequality involves " $\le$ ". Test $(0, 0)$, which yields $0 \le 4$, a true statement. Shade the region that includes $(0, 0)$.

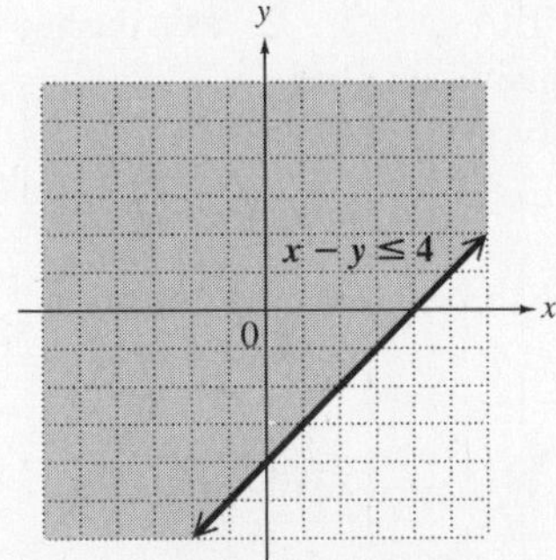

Graph $x = -2$ as a solid vertical line through $(-2, 0)$. Shade the region to the right of $x = -2$.

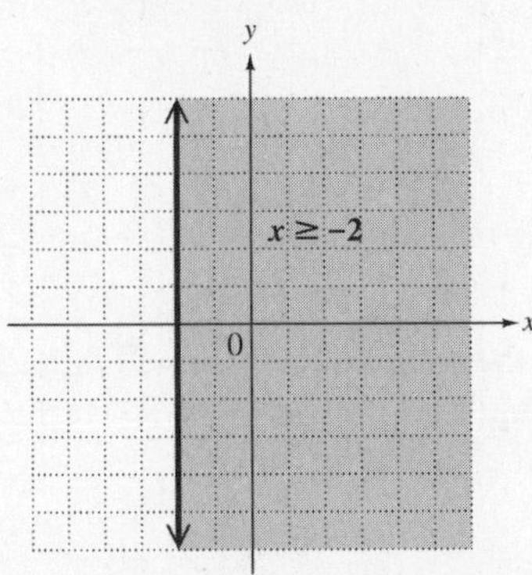

The graph of the intersection is the region common to both graphs.

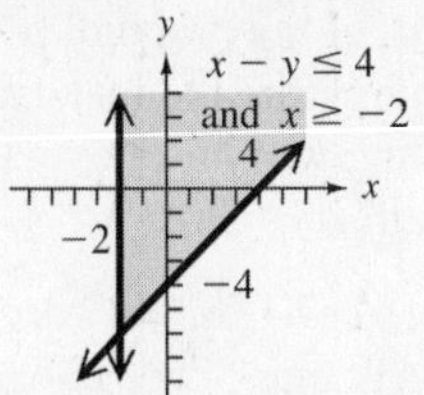

4. $7x - 3y < 21$ or $x > 2$
Graph $7x - 3y = 21$ as a dashed line through its intercepts $(3, 0)$ and $(0, -7)$. Test $(0, 0)$, which yields $0 < 21$, a true statement. Shade the region that includes $(0, 0)$.

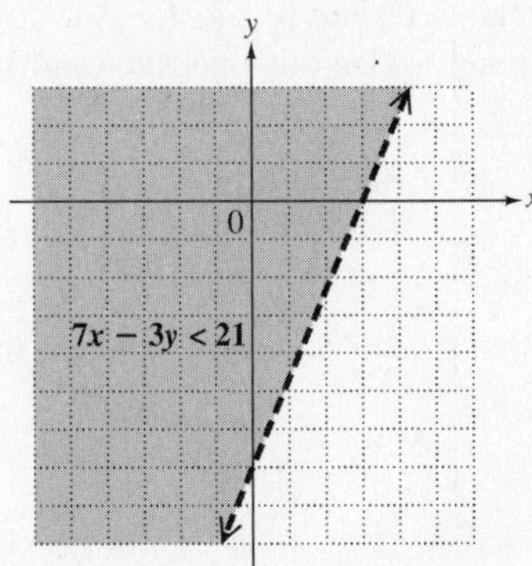

Graph $x = 2$ as a dashed vertical line through $(2, 0)$. Shade the region to the right of $x = 2$.

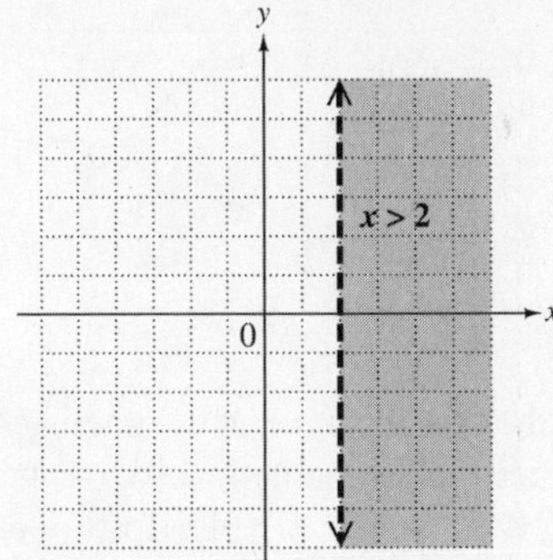

The graph of the union is the region that includes all the points in both graphs.

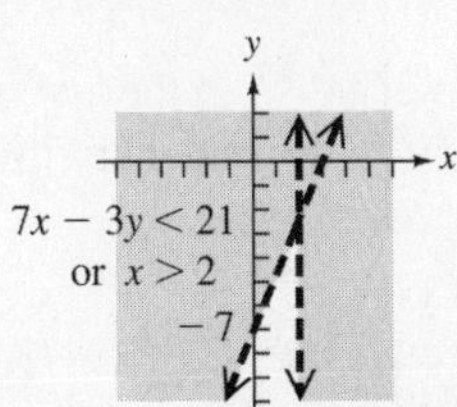

4.4 Section Exercises

1. The boundary of the graph of $y \le -x + 2$ will be a <u>solid</u> line (since the inequality involves $\le$), and the shading will be <u>below</u> the line (since the inequality sign is $\le$ or $<$).

3. The boundary of the graph of $y > -x + 2$ will be a dashed line (since the inequality involves $>$), and the shading will be above the line (since the inequality sign is $\geq$ or $>$).

5. The graph of $Ax + By = C$ divides the plane into two regions. In one of these regions, the ordered pairs satisfy $Ax + By < C$; in the other, they satisfy $Ax + By > C$.

7. $x + y \leq 2$

Graph the line $x + y = 2$ by drawing a solid line (since the inequality involves $\leq$) through the intercepts $(2, 0)$ and $(0, 2)$.
Test a point not on this line, such as $(0, 0)$.

$$x + y \leq 2$$
$$0 + 0 \overset{?}{\leq} 2$$
$$0 \leq 2 \quad \textit{True}$$

Shade that side of the line containing the test point $(0, 0)$.

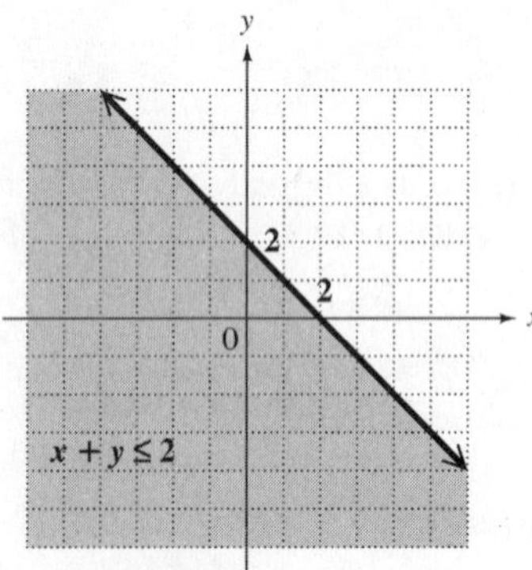

9. $4x - y < 4$

Graph the line $4x - y = 4$ by drawing a dashed line (since the inequality involves $<$) through the intercepts $(1, 0)$ and $(0, -4)$. Instead of using a test point, we will solve the inequality for y.

$$-y < -4x + 4$$
$$y > 4x - 4$$

Since we have "$y >$ " in the last inequality, shade the region *above* the boundary line.

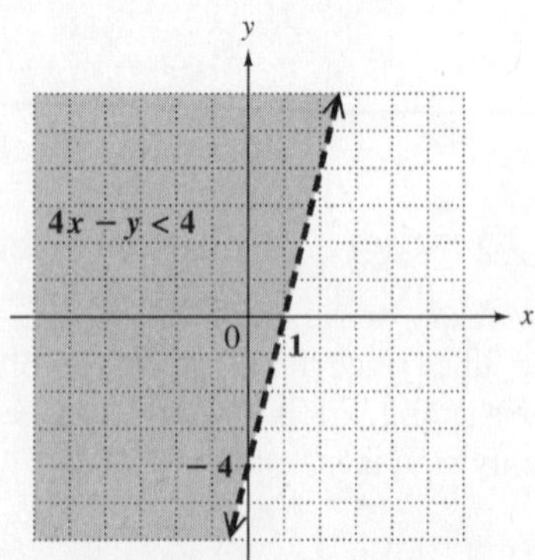

11. $x + 3y \geq -2$

Graph the solid line $x + 3y = -2$ (since the inequality involves $\geq$) through the intercepts $(-2, 0)$ and $(0, -\frac{2}{3})$.
Test a point not on this line such as $(0, 0)$.

$$0 + 3(0) \overset{?}{\geq} -2$$
$$0 \geq -2 \quad \textit{True}$$

Shade that side of the line containing the test point $(0, 0)$.

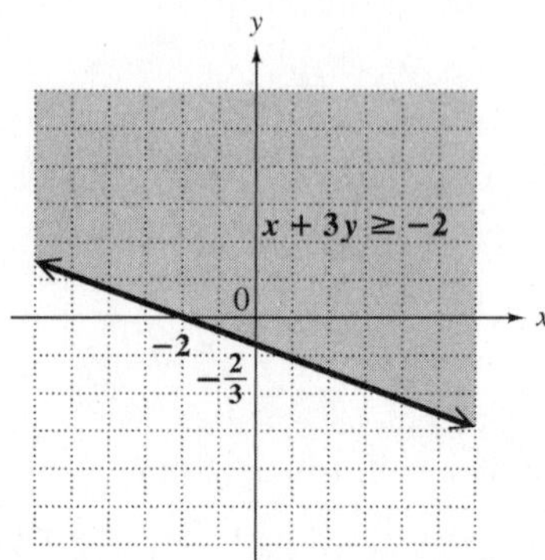

13. $x + y > 0$

Graph the line $x + y = 0$, which includes the points $(0, 0)$ and $(2, -2)$, as a dashed line (since the inequality involves $>$). Solving the inequality for y gives us

$$y > -x,$$

So shade the region above the boundary line.

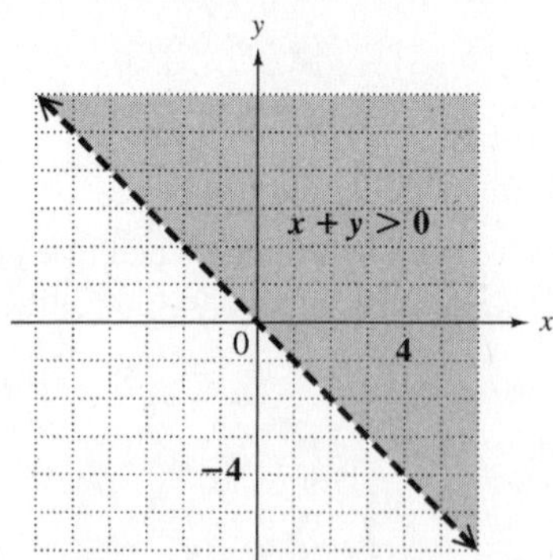

15. $x - 3y \leq 0$

Graph the solid line $x - 3y = 0$ through the points $(0, 0)$ and $(3, 1)$.
Solve the inequality for y.

$$-3y \leq -x$$
$$y \geq \tfrac{1}{3}x$$

Shade the region above the boundary line.

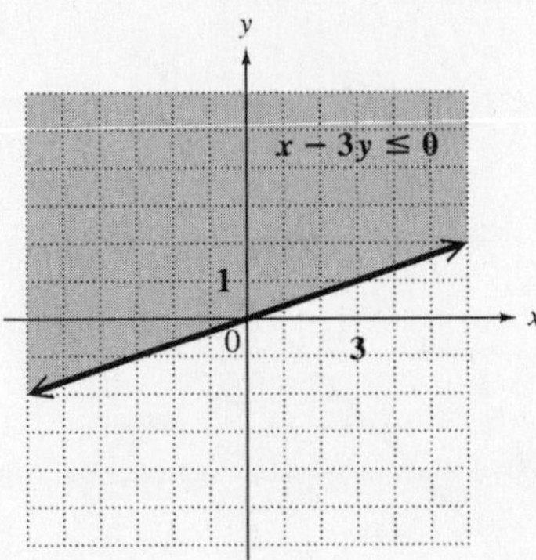

17. $y < x$

Graph the dashed line $y = x$ through $(0, 0)$ and $(2, 2)$. Since we have "$y <$ " in the inequality, shade the region *below* the boundary line.

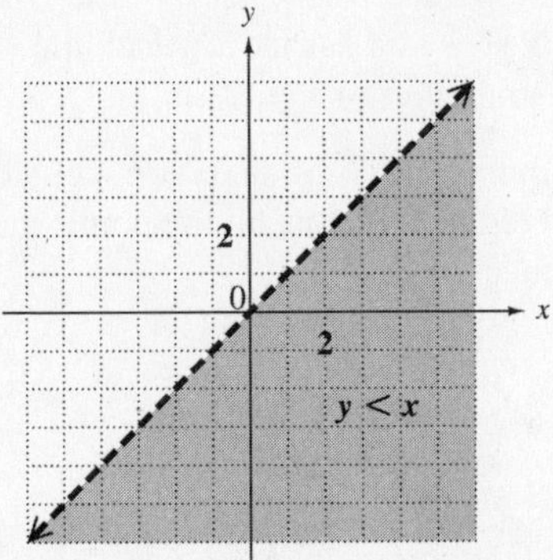

19. $x + y \leq 1$ and $x \geq 1$

Graph the solid line $x + y = 1$ through $(0, 1)$ and $(1, 0)$. The inequality $x + y \leq 1$ can be written as $y \leq -x + 1$, so shade the region below the boundary line.
Graph the solid vertical line $x = 1$ through $(1, 0)$ and shade the region to the right. The required graph is the common shaded area as well as the portions of the lines that bound it.

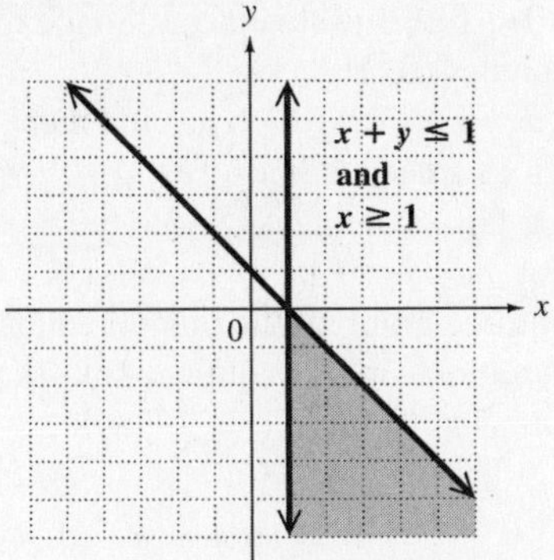

21. $2x - y \geq 2$ and $y < 4$

Graph the solid line $2x - y = 2$ through the intercepts $(1, 0)$ and $(0, -2)$. Test $(0, 0)$ to get $0 \geq 2$, a false statement. Shade that side of the graph not containing $(0, 0)$. To graph $y < 4$ on the same axes, graph the dashed horizontal line through $(0, 4)$. Test $(0, 0)$ to get $0 < 4$, a true statement. Shade that side of the dashed line containing $(0, 0)$.
The word "and" indicates the intersection of the two graphs. The final solution set consists of the region where the two shaded regions overlap.

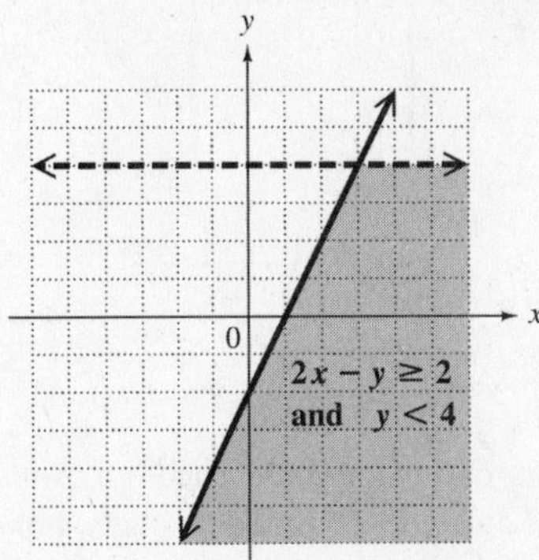

23. $x + y > -5$ and $y < -2$

Graph $x + y = -5$, which has intercepts $(-5, 0)$ and $(0, -5)$, as a dashed line. Test $(0, 0)$, which yields $0 > -5$, a true statement. Shade the region that includes $(0, 0)$.
Graph $y = -2$ as a dashed horizontal line. Shade the region below $y = -2$. The required graph of the intersection is the region common to both graphs.

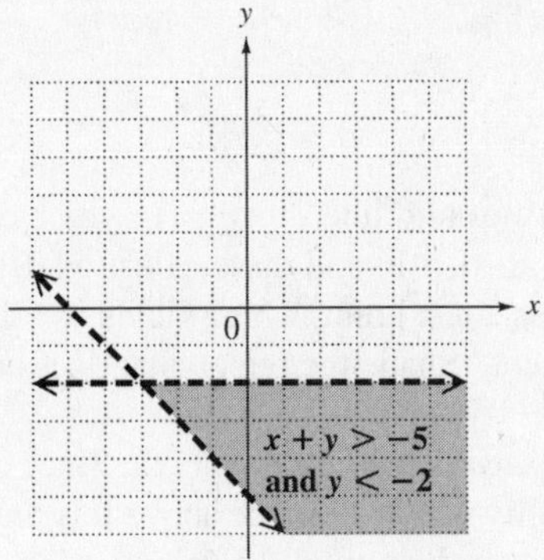

25. $|x| \geq 3$

Rewrite $|x| \geq 3$ as $x \geq 3$ or $x \leq -3$. The graph consists of the region to the left of the solid vertical line $x = -3$ and to the right of the solid vertical line $x = 3$.

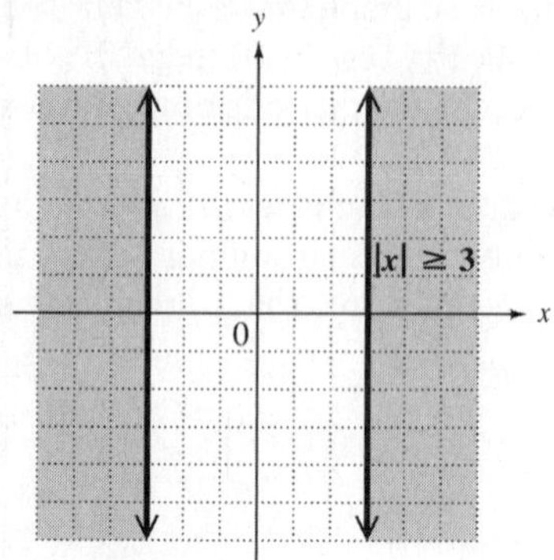

27. $|y + 1| < 2$ can be rewritten as

$$-2 < y + 1 < 2$$
$$-3 < y < 1.$$

The boundaries are the dashed horizontal lines $y = -3$ and $y = 1$. Since y is between -3 and 1, the graph includes all points between the lines.

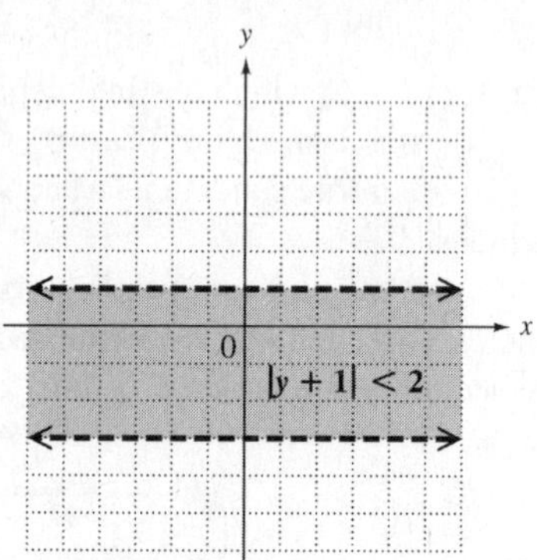

29. $x - y \geq 1$ or $y \geq 2$

Graph the solid line $x - y = 1$, which crosses the y-axis at $(0, -1)$ and the x-axis at $(1, 0)$. Use $(0, 0)$ as a test point, which yields $0 \geq 1$, a false statement. Shade the region that does not include $(0, 0)$.
Now graph the solid line $y = 2$. Since the inequality is $y \geq 2$, shade above this line.
The required graph of the union includes all the shaded regions, that is, all the points that satisfy either inequality.

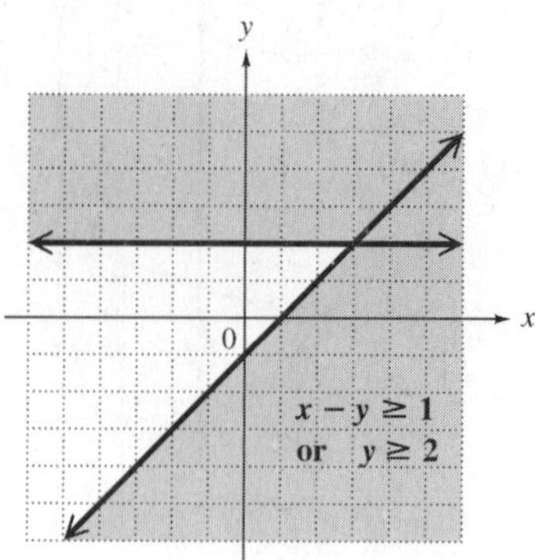

31. $x - 2 > y$ or $x < 1$

Graph $x - 2 = y$, which has intercepts $(2, 0)$ and $(0, -2)$, as a dashed line. Test $(0, 0)$, which yields $-2 > 0$, a false statement. Shade the region that does not include $(0, 0)$.
Graph $x = 1$ as a dashed vertical line. Shade the region to the left of $x = 1$.

The required graph of the union includes all the shaded regions, that is, all the points that satisfy either inequality.

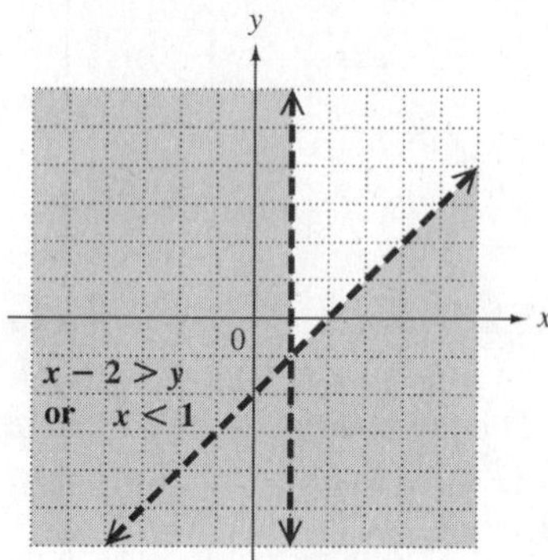

33. $3x + 2y < 6$ or $x - 2y > 2$

Graph $3x + 2y = 6$, which has intercepts $(2, 0)$ and $(0, 3)$, as a dashed line. Test $(0, 0)$, which yields $0 < 6$, a true statement. Shade the region that includes $(0, 0)$.
Graph $x - 2y = 2$, which has intercepts $(2, 0)$ and $(0, -1)$, as a dashed line. Test $(0, 0)$, which yields $0 > 2$, a false statement. Shade the region that does not include $(0, 0)$.
The required graph of the union includes all the shaded regions, that is, all the points that satisfy either inequality.

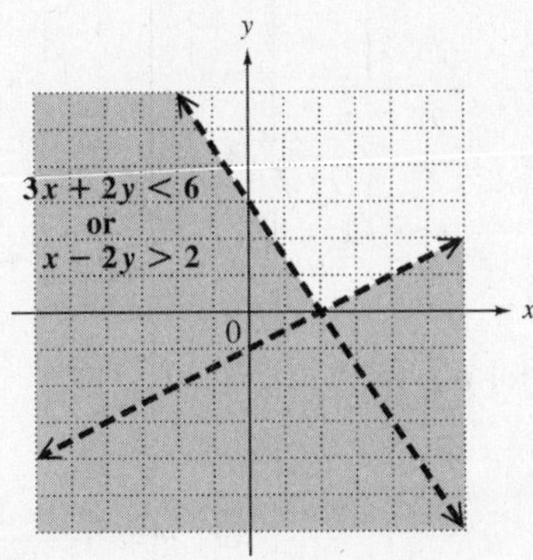

4.5 Introduction to Functions

4.5 Margin Exercises

1. The ordered pair $(40, 320)$ indicates that when you work 40 hours, your paycheck is \$320.

2. **(a)** $\{(0,3), (-1,2), (-1,3)\}$

   The last two ordered pairs have the *same* x-value paired with *two different* y-values (-1 is paired with both 2 and 3), so this relation *is not a function.*

   (b) $\{(2,-2), (4,-4), (6,-6)\}$

   The relation *is a function* because for each different x-value there is exactly one y-value.

   (c) $\{(-1,5), (0,5)\}$

   The relation *is a function* because for each different x-value there is exactly one y-value. It is acceptable to have different x-values paired with the same y-value.

3. **(a)** $\{(4,0), (4,1), (4,2)\}$

   The domain of this relation is the set of x-values, that is, $\{4\}$. The range of this relation is the set of y-values, that is, $\{0, 1, 2\}$. This relation *is not a function* because the same x-value 4 is paired with more than one y-value.

   (b) The domain of this relation is the set of all first components, that is, $\{-1, 4, 7\}$. The range of this relation is the set of all second components, that is, $\{0, -2, 3, 7\}$. This relation *is not a function* because the same x-value 4 is paired with more than one y-value.

   (c) The domain of this relation is the set of all first components, that is, $\{2002, 2003, 2004, 2005, 2006\}$. The range of this relation is the set of all second components, that is, $\{140{,}766, 158{,}722, 182{,}140, 207{,}896, 233{,}041\}$. This relation *is a function* because for each different first component, there is exactly one second component.

4. **(a)** The domain is the set of x-values, $\{-3, -2, 2, 3\}$. The range is the set of y-values, $\{-2, -1, 2, 3\}$.

   (b) The arrowheads indicate that the graph extends indefinitely right, as well as up and down. Because there is a smallest x-value, -2, the domain includes all numbers greater than or equal to -2, $[-2, \infty)$. The range includes all real numbers, written $(-\infty, \infty)$.

   (c) The arrowheads indicate that the graph extends indefinitely left and right, as well as downward. The domain includes all real numbers, written $(-\infty, \infty)$. Because there is a greatest y-value, 0, the range includes all numbers less than or equal to 0, $(-\infty, 0]$.

5. Any vertical line would intersect the graph at most once on graphs **A** and **C**, so graphs **A** and **C** represent functions.

   Graph **B** does not represent a function since a vertical line, such as the y-axis, intersects the graph in more than one point.

6. **(a)** $y = 6x + 12$ is a function because each value of x corresponds to exactly one value of y. Its domain is the set of all real numbers, $(-\infty, \infty)$.

   (b) $y \leq 4x$ is not a function because if $x = 0$, then $y \leq 0$. Thus, the x-value 0 corresponds to many y-values. Its domain is the set of all real numbers, $(-\infty, \infty)$.

   (c) $y = -\sqrt{3x - 2}$ is a function because each value of x corresponds to exactly one value of y. Since the quantity under the radical must be nonnegative, the domain is the set of real numbers that satisfy the condition

   $$3x - 2 \geq 0$$
   $$x \geq \tfrac{2}{3}.$$

   Therefore, the domain is $[\frac{2}{3}, \infty)$.

   (d) $y^2 = 25x$ is not a function. If $x = 1$, for example, $y^2 = 25$ and $y = 5$ or $y = -5$. Since y^2 must be nonnegative, the domain is the set of nonnegative real numbers, $[0, \infty)$.

7. **(a)**

$$\begin{aligned} f(x) &= 6x - 2 \\ f(-3) &= 6(-3) - 2 \quad \textit{Replace x with } -3. \\ &= -18 - 2 \\ &= -20 \\ f(p) &= 6p - 2 \quad \textit{Replace x with p.} \\ f(m+1) &= 6(m+1) - 2 \quad \textit{Replace x with } m+1. \\ &= 6m + 6 - 2 \\ &= 6m + 4 \end{aligned}$$

(b)

$$\begin{aligned} f(x) &= \frac{-3x+5}{2} \\ f(-3) &= \frac{-3(-3)+5}{2} \quad \textit{Replace x with } -3. \\ &= \frac{14}{2} = 7 \\ f(p) &= \frac{-3p+5}{2} \quad \textit{Replace x with p.} \\ f(m+1) &= \frac{-3(m+1)+5}{2} \quad \textit{Replace x with } m+1. \\ &= \frac{-3m-3+5}{2} \\ &= \frac{-3m+2}{2} \end{aligned}$$

(c)

$$\begin{aligned} f(x) &= \tfrac{1}{6}x - 1 \\ f(-3) &= \tfrac{1}{6}(-3) - 1 \quad \textit{Replace x with } -3. \\ &= -\tfrac{1}{2} - 1 = -\tfrac{3}{2} \\ f(p) &= \tfrac{1}{6}p - 1 \quad \textit{Replace x with p.} \\ f(m+1) &= \tfrac{1}{6}(m+1) - 1 \quad \textit{Replace x with } m+1. \\ &= \tfrac{1}{6}m + \tfrac{1}{6} - 1 \\ &= \tfrac{1}{6}m - \tfrac{5}{6} \end{aligned}$$

8. **(a)**

$$\begin{aligned} f(x) &= -4x - 8 \\ f(-2) &= -4(-2) - 8 \quad \textit{Replace x with } -2. \\ &= 8 - 8 \\ &= 0 \end{aligned}$$

(b) $f = \{(0,5), (-1,3), (-2,1)\}$

We want $f(-2)$, the y-value of the ordered pair where $x = -2$. As indicated by the ordered pair $(-2, 1)$, when $x = -2$, $y = 1$, so $f(-2) = 1$.

(c) Ordered pairs of a function can be written as $(x, f(x))$. In this case, we have $(-2, f(-2)) = (-2, 4)$, so $f(-2) = 4$.

9. **(a)**

$$\begin{aligned} y &= \sqrt{x+2} \\ f(x) &= \sqrt{x+2} \quad \textit{Replace y with f(x).} \\ f(-1) &= \sqrt{-1+2} \quad \textit{Let } x = -1. \\ &= \sqrt{1} = 1 \end{aligned}$$

(b) $x^2 - 4y = 3$

Solve for y.

$$\begin{aligned} -4y &= -x^2 + 3 \\ y &= \frac{1}{4}x^2 - \frac{3}{4}, \text{ or } y = \frac{x^2-3}{4} \end{aligned}$$

$$\begin{aligned} f(x) &= \frac{x^2-3}{4} \quad \textit{Replace y with f(x).} \\ f(-1) &= \frac{(-1)^2-3}{4} \quad \textit{Let } x = -1. \\ &= \frac{1-3}{4} \\ &= \frac{-2}{4} = -\frac{1}{2} \end{aligned}$$

10. **(a)** For $f(x) = \frac{3}{4}x - 2$, the y-intercept is $(0, -2)$ and the slope is $\frac{3}{4}$. Plot the point $(0, -2)$, then move up 3 units and right 4 units to plot another point, namely, $(4, 1)$. Draw a line through the points.

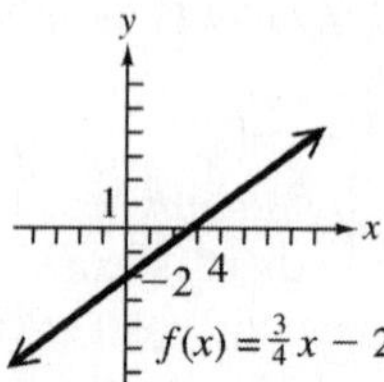

The domain and range are each $(-\infty, \infty)$.

(b) $g(x) = 3$ is a constant function. Its graph is a horizontal line.

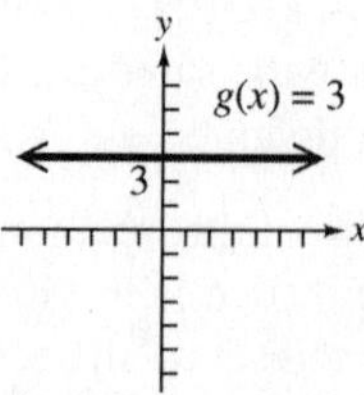

The domain of a constant function is $(-\infty, \infty)$. The range consists of the only y-value, 3, so it is $\{3\}$.

4.5 Section Exercises

1. In an ordered pair of a relation, the first element is the independent variable.

3. **(a)** A relation is a set of ordered pairs.

(b) The domain is the set of all first components (x-values).

(c) The range is the set of all second components (y-values).

(d) A function is a relation in which each domain element is paired with one and only one range element.

5. $\{(5,1),(3,2),(4,9),(7,3)\}$

The relation is a function because for each x-value, there is only one y-value.

The domain, the set of x-values, is $\{5,3,4,7\}$. The range, the set of y-values, is $\{1,2,9,3\}$.

7. $\{(2,4),(0,2),(2,6)\}$

The relation is not a function since the x-value 2 is paired with two different y-values, 4 and 6.

The domain, the set of x-values, is $\{0,2\}$. The range, the set of y-values, is $\{2,4,6\}$.

9. $\{(-3,1),(4,1),(-2,7)\}$

The relation is a function since for each x-value, there is only one y-value.

The domain is the set of x-values: $\{-3,4,-2\}$
The range is the set of y-values: $\{1,7\}$

11. $\{(1,1),(1,-1),(0,0),(2,4),(2,-4)\}$

The relation is not a function since the x-value 1 has two different y-values associated with it, 1 and -1. (A similar statement can be made for $x = 2$.)

The domain is the set of x-values: $\{0,1,2\}$.
The range is the set of y-values: $\{-4,-1,0,1,4\}$.

13. The relation can be described by the set of ordered pairs

$$\{(2,1),(5,1),(11,7),(17,20),(3,20)\}.$$

The relation is a function since for each x-value, there is only one y-value.

The domain is the set of x-values: $\{2,3,5,11,17\}$.
The range is the set of y-values: $\{1,7,20\}$.

15. The relation can be described by the set of ordered pairs

$$\{(1,5),(1,2),(1,-1),(1,-4)\}.$$

The relation is not a function since the x-value 1 has four different y-values associated with it, 5, 2, -1, and -4.

The domain is the set of x-values: $\{1\}$.
The range is the set of y-values: $\{5,2,-1,-4\}$.

17. Using the vertical line test, we find any vertical line will intersect the graph at most once. This indicates that the graph represents a function. This graph extends indefinitely to the left $(-\infty)$ and indefinitely to the right (∞). Therefore, the domain is $(-\infty,\infty)$. This graph extends indefinitely downward $(-\infty)$, and indefinitely upward (∞). Thus, the range is $(-\infty,\infty)$.

19. Using the vertical line test, we find any vertical line will intersect the graph at most once. This indicates that the graph represents a function. This graph extends indefinitely to the left $(-\infty)$ and indefinitely to the right (∞). Therefore, the domain is $(-\infty,\infty)$. This graph extends indefinitely downward $(-\infty)$, and reaches a high point at $y = 4$. Therefore, the range is $(-\infty,4]$.

21. Since a vertical line, such as $x = 4$, intersects the graph in two points, the relation is not a function. The domain is $[3,\infty)$, and the range is $(-\infty,\infty)$.

23. $y = x^2$

Each value of x corresponds to one y-value. For example, if $x = 3$, then $y = 3^2 = 9$. Therefore, $y = x^2$ defines y as a function of x.
Since any x-value, positive, negative, or zero, can be squared, the domain is $(-\infty,\infty)$.

25. $x = y^6$

The ordered pairs $(64,2)$ and $(64,-2)$ both satisfy the equation. Since one value of x, 64, corresponds to two values of y, 2 and -2, the relation does not define a function. Because x is equal to the sixth power of y, the values of x must always be nonnegative. The domain is $[0,\infty)$.

27. $y = 2x - 6$

For any value of x, there is exactly one value of y, so this equation defines a function. The domain is the set of all real numbers, $(-\infty,\infty)$.

29. $x + y < 4$

For a particular x-value, more than one y-value can be selected to satisfy $x + y < 4$. For example, if $x = 2$ and $y = 0$, then

$$2 + 0 < 4. \quad \textit{True}$$

Now, if $x = 2$ and $y = 1$, then

$$2 + 1 < 4. \quad \textit{Also true}$$

Therefore, $x + y < 4$ does not define y as a function of x.
The graph of $x + y < 4$ consists of the shaded region below the dashed line $x + y = 4$, which extends indefinitely from left to right. Therefore, the domain is $(-\infty,\infty)$.

31. $y = \sqrt{x}$

For any value of x, there is exactly one corresponding value for y, so this relation defines a function. Since the radicand must be a nonnegative number, x must always be nonnegative. The domain is $[0, \infty)$.

33. $xy = 1$

Rewrite $xy = 1$ as $y = \dfrac{1}{x}$. Note that x can never equal 0, otherwise the denominator would equal 0. The domain is $(-\infty, 0) \cup (0, \infty)$.
Each nonzero x-value gives exactly one y-value. Therefore, $xy = 1$ defines y as a function of x.

35. $y = \sqrt{4x + 2}$

To determine the domain of $y = \sqrt{4x+2}$, recall that the radicand must be nonnegative. Solve the inequality $4x + 2 \geq 0$, which gives us $x \geq -\frac{1}{2}$. Therefore, the domain is $[-\frac{1}{2}, \infty)$.
Each x-value from the domain produces exactly one y-value. Therefore, $y = \sqrt{4x+2}$ defines a function.

37. $y = \dfrac{2}{x - 9}$

Given any value of x, y is found by subtracting 9, then dividing the result into 2. This process produces exactly one value of y for each x-value, so the relation represents a function. The domain includes all real numbers except those that make the denominator 0, namely 9. The domain is $(-\infty, 9) \cup (9, \infty)$.

39. **(a)** The independent variable is t, the number of hours, and the possible values are in the set $[0, 100]$. The dependent variable is g, the number of gallons, and the possible values are in the set $[0, 3000]$.

(b) The graph rises for the first 25 hours, so the water level increases for 25 hours. The graph falls for $t = 50$ to $t = 75$, so the water level decreases for 25 hours.

(c) There are 2000 gallons in the pool when $t = 90$.

(d) $g(0)$ is the number of gallons in the pool at time $t = 0$. Here, $g(0) = 0$, which means the pool is empty at time 0.

41. Here are two examples.

The cost of gasoline depends on the number of gallons purchased, so cost is a function of the number of gallons purchased.

The amount of income tax you pay depends on your taxable income, so income tax is a function of taxable income.

43.
$$\begin{aligned} f(x) &= -3x + 4 \\ f(0) &= -3(0) + 4 \quad \textit{Replace x with 0.} \\ &= 0 + 4 \\ &= 4 \end{aligned}$$

45.
$$\begin{aligned} g(x) &= -x^2 + 4x + 1 \\ g(-2) &= -(-2)^2 + 4(-2) + 1 \\ &\quad \textit{Replace x with } -2. \\ &= -(4) - 8 + 1 \\ &= -11 \end{aligned}$$

47.
$$\begin{aligned} f(x) &= -3x + 4 \\ f(p) &= -3(p) + 4 \quad \textit{Replace x with p.} \\ &= -3p + 4 \end{aligned}$$

49.
$$\begin{aligned} f(x) &= -3x + 4 \\ f(-x) &= -3(-x) + 4 \quad \textit{Replace x with } -x. \\ &= 3x + 4 \end{aligned}$$

51.
$$\begin{aligned} f(x) &= -3x + 4 \\ f(x+2) &= -3(x+2) + 4 \\ &\quad \textit{Replace x with x + 2.} \\ &= -3x - 6 + 4 \\ &= -3x - 2 \end{aligned}$$

53.
$$\begin{aligned} g(x) &= -x^2 + 4x + 1 \\ g\left(\frac{p}{3}\right) &= -\left(\frac{p}{3}\right)^2 + 4\left(\frac{p}{3}\right) + 1 \\ &\quad \textit{Replace x with p/3.} \\ &= -\frac{p^2}{9} + \frac{4p}{3} + 1 \end{aligned}$$

55. **(a)** When $x = 2$, $y = 2$, so $f(2) = 2$.

(b) When $x = -1$, $y = 3$, so $f(-1) = 3$.

57. **(a)** When $x = 2$, $y = 15$, so $f(2) = 15$.

(b) When $x = -1$, $y = 10$, so $f(-1) = 10$.

59. **(a)** The point $(2, 3)$ is on the graph of f, so $f(2) = 3$.

(b) The point $(-1, -3)$ is on the graph of f, so $f(-1) = -3$.

61. The equation $2x + y = 4$ has a straight <u>line</u> as its graph. To find y in $(3, y)$, let $x = 3$ in the equation.

$$\begin{aligned} 2x + y &= 4 \\ 2(3) + y &= 4 \\ 6 + y &= 4 \\ y &= -2 \end{aligned}$$

To use functional notation for $2x + y = 4$, solve for y to get

$$y = -2x + 4.$$

Replace y with $f(x)$ to get the <u>linear</u> function

$$\begin{aligned} f(x) &= -2x + 4. \\ f(3) &= -2(3) + 4 = -2 \end{aligned}$$

Because $y = -2$ when $x = 3$, the point $(3, -2)$ lies on the graph of the function.

63. **(a)** Solve the equation for y.

$$\begin{aligned} x + 3y &= 12 \\ 3y &= 12 - x \\ y &= \frac{12 - x}{3} \end{aligned}$$

Since $y = f(x)$,

$$f(x) = \frac{12 - x}{3}.$$

(b) $f(3) = \dfrac{12 - 3}{3} = \dfrac{9}{3} = 3$

65. **(a)** Solve the equation for y.

$$\begin{aligned} y + 2x^2 &= 3 \\ y &= 3 - 2x^2 \end{aligned}$$

Since $y = f(x)$,

$$f(x) = 3 - 2x^2.$$

(b) $$\begin{aligned} f(3) &= 3 - 2(3)^2 \\ &= 3 - 2(9) \\ &= -15 \end{aligned}$$

67. **(a)** Solve the equation for y.

$$\begin{aligned} 4x - 3y &= 8 \\ -3y &= 8 - 4x \\ y &= \frac{8 - 4x}{-3} \end{aligned}$$

Since $y = f(x)$,

$$f(x) = \frac{8 - 4x}{-3}.$$

(b) $$\begin{aligned} f(3) &= \frac{8 - 4(3)}{-3} \\ &= \frac{8 - 12}{-3} = \frac{-4}{-3} = \frac{4}{3} \end{aligned}$$

69. $f(x) = -2x + 5$

The graph will be a line. The intercepts are $(0, 5)$ and $(\frac{5}{2}, 0)$.
The domain is $(-\infty, \infty)$. The range is $(-\infty, \infty)$.

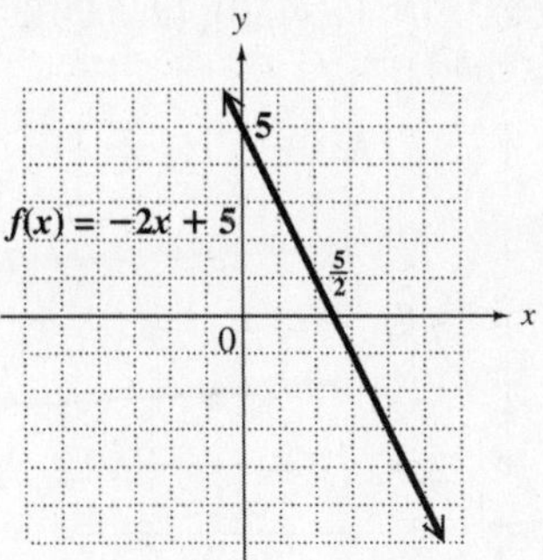

71. $h(x) = \frac{1}{2}x + 2$

The graph will be a line. The intercepts are $(0, 2)$ and $(-4, 0)$.
The domain is $(-\infty, \infty)$. The range is $(-\infty, \infty)$.

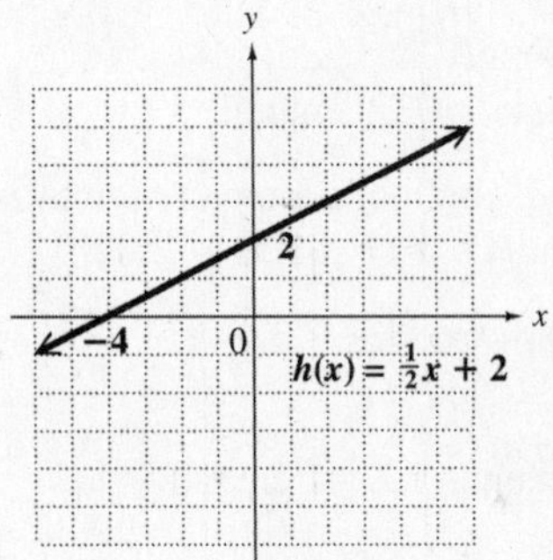

73. $g(x) = -4$

Using a y-intercept of $(0, -4)$ and a slope of $m = 0$, we graph the horizontal line.
On the line, the value of x can be any real number, so the domain is $(-\infty, \infty)$. The range is $\{-4\}$.

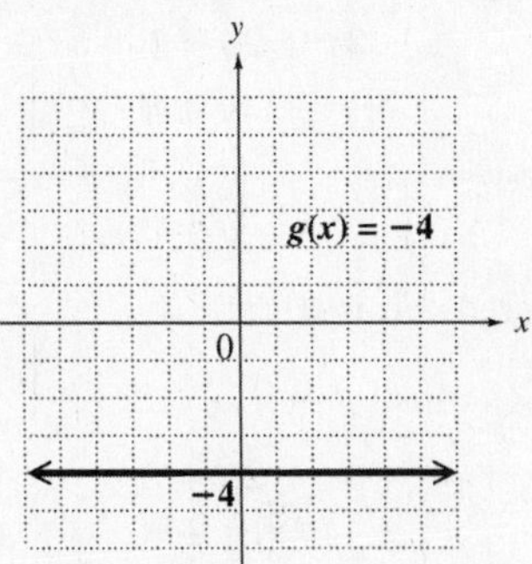

75. **(a)**

x	$f(x)$
0	\$0
1	\$2.50
2	\$5.00
3	\$7.50

(b) Since the charge equals the cost per mile, \$2.50, times the number of miles, x, $f(x) = \underline{\$2.50x}$.

(c) To graph $y = f(x)$ for x in $\{0, 1, 2, 3\}$, plot the points $(0, 0)$, $(1, 2.50)$, $(2, 5.00)$, and $(3, 7.50)$ from the chart.

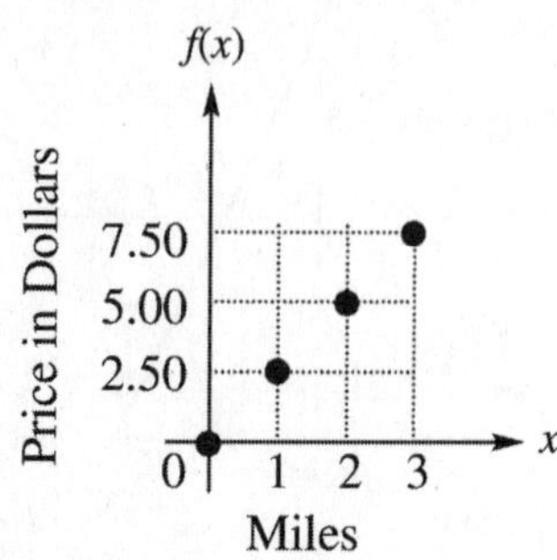

77. Since the length of a man's femur is given, use the formula $h(r) = 69.09 + 2.24r$.

$$\begin{aligned} h(56) &= 69.09 + 2.24(56) \quad \textit{Let } r = 56. \\ &= 194.53 \end{aligned}$$

The man is 194.53 cm tall.

79. Since the length of a woman's femur is given, use the formula $h(r) = 61.41 + 2.32r$.

$$\begin{aligned} h(50) &= 61.41 + 2.32(50) \quad \textit{Let } r = 50. \\ &= 177.41 \end{aligned}$$

The woman is 177.41 cm tall.

Chapter 4 Review Exercises

1. $3x + 2y = 6$

To complete the ordered pairs, substitute the given values for x or y in the equation.

For (0, ____), let $x = 0$.

$$\begin{aligned} 3x + 2y &= 6 \\ 3(0) + 2y &= 6 \\ 2y &= 6 \\ y &= 3 \end{aligned}$$

The ordered pair is $(0, 3)$.

For (____, 0), let $y = 0$.

$$\begin{aligned} 3x + 2y &= 6 \\ 3x + 2(0) &= 6 \\ 3x &= 6 \\ x &= 2 \end{aligned}$$

The ordered pair is $(2, 0)$.

For (____, −2), let $y = -2$.

$$\begin{aligned} 3x + 2y &= 6 \\ 3x + 2(-2) &= 6 \\ 3x - 4 &= 6 \\ 3x &= 10 \\ x &= \tfrac{10}{3} \end{aligned}$$

The ordered pair is $(\frac{10}{3}, -2)$.

Plot the ordered pairs, and draw the line through them.

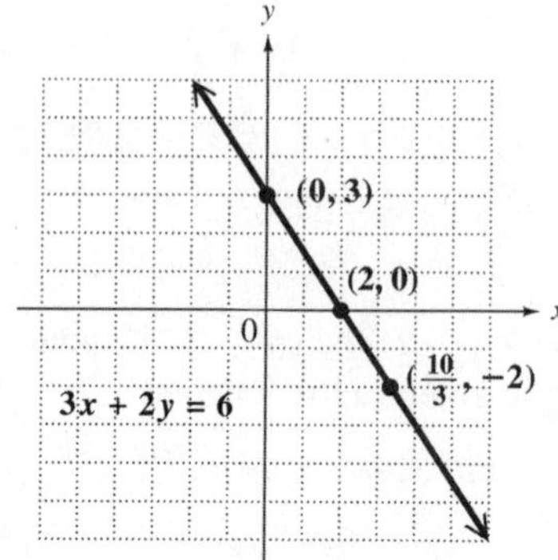

2. $x - y = 6$

To complete the ordered pairs, substitute the given values for x or y in the equation.

For (2, ____), let $x = 2$.

$$\begin{aligned} x - y &= 6 \\ 2 - y &= 6 \\ -y &= 4 \\ y &= -4 \end{aligned}$$

The ordered pair is $(2, -4)$.

For (____, −3), let $y = -3$.

$$\begin{aligned} x - y &= 6 \\ x - (-3) &= 6 \\ x + 3 &= 6 \\ x &= 3 \end{aligned}$$

The ordered pair is $(3, -3)$.

For (1, ____), let $x = 1$.

$$\begin{aligned} x - y &= 6 \\ 1 - y &= 6 \\ -y &= 5 \\ y &= -5 \end{aligned}$$

The ordered pair is $(1, -5)$.

For (____, −2), let $y = -2$.

$$\begin{aligned} x - y &= 6 \\ x - (-2) &= 6 \\ x + 2 &= 6 \\ x &= 4 \end{aligned}$$

The ordered pair is $(4, -2)$.

Plot the ordered pairs, and draw the line through them.

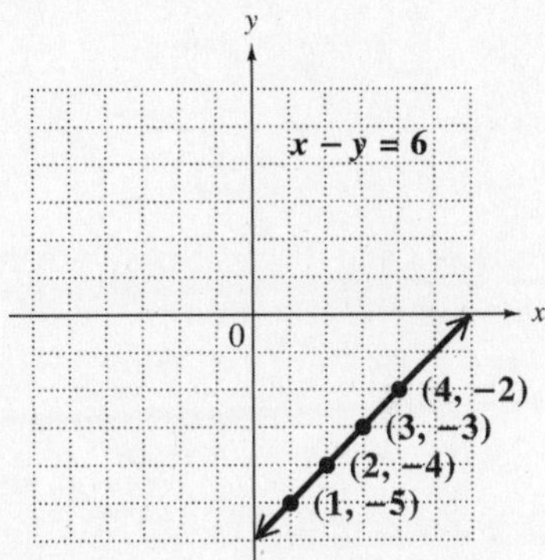

3. $4x + 3y = 12$

To find the x-intercept, let $y = 0$.

$$\begin{aligned} 4x + 3y &= 12 \\ 4x + 3(0) &= 12 \\ 4x &= 12 \\ x &= 3 \end{aligned}$$

The x-intercept is $(3, 0)$.

To find the y-intercept, let $x = 0$.

$$\begin{aligned} 4x + 3y &= 12 \\ 4(0) + 3y &= 12 \\ 3y &= 12 \\ y &= 4 \end{aligned}$$

The y-intercept is $(0, 4)$.

Plot the intercepts and draw the line through them.

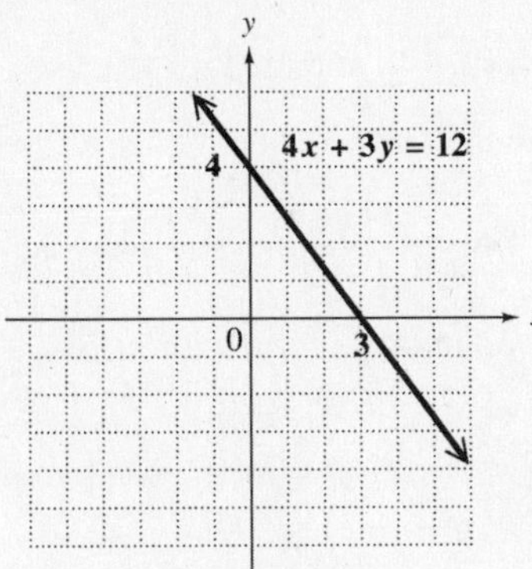

4. $5x + 7y = 15$

To find the x-intercept, let $y = 0$.

$$\begin{aligned} 5x + 7y &= 15 \\ 5x + 7(0) &= 15 \\ 5x &= 15 \\ x &= 3 \end{aligned}$$

The x-intercept is $(3, 0)$.

To find the y-intercept, let $x = 0$.

$$\begin{aligned} 5x + 7y &= 15 \\ 5(0) + 7y &= 15 \\ 7y &= 15 \\ y &= \tfrac{15}{7} \end{aligned}$$

The y-intercept is $(0, \frac{15}{7})$.

Plot the intercepts and draw the line through them.

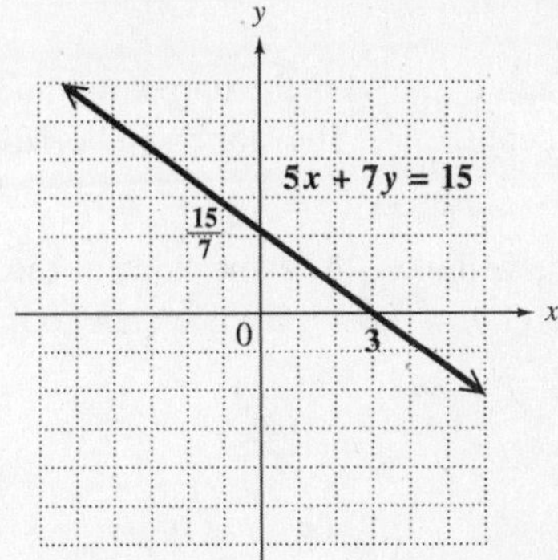

5. By the midpoint formula, the midpoint of the segment with endpoints $(-8, -12)$ and $(8, 16)$ is

$$\left(\frac{-8+8}{2}, \frac{-12+16}{2}\right) = \left(\frac{0}{2}, \frac{4}{2}\right) = (0, 2).$$

6. By the midpoint formula, the midpoint of the segment with endpoints $(0, -5)$ and $(-9, 8)$ is

$$\left(\frac{0+(-9)}{2}, \frac{-5+8}{2}\right) = \left(\frac{-9}{2}, \frac{3}{2}\right) = \left(-\frac{9}{2}, \frac{3}{2}\right).$$

7. By the midpoint formula, the midpoint of the segment with endpoints $(3.8, 8.6)$ and $(1.4, 15.2)$ is

$$\left(\frac{3.8+1.4}{2}, \frac{8.6+15.2}{2}\right) = \left(\frac{5.2}{2}, \frac{23.8}{2}\right) = (2.6, 11.9).$$

8. By the midpoint formula, the midpoint of the segment with endpoints $(15.5, -6.3)$ and $(-6.5, -12.7)$ is

$$\left(\frac{15.5+(-6.5)}{2}, \frac{-6.3+(-12.7)}{2}\right) = \left(\frac{9.0}{2}, \frac{-19.0}{2}\right) = (4.5, -9.5).$$

9. Let $(x_1, y_1) = (-1, 2)$ and $(x_2, y_2) = (4, -6)$. Then

$$m = \frac{y_2 - y_1}{x_2 - x_1} = \frac{-6-2}{4-(-1)} = \frac{-8}{5} = -\frac{8}{5}.$$

The slope is $-\frac{8}{5}$.

10. The equation $y = 2x + 3$ is in slope-intercept form, $y = mx + b$, so the slope, m, is 2.

11. $-3x + 4y = 5$

Write the equation in slope-intercept form by solving for y.

$$\begin{aligned} -3x + 4y &= 5 \\ 4y &= 3x + 5 \\ y &= \tfrac{3}{4}x + \tfrac{5}{4} \end{aligned}$$

The slope is $\frac{3}{4}$.

12. The graph of $y = 4$ is the horizontal line with y-intercept $(0, 4)$. The slope of a horizontal line is 0.

13. The line will have the same slope as the line with equation $3y = -2x + 5$ since the two lines are parallel.

$$\begin{aligned} 3y &= -2x + 5 \\ y &= -\tfrac{2}{3}x + \tfrac{5}{3} \end{aligned}$$

The slope of the line with equation $3y = -2x + 5$ (and any line parallel to it) is $-\frac{2}{3}$.

14. The slope of the line will be the negative reciprocal of the slope of the line with equation $3x - y = 6$ since the two lines are perpendicular.

$$\begin{aligned} 3x - y &= 6 \\ -y &= -3x + 6 \\ y &= 3x - 6 \end{aligned}$$

The slope of the line with equation $3x - y = 6$ is 3. The negative reciprocal of 3 is $-\frac{1}{3}$.

The slope of a line perpendicular to the line with equation $3x - y = 6$ is $-\frac{1}{3}$.

15. The line goes up from left to right, so it has positive slope.

16. The line goes down from left to right, so it has negative slope.

17. The line is horizontal, so it has 0 slope.

18. The line is vertical, so it has *undefined* slope.

19. To rise 1 foot, we must move 4 feet in the horizontal direction. To rise 3 feet, we must move $3(4) = 12$ feet in the horizontal direction.

20. Use $(1980, 21{,}000)$ and $(2005, 56{,}200)$.

$$\begin{aligned} \text{average rate of change} &= \frac{56{,}200 - 21{,}000}{2005 - 1980} \\ &= \frac{35{,}200}{25} \\ &= 1408 \end{aligned}$$

The average rate of change is \$1408 per year.

21. Slope $\frac{3}{5}$; y-intercept $(0, -8)$

Here, $m = \frac{3}{5}$ and $b = -8$. Substitute these values in the slope-intercept form.

$$\begin{aligned} y &= mx + b \\ y &= \tfrac{3}{5}x - 8 \end{aligned}$$

22. Slope $-\frac{1}{3}$; y-intercept $(0, 5)$

Here, $m = -\frac{1}{3}$ and $b = 5$. Substitute these values in the slope-intercept form.

$$\begin{aligned} y &= mx + b \\ y &= -\tfrac{1}{3}x + 5 \end{aligned}$$

23. Slope 0; y-intercept $(0, 12)$

A line with slope 0 is horizontal. The horizontal line through the point (c, d) has equation $y = d$. Here $d = 12$, so the line has equation $y = 12$, or $y = 0x + 12$.

24. Undefined slope; through $(2, 7)$

A line with undefined slope is vertical. The vertical line through the point (c, d) has equation $x = c$. Here $c = 2$, so the line has equation $x = 2$.

25. Horizontal; through $(-1, 4)$

The horizontal line through the point (c, d) has equation $y = d$. Here $d = 4$, so the line has equation $y = 4$.

26. Vertical; through $(0.3, 0.6)$

The vertical line through the point (c, d) has equation $x = c$. Here $c = 0.3$, so the line has equation $x = 0.3$.

27. Through $(2, -5)$ and $(1, 4)$

(a) Find the slope.

$$m = \frac{4 - (-5)}{1 - 2} = \frac{9}{-1} = -9$$

Use the point-slope form with $m = -9$ and $(x_1, y_1) = (2, -5)$.

$$\begin{aligned} y - y_1 &= m(x - x_1) \\ y - (-5) &= -9(x - 2) \\ y + 5 &= -9x + 18 \\ y &= -9x + 13 \end{aligned}$$

(b)
$$\begin{aligned} y &= -9x + 13 \\ 9x + y &= 13 \end{aligned}$$

28. Through $(-3,-1)$ and $(2,6)$

(a) Find the slope.

$$m = \frac{6-(-1)}{2-(-3)} = \frac{7}{5}$$

Use the point-slope form with $m = \frac{7}{5}$ and $(x_1, y_1) = (2, 6)$.

$$\begin{aligned} y - y_1 &= m(x - x_1) \\ y - 6 &= \tfrac{7}{5}(x-2) \\ y - 6 &= \tfrac{7}{5}x - \tfrac{14}{5} \\ y &= \tfrac{7}{5}x + \tfrac{16}{5} \end{aligned}$$

(b)
$$\begin{aligned} y &= \tfrac{7}{5}x + \tfrac{16}{5} \\ 5y &= 7x + 16 \quad \textit{Multiply by 5.} \\ -7x + 5y &= 16 \\ 7x - 5y &= -16 \end{aligned}$$

29. Parallel to $4x - y = 3$ and through $(6, -2)$

(a) Writing $4x - y = 3$ in slope-intercept form gives us $y = 4x - 3$, which has slope 4. Lines parallel to it will also have slope 4. The line with slope 4 through $(6, -2)$ is:

$$\begin{aligned} y - y_1 &= m(x - x_1) \\ y - (-2) &= 4(x - 6) \\ y + 2 &= 4x - 24 \\ y &= 4x - 26 \end{aligned}$$

(b)
$$\begin{aligned} y &= 4x - 26 \\ -4x + y &= -26 \\ 4x - y &= 26 \end{aligned}$$

30. Perpendicular to $2x - 5y = 7$ and through $(0, 1)$

(a) Write the equation in slope-intercept form.

$$\begin{aligned} 2x - 5y &= 7 \\ -5y &= -2x + 7 \\ y &= \tfrac{2}{5}x - \tfrac{7}{5} \end{aligned}$$

$y = \frac{2}{5}x - \frac{7}{5}$ has slope $\frac{2}{5}$ and is perpendicular to lines with slope $-\frac{5}{2}$.
The line with slope $-\frac{5}{2}$ through $(0, 1)$ is

$$\begin{aligned} y - y_1 &= m(x - x_1) \\ y - 1 &= -\tfrac{5}{2}(x - 0) \\ y &= -\tfrac{5}{2}x + 1 \end{aligned}$$

(b)
$$\begin{aligned} y &= -\tfrac{5}{2}x + 1 \\ 2y &= -5x + 2 \quad \textit{Multiply by 2.} \\ 5x + 2y &= 2 \end{aligned}$$

31. **(a)** The fixed cost is \$159, so that is the value of b. The variable cost is \$57, so

$$y = mx + b = 57x + 159.$$

The cost of a 1-year membership can be found by substituting 12 for x.

$$\begin{aligned} y &= 57(12) + 159 \\ &= 684 + 159 = 843 \end{aligned}$$

The cost is \$843.

(b) As in part (a),

$$\begin{aligned} y &= 47x + 159. \\ y &= 47(12) + 159 \\ &= 564 + 159 = 723 \end{aligned}$$

The cost is \$723.

32. $3x - 2y \le 12$

Graph $3x - 2y = 12$ as a solid line through $(0, -6)$ and $(4, 0)$. Use $(0, 0)$ as a test point. Since $(0, 0)$ satisfies the inequality, shade the region on the side of the line containing $(0, 0)$.

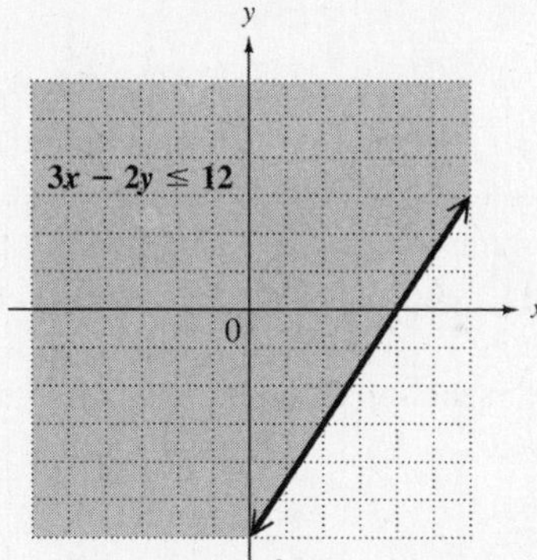

33. $5x - y > 6$

Graph $5x - y = 6$ as a dashed line through $(0, -6)$ and $(\frac{6}{5}, 0)$. Use $(0, 0)$ as a test point. Since $(0, 0)$ does not satisfy the inequality, shade the region on the side of the line that does not contain $(0, 0)$.

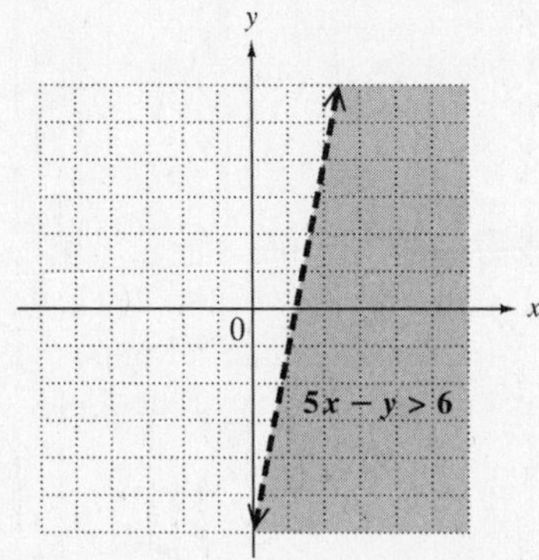

34. $x \geq 2$
Graph $x = 2$ as a solid vertical line through $(2, 0)$. Shade the region to the right of $x = 2$.

$y \geq 2$
Graph $y = 2$ as a solid horizontal line through $(0, 2)$. Shade the region above $y = 2$. The graph of

$$x \geq 2 \quad \text{or} \quad y \geq 2$$

includes all the shaded regions.

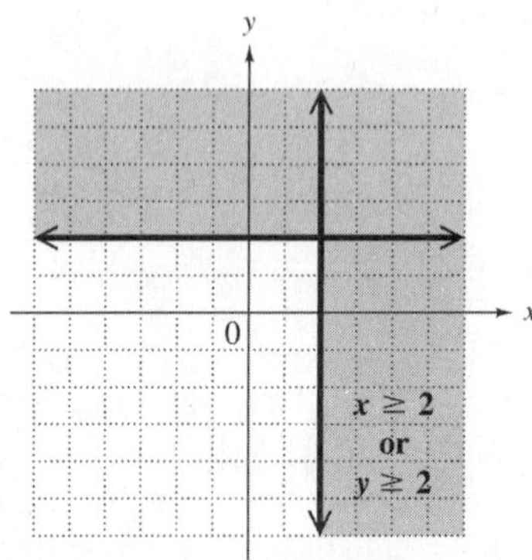

35. $2x + y \leq 1$ and $x \geq 2y$

Graph $2x + y = 1$ as a solid line through $(\frac{1}{2}, 0)$ and $(0, 1)$, and shade the region on the side containing $(0, 0)$ since it satisfies the inequality. Next, graph $x = 2y$ as a solid line through $(0, 0)$ and $(2, 1)$, and shade the region on the side containing $(2, 0)$ since $2 > 2(0)$ or $2 > 0$ is true. The intersection is the region where the graphs overlap.

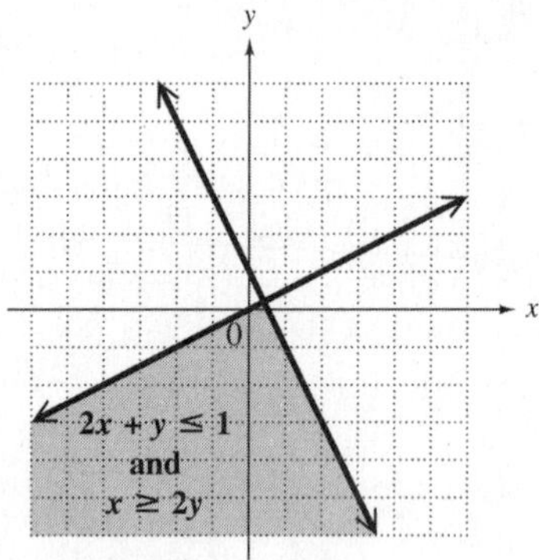

36. $\{(-4, 2), (-4, -2), (1, 5), (1, -5)\}$

The domain, the set of x-values, is $\{-4, 1\}$. The range, the set of y-values, is $\{2, -2, 5, -5\}$. Since at least one x-value has more than one y-value, the relation is not a function.

37. The relation can be described by the set of ordered pairs

$$\{(9, 32), (11, 47), (4, 47), (17, 69), (25, 14)\}.$$

The relation is a function since for each x-value, there is only one y-value.

The domain is the set of x-values:

$$\{9, 11, 4, 17, 25\}.$$

The range is the set of y-values: $\{32, 47, 69, 14\}$.

38. The domain, the x-values of the points on the graph, is $[-4, 4]$. The range, the y-values of the points on the graph, is $[0, 2]$. Since any vertical line intersects the graph of the relation in at most one point, the relation is a function.

39. $y = 3x - 3$
For any value of x, there is exactly one value of y, so the equation defines a function, actually a linear function. The domain is the set of all real numbers, $(-\infty, \infty)$.

40. $y < x + 2$
For any value of x, there are many values of y. For example, $(1, 0)$ and $(1, 1)$ are both solutions of the inequality that have the same x-value but different y-values. The inequality does not define a function. The domain is the set of all real numbers, $(-\infty, \infty)$.

41. $y = |x - 4|$
Given any value of x, y is found by subtracting 4, and taking the absolute value of the result. This process produces exactly one value of y for each x-value, so the equation defines a function. The domain is the set of all real numbers, $(-\infty, \infty)$.

42. $y = \sqrt{4x + 7}$
Given any value of x, y is found by multiplying x by 4, adding 7, and taking the square root of the result. This process produces exactly one value of y for each x-value, so the equation defines a function. Since the radicand must be nonnegative,

$$\begin{aligned} 4x + 7 &\geq 0 \\ 4x &\geq -7 \\ x &\geq -\tfrac{7}{4}. \end{aligned}$$

The domain is $[-\frac{7}{4}, \infty)$.

43. $x = y^2$
The ordered pairs $(4, 2)$ and $(4, -2)$ both satisfy the equation. Since one value of x, 4, corresponds to two values of y, 2 and -2, the equation does not define a function. Because x is equal to the square of y, the values of x must always be nonnegative. The domain is $[0, \infty)$.

44. $y = \dfrac{7}{x - 36}$
Given any value of x, y is found by subtracting 36, then dividing the result into 7. This process produces exactly one value of y for each x-value, so the equation defines a function. The domain includes all real numbers except those that make the denominator 0, namely 36. The domain is $(-\infty, 36) \cup (36, \infty)$.

45. **(a)** For each year, there is exactly one life expectancy associated with the year, so the table defines a function.

(b) The domain is the set of years, that is, $\{1943, 1953, 1963, 1973, 1983, 1993, 2003\}$. The range is the set of life expectancies, that is, $\{63.3, 68.8, 69.9, 71.4, 74.6, 75.5, 77.6\}$.

(c) Answers will vary. Two possible ordered pairs are $(1953, 68.8)$ and $(1973, 71.4)$.

(d) $f(2003) = 77.6$. In 2003, life expectancy at birth was 77.6 yr.

(e) Since $f(1993) = 75.5$, $x = 1993$.

In Exercises 46–49, use

$$f(x) = -2x^2 + 3x - 6.$$

46. $f(0) = -2(0)^2 + 3(0) - 6 = -6$

47. $f(3) = -2(3)^2 + 3(3) - 6$
$= -18 + 9 - 6 = -15$

48. $f(p) = -2p^2 + 3p - 6$

49. $f(-k) = -2(-k)^2 + 3(-k) - 6$
$= -2k^2 - 3k - 6$

50.
$$\begin{aligned} 2x^2 - y &= 0 \\ -y &= -2x^2 \\ y &= 2x^2 \end{aligned}$$
Since $y = f(x)$,
$f(x) = 2x^2$,
and $f(3) = 2(3)^2 = 2(9) = 18$.

51. Solve for y in terms of x.
$$\begin{aligned} 2x - 5y &= 7 \\ 2x - 7 &= 5y \\ \frac{2x - 7}{5} &= y \end{aligned}$$
This is the same as choice **C**,
$$f(x) = \frac{-7 + 2x}{5}.$$

52. The graph of a constant function is the graph of a horizontal line.

Chapter 4 Test

1. Let $(x_1, y_1) = (6, 4)$ and $(x_2, y_2) = (-4, -1)$. Then
$$m = \frac{y_2 - y_1}{x_2 - x_1} = \frac{-1 - 4}{-4 - 6} = \frac{-5}{-10} = \frac{1}{2}.$$
The slope is $\frac{1}{2}$.

2. To find the slope and y-intercept of
$$3x - 2y = 13,$$
write the equation in slope-intercept form by solving for y.
$$\begin{aligned} 3x - 2y &= 13 \\ -2y &= -3x + 13 \\ y &= \tfrac{3}{2}x - \tfrac{13}{2} \end{aligned}$$
The slope is $\frac{3}{2}$, and the y-intercept is $(0, -\frac{13}{2})$. To find the x-intercept, substitute 0 for y in the original equation and solve for x.
$$\begin{aligned} 3x - 2y &= 13 \\ 3x - 2(0) &= 13 \\ 3x &= 13 \\ x &= \tfrac{13}{3} \end{aligned}$$
The x-intercept is $(\frac{13}{3}, 0)$.

3. The graph of $y = 5$ is the horizontal line with slope 0 and y-intercept $(0, 5)$. There is no x-intercept.

4. The graph of a line with undefined slope is the graph of a vertical line.

5. $4x - 3y = -12$

The x-intercept is $(-3, 0)$ and the y-intercept is $(0, 4)$. To sketch the graph, plot the intercepts and draw the line through them.

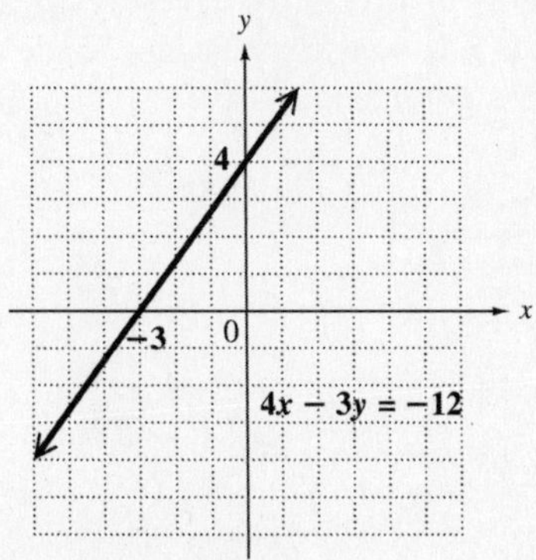

6. $y - 2 = 0$

$y = 2$

The graph of $y = 2$ is the horizontal line with y-intercept $(0, 2)$. There is no x-intercept.

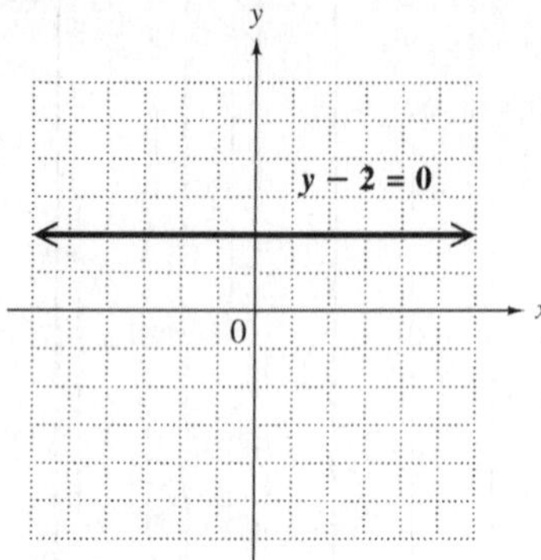

7. $y = -2x$

There is only one intercept, $(0, 0)$. Another ordered pair that satisfies the equation is $(1, -2)$. To sketch the graph, plot the two points and draw the line through them.

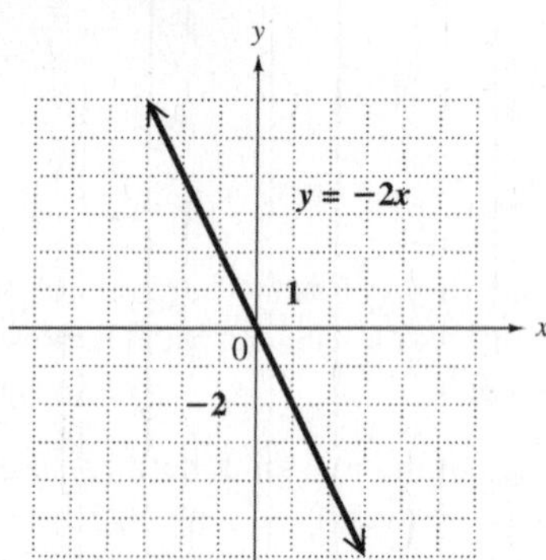

8. $3x - 2y > 6$

Graph the line $3x - 2y = 6$, which has intercepts $(2, 0)$ and $(0, -3)$, as a dashed line since the inequality involves " $>$ ". Test $(0, 0)$, which yields $0 > 6$, a false statement. Shade the region that does not include $(0, 0)$.

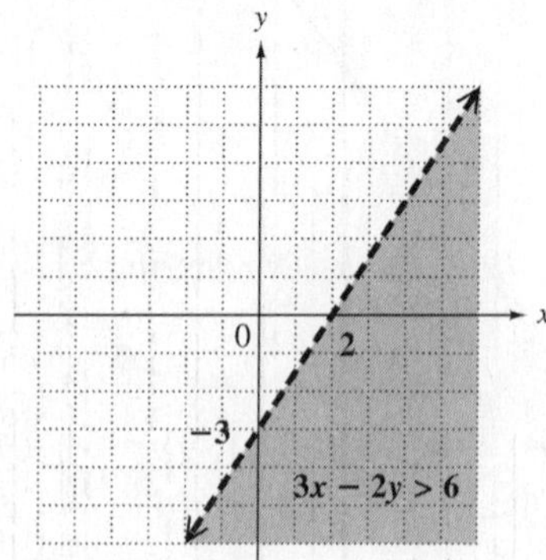

9. Through $(-3, 14)$ and $(-6, 9)$

Find the slope.

$$m = \frac{9 - 14}{-6 - (-3)} = \frac{-5}{-3} = \frac{5}{3}$$

Use the point-slope form with $(x_1, y_1) = (-3, 14)$ and $m = \frac{5}{3}$.

$$\begin{aligned} y - 14 &= \tfrac{5}{3}[x - (-3)] \\ 3(y - 14) &= 5(x + 3) \\ 3y - 42 &= 5x + 15 \\ -5x + 3y &= 57 \\ 5x - 3y &= -57 \end{aligned}$$

10. Through $(4, -1)$; $m = -5$

Let $m = -5$ and $(x_1, y_1) = (4, -1)$ in the point-slope form.

$$\begin{aligned} y - y_1 &= m(x - x_1) \\ y - (-1) &= -5(x - 4) \\ y + 1 &= -5x + 20 \\ 5x + y &= 19 \end{aligned}$$

11. **(a)** Through $(-7, 2)$ and parallel to $3x + 5y = 6$

To find the slope of the line with equation $3x + 5y = 6$, write the equation in slope-intercept form by solving for y.

$$\begin{aligned} 3x + 5y &= 6 \\ 5y &= -3x + 6 \\ y &= -\tfrac{3}{5}x + \tfrac{6}{5} \end{aligned}$$

The slope is $-\frac{3}{5}$, so a line parallel to it also has slope $-\frac{3}{5}$. Let $m = -\frac{3}{5}$ and $(x_1, y_1) = (-7, 2)$ in the point-slope form.

$$\begin{aligned} y - y_1 &= m(x - x_1) \\ y - 2 &= -\tfrac{3}{5}[x - (-7)] \\ y - 2 &= -\tfrac{3}{5}(x + 7) \\ y - 2 &= -\tfrac{3}{5}x - \tfrac{21}{5} \\ y &= -\tfrac{3}{5}x - \tfrac{11}{5} \end{aligned}$$

(b) Through $(-7, 2)$ and perpendicular to $y = 2x$

Since $y = 2x$ is in slope-intercept form $(b = 0)$, the slope, m, of the line with equation $y = 2x$ is 2. A line perpendicular to it has a slope that is the negative reciprocal of 2, that is, $-\frac{1}{2}$. Let $m = -\frac{1}{2}$ and $(x_1, y_1) = (-7, 2)$ in the point-slope form.

$$\begin{aligned} y - y_1 &= m(x - x_1) \\ y - 2 &= -\tfrac{1}{2}(x + 7) \\ y - 2 &= -\tfrac{1}{2}x - \tfrac{7}{2} \\ y &= -\tfrac{1}{2}x - \tfrac{3}{2} \end{aligned}$$

12. Choice **D** is the only graph that passes the vertical line test. Since x can be any value, the domain is $(-\infty, \infty)$. Since $y \geq 0$, the range is $[0, \infty)$.

13. Choice **D** does not define a function, since its domain (input) element 0 is paired with two different range (output) elements, 1 and 2.

The domain is the set of x-values: $\{0, 3, 6\}$
The range is the set of y-values: $\{1, 2, 3\}$

14.
$$\begin{aligned} f(x) &= -x^2 + 2x - 1 \\ f(1) &= -(1)^2 + 2(1) - 1 \\ &= -1 + 2 - 1 \\ &= 0 \\ f(a) &= -a^2 + 2a - 1 \end{aligned}$$

15. Use the points $(1980, 119{,}000)$ and $(2005, 89{,}000)$.

average rate of change
$$\begin{aligned} &= \frac{\text{change in } y}{\text{change in } x} = \frac{89{,}000 - 119{,}000}{2005 - 1980} \\ &= \frac{-30{,}000}{25} = -1200 \end{aligned}$$

The average rate of change is about -1200 farms per year, that is, the number of farms decreased by about 1200 each year from 1980 to 2005.

Cumulative Review Exercises (Chapters 1–4)

1. The absolute value of a negative number is a positive number and the additive inverse of the same negative number is the same positive number. For example, suppose the negative number is -5:

$$\begin{aligned} &|-5| = -(-5) = 5 \\ \text{and} \quad &-(-5) = 5 \end{aligned}$$

The statement is *always true*.

2. The statement is *always true*; in fact, it is the definition of a rational number.

3. The sum of two negative numbers is another negative number, so the statement is *never true.*

4. The statement is *sometimes true*. For example,

$$\begin{aligned} &3 + (-3) = 0, \\ \text{but} \quad &3 + (-1) = 2 \neq 0. \end{aligned}$$

5.
$$\begin{aligned} &-|-2| - 4 + |-3| + 7 \\ &= -2 - 4 + 3 + 7 \\ &= -6 + 3 + 7 \\ &= -3 + 7 \\ &= 4 \end{aligned}$$

6. $(-0.8)^2 = (-0.8)(-0.8) = 0.64$

7. $\sqrt{-64}$ is not a real number.

8. The product of two numbers that have the same sign is positive.

$$-\frac{2}{3}\left(-\frac{12}{5}\right) = \frac{2 \cdot 12}{3 \cdot 5} = \frac{2 \cdot 4}{5} = \frac{8}{5}$$

9.
$$\begin{aligned} -(-4m + 3) &= -(-4m) - 3 \\ &= 4m - 3 \end{aligned}$$

10.
$$\begin{aligned} &3x^2 - 4x + 4 + 9x - x^2 \\ &= 3x^2 - x^2 - 4x + 9x + 4 \\ &= 2x^2 + 5x + 4 \end{aligned}$$

11.
$$\begin{aligned} \frac{3\sqrt{16} - (-1)(7)}{4 + (-6)} &= \frac{3(4) - (-7)}{-2} \\ &= \frac{12 + 7}{-2} = -\frac{19}{2} \end{aligned}$$

12. $-3 < x \leq 5$
This is the set of numbers between -3 and 5, not including -3 (use a parenthesis), but including 5 (use a bracket). In interval notation, the set is $(-3, 5]$.

13.
$$\sqrt{\frac{-2 + 4}{-5}} = \sqrt{\frac{2}{-5}} = \sqrt{-\frac{2}{5}}$$

This is not a real number since the number under the radical sign is negative.

For Exercises 14–16, let $p = -4$, $q = -2$, and $r = 5$.

14.
$$\begin{aligned} -3(2q - 3p) &= -3[2(-2) - 3(-4)] \\ &= -3(-4 + 12) \\ &= -3(8) \\ &= -24 \end{aligned}$$

15.
$$\begin{aligned} |p|^3 - |q^3| &= |-4|^3 - |(-2)^3| \\ &= 4^3 - |-8| \\ &= 64 - 8 \\ &= 56 \end{aligned}$$

16.
$$\begin{aligned} \frac{\sqrt{r}}{-p + 2q} &= \frac{\sqrt{5}}{-(-4) + 2(-2)} \\ &= \frac{\sqrt{5}}{4 - 4} \\ &= \frac{\sqrt{5}}{0} \end{aligned}$$

This expression is *undefined* since the denominator is zero.

17.
$$\begin{aligned} 2z - 5 + 3z &= 4 - (z + 2) \\ 5z - 5 &= 4 - z - 2 \\ 5z - 5 &= 2 - z \\ 6z &= 7 \\ z &= \tfrac{7}{6} \end{aligned}$$

Solution set: $\{\frac{7}{6}\}$

18. $\frac{3a-1}{5}+\frac{a+2}{2}=-\frac{3}{10}$

Multiply each side by the LCD, 10.

$$10\left(\frac{3a-1}{5}+\frac{a+2}{2}\right)=10\left(-\frac{3}{10}\right)$$
$$2(3a-1)+5(a+2)=-3$$
$$6a-2+5a+10=-3$$
$$11a+8=-3$$
$$11a=-11$$
$$a=-1$$

Solution set: $\{-1\}$

19. Solve $V=\frac{1}{3}\pi r^2 h$ for h.

$3V=\pi r^2 h$ *Multiply by 3.*

$\frac{3V}{\pi r^2}=h$ *Divide by πr^2.*

20. Let $x=$ the time it takes for the planes to be 2100 miles apart.
Make a table. Use the formula $d=rt$.

	r	*t*	*d*
Eastbound Plane	550	x	$550x$
Westbound Plane	500	x	$500x$

The total distance is 2100 miles.

$$550x+500x=2100$$
$$1050x=2100$$
$$x=2$$

It will take 2 hr for the planes to be 2100 mi apart.

21. Let x represent the number of white pills and $2x$ the number of yellow pills.

Make a table.

	Strength (in Units)	**Number of Pills**	**Amount of Medication**
White	3	x	$3x$
Yellow	3	$2x$	$3(2x)$

The total amount of medication must be at least 30 units. Solve the inequality.

$$3x+3(2x)\geq 30$$
$$3x+6x\geq 30$$
$$9x\geq 30$$
$$x\geq \tfrac{30}{9}$$
$$x\geq \tfrac{10}{3},\text{ or } 3\tfrac{1}{3}$$

Since Ms. Bell must take a whole number of pills, she must take at least 4 white pills.

22. Let x denote the side of the original square and $4x$ the perimeter. Now $x+4$ is the side of the new square and $4(x+4)$ is its perimeter.
"The perimeter would be 8 inches less than twice the perimeter of the original square " translates as

$$4(x+4)=2(4x)-8.$$
$$4x+16=8x-8$$
$$24=4x$$
$$6=x$$

The length of a side of the original square is 6 inches.

23. The union of the three solution sets is $(-\infty,\infty)$; that is, the set of all real numbers.

24. $3-2(m+3)<4m$

$$3-2m-6<4m$$
$$-2m-3<4m$$
$$-6m<3$$

Multiply by $-\frac{1}{6}$; reverse the direction of the inequality symbol.

$$-\tfrac{1}{6}(-6m)>-\tfrac{1}{6}(3)$$
$$m>-\tfrac{1}{2}$$

Solution set: $(-\frac{1}{2},\infty)$

25. $2k+4<10$ and $3k-1>5$

$2k<6$ and $3k>6$

$k<3$ and $k>2$

The solution set of these inequalities is the set of numbers that are both less than 3 and greater than 2. This is the set of all numbers between 2 and 3.

Solution set: $(2,3)$

26. $2k+4>10$ or $3k-1<5$

$2k>6$ or $3k<6$

$k>3$ or $k<2$

The solution set is the set of numbers that are either greater than 3 or less than 2.

Solution set: $(-\infty,2)\cup(3,\infty)$

27. $|5x+3|=13$

$5x+3=13$ or $5x+3=-13$

$5x=10$ or $5x=-16$

$x=2$ or $x=-\frac{16}{5}$

Solution set: $\{-\frac{16}{5},2\}$

28. $|x+2|<9$

$-9<x+2<9$

Subtract 2 from each part of the inequality.

$-11<x<7$

Solution set: $(-11,7)$

29. $|2x-5|\geq 9$

$$\begin{aligned} 2x-5\geq 9 &\quad \text{or} \quad 2x-5\leq -9 \\ 2x\geq 14 &\quad \text{or} \quad 2x\leq -4 \\ x\geq 7 &\quad \text{or} \quad x\leq -2 \end{aligned}$$

Solution set: $(-\infty,-2]\cup[7,\infty)$

30. $3x-4y=12$

To complete the ordered pairs, substitute the given values for x or y in the equation.

For (0,____), let $x=0$.

$$\begin{aligned} 3x-4y&=12 \\ 3(0)-4y&=12 \\ -4y&=12 \\ y&=-3 \end{aligned}$$

The ordered pair is $(0,-3)$.

For (____, 0), let $y=0$.

$$\begin{aligned} 3x-4y&=12 \\ 3x-4(0)&=12 \\ 3x&=12 \\ x&=4 \end{aligned}$$

The ordered pair is $(4,0)$.

For (2,____), let $x=2$.

$$\begin{aligned} 3x-4y&=12 \\ 3(2)-4y&=12 \\ 6-4y&=12 \\ -4y&=6 \\ y&=-\tfrac{6}{4}=-\tfrac{3}{2} \end{aligned}$$

The ordered pair is $(2,-\frac{3}{2})$.

31. $-4x+2y=8$

To find the x-intercept of the equation, let $y=0$.

$$\begin{aligned} -4x+2y&=8 \\ -4x+2(0)&=8 \\ -4x&=8 \\ x&=-2 \end{aligned}$$

The x-intercept is $(-2,0)$.

To find the y-intercept of the equation, let $x=0$.

$$\begin{aligned} -4x+2y&=8 \\ -4(0)+2y&=8 \\ 2y&=8 \\ y&=4 \end{aligned}$$

The y-intercept is $(0,4)$.

Plot the intercepts and draw the line through them.

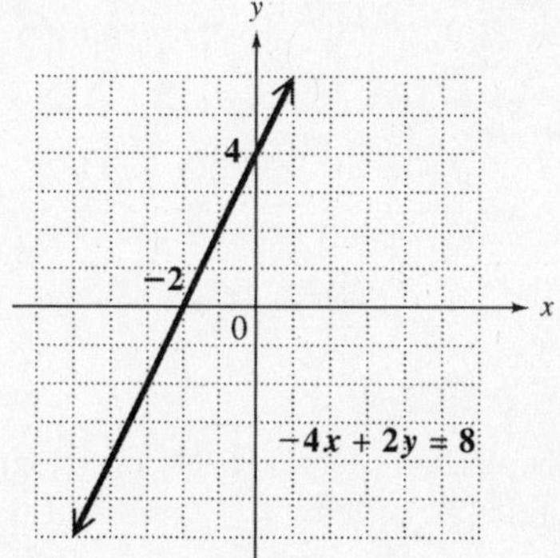

32. Through $(-5,8)$ and $(-1,2)$

Let $(x_1,y_1)=(-5,8)$ and $(x_2,y_2)=(-1,2)$.

$$m=\frac{y_2-y_1}{x_2-x_1}=\frac{2-8}{-1-(-5)}=\frac{-6}{4}=-\frac{3}{2}$$

The slope is $-\frac{3}{2}$.

33. The equation is in the form $y=mx+b$. The slope, m, of the line with equation $y=-\frac{1}{2}x+5$ is $-\frac{1}{2}$. Thus, the slope of any line parallel to it is $-\frac{1}{2}$.

34. Perpendicular to $4x-3y=12$

Solve for y to write the equation in slope-intercept form.

$$\begin{aligned} 4x-3y&=12 \\ -3y&=-4x+12 \\ y&=\tfrac{4}{3}x-4 \end{aligned}$$

The slope of the given line is $\frac{4}{3}$. Perpendicular lines have slopes that are negative reciprocals of each other. The negative reciprocal of $\frac{4}{3}$ is $-\frac{3}{4}$. The slope of a line perpendicular to the given line is $-\frac{3}{4}$.

35. Slope $-\frac{3}{4}$; y-intercept $(0, -1)$

To write an equation of this line, let $m = -\frac{3}{4}$ and $b = -1$ in the slope-intercept form.

$$y = mx + b$$
$$y = -\tfrac{3}{4}x - 1$$

36. Horizontal; through $(2, -2)$

A horizontal line through the point (c, d) has equation $y = d$. Here $d = -2$, so the equation of the line is $y = -2$.

37. Through $(4, -3)$ and $(1, 1)$

First find the slope of the line.

$$m = \frac{1-(-3)}{1-4} = \frac{4}{-3} = -\frac{4}{3}$$

Now substitute $(x_1, y_1) = (4, -3)$ and $m = -\frac{4}{3}$ in the point-slope form. Then solve for y.

$$\begin{aligned} y - y_1 &= m(x - x_1) \\ y - (-3) &= -\tfrac{4}{3}(x - 4) \\ y + 3 &= -\tfrac{4}{3}x + \tfrac{16}{3} \\ y &= -\tfrac{4}{3}x + \tfrac{7}{3} \end{aligned}$$

38. $f(x) = -4x + 10$

(a) The variable x can be any real number, so the domain is $(-\infty, \infty)$.

(b) $f(-3) = -4(-3) + 10 = 12 + 10 = 22$

39. Use (1997, 12,000) and (2004, 86,000).

average rate of change

$$= \frac{\Delta y}{\Delta x} = \frac{86{,}000 - 12{,}000}{2004 - 1997} = \frac{74{,}000}{7} \approx 10{,}571$$

So the average rate of change is 10,571 per year; that is, the number of motor scooters sold in the United States increased by an average of 10,571 per year from 1997 to 2004.

40. The slope (in thousands) is 10.571 and the y-intercept is $(0, 12)$, so the slope-intercept form of the model is $y = 10.571x + 12$.

CHAPTER 5 SYSTEMS OF LINEAR EQUATIONS

5.1 Systems of Linear Equations in Two Variables

5.1 Margin Exercises

1. **(a)** $x - y = 3$ (1)
$2x - y = 4$ (2)

When the equations are graphed, the point of intersection appears to be $(1, -2)$. To check, substitute 1 for x and -2 for y in each equation of the system.

$x - y = 3$ (1)	$2x - y = 4$ (2)
$1 - (-2) \stackrel{?}{=} 3$	$2(1) - (-2) \stackrel{?}{=} 4$
$3 = 3$ *True*	$4 = 4$ *True*

Since $(1, -2)$ makes both equations true, the solution set of the system is $\{(1, -2)\}$.

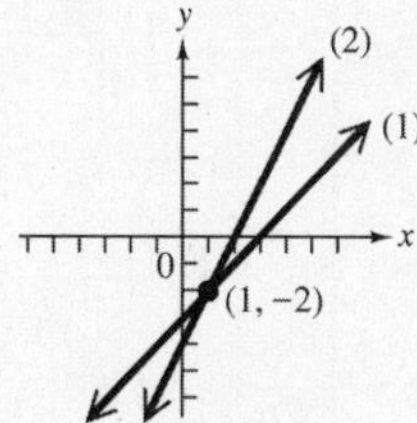

(b) $2x + y = -5$ (1)
$-x + 3y = 6$ (2)

When the equations are graphed, the point of intersection appears to be $(-3, 1)$. To check, substitute -3 for x and 1 for y in each equation of the system.

$2x + y = -5$ (1)	$-x + 3y = 6$ (2)
$2(-3) + 1 \stackrel{?}{=} -5$	$-(-3) + 3(1) \stackrel{?}{=} 6$
$-5 = -5$ *True*	$6 = 6$ *True*

Since $(-3, 1)$ makes both equations true, the solution set is $\{(-3, 1)\}$.

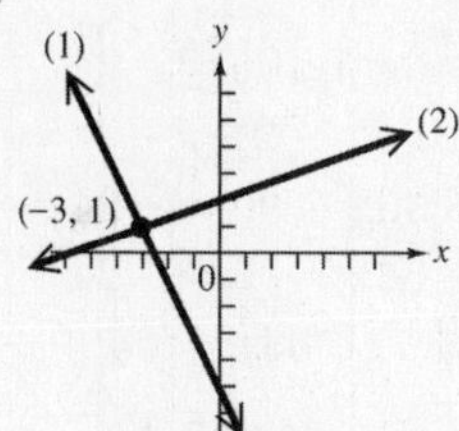

2. **(a)** To determine whether $(-4, 2)$ is a solution of the system

$$2x + y = -6 \quad (1)$$
$$x + 3y = 2, \quad (2)$$

replace x with -4 and y with 2 in each equation.

$2x + y = -6$ (1)	$x + 3y = 2$ (2)
$2(-4) + 2 \stackrel{?}{=} -6$	$-4 + 3(2) \stackrel{?}{=} 2$
$-8 + 2 \stackrel{?}{=} -6$	$-4 + 6 \stackrel{?}{=} 2$
$-6 = -6$ *True*	$2 = 2$ *True*

Since $(-4, 2)$ makes both equations true, $(-4, 2)$ is a solution of the system.

(b) To determine whether $(-1, 5)$ is a solution of the system

$$9x - y = -4$$
$$4x + 3y = 11,$$

replace x with -1 and y with 5 in each equation.

$$9x - y = -4$$
$$9(-1) - 5 \stackrel{?}{=} -4$$
$$-9 - 5 \stackrel{?}{=} -4$$
$$-14 = -4 \quad \textit{False}$$

Since $(-1, 5)$ does not satisfy the first equation, it is not necessary to check whether it satisfies the second equation. The ordered pair $(-1, 5)$ is not a solution of the system.

3. **(a)** $7x - 2y = -2$ (1)
$y = 3x$ (2)

Since equation (2) is given in terms of y, substitute $3x$ for y in equation (1).

$$7x - 2y = -2 \quad (1)$$
$$7x - 2(3x) = -2$$
$$7x - 6x = -2$$
$$x = -2$$

Since $y = 3x$ and $x = -2$,

$$y = 3(-2) = -6.$$

Check $(-2, -6)$: $-2 = -2$; $-6 = -6$

Solution set: $\{(-2, -6)\}$

(b) $5x - 3y = -6$ (1)
$x = 2 - y$ (2)

Since equation (2) is given in terms of x, substitute $2 - y$ for x in equation (1).

$$5x - 3y = -6 \quad (1)$$
$$5(2 - y) - 3y = -6$$
$$10 - 5y - 3y = -6$$
$$10 - 8y = -6$$
$$-8y = -16$$
$$y = 2$$

Since $x = 2 - y$ and $y = 2$,

$$x = 2 - 2 = 0.$$

Check $(0, 2)$: $-6 = -6$; $0 = 0$

Solution set: $\{(0, 2)\}$

4. **(a)** $3x - y = 10$ (1)
$2x + 5y = 1$ (2)

Step 1
To use the substitution method, first solve one of the equations for x or y. Since the coefficient of y in equation (1) is -1, it is easiest to solve for y in this equation.

$$\begin{aligned} 3x - y &= 10 \quad (1) \\ -y &= -3x + 10 \\ y &= 3x - 10 \end{aligned}$$

Step 2
Substitute $3x - 10$ for y in equation (2) and solve for x.

$$\begin{aligned} 2x + 5y &= 1 \quad (2) \\ 2x + 5(3x - 10) &= 1 \end{aligned}$$

Step 3 $2x + 15x - 50 = 1$
$$\begin{aligned} 17x &= 51 \\ x &= 3 \end{aligned}$$

Step 4
Since $y = 3x - 10$ and $x = 3$,

$$y = 3(3) - 10 = 9 - 10 = -1.$$

Step 5
Check $(3, -1)$: $10 = 10$; $1 = 1$

Solution set: $\{(3, -1)\}$

(b) $4x - 5y = -11$ (1)
$x + 2y = 7$ (2)

Step 1
Solve equation (2) for x.

$$\begin{aligned} x + 2y &= 7 \quad (2) \\ x &= 7 - 2y \end{aligned}$$

Step 2
Substitute $7 - 2y$ for x in equation (1).

$$\begin{aligned} 4x - 5y &= -11 \quad (1) \\ 4(7 - 2y) - 5y &= -11 \end{aligned}$$

Step 3 $28 - 8y - 5y = -11$
$$\begin{aligned} 28 - 13y &= -11 \\ -13y &= -39 \\ y &= 3 \end{aligned}$$

Step 4
Since $x = 7 - 2y$ and $y = 3$,

$$x = 7 - 2(3) = 7 - 6 = 1.$$

Step 5
Check $(1, 3)$: $-11 = -11$; $7 = 7$

Solution set: $\{(1, 3)\}$

5. **(a)** $-2x + 5y = 22$ (1)
$\frac{1}{2}x + \frac{1}{4}y = \frac{1}{2}$ (2)

Multiply equation (2) by the LCD, 4. The new system is

$$\begin{aligned} -2x + 5y &= 22 \quad (1) \\ 2x + y &= 2. \quad (3) \end{aligned}$$

Solve equation (3) for y.

$$y = -2x + 2$$

Substitute $-2x + 2$ for y in equation (1) and solve for x.

$$\begin{aligned} -2x + 5y &= 22 \\ -2x + 5(-2x + 2) &= 22 \\ -2x - 10x + 10 &= 22 \\ -12x + 10 &= 22 \\ -12x &= 12 \\ x &= -1 \end{aligned}$$

Since $y = -2x + 2$ and $x = -1$,

$$y = -2(-1) + 2 = 2 + 2 = 4.$$

Check $(-1, 4)$: $2 + 20 = 22$; $-\frac{1}{2} + 1 = \frac{1}{2}$

The solution set is $\{(-1, 4)\}$.

(b) $\frac{1}{5}x + \frac{2}{3}y = -\frac{8}{5}$ (1)
$3x - y = 9$ (2)

To eliminate the fractions in equation (1), multiply by 15. The new system is

$$\begin{aligned} 3x + 10y &= -24 \quad (3) \\ 3x - y &= 9. \quad (2) \end{aligned}$$

Solve equation (2) for y.

$$\begin{aligned} 3x - y &= 9 \quad (2) \\ -y &= -3x + 9 \\ y &= 3x - 9 \end{aligned}$$

Substitute $3x - 9$ for y in equation (3) and solve for x.

$$\begin{aligned} 3x + 10y &= -24 \quad (3) \\ 3x + 10(3x - 9) &= -24 \\ 3x + 30x - 90 &= -24 \\ 33x - 90 &= -24 \\ 33x &= 66 \\ x &= 2 \end{aligned}$$

Since $y = 3x - 9$ and $x = 2$,

$$y = 3(2) - 9 = 6 - 9 = -3.$$

Check $(2, -3)$: $\frac{2}{5} - 2 = -\frac{8}{5}$; $9 = 9$

Solution set: $\{(2, -3)\}$

6. **(a)** $3x - y = -7$ (1)
$2x + y = -3$ (2)

Eliminate y by adding equations (1) and (2).

$$\begin{array}{rcll} 3x - y &=& -7 & (1) \\ 2x + y &=& -3 & (2) \\ \hline 5x &=& -10 & \textit{Add.} \\ x &=& -2 & \textit{Divide by 5.} \end{array}$$

To find y, replace x with -2 in either equation (1) or (2).

$$\begin{aligned} 3x - y &= -7 \quad (1) \\ 3(-2) - y &= -7 \\ -6 - y &= -7 \\ -y &= -1 \\ y &= 1 \end{aligned}$$

Check $(-2, 1)$: $-7 = -7$; $-3 = -3$

Solution set: $\{(-2, 1)\}$

(b) $-2x + 3y = -10$ (1)
$2x + 2y = 5$ (2)

Eliminate x by adding equations (1) and (2).

$$\begin{array}{rcll} -2x + 3y &=& -10 & (1) \\ 2x + 2y &=& 5 & (2) \\ \hline 5y &=& -5 & \textit{Add.} \\ y &=& -1 & \textit{Divide by 5.} \end{array}$$

To find x, replace y with -1 in either equation (1) or (2).

$$\begin{aligned} -2x + 3y &= -10 \quad (1) \\ -2x + 3(-1) &= -10 \\ -2x - 3 &= -10 \\ -2x &= -7 \\ x &= \tfrac{7}{2} \end{aligned}$$

Check $(\frac{7}{2}, -1)$: $-10 = -10$; $5 = 5$

Solution set: $\{(\frac{7}{2}, -1)\}$

7. **(a)** $x + 3y = 8$ (1)
$2x - 5y = -17$ (2)

Step 1
Both equations are in standard form.

Step 2
To eliminate x, multiply equation (1) by -2 and add the result to equation (2).

$$\begin{array}{lrcll} & -2x - 6y &=& -16 & -2 \times (1) \\ & 2x - 5y &=& -17 & (2) \\ \hline \textit{Step 3} & -11y &=& -33 & \textit{Add.} \\ \textit{Step 4} & y &=& 3 & \textit{Div. } -11. \end{array}$$

Step 5
To find x, substitute 3 for y in equation (1) or (2).

$$\begin{aligned} x + 3y &= 8 \quad (1) \\ x + 3(3) &= 8 \\ x + 9 &= 8 \\ x &= -1 \end{aligned}$$

Step 6
Check $(-1, 3)$: $8 = 8$; $-17 = -17$

Solution set: $\{(-1, 3)\}$

(b) $6x - 2y = -21$ (1)
$-3x + 4y = 36$ (2)

To eliminate x, multiply equation (2) by 2 and add the result to equation (1).

$$\begin{array}{rcll} 6x - 2y &=& -21 & (1) \\ -6x + 8y &=& 72 & 2 \times (2) \\ \hline 6y &=& 51 & \textit{Add.} \\ y &=& \frac{51}{6} = \frac{17}{2} & \textit{Divide by 6.} \end{array}$$

To find x, substitute $\frac{17}{2}$ for y in equation (1) or (2).

$$\begin{aligned} 6x - 2y &= -21 \quad (1) \\ 6x - 2(\tfrac{17}{2}) &= -21 \\ 6x - 17 &= -21 \\ 6x &= -4 \\ x &= -\tfrac{2}{3} \end{aligned}$$

Check $(-\frac{2}{3}, \frac{17}{2})$: $-4 - 17 = -21$; $2 + 34 = 36$

Solution set: $\{(-\frac{2}{3}, \frac{17}{2})\}$

(c) $2x + 3y = 19$ (1)
$3x - 7y = -6$ (2)

To eliminate x, multiply equation (1) by -3 and add the result to equation (2) multiplied by 2.

$$\begin{array}{rcll} -6x - 9y &=& -57 & -3 \times (1) \\ 6x - 14y &=& -12 & 2 \times (2) \\ \hline -23y &=& -69 & \textit{Add.} \\ y &=& 3 & \textit{Divide by } -23. \end{array}$$

To find x, substitute 3 for y in equation (1) or (2).

$$\begin{aligned} 2x + 3y &= 19 \quad (1) \\ 2x + 3(3) &= 19 \\ 2x + 9 &= 19 \\ 2x &= 10 \\ x &= 5 \end{aligned}$$

Check $(5, 3)$: $19 = 19$; $-6 = -6$

Solution set: $\{(5, 3)\}$

8. $2x + y = 6$ (1)
$-8x - 4y = -24$ (2)

Multiply equation (1) by 4, and add the result to equation (2).

$$\begin{array}{rcrl} 8x + 4y & = & 24 & 4 \times (1) \\ -8x - 4y & = & -24 & (2) \\ \hline 0 & = & 0 & \textit{True} \end{array}$$

The equations are dependent.

Solution set: $\{(x, y) \mid 2x + y = 6\}$

Equations (1) and (2) have the same graph.

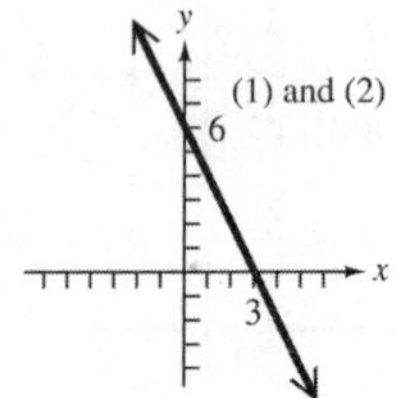

9. $4x - 3y = 8$ (1)
$8x - 6y = 14$ (2)

Multiply equation (1) by -2 and add the result to equation (2).

$$\begin{array}{rcrl} -8x + 6y & = & -16 & -2 \times (1) \\ 8x - 6y & = & 14 & (2) \\ \hline 0 & = & -2 & \textit{False} \end{array}$$

The system is inconsistent.

Solution set: $\emptyset$

The graphs of the equations are parallel lines.

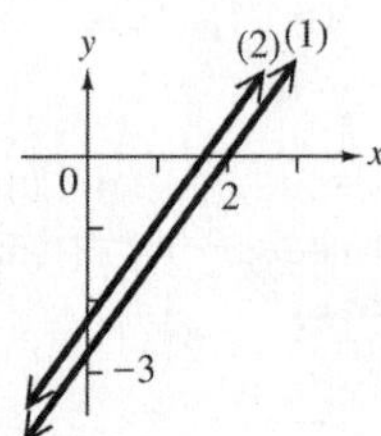

10. Since the equations are equivalent, solve one equation for y.

$$\begin{aligned} 2x - y &= 3 \quad (1) \\ -y &= -2x + 3 \\ y &= 2x - 3 \end{aligned}$$

Both equations are $f(x) = 2x - 3$.

11. $x + 3y = 4$ (1)

$$\begin{aligned} 3y &= -x + 4 \\ y &= -\tfrac{1}{3}x + \tfrac{4}{3} \\ f(x) &= -\tfrac{1}{3}x + \tfrac{4}{3} \end{aligned}$$

$$\begin{aligned} -2x - 6y &= 3 \quad (2) \\ -6y &= 2x + 3 \\ y &= -\tfrac{1}{3}x - \tfrac{1}{2} \\ f(x) &= -\tfrac{1}{3}x - \tfrac{1}{2} \end{aligned}$$

5.1 Section Exercises

1. If $(3, -6)$ is a solution of a linear system in two variables, then substituting <u>3</u> for x and <u>-6</u> for y leads to true statements in *both* equations.

3. **D**; The ordered pair solution must be in quadrant IV, since that is where the graphs of the equations intersect.

5. $x - y = 0$ implies that $x = y$, so the coordinates of the solution must be equal—this limits our choice to **B** or **C**. Since $x + y = 6$, graph **B** is correct.

7. $x + y = 0$ implies that $x = -y$, so the coordinates of the solution must be opposites—this limits our choice to **A** or **D**. Since $x - y = -6$, graph **A** is correct.

9. $x + y = -5$
$-2x + y = 1$
Graph the line $x + y = -5$ through its intercepts, $(-5, 0)$ and $(0, -5)$, and the line $-2x + y = 1$ through its intercepts, $(-\tfrac{1}{2}, 0)$ and $(0, 1)$. The lines appear to intersect at $(-2, -3)$.

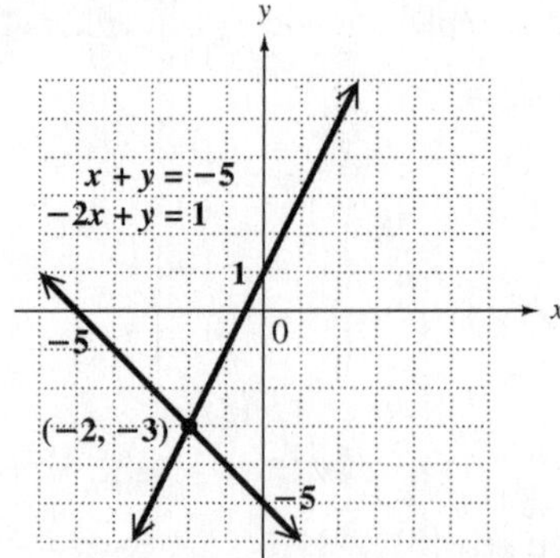

Check this ordered pair in the system.

$$\begin{aligned} (-2) + (-3) &\stackrel{?}{=} -5 \\ -5 &= -5 \quad \textit{True} \end{aligned}$$

$$\begin{aligned} -2(-2) + (-3) &\stackrel{?}{=} 1 \\ 4 + (-3) &= 1 \quad \textit{True} \end{aligned}$$

Solution set: $\{(-2, -3)\}$

11. $x - 4y = -4$
$3x + y = 1$

Graph the line $x - 4y = -4$ through its intercepts, $(-4, 0)$ and $(0, 1)$, and the line $3x + y = 1$ through its intercepts, $(\frac{1}{3}, 0)$ and $(0, 1)$. The lines have the same y-intercept, $(0, 1)$, so $\{(0, 1)\}$ is the solution set.

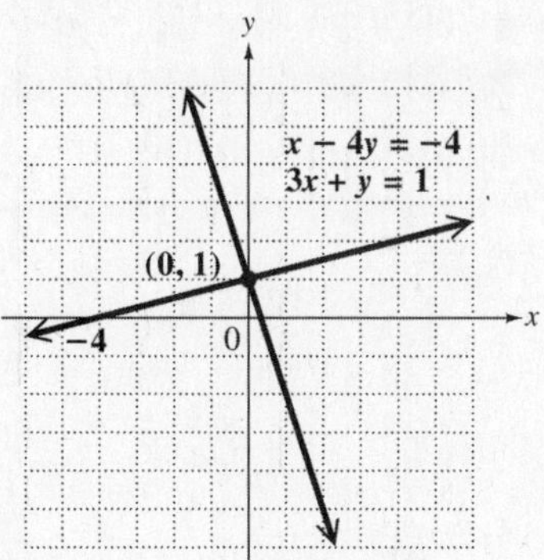

13. $2x + 3y = -6$
$x - 3y = -3$

Graph the line $2x + 3y = -6$ through its intercepts, $(-3, 0)$ and $(0, -2)$, and the line $x - 3y = -3$ through its intercepts, $(-3, 0)$ and $(0, 1)$. The lines have the same x-intercept, $(-3, 0)$, so $\{(-3, 0)\}$ is the solution set.

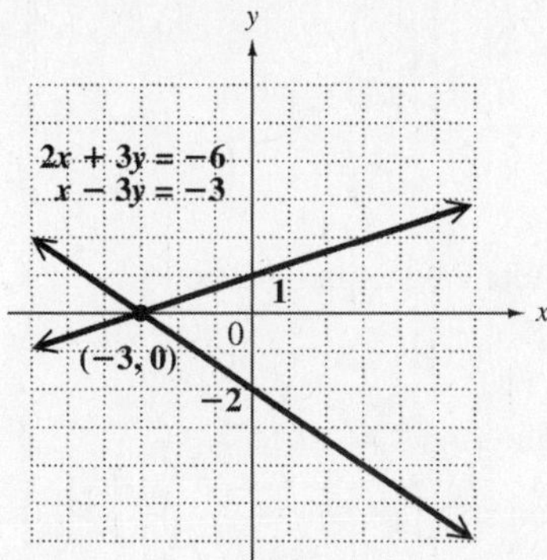

15. To determine whether $(5, 1)$ is a solution of the system

$$x + y = 6$$
$$x - y = 4,$$

replace x with 5 and y with 1 in each equation.

$$x + y = 6$$
$$5 + 1 \stackrel{?}{=} 6$$
$$6 = 6 \quad \textit{True}$$

$$x - y = 4$$
$$5 - 1 \stackrel{?}{=} 4$$
$$4 = 4 \quad \textit{True}$$

Since $(5, 1)$ makes both equations true, $(5, 1)$ is a solution of the system.

17. To determine whether $(5, 2)$ is a solution of the system

$$2x - y = 8$$
$$3x + 2y = 20,$$

replace x with 5 and y with 2 in each equation.

$$2x - y = 8$$
$$2(5) - 2 \stackrel{?}{=} 8$$
$$10 - 2 = 8 \quad \textit{True}$$

$$3x + 2y = 20$$
$$3(5) + 2(2) \stackrel{?}{=} 20$$
$$15 + 4 = 20 \quad \textit{False}$$

The ordered pair $(5, 2)$ is *not* a solution of the system since it does not make both equations true.

19. To determine whether $(-1, 1)$ is a solution of the system

$$4x + 3y = -1$$
$$-2x + 5y = 3,$$

replace x with -1 and y with 1 in each equation.

$$4x + 3y = -1$$
$$4(-1) + 3(1) \stackrel{?}{=} -1$$
$$-4 + 3 = -1 \quad \textit{True}$$

$$-2x + 5y = 3$$
$$-2(-1) + 5(1) \stackrel{?}{=} 3$$
$$2 + 5 = 3 \quad \textit{False}$$

The ordered pair $(-1, 1)$ is *not* a solution of the system since it does not make both equations true.

21. $4x + y = 6 \quad (1)$
$y = 2x \quad (2)$

Since equation (2) is already solved for y, substitute $2x$ for y in equation (1).

$$4x + y = 6 \quad (1)$$
$$4x + 2x = 6 \quad \textit{Let } y = 2x.$$
$$6x = 6$$
$$x = 1$$

Substitute 1 for x in (2).

$$y = 2(1) = 2$$

The solution $(1, 2)$ checks.

Solution set: $\{(1, 2)\}$

23. $-x - 4y = -14$ (1)
$y = 2x - 1$ (2)

Substitute $2x - 1$ for y in equation (1) and solve for x.

$$\begin{aligned} -x - 4y &= -14 \quad (1) \\ -x - 4(2x - 1) &= -14 \\ -x - 8x + 4 &= -14 \\ -9x &= -18 \\ x &= 2 \end{aligned}$$

Substitute 2 for x in (2).

$$y = 2(2) - 1 = 3$$

The solution $(2, 3)$ checks.

Solution set: $\{(2, 3)\}$

25. $3x - 4y = -22$ (1)
$-3x + y = 0$ (2)

Solve equation (2) for y to get

$$y = 3x. \quad (3)$$

Substitute $3x$ for y in equation (1).

$$\begin{aligned} 3x - 4y &= -22 && (1) \\ 3x - 4(3x) &= -22 && \textit{Let } y = 3x. \\ 3x - 12x &= -22 \\ -9x &= -22 \\ x &= \tfrac{-22}{-9} = \tfrac{22}{9} \end{aligned}$$

Substitute $\frac{22}{9}$ for x in equation (3) to get

$$y = 3x = 3(\tfrac{22}{9}) = \tfrac{22}{3}.$$

The solution $(\frac{22}{9}, \frac{22}{3})$ checks.

Solution set: $\{(\frac{22}{9}, \frac{22}{3})\}$

27. $5x - 4y = 9$ (1)
$3 - 2y = -x$ (2)

Solve equation (2) for x.

$$\begin{aligned} 3 - 2y &= -x && (2) \\ 2y - 3 &= x && (3) \quad -1 \times (2) \end{aligned}$$

Substitute $2y - 3$ for x in equation (1).

$$\begin{aligned} 5x - 4y &= 9 && (1) \\ 5(2y - 3) - 4y &= 9 && \textit{Let } x = 2y - 3. \\ 10y - 15 - 4y &= 9 \\ 6y - 15 &= 9 \\ 6y &= 24 \\ y &= 4 \end{aligned}$$

Substitute 4 for y in equation (3).

$$x = 2y - 3 = 2(4) - 3 = 5$$

The solution $(5, 4)$ checks.

Solution set: $\{(5, 4)\}$

29. $x = 3y + 5$ (1)
$x = \frac{3}{2}y$ (2)

Both equations are given in terms of x. Choose equation (2), and substitute $\frac{3}{2}y$ for x in equation (1).

$$\begin{aligned} x &= 3y + 5 && (1) \\ \tfrac{3}{2}y &= 3y + 5 && \textit{Let } x = \tfrac{3}{2}y. \\ 3y &= 6y + 10 && \textit{Multiply by 2.} \\ -3y &= 10 \\ y &= -\tfrac{10}{3} \end{aligned}$$

Since $x = \frac{3}{2}y$ and $y = -\frac{10}{3}$,

$$x = \tfrac{3}{2}(-\tfrac{10}{3}) = -\tfrac{10}{2} = -5.$$

The solution $(-5, -\frac{10}{3})$ checks.

Solution set: $\{(-5, -\frac{10}{3})\}$

31. $\frac{1}{2}x + \frac{1}{3}y = 3$ (1)
$-3x + y = 0$ (2)

Multiply (1) by its LCD, 6, to eliminate fractions

$$3x + 2y = 18 \quad (3)$$

From (2), we have $y = 3x$ (4), so substitute $3x$ for y in (3).

$$\begin{aligned} 3x + 2(3x) &= 18 \\ 3x + 6x &= 18 \\ 9x &= 18 \\ x &= 2 \end{aligned}$$

Substitute 2 for x in equation (4).

$$y = 3x = 3(2) = 6$$

The solution $(2, 6)$ checks.

Solution set: $\{(2, 6)\}$

33. $y = 2x$ (1)
$4x - 2y = 0$ (2)

From equation (1), substitute $2x$ for y in equation (2).

$$\begin{aligned} 4x - 2y &= 0 && (2) \\ 4x - 2(2x) &= 0 \\ 4x - 4x &= 0 \\ 0 &= 0 && \textit{True} \end{aligned}$$

The equations are dependent, and the solution is the set of all points on the line.
The solution set is $\{(x, y) \mid 2x - y = 0\}$.

35. $5x - 25y = 5$ (1)
$x = 5y$ (2)

From equation (2), substitute $5y$ for x in equation (1).

$$\begin{aligned} 5x - 25y &= 5 \quad (1) \\ 5(5y) - 25y &= 5 \\ 25y - 25y &= 5 \\ 0 &= 5 \quad \textit{False} \end{aligned}$$

The system is *inconsistent*. Since the graphs of the equations are parallel lines, there are no ordered pairs that satisfy both equations.
The solution set is $\emptyset$.

37. $-2x + 3y = -16$ (1)
$2x - 5y = 24$ (2)
To eliminate x, add the equations.

$$\begin{array}{rcrl} -2x + 3y &=& -16 & (1) \\ 2x - 5y &=& 24 & (2) \\ \hline -2y &=& 8 & \\ y &=& -4 & \end{array}$$

To find x, substitute -4 for y in equation (2).

$$\begin{aligned} 2x - 5y &= 24 \quad (2) \\ 2x - 5(-4) &= 24 \\ 2x + 20 &= 24 \\ 2x &= 4 \\ x &= 2 \end{aligned}$$

The ordered pair $(2, -4)$ satisfies both equations, so it checks.
The solution set is $\{(2, -4)\}$.

39. $2x - 5y = 11$ (1)
$3x + y = 8$ (2)

To eliminate y, multiply equation (2) by 5 and add the result to equation (1).

$$\begin{array}{rcrl} 2x - 5y &=& 11 & (1) \\ 15x + 5y &=& 40 & 5 \times (2) \\ \hline 17x &=& 51 & \\ x &=& 3 & \end{array}$$

To find y, substitute 3 for x in equation (2).

$$\begin{aligned} 3x + y &= 8 \quad (2) \\ 3(3) + y &= 8 \\ 9 + y &= 8 \\ y &= -1 \end{aligned}$$

The ordered pair $(3, -1)$ satisfies both equations, so it checks.

Solution set: $\{(3, -1)\}$

41. $3x + 4y = -6$ (1)
$5x + 3y = 1$ (2)

To eliminate x, multiply equation (1) by 5 and equation (2) by -3. Then add the results.

$$\begin{array}{rcrl} 15x + 20y &=& -30 & 5 \times (1) \\ -15x - 9y &=& -3 & -3 \times (2) \\ \hline 11y &=& -33 & \\ y &=& -3 & \end{array}$$

To find x, substitute -3 for y in equation (2).

$$\begin{aligned} 5x + 3y &= 1 \quad (2) \\ 5x + 3(-3) &= 1 \\ 5x - 9 &= 1 \\ 5x &= 10 \\ x &= 2 \end{aligned}$$

The ordered pair $(2, -3)$ satisfies both equations, so it checks.

Solution set: $\{(2, -3)\}$

43. $3x + 3y = 0$ (1)
$4x + 2y = 3$ (2)

To eliminate y, multiply equation (1) by 2 and equation (2) by -3. Then add the results.

$$\begin{array}{rcrl} 6x + 6y &=& 0 & 2 \times (1) \\ -12x - 6y &=& -9 & -3 \times (2) \\ \hline -6x &=& -9 & \\ x &=& \frac{-9}{-6} = \frac{3}{2} & \end{array}$$

To find y, substitute $\frac{3}{2}$ for x in equation (1).

$$\begin{aligned} 3x + 3y &= 0 \quad (1) \\ 3(\tfrac{3}{2}) + 3y &= 0 \\ \tfrac{9}{2} + 3y &= 0 \\ 3y &= -\tfrac{9}{2} \\ y &= \tfrac{1}{3}(-\tfrac{9}{2}) = -\tfrac{3}{2} \end{aligned}$$

The solution $(\frac{3}{2}, -\frac{3}{2})$ checks.

Solution set: $\{(\frac{3}{2}, -\frac{3}{2})\}$

When you get a solution that has non-integer components, it is sometimes more difficult to check the problem than it was to solve it. A graphing calculator can be very helpful in this case. Just store the values for x and y in their respective memory locations, and then type the expressions as shown in the following screen. The results 0 and 3 (the right sides of the equations) indicate that we have found the correct solution.

continued

```
3/2→X: -3/2→Y
                -1.5
3X+3Y
                   0
4X+2Y
                   3
```

45. $7x + 2y = 6$ (1)
$-14x - 4y = -12$ (2)

To eliminate y, multiply equation (1) by 2 and add the result to equation (2).

$$\begin{array}{rcrl} 14x + 4y & = & 12 & 2 \times (1) \\ -14x - 4y & = & -12 & (2) \\ \hline 0 & = & 0 & \textit{True} \end{array}$$

Multiplying equation (1) by -2 gives equation (2). The equations are dependent, and the solution is the set of all points on the line.

Solution set: $\{(x, y) \mid 7x + 2y = 6\}$

47. $\frac{x}{2} + \frac{y}{3} = -\frac{1}{3}$ (1)
$\frac{x}{2} + 2y = -7$ (2)

Eliminate the fractions by multiplying equation (1) by -6 and equation (2) by 6. Then add the results to eliminate x.

$$\begin{array}{rcrl} -3x - 2y & = & 2 & (3) \quad -6 \times (1) \\ 3x + 12y & = & -42 & 6 \times (2) \\ \hline 10y & = & -40 & \\ y & = & -4 & \end{array}$$

To find x, substitute -4 for y in equation (3).

$$\begin{aligned} -3x - 2y &= 2 \quad (3) \\ -3x - 2(-4) &= 2 \\ -3x + 8 &= 2 \\ -3x &= -6 \\ x &= 2 \end{aligned}$$

The solution $(2, -4)$ checks.

Solution set: $\{(2, -4)\}$

49. $5x - 5y = 3$ (1)
$x - y = 12$ (2)

To eliminate x, multiply equation (2) by -5 and add the result to equation (1).

$$\begin{array}{rcrl} 5x - 5y & = & 3 & (1) \\ -5x + 5y & = & -60 & -5 \times (2) \\ \hline 0 & = & -57 & \textit{False} \end{array}$$

The system is *inconsistent*. Since the graphs of the equations are parallel lines, there are no ordered pairs that satisfy both equations.

Solution set: $\emptyset$

51. $3x + 7y = 4$ (1)
$6x + 14y = 3$ (2)

Write each equation in slope-intercept form by solving for y.

$$\begin{aligned} 3x + 7y &= 4 \qquad (1) \\ 7y &= -3x + 4 \\ y &= -\tfrac{3}{7}x + \tfrac{4}{7} \end{aligned}$$

$$\begin{aligned} 6x + 14y &= 3 \qquad (2) \\ 14y &= -6x + 3 \\ y &= -\tfrac{6}{14}x + \tfrac{3}{14} \\ y &= -\tfrac{3}{7}x + \tfrac{3}{14} \end{aligned}$$

Since the equations have the same slope, $-\frac{3}{7}$, but different y-intercepts, $\frac{4}{7}$ and $\frac{3}{14}$, the lines when graphed are parallel. The system is inconsistent and has no solution.

53. $2x = -3y + 1$ (1)
$6x = -9y + 3$ (2)

Write each equation in slope-intercept form by solving for y.

$$\begin{aligned} 2x &= -3y + 1 \qquad (1) \\ 3y &= -2x + 1 \\ y &= -\tfrac{2}{3}x + \tfrac{1}{3} \end{aligned}$$

$$\begin{aligned} 6x &= -9y + 3 \qquad (2) \\ 9y &= -6x + 3 \\ y &= -\tfrac{6}{9}x + \tfrac{3}{9} \\ y &= -\tfrac{2}{3}x + \tfrac{1}{3} \end{aligned}$$

Since both equations are the same, the solution set is all points on the line $y = -\frac{2}{3}x + \frac{1}{3}$. The system has infinitely many solutions.

55. (a) $6x - y = 5$ (1)
$y = 11x$ (2)

Use substitution since the second equation is solved for y.

(b) $3x + y = -7$ (1)
$x - y = -5$ (2)

Use elimination since the coefficients of the y-terms are opposites.

(c) $3x - 2y = 0$ (1)
$9x + 8y = 7$ (2)

Use elimination since the equations are in standard form with no coefficients of 1 and -1. Solving by substitution would involve fractions.

57. $3x + y = -7$ (1)
$x - y = -5$ (2)

Add the equations to eliminate y.

$$\begin{aligned} 3x + y &= -7 \\ x - y &= -5 \\ \hline 4x \quad &= -12 \\ x &= -3 \end{aligned}$$

Substitute -3 for x in (2).

$$\begin{aligned} -3 - y &= -5 \\ -y &= -2 \\ y &= 2 \end{aligned}$$

The solution $(-3, 2)$ checks.

Solution set: $\{(-3, 2)\}$

59. $2x + 3y = 10$ (1)
$-3x + y = 18$ (2)

Solve equation (2) for y.

$$y = 3x + 18 \qquad (3)$$

Substitute $3x + 18$ for y in (1).

$$\begin{aligned} 2x + 3(3x + 18) &= 10 \\ 2x + 9x + 54 &= 10 \\ 11x + 54 &= 10 \\ 11x &= -44 \\ x &= -4 \end{aligned}$$

Substitute -4 for x in (3).

$$y = 3(-4) + 18 = 6$$

The solution $(-4, 6)$ checks.

Solution set: $\{(-4, 6)\}$

61. $\frac{1}{2}x - \frac{1}{8}y = -\frac{1}{4}$ (1)
$-4x + y = 2$ (2)

Multiply equation (1) by 8 and add to equation (2).

$$\begin{aligned} 4x - y &= -2 \quad (3) \quad 8 \times (1) \\ -4x + y &= 2 \quad (2) \\ \hline 0 &= 0 \end{aligned}$$

The equations are dependent.

Solution set: $\{(x, y) \mid 4x - y = -2\}$

63. $0.3x + 0.2y = 0.4$ (1)
$0.5x + 0.4y = 0.7$ (2)

To eliminate y, multiply equation (1) by -20 and equation (2) by 10, then add.

$$\begin{aligned} -6x - 4y &= -8 \qquad\quad -20 \times (1) \\ 5x + 4y &= 7 \quad (3) \quad 10 \times (2) \\ \hline -x \quad &= -1 \\ x \quad &= 1 \end{aligned}$$

Substitute 1 for x in equation (3).

$$\begin{aligned} 5(1) + 4y &= 7 \\ 4y &= 2 \\ y &= \tfrac{1}{2} \end{aligned}$$

Check $(1, \frac{1}{2})$: $0.3 + 0.1 = 0.4$; $0.5 + 0.2 = 0.7$

Solution set: $\{(1, \frac{1}{2})\}$

65. **(a)** The supply and demand graphs intersect at 4, so supply equals demand at a price of \$4 per half-gallon.

(b) At a price of \$4 per half-gallon, the supply and demand are both about 300 half-gallons.

(c) At a price of \$2 per half-gallon, the supply is 200 half-gallons and the demand is 400 half-gallons.

67. **(a)** Rising graphs indicate population growth, so Houston, Phoenix, and Dallas will experience population growth.

(b) Philadelphia's graph indicates that it will experience population decline.

(c) In the year 2020, the city populations from least to greatest are Philadelphia, Dallas, Phoenix, and Houston.

(d) The graphs for Dallas and Philadelphia intersect in the year 2010. The population for each city will be about 1.45 million.

(e) The graphs for Houston and Phoenix appear to intersect in the year 2025 with about 2.8 million. This can be represented by the ordered pair $(2025, 2.8)$.

69. The graphs intersect at about 3.5 (years since 2000). So the sales of digital cameras were less than the sales of conventional cameras for 2000, 2001, 2002, and the first half of 2003.

71. $2.5x + y = 19.4 \quad (1)$
$-1.7x + y = 4.4 \quad (2)$

Solve equation (2) for y.

$$y = 1.7x + 4.4 \quad (3)$$

Substitute $1.7x + 4.4$ for y in equation (1).

$$\begin{aligned} 2.5x + (1.7x + 4.4) &= 19.4 \\ 4.2x + 4.4 &= 19.4 \\ 4.2x &= 15 \\ x = \tfrac{15}{4.2} = \tfrac{150}{42} &= \tfrac{25}{7} \approx 3.6 \end{aligned}$$

Substitute $\frac{25}{7}$ for x in (3).

$$y = 1.7(\tfrac{25}{7}) + 4.4 = \tfrac{17}{10}(\tfrac{25}{7}) + \tfrac{44}{10} \cdot \tfrac{7}{7} = \tfrac{733}{70}$$

Written as an ordered pair, the solution is $(\frac{25}{7}, \frac{733}{70})$.

Approximating these values to the nearest tenth, we get $\frac{25}{7} \approx 3.6$ and $\frac{733}{70} \approx 10.5$. Written as an ordered pair, the solution is $(3.6, 10.5)$. (Values may vary slightly based on the method of solution used.)

73. $3x + y = 6 \quad (1)$
$-2x + 3y = 7 \quad (2)$

Multiply equation (1) by -3 and add the result to equation (2).

$$\begin{array}{rcrl} -9x - 3y &=& -18 & -3 \times (1) \\ -2x + 3y &=& 7 & (2) \\ \hline -11x &=& -11 & \\ x &=& 1 & \end{array}$$

To find y, substitute 1 for x in equation (1).

$$\begin{aligned} 3x + y &= 6 \quad (1) \\ 3(1) + y &= 6 \\ y &= 3 \end{aligned}$$

The solution $(1, 3)$ checks.

Solution set: $\{(1, 3)\}$

74. $3x + y = 6$
$y = -3x + 6$

Replace y with $f(x)$.

$$f(x) = -3x + 6$$

Since f is in the form $f(x) = mx + b$, it is a linear function.

75. $-2x + 3y = 7$
$3y = 2x + 7$
$y = \frac{2}{3}x + \frac{7}{3}$

Replace y with $g(x)$.

$$g(x) = \tfrac{2}{3}x + \tfrac{7}{3}$$

Since g is in the form $g(x) = mx + b$, it is a linear function.

76. Because the graphs of f and g are straight lines that are neither parallel nor coincide, they intersect in exactly <u>one</u> point. The coordinates of the point are (<u>1</u>, <u>3</u>). Using functional notation, this is given by $f(\underline{1}) = \underline{3}$ and $g(\underline{1}) = \underline{3}$.

5.2 Systems of Linear Equations in Three Variables

5.2 Margin Exercises

1. (a) $x + 7y - 3z = -14 \quad (2)$

Let $x = -3$, $y = 1$, $z = 6$.

$$\begin{aligned} -3 + 7(1) - 3(6) &\stackrel{?}{=} -14 \\ -3 + 7 - 18 &\stackrel{?}{=} -14 \\ -14 &= -14 \quad \textit{True} \end{aligned}$$

Yes, the solution satisfies equation (2).

(b) $2x - 3y + 2z = 3 \quad (3)$

Let $x = -3$, $y = 1$, $z = 6$.

$$\begin{aligned} 2(-3) - 3(1) + 2(6) &\stackrel{?}{=} 3 \\ -6 - 3 + 12 &\stackrel{?}{=} 3 \\ 3 &= 3 \quad \textit{True} \end{aligned}$$

Yes, the solution satisfies equation (3).

2. (a) $x + y + z = 2 \quad (1)$
$x - y + 2z = 2 \quad (2)$
$-x + 2y - z = 1 \quad (3)$

Step 1
We'll select y as the focus variable and equation (2) as the working equation.

Step 2
Eliminate y by adding equations (1) and (2).

$$\begin{array}{rcrl} x + y + z &=& 2 & (1) \\ x - y + 2z &=& 2 & (2) \\ \hline 2x \quad + 3z &=& 4 & (4) \end{array}$$

Step 3
To eliminate y again, multiply equation (2) by 2 and add the result to equation (3).

$$\begin{array}{rcrl} 2x - 2y + 4z &=& 4 & 2 \times (2) \\ -x + 2y - z &=& 1 & (3) \\ \hline x \quad + 3z &=& 5 & (5) \end{array}$$

Step 4
Use equations (4) and (5) to eliminate z. Multiply equation (5) by -1 and add the result to equation (4).

$$\begin{array}{rcrl} -x - 3z &=& -5 & -1 \times (5) \\ 2x + 3z &=& 4 & (4) \\ \hline x &=& -1 & \end{array}$$

Now substitute -1 for x in equation (5) to find z.

$$\begin{aligned} x + 3z &= 5 && (5) \\ -1 + 3z &= 5 && \textit{Let } x = -1. \\ 3z &= 6 \\ z &= 2 \end{aligned}$$

Step 5
Substitute -1 for x and 2 for z in equation (1) to find y.

$$\begin{aligned} x + y + z &= 2 && (1) \\ -1 + y + 2 &= 2 && \textit{Let } x = -1, z = 2. \\ y + 1 &= 2 \\ y &= 1 \end{aligned}$$

Step 6
Check $(-1, 1, 2)$

Equation (1): $-1 + 1 + 2 = 2$ *True*
Equation (2): $-1 - 1 + 4 = 2$ *True*
Equation (3): $1 + 2 - 2 = 1$ *True*

Solution set: $\{(-1, 1, 2)\}$

(b)
$$\begin{aligned} 2x + y + z &= 9 && (1) \\ -x - y + z &= 1 && (2) \\ 3x - y + z &= 9 && (3) \end{aligned}$$

We'll select y as the focus variable and equation (1) as the working equation. Eliminate y by adding equations (1) and (2).

$$\begin{array}{rcl} 2x + y + z &=& 9 \quad (1) \\ -x - y + z &=& 1 \quad (2) \\ \hline x \qquad + 2z &=& 10 \quad (4) \end{array}$$

To eliminate y again, add equations (1) and (3).

$$\begin{array}{rcl} 2x + y + z &=& 9 \quad (1) \\ 3x - y + z &=& 9 \quad (3) \\ \hline 5x \qquad + 2z &=& 18 \quad (5) \end{array}$$

Multiply equation (4) by -1 and add to equation (5) to eliminate z.

$$\begin{array}{rcl} -x - 2z &=& -10 \qquad -1 \times (4) \\ 5x + 2z &=& 18 \quad (5) \\ \hline 4x &=& 8 \\ x &=& 2 \end{array}$$

Substitute 2 for x in equation (4) to find z.

$$\begin{aligned} x + 2z &= 10 && (4) \\ 2 + 2z &= 10 && \textit{Let } x = 2. \\ 2z &= 8 \\ z &= 4 \end{aligned}$$

Substitute 2 for x and 4 for z in equation (1) to find y.

$$\begin{aligned} 2x + y + z &= 9 && (1) \\ 2(2) + y + 4 &= 9 && \textit{Let } x = 2, z = 4. \\ 4 + y + 4 &= 9 \\ 8 + y &= 9 \\ y &= 1 \end{aligned}$$

Check $(2, 1, 4)$

Equation (1): $4 + 1 + 4 = 9$ *True*
Equation (2): $-2 - 1 + 4 = 1$ *True*
Equation (3): $6 - 1 + 4 = 9$ *True*

Solution set: $\{(2, 1, 4)\}$

3. **(a)**
$$\begin{aligned} x - y \qquad &= 6 && (1) \\ 2y + 5z &= 1 && (2) \\ 3x \qquad - 4z &= 8 && (3) \end{aligned}$$

Since equation (3) is missing y, eliminate y again from equations (1) and (2). Multiply equation (1) by 2 and add the result to equation (2).

$$\begin{array}{rcl} 2x - 2y \qquad &=& 12 \qquad 2 \times (1) \\ 2y + 5z &=& 1 \quad (2) \\ \hline 2x \qquad + 5z &=& 13 \quad (4) \end{array}$$

Use equation (4) together with equation (3) to eliminate x. Multiply equation (4) by 3 and equation (3) by -2. Then add the results.

$$\begin{array}{rcl} 6x + 15z &=& 39 \qquad 3 \times (4) \\ -6x + 8z &=& -16 \qquad -2 \times (3) \\ \hline 23z &=& 23 \\ z &=& 1 \end{array}$$

Substitute 1 for z in equation (2) to find y.

$$\begin{aligned} 2y + 5z &= 1 && (2) \\ 2y + 5(1) &= 1 && \textit{Let } z = 1. \\ 2y + 5 &= 1 \\ 2y &= -4 \\ y &= -2 \end{aligned}$$

Substitute -2 for y in equation (1) to find x.

$$\begin{aligned} x - y &= 6 && (1) \\ x - (-2) &= 6 && \textit{Let } y = -2. \\ x + 2 &= 6 \\ x &= 4 \end{aligned}$$

Check $(4, -2, 1)$

Equation (1): $4 + 2 = 6$ *True*
Equation (2): $-4 + 5 = 1$ *True*
Equation (3): $12 - 4 = 8$ *True*

Solution set: $\{(4, -2, 1)\}$

(b)
$$\begin{aligned} 5x - y &= 26 && (1)\\ 4y + 3z &= -4 && (2)\\ x + z &= 5 && (3)\end{aligned}$$

Since equation (1) is missing z, eliminate z again from equations (2) and (3). Multiply equation (3) by -3 and add to equation (2).

$$\begin{array}{rrrrl} & 4y & +\,3z & = -4 & (2)\\ -3x & & -\,3z & = -15 & -3\times(3)\\ \hline -3x & +\,4y & & = -19 & (4)\end{array}$$

Multiply equation (1) by 4 and add to equation (4).

$$\begin{array}{rrrl} 20x & -\,4y & = 104 & 4\times(1)\\ -3x & +\,4y & = -19 & (4)\\ \hline 17x & & = 85 & \\ x & & = 5 & \end{array}$$

Substitute 5 for x in equation (1) to find y.

$$\begin{aligned} 5x - y &= 26 && (1)\\ 5(5) - y &= 26 && \textit{Let } x = 5.\\ 25 - y &= 26 &&\\ y &= -1 &&\end{aligned}$$

Substitute 5 for x in equation (3) to find z.

$$\begin{aligned} x + z &= 5 && (3)\\ 5 + z &= 5 && \textit{Let } x = 5.\\ z &= 0 &&\end{aligned}$$

Check $(5, -1, 0)$

Equation (1): $25 + 1 = 26$ *True*
Equation (2): $-4 + 0 = -4$ *True*
Equation (3): $5 + 0 = 5$ *True*

Solution set: $\{(5, -1, 0)\}$

4. **(a)**
$$\begin{aligned} 3x - 5y + 2z &= 1 && (1)\\ 5x + 8y - z &= 4 && (2)\\ -6x + 10y - 4z &= 5 && (3)\end{aligned}$$

Multiply equation (1) by 2 and add the result to equation (3).

$$\begin{array}{rrrrl} 6x & -\,10y & +\,4z & = 2 & 2\times(1)\\ -6x & +\,10y & -\,4z & = 5 & (3)\\ \hline & & 0 & = 7 & \textit{False}\end{array}$$

Since a false statement results, the system is inconsistent.

Solution set: $\emptyset$

(b)
$$\begin{aligned} 7x - 9y + 2z &= 0 && (1)\\ y + z &= 0 && (2)\\ 8x - z &= 0 && (3)\end{aligned}$$

Since equation (3) is missing y, eliminate y again from equations (1) and (2). Multiply equation (2) by 9 and add to equation (1).

$$\begin{array}{rrrrl} 7x & -\,9y & +\,2z & = 0 & (1)\\ & 9y & +\,9z & = 0 & 9\times(2)\\ \hline 7x & & +\,11z & = 0 & (4)\end{array}$$

Use equations (3) and (4) to eliminate z. Multiply equation (3) by 11 and add to equation (4).

$$\begin{array}{rrrl} 88x & -\,11z & = 0 & 11\times(3)\\ 7x & +\,11z & = 0 & (4)\\ \hline 95x & & = 0 & \\ & x & = 0 & \end{array}$$

Substitute 0 for x in equation (3) to find z.

$$\begin{aligned} 8x - z &= 0 && (3)\\ 8(0) - z &= 0 && \textit{Let } x = 0.\\ -z &= 0 &&\\ z &= 0 &&\end{aligned}$$

Substitute 0 for z in equation (2) to find y.

$$\begin{aligned} y + z &= 0 && (2)\\ y + 0 &= 0 && \textit{Let } z = 0.\\ y &= 0 &&\end{aligned}$$

Check The check is trivial since all the terms are equal to 0 when $x = 0$, $y = 0$, and $z = 0$.

Solution set: $\{(0, 0, 0)\}$

5.
$$\begin{aligned} x - y + z &= 4 && (1)\\ -3x + 3y - 3z &= -12 && (2)\\ 2x - 2y + 2z &= 8 && (3)\end{aligned}$$

Since equation (2) is -3 times equation (1) and equation (3) is 2 times equation (1), the three equations are dependent. All three have the same graph.

Solution set: $\{(x, y, z) \mid x - y + z = 4\}$

6.
$$\begin{aligned} 2x + 3y - z &= 8 && (1)\\ \tfrac{1}{2}x + \tfrac{3}{4}y - \tfrac{1}{4}z &= 2 && (2)\\ x + \tfrac{3}{2}y - \tfrac{1}{2}z &= -6 && (3)\end{aligned}$$

Eliminate the fractions in equations (2) and (3).

$$\begin{aligned} 2x + 3y - z &= 8 && (1)\\ 2x + 3y - z &= 8 && (4)\quad 4\times(2)\\ 2x + 3y - z &= -12 && (5)\quad 2\times(3)\end{aligned}$$

Equations (1) and (4) are dependent (they have the same graph). Equations (1) and (5) are not equivalent. Since they have the same coefficients, but a different constant term, their graphs have no points in common (the planes are parallel). Thus, the system is inconsistent and the solution set is $\emptyset$.

5.2 Section Exercises

1. The statement means that when -1 is substituted for x, 2 is substituted for y, and 3 is substituted for z in the three equations, the resulting three statements are true.

3.

$$\begin{aligned} 2x - 5y + 3z &= -1 \quad (1) \\ x + 4y - 2z &= 9 \quad (2) \\ x - 2y - 4z &= -5 \quad (3) \end{aligned}$$

We'll select x as the focus variable and equation (2) as the working equation. Eliminate x by adding equation (1) to -2 times equation (2).

$$\begin{array}{rcll} 2x - 5y + 3z &=& -1 & (1) \\ -2x - 8y + 4z &=& -18 & -2 \times (2) \\ \hline -13y + 7z &=& -19 & (4) \end{array}$$

Now eliminate x by adding equation (2) to -1 times equation (3).

$$\begin{array}{rcll} x + 4y - 2z &=& 9 & (2) \\ -x + 2y + 4z &=& 5 & -1 \times (3) \\ \hline 6y + 2z &=& 14 & (5) \end{array}$$

Use equations (4) and (5) to eliminate z. Multiply equation (4) by -2 and add the result to 7 times equation (5).

$$\begin{array}{rcll} 26y - 14z &=& 38 & -2 \times (4) \\ 42y + 14z &=& 98 & 7 \times (5) \\ \hline 68y &=& 136 & \\ y &=& 2 & \end{array}$$

Substitute 2 for y in equation (5) to find z.

$$\begin{aligned} 6y + 2z &= 14 \quad (5) \\ 6(2) + 2z &= 14 \\ 12 + 2z &= 14 \\ 2z &= 2 \\ z &= 1 \end{aligned}$$

Substitute 2 for y and 1 for z in equation (3) to find x.

$$\begin{aligned} x - 2y - 4z &= -5 \quad (3) \\ x - 2(2) - 4(1) &= -5 \\ x - 4 - 4 &= -5 \\ x - 8 &= -5 \\ x &= 3 \end{aligned}$$

Check $(3, 2, 1)$

Equation (1): $6 - 10 + 3 = -1$ *True*
Equation (2): $3 + 8 - 2 = 9$ *True*
Equation (3): $3 - 4 - 4 = -5$ *True*

The solution $(3, 2, 1)$ checks in all three of the original equations.

Solution set: $\{(3, 2, 1)\}$

5.

$$\begin{aligned} 3x + 2y + z &= 8 \quad (1) \\ 2x - 3y + 2z &= -16 \quad (2) \\ x + 4y - z &= 20 \quad (3) \end{aligned}$$

We'll select z as the focus variable and equation (3) as the working equation. Eliminate z by adding equations (1) and (3).

$$\begin{array}{rcll} 3x + 2y + z &=& 8 & (1) \\ x + 4y - z &=& 20 & (3) \\ \hline 4x + 6y &=& 28 & (4) \end{array}$$

To get another equation without z, multiply equation (3) by 2 and add the result to equation (2).

$$\begin{array}{rcll} 2x + 8y - 2z &=& 40 & 2 \times (3) \\ 2x - 3y + 2z &=& -16 & (2) \\ \hline 4x + 5y &=& 24 & (5) \end{array}$$

Use equations (4) and (5) to eliminate x. Multiply equation (4) by -1 and add the result to equation (5).

$$\begin{array}{rcll} -4x - 6y &=& -28 & -1 \times (4) \\ 4x + 5y &=& 24 & (5) \\ \hline -y &=& -4 & \\ y &=& 4 & \end{array}$$

Substitute 4 for y in equation (5) to find x.

$$\begin{aligned} 4x + 5y &= 24 \quad (5) \\ 4x + 5(4) &= 24 \\ 4x + 20 &= 24 \\ 4x &= 4 \\ x &= 1 \end{aligned}$$

Substitute 1 for x and 4 for y in equation (3) to find z.

$$\begin{aligned} x + 4y - z &= 20 \quad (3) \\ 1 + 4(4) - z &= 20 \\ 1 + 16 - z &= 20 \\ 17 - z &= 20 \\ -z &= 3 \\ z &= -3 \end{aligned}$$

Check $(1, 4, -3)$

Equation (1): $3 + 8 - 3 = 8$ *True*
Equation (2): $2 - 12 - 6 = -16$ *True*
Equation (3): $1 + 16 + 3 = 20$ *True*

The solution $(1, 4, -3)$ checks in all three of the original equations.

Solution set: $\{(1, 4, -3)\}$

7.
$$\begin{aligned} x + 2y + z &= 4 \quad (1)\\ 2x + y - z &= -1 \quad (2)\\ x - y - z &= -2 \quad (3) \end{aligned}$$

Add (1) and (2) to eliminate z.

$$\begin{aligned} x + 2y + z &= 4 \quad (1)\\ 2x + y - z &= -1 \quad (2)\\ \hline 3x + 3y &= 3 \quad (4) \end{aligned}$$

Add (1) and (3) to eliminate z.

$$\begin{aligned} x + 2y + z &= 4 \quad (1)\\ x - y - z &= -2 \quad (3)\\ \hline 2x + y &= 2 \quad (5) \end{aligned}$$

Multiply equation (5) by -3 and add to (4).

$$\begin{aligned} -6x - 3y &= -6 \quad (6)\ -3 \times (5)\\ 3x + 3y &= 3 \quad (4)\\ \hline -3x &= -3\\ x &= 1 \end{aligned}$$

Substitute 1 for x in equation (5) and solve for y.

$$\begin{aligned} 2(1) + y &= 2\\ y &= 0 \end{aligned}$$

Substitute 1 for x and 0 for y in (1).

$$\begin{aligned} (1) + 2(0) + z &= 4\\ z &= 3 \end{aligned}$$

The solution $(1, 0, 3)$ checks in all three of the original equations.

Solution set: $\{(1, 0, 3)\}$

9.
$$\begin{aligned} -x + 2y + 6z &= 2 \quad (1)\\ 3x + 2y + 6z &= 6 \quad (2)\\ x + 4y - 3z &= 1 \quad (3) \end{aligned}$$

Eliminate y and z by adding equation (1) to -1 times equation (2).

$$\begin{aligned} -x + 2y + 6z &= 2 \quad (1)\\ -3x - 2y - 6z &= -6 \quad -1 \times (2)\\ \hline -4x &= -4\\ x &= 1 \end{aligned}$$

Eliminate z by adding equation (2) to 2 times equation (3).

$$\begin{aligned} 3x + 2y + 6z &= 6 \quad (2)\\ 2x + 8y - 6z &= 2 \quad 2 \times (3)\\ \hline 5x + 10y &= 8 \quad (4) \end{aligned}$$

Substitute 1 for x in equation (4).

$$\begin{aligned} 5x + 10y &= 8 \quad (4)\\ 5(1) + 10y &= 8\\ 10y &= 3\\ y &= \tfrac{3}{10} \end{aligned}$$

Substitute 1 for x and $\frac{3}{10}$ for y in equation (1).

$$\begin{aligned} -x + 2y + 6z &= 2 \quad (1)\\ -1 + 2(\tfrac{3}{10}) + 6z &= 2\\ \tfrac{3}{5} + 6z &= 3\\ 6z &= \tfrac{12}{5}\\ z &= \tfrac{2}{5} \end{aligned}$$

Solution set: $\{(1, \frac{3}{10}, \frac{2}{5})\}$

11.
$$\begin{aligned} 2x + 5y + 2z &= 0 \quad (1)\\ 4x - 7y - 3z &= 1 \quad (2)\\ 3x - 8y - 2z &= -6 \quad (3) \end{aligned}$$

Add equations (1) and (3) to eliminate z.

$$\begin{aligned} 2x + 5y + 2z &= 0 \quad (1)\\ 3x - 8y - 2z &= -6 \quad (3)\\ \hline 5x - 3y &= -6 \quad (4) \end{aligned}$$

Multiply equation (1) by 3 and equation (2) by 2. Then add the results to eliminate z again.

$$\begin{aligned} 6x + 15y + 6z &= 0 \quad 3 \times (1)\\ 8x - 14y - 6z &= 2 \quad 2 \times (2)\\ \hline 14x + y &= 2 \quad (5) \end{aligned}$$

Multiply equation (5) by 3 then add this result to (4).

$$\begin{aligned} 5x - 3y &= -6 \quad (4)\\ 42x + 3y &= 6 \quad 3 \times (5)\\ \hline 47x &= 0\\ x &= 0 \end{aligned}$$

To find y, substitute $x = 0$ into equation (4).

$$\begin{aligned} 5x - 3y &= -6 \quad (4)\\ 5(0) - 3y &= -6\\ y &= 2 \end{aligned}$$

To find z, substitute $x = 0$ and $y = 2$ in equation (1).

$$\begin{aligned} 2x + 5y + 2z &= 0 \quad (1)\\ 2(0) + 5(2) + 2z &= 0\\ 10 + 2z &= 0\\ 2z &= -10\\ z &= -5 \end{aligned}$$

Solution set: $\{(0, 2, -5)\}$

13.
$$\begin{aligned} x + 2y + 3z &= 1 \quad (1)\\ -x - y + 3z &= 2 \quad (2)\\ -6x + y + z &= -2 \quad (3) \end{aligned}$$

Add (1) and (2).

$$y + 6z = 3 \quad (4)$$

Multiply (1) by 6 and add to (3).

$$\begin{array}{rcrcrcrl} 6x & + & 12y & + & 18z & = & 6 & 6\times(1) \\ -6x & + & y & + & z & = & -2 & (3) \\ \hline & & 13y & + & 19z & = & 4 & (5) \end{array}$$

Multiply (4) by -13 and add to (5).

$$\begin{array}{rcrcrl} -13y & - & 78z & = & -39 & -13\times(4) \\ 13y & + & 19z & = & 4 & (5) \\ \hline & & -59z & = & -35 & \\ & & z & = & \frac{35}{59} & \end{array}$$

Substitute $\frac{35}{59}$ for z in (4).

$$\begin{aligned} y+6(\tfrac{35}{59}) &= 3 \\ y+\tfrac{210}{59} &= 3 \\ y &= \tfrac{177}{59}-\tfrac{210}{59} \\ y &= -\tfrac{33}{59} \end{aligned}$$

Substitute $-\frac{33}{59}$ for y and $\frac{35}{59}$ for z in (1).

$$\begin{aligned} x+2(-\tfrac{33}{59})+3(\tfrac{35}{59}) &= 1 \\ x-\tfrac{66}{59}+\tfrac{105}{59} &= 1 \\ x &= \tfrac{59}{59}-\tfrac{39}{59} \\ x &= \tfrac{20}{59} \end{aligned}$$

Solution set: $\{(\frac{20}{59}, -\frac{33}{59}, \frac{35}{59})\}$

A calculator check reduces the probability of making any arithmetic errors and is highly recommended. The following screen shows the substitution of the solution for x, y, and z.

The following screen shows the left sides of the three original equations. The evaluation of the three expressions yields 1, 2, and -2 (the right sides of the three equations), indicating that we have found the correct solution.

```
X+2Y+3Z
                1
-X-Y+3Z
                2
-6X+Y+Z
               -2
```

15. $$\begin{array}{rcrcrcrl} 2x & - & 3y & + & 2z & = & -1 & (1) \\ x & + & 2y & + & z & = & 17 & (2) \\ & & 2y & - & z & = & 7 & (3) \end{array}$$

Multiply equation (2) by -2, and add the result to equation (1).

$$\begin{array}{rcrcrcrl} 2x & - & 3y & + & 2z & = & -1 & (1) \\ -2x & - & 4y & - & 2z & = & -34 & -2\times(2) \\ \hline & & -7y & & & = & -35 & \\ & & y & & & = & 5 & \end{array}$$

To find z, substitute 5 for y in equation (3).

$$\begin{aligned} 2y-z &= 7 \quad (3) \\ 2(5)-z &= 7 \\ 10-z &= 7 \\ -z &= -3 \\ z &= 3 \end{aligned}$$

To find x, substitute $y=5$ and $z=3$ into equation (1).

$$\begin{aligned} 2x-3y+2z &= -1 \quad (1) \\ 2x-3(5)+2(3) &= -1 \\ 2x-9 &= -1 \\ 2x &= 8 \\ x &= 4 \end{aligned}$$

Solution set: $\{(4,5,3)\}$

17. $$\begin{array}{rcrcrcrl} 4x & + & 2y & - & 3z & = & 6 & (1) \\ x & - & 4y & + & z & = & -4 & (2) \\ -x & & & + & 2z & = & 2 & (3) \end{array}$$

Equation (3) is missing y. Eliminate y again by multiplying equation (1) by 2 and adding the result to equation (2).

$$\begin{array}{rcrcrcrl} 8x & + & 4y & - & 6z & = & 12 & 2\times(1) \\ x & - & 4y & + & z & = & -4 & (2) \\ \hline 9x & & & - & 5z & = & 8 & (4) \end{array}$$

Use equations (3) and (4) to eliminate x. Multiply equation (3) by 9 and add the result to equation (4).

$$\begin{array}{rcrcrl} -9x & + & 18z & = & 18 & 9\times(3) \\ 9x & - & 5z & = & 8 & (4) \\ \hline & & 13z & = & 26 & \\ & & z & = & 2 & \end{array}$$

Substitute 2 for z in equation (3) to find x.

$$\begin{aligned} -x+2z &= 2 \quad (3) \\ -x+2(2) &= 2 \\ -x+4 &= 2 \\ -x &= -2 \\ x &= 2 \end{aligned}$$

continued

Substitute 2 for x and 2 for z in equation (2) to find y.

$$\begin{aligned} x - 4y + z &= -4 \quad (2) \\ 2 - 4y + 2 &= -4 \\ -4y + 4 &= -4 \\ -4y &= -8 \\ y &= 2 \end{aligned}$$

Solution set: $\{(2, 2, 2)\}$

19. $$\begin{aligned} -5x + 2y + z &= 5 \quad (1) \\ -3x - 2y - z &= 3 \quad (2) \\ -x + 6y &= 1 \quad (3) \end{aligned}$$

Add (1) and (2) to eliminate y and z.

$$\begin{aligned} -8x &= 8 \\ x &= -1 \end{aligned}$$

Substitute -1 for x in equation (3).

$$\begin{aligned} -x + 6y &= 1 \quad (3) \\ -(-1) + 6y &= 1 \\ 6y &= 0 \\ y &= 0 \end{aligned}$$

Substitute -1 for x and 0 for y in (1).

$$\begin{aligned} -5x + 2y + z &= 5 \quad (1) \\ -5(-1) + 2(0) + z &= 5 \\ 5 + z &= 5 \\ z &= 0 \end{aligned}$$

The solution set is $\{(-1, 0, 0)\}$.

21. $$\begin{aligned} 2x + y &= 6 \quad (1) \\ 3y - 2z &= -4 \quad (2) \\ 3x - 5z &= -7 \quad (3) \end{aligned}$$

To eliminate y, multiply equation (1) by -3 and add the result to equation (2).

$$\begin{aligned} -6x - 3y &= -18 \quad -3 \times (1) \\ 3y - 2z &= -4 \quad (2) \\ \hline -6x - 2z &= -22 \quad (4) \end{aligned}$$

Since equation (3) does not have a y-term, we can multiply equation (3) by 2 and add the result to equation (4) to eliminate x and solve for z.

$$\begin{aligned} 6x - 10z &= -14 \quad 2 \times (3) \\ -6x - 2z &= -22 \quad (4) \\ \hline -12z &= -36 \\ z &= 3 \end{aligned}$$

To find x, substitute 3 for z into equation (3).

$$\begin{aligned} 3x - 5z &= -7 \quad (3) \\ 3x - 5(3) &= -7 \\ 3x &= 8 \\ x &= \tfrac{8}{3} \end{aligned}$$

To find y, substitute 3 for z into equation (2).

$$\begin{aligned} 3y - 2z &= -4 \quad (2) \\ 3y - 2(3) &= -4 \\ 3y &= 2 \\ y &= \tfrac{2}{3} \end{aligned}$$

Solution set: $\{(\frac{8}{3}, \frac{2}{3}, 3)\}$

23. Answers will vary.

(a) One example is two perpendicular walls and the ceiling (or floor) of a room, which meet at one point.

(b) One example is two opposite walls and the floor, which have no points in common. Another is the floors of three different levels of an office building.

(c) One example is the plane through the ceiling of a house, the plane through an outside wall, and the plane through one side of a slanted roof, which all meet in one line; therefore, they intersect in infinitely many points. Another example is three pages of this book, which intersect in the spine.

25. $$\begin{aligned} 2x + 2y - 6z &= 5 \quad (1) \\ -3x + y - z &= -2 \quad (2) \\ -x - y + 3z &= 4 \quad (3) \end{aligned}$$

Multiply equation (3) by 2 and add the result to equation (1).

$$\begin{aligned} 2x + 2y - 6z &= 5 \quad (1) \\ -2x - 2y + 6z &= 8 \quad 2 \times (3) \\ \hline 0 &= 13 \quad \textit{False} \end{aligned}$$

Solution set: $\emptyset$; inconsistent system

27. $$\begin{aligned} -5x + 5y - 20z &= -40 \quad (1) \\ x - y + 4z &= 8 \quad (2) \\ 3x - 3y + 12z &= 24 \quad (3) \end{aligned}$$

Dividing equation (1) by -5 gives equation (2). Dividing equation (3) by 3 also gives equation (2). The resulting equations are the same, so the three equations are dependent.

Solution set: $\{(x, y, z) \mid x - y + 4z = 8\}$

29. $$\begin{aligned} 2x + y - z &= 6 \quad (1) \\ 4x + 2y - 2z &= 12 \quad (2) \\ -x - \tfrac{1}{2}y + \tfrac{1}{2}z &= -3 \quad (3) \end{aligned}$$

Multiplying equation (1) by 2 gives equation (2). Multiplying equation (3) by -4 also gives equation (2). The resulting equations are the same, so the three equations are dependent.

Solution set: $\{(x, y, z) \mid 2x + y - z = 6\}$

31. $x + y - 2z = 0$ (1)
$3x - y + z = 0$ (2)
$4x + 2y - z = 0$ (3)

Eliminate z by adding equations (2) and (3).

$$\begin{array}{rcll} 3x - y + z &=& 0 & (2) \\ 4x + 2y - z &=& 0 & (3) \\ \hline 7x + y &=& 0 & (4) \end{array}$$

To get another equation without z, multiply equation (2) by 2 and add the result to equation (1).

$$\begin{array}{rcll} 6x - 2y + 2z &=& 0 & 2 \times (2) \\ x + y - 2z &=& 0 & (1) \\ \hline 7x - y &=& 0 & (5) \end{array}$$

Add equations (4) and (5) to find x.

$$\begin{array}{rcll} 7x + y &=& 0 & (4) \\ 7x - y &=& 0 & (5) \\ \hline 14x &=& 0 & \\ x &=& 0 & \end{array}$$

Substitute 0 for x in equation (4) to find y.

$$\begin{aligned} 7x + y &= 0 \quad (4) \\ 7(0) + y &= 0 \\ 0 + y &= 0 \\ y &= 0 \end{aligned}$$

Substitute 0 for x and 0 for y in equation (1) to find z.

$$\begin{aligned} x + y - 2z &= 0 \quad (1) \\ 0 + 0 - 2z &= 0 \\ -2z &= 0 \\ z &= 0 \end{aligned}$$

Solution set: $\{(0, 0, 0)\}$

33. $x - 2y + \frac{1}{3}z = 4$ (1)
$3x - 6y + z = 12$ (2)
$-6x + 12y - 2z = -3$ (3)

The coefficients of z are $\frac{1}{3}$, 1, and -2. We'll multiply the equations by values that will eliminate fractions and make the coefficients of z easy to compare.

$$\begin{array}{rcrll} -6x + 12y - 2z &=& -24 & (4) & -6 \times (1) \\ -6x + 12y - 2z &=& -24 & (5) & -2 \times (2) \\ -6x + 12y - 2z &=& -3 & (3) & \end{array}$$

We can now easily see that (4) and (5) are dependent equations (they have the same graph—in fact, their graph is the same plane). Equation (3) has the same coefficients, but a different constant term, so its graph is a plane parallel to the other plane—that is, there are no points in common. Thus, the system is inconsistent and the solution set is $\emptyset$.

35. $x + 5y - 2z = -1$ (1)
$-2x + 8y + z = -4$ (2)
$3x - y + 5z = 19$ (3)

Eliminate x by adding (2) to 2 times (1).

$$\begin{array}{rcrll} 2x + 10y - 4z &=& -2 & & 2 \times (1) \\ -2x + 8y + z &=& -4 & (2) & \\ \hline 18y - 3z &=& -6 & (4) & \end{array}$$

Eliminate x by adding (3) to -3 times (1).

$$\begin{array}{rcrll} -3x - 15y + 6z &=& 3 & & -3 \times (1) \\ 3x - y + 5z &=& 19 & (3) & \\ \hline -16y + 11z &=& 22 & (5) & \end{array}$$

To eliminate z from equations (4) and (5), we could first divide equation (4) by 3 and then multiply the resulting equation by 11, or we could simply multiply equation (4) by $\frac{11}{3}$ and add it to equation (5).

$$\begin{array}{rcrll} 66y - 11z &=& -22 & & \frac{11}{3} \times (4) \\ -16y + 11z &=& 22 & (5) & \\ \hline 50y &=& 0 & & \\ y &=& 0 & & \end{array}$$

Substitute 0 for y in equation (5).

$$\begin{aligned} -16y + 11z &= 22 \quad (5) \\ -16(0) + 11z &= 22 \\ 11z &= 22 \\ z &= 2 \end{aligned}$$

Substitute 0 for y and 2 for z in equation (1).

$$\begin{aligned} x + 5y - 2z &= -1 \quad (1) \\ x + 5(0) - 2(2) &= -1 \\ x - 4 &= -1 \\ x &= 3 \end{aligned}$$

Solution set: $\{(3, 0, 2)\}$

37. $f(1) = a(1)^2 + b(1) + c$
$= a + b + c$

Since $f(1) = 128$, the first equation is

$$a + b + c = 128.$$

38. $f(1.5) = a(1.5)^2 + b(1.5) + c$
$= 2.25a + 1.5b + c$

Since $f(1.5) = 140$, the second equation is

$$2.25a + 1.5b + c = 140.$$

39. $f(3) = a(3)^2 + b(3) + c$
$= 9a + 3b + c$

Since $f(3) = 80$, the third equation is

$$9a + 3b + c = 80.$$

40. Using the three equations from Exercises 37–39, the system is

$$\begin{aligned} a + b + c &= 128 \quad (1) \\ 2.25a + 1.5b + c &= 140 \quad (2) \\ 9a + 3b + c &= 80. \quad (3) \end{aligned}$$

Multiply equation (2) by -1 and add the result to equation (1).

$$\begin{array}{rcll} a + b + c &=& 128 & (1) \\ -2.25a - 1.5b - c &=& -140 & -1 \times (2) \\ \hline -1.25a - 0.5b &=& -12 & (4) \end{array}$$

Multiply equation (3) by -1 and add the result to equation (1).

$$\begin{array}{rcll} a + b + c &=& 128 & (1) \\ -9a - 3b - c &=& -80 & -1 \times (3) \\ \hline -8a - 2b &=& 48 & (5) \end{array}$$

Use equations (4) and (5) to eliminate b. Multiply equation (4) by -4 and add the result to equation (5).

$$\begin{array}{rcll} 5a + 2b &=& 48 & -4 \times (4) \\ -8a - 2b &=& 48 & (5) \\ \hline -3a &=& 96 & \\ a &=& -32 & \end{array}$$

To find b, substitute $a = -32$ into equation (5).

$$\begin{aligned} -8a - 2b &= 48 \quad (5) \\ -8(-32) - 2b &= 48 \\ 256 - 2b &= 48 \\ -2b &= -208 \\ b &= 104 \end{aligned}$$

To find c, substitute $a = -32$ and $b = 104$ into equation (1).

$$\begin{aligned} a + b + c &= 128 \quad (1) \\ -32 + 104 + c &= 128 \\ 72 + c &= 128 \\ c &= 56 \end{aligned}$$

Solution set: $\{(-32, 104, 56)\}$

41. If $(a, b, c) = (-32, 104, 56)$, then
$f(x) = -32x^2 + 104x + 56.$

42. In the function f written in Exercise 41, the height of the projectile is a function of the time elapsed since it was projected.

43. $f(x) = -32x^2 + 104x + 56$
$f(0) = -32(0)^2 + 104(0) + 56$
$= 56$
The initial height is 56 ft.

44. $f(1.625) = -32(1.625)^2 + 104(1.625) + 56$
$= -84.5 + 169 + 56$
$= 140.5$
The maximum height is 140.5 ft.

5.3 Applications of Systems of Linear Equations

5.3 Margin Exercises

1. *Step 2*
Let L = the length and W = the width of the foundation of the rectangular house.

Step 3
Using the fact that the length is 6 m more than the width and the perimeter formula, we get the following system.

$$\begin{aligned} L &= 6 + W \quad (1) \\ 48 &= 2L + 2W \quad (2) \end{aligned}$$

Step 4
Substitute $6 + W$ for L in equation (2).

$$\begin{aligned} 48 &= 2L + 2W \\ 48 &= 2(6 + W) + 2W \\ 48 &= 12 + 2W + 2W \\ 36 &= 4W \\ 9 &= W \end{aligned}$$

Since $L = 6 + W$ and $W = 9$,

$$L = 6 + 9 = 15.$$

Step 5
The length of the house is 15 m and the width is 9 m.

Step 6
Check: 15 m is 6 m more than 9 m and $2(15) + 2(9) = 48$.

2. *Step 2*
Let x = the average cost of a baseball ticket, and y = the average cost of a football ticket.

Step 3
From the given information,

$$\begin{aligned} 3x + 2y &= 181.41 \quad (1) \\ 2x + y &= 101.29. \quad (2) \end{aligned}$$

Step 4
Multiply equation (2) by -2 and add to equation (1).

$$\begin{array}{rcll} 3x + 2y &=& 181.41 & (1) \\ -4x - 2y &=& -202.58 & -2 \times (2) \\ \hline -x &=& -21.17 & \\ x &=& 21.17 & \end{array}$$

Let $x = 21.17$ in equation (2).

$$2(21.17) + y = 101.29$$
$$42.34 + y = 101.29$$
$$y = 58.95$$

Step 5
The average cost of a baseball ticket is \$21.17 and the average cost of a football ticket is \$58.95.

Step 6
Check: $3(21.17) + 2(58.95) = 181.41$ and $2(21.17) + 58.95 = 101.29$.

3. **(a)** *Step 2*
Let x represent the number of pounds of the \$4 per lb coffee and y represent the number of pounds of the \$8 per lb coffee.

Number of Pounds	Price per Pound	Value of Coffee
x	\$4	$4x$
y	\$8	$8y$
50	\$5.60	$5.6(50) = 280$

Step 3
Write the system from the columns.

$$x + y = 50 \qquad (1)$$
$$4x + 8y = 280 \qquad (2)$$

Step 4
To eliminate x, multiply equation (1) by -4 and add the result to equation (2).

$$\begin{array}{rcrl} -4x - 4y &=& -200 & -4 \times (1) \\ 4x + 8y &=& 280 & (2) \\ \hline 4y &=& 80 & \\ y &=& 20 & \end{array}$$

Since $y = 20$ and $x + y = 50$, $x = 30$.

Step 5
To mix the coffee, 30 lb of \$4 per lb coffee and 20 lb of \$8 per lb coffee should be used.

Step 6
Check: $30 + 20 = 50$ and $4(30) + 8(20) = 280$.

(b) *Step 2*
Let $x =$ the amount of 40% alcohol solution. and $y =$ the amount of 80% alcohol solution.

Liters of Solution	Percent (as a Decimal)	Liters of Pure Alcohol
x	$40\% = 0.40$	$0.4x$
y	$80\% = 0.80$	$0.8y$
200	$50\% = 0.50$	$0.50(200) = 100$

Step 3
Write the system from the columns.

$$x + y = 200 \qquad (1)$$
$$0.4x + 0.8y = 100 \qquad (2)$$

Step 4
To eliminate x, multiply equation (1) by -4 and equation (2) by 10. Then add the results.

$$\begin{array}{rcrl} -4x - 4y &=& -800 & -4 \times (1) \\ 4x + 8y &=& 1000 & 10 \times (2) \\ \hline 4y &=& 200 & \\ y &=& 50 & \end{array}$$

Since $y = 50$ and $x + y = 200$, $x = 150$.

Step 5
The mixture should contain 150 L of 40% solution and 50 L of 80% solution.

Step 6
Check: $150 + 50 = 200$ and $0.4(150) + 0.8(50) = 100$.

4. *Step 2*
Let $x =$ the train's speed and $y =$ the truck's speed.

Make a table. Use $t = \frac{d}{r}$.

	Distance	Rate	Time
Train	600	x	$\frac{600}{x}$
Truck	520	y	$\frac{520}{y}$

Step 3
Since the times are the same,

$$\frac{600}{x} = \frac{520}{y}$$
$$600y = 520x \qquad \textit{Multiply by xy.}$$
$$-520x + 600y = 0. \qquad (1)$$

Since the train's average speed, x, is 8 mph faster than the truck's speed, y,

$$x = y + 8. \qquad (2)$$

Step 4
Substitute $y + 8$ for x in equation (1) to find y.

$$-520x + 600y = 0 \qquad (1)$$
$$-520(y + 8) + 600y = 0$$
$$-520y - 4160 + 600y = 0$$
$$80y = 4160$$
$$y = 52$$

Since $y = 52$ and $x = y + 8$, $x = 60$.

Step 5
The train's speed is 60 mph, and the truck's speed is 52 mph.

Step 6
Check: 60 mph is 8 mph faster than 52 mph. It would take the train 10 hours to travel 600 miles at 60 mph, which is the same amount of time it would take the truck to travel 520 miles at 52 mph.

5.
$$\begin{aligned} x - 3y &= 0 && (1) \\ x - z &= 5 && (2) \\ 295x + 299y + 579z &= 8789 && (3) \end{aligned}$$

Eliminate z by adding equation (3) to 579 times equation (2).

$$\begin{array}{rcrcrcrl} 579x & & & - & 579z & = & 2895 & 579 \times (2) \\ 295x & + & 299y & + & 579z & = & 8789 & (3) \\ \hline 874x & + & 299y & & & = & 11{,}684 & (4) \end{array}$$

Eliminate x by adding equation (4) to -874 times equation (1).

$$\begin{array}{rcrcrl} -874x & + & 2622y & = & 0 & -874 \times (1) \\ 874x & + & 299y & = & 11{,}684 & (4) \\ \hline & & 2921y & = & 11{,}684 & \\ & & y & = & 4 & \end{array}$$

Substitute 4 for y in equation (1) and solve for x.

$$\begin{aligned} x - 3(4) &= 0 \\ x &= 12 \end{aligned}$$

Substitute 12 for x in equation (2) and solve for z.

$$\begin{aligned} 12 - z &= 5 \\ 7 &= z \end{aligned}$$

Solution set: $\{(12, 4, 7)\}$

6. Let $x =$ the number of bottles of Felice, at $8,
$y =$ the number of bottles of Vivid, at $15, and
$z =$ the number of bottles of Joy, at $32.

There are 10 more bottles of Felice than Vivid, so

$$x = y + 10. \qquad (1)$$

There are 3 fewer bottles of Joy than Vivid, so

$$z = y - 3. \qquad (2)$$

The total value is $589, so

$$8x + 15y + 32z = 589. \qquad (3)$$

Substitute $y + 10$ for x and $y - 3$ for z in equation (3) to find y.

$$\begin{aligned} 8(y + 10) + 15y + 32(y - 3) &= 589 \\ 8y + 80 + 15y + 32y - 96 &= 589 \\ 55y - 16 &= 589 \\ 55y &= 605 \\ y &= 11 \end{aligned}$$

Since $y = 11$,

$$x = y + 10 = 21 \quad \text{and} \quad z = y - 3 = 8.$$

There are 21 bottles of Felice, 11 of Vivid, and 8 of Joy.

7. Let $x =$ the number of tons of newsprint,
$y =$ the number of tons of bond, and
$z =$ the number of tons of copy machine paper.

The amount of recycled paper used to make all three items is 4200 tons, so

$$3x + 2y + 2z = 4200. \qquad (1)$$

The amount of wood pulp used to make all three items is 5800 tons, so

$$x + 4y + 3z = 5800. \qquad (2)$$

The amount of rags used to make all three items is 3900 tons, so

$$3y + 2z = 3900. \qquad (3)$$

Solve the system of equations (1), (2), and (3). To eliminate z, multiply equation (3) by -1 and add the result to equation (1).

$$\begin{array}{rcrcrcrl} & & -3y & - & 2z & = & -3900 & -1 \times (3) \\ 3x & + & 2y & + & 2z & = & 4200 & (1) \\ \hline 3x & - & y & & & = & 300 & (4) \end{array}$$

To eliminate z again, multiply equation (2) by -2 and equation (3) by 3 and add the results.

$$\begin{array}{rcrcrcrl} -2x & - & 8y & - & 6z & = & -11{,}600 & -2 \times (2) \\ & & 9y & + & 6z & = & 11{,}700 & 3 \times (3) \\ \hline -2x & + & y & & & = & 100 & (5) \end{array}$$

Add equations (4) and (5) to eliminate y.

$$\begin{array}{rcrcrl} 3x & - & y & = & 300 & (4) \\ -2x & + & y & = & 100 & (5) \\ \hline x & & & = & 400 & \end{array}$$

Substitute 400 for x in equation (4).

$$\begin{aligned} 3x - y &= 300 && (4) \\ 3(400) - y &= 300 \\ 1200 - y &= 300 \\ 900 &= y \end{aligned}$$

Substitute 900 for y in equation (3).

$$\begin{aligned} 3y + 2z &= 3900 && (3) \\ 3(900) + 2z &= 3900 \\ 2700 + 2z &= 3900 \\ 2z &= 1200 \\ z &= 600 \end{aligned}$$

The paper mill can make 400 tons of newsprint, 900 tons of bond, and 600 tons of copy machine paper.

5.3 Section Exercises

1. *Step 2*
Let $x =$ the number of games that the Indians won and let $y =$ the number of games that they lost.

Step 3
They played 162 games, so

$$x + y = 162. \quad (1)$$

They won 30 more games than they lost, so

$$x = 30 + y. \quad (2)$$

Step 4
Substitute $30 + y$ for x in (1).

$$\begin{aligned} (30 + y) + y &= 162 \\ 30 + 2y &= 162 \\ 2y &= 132 \\ y &= 66 \end{aligned}$$

Substitute 66 for y in (2).

$$x = 30 + 66 = 96$$

Step 5
The Indians' win-loss record was 96 wins and 66 losses.

Step 6
96 is 30 more than 66 and the sum of 96 and 66 is 162.

3. Let $W =$ the width of the tennis court and $L =$ the length of the court.

Since the length is 42 ft more than the width,

$$L = W + 42. \quad (1)$$

The perimeter of a rectangle is given by

$$2W + 2L = P.$$

With perimeter $P = 228$ ft,

$$2W + 2L = 228. \quad (2)$$

Substitute $W + 42$ for L in equation (2).

$$\begin{aligned} 2W + 2(W + 42) &= 228 \\ 2W + 2W + 84 &= 228 \\ 4W &= 144 \\ W &= 36 \end{aligned}$$

Substitute $W = 36$ into equation (1).

$$\begin{aligned} L &= W + 42 \qquad (1) \\ L &= 36 + 42 \\ &= 78 \end{aligned}$$

The length is 78 ft; the width is 36 ft.

5. Let $x =$ the revenue for Wal-Mart and $y =$ the revenue for ExxonMobil (both in billions of dollars).

The total revenue was \$656 billion.

$$x + y = 656 \quad (1)$$

ExxonMobil's revenue was \$24 billion more than that of Wal-Mart.

$$y = x + 24 \quad (2)$$

Substitute $x + 24$ for y in equation (1).

$$\begin{aligned} x + (x + 24) &= 656 \\ 2x + 24 &= 656 \\ 2x &= 632 \\ x &= 316 \end{aligned}$$

Substitute 316 for x in equation (2).

$$y = x + 24 = 316 + 24 = 340$$

Wal-Mart's revenue was \$316 billion and ExxonMobil's revenue was \$340 billion.

7. From the figure in the text, the angles marked y and $3x + 10$ are supplementary, so

$$(3x + 10) + y = 180. \quad (1)$$

Also, the angles x and y are complementary, so

$$x + y = 90. \quad (2)$$

Solve equation (2) for y to get

$$y = 90 - x. \quad (3)$$

Substitute $90 - x$ for y in equation (1).

$$\begin{aligned} (3x + 10) + (90 - x) &= 180 \\ 2x + 100 &= 180 \\ 2x &= 80 \\ x &= 40 \end{aligned}$$

Substitute $x = 40$ into equation (3) to get

$$y = 90 - x = 90 - 40 = 50.$$

The angles measure 40° and 50°.

9. Let $x =$ the hockey FCI and $y =$ the basketball FCI.

The sum is \$514.69, so

$$x + y = 514.69. \quad (1)$$

The hockey FCI was \$20.05 less than the basketball FCI, so

$$x = y - 20.05. \quad (2)$$

continued

From (2), substitute $y - 20.05$ for x in (1).

$$\begin{aligned}(y - 20.05) + y &= 514.69\\ 2y - 20.05 &= 514.69\\ 2y &= 534.74\\ y &= 267.37\end{aligned}$$

From (2),

$$x = y - 20.05 = 267.37 - 20.05 = 247.32.$$

The hockey FCI was \$247.32 and the basketball FCI was \$267.37.

11. Let $x =$ the cost of a single Junior Roast Beef sandwich, and $y =$ the cost of a single Big Montana sandwich.

15 Junior Roast Beef sandwiches and 10 Big Montana sandwiches cost \$75.25.

$$15x + 10y = 75.25 \quad (1)$$

30 Junior Roast Beef sandwiches and 5 Big Montana sandwiches cost \$84.65.

$$30x + 5y = 84.65 \quad (2)$$

Multiply equation (1) by -2 and add to equation (2).

$$\begin{array}{rcrl} -30x - 20y &=& -150.50 & -2 \times (1)\\ 30x + 5y &=& 84.65 & (2)\\ \hline -15y &=& -65.85 & \\ y &=& 4.39 & \end{array}$$

Substitute 4.39 for y in equation (1).

$$\begin{aligned}15x + 10y &= 75.25\\ 15x + 10(4.39) &= 75.25\\ 15x + 43.90 &= 75.25\\ 15x &= 31.35\\ x &= 2.09\end{aligned}$$

A single Junior Roast Beef sandwich costs \$2.09 and a single Big Montana sandwich costs \$4.39.

13. Use the formula (rate of percent) • (base amount) = amount (percentage) of pure acid to compute parts (a) – (d).

(a) $0.10(60) = 6$ oz

(b) $0.25(60) = 15$ oz

(c) $0.40(60) = 24$ oz

(d) $0.50(60) = 30$ oz

15. The cost is the price per pound, \$1.29, times the number of pounds, x, or $\$1.29x$.

17. Let $x =$ the amount of 25% alcohol solution and $y =$ the amount of 35% alcohol solution.

Make a table. The amount of solution times the percent gives the amount of pure alcohol in the third column.

Gallons of Solution	Percent (as a Decimal)	Gallons of Pure Alcohol
x	$25\% = 0.25$	$0.25x$
y	$35\% = 0.35$	$0.35y$
20	$32\% = 0.32$	$0.32(20) = 6.4$

The third row gives the total amounts of solution and pure alcohol. From the columns in the table, write a system of equations.

$$\begin{aligned}x + y &= 20 \quad (1)\\ 0.25x + 0.35y &= 6.4 \quad (2)\end{aligned}$$

Solve the system. Multiply equation (1) by -25 and equation (2) by 100. Then add the results.

$$\begin{array}{rcrl} -25x - 25y &=& -500 & -25 \times (1)\\ 25x + 35y &=& 640 & 100 \times (2)\\ \hline 10y &=& 140 & \\ y &=& 14 & \end{array}$$

Substitute $y = 14$ into equation (1).

$$\begin{aligned}x + y &= 20 \quad (1)\\ x + 14 &= 20\\ x &= 6\end{aligned}$$

Mix 6 gal of 25% solution and 14 gal of 35% solution.

19. Let $x =$ the amount of pure acid and $y =$ the amount of 10% acid.

Make a table.

Liters of Solution	Percent (as a Decimal)	Liters of Pure Acid
x	$100\% = 1$	$1.00x = x$
y	$10\% = 0.10$	$0.10y$
54	$20\% = 0.20$	$0.20(54) = 10.8$

Solve the following system.

$$\begin{aligned}x + y &= 54 \quad (1)\\ x + 0.10y &= 10.8 \quad (2)\end{aligned}$$

Multiply equation (2) by 10 to clear the decimals.

$$10x + y = 108 \quad (3)$$

To eliminate y, multiply equation (1) by -1 and add the result to equation (3).

$$\begin{array}{rcrl} -x - y &=& -54 & -1 \times (1) \\ 10x + y &=& 108 & (3) \\ \hline 9x &=& 54 & \\ x &=& 6 & \end{array}$$

Substitute $x = 6$ into equation (1).

$$\begin{aligned} x + y &= 54 \quad (1) \\ 6 + y &= 54 \\ y &= 48 \end{aligned}$$

Use 6 L of pure acid and 48 L of 10% acid.

21. Complete the table.

	Number of Kilograms	Price per Kilogram	Value
Nuts	x	2.50	$2.50x$
Cereal	y	1.00	$1.00y$
Mixture	30	1.70	$1.70(30)$

From the "Number of Kilograms" column,

$$x + y = 30. \quad (1)$$

From the "Value" column,

$$2.50x + 1.00y = 1.70(30) = 51. \quad (2)$$

Solve the system.

$$\begin{array}{rcrl} -10x - 10y &=& -300 & -10 \times (1) \\ 25x + 10y &=& 510 & 10 \times (2) \\ \hline 15x &=& 210 & \\ x &=& 14 & \end{array}$$

From (1), $14 + y = 30$, so $y = 16$.
The party mix should be made from 14 kg of nuts and 16 kg of cereal.

23. From the "Principal" column in the text,

$$x + y = 3000. \quad (1)$$

From the "Interest" column in the text,

$$0.02x + 0.04y = 100. \quad (2)$$

Multiply equation (2) by 100 to clear the decimals.

$$2x + 4y = 10{,}000 \quad (3)$$

To eliminate x, multiply equation (1) by -2 and add the result to equation (3).

$$\begin{array}{rcrl} -2x - 2y &=& -6000 & -2 \times (1) \\ 2x + 4y &=& 10{,}000 & (3) \\ \hline 2y &=& 4000 & \\ y &=& 2000 & \end{array}$$

From (1), $x + 2000 = 3000$, so $x = 1000$.
\$1000 is invested at 2%, and \$2000 is invested at 4%.

25. **(a)** The speed of the boat going upstream is *decreased* by the speed of the current, so it is $(10 - x)$ mph.

(b) The speed of the boat going downstream is *increased* by the speed of the current, so it is $(10 + x)$ mph.

27. Let $x =$ speed of the bicycle and $y =$ the speed of the scooter.

Make a table. Use $t = \frac{d}{r}$.

	r	t	d
Bicycle	x	$\frac{8}{x}$	8
Scooter	y	$\frac{20}{y}$	20

The times are equal, so

$$\frac{8}{x} = \frac{20}{y}. \quad (1)$$

The speed of the scooter is 5 mph more than twice the speed of the bicycle, so

$$y = 2x + 5. \quad (2)$$

Multiply (1) by xy (or use cross products).

$$8y = 20x \quad (3)$$

From (2), substitute $2x + 5$ for y in (3).

$$\begin{aligned} 8(2x + 5) &= 20x \\ 16x + 40 &= 20x \\ 40 &= 4x \\ 10 &= x \end{aligned}$$

From (2), $y = 2(10) + 5 = 25$.
The speed of the scooter is 25 mph, and the speed of the bicycle is 10 mph.

29. Let $x =$ the speed of the boat in still water and $y =$ the speed of the current.

Furthermore,

$$\text{rate upstream} = x - y$$
$$\text{and} \quad \text{rate downstream} = x + y.$$

Use these rates and the information in the problem to make a table.

	r	t	d
Upstream	$x - y$	2	36
Downstream	$x + y$	1.5	36

From the table, use the formula $d = rt$ to write a system of equations.

$$\begin{aligned} 36 &= 2(x - y) \\ 36 &= 1.5(x + y) \end{aligned}$$

continued

Remove the parentheses and move the variables to the left side.

$$\begin{aligned} 2x - 2y &= 36 \quad (1) \\ 1.5x + 1.5y &= 36 \quad (2) \end{aligned}$$

Solve the system. Multiply equation (1) by -3 and equation (2) by 4. Then add the results.

$$\begin{array}{rcrl} -6x + 6y &=& -108 & -3 \times (1) \\ 6x + 6y &=& 144 & 4 \times (2) \\ \hline 12y &=& 36 & \\ y &=& 3 & \end{array}$$

Substitute $y = 3$ into equation (1).

$$\begin{aligned} 2x - 2y &= 36 \quad (1) \\ 2x - 2(3) &= 36 \\ 2x &= 42 \\ x &= 21 \end{aligned}$$

The speed of the boat is 21 mph, and the speed of the current is 3 mph.

31. Let $x =$ the number of pounds of the \$0.75-per-lb candy and $y =$ the number of pounds of the \$1.25-per-lb candy.

Make a table.

	Price per Pound	Number of Pounds	Value
Less Expensive Candy	\$0.75	x	$\$0.75x$
More Expensive Candy	\$1.25	y	$\$1.25y$
Mixture	\$0.96	9	$\$0.96(9) = \$8.64$

From the "Number of Pounds" column,

$$x + y = 9. \quad (1)$$

From the "Value" column,

$$0.75x + 1.25y = 8.64. \quad (2)$$

Solve the system.

$$\begin{array}{rcrl} -75x - 75y &=& -675 & -75 \times (1) \\ 75x + 125y &=& 864 & 100 \times (2) \\ \hline 50y &=& 189 & \\ y &=& 3.78 & \end{array}$$

From (1), $x + 3.78 = 9$, so $x = 5.22$.
Mix 5.22 pounds of the \$0.75-per-lb candy with 3.78 pounds of the \$1.25-per-lb candy to obtain 9 pounds of a mixture that sells for \$0.96 per pound.

33. Let $x =$ the number of general admission tickets and $y =$ the number of student tickets.

Make a table.

Ticket	Number	Value of Tickets
General	x	$5 \cdot x = 5x$
Student	y	$4 \cdot y = 4y$
Totals	184	812

Solve the system.

$$\begin{aligned} x + y &= 184 \quad (1) \\ 5x + 4y &= 812 \quad (2) \end{aligned}$$

To eliminate y, multiply equation (1) by -4 and add the result to equation (2).

$$\begin{array}{rcrl} -4x - 4y &=& -736 & -4 \times (1) \\ 5x + 4y &=& 812 & (2) \\ \hline x &=& 76 & \end{array}$$

From (1), $76 + y = 184$, so $y = 108$.

76 general admission tickets and 108 student tickets were sold.

35. Let $x =$ the price for a citron and let $y =$ the price for a wood apple.

"9 citrons and 7 fragrant wood apples is 107" gives us

$$9x + 7y = 107. \quad (1)$$

"7 citrons and 9 fragrant wood apples is 101" gives us

$$7x + 9y = 101. \quad (2)$$

Multiply equation (1) by -7 and equation (2) by 9. Then add.

$$\begin{array}{rcrl} -63x - 49y &=& -749 & -7 \times (1) \\ 63x + 81y &=& 909 & 9 \times (2) \\ \hline 32y &=& 160 & \\ y &=& 5 & \end{array}$$

Substitute 5 for y in equation (1).

$$\begin{aligned} 9x + 7(5) &= 107 \\ 9x + 35 &= 107 \\ 9x &= 72 \\ x &= 8 \end{aligned}$$

The prices are 8 for a citron and 5 for a wood apple.

37. Let $x =$ the measure of one angle,
$y =$ the measure of another angle,
and $z =$ the measure of the last angle.

Two equations are given, so

$$z = x + 10$$
$$\text{or} \quad -x + z = 10 \qquad (1)$$
$$\text{and} \quad x + y = 100. \qquad (2)$$

Since the sum of the measures of the angles of a triangle is 180°, the third equation of the system is

$$x + y + z = 180. \qquad (3)$$

Equation (1) is missing y. To eliminate y again, multiply equation (2) by -1 and add the result to equation (3).

$$\begin{array}{rcrl} -x - y & = & -100 & -1 \times (2) \\ x + y + z & = & 180 & (3) \\ \hline z & = & 80 & \end{array}$$

Since $z = 80$,

$$\begin{aligned} -x + z &= 10 \qquad (1) \\ -x + 80 &= 10 \\ -x &= -70 \\ x &= 70. \end{aligned}$$

From (2), $70 + y = 100$, so $y = 30$.
The measures of the angles are 70°, 30°, and 80°.

39. Let $x =$ the measure of the first angle,
$y =$ the measure of the second angle, and
$z =$ the measure of the third angle.

The sum of the angles in a triangle equals 180°, so

$$x + y + z = 180. \qquad (1)$$

The measure of the second angle is 10° more than 3 times that of the first angle, so

$$y = 3x + 10. \qquad (2)$$

The third angle is equal to the sum of the other two, so

$$z = x + y. \qquad (3)$$

Solve the system. Substitute z for $x + y$ in equation (1).

$$\begin{aligned} (x + y) + z &= 180 \qquad (1) \\ z + z &= 180 \\ 2z &= 180 \\ z &= 90 \end{aligned}$$

Substitute 90 for z and $3x + 10$ for y in equation (3).

$$\begin{aligned} z &= x + y \qquad (3) \\ 90 &= x + (3x + 10) \\ 80 &= 4x \\ 20 &= x \end{aligned}$$

Substitute $x = 20$ and $z = 90$ into equation (3).

$$\begin{aligned} z &= x + y \qquad (3) \\ 90 &= 20 + y \\ 70 &= y \end{aligned}$$

The three angles have measures of 20°, 70°, and 90°.

41. Let $x =$ the length of the longest side,
$y =$ the length of the middle side,
and $z =$ the length of the shortest side.

Perimeter is the sum of the measures of the sides, so

$$x + y + z = 70. \qquad (1)$$

The longest side is 4 cm less than the sum of the other sides, so

$$x = y + z - 4$$
$$\text{or} \quad x - y - z = -4. \qquad (2)$$

Twice the shortest side is 9 cm less than the longest side, so

$$2z = x - 9$$
$$\text{or} \quad -x + 2z = -9 \qquad (3)$$

Add equations (1) and (2) to eliminate y and z.

$$\begin{array}{rcrl} x + y + z & = & 70 & (1) \\ x - y - z & = & -4 & (2) \\ \hline 2x & = & 66 & \\ x & = & 33 & \end{array}$$

Substitute 33 for x in (3).

$$\begin{aligned} -x + 2z &= -9 \qquad (3) \\ -33 + 2z &= -9 \\ 2z &= 24 \\ z &= 12 \end{aligned}$$

Substitute 33 for x and 12 for z in (1).

$$\begin{aligned} x + y + z &= 70 \qquad (1) \\ 33 + y + 12 &= 70 \\ y + 45 &= 70 \\ y &= 25 \end{aligned}$$

The shortest side is 12 cm long, the middle side is 25 cm long, and the longest side is 33 cm long.

43. Let $x =$ the number of Republicans, $y =$ the number of Democrats, and $z =$ the number of Independents. There are 100 Americans in the sample, so

$$x + y + z = 100. \quad (1)$$

10 more Americans identify themselves as Independents than Republicans, so

$$z = x + 10. \quad (2)$$

Six fewer Americans identify themselves as Republicans than Democrats, so

$$y - 6 = x. \quad (3)$$

Substitute $x + 6$ for y and $x + 10$ for z in (1).

$$\begin{aligned} x + (x + 6) + (x + 10) &= 100 \\ 3x + 16 &= 100 \\ 3x &= 84 \\ x &= 28 \end{aligned}$$

From (3), $y - 6 = 28$, so $y = 34$.
From (2), $z = 28 + 10 = 38$.

There are 28 Republicans, 34 Democrats, and 38 Independents.

45. Let $x =$ the number of \$14 tickets, $y =$ the number of \$20 tickets, and $z =$ the number of VIP \$50 tickets.

Five times as many \$14 tickets have been sold as VIP tickets, so

$$x = 5z. \quad (1)$$

The number of \$14 tickets is 15 more than the sum of the number of \$20 tickets and the number of VIP tickets, so

$$x = 15 + y + z. \quad (2)$$

Since x is in terms of z in (1), we'll substitute $5z$ for x in (2) and then get y in terms of z.

$$\begin{aligned} 5z &= 15 + y + z \\ 4z - 15 &= y \qquad (3) \end{aligned}$$

Sales of these tickets totaled \$11,700.

$$\begin{aligned} 14x + 20y + 50z &= 11{,}700 \\ 14(5z) + 20(4z - 15) + 50z &= 11{,}700 \\ 70z + 80z - 300 + 50z &= 11{,}700 \\ 200z &= 12{,}000 \\ z &= 60 \end{aligned}$$

From (1), $x = 5(60) = 300$.
From (3), $y = 4(60) - 15 = 225$.

There were 300 \$14 tickets, 225 \$20 tickets, and 60 \$50 tickets sold.

47. Let $x =$ the number of T-shirts shipped to bookstore A,
$y =$ the number of T-shirts shipped to bookstore B,
and $z =$ the number of T-shirts shipped to bookstore C.

Twice as many T-shirts were shipped to bookstore B as to bookstore A, so

$$y = 2x. \quad (1)$$

The number shipped to bookstore C was 40 less than the sum of the numbers shipped to the other two bookstores, so

$$z = x + y - 40. \quad (2)$$

Substitute $2x$ for y [from (1)] into equation (2) to get z in terms of x.

$$\begin{aligned} z &= x + y - 40 \qquad (2) \\ z &= x + (2x) - 40 \\ z &= 3x - 40 \qquad (3) \end{aligned}$$

The total number of T-shirts shipped was 800, so

$$x + y + z = 800. \quad (4)$$

Substitute $2x$ for y and $3x - 40$ for z in (4).

$$\begin{aligned} x + (2x) + (3x - 40) &= 800 \\ 6x - 40 &= 800 \\ 6x &= 840 \\ x &= 140 \end{aligned}$$

From (1), $y = 2(140) = 280$.
From (3), $z = 3(140) - 40 = 380$.
The number of T-shirts shipped to bookstores A, B, and C was 140, 280, and 380, respectively.

49. *Step 2*
Let $x =$ the number of wins,
$y =$ the number of losses,
and $z =$ the number of ties.

Step 3
They played 82 games, so

$$x + y + z = 82. \quad (1)$$

Their wins and losses totaled 71, so

$$x + y = 71. \quad (2)$$

They tied 14 fewer games than they lost, so

$$z = y - 14. \quad (3)$$

Step 4
Multiply (2) by -1 and add to (1).

$$\begin{array}{rcll} x + y + z &=& 82 & (1) \\ -x - y &=& -71 & -1\times(2) \\ \hline z &=& 11 & \end{array}$$

Substitute 11 for z in (3).

$$\begin{aligned} 11 &= y - 14 \\ 25 &= y \end{aligned}$$

Substitute 25 for y in (2).

$$\begin{aligned} x + 25 &= 71 \\ x &= 46 \end{aligned}$$

Step 5
The Flames won 46 games, lost 25 games, and tied 11 games.

Step 6
Adding 46, 25, and 11 gives 82 total games. The wins and losses add up to 71, and there were 14 fewer ties than losses. The solution is correct.

5.4 Solving Systems of Linear Equations by Matrix Methods

5.4 Margin Exercises

1. $x - 2y = 9$
$3x + y = 13$

Write the augmented matrix for this system.

$$\left[\begin{array}{rr|r} 1 & -2 & 9 \\ 3 & 1 & 13 \end{array}\right]$$

$$\left[\begin{array}{rr|r} 1 & -2 & 9 \\ 0 & 7 & -14 \end{array}\right] \quad -3R_1 + R_2$$

$$\left[\begin{array}{rr|r} 1 & -2 & 9 \\ 0 & 1 & -2 \end{array}\right] \quad \tfrac{1}{7}R_2$$

This matrix gives the system

$$\begin{aligned} x - 2y &= 9 \\ y &= -2. \end{aligned}$$

Substitute -2 for y in the first equation.

$$\begin{aligned} x - 2(-2) &= 9 \\ x + 4 &= 9 \\ x &= 5 \end{aligned}$$

Solution set: $\{(5, -2)\}$

2. $2x - y + z = 7$ $x - 3y - z = 7$
$x - 3y - z = 7$ **OR** $2x - y + z = 7$
$-x + y - 5z = -9$ $-x + y - 5z = -9$

Write the augmented matrix for this system.

$$\left[\begin{array}{rrr|r} 1 & -3 & -1 & 7 \\ 2 & -1 & 1 & 7 \\ -1 & 1 & -5 & -9 \end{array}\right]$$

$$\left[\begin{array}{rrr|r} 1 & -3 & -1 & 7 \\ 0 & 5 & 3 & -7 \\ -1 & 1 & -5 & -9 \end{array}\right] \quad -2R_1 + R_2$$

$$\left[\begin{array}{rrr|r} 1 & -3 & -1 & 7 \\ 0 & 5 & 3 & -7 \\ 0 & -2 & -6 & -2 \end{array}\right] \quad R_1 + R_3$$

$$\left[\begin{array}{rrr|r} 1 & -3 & -1 & 7 \\ 0 & 1 & \frac{3}{5} & -\frac{7}{5} \\ 0 & -2 & -6 & -2 \end{array}\right] \quad \tfrac{1}{5}R_2$$

$$\left[\begin{array}{rrr|r} 1 & -3 & -1 & 7 \\ 0 & 1 & \frac{3}{5} & -\frac{7}{5} \\ 0 & 0 & -\frac{24}{5} & -\frac{24}{5} \end{array}\right] \quad 2R_2 + R_3$$

$$\left[\begin{array}{rrr|r} 1 & -3 & -1 & 7 \\ 0 & 1 & \frac{3}{5} & -\frac{7}{5} \\ 0 & 0 & 1 & 1 \end{array}\right] \quad -\tfrac{5}{24}R_3$$

This matrix gives the system

$$\begin{aligned} x - 3y - z &= 7 \\ y + \tfrac{3}{5}z &= -\tfrac{7}{5} \\ z &= 1. \end{aligned}$$

Substitute 1 for z in the second equation.

$$\begin{aligned} y + \tfrac{3}{5}(1) &= -\tfrac{7}{5} \\ y &= -2 \end{aligned}$$

Substitute -2 for y and 1 for z in the first equation.

$$\begin{aligned} x - 3(-2) - 1 &= 7 \\ x + 5 &= 7 \\ x &= 2 \end{aligned}$$

Solution set: $\{(2, -2, 1)\}$

3. **(a)** $x - y = 2$
$-2x + 2y = 2$

Write the augmented matrix for this system.

$$\left[\begin{array}{rr|r} 1 & -1 & 2 \\ -2 & 2 & 2 \end{array}\right]$$

$$\left[\begin{array}{rr|r} 1 & -1 & 2 \\ 0 & 0 & 6 \end{array}\right] \quad 2R_1 + R_2$$

This matrix gives the system

$$\begin{aligned} x - y &= 2 \\ 0 &= 6. \quad \textit{False} \end{aligned}$$

The false statement indicates that the system is inconsistent and has no solution.

Solution set: $\emptyset$

(b) $x - y = 2$
$-2x + 2y = -4$

Write the augmented matrix for this system.

$$\left[\begin{array}{rr|r} 1 & -1 & 2 \\ -2 & 2 & -4 \end{array}\right]$$

$$\left[\begin{array}{rr|r} 1 & -1 & 2 \\ 0 & 0 & 0 \end{array}\right] \quad 2R_1 + R_2$$

This matrix gives the system

$$x - y = 2$$
$$0 = 0. \quad \textit{True}$$

The true statement indicates that the system has dependent equations.

Solution set: $\{(x, y) \mid x - y = 2\}$

5.4 Section Exercises

1. $\begin{bmatrix} -2 & 3 & 1 \\ 0 & 5 & -3 \\ 1 & 4 & 8 \end{bmatrix}$

(a) The elements of the second row are 0, 5, and −3.

(b) The elements of the third column are 1, −3, and 8.

(c) The matrix is square since the number of rows (three) is the same as the number of columns.

(d) The matrix obtained by interchanging the first and third rows is

$$\begin{bmatrix} 1 & 4 & 8 \\ 0 & 5 & -3 \\ -2 & 3 & 1 \end{bmatrix}.$$

(e) The matrix obtained by multiplying the first row by $-\frac{1}{2}$ is

$$\begin{bmatrix} -2(-\frac{1}{2}) & 3(-\frac{1}{2}) & 1(-\frac{1}{2}) \\ 0 & 5 & -3 \\ 1 & 4 & 8 \end{bmatrix} = \begin{bmatrix} 1 & -\frac{3}{2} & -\frac{1}{2} \\ 0 & 5 & -3 \\ 1 & 4 & 8 \end{bmatrix}.$$

(f) The matrix obtained by multiplying the third row by 3 and adding to the first row is

$$\begin{bmatrix} -2+3(1) & 3+3(4) & 1+3(8) \\ 0 & 5 & -3 \\ 1 & 4 & 8 \end{bmatrix} = \begin{bmatrix} 1 & 15 & 25 \\ 0 & 5 & -3 \\ 1 & 4 & 8 \end{bmatrix}.$$

3. $4x + 8y = 44$
$2x - y = -3$

$$\left[\begin{array}{rr|r} 4 & 8 & 44 \\ 2 & -1 & -3 \end{array}\right]$$

$$\left[\begin{array}{rr|r} 1 & \underline{2} & \underline{11} \\ 2 & -1 & -3 \end{array}\right] \quad \tfrac{1}{4}R_1$$

$$\left[\begin{array}{rr|r} 1 & 2 & 11 \\ 0 & \underline{-5} & \underline{-25} \end{array}\right] \quad -2R_1 + R_2$$

Note: $\begin{cases} -2(2) + (-1) = -5 \\ -2(11) + (-3) = -25 \end{cases}$

$$\left[\begin{array}{rr|r} 1 & 2 & 11 \\ 0 & 1 & \underline{5} \end{array}\right] \quad -\tfrac{1}{5}R_2$$

This represents the system

$$x + 2y = 11$$
$$y = 5.$$

Substitute $y = 5$ in the first equation.

$$x + 2y = 11$$
$$x + 2(5) = 11$$
$$x + 10 = 11$$
$$x = 1$$

Solution set: $\{(1, 5)\}$

5. $x + y = 5$
$x - y = 3$

Write the augmented matrix for this system.

$$\left[\begin{array}{rr|r} 1 & 1 & 5 \\ 1 & -1 & 3 \end{array}\right]$$

$$\left[\begin{array}{rr|r} 1 & 1 & 5 \\ 0 & -2 & -2 \end{array}\right] \quad -1R_1 + R_2$$

$$\left[\begin{array}{rr|r} 1 & 1 & 5 \\ 0 & 1 & 1 \end{array}\right] \quad -\tfrac{1}{2}R_2$$

This matrix gives the system

$$x + y = 5$$
$$y = 1.$$

Substitute $y = 1$ in the first equation.

$$x + y = 5$$
$$x + 1 = 5$$
$$x = 4$$

Solution set: $\{(4, 1)\}$

7. $2x + 4y = 6$
$3x - y = 2$

Write the augmented matrix.

$$\left[\begin{array}{rr|r} 2 & 4 & 6 \\ 3 & -1 & 2 \end{array}\right]$$

The easiest way to get a 1 in the first row, first column position is to multiply the elements in the first row by $\frac{1}{2}$.

$$\left[\begin{array}{cc|c} 1 & 2 & 3 \\ 3 & -1 & 2 \end{array}\right] \quad \tfrac{1}{2}\text{R}_1$$

To get a 0 in row two, column 1, we need to subtract 3 from the 3 that is in that position. To do this we will multiply row 1 by -3 and add the result to row 2.

$$\left[\begin{array}{cc|c} 1 & 2 & 3 \\ 0 & -7 & -7 \end{array}\right] \quad -3\text{R}_1 + \text{R}_2$$

$$\left[\begin{array}{cc|c} 1 & 2 & 3 \\ 0 & 1 & 1 \end{array}\right] \quad -\tfrac{1}{7}\text{R}_2$$

This matrix gives the system

$$\begin{aligned} x + 2y &= 3 \\ y &= 1. \end{aligned}$$

Substitute $y = 1$ in the first equation.

$$\begin{aligned} x + 2y &= 3 \\ x + 2(1) &= 3 \\ x &= 1 \end{aligned}$$

Solution set: $\{(1, 1)\}$

9. $\begin{aligned} 3x + 4y &= 13 \\ 2x - 3y &= -14 \end{aligned}$

Write the augmented matrix.

$$\left[\begin{array}{cc|c} 3 & 4 & 13 \\ 2 & -3 & -14 \end{array}\right]$$

$$\left[\begin{array}{cc|c} 1 & 7 & 27 \\ 2 & -3 & -14 \end{array}\right] \quad -1\text{R}_2 + \text{R}_1$$

$$\left[\begin{array}{cc|c} 1 & 7 & 27 \\ 0 & -17 & -68 \end{array}\right] \quad -2\text{R}_1 + \text{R}_2$$

$$\left[\begin{array}{cc|c} 1 & 7 & 27 \\ 0 & 1 & 4 \end{array}\right] \quad -\tfrac{1}{17}\text{R}_2$$

This matrix gives the system

$$\begin{aligned} x + 7y &= 27 \\ y &= 4. \end{aligned}$$

Substitute $y = 4$ in the first equation.

$$\begin{aligned} x + 7y &= 27 \\ x + 7(4) &= 27 \\ x + 28 &= 27 \\ x &= -1 \end{aligned}$$

Solution set: $\{(-1, 4)\}$

11. $\begin{aligned} -4x + 12y &= 36 \\ x - 3y &= 9 \end{aligned}$

Write the augmented matrix.

$$\left[\begin{array}{cc|c} -4 & 12 & 36 \\ 1 & -3 & 9 \end{array}\right]$$

$$\left[\begin{array}{cc|c} -1 & 3 & 9 \\ 1 & -3 & 9 \end{array}\right] \quad \tfrac{1}{4}\text{R}_1$$

$$\left[\begin{array}{cc|c} -1 & 3 & 9 \\ 0 & 0 & 18 \end{array}\right] \quad \text{R}_1 + \text{R}_2$$

The corresponding system is

$$\begin{aligned} -x + 3y &= 9 \\ 0 &= 18 \quad \textit{False} \end{aligned}$$

which is inconsistent and has no solution.

Solution set: $\emptyset$

13. $\begin{aligned} 2x + y &= 4 \\ 4x + 2y &= 8 \end{aligned}$

Write the augmented matrix.

$$\left[\begin{array}{cc|c} 2 & 1 & 4 \\ 4 & 2 & 8 \end{array}\right]$$

$$\left[\begin{array}{cc|c} 1 & \frac{1}{2} & 2 \\ 4 & 2 & 8 \end{array}\right] \quad \tfrac{1}{2}\text{R}_1$$

$$\left[\begin{array}{cc|c} 1 & \frac{1}{2} & 2 \\ 0 & 0 & 0 \end{array}\right] \quad -4\text{R}_1 + \text{R}_2$$

Row 2, $0 = 0$, indicates that the system has dependent equations.

Solution set: $\{(x, y) \mid 2x + y = 4\}$

15. $\begin{aligned} \tfrac{1}{2}x + \tfrac{1}{3}y &= 0 \\ \tfrac{2}{3}x + \tfrac{3}{4}y &= 0 \end{aligned}$

Write the augmented matrix.

$$\left[\begin{array}{cc|c} \frac{1}{2} & \frac{1}{3} & 0 \\ \frac{2}{3} & \frac{3}{4} & 0 \end{array}\right]$$

$$\left[\begin{array}{cc|c} 1 & \frac{2}{3} & 0 \\ \frac{2}{3} & \frac{3}{4} & 0 \end{array}\right] \quad 2\text{R}_1$$

$$\left[\begin{array}{cc|c} 1 & \frac{2}{3} & 0 \\ 0 & \frac{11}{36} & 0 \end{array}\right] \quad -\tfrac{2}{3}\text{R}_1 + \text{R}_2$$

$$\left[\begin{array}{cc|c} 1 & \frac{2}{3} & 0 \\ 0 & 1 & 0 \end{array}\right] \quad \tfrac{36}{11}\text{R}_2$$

This matrix gives the system

$$\begin{aligned} x + \tfrac{2}{3}y &= 0 \\ y &= 0. \end{aligned}$$

Substitute $y = 0$ in the first equation.

$$\begin{aligned} x + \tfrac{2}{3}(0) &= 0 \\ x &= 0 \end{aligned}$$

Solution set: $\{(0, 0)\}$

17. $\begin{aligned} x + y - z &= -3 \\ 2x + y + z &= 4 \\ 5x - y + 2z &= 23 \end{aligned}$

Write the augmented matrix.

$$\left[\begin{array}{ccc|c} 1 & 1 & -1 & -3 \\ 2 & 1 & 1 & 4 \\ 5 & -1 & 2 & 23 \end{array}\right]$$

$$\left[\begin{array}{ccc|c} 1 & 1 & -1 & -3 \\ 0 & \underline{-1} & \underline{3} & \underline{10} \\ 0 & \underline{-6} & \underline{7} & \underline{38} \end{array}\right] \quad \begin{array}{l} -2R_1 + R_2 \\ -5R_1 + R_3 \end{array}$$

$$\left[\begin{array}{ccc|c} 1 & 1 & -1 & -3 \\ 0 & 1 & \underline{-3} & \underline{-10} \\ 0 & -6 & 7 & 38 \end{array}\right] \quad -1R_2$$

$$\left[\begin{array}{ccc|c} 1 & 1 & -1 & -3 \\ 0 & 1 & -3 & -10 \\ 0 & 0 & \underline{-11} & \underline{-22} \end{array}\right] \quad 6R_2 + R_3$$

$$\left[\begin{array}{ccc|c} 1 & 1 & -1 & -3 \\ 0 & 1 & -3 & -10 \\ 0 & 0 & 1 & \underline{2} \end{array}\right] \quad -\tfrac{1}{11}R_3$$

This matrix gives the system

$$\begin{aligned} x + y - z &= -3 \\ y - 3z &= -10 \\ z &= 2. \end{aligned}$$

Substitute $z = 2$ in the second equation.

$$\begin{aligned} y - 3z &= -10 \\ y - 3(2) &= -10 \\ y - 6 &= -10 \\ y &= -4 \end{aligned}$$

Substitute $y = -4$ and $z = 2$ in the first equation.

$$\begin{aligned} x + y - z &= -3 \\ x - 4 - 2 &= -3 \\ x - 6 &= -3 \\ x &= 3 \end{aligned}$$

Solution set: $\{(3, -4, 2)\}$

19. $\begin{aligned} x + y - 3z &= 1 \\ 2x - y + z &= 9 \\ 3x + y - 4z &= 8 \end{aligned}$

Write the augmented matrix.

$$\left[\begin{array}{ccc|c} 1 & 1 & -3 & 1 \\ 2 & -1 & 1 & 9 \\ 3 & 1 & -4 & 8 \end{array}\right]$$

$$\left[\begin{array}{ccc|c} 1 & 1 & -3 & 1 \\ 0 & -3 & 7 & 7 \\ 0 & -2 & 5 & 5 \end{array}\right] \quad \begin{array}{l} -2R_1 + R_2 \\ -3R_1 + R_3 \end{array}$$

$$\left[\begin{array}{ccc|c} 1 & 1 & -3 & 1 \\ 0 & 1 & -\frac{7}{3} & -\frac{7}{3} \\ 0 & -2 & 5 & 5 \end{array}\right] \quad -\tfrac{1}{3}R_2$$

$$\left[\begin{array}{ccc|c} 1 & 1 & -3 & 1 \\ 0 & 1 & -\frac{7}{3} & -\frac{7}{3} \\ 0 & 0 & \frac{1}{3} & \frac{1}{3} \end{array}\right] \quad 2R_2 + R_3$$

$$\left[\begin{array}{ccc|c} 1 & 1 & -3 & 1 \\ 0 & 1 & -\frac{7}{3} & -\frac{7}{3} \\ 0 & 0 & 1 & 1 \end{array}\right] \quad 3R_3$$

This matrix gives the system

$$\begin{aligned} x + y - 3z &= 1 \\ y - \tfrac{7}{3}z &= -\tfrac{7}{3} \\ z &= 1. \end{aligned}$$

Substitute $z = 1$ in the second equation.

$$\begin{aligned} y - \tfrac{7}{3}z &= -\tfrac{7}{3} \\ y - \tfrac{7}{3}(1) &= -\tfrac{7}{3} \\ y &= 0 \end{aligned}$$

Substitute $y = 0$ and $z = 1$ in the first equation.

$$\begin{aligned} x + y - 3z &= 1 \\ x + 0 - 3(1) &= 1 \\ x - 3 &= 1 \\ x &= 4 \end{aligned}$$

Solution set: $\{(4, 0, 1)\}$

21. $\begin{aligned} x + y - z &= 6 \\ 2x - y + z &= -9 \\ x - 2y + 3z &= 1 \end{aligned}$

Write the augmented matrix.

$$\left[\begin{array}{ccc|c} 1 & 1 & -1 & 6 \\ 2 & -1 & 1 & -9 \\ 1 & -2 & 3 & 1 \end{array}\right]$$

$$\left[\begin{array}{ccc|c} 1 & 1 & -1 & 6 \\ 0 & -3 & 3 & -21 \\ 0 & -3 & 4 & -5 \end{array}\right] \quad \begin{array}{l} -2R_1 + R_2 \\ -1R_1 + R_3 \end{array}$$

$$\left[\begin{array}{ccc|c} 1 & 1 & -1 & 6 \\ 0 & 1 & -1 & 7 \\ 0 & -3 & 4 & -5 \end{array}\right] \quad -\tfrac{1}{3}R_2$$

$$\left[\begin{array}{ccc|c} 1 & 1 & -1 & 6 \\ 0 & 1 & -1 & 7 \\ 0 & 0 & 1 & 16 \end{array}\right] \quad 3R_2 + R_3$$

This matrix gives the system

$$\begin{aligned} x + y - z &= 6 \\ y - z &= 7 \\ z &= 16. \end{aligned}$$

Substitute $z = 16$ in the second equation.

$$\begin{aligned} y - 16 &= 7 \\ y &= 23 \end{aligned}$$

Substitute $y = 23$ and $z = 16$ in the first equation.

$$\begin{aligned} x + 23 - 16 &= 6 \\ x + 7 &= 6 \\ x &= -1 \end{aligned}$$

Solution set: $\{(-1, 23, 16)\}$

23. $$\begin{aligned} x - y \quad &= 1 \\ y - z &= 6 \\ x + \quad z &= -1 \end{aligned}$$

Write the augmented matrix.

$$\left[\begin{array}{rrr|r} 1 & -1 & 0 & 1 \\ 0 & 1 & -1 & 6 \\ 1 & 0 & 1 & -1 \end{array}\right]$$

$$\left[\begin{array}{rrr|r} 1 & -1 & 0 & 1 \\ 0 & 1 & -1 & 6 \\ 0 & 1 & 1 & -2 \end{array}\right] \quad -1R_1 + R_3$$

$$\left[\begin{array}{rrr|r} 1 & -1 & 0 & 1 \\ 0 & 1 & -1 & 6 \\ 0 & 0 & 2 & -8 \end{array}\right] \quad -1R_2 + R_3$$

$$\left[\begin{array}{rrr|r} 1 & -1 & 0 & 1 \\ 0 & 1 & -1 & 6 \\ 0 & 0 & 1 & -4 \end{array}\right] \quad \tfrac{1}{2}R_3$$

This matrix gives the system

$$\begin{aligned} x - y \quad &= 1 \\ y - z &= 6 \\ z &= -4. \end{aligned}$$

Substitute $z = -4$ in the second equation.

$$\begin{aligned} y - z &= 6 \\ y - (-4) &= 6 \\ y &= 2 \end{aligned}$$

Substitute $y = 2$ in the first equation.

$$\begin{aligned} x - y &= 1 \\ x - 2 &= 1 \\ x &= 3 \end{aligned}$$

Solution set: $\{(3, 2, -4)\}$

25. $$\begin{aligned} 4x + 8y + 4z &= 9 \\ x + 3y + 4z &= 10 \\ 5x + 10y + 5z &= 12 \end{aligned}$$

Write the augmented matrix.

$$\left[\begin{array}{rrr|r} 4 & 8 & 4 & 9 \\ 1 & 3 & 4 & 10 \\ 5 & 10 & 5 & 12 \end{array}\right]$$

$$\left[\begin{array}{rrr|r} 1 & 3 & 4 & 10 \\ 4 & 8 & 4 & 9 \\ 5 & 10 & 5 & 12 \end{array}\right] \quad R_1 \leftrightarrow R_2$$

$$\left[\begin{array}{rrr|r} 1 & 3 & 4 & 10 \\ 0 & -4 & -12 & -31 \\ 0 & -5 & -15 & -38 \end{array}\right] \quad \begin{array}{l} -4R_1 + R_2 \\ -5R_1 + R_3 \end{array}$$

$$\left[\begin{array}{rrr|r} 1 & 3 & 4 & 10 \\ 0 & 1 & 3 & 7 \\ 0 & -5 & -15 & -38 \end{array}\right] \quad -R_3 + R_2$$

$$\left[\begin{array}{rrr|r} 1 & 3 & 4 & 10 \\ 0 & 1 & 3 & 7 \\ 0 & 0 & 0 & -3 \end{array}\right] \quad 5R_2 + R_3$$

From the last row, $0 = -3$, we see that the system is inconsistent.

Solution set: $\emptyset$

27. $$\begin{aligned} x - 2y + z &= 4 \\ 3x - 6y + 3z &= 12 \\ -2x + 4y - 2z &= -8 \end{aligned}$$

Write the augmented matrix.

$$\left[\begin{array}{rrr|r} 1 & -2 & 1 & 4 \\ 3 & -6 & 3 & 12 \\ -2 & 4 & -2 & -8 \end{array}\right]$$

$$\left[\begin{array}{rrr|r} 1 & -2 & 1 & 4 \\ 1 & -2 & 1 & 4 \\ -1 & 2 & -1 & -4 \end{array}\right] \quad \begin{array}{l} \tfrac{1}{3}R_2 \\ \tfrac{1}{2}R_3 \end{array}$$

$$\left[\begin{array}{rrr|r} 1 & -2 & 1 & 4 \\ 0 & 0 & 0 & 0 \\ 0 & 0 & 0 & 0 \end{array}\right] \quad \begin{array}{r} -1R_1 + R_2 \\ R_1 + R_3 \end{array}$$

This augmented matrix represents a system of dependent equations.

Solution set: $\{(x, y, z) \,|\, x - 2y + z = 4\}$

29. $5x + 3y - z = 0$
$2x - 3y + z = 0$
$x + 4y - 2z = 0$

Write the augmented matrix.

$$\left[\begin{array}{rrr|r} 5 & 3 & -1 & 0 \\ 2 & -3 & 1 & 0 \\ 1 & 4 & -2 & 0 \end{array}\right]$$

$$\left[\begin{array}{rrr|r} 1 & 4 & -2 & 0 \\ 2 & -3 & 1 & 0 \\ 5 & 3 & -1 & 0 \end{array}\right] \quad R_1 \leftrightarrow R_3$$

$$\left[\begin{array}{rrr|r} 1 & 4 & -2 & 0 \\ 0 & -11 & 5 & 0 \\ 0 & -17 & 9 & 0 \end{array}\right] \quad \begin{array}{l} -2R_1 + R_2 \\ -5R_1 + R_3 \end{array}$$

$$\left[\begin{array}{rrr|r} 1 & 4 & -2 & 0 \\ 0 & 1 & -\frac{5}{11} & 0 \\ 0 & -17 & 9 & 0 \end{array}\right] \quad -\tfrac{1}{11}R_2$$

$$\left[\begin{array}{rrr|r} 1 & 4 & -2 & 0 \\ 0 & 1 & -\frac{5}{11} & 0 \\ 0 & 0 & \frac{14}{11} & 0 \end{array}\right] \quad 17R_2 + R_3$$

$$\left[\begin{array}{rrr|r} 1 & 4 & -2 & 0 \\ 0 & 1 & -\frac{5}{11} & 0 \\ 0 & 0 & 1 & 0 \end{array}\right] \quad \tfrac{11}{14}R_3$$

This matrix gives the system

$$\begin{aligned} x + 4y - 2z &= 0 \\ y - \tfrac{5}{11}z &= 0 \\ z &= 0. \end{aligned}$$

Substituting $z = 0$ in the second equation gives $y = 0$ and substituting $y = 0$ and $z = 0$ in the first equation gives $x = 0$.

Solution set: $\{(0, 0, 0)\}$

Chapter 5 Review Exercises

When solving systems, check each solution in the equations of the original system.

1. $x + 3y = 8 \quad (1)$
$2x - \;\; y = 2 \quad (2)$

Graph the two lines. They appear to intersect at the point $(2, 2)$. Check $(2, 2)$ in both equations.

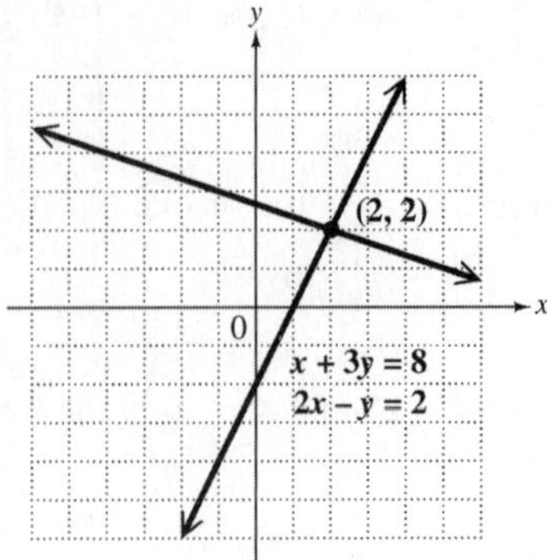

$$\begin{aligned} x + 3y &= 8 \qquad (1) \\ 2 + 3(2) &\stackrel{?}{=} 8 \\ 8 &= 8 \quad \textit{True} \end{aligned}$$

$$\begin{aligned} 2x - y &= 2 \qquad (2) \\ 2(2) - 2 &\stackrel{?}{=} 2 \\ 2 &= 2 \quad \textit{True} \end{aligned}$$

Solution set: $\{(2, 2)\}$

2. Checking the ordered pairs in choices **A**, **B**, and **C** yields true statements.

For choice **D**, we have:

$$\begin{aligned} 3x + 2y &= 6 \\ 3(3) + 2(-2) &\stackrel{?}{=} 6 \\ 9 - 4 &\stackrel{?}{=} 6 \\ 5 &= 6 \quad \textit{False} \end{aligned}$$

So **D** is the answer.

3. **(a)** The graphs meet between the years of 1980 and 1985. Therefore, the number of degrees for men equal the number of degrees for women between 1980 and 1985.

(b) The number was just less than 500,000.

4. $3x + y = -4 \quad (1)$
$x = \frac{2}{3}y \quad (2)$

Substitute $\frac{2}{3}y$ for x in equation (1) and solve for y.

$$\begin{aligned} 3x + y &= -4 \qquad (1) \\ 3(\tfrac{2}{3}y) + y &= 4 \\ 2y + y &= -4 \\ 3y &= -4 \\ y &= -\tfrac{4}{3} \end{aligned}$$

Since $x = \frac{2}{3}y$ and $y = -\frac{4}{3}$,

$$x = \tfrac{2}{3}(-\tfrac{4}{3}) = -\tfrac{8}{9}.$$

Solution set: $\{(-\frac{8}{9}, -\frac{4}{3})\}$

5. $9x - y = -4$ (1)
$y = x + 4$ (2)

Substitute $x + 4$ for y in equation (1) and solve for x.

$$\begin{aligned} 9x - y &= -4 \quad (1) \\ 9x - (x+4) &= -4 \\ 9x - x - 4 &= -4 \\ 8x &= 0 \\ x &= 0 \end{aligned}$$

Since $x = 0$, $y = x + 4 = 0 + 4 = 4$.

Solution set: $\{(0, 4)\}$

6. $-5x + 2y = -2$ (1)
$x + 6y = 26$ (2)

Solve equation (2) for x.

$$x = 26 - 6y$$

Substitute $26 - 6y$ for x in equation (1).

$$\begin{aligned} -5x + 2y &= -2 \quad (1) \\ -5(26 - 6y) + 2y &= -2 \\ -130 + 30y + 2y &= -2 \\ -130 + 32y &= -2 \\ 32y &= 128 \\ y &= 4 \end{aligned}$$

Since $x = 26 - 6y$ and $y = 4$,

$$x = 26 - 6(4) = 26 - 24 = 2.$$

Solution set: $\{(2, 4)\}$

7. $6x + 5y = 4$ (1)
$-4x + 2y = 8$ (2)

To eliminate x, multiply equation (1) by 2 and equation (2) by 3. Then add the results.

$$\begin{array}{rcrcrl} 12x & + & 10y & = & 8 & 2 \times (1) \\ -12x & + & 6y & = & 24 & 3 \times (2) \\ \hline & & 16y & = & 32 & \\ & & y & = & 2 & \end{array}$$

Since $y = 2$,

$$\begin{aligned} -4x + 2y &= 8 \quad (2) \\ -4x + 2(2) &= 8 \\ -4x + 4 &= 8 \\ -4x &= 4 \\ x &= -1. \end{aligned}$$

Solution set: $\{(-1, 2)\}$

8. $\frac{x}{6} + \frac{y}{6} = -\frac{1}{2}$ (1)
$x - y = -9$ (2)

Multiply equation (1) by 6 to clear the fractions. Add the result to equation (2) to eliminate y.

$$\begin{array}{rcrcrl} x & + & y & = & -3 & 6 \times (1) \\ x & - & y & = & -9 & (2) \\ \hline 2x & & & = & -12 & \\ x & & & = & -6 & \end{array}$$

Substitute -6 for x in equation (2).

$$\begin{aligned} x - y &= -9 \quad (2) \\ -6 - y &= -9 \\ -y &= -3 \\ y &= 3 \end{aligned}$$

Solution set: $\{(-6, 3)\}$

9. $4x + 5y = 9$ (1)
$3x + 7y = -1$ (2)

To eliminate x, multiply equation (1) by 3 and equation (2) by -4. Then add the results.

$$\begin{array}{rcrcrl} 12x & + & 15y & = & 27 & 3 \times (1) \\ -12x & - & 28y & = & 4 & -4 \times (2) \\ \hline & & -13y & = & 31 & \\ & & y & = & -\frac{31}{13} & \end{array}$$

To eliminate y, multiply equation (1) by 7 and equation (2) by -5. Then add the results.

$$\begin{array}{rcrcrl} 28x & + & 35y & = & 63 & 7 \times (1) \\ -15x & - & 35y & = & 5 & -5 \times (2) \\ \hline 13x & & & = & 68 & \\ & & x & = & \frac{68}{13} & \end{array}$$

Solution set: $\{(\frac{68}{13}, -\frac{31}{13})\}$

10. $-3x + y = 6$ (1)
$2y = 12 + 6x$ (2)

Since equation (2) can be rewritten as

$$y = 6 + 3x$$
$$\text{or} \quad -3x + y = 6,$$

the two equations are the same, and hence, dependent.

Solution set: $\{(x, y) \mid 3x - y = -6\}$

11.
$$\begin{aligned} 5x - 4y &= 2 \quad (1) \\ -10x + 8y &= 7 \quad (2) \end{aligned}$$

Multiply equation (1) by 2 and add the result to equation (2).

$$\begin{array}{rcrl} 10x - 8y &=& 4 & 2 \times (1) \\ -10x + 8y &=& 7 & (2) \\ \hline 0 &=& 11 & \textit{False} \end{array}$$

Since a false statement results, the system is *inconsistent*. The solution set is $\emptyset$.

12.
$$\begin{aligned} 3x + 3y &= 0 \quad (1) \\ -2x - y &= 0 \quad (2) \end{aligned}$$

To eliminate y, multiply equation (2) by 3 and add the result to equation (1).

$$\begin{array}{rcrl} 3x + 3y &=& 0 & (1) \\ -6x - 3y &=& 0 & 3 \times (2) \\ \hline -3x &=& 0 & \\ x &=& 0 & \end{array}$$

Since $x = 0$,

$$\begin{aligned} 3x + 3y &= 0 \quad (1) \\ 3(0) + 3y &= 0 \\ 0 + 3y &= 0 \\ 3y &= 0 \\ y &= 0. \end{aligned}$$

Solution set: $\{(0,0)\}$

13. If the system has a single solution, the graphs of the two linear equations of the system will intersect in one point.

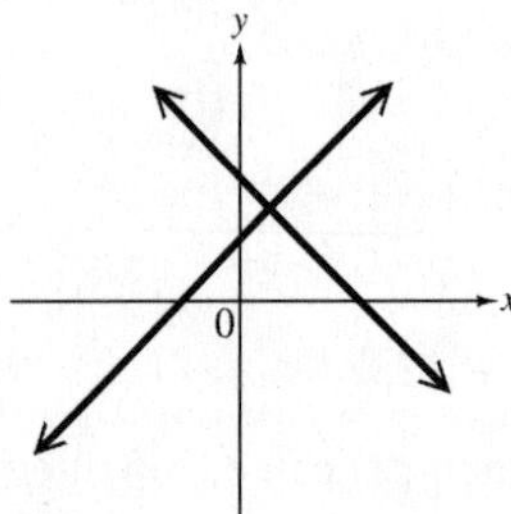

14. If the system has no solution, the graphs of the two linear equations are two parallel lines and do not intersect.

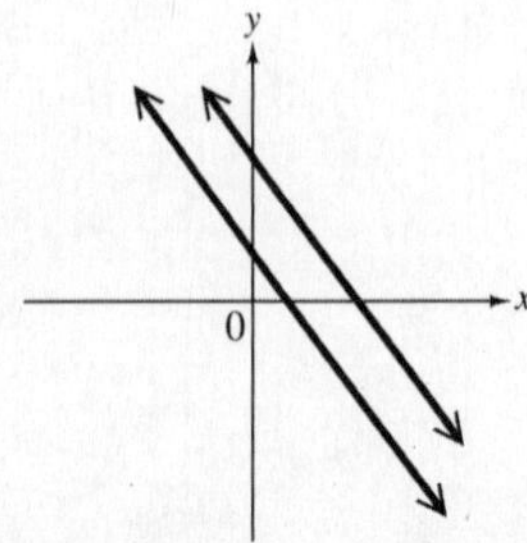

15. If the system has infinitely many solutions, the graphs of the two linear equations are the same line.

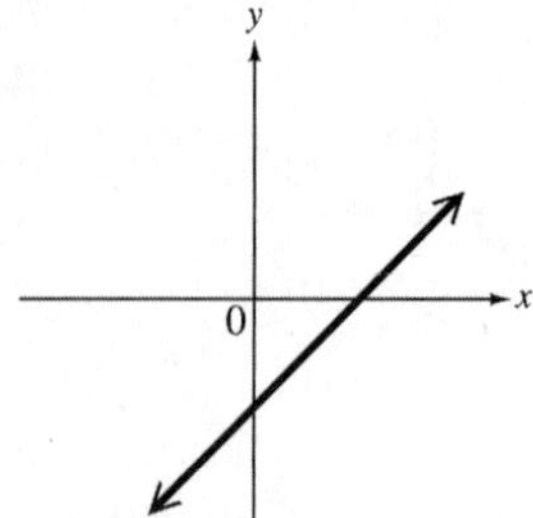

16.
$$\begin{aligned} y &= 3x + 2 \\ y &= 3x - 4 \end{aligned}$$

Since both equations are in slope-intercept form, their slopes and y-intercepts can be easily determined. Since they have the same slope (3), but different y-intercepts ($(0,2)$ and $(0,-4)$), they are parallel lines. Thus, there is no point common to both lines, and the system has $\emptyset$ as its solution set.

17.
$$\begin{array}{rcrl} 2x + 3y - z &=& -16 & (1) \\ x + 2y + 2z &=& -3 & (2) \\ -3x + y + z &=& -5 & (3) \end{array}$$

To eliminate z, add equations (1) and (3).

$$\begin{array}{rcrl} 2x + 3y - z &=& -16 & (1) \\ -3x + y + z &=& -5 & (3) \\ \hline -x + 4y &=& -21 & (4) \end{array}$$

To eliminate z again, multiply equation (1) by 2 and add the result to equation (2).

$$\begin{array}{rcrl} 4x + 6y - 2z &=& -32 & 2 \times (1) \\ x + 2y + 2z &=& -3 & (2) \\ \hline 5x + 8y &=& -35 & (5) \end{array}$$

Use equations (4) and (5) to eliminate x. Multiply equation (4) by 5 and add the result to equation (5).

$$\begin{array}{rcrl} -5x + 20y &=& -105 & 5 \times (4) \\ 5x + 8y &=& -35 & (5) \\ \hline 28y &=& -140 & \\ y &=& -5 & \end{array}$$

Substitute -5 for y in equation (4) to find x.

$$\begin{aligned} -x + 4y &= -21 \quad (4) \\ -x + 4(-5) &= -21 \\ -x - 20 &= -21 \\ -x &= -1 \\ x &= 1 \end{aligned}$$

Substitute 1 for x and -5 for y in equation (2) to find z.

$$\begin{aligned} x+2y+2z &= -3 \quad (2) \\ 1+2(-5)+2z &= -3 \\ 1-10+2z &= -3 \\ 2z &= 6 \\ z &= 3 \end{aligned}$$

The solution set is $\{(1,-5,3)\}$.

18.

$$\begin{aligned} 3x - y - z &= -8 \quad (1) \\ 4x + 2y + 3z &= 15 \quad (2) \\ -6x + 2y + 2z &= 10 \quad (3) \end{aligned}$$

To eliminate y, multiply equation (1) by 2 and add the result to equation (3).

$$\begin{aligned} 6x - 2y - 2z &= -16 \quad 2\times(1) \\ -6x + 2y + 2z &= 10 \quad (3) \\ \hline 0 &= -6 \quad \textit{False} \end{aligned}$$

Since a false statement results, equations (1) and (3) have no common solution. The system is *inconsistent*. The solution set is $\emptyset$.

19.

$$\begin{aligned} 4x - y &= 2 \quad (1) \\ 3y + z &= 9 \quad (2) \\ x + 2z &= 7 \quad (3) \end{aligned}$$

To eliminate y, multiply equation (1) by 3 and add the result to equation (2).

$$\begin{aligned} 12x - 3y &= 6 \quad 3\times(1) \\ 3y + z &= 9 \quad (2) \\ \hline 12x + z &= 15 \quad (4) \end{aligned}$$

To eliminate z, multiply equation (4) by -2 and add the result to equation (3).

$$\begin{aligned} -24x - 2z &= -30 \quad -2\times(4) \\ x + 2z &= 7 \quad (3) \\ \hline -23x &= -23 \\ x &= 1 \end{aligned}$$

Substitute 1 for x in equation (3) to find z.

$$\begin{aligned} x+2z &= 7 \quad (3) \\ 1+2z &= 7 \\ 2z &= 6 \\ z &= 3 \end{aligned}$$

Substitute 1 for x in equation (1) to find y.

$$\begin{aligned} 4x-y &= 2 \quad (1) \\ 4(1)-y &= 2 \\ 4-y &= 2 \\ -y &= -2 \\ y &= 2 \end{aligned}$$

The solution set is $\{(1,2,3)\}$.

20. Let $x =$ the width of the rink and $y =$ the length of the rink.

The length is 30 ft longer than twice the width.

$$y = 2x + 30 \quad (1)$$

The perimeter is 570 ft.

$$2x + 2y = 570 \quad (2)$$

Substitute $2x + 30$ for y in equation (2).

$$\begin{aligned} 2x+2y &= 570 \quad (2) \\ 2x+2(2x+30) &= 570 \\ 2x+4x+60 &= 570 \\ 6x &= 510 \\ x &= 85 \end{aligned}$$

From (1), $y = 2(85) + 30 = 200$.

The width is 85 ft and the length is 200 ft.

21. Let $x =$ the average price for a Red Sox ticket and $y =$ the average price for a Cubs ticket.

From the given information, we get the following system of equations:

$$\begin{aligned} 4x+4y &= 276.88 \quad (1) \\ 2x+6y &= 252.24 \quad (2) \end{aligned}$$

Divide (1) by 4 and (2) by -2.

$$\begin{aligned} x + y &= 69.22 \quad (3)\ (1)\div 4 \\ -x - 3y &= -126.12 \quad (4)\ (2)\div(-2) \\ \hline -2y &= -56.90 \quad \textit{Add (3) and (4).} \\ y &= 28.45 \end{aligned}$$

From (3), $x + 28.45 = 69.22$, so $x = 40.77$. The average price for a Red Sox ticket was \$40.77 and the average price for a Cubs ticket was \$28.45.

22. Let $x =$ the speed of the plane and $y =$ the speed of the wind.

Complete the chart.

	r	t	d
With Wind	$x+y$	1.75	$1.75(x+y)$
Against Wind	$x-y$	2	$2(x-y)$

The distance each way is 560 miles. From the chart,

$$1.75(x+y) = 560.$$

Divide by 1.75.

$$x + y = 320 \quad (1)$$

continued

From the chart,

$$2(x - y) = 560$$
$$x - y = 280. \qquad (2)$$

Solve the system by adding equations (1) and (2) to eliminate y.

$$\begin{array}{rcl} x + y &=& 320 \quad (1) \\ x - y &=& 280 \quad (2) \\ \hline 2x &=& 600 \\ x &=& 300 \end{array}$$

From (1), $300 + y = 320$, so $y = 20$.

The speed of the plane was 300 mph, and the speed of the wind was 20 mph.

23. Let $x =$ the amount of \$2 per pound nuts and $y =$ the amount of \$1 per pound chocolate candy.

Complete the chart.

	Number of Pounds	Price per Pound	Value
Nuts	x	2	$2x$
Chocolate	y	1	$1y = y$
Mixture	100	1.30	$1.30(100) = 130$

Solve the system formed from the second and fourth columns.

$$x + y = 100 \quad (1)$$
$$2x + y = 130 \quad (2)$$

Solve equation (1) for y.

$$y = 100 - x \quad (3)$$

Substitute $100 - x$ for y in equation (2).

$$2x + (100 - x) = 130$$
$$x = 30$$

From (3), $y = 100 - 30 = 70$.

She should use 30 lb of \$2 per pound nuts and 70 lb of \$1 per pound candy.

24. Let $g =$ the number of vats of green algae to be grown and $b =$ the number of vats of brown algae to be grown.

Arrange the information in a table. The table shows the amount of each nutrient needed to grow the algae (in kg).

	Amount of Nutrient X Needed	Amount of Nutrient Y Needed
Green Algae	$2g$	$3g$
Brown Algae	$1b$	$2b$
Total Amount of Nutrient Available	15	26

From the last two columns, we have the following system.

$$2g + b = 15 \quad (1)$$
$$3g + 2b = 26 \quad (2)$$

To solve, multiply (1) by 2 and (2) by -1 to get

$$\begin{array}{rcl} 4g + 2b &=& 30 \qquad 2 \times (1) \\ -3g - 2b &=& -26 \qquad -1 \times (2) \\ \hline g &=& 4. \end{array}$$

Substitute $g = 4$ in (1).

$$2(4) + b = 15$$
$$b = 7$$

She should grow 4 vats of green algae and 7 vats of brown algae in order to use all the nutrients.

25. Let $x =$ the measure of the largest angle, $y =$ the measure of the middle-sized angle, and $z =$ the measure of the smallest angle.

Since the sum of the measures of the angles of a triangle is 180°,

$$x + y + z = 180. \qquad (1)$$

Since the largest angle measures 10° less than the sum of the other two,

$$x = y + z - 10$$
$$\text{or} \quad x - y - z = -10. \qquad (2)$$

Since the measure of the middle-sized angle is the average of the other two,

$$y = \frac{x + z}{2}$$
$$2y = x + z$$
$$-x + 2y - z = 0. \qquad (3)$$

Solve the system.

$$\begin{array}{rcl} x + y + z &=& 180 \quad (1) \\ x - y - z &=& -10 \quad (2) \\ -x + 2y - z &=& 0 \quad (3) \end{array}$$

Add equations (1) and (2) to find x.

$$\begin{array}{rcl} x + y + z &=& 180 \quad (1) \\ x - y - z &=& -10 \quad (2) \\ \hline 2x &=& 170 \\ x &=& 85 \end{array}$$

Add equations (1) and (3), to find y.

$$\begin{array}{rl} x + y + z = 180 & (1) \\ -x + 2y - z = 0 & (3) \\ \hline 3y = 180 & \\ y = 60 & \end{array}$$

Substitute 85 for x and 60 for y in equation (1) to find z.

$$\begin{aligned} x + y + z &= 180 \quad (1) \\ 85 + 60 + z &= 180 \\ 145 + z &= 180 \\ z &= 35 \end{aligned}$$

The three angles are 85°, 60°, and 35°.

26. Let $x =$ the number of liters of 8% solution,
$y =$ the number of liters of 10% solution,
and $z =$ the number of liters of 20% solution.

Since the amount of the mixture will be 8 L,

$$x + y + z = 8. \qquad (1)$$

Since the final solution will be 12.5% hydrogen peroxide,

$$0.08x + 0.10y + 0.20z = 0.125(8).$$

Multiply by 100 to clear the decimals.

$$8x + 10y + 20z = 100 \qquad (2)$$

Since the amount of 8% solution used must be 2 L more than the amount of 20% solution,

$$x = z + 2. \qquad (3)$$

Solve the system.

$$\begin{array}{rl} x + y + z = 8 & (1) \\ 8x + 10y + 20z = 100 & (2) \\ x = z + 2 & (3) \end{array}$$

Since equation (3) is given in terms of x, substitute $z + 2$ for x in equations (1) and (2).

$$\begin{aligned} x + y + z &= 8 \quad (1) \\ (z + 2) + y + z &= 8 \\ y + 2z &= 6 \quad (4) \end{aligned}$$

$$\begin{aligned} 8x + 10y + 20z &= 100 \quad (2) \\ 8(z + 2) + 10y + 20z &= 100 \\ 8z + 16 + 10y + 20z &= 100 \\ 10y + 28z &= 84 \quad (5) \end{aligned}$$

To eliminate y, multiply equation (4) by -10 and add the result to equation (5).

$$\begin{array}{rl} -10y - 20z = -60 & -10 \times (4) \\ 10y + 28z = 84 & (5) \\ \hline 8z = 24 & \\ z = 3 & \end{array}$$

From (3), $x = z + 2 = 3 + 2 = 5$.
From (4), $y = 6 - 2z = 6 - 2(3) = 0$.

Mix 5 L of 8% solution, none of 10% solution, and 3 L of 20% solution.

27. Let $x =$ the number of home runs hit by Mantle,
$y =$ the number of home runs hit by Maris,
and $z =$ the number of home runs hit by Blanchard.

They combined for 136 home runs, so

$$x + y + z = 136. \qquad (1)$$

Mantle hit 7 fewer than Maris, so

$$x = y - 7. \qquad (2)$$

Maris hit 40 more than Blanchard, so

$$y = z + 40 \quad \text{or} \quad z = y - 40. \qquad (3)$$

Substitute $y - 7$ for x and $y - 40$ for z in (1).

$$\begin{aligned} x + y + z &= 136 \quad (1) \\ (y - 7) + y + (y - 40) &= 136 \\ 3y - 47 &= 136 \\ 3y &= 183 \\ y &= 61 \end{aligned}$$

From (2), $x = y - 7 = 61 - 7 = 54$.
From (3), $z = y - 40 = 61 - 40 = 21$.

Mantle hit 54 home runs, Maris hit 61 home runs, and Blanchard hit 21 home runs.

28. $2x + 5y = -4$
$4x - y = 14$

Write the augmented matrix.

$$\left[\begin{array}{rr|r} 2 & 5 & -4 \\ 4 & -1 & 14 \end{array}\right]$$

$$\left[\begin{array}{rr|r} 2 & 5 & -4 \\ 0 & -11 & 22 \end{array}\right] \quad -2R_1 + R_2$$

$$\left[\begin{array}{rr|r} 2 & 5 & -4 \\ 0 & 1 & -2 \end{array}\right] \quad -\tfrac{1}{11}R_2$$

This matrix gives the system

$$\begin{aligned} 2x + 5y &= -4 \\ y &= -2. \end{aligned}$$

Substitute $y = -2$ in the first equation.

$$\begin{aligned} 2x + 5y &= -4 \\ 2x + 5(-2) &= -4 \\ 2x - 10 &= -4 \\ 2x &= 6 \\ x &= 3 \end{aligned}$$

Solution set: $\{(3, -2)\}$

29. $$\begin{aligned} 6x + 3y &= 9 \\ -7x + 2y &= 17 \end{aligned}$$

Write the augmented matrix.

$$\left[\begin{array}{rr|r} 6 & 3 & 9 \\ -7 & 2 & 17 \end{array}\right]$$

$$\left[\begin{array}{rr|r} -1 & 5 & 26 \\ -7 & 2 & 17 \end{array}\right] \quad R_2 + R_1$$

$$\left[\begin{array}{rr|r} 1 & -5 & -26 \\ -7 & 2 & 17 \end{array}\right] \quad -R_1$$

$$\left[\begin{array}{rr|r} 1 & -5 & -26 \\ 0 & -33 & -165 \end{array}\right] \quad 7R_1 + R_2$$

$$\left[\begin{array}{rr|r} 1 & -5 & -26 \\ 0 & 1 & 5 \end{array}\right] \quad -\tfrac{1}{33}R_2$$

This matrix gives the system

$$\begin{aligned} x - 5y &= -26 \\ y &= 5. \end{aligned}$$

Substitute $y = 5$ in the first equation.

$$\begin{aligned} x - 5y &= -26 \\ x - 5(5) &= -26 \\ x - 25 &= -26 \\ x &= -1 \end{aligned}$$

Solution set: $\{(-1, 5)\}$

30. $$\begin{aligned} x + 2y - z &= 1 \\ 3x + 4y + 2z &= -2 \\ -2x - y + z &= -1 \end{aligned}$$

$$\left[\begin{array}{rrr|r} 1 & 2 & -1 & 1 \\ 3 & 4 & 2 & -2 \\ -2 & -1 & 1 & -1 \end{array}\right]$$

$$\left[\begin{array}{rrr|r} 1 & 2 & -1 & 1 \\ 0 & -2 & 5 & -5 \\ 0 & 3 & -1 & 1 \end{array}\right] \quad \begin{array}{r} -3R_1 + R_2 \\ 2R_1 + R_3 \end{array}$$

$$\left[\begin{array}{rrr|r} 1 & 2 & -1 & 1 \\ 0 & 1 & 4 & -4 \\ 0 & 3 & -1 & 1 \end{array}\right] \quad R_3 + R_2$$

$$\left[\begin{array}{rrr|r} 1 & 2 & -1 & 1 \\ 0 & 1 & 4 & -4 \\ 0 & 0 & -13 & 13 \end{array}\right] \quad -3R_2 + R_3$$

$$\left[\begin{array}{rrr|r} 1 & 2 & -1 & 1 \\ 0 & 1 & 4 & -4 \\ 0 & 0 & 1 & -1 \end{array}\right] \quad -\tfrac{1}{13}R_3$$

This matrix gives the system

$$\begin{aligned} x + 2y - z &= 1 \\ y + 4z &= -4 \\ z &= -1. \end{aligned}$$

Substitute $z = -1$ in the second equation.

$$\begin{aligned} y + 4z &= -4 \\ y + 4(-1) &= -4 \\ y &= 0 \end{aligned}$$

Substitute $y = 0$ and $z = -1$ in the first equation.

$$\begin{aligned} x + 2y - z &= 1 \\ x + 2(0) - (-1) &= 1 \\ x + 1 &= 1 \\ x &= 0 \end{aligned}$$

Solution set: $\{(0, 0, -1)\}$

31. **[5.1]** System **B** would be easier to solve using the substitution method because the second equation is already solved for y.

32. **[5.1]** $$\begin{aligned} \tfrac{2}{3}x + \tfrac{1}{6}y &= \tfrac{19}{2} && (1) \\ \tfrac{1}{3}x - \tfrac{2}{9}y &= 2 && (2) \end{aligned}$$

Multiply equation (1) by 6 and equation (2) by 9 to clear the fractions.

$$\begin{aligned} 4x + y &= 57 && (3) \quad 6 \times (1) \\ 3x - 2y &= 18 && (4) \quad 9 \times (2) \end{aligned}$$

To eliminate y, multiply equation (3) by 2 and add the result to equation (4).

$$\begin{array}{rcll} 8x + 2y &=& 114 & 2 \times (3) \\ 3x - 2y &=& 18 & (4) \\ \hline 11x &=& 132 & \\ x &=& 12 & \end{array}$$

Substitute 12 for x in equation (3) to find y.

$$\begin{aligned} 4x + y &= 57 && (3) \\ 4(12) + y &= 57 \\ 48 + y &= 57 \\ y &= 9 \end{aligned}$$

Solution set: $\{(12, 9)\}$

33. **[5.1]** $$\begin{aligned} 2x - 5y &= 8 && (1) \\ 3x + 4y &= 10 && (2) \end{aligned}$$

To eliminate y, multiply equation (1) by 4 and equation (2) by 5 and add the results.

$$\begin{array}{rcll} 8x - 20y &=& 32 & 4 \times (1) \\ 15x + 20y &=& 50 & 5 \times (2) \\ \hline 23x &=& 82 & \\ x &=& \frac{82}{23} & \end{array}$$

Instead of substituting to find y, we'll choose different multipliers and eliminate x from the original system.

$$\begin{array}{rcll} 6x - 15y &=& 24 & 3 \times (1) \\ -6x - 8y &=& -20 & -2 \times (2) \\ \hline -23y &=& 4 & \\ y &=& -\frac{4}{23} & \end{array}$$

Solution set: $\{(\frac{82}{23}, -\frac{4}{23})\}$

34. **[5.1]** $x = 7y + 10$ (1)
$2x + 3y = 3$ (2)

Since equation (1) is given in terms of x, substitute $7y + 10$ for x in equation (2) and solve for y.

$$\begin{aligned} 2(7y+10) + 3y &= 3 \\ 14y + 20 + 3y &= 3 \\ 17y &= -17 \\ y &= -1 \end{aligned}$$

From (1), $x = 7(-1) + 10 = 3$.

Solution set: $\{(3, -1)\}$

35. **[5.1]** $x + 4y = 17$ (1)
$-3x + 2y = -9$ (2)

To eliminate x, multiply equation (1) by 3 and add the result to equation (2).

$$\begin{array}{rcll} 3x + 12y &=& 51 & 3 \times (1) \\ -3x + 2y &=& -9 & (2) \\ \hline 14y &=& 42 & \\ y &=& 3 & \end{array}$$

Substitute 3 for y in equation (1) to find x.

$$\begin{aligned} x + 4y &= 17 \quad (1) \\ x + 4(3) &= 17 \\ x + 12 &= 17 \\ x &= 5 \end{aligned}$$

Solution set: $\{(5, 3)\}$

36. **[5.1]** $-7x + 3y = 12$ (1)
$5x + 2y = 8$ (2)

To eliminate y, multiply equation (1) by 2 and equation (2) by -3. Then add the results.

$$\begin{array}{rcll} -14x + 6y &=& 24 & 2 \times (1) \\ -15x - 6y &=& -24 & -3 \times (2) \\ \hline -29x &=& 0 & \\ x &=& 0 & \end{array}$$

Substitute 0 for x in equation (1) to find y.

$$\begin{aligned} -7x + 3y &= 12 \quad (1) \\ -7(0) + 3y &= 12 \\ 3y &= 12 \\ y &= 4 \end{aligned}$$

Solution set: $\{(0, 4)\}$

37. **[5.2]** $2x + 5y - z = 12$ (1)
$-x + y - 4z = -10$ (2)
$-8x - 20y + 4z = 31$ (3)

Multiply equation (1) by 4 and add the result to equation (3).

$$\begin{array}{rcll} 8x + 20y - 4z &=& 48 & 4 \times (1) \\ -8x - 20y + 4z &=& 31 & (3) \\ \hline 0 &=& 79 & \textit{False} \end{array}$$

Since a false statement results, the system is *inconsistent*. The solution set is $\emptyset$.

38. **[5.3]** Let x = the number of liters of 5% solution and y = the number of liters of 10% solution.

Liters of Solution	Percent (as a Decimal)	Liters of Pure Acid
x	0.05	$0.05x$
10	0.20	$0.20(10) = 2$
y	0.10	$0.10y$

Solve the system formed from the first and third columns.

$$\begin{aligned} x + 10 &= y \quad (1) \\ 0.05x + 2 &= 0.10y \quad (2) \end{aligned}$$

Multiply equation (2) by 100 to clear the decimals.

$$5x + 200 = 10y \quad (3)$$

Substitute $x + 10$ for y in equation (3) and solve for x.

$$\begin{aligned} 5x + 200 &= 10y \quad (3) \\ 5x + 200 &= 10(x + 10) \\ 5x + 200 &= 10x + 100 \\ 100 &= 5x \\ 20 &= x \end{aligned}$$

He should use 20 L of 5% solution.

39. **[5.3]** Let x, y, and z denote the number of medals won by Germany, the United States, and Canada, respectively.

The total number of medals won was 78, so

$$x + y + z = 78. \quad (1)$$

Germany won four more medals than the United States, so

$$x = y + 4. \quad (2)$$

continued

Canada won one fewer medal than the United States, so

$$z = y - 1. \quad (3)$$

Substitute $y + 4$ for x and $y - 1$ for z in (1).

$$\begin{aligned}(y+4) + y + (y-1) &= 78\\ 3y + 3 &= 78\\ 3y &= 75\\ y &= 25\end{aligned}$$

From (2), $x = 25 + 4 = 29$.
From (3), $z = 25 - 1 = 24$.

Germany won 29 medals, the United States won 25 medals, and Canada won 24 medals.

40. $x^2 + y^2 + ax + by + c = 0$
Let $x = 2$ and $y = 1$.

$$\begin{aligned}2^2 + 1^2 + a(2) + b(1) + c &= 0\\ 4 + 1 + 2a + b + c &= 0\\ 2a + b + c &= -5\end{aligned}$$

41. $x^2 + y^2 + ax + by + c = 0$
Let $x = -1$ and $y = 0$.

$$\begin{aligned}(-1)^2 + 0^2 + a(-1) + b(0) + c &= 0\\ 1 - a + c &= 0\\ -a + c &= -1\\ a - c &= 1\end{aligned}$$

42. $x^2 + y^2 + ax + by + c = 0$
Let $x = 3$ and $y = 3$.

$$\begin{aligned}3^2 + 3^2 + a(3) + b(3) + c &= 0\\ 9 + 9 + 3a + 3b + c &= 0\\ 3a + 3b + c &= -18\end{aligned}$$

43. Use the equations from Exercises 40–42 to form a system of equations.

$$\begin{aligned}2a + b + c &= -5 \quad (1)\\ a \qquad - c &= 1 \quad (2)\\ 3a + 3b + c &= -18 \quad (3)\end{aligned}$$

Add equations (1) and (2).

$$\begin{array}{rcl}2a + b + c &=& -5 \quad (1)\\ a \qquad - c &=& 1 \quad (2)\\ \hline 3a + b &=& -4 \quad (4)\end{array}$$

Add equations (2) and (3).

$$\begin{array}{rcl}a \qquad - c &=& 1 \quad (2)\\ 3a + 3b + c &=& -18 \quad (3)\\ \hline 4a + 3b &=& -17 \quad (5)\end{array}$$

Multiply equation (4) by -3 and add the result to equation (5).

$$\begin{array}{rcl}-9a - 3b &=& 12 \qquad -3 \times (4)\\ 4a + 3b &=& -17 \quad (5)\\ \hline -5a &=& -5\\ a &=& 1\end{array}$$

Substitute $a = 1$ into equation (4).

$$\begin{aligned}3a + b &= -4 \quad (4)\\ 3(1) + b &= -4\\ b &= -7\end{aligned}$$

Substitute $a = 1$ into equation (2).

$$\begin{aligned}a - c &= 1 \quad (2)\\ 1 - c &= 1\\ c &= 0\end{aligned}$$

Since $a = 1$, $b = -7$, and $c = 0$, the equation of the circle is

$$\begin{aligned}x^2 + y^2 + ax + by + c &= 0\\ x^2 + y^2 + x - 7y &= 0.\end{aligned}$$

44. The graph of the circle is not the graph of a function because it fails the vertical line test.

Chapter 5 Test

1. The graphs intersect to the right of 1997 and to the left of 1998, so the numbers were the same in the year 1997.

2. The graphs intersect about halfway between 100 and 120, so about 110 thousand people of each race were living with AIDS in 1997.

3. When each equation of the system

$$\begin{aligned}x + y &= 7\\ x - y &= 5\end{aligned}$$

is graphed, the point of intersection appears to be $(6, 1)$. To check, substitute 6 for x and 1 for y in each of the equations. Since $(6, 1)$ makes both equations true, the solution set of the system is $\{(6, 1)\}$.

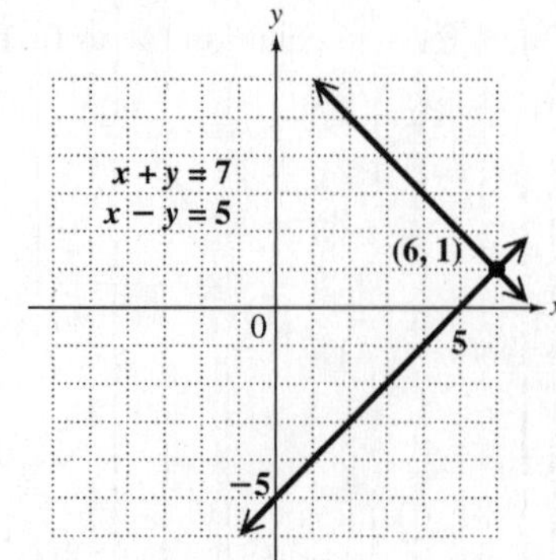

4. $$\begin{aligned} 2x - 3y &= 24 \quad (1) \\ y &= -\tfrac{2}{3}x \quad (2) \end{aligned}$$

Since equation (2) is solved for y, substitute $-\frac{2}{3}x$ for y in equation (1) and solve for x.

$$\begin{aligned} 2x - 3y &= 24 \quad (1) \\ 2x - 3(-\tfrac{2}{3}x) &= 24 \\ 2x + 2x &= 24 \\ 4x &= 24 \\ x &= 6 \end{aligned}$$

From (2), $y = -\frac{2}{3}(6) = -4$.

Solution set: $\{(6, -4)\}$

5. $$\begin{aligned} 12x - 5y &= 8 \quad (1) \\ 3x &= \tfrac{5}{4}y + 2 \\ \text{or} \quad x &= \tfrac{5}{12}y + \tfrac{2}{3} \quad (2) \end{aligned}$$

Substitute $\frac{5}{12}y + \frac{2}{3}$ for x in equation (1) and solve for y.

$$\begin{aligned} 12x - 5y &= 8 \quad (1) \\ 12(\tfrac{5}{12}y + \tfrac{2}{3}) - 5y &= 8 \\ 5y + 8 - 5y &= 8 \\ 8 &= 8 \quad \textit{True} \end{aligned}$$

Equations (1) and (2) are dependent.

Solution set: $\{(x, y) \mid 12x - 5y = 8\}$

6. $$\begin{aligned} 3x - y &= -8 \quad (1) \\ 2x + 6y &= 3 \quad (2) \end{aligned}$$

Solve equation (1) for y.

$$\begin{aligned} 3x - y &= -8 \\ -y &= -3x - 8 \\ y &= 3x + 8 \end{aligned}$$

Substitute $3x + 8$ for y in equation (2).

$$\begin{aligned} 2x + 6(3x + 8) &= 3 \\ 2x + 18x + 48 &= 3 \\ 20x &= -45 \\ x &= -\tfrac{9}{4} \end{aligned}$$

Since $y = 3x + 8$ and $x = -\frac{9}{4}$,

$$y = 3(-\tfrac{9}{4}) + 8 = -\tfrac{27}{4} + \tfrac{32}{4} = \tfrac{5}{4}.$$

Solution set: $\{(-\frac{9}{4}, \frac{5}{4})\}$

7. $$\begin{aligned} 3x + y &= 12 \quad (1) \\ 2x - y &= 3 \quad (2) \end{aligned}$$

To eliminate y, add equations (1) and (2).

$$\begin{aligned} 3x + y &= 12 \quad (1) \\ 2x - y &= 3 \quad (2) \\ \hline 5x &= 15 \\ x &= 3 \end{aligned}$$

Substitute 3 for x in equation (1) to find y.

$$\begin{aligned} 3x + y &= 12 \quad (1) \\ 3(3) + y &= 12 \\ 9 + y &= 12 \\ y &= 3 \end{aligned}$$

Solution set: $\{(3, 3)\}$

8. $$\begin{aligned} -5x + 2y &= -4 \quad (1) \\ 6x + 3y &= -6 \quad (2) \end{aligned}$$

To eliminate x, multiply equation (1) by 6 and equation (2) by 5. Then add the results.

$$\begin{aligned} -30x + 12y &= -24 \quad 6 \times (1) \\ 30x + 15y &= -30 \quad 5 \times (2) \\ \hline 27y &= -54 \\ y &= -2 \end{aligned}$$

Substitute -2 for y in equation (1) to find x.

$$\begin{aligned} -5x + 2y &= -4 \quad (1) \\ -5x + 2(-2) &= -4 \\ -5x - 4 &= -4 \\ -5x &= 0 \\ x &= 0 \end{aligned}$$

Solution set: $\{(0, -2)\}$

9. $$\begin{aligned} 3x + 4y &= 8 \quad (1) \\ 8y &= 7 - 6x \\ \text{or} \quad 6x + 8y &= 7 \quad (2) \end{aligned}$$

Multiply equation (1) by -2 and add the result to equation (2).

$$\begin{aligned} -6x - 8y &= -16 \quad -2 \times (1) \\ 6x + 8y &= 7 \quad (2) \\ \hline 0 &= -9 \quad \textit{False} \end{aligned}$$

Since a false statement results, the system is *inconsistent*. The solution set is $\emptyset$.

10. $$\begin{aligned} 3x + 5y + 3z &= 2 \quad (1) \\ 6x + 5y + z &= 0 \quad (2) \\ 3x + 10y - 2z &= 6 \quad (3) \end{aligned}$$

To eliminate x, multiply equation (1) by -1 and add the result to equation (3).

continued

$$\begin{array}{rcrcrcrl} -3x & - & 5y & - & 3z & = & -2 & \quad -1 \times (1) \\ 3x & + & 10y & - & 2z & = & 6 & \quad (3) \\ \hline & & 5y & - & 5z & = & 4 & \quad (4) \end{array}$$

To eliminate x again, multiply equation (1) by -2 and add the result to equation (2).

$$\begin{array}{rcrcrcrl} -6x & - & 10y & - & 6z & = & -4 & \quad -2 \times (1) \\ 6x & + & 5y & + & z & = & 0 & \quad (2) \\ \hline & & -5y & - & 5z & = & -4 & \quad (5) \end{array}$$

To eliminate y, add equations (4) and (5).

$$\begin{array}{rcrcrl} 5y & - & 5z & = & 4 & \quad (4) \\ -5y & - & 5z & = & -4 & \quad (5) \\ \hline & & -10z & = & 0 & \\ & & z & = & 0 & \end{array}$$

Substitute 0 for z in equation (4) to find y.

$$\begin{aligned} 5y - 5z &= 4 \qquad (4) \\ 5y - 5(0) &= 4 \\ 5y - 0 &= 4 \\ 5y &= 4 \\ y &= \tfrac{4}{5} \end{aligned}$$

Substitute $\frac{4}{5}$ for y and 0 for z in equation (1) to find x.

$$\begin{aligned} 3x + 5y + 3z &= 2 \qquad (1) \\ 3x + 5(\tfrac{4}{5}) + 3(0) &= 2 \\ 3x + 4 + 0 &= 2 \\ 3x &= -2 \\ x &= -\tfrac{2}{3} \end{aligned}$$

Solution set: $\{(-\frac{2}{3}, \frac{4}{5}, 0)\}$

11.

$$\begin{array}{rcrcrcrl} 4x & + & y & + & z & = & 11 & \quad (1) \\ x & - & y & - & z & = & 4 & \quad (2) \\ & & y & + & 2z & = & 0 & \quad (3) \end{array}$$

To eliminate x, multiply equation (2) by -4 and add the result to equation (1).

$$\begin{array}{rcrcrcrl} 4x & + & y & + & z & = & 11 & \quad (1) \\ -4x & + & 4y & + & 4z & = & -16 & \quad -4 \times (2) \\ \hline & & 5y & + & 5z & = & -5 & \quad (4) \end{array}$$

To eliminate y, multiply equation (3) by -5 and add the result to equation (4).

$$\begin{array}{rcrcrl} -5y & - & 10z & = & 0 & \quad -5 \times (3) \\ 5y & + & 5z & = & -5 & \quad (4) \\ \hline & & -5z & = & -5 & \\ & & z & = & 1 & \end{array}$$

From (3), $y + 2(1) = 0$, so $y = -2$.
From (2), $x - (-2) - 1 = 4$, so $x = 3$.

Solution set: $\{(3, -2, 1)\}$

12. Let $x =$ the gross (in millions of dollars) for *Ocean's Eleven*, and $y =$ the gross (in millions of dollars) for *Runaway Bride*.

Together the movies grossed \$335.5 million, so

$$x + y = 335.5. \qquad (1)$$

Runaway Bride grossed \$31.3 million less than *Ocean's Eleven*, so

$$y = x - 31.3. \qquad (2)$$

Substitute $x - 31.3$ for y in equation (1).

$$\begin{aligned} x + y &= 335.5 \qquad (1) \\ x + (x - 31.3) &= 335.5 \\ 2x &= 366.8 \\ x &= 183.4 \end{aligned}$$

From (2), $y = 183.4 - 31.3 = 152.1$.

Ocean's Eleven grossed \$183.4 million and *Runaway Bride* grossed \$152.1 million.

13. Let $x =$ the speed of the faster car and $y =$ the speed of the slower car.

Make a table. Use the formula $d = rt$.

	r	t	d
Faster Car	x	3.5	$3.5x$
Slower Car	y	3.5	$3.5y$

Since the fast car travels 30 mph faster than the slow car,

$$x - y = 30. \qquad (1)$$

Since the cars travel a total of 420 miles,

$$3.5x + 3.5y = 420.$$

Multiply by 10 to clear the decimals.

$$35x + 35y = 4200 \qquad (2)$$

To eliminate y, multiply equation (1) by 35 and add the result to equation (2).

$$\begin{array}{rcrcrl} 35x & - & 35y & = & 1050 & \quad 35 \times (1) \\ 35x & + & 35y & = & 4200 & \quad (2) \\ \hline 70x & & & = & 5250 & \\ & & x & = & 75 & \end{array}$$

Substitute 75 for x in equation (1) to find y.

$$\begin{aligned} x - y &= 30 \qquad (1) \\ 75 - y &= 30 \\ -y &= -45 \\ y &= 45 \end{aligned}$$

The faster car is traveling at 75 mph, and the slower car is traveling at 45 mph.

14. Let $x =$ the number of liters of 20% solution and $y =$ the number of liters of 50% solution.

Make a table.

Liters of Solution	Percent (as a Decimal)	Liters of Pure Alcohol
x	0.20	$0.20x$
y	0.50	$0.50y$
12	0.40	$0.40(12) = 4.8$

Since 12 L of the mixture are needed,

$$x + y = 12. \quad (1)$$

Since the amount of pure alcohol in the 20% solution plus the amount of pure alcohol in the 50% solution must equal the amount of alcohol in the mixture,

$$0.2x + 0.5y = 4.8.$$

Multiply by 10 to clear the decimals.

$$2x + 5y = 48 \quad (2)$$

Multiply equation (1) by -2 and add the result to equation (2).

$$\begin{array}{rcrl} -2x - 2y &=& -24 & \quad -2 \times (1) \\ 2x + 5y &=& 48 & \quad (2) \\ \hline 3y &=& 24 & \\ y &=& 8 & \end{array}$$

From (1), $x + 8 = 12$, so $x = 4$.

4 L of 20% solution and 8 L of 50% solution are needed.

15. Let $x =$ the price of an AC adaptor and $y =$ the price of a rechargeable flashlight.

Since 7 AC adaptors and 2 rechargeable flashlights cost \$86,

$$7x + 2y = 86. \quad (1)$$

Since 3 AC adaptors and 4 rechargeable flashlights cost \$84,

$$3x + 4y = 84. \quad (2)$$

Solve the system.

$$\begin{array}{rcrl} 7x + 2y &=& 86 & \quad (1) \\ 3x + 4y &=& 84 & \quad (2) \end{array}$$

To eliminate y, multiply equation (1) by -2 and add the result to equation (2).

$$\begin{array}{rcrl} -14x - 4y &=& -172 & \quad -2 \times (1) \\ 3x + 4y &=& 84 & \quad (2) \\ \hline -11x &=& -88 & \\ x &=& 8 & \end{array}$$

Substitute 8 for x in equation (1) to find y.

$$\begin{aligned} 7x + 2y &= 86 \quad (1) \\ 7(8) + 2y &= 86 \\ 56 + 2y &= 86 \\ 2y &= 30 \\ y &= 15 \end{aligned}$$

An AC adaptor costs \$8, and a rechargeable flashlight costs \$15.

16. Let $x =$ the amount of Orange Pekoe, $y =$ the amount of Irish Breakfast, and $z =$ the amount of Earl Grey.

The owner wants 100 oz of tea, so

$$x + y + z = 100. \quad (1)$$

An equation which relates the prices of the tea is

$$0.80x + 0.85y + 0.95z = 0.83(100).$$

Multiply by 100 to clear the decimals.

$$80x + 85y + 95z = 8300 \quad (2)$$

The mixture must use twice as much Orange Pekoe as Irish Breakfast, so

$$x = 2y. \quad (3)$$

To eliminate z, multiply equation (1) by -95 and add the result to equation (2).

$$\begin{array}{rcrl} -95x - 95y - 95z &=& -9500 & \quad -95 \times (1) \\ 80x + 85y + 95z &=& 8300 & \quad (2) \\ \hline -15x - 10y &=& -1200 & \quad (4) \end{array}$$

Divide equation (4) by -5.

$$3x + 2y = 240 \quad (5)$$

Substitute $2y$ for x in equation (5) to find y.

$$\begin{aligned} 3x + 2y &= 240 \quad (5) \\ 3(2y) + 2y &= 240 \\ 8y &= 240 \\ y &= 30 \end{aligned}$$

From (3), $x = 2(30) = 60$.
Substitute 60 for x and 30 for y in equation (1) to find z.

$$\begin{aligned} x + y + z &= 100 \quad (1) \\ 60 + 30 + z &= 100 \\ z &= 10 \end{aligned}$$

He should use 60 oz of Orange Pekoe, 30 oz of Irish Breakfast, and 10 oz of Earl Grey.

17. $3x + 2y = 4$
$5x + 5y = 9$

Write the augmented matrix.

$$\left[\begin{array}{rr|r} 3 & 2 & 4 \\ 5 & 5 & 9 \end{array}\right]$$

$$\left[\begin{array}{rr|r} 6 & 4 & 8 \\ 5 & 5 & 9 \end{array}\right] \quad 2R_1$$

$$\left[\begin{array}{rr|r} 1 & -1 & -1 \\ 5 & 5 & 9 \end{array}\right] \quad -R_2 + R_1$$

$$\left[\begin{array}{rr|r} 1 & -1 & -1 \\ 0 & 10 & 14 \end{array}\right] \quad -5R_1 + R_2$$

$$\left[\begin{array}{rr|r} 1 & -1 & -1 \\ 0 & 1 & \frac{7}{5} \end{array}\right] \quad \tfrac{1}{10}R_2$$

This matrix gives the system

$$x - y = -1$$
$$y = \tfrac{7}{5}.$$

Substitute $y = \frac{7}{5}$ in the first equation.

$$x - \tfrac{7}{5} = -1$$
$$x = -1 + \tfrac{7}{5}$$
$$= -\tfrac{5}{5} + \tfrac{7}{5} = \tfrac{2}{5}$$

Solution set: $\{(\frac{2}{5}, \frac{7}{5})\}$

18. $x + 3y + 2z = 11$
$3x + 7y + 4z = 23$
$5x + 3y - 5z = -14$

Write the augmented matrix.

$$\left[\begin{array}{rrr|r} 1 & 3 & 2 & 11 \\ 3 & 7 & 4 & 23 \\ 5 & 3 & -5 & -14 \end{array}\right]$$

$$\left[\begin{array}{rrr|r} 1 & 3 & 2 & 11 \\ 0 & -2 & -2 & -10 \\ 0 & -12 & -15 & -69 \end{array}\right] \quad \begin{array}{l} \\ -3R_1 + R_2 \\ -5R_1 + R_3 \end{array}$$

$$\left[\begin{array}{rrr|r} 1 & 3 & 2 & 11 \\ 0 & 1 & 1 & 5 \\ 0 & -12 & -15 & -69 \end{array}\right] \quad -\tfrac{1}{2}R_2$$

$$\left[\begin{array}{rrr|r} 1 & 3 & 2 & 11 \\ 0 & 1 & 1 & 5 \\ 0 & 0 & -3 & -9 \end{array}\right] \quad 12R_2 + R_3$$

$$\left[\begin{array}{rrr|r} 1 & 3 & 2 & 11 \\ 0 & 1 & 1 & 5 \\ 0 & 0 & 1 & 3 \end{array}\right] \quad -\tfrac{1}{3}R_3$$

This matrix gives the system

$$x + 3y + 2z = 11$$
$$y + z = 5$$
$$z = 3.$$

Substitute $z = 3$ in the second equation.

$$y + z = 5$$
$$y + 3 = 5$$
$$y = 2$$

Substitute $y = 2$ and $z = 3$ in the first equation.

$$x + 3y + 2z = 11$$
$$x + 3(2) + 2(3) = 11$$
$$x + 6 + 6 = 11$$
$$x = -1$$

Solution set: $\{(-1, 2, 3)\}$

Cumulative Review Exercises (Chapters 1–5)

1. $(-3)^4 = (-3)(-3)(-3)(-3) = 81$

2. $-3^4 = -(3)(3)(3)(3) = -81$

3. $-(-3)^4 = -(-3)(-3)(-3)(-3) = -81$

4. $\sqrt{0.49} = 0.7$, since 0.7 is positive and $(0.7)^2 = 0.49$.

5. $-\sqrt{0.49} = -0.7$, since $(0.7)^2 = 0.49$ and the negative sign in front of the radical must be applied.

6. $\sqrt{-0.49}$ is not a real number because of the negative sign under the radical. No real number squared is negative.

In Exercises 7–9, let $x = -4$, $y = 3$, and $z = 6$.

7. $|2x| + y^2 - z^3 = |(2)(-4)| + 3^2 - (6)^3$
$= |-8| + 9 - 216$
$= 8 + 9 - 216$
$= -199$

8. $-5(x^3 - y^3) = -5[(-4)^3 - (3)^3]$
$= -5(-64 - 27)$
$= -5(-91)$
$= 455$

9. $$\frac{2x^2 - x + z}{y^2 - z} = \frac{2(-4)^2 - (-4) + 6}{3^2 - 6} = \frac{2(16) + 4 + 6}{9 - 6} = \frac{32 + 4 + 6}{3} = \frac{42}{3} = 14$$

10. $7(2x + 3) - 4(2x + 1) = 2(x + 1)$
$14x + 21 - 8x - 4 = 2x + 2$
$6x + 17 = 2x + 2$
$4x = -15$
$x = -\frac{15}{4}$

Solution set: $\{-\frac{15}{4}\}$

11. $$0.04x + 0.06(x-1) = 1.04$$

Multiply each side by 100 to clear the decimals.

$$\begin{aligned} 4x + 6(x-1) &= 104 \\ 4x + 6x - 6 &= 104 \\ 10x - 6 &= 104 \\ 10x &= 110 \\ x &= 11 \end{aligned}$$

Solution set: $\{11\}$

12. Solve $ax + by = c$ for x.

$$\begin{aligned} ax &= c - by \\ x &= \frac{c - by}{a} \end{aligned}$$

13. $|6x - 8| = 4$

$$\begin{aligned} 6x - 8 &= 4 \quad \text{or} \quad 6x - 8 = -4 \\ 6x &= 12 \quad \text{or} \quad 6x = 4 \\ x &= 2 \quad \text{or} \quad x = \tfrac{4}{6} = \tfrac{2}{3} \end{aligned}$$

Solution set: $\{\frac{2}{3}, 2\}$

14. $$\tfrac{2}{3}x + \tfrac{5}{12}x \le 20$$

$$12(\tfrac{2}{3}x + \tfrac{5}{12}x) \le 12(20) \quad \textit{Multiply by the LCD, 12.}$$

$$\begin{aligned} 8x + 5x &\le 240 \\ 13x &\le 240 \\ x &\le \tfrac{240}{13} \end{aligned}$$

Solution set: $(-\infty, \frac{240}{13}]$

15. $|3x + 2| \le 4$

$$\begin{aligned} -4 \le 3x + 2 &\le 4 \\ -6 \le 3x &\le 2 \quad \textit{Subtract 2.} \\ -2 \le x &\le \tfrac{2}{3} \quad \textit{Divide by 3.} \end{aligned}$$

Solution set: $[-2, \frac{2}{3}]$

16. $|12t + 7| \ge 0$

The solution set is $(-\infty, \infty)$ since the absolute value of any number is greater than or equal to 0.

17. 80.4% of 2500 is $0.804(2500) = 2010$
72.5% of 2500 is $0.725(2500) = 1812.5 \approx 1813$

$$\frac{1570}{2500} = 0.628 \text{ or } 62.8\%; \quad \frac{1430}{2500} = 0.572 \text{ or } 57.2\%$$

Product or Company	Percent	Actual Number
Charmin	80.4%	*2010*
Wheaties	72.5%	*1813*
Budweiser	*62.8%*	1570
State Farm	*57.2%*	1430

18. Let $x =$ the number of nickels,
$x + 1 =$ the number of dimes, and
$x + 6 =$ the number of pennies.

The total value is \$4.80, so

$$0.05x + 0.10(x+1) + 0.01(x+6) = 4.80.$$

Multiply each side by 100 to clear the decimals.

$$\begin{aligned} 5x + 10(x+1) + 1(x+6) &= 480 \\ 5x + 10x + 10 + x + 6 &= 480 \\ 16x + 16 &= 480 \\ 16x &= 464 \\ x &= 29 \end{aligned}$$

Since $x = 29$, $x + 1 = 29 + 1 = 30$, and $x + 6 = 29 + 6 = 35$.

There are 35 pennies, 29 nickels, and 30 dimes.

19. Let $x =$ the measure of the equal angles and
$2x - 4 =$ the measure of the third angle.

The sum of the measures of the angles in a triangle is 180°, so

$$\begin{aligned} x + x + (2x - 4) &= 180 \\ 4x - 4 &= 180 \\ 4x &= 184 \\ x &= 46. \end{aligned}$$

So, $2x - 4 = 2(46) - 4 = 92 - 4 = 88$.

The measures of the angles are 46°, 46°, and 88°.

20. A horizontal line through the point (x, k) has equation $y = k$. Since point A has coordinates $(-2, 6)$, $k = 6$. The equation of the horizontal line through A is $y = 6$.

21. A vertical line through the point (k, y) has equation $x = k$. Since point B has coordinates $(4, -2)$, $k = 4$. The equation of the vertical line through B is $x = 4$.

22. Let $(x_1, y_1) = (-2, 6)$ and $(x_2, y_2) = (4, -2)$. Then,

$$m = \frac{y_2 - y_1}{x_2 - x_1} = \frac{-2 - 6}{4 - (-2)} = \frac{-8}{6} = -\frac{4}{3}.$$

The slope is $-\frac{4}{3}$.

23. Perpendicular lines have slopes that are negative reciprocals of each other. The slope of line *AB* is $-\frac{4}{3}$ (from Exercise 22). The negative reciprocal of $-\frac{4}{3}$ is $\frac{3}{4}$, so the slope of a line perpendicular to line *AB* is $\frac{3}{4}$.

24. Let $m = -\frac{4}{3}$ and $(x_1, y_1) = (4, -2)$ in the point-slope form.

$$\begin{aligned} y - y_1 &= m(x - x_1) \\ y - (-2) &= -\tfrac{4}{3}(x - 4) \\ y + 2 &= -\tfrac{4}{3}x + \tfrac{16}{3} \end{aligned}$$

Multiply by 3 to clear the fractions, and then write the equation in standard form, $Ax + By = C$.

$$\begin{aligned} 3y + 6 &= -4x + 16 \\ 4x + 3y &= 10 \end{aligned}$$

25. Write the equation of the line in slope-intercept form.

$$\begin{aligned} 4x + 3y &= 10 \\ 3y &= -4x + 10 \\ y &= -\tfrac{4}{3}x + \tfrac{10}{3} \end{aligned}$$

Replace y with $f(x)$.

$$f(x) = -\tfrac{4}{3}x + \tfrac{10}{3}$$

26. First locate the point $(-1, -3)$ on a graph. Then use the definition of slope to find a second point on the line.

$$m = \frac{\text{change in } y}{\text{change in } x} = \frac{2}{3}$$

From $(-1, -3)$, move 2 units up and 3 units to the right to get to $(2, -1)$. The line through $(-1, -3)$ and the new point, $(2, -1)$, is the graph.

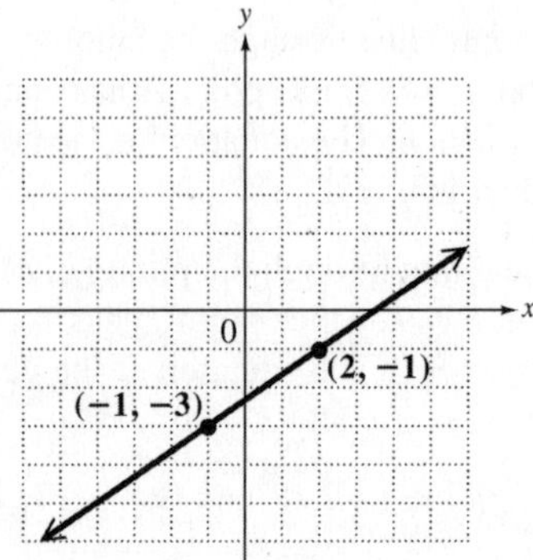

27. $-3x - 2y \le 6$

Graph the line $-3x - 2y = 6$ through its intercepts, $(-2, 0)$ and $(0, -3)$, as a solid line, since the inequality involves $\le$.
To determine the region that belongs to the graph, test $(0, 0)$.

$$\begin{aligned} -3x - 2y &\le 6 \\ -3(0) - 2(0) &\overset{?}{\le} 6 \\ 0 &\le 6 \quad \textit{True} \end{aligned}$$

Since the result is true, shade the region that includes $(0, 0)$.

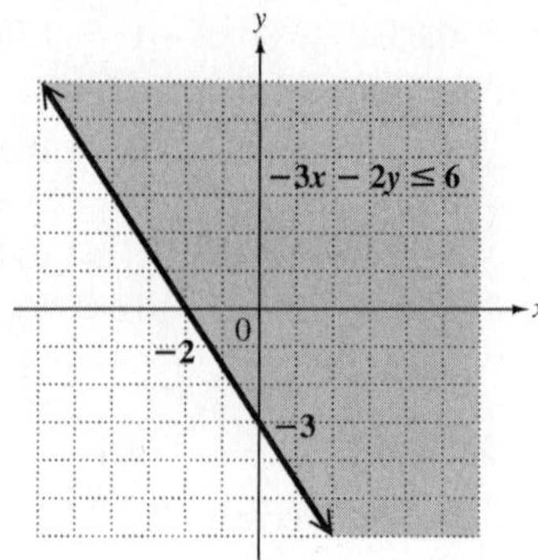

28.
$$\begin{aligned} -2x + 3y &= -15 \quad (1) \\ 4x - y &= 15 \quad (2) \end{aligned}$$

To eliminate x, multiply equation (1) by 2 and add the result to equation (2).

$$\begin{aligned} -4x + 6y &= -30 \quad 2 \times (1) \\ 4x - y &= 15 \quad (2) \\ \hline 5y &= -15 \\ y &= -3 \end{aligned}$$

Substitute -3 for y in equation (2) to find x.

$$\begin{aligned} 4x - y &= 15 \quad (2) \\ 4x - (-3) &= 15 \\ 4x + 3 &= 15 \\ 4x &= 12 \\ x &= 3 \end{aligned}$$

Solution set: $\{(3, -3)\}$

29.
$$\begin{aligned} x + y + z &= 10 \quad (1) \\ x - y - z &= 0 \quad (2) \\ -x + y - z &= -4 \quad (3) \end{aligned}$$

Add equations (1) and (2) to eliminate y and z. The result is

$$\begin{aligned} 2x &= 10 \\ x &= 5. \end{aligned}$$

Add equations (2) and (3) to eliminate x and y. The result is

$$\begin{aligned} -2z &= -4 \\ z &= 2. \end{aligned}$$

Substitute 5 for x and 2 for z in equation (1) to find y.

$$\begin{aligned} x + y + z &= 10 \quad (1) \\ 5 + y + 2 &= 10 \\ y + 7 &= 10 \\ y &= 3 \end{aligned}$$

Solution set: $\{(5, 3, 2)\}$

30. Let x = the number of pounds of \$1.20/lb candy, and y = the number of pounds of \$2.40/lb candy.

Number of Pounds	Price per Pound	Value
x	\$1.20	\$1.20x
y	\$2.40	\$2.40y
80	\$1.65	\$1.65(80)

From the first and third columns, we have:

$$\begin{aligned} x + y &= 80 \\ 1.2x + 2.4y &= 1.65(80) \end{aligned}$$

or

$$\begin{aligned} x &= 80 - y \quad (1) \\ 1.2x + 2.4y &= 132 \quad (2) \end{aligned}$$

Substitute $80 - y$ for x in equation (2).

$$\begin{aligned} 1.2x + 2.4y &= 132 \quad (2) \\ 1.2(80 - y) + 2.4y &= 132 \\ 96 - 1.2y + 2.4y &= 132 \\ 1.2y &= 36 \\ y &= 30 \end{aligned}$$

Since $x = 80 - y$ and $y = 30$,

$$x = 80 - 30 = 50.$$

He should mix 50 lb of \$1.20/lb candy with 30 lb of \$2.40/lb candy.

31. Let x = the amount borrowed at 8%, y = the amount borrowed at 9%, and z = the amount borrowed at 10%.

Amount Borrowed	Rate (as a Decimal)	Annual Interest
x	0.08	$0.08x$
y	0.09	$0.09y$
z	0.10	$0.10z$
25,000	← Totals →	2220

$$\begin{aligned} x + y + z &= 25{,}000 \\ z &= \tfrac{1}{2}x + 2000 \\ 0.08x + 0.09y + 0.10z &= 2220 \end{aligned}$$

or

$$\begin{aligned} x + y + z &= 25{,}000 \\ -x + 2z &= 4000 \\ 8x + 9y + 10z &= 222{,}000 \end{aligned}$$

Write the augmented matrix.

$$\left[\begin{array}{ccc|c} 1 & 1 & 1 & 25{,}000 \\ -1 & 0 & 2 & 4000 \\ 8 & 9 & 10 & 222{,}000 \end{array}\right]$$

$$\left[\begin{array}{ccc|c} 1 & 1 & 1 & 25{,}000 \\ 0 & 1 & 3 & 29{,}000 \\ 0 & 1 & 2 & 22{,}000 \end{array}\right] \quad \begin{array}{l} \\ R_1 + R_2 \\ -8R_1 + R_3 \end{array}$$

$$\left[\begin{array}{ccc|c} 1 & 1 & 1 & 25{,}000 \\ 0 & 1 & 3 & 29{,}000 \\ 0 & 0 & -1 & -7000 \end{array}\right] \quad \begin{array}{l} \\ \\ -R_2 + R_3 \end{array}$$

$$\left[\begin{array}{ccc|c} 1 & 1 & 1 & 25{,}000 \\ 0 & 1 & 3 & 29{,}000 \\ 0 & 0 & 1 & 7000 \end{array}\right] \quad \begin{array}{l} \\ \\ -R_3 \end{array}$$

This matrix represents the system

$$\begin{aligned} x + y + z &= 25{,}000 \\ y + 3z &= 29{,}000 \\ z &= 7000. \end{aligned}$$

Substitute 7000 for z in the second equation.

$$\begin{aligned} y + 3(7000) &= 29{,}000 \\ y + 21{,}000 &= 29{,}000 \\ y &= 8000 \end{aligned}$$

Substitute 8000 for y and 7000 for z in the first equation.

$$\begin{aligned} x + 8000 + 7000 &= 25{,}000 \\ x &= 10{,}000 \end{aligned}$$

The company borrowed \$10,000 at 8%, \$8000 at 9% and \$7000 at 10%.

32. The lines intersect at $(8, 3000)$, so the cost equals the revenue at $x = 8$ (which is 800 parts). The revenue is \$3000.

33. On the sale of 1100 parts ($x = 11$), the revenue is about \$4100 and the cost is about \$3700.

$$\begin{aligned} \text{Profit} &= \text{Revenue} - \text{Cost} \\ &\approx 4100 - 3700 \\ &= 400 \end{aligned}$$

The profit is about \$400.

CHAPTER 6 EXPONENTS, POLYNOMIALS, AND POLYNOMIAL FUNCTIONS

6.1 Integer Exponents and Scientific Notation

6.1 Margin Exercises

1. Use the product rule for exponents, if possible.

(a) $m^8 \cdot m^6 = m^{8+6} = m^{14}$

(b) $r^7 \cdot r = r^7 \cdot r^1 = r^{7+1} = r^8$

(c) $k^4k^3k^6 = k^{4+3+6} = k^{13}$

(d) $m^5 \cdot p^4$ cannot be simplified further because the bases m and p are not the same. The product rule does not apply.

(e) Use the associative and commutative properties to group constants together and variables together.

$$\begin{aligned}(-4a^3)(6a^2) &= (-4)(6)(a^3a^2)\\ &= -24a^{3+2}\\ &= -24a^5\end{aligned}$$

(f) $$\begin{aligned}(-5p^4)(-9p^5) &= (-5)(-9)(p^4p^5)\\ &= 45p^{4+5}\\ &= 45p^9\end{aligned}$$

2. **(a)** $29^0 = 1$

Any real number (except 0) raised to the power zero is equal to 1.

(b) $(-29)^0 = 1$

(c) $-(-29)^0 = -(1) = -1$

(d) $-29^0 = -(29^0) = -1$

(e) $8^0 - 15^0 = 1 - 1 = 0$

(f) $(-15p^5)^0 = 1$

Since $p \neq 0$, $-15p^5$ will not equal zero. Therefore, $-15p^5$ raised to the power zero will equal 1.

3. **(a)** $6^{-3} = \frac{1}{6^3}$

(b) $8^{-1} = \frac{1}{8^1} = \frac{1}{8}$

(c) $(2x)^{-4} = \frac{1}{(2x)^4}, x \neq 0$

(d) $7r^{-6} = 7\left(\frac{1}{r^6}\right) = \frac{7}{r^6}, r \neq 0$

(e) $-q^{-4} = -\left(\frac{1}{q^4}\right) = -\frac{1}{q^4}, q \neq 0$

(f) $(-q)^{-4} = \frac{1}{(-q)^4}, q \neq 0$

(g) $3^{-1} + 5^{-1} = \frac{1}{3} + \frac{1}{5} = \frac{5}{15} + \frac{3}{15} = \frac{8}{15}$

(h) $4^{-1} - 2^{-1} = \frac{1}{4} - \frac{1}{2} = \frac{1}{4} - \frac{2}{4} = -\frac{1}{4}$

4. **(a)** $\frac{1}{4^{-3}} = \frac{1}{\frac{1}{4^3}} = 1 \div \frac{1}{4^3} = 1 \cdot \frac{4^3}{1} = 4^3 = 64$

(b) $$\begin{aligned}\frac{3^{-3}}{9^{-1}} &= \frac{\frac{1}{3^3}}{\frac{1}{9}} = \frac{1}{3^3} \div \frac{1}{9} = \frac{1}{3^3} \cdot \frac{9}{1}\\ &= \frac{1}{27} \cdot \frac{9}{1} = \frac{9}{27} = \frac{1}{3}\end{aligned}$$

5. **(a)** $\frac{4^8}{4^6} = 4^{8-6} = 4^2$

(b) $\frac{x^{12}}{x^3} = x^{12-3} = x^9, x \neq 0$

(c) $\frac{r^5}{r^8} = r^{5-8} = r^{-3} = \frac{1}{r^3}, r \neq 0$

(d) $\frac{2^8}{2^{-4}} = 2^{8-(-4)} = 2^{8+4} = 2^{12}$

(e) $\frac{6^{-3}}{6^4} = 6^{-3-4} = 6^{-7} = \frac{1}{6^7}$

(f) $\frac{8}{8^{-1}} = \frac{8^1}{8^{-1}} = 8^{1-(-1)} = 8^{1+1} = 8^2$

(g) $\frac{t^{-4}}{t^{-6}} = t^{-4-(-6)} = t^{-4+6} = t^2, t \neq 0$

(h) $\frac{x^3}{y^5}, y \neq 0$ cannot be simplified because the quotient rule does not apply. The bases x and y are different.

6. **(a)** $(r^5)^4 = r^{5\cdot 4} = r^{20}$

(b) $\left(\frac{3}{4}\right)^3 = \frac{3^3}{4^3} = \frac{27}{64}$

(c) $(9x)^3 = 9^3x^3 = 729x^3$

(d) $(5r^6)^3 = 5^3r^{6\cdot 3} = 5^3r^{18} = 125r^{18}$

(e) $$\begin{aligned}\left(\frac{-3n^4}{m}\right)^3 &= \frac{(-3)^3n^{4\cdot 3}}{m^3} = \frac{(-3)^3n^{12}}{m^3}\\ &= \frac{-27n^{12}}{m^3}, m \neq 0\end{aligned}$$

7. **(a)** $\left(\frac{3}{4}\right)^{-3} = \left(\frac{4}{3}\right)^3 = \frac{4^3}{3^3} = \frac{64}{27}$

(b) $\left(\frac{5}{6}\right)^{-2} = \left(\frac{6}{5}\right)^{2} = \frac{6^2}{5^2} = \frac{36}{25}$

8. (a) $5^4 \cdot 5^{-6} = 5^{4+(-6)} = 5^{-2} = \frac{1}{5^2}$, or $\frac{1}{25}$

(b) $x^{-4} \cdot x^{-6} \cdot x^8 = x^{-4+(-6)+8} = x^{-2} = \frac{1}{x^2}$

(c) $(5^{-3})^{-2} = 5^{(-3)(-2)} = 5^6$

(d) $(y^{-2})^7 = y^{(-2)7} = y^{-14} = \frac{1}{y^{14}}$

(e) $\frac{a^{-3}b^5}{a^4b^{-2}} = \frac{a^{-3}}{a^4} \cdot \frac{b^5}{b^{-2}} = a^{-3-4} \cdot b^{5-(-2)}$

$= a^{-7}b^7 = \frac{b^7}{a^7}$

(f) $(3^2k^{-4})^{-1} = (3^2)^{-1} \cdot (k^{-4})^{-1} = 3^{-2}k^4$

$= \frac{k^4}{3^2}$, or $\frac{k^4}{9}$

(g) $\left(\frac{2y}{x^3}\right)^2 \left(\frac{4y}{x}\right)^{-1}$

$= \frac{2^2y^2}{x^6} \cdot \frac{4^{-1}y^{-1}}{x^{-1}}$ *Combination of rules*

$= \frac{2^2 4^{-1} y^1}{x^5}$

$= \frac{2^2 y}{4x^5} = \frac{y}{x^5}$

(h) $\left(\frac{-28a^3b^{-5}}{7a^{-7}b^3}\right)^{-3}$

$= \left(\frac{-4a^{3-(-7)}b^{-5-3}}{1}\right)^{-3}$

$= (-4a^{10}b^{-8})^{-3}$

$= (-4)^{-3}(a^{10})^{-3}(b^{-8})^{-3}$

$= \frac{1}{(-4)^3}(a^{-30})(b^{24})$

$= \frac{b^{24}}{-64a^{30}} = -\frac{b^{24}}{64a^{30}}$

9. (a) $400{,}000 = 4_\wedge 00{,}000.$ ← Decimal point

Place a caret after the first nonzero digit, 4. Count 5 places from the decimal point to the caret (understood to be after the last 0). Use a positive exponent on 10 since $400{,}000 > 4$.

$$400{,}000 = 4 \times 10^5$$

(b) $29{,}800{,}000 = 2_\wedge 9{,}800{,}000.$

Count 7 places. Use a positive exponent on 10 since $29{,}800{,}000 > 2.98$.

$$29{,}800{,}000 = 2.98 \times 10^7$$

(c) $-6083 = -6_\wedge 083.$

Count 3 places. Use a positive exponent on 10 since $6083 > 6.083$.

$$-6083 = -6.083 \times 10^3$$

(d) $0.00172 = 0.001_\wedge 72$

Count 3 places. Use a negative exponent on 10 since $0.00172 < 1.72$.

$$0.00172 = 1.72 \times 10^{-3}$$

(e) $0.000\,000\,0503 = 0.000\,000\,05_\wedge 03$

Count 8 places. Use a negative exponent on 10 since $0.000\,000\,0503 < 5.03$.

$$0.000\,000\,0503 = 5.03 \times 10^{-8}$$

(f) $-0.0031 = -0.003_\wedge 1$

Count 3 places. Use a negative exponent on 10 since $0.0031 < 3.1$.

$$-0.0031 = -3.1 \times 10^{-3}$$

10. (a) $4.98 \times 10^5 = 4.98\,000. = 498{,}000$

Move the decimal 5 places to the right.

(b) $6.8 \times 10^{-7} = 0.000\,000\,6.8 = 0.000\,000\,68$

Move the decimal 7 places to the left.

(c) $-5.372 \times 10^0 = -5.372 \times 1 = -5.372$

11. $\frac{200{,}000 \times 0.0003}{0.06 \times 4{,}000{,}000} = \frac{2 \times 10^5 \times 3 \times 10^{-4}}{6 \times 10^{-2} \times 4 \times 10^6}$

$= \frac{2 \times 3 \times 10^5 \times 10^{-4}}{6 \times 4 \times 10^{-2} \times 10^6}$

$= \frac{2 \times 3 \times 10^1}{6 \times 4 \times 10^4}$

$= \frac{2 \times 3}{6 \times 4} \times 10^{-3}$

$= 0.25 \times 10^{-3}$

$= (2.5 \times 10^{-1}) \times 10^{-3}$

$= 2.5 \times 10^{-4}$, or 0.00025

12. $d = rt$, so

$t = \frac{d}{r}$

$= \frac{9.3 \times 10^7}{3.2 \times 10^3} = \frac{9.3}{3.2} \times 10^{7-3} \approx 2.9 \times 10^4.$

It would take approximately 2.9×10^4 hours.

6.1 Section Exercises

1. $(ab)^2 = a^2b^2$ by a power rule. Since $a^2b^2 \neq ab^2$, the expression $(ab)^2 = ab^2$ has been simplified incorrectly. The exponent should apply to both a and b.

3. $\left(\frac{4}{a}\right)^3 = \frac{4^3}{a^3} \quad (a \neq 0)$

Since $\frac{4^3}{a^3} \neq \frac{4^3}{a}$, the expression $\left(\frac{4}{a}\right)^3 = \frac{4^3}{a}$ has been simplified *incorrectly*.

5. $x^3 \cdot x^4 = x^{3+4} = x^7$ is correct.

7. $13^4 \cdot 13^8 = 13^{4+8} = 13^{12}$

9. $8^9 \cdot 8 = 8^9 \cdot 8^1 = 8^{9+1} = 8^{10}$

11. $x^3 \cdot x^5 \cdot x^9 = x^{3+5+9} = x^{17}$

13. $(-3w^5)(9w^3) = (-3)(9)w^{5+3} = -27w^8$

15. $(2x^2y^5)(9xy^3) = (2)(9)x^{2+1}y^{5+3} = 18x^3y^8$

17. $r^2 \cdot s^4$ Because the bases are not the same, the product rule does not apply.

19. **(a)** $9^0 = 1$ **(B)**

(b) $-9^0 = -(9^0) = -(1) = -1$ **(C)**

(c) $(-9)^0 = 1$ **(B)**

(d) $-(-9)^0 = -1$ **(C)**

21. $17^0 = 1$, since $a^0 = 1$ for any nonzero base a.

23. $-5^0 = -(5^0) = -(1) = -1$

25. $(-15)^0 = 1$, since $a^0 = 1$ for any nonzero base a.

27. $-4^0 - m^0 = -1 - 1 = -2$

29. **(a)** $4^{-2} = \frac{1}{4^2} = \frac{1}{16}$ **(B)**

(b) $-4^{-2} = -\frac{1}{4^2} = -\frac{1}{16}$ **(D)**

(c) $(-4)^{-2} = \frac{1}{(-4)^2} = \frac{1}{16}$ **(B)**

(d) $-(-4)^{-2} = -\frac{1}{(-4)^2} = -\frac{1}{16}$ **(D)**

31. $5^{-4} = \frac{1}{5^4}$, or $\frac{1}{625}$

33. $8^{-1} = \frac{1}{8^1} = \frac{1}{8}$

35. $(4x)^{-2} = \frac{1}{(4x)^2} = \frac{1}{4^2x^2} = \frac{1}{16x^2}$

37. $4x^{-2} = \frac{4}{x^2}$

39. $-a^{-3} = -\frac{1}{a^3}$

41. $(-a)^{-4} = \frac{1}{(-a)^4}$, or $\frac{1}{a^4}$

43. $5^{-1} + 6^{-1} = \frac{1}{5} + \frac{1}{6} = \frac{6}{30} + \frac{5}{30} = \frac{11}{30}$

45. $8^{-1} - 3^{-1} = \frac{1}{8} - \frac{1}{3} = \frac{3}{24} - \frac{8}{24} = -\frac{5}{24}$

47. $\frac{1}{4^{-2}} = 4^2 = 16$

49. $\frac{2^{-2}}{3^{-3}} = \frac{3^3}{2^2} = \frac{27}{4}$

51. $\left(\frac{2}{3}\right)^{-3} = \left(\frac{3}{2}\right)^3 = \frac{3^3}{2^3} = \frac{27}{8}$

53. $\left(\frac{4}{5}\right)^{-2} = \left(\frac{5}{4}\right)^2 = \frac{5^2}{4^2} = \frac{25}{16}$

55. **(a)** $\left(\frac{1}{3}\right)^{-1} = \left(\frac{3}{1}\right)^1 = 3$ **(B)**

(b) $\left(-\frac{1}{3}\right)^{-1} = \left(-\frac{3}{1}\right)^1 = -3$ **(D)**

(c) $-\left(\frac{1}{3}\right)^{-1} = -\left(\frac{3}{1}\right)^1 = -3$ **(D)**

(d) $-\left(-\frac{1}{3}\right)^{-1} = -\left(-\frac{3}{1}\right)^1 = -(-3) = 3$ **(B)**

57. $\frac{4^8}{4^6} = 4^{8-6} = 4^2$, or 16

59. $\frac{x^{12}}{x^8} = x^{12-8} = x^4$

61. $\frac{r^7}{r^{10}} = r^{7-10} = r^{-3} = \frac{1}{r^3}$

63. $\frac{6^4}{6^{-2}} = 6^{4-(-2)} = 6^{4+2} = 6^6$

65. $\frac{6^{-3}}{6^7} = 6^{-3-7} = 6^{-10} = \frac{1}{6^{10}}$

67. $\frac{7}{7^{-1}} = 7^{1-(-1)} = 7^2$, or 49

69. $\frac{r^{-3}}{r^{-6}} = r^{-3-(-6)} = r^{-3+6} = r^3$

71. $\dfrac{x^3}{y^2}$ Because the bases are not the same, the quotient rule does not apply.

73. $(x^3)^6 = x^{3\cdot 6} = x^{18}$

75. $\left(\dfrac{3}{5}\right)^3 = \dfrac{3^3}{5^3} = \dfrac{27}{125}$

77. $(4t)^3 = 4^3t^3 = 64t^3$

79. $(-6x^2)^3 = (-6)^3x^{2\cdot 3} = -216x^6$

81. $\left(\dfrac{-4m^2}{t}\right)^3 = \dfrac{(-4)^3m^{2\cdot 3}}{t^3} = \dfrac{-64m^6}{t^3} = -\dfrac{64m^6}{t^3}$

83. $\left(\dfrac{-s^3}{t^5}\right)^4 = \dfrac{(-1)^4s^{3\cdot 4}}{t^{5\cdot 4}} = \dfrac{s^{12}}{t^{20}}$

85. $3^5 \cdot 3^{-6} = 3^{5+(-6)} = 3^{-1} = \dfrac{1}{3^1} = \dfrac{1}{3}$

87. $a^{-3}a^2a^{-4} = a^{-3+2+(-4)} = a^{-5} = \dfrac{1}{a^5}$

89. $(k^2)^{-3}k^4 = k^{2(-3)}k^4$
$= k^{-6}k^4$
$= k^{-6+4}$
$= k^{-2}$
$= \dfrac{1}{k^2}$

91. $-4r^{-2}(r^4)^2 = -4r^{-2}(r^8)$
$= -4r^{-2+8}$
$= -4r^6$

93. $(5a^{-1})^4(a^2)^{-3} = 5^4a^{-1\cdot 4}a^{2(-3)}$
$= 5^4a^{-4}a^{-6}$
$= 5^4a^{-4-6}$
$= 5^4a^{-10}$
$= \dfrac{5^4}{a^{10}} = \dfrac{625}{a^{10}}$

95. $(z^{-4}x^3)^{-1} = z^{-4(-1)}x^{3(-1)}$
$= z^4x^{-3}$
$= \dfrac{z^4}{x^3}$

97. $\dfrac{(p^{-2})^3}{5p^4} = \dfrac{p^{-6}}{5p^4} = \dfrac{p^{-6-4}}{5} = \dfrac{p^{-10}}{5} = \dfrac{1}{5p^{10}}$

99. $\dfrac{4a^5(a^{-1})^3}{(a^{-2})^{-2}} = \dfrac{4a^5a^{-1\cdot 3}}{a^{-2(-2)}} = \dfrac{4a^5a^{-3}}{a^4}$
$= 4a^{5+(-3)-4}$
$= 4a^{-2}$
$= \dfrac{4}{a^2}$

101. $\dfrac{(2k)^2m^{-5}}{(km)^{-3}} = \dfrac{2^2k^2m^{-5}}{k^{-3}m^{-3}}$
$= 2^2k^{2-(-3)}m^{-5-(-3)}$
$= 2^2k^5m^{-2}$
$= \dfrac{2^2k^5}{m^2}$, or $\dfrac{4k^5}{m^2}$

103. $\left(\dfrac{3k^{-2}}{k^4}\right)^{-1} \cdot \dfrac{2}{k} = \dfrac{(3k^{-2})^{-1}}{(k^4)^{-1}} \cdot \dfrac{2}{k}$
$= \dfrac{3^{-1}k^2 \cdot 2}{k^{-4}k^1}$
$= \dfrac{3^{-1} \cdot 2k^2}{k^{-3}}$
$= \dfrac{2k^2k^3}{3}$
$= \dfrac{2k^5}{3}$

105. $\left(\dfrac{2p}{q^2}\right)^3\left(\dfrac{3p^4}{q^{-4}}\right)^{-1} = \dfrac{(2p)^3}{(q^2)^3} \cdot \dfrac{(3p^4)^{-1}}{(q^{-4})^{-1}}$
$= \dfrac{2^3p^3}{q^6} \cdot \dfrac{3^{-1}p^{-4}}{q^4}$
$= \dfrac{2^3p^{3+(-4)}}{3^1q^{6+4}}$
$= \dfrac{8p^{-1}}{3q^{10}} = \dfrac{8}{3pq^{10}}$

107. $\left(\dfrac{3a^{-4}b^6}{15a^2b^{-4}}\right)^{-2} = \left(\dfrac{a^{-4-2}b^{6-(-4)}}{5}\right)^{-2}$
$= \left(\dfrac{a^{-6}b^{10}}{5}\right)^{-2}$
$= \left(\dfrac{5}{a^{-6}b^{10}}\right)^2$
$= \dfrac{5^2}{(a^{-6})^2(b^{10})^2}$
$= \dfrac{25}{a^{-12}b^{20}}$
$= \dfrac{25a^{12}}{b^{20}}$

109. $530 = 5_{\wedge}30.$ ← Decimal point

Count 2 places.

Since the number 5.3 is to be made greater, the exponent on 10 is positive.

$$530 = 5.3 \times 10^2$$

111. $0.830 = 0.8_{\wedge}30$

Count 1 place.

Since the number 8.3 is to be made less, the exponent on 10 is negative.

$$0.830 = 8.3 \times 10^{-1}$$

113. $0.000\,006\,92 = 0.00\,0\,0\,0\,6_{\wedge}92$

Count 6 places.

Since the number 6.92 is to be made less, the exponent on 10 is negative.

$$0.000\,006\,92 = 6.92 \times 10^{-6}$$

115. $-38{,}500 = -3_{\wedge}85\,00.$

Count 4 places.

Since the number 3.85 is to be made greater, the exponent on 10 is positive. Also, affix a negative sign in front of the number.

$$-38{,}500 = -3.85 \times 10^4$$

117. $7.2 \times 10^4 = 7.2000. = 72{,}000$

Move the decimal point 4 places to the *right* because of the *positive* exponent. Attach extra zeros.

119. $2.54 \times 10^{-3} = 0.002.54 = 0.00254$

Since the exponent is *negative*, move the decimal point 3 places to the *left.*

121. $-6 \times 10^4 = -6.0000. = -60{,}000$

Move the decimal point 4 places to the *right* because of the *positive* exponent. Attach extra zeros.

123. $1.2 \times 10^{-5} = 0.00001.2 = 0.000\,012$

Since the exponent is *negative*, move the decimal point 5 places to the *left.*

125.
$$\begin{aligned}\frac{3 \times 10^{-2}}{12 \times 10^3} &= \frac{3 \times 10^{-2}}{1.2 \times 10^4}\\ &= \frac{3}{1.2} \times \frac{10^{-2}}{10^4}\\ &= 2.5 \times 10^{-6}\\ &= 0.000\,002\,5\end{aligned}$$

127.
$$\begin{aligned}\frac{0.05 \times 1600}{0.0004} &= \frac{5 \times 10^{-2} \times 1.6 \times 10^3}{4 \times 10^{-4}}\\ &= \frac{5(1.6)}{4} \times \frac{10^{-2} \times 10^3}{10^{-4}}\\ &= 2 \times 10^{-2+3-(-4)}\\ &= 2 \times 10^5\\ &= 200{,}000\end{aligned}$$

129.
$$\begin{aligned}\$1{,}000{,}000{,}000 &= \$1 \times 10^9\\ \$1{,}000{,}000{,}000{,}000 &= \$1 \times 10^{12}\\ \$3{,}100{,}000{,}000{,}000 &= \$3.1 \times 10^{12}\\ 210{,}385 &= 2.10385 \times 10^5\end{aligned}$$

131. (a) 304.1 million
$$\begin{aligned}&= 304{,}100{,}000\\ &= 3.041 \times 10^8\end{aligned}$$

(b) $\$1{,}000{,}000{,}000{,}000 = \$1 \times 10^{12}$

(c) Divide the amount by the number of people to determine how much each person would have to contribute.

$$\begin{aligned}\frac{\$1 \times 10^{12}}{3.041 \times 10^8} &\approx \$0.3288 \times 10^4\\ &= \$3288\end{aligned}$$

In 2008, each person in the United States would have had to contribute about \$3288 in order to make someone a trillionaire.

133. If x represents the population of Monaco in 2006, then $38.64x$ represents the population of Japan.

$$\begin{aligned}38.64x &= 1.275 \times 10^6.\\ x &= \frac{1.275 \times 10^6}{3.864 \times 10^1}\\ &= \frac{1.275}{3.864} \times \frac{10^6}{10^1}\\ &\approx 0.33 \times 10^5\\ &= 33{,}000\end{aligned}$$

In 2006, the population of Monaco was about 33,000, or 3.3×10^4.

135. Use $d = rt$, or $\frac{d}{r} = t$, where $d = 9.3 \times 10^7$ and $r = 2.9 \times 10^3$.

$$\begin{aligned}\frac{9.3 \times 10^7}{2.9 \times 10^3} &= \frac{9.3}{2.9} \times 10^{7-3}\\ &\approx 3.2 \times 10^4\end{aligned}$$

It would take about 3.2×10^4 or 32,000 hours. Note that there are

$$24 \times 365 = 8760$$

hours in one year.

$$\frac{32{,}000 \text{ hours}}{8760 \text{ hours/year}} \approx 3.7 \text{ years}$$

137. (a) The distance from Mercury to the sun is 3.6×10^7 mi and the distance from Venus to the sun is 6.7×10^7 mi, so the distance between Mercury and Venus in miles is

$$\begin{aligned}&(6.7 \times 10^7) - (3.6 \times 10^7)\\&= (6.7 - 3.6) \times 10^7\\&= 3.1 \times 10^7.\end{aligned}$$

Use $d = rt$, or $d/r = t$, where $d = 3.1 \times 10^7$ and $r = 1.55 \times 10^3$.

$$\begin{aligned}\frac{3.1 \times 10^7}{1.55 \times 10^3} &= \frac{3.1}{1.55} \times 10^{7-3}\\&= 2 \times 10^4\\&= 20{,}000\end{aligned}$$

It would take 20,000 hr.

(b) From part (a), it takes 20,000 hr for a spacecraft to travel from Venus to Mercury. Convert this to days (24 hours = 1 day).

$$\begin{aligned}20{,}000 \text{ hr} &= \tfrac{20{,}000}{24} \text{ days}\\&\approx 833 \text{ days}\end{aligned}$$

It would take about 833 days.

139. Consider $-a^n$ and $(-a)^n$. Let $a = 2$.

If $n = 2$, then
$-a^n = -2^2 = -4$ and $(-a)^n = (-2)^2 = 4$.
If $n = 3$, then
$-a^n = -2^3 = -8$ and $(-a)^n = (-2)^3 = -8$.
If $n = 4$, then
$-a^n = -2^4 = -16$ and $(-a)^n = (-2)^4 = 16$.
If $n = 5$, then
$-a^n = -2^5 = -32$ and $(-a)^n = (-2)^5 = -32$.
If $n = 6$, then
$-a^n = -2^6 = -64$ and $(-a)^n = (-2)^6 = 64$.

Based on these cases, when n is even, the expressions are opposites. When n is odd, the expressions are equal.

141. Write the fraction as its reciprocal raised to the opposite of the negative power.

6.2 Adding and Subtracting Polynomials

6.2 Margin Exercises

1. **(a)** In the term $-9m^5$, the coefficient is -9.

(b) In $12y^2x$, the coefficient is 12.

(c) In x, the coefficient is 1 since $x = 1 \cdot x$.

(d) In $-y$, the coefficient is -1 since $-y = -1 \cdot y$.

(e) In $\frac{z}{4}$, the coefficient is $\frac{1}{4}$ since $\frac{z}{4} = \frac{1}{4}z$.

2. **(a)** $-4 + 9y + y^3$ is written in descending powers as $y^3 + 9y - 4$.

(b) $-3z^4 + 2z^3 + z^5 - 6z$ is written in descending powers as $z^5 - 3z^4 + 2z^3 - 6z$.

(c) $-12m^{10} + 8m^9 + 10m^{12}$ is written in descending powers as $10m^{12} - 12m^{10} + 8m^9$.

3. **(a)** $12m^4 - 6m^2$ is a binomial since it has two terms.

(b) $-6y^3 + 2y^2 - 8y$ is a trinomial since it has three terms.

(c) $3a^5$ is a monomial since it has one term.

(d) $-2k^{10} + 2k^9 - 8k^5 + 2k$ has four terms, so it is none of the choices.

4. The degree of a polynomial is the highest degree of any of the terms.

(a) $9y^4 + 8y^3 - 6$ has degree 4 since the largest exponent on y is 4.

(b) $-12m^7 + 11m^3 + m^9$ has degree 9.

(c) $-2k$ has degree 1 since $-2k = -2k^1$.

(d) 10 has degree 0 since $10 = 10x^0$.

(e) $3mn^2 + 2m^3n$ has degree 4.

Since the degree of $3mn^2$, or $3m^1n^2$, is $1 + 2 = 3$, and the degree of $2m^3n$, or $2m^3n^1$, is $3 + 1 = 4$, the degree of the polynomial is the greatest degree, 4.

5. **(a)**
$$\begin{aligned}&11x + 12x - 7x - 3x\\&= (11 + 12 - 7 - 3)x = 13x\end{aligned}$$

(b)
$$\begin{aligned}&11p^5 + 4p^5 - 6p^3 + 8p^3\\&= (11 + 4)p^5 + (-6 + 8)p^3\\&= 15p^5 + 2p^3\end{aligned}$$

(c)
$$\begin{aligned}&2y^2z^4 + 3y^4 + 5y^4 - 9y^4z^2\\&= 2y^2z^4 + (3 + 5)y^4 - 9y^4z^2\\&= 2y^2z^4 + 8y^4 - 9y^4z^2\end{aligned}$$

6. **(a)** Add using the horizontal method.

$$\begin{aligned}&(12y^2 - 7y + 9) + (-4y^2 - 11y + 5)\\&= 12y^2 - 4y^2 - 7y - 11y + 9 + 5\\&= 8y^2 - 18y + 14\end{aligned}$$

Add using the vertical method.

$$\begin{array}{rcrcr} 12y^2 & - & 7y & + & 9\\ -4y^2 & - & 11y & + & 5\\ \hline 8y^2 & - & 18y & + & 14\end{array}$$

(b) Add using the vertical method.

$$\begin{array}{r} -6r^5 + 2r^3 - r^2 \\ 8r^5 - 2r^3 + 5r^2 \\ \hline 2r^5 + 0r^3 + 4r^2 \end{array} = 2r^5 + 4r^2$$

Add using the horizontal method.

$$(-6r^5 + 2r^3 - r^2) + (8r^5 - 2r^3 + 5r^2)$$
$$= -6r^5 + 8r^5 + 2r^3 - 2r^3 - r^2 + 5r^2$$
$$= 2r^5 + 4r^2$$

7. **(a)** Subtract using the horizontal method.

$$(6y^3 - 9y^2 + 8) - (2y^3 + y^2 + 5)$$
$$= 6y^3 - 9y^2 + 8 - 2y^3 - y^2 - 5$$
$$= 6y^3 - 2y^3 - 9y^2 - y^2 + 8 - 5$$
$$= 4y^3 - 10y^2 + 3$$

Subtract using the vertical method.

$$\begin{array}{r} 6y^3 - 9y^2 + 8 \\ \underline{2y^3 + y^2 + 5} \end{array}$$

Change all the signs in the second polynomial, and add.

$$\begin{array}{rl} 6y^3 - 9y^2 + 8 & \\ -2y^3 - y^2 - 5 & \textit{Change all signs.} \\ \hline 4y^3 - 10y^2 + 3 & \textit{Add in columns.} \end{array}$$

(b) Subtract using the vertical method.

$$\begin{array}{r} 6y^3 - 2y^2 + 5y \\ \underline{-2y^3 + 8y^2 - 11y} \end{array}$$

Change all the signs in the second polynomial, and add.

$$\begin{array}{rl} 6y^3 - 2y^2 + 5y & \\ 2y^3 - 8y^2 + 11y & \textit{Change all signs.} \\ \hline 8y^3 - 10y^2 + 16y & \textit{Add in columns.} \end{array}$$

Subtract using the horizontal method.

$$(6y^3 - 2y^2 + 5y) - (-2y^3 + 8y^2 - 11y)$$
$$= 6y^3 - 2y^2 + 5y + 2y^3 - 8y^2 + 11y$$
$$= 6y^3 + 2y^3 - 2y^2 - 8y^2 + 5y + 11y$$
$$= 8y^3 - 10y^2 + 16y$$

6.2 Section Exercises

1. In $2x^3 + x - 3x^2$, the degrees of the terms are 3, 1, and 2 from left to right. Therefore, this polynomial is written in neither ascending nor descending powers.

3. The polynomial $4p^3 - 8p^5 + p^7$ is written in ascending powers, since the exponents are increasing from left to right.

5. In $-m^3 + 5m^2 + 3m + 10$, the degrees of the terms are 3, 2, 1, and 0 from left to right. Therefore, this polynomial is written in descending powers.

7. In $7z$, the coefficient is 7 and, since $7z = 7z^1$, the degree is 1.

9. In $-15p^2$, the coefficient is -15 and the degree is 2.

11. In x^4, since $x^4 = 1x^4$, the coefficient is 1 and the degree is 4.

13. In $-mn^5$, the coefficient is -1, since $-mn^5 = -1mn^5$, and the degree is 6, since the sum of the exponents on m and n is $1 + 5 = 6$.

15. 24 is one term, so it's a *monomial.* 24 is a nonzero constant, so it has degree zero.

17. $7m - 21$ has two terms, so it's a *binomial.* The exponent on m is 1, so $7m - 21$ has degree 1.

19. $2r^3 + 3r^2 + 5r$ has three terms, so it's a *trinomial.* The greatest exponent is 3, so the degree is 3.

21. $-6p^4q - 3p^3q^2 + 2pq^3 - q^4$ has four terms, so it is classified as *none of these.* The greatest sum of exponents on any term is 5, so the polynomial has degree 5.

23. $5z^4 + 3z^4 = (5 + 3)z^4 = 8z^4$

25. $-m^3 + 2m^3 + 6m^3$
$$= (-1 + 2 + 6)m^3$$
$$= 7m^3$$

27. $x + x + x + x + x$
$$= (1 + 1 + 1 + 1 + 1)x$$
$$= 5x$$

29. $m^4 - 3m^2 + m$ is *already simplified* since there are no like terms to be combined.

31. $y^2 + 7y - 4y^2 = (y^2 - 4y^2) + 7y$
$$= (1 - 4)y^2 + 7y$$
$$= -3y^2 + 7y$$

33. $2k + 3k^2 + 5k^2 - 7$
$$= (3k^2 + 5k^2) + 2k - 7$$
$$= (3 + 5)k^2 + 2k - 7$$
$$= 8k^2 + 2k - 7$$

35. $n^4 - 2n^3 + n^2 - 3n^4 + n^3$
$$= n^4 - 3n^4 - 2n^3 + n^3 + n^2$$
$$= (1 - 3)n^4 + (-2 + 1)n^3 + n^2$$
$$= -2n^4 - n^3 + n^2$$

37.
$$\begin{array}{r} -12p^2 + 4p - 1 \\ 3p^2 + 7p - 8 \\ \hline -9p^2 + 11p - 9 \end{array}$$
Add vertically.

39.
$$\begin{array}{r} 12a + 15 \\ 7a - 3 \\ \hline \end{array}$$

To subtract, change all the signs in the second polynomial, and add.

$$\begin{array}{r} 12a + 15 \\ -7a + 3 \\ \hline 5a + 18 \end{array}$$

41.
$$\begin{array}{r} 6m^2 - 11m + 5 \\ -8m^2 + 2m - 1 \\ \hline \end{array}$$

To subtract, change all the signs in the second polynomial, and add.

$$\begin{array}{r} 6m^2 - 11m + 5 \\ 8m^2 - 2m + 1 \\ \hline 14m^2 - 13m + 6 \end{array}$$

43. Add column by column to obtain the result on the bottom line.

$$\begin{array}{r} 12z^2 - 11z + 8 \\ 5z^2 + 16z - 2 \\ -4z^2 + 5z - 9 \\ \hline 13z^2 + 10z - 3 \end{array}$$

45.
$$\begin{array}{r} 6y^3 - 9y^2 \qquad + 8 \\ 4y^3 + 2y^2 + 5y \qquad \\ \hline 10y^3 - 7y^2 + 5y + 8 \end{array}$$

47.
$$\begin{array}{r} -5a^4 \qquad + 8a^2 - 9 \\ 6a^3 - a^2 + 2 \\ \hline \end{array}$$

To subtract, change all the signs in the second polynomial, and add.

$$\begin{array}{r} -5a^4 \qquad + 8a^2 - 9 \\ - 6a^3 + a^2 - 2 \\ \hline -5a^4 - 6a^3 + 9a^2 - 11 \end{array}$$

49. $(3r + 8) - (2r - 5)$

Change all signs in the second polynomial and add.

$$\begin{aligned} &= (3r + 8) + (-2r + 5) \\ &= 3r + 8 - 2r + 5 \\ &= 3r - 2r + 8 + 5 \\ &= r + 13 \end{aligned}$$

51.
$$\begin{aligned} &(5x^2 + 7x - 4) + (3x^2 - 6x + 2) \\ &= 5x^2 + 3x^2 + 7x - 6x - 4 + 2 \\ &= 8x^2 + x - 2 \end{aligned}$$

53.
$$\begin{aligned} &(2a^2 + 3a - 1) - (4a^2 + 5a + 6) \\ &= (2a^2 + 3a - 1) + (-4a^2 - 5a - 6) \\ &= 2a^2 - 4a^2 + 3a - 5a - 1 - 6 \\ &= -2a^2 - 2a - 7 \end{aligned}$$

55.
$$\begin{aligned} &(z^5 + 3z^2 + 2z) - (4z^5 + 2z^2 - 5z) \\ &= z^5 + 3z^2 + 2z - 4z^5 - 2z^2 + 5z \\ &= z^5 - 4z^5 + 3z^2 - 2z^2 + 2z + 5z \\ &= -3z^5 + z^2 + 7z \end{aligned}$$

6.3 Polynomial Functions

6.3 Margin Exercises

1. $f(x) = -x^2 + 5x - 11$

(a)
$$\begin{aligned} f(1) &= -1^2 + 5(1) - 11 \\ &= -1 + 5 - 11 \\ &= -7 \end{aligned}$$

(b)
$$\begin{aligned} f(-4) &= -(-4)^2 + 5(-4) - 11 \\ &= -16 - 20 - 11 \\ &= -47 \end{aligned}$$

(c)
$$\begin{aligned} f(0) &= -0^2 + 5(0) - 11 \\ &= 0 + 0 - 11 \\ &= -11 \end{aligned}$$

2. For the year 2000, $x = 2000 - 1990 = 10$.

$$\begin{aligned} P(x) &= -0.01085x^2 + 0.4572x + 12.70 \\ P(10) &= -0.01085(10)^2 + 0.4572(10) + 12.70 \\ &= 16.187 \end{aligned}$$

Thus, in 2000, there were about 16.2 million high school students in the United States.

3. $f(x) = 3x^2 + 8x - 6$, $g(x) = -4x^2 + 4x - 8$

(a)
$$\begin{aligned} &(f + g)(x) \\ &= f(x) + g(x) \\ &= (3x^2 + 8x - 6) + (-4x^2 + 4x - 8) \\ &= 3x^2 - 4x^2 + 8x + 4x - 6 - 8 \\ &= -x^2 + 12x - 14 \end{aligned}$$

(b)
$$\begin{aligned} &(f - g)(x) \\ &= f(x) - g(x) \\ &= (3x^2 + 8x - 6) - (-4x^2 + 4x - 8) \\ &= 3x^2 + 8x - 6 + 4x^2 - 4x + 8 \\ &= 7x^2 + 4x + 2 \end{aligned}$$

4. **(a)** $(f+g)(x)$
$= f(x) + g(x)$
$= (18x^2 - 24x) + (3x)$
$= 18x^2 - 21x$

$(f+g)(-1)$
$= 18(-1)^2 - 21(-1)$
$= 18 + 21$
$= 39$

(b) $(f-g)(x)$
$= f(x) - g(x)$
$= (18x^2 - 24x) - (3x)$
$= 18x^2 - 27x$

$(f-g)(1)$
$= 18(1)^2 - 27(1)$
$= 18 - 27$
$= -9$

5.

x	x^2	$f(x) = -2x^2$
−2	4	−8
−1	1	−2
0	0	0
1	1	−2
2	4	−8

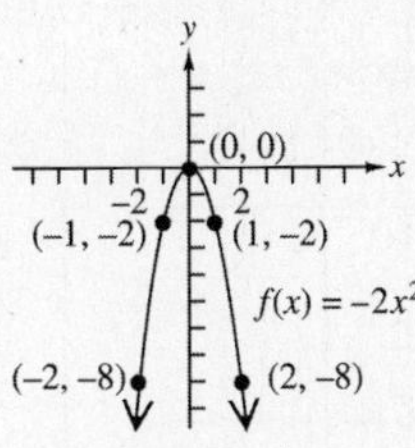

Any value of x can be used, so the domain is $(-\infty, \infty)$. The maximum y-value is 0 and there is no minimum y-value, so the range is $(-\infty, 0]$.

6.3 Section Exercises

1. $f(x) = 6x - 4$

(a) $f(-1) = 6(-1) - 4$
$= -6 - 4 = -10$

(b) $f(2) = 6(2) - 4$
$= 12 - 4 = 8$

3. $f(x) = x^2 - 3x + 4$

(a) $f(-1) = (-1)^2 - 3(-1) + 4$
$= 1 + 3 + 4 = 8$

(b) $f(2) = (2)^2 - 3(2) + 4$
$= 4 - 6 + 4 = 2$

5. $f(x) = 5x^4 - 3x^2 + 6$

(a) $f(-1) = 5(-1)^4 - 3(-1)^2 + 6$
$= 5 \cdot 1 - 3 \cdot 1 + 6$
$= 5 - 3 + 6 = 8$

(b) $f(2) = 5(2)^4 - 3(2)^2 + 6$
$= 5 \cdot 16 - 3 \cdot 4 + 6$
$= 80 - 12 + 6 = 74$

7. $f(x) = -x^2 + 2x^3 - 8$

(a) $f(-1) = -(-1)^2 + 2(-1)^3 - 8$
$= -(1) + 2(-1) - 8$
$= -1 - 2 - 8 = -11$

(b) $f(2) = -(2)^2 + 2(2)^3 - 8$
$= -(4) + 2 \cdot 8 - 8$
$= -4 + 16 - 8 = 4$

9. $f(x) = -2.3425x^2 + 248.04x + 15{,}160$

(a) $x = 1980 - 1980 = 0$

$f(0) = -2.3425(0)^2 + 248.04(0) + 15{,}160$
$= 15{,}160$ airports

(b) $x = 1995 - 1980 = 15$

$f(15) = -2.3425(15)^2 + 248.04(15) + 15{,}160$
$= 18{,}353.5375 \approx 18{,}354$ airports

(c) $x = 2005 - 1980 = 25$

$f(25) = -2.3425(25)^2 + 248.04(25) + 15{,}160$
$= 19{,}896.9375 \approx 19{,}897$ airports

11. $P(x) =$
$-0.00189x^3 + 0.1193x^2 + 2.027x + 28.19$

(a) $x = 1985 - 1985 = 0$

$P(0) = -0.00189(0)^3 + 0.1193(0)^2$
$+ 2.027(0) + 28.19$
$= 28.19$ ($28.2 billion)

(b) $x = 2000 - 1985 = 15$

$P(15) = -0.00189(15)^3 + 0.1193(15)^2$
$+ 2.027(15) + 28.19$
$= 79.05875$ ($79.1 billion)

(c) $x = 2006 - 1985 = 21$

$P(21) = -0.00189(21)^3 + 0.1193(21)^2$
$+ 2.027(21) + 28.19$
$= 105.86501$ ($105.9 billion)

13. **(a)** $(f+g)(x) = f(x) + g(x)$
$= (5x - 10) + (3x + 7)$
$= 8x - 3$

(b) $(f-g)(x) = f(x) - g(x)$
$= (5x-10) - (3x+7)$
$= (5x-10) + (-3x-7)$
$= 2x - 17$

15. **(a)** $(f+g)(x)$
$= f(x) + g(x)$
$= (4x^2+8x-3) + (-5x^2+4x-9)$
$= -x^2 + 12x - 12$

(b) $(f-g)(x)$
$= f(x) - g(x)$
$= (4x^2+8x-3) - (-5x^2+4x-9)$
$= (4x^2+8x-3) + (5x^2-4x+9)$
$= 9x^2 + 4x + 6$

For Exercises 17–28, let $f(x) = x^2 - 9$, $g(x) = 2x$, and $h(x) = x - 3$.

17. $(f+g)(x) = f(x) + g(x)$
$= (x^2-9) + (2x)$
$= x^2 + 2x - 9$

19. $(f+g)(3) = f(3) + g(3)$
$= (3^2-9) + 2(3)$
$= 0 + 6 = 6$

Alternatively, we could evaluate the polynomial in Exercise 17, $x^2 + 2x - 9$, using $x = 3$.

21. $(f-h)(x) = f(x) - h(x)$
$= (x^2-9) - (x-3)$
$= x^2 - 9 - x + 3$
$= x^2 - x - 6$

23. $(f-h)(-3) = f(-3) - h(-3)$
$= [(-3)^2 - 9] - [(-3) - 3]$
$= (9-9) - (-6)$
$= 0 + 6 = 6$

Alternatively, we could evaluate the polynomial in Exercise 21, $x^2 - x - 6$, using $x = -3$.

25. $(g+h)(-10) = g(-10) + h(-10)$
$= 2(-10) + [(-10) - 3]$
$= -20 + (-13)$
$= -33$

27. $(g-h)(-3) = g(-3) - h(-3)$
$= 2(-3) - [(-3) - 3]$
$= -6 - (-6)$
$= -6 + 6 = 0$

29.

x	$f(x) = -2x + 1$
-2	$-2(-2) + 1 = 5$
-1	$-2(-1) + 1 = 3$
0	$-2(0) + 1 = 1$
1	$-2(1) + 1 = -1$
2	$-2(2) + 1 = -3$

This is a linear function, so plot the points and draw a line through them.

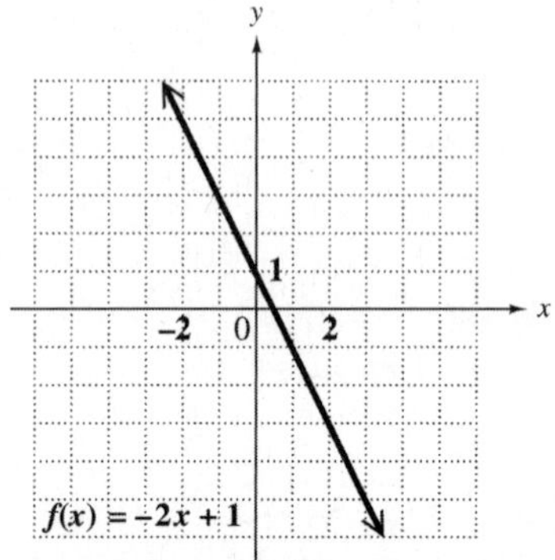

Any x-value can be used, so the domain is $(-\infty, \infty)$. From the graph, we see that any y-value can be obtained from the function, so the range is $(-\infty, \infty)$.

31.

x	$f(x) = -3x^2$
-2	$-3(-2)^2 = -12$
-1	$-3(-1)^2 = -3$
0	$-3(0)^2 = 0$
1	$-3(1)^2 = -3$
2	$-3(2)^2 = -12$

Since the greatest exponent is 2, the graph of f is a parabola.

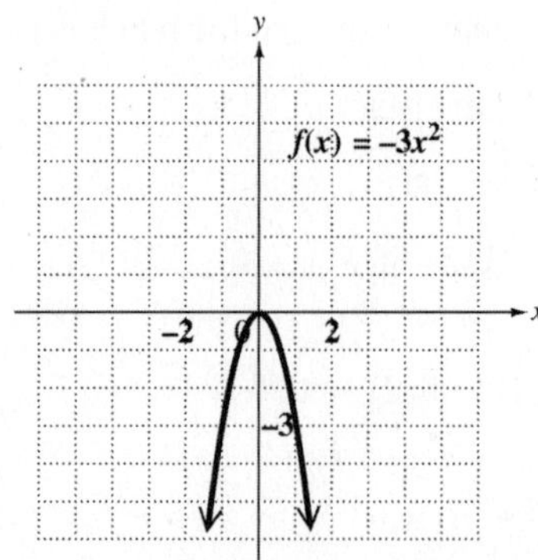

Any x-value can be used, so the domain is $(-\infty, \infty)$. From the graph, we see that the y-values are at most 0, so the range is $(-\infty, 0]$.

33.

x	$f(x) = x^3 + 1$
-2	$(-2)^3 + 1 = -7$
-1	$(-1)^3 + 1 = 0$
0	$(0)^3 + 1 = 1$
1	$(1)^3 + 1 = 2$
2	$(2)^3 + 1 = 9$

The greatest exponent is 3, so the graph of f is s-shaped.

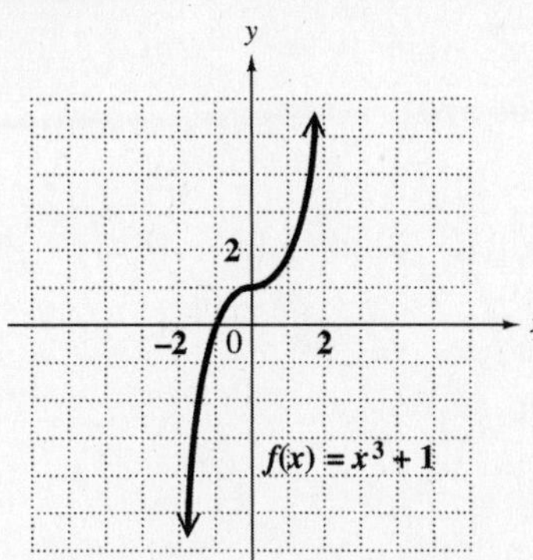

Any x-value can be used, so the domain is $(-\infty, \infty)$. From the graph, we see that any y-value can be obtained from the function, so the range is $(-\infty, \infty)$.

6.4 Multiplying Polynomials

6.4 Margin Exercises

1. **(a)** $-6m^5(2m^4) = (-6)(2)m^5 \cdot m^4$
$= -12m^{5+4}$
$= -12m^9$

(b) $8k^3y(9ky^3) = (8)(9)k^3 \cdot k^1 \cdot y^1 \cdot y^3$
$= 72k^{3+1}y^{1+3}$
$= 72k^4y^4$

2. **(a)** $-2r(9r - 5) = -2r(9r) - 2r(-5)$
$= -18r^2 + 10r$

(b) $3p^2(5p^3 + 2p^2 - 7)$
$= 3p^2(5p^3) + 3p^2(2p^2) + 3p^2(-7)$
$= 15p^5 + 6p^4 - 21p^2$

(c) $(4a - 5)(3a + 6)$
$= (4a - 5)(3a) + (4a - 5)(6)$
$= (4a)(3a) + (-5)(3a) + (4a)(6) + (-5)(6)$
$= 12a^2 - 15a + 24a - 30$
$= 12a^2 + 9a - 30$

(d) $3x^3(x + 4)(x - 6)$
$= 3x^3[(x + 4)(x) + (x + 4)(-6)]$
$= 3x^3[x^2 + 4x - 6x - 24]$
$= 3x^3(x^2 - 2x - 24)$
$= 3x^5 - 6x^4 - 72x^3$

3. **(a)**

$$\begin{array}{rrrl} & 2m & -\ 5 & \\ & 3m & +\ 4 & \\ \hline & 8m & -\ 20 & \leftarrow 4(2m-5) \\ 6m^2 & -\ 15m & & \leftarrow 3m(2m-5) \\ \hline 6m^2 & -\ 7m & -\ 20 & \textit{Combine like terms.} \end{array}$$

(b)

$$\begin{array}{rrrrr} & 5a^3 & -\ 6a^2 & +\ 2a & -\ 3 \\ & & & 2a & -\ 5 \\ \hline & -\ 25a^3 & +\ 30a^2 & -\ 10a & +\ 15 \\ 10a^4 & -\ 12a^3 & +\ 4a^2 & -\ 6a & \\ \hline 10a^4 & -\ 37a^3 & +\ 34a^2 & -\ 16a & +\ 15 \end{array}$$

4. **(a)** $(3z + 2)(z + 1)$
F O I L
$= 3z \cdot z + 3z \cdot 1 + 2 \cdot z + 2 \cdot 1$
$= 3z^2 + 3z + 2z + 2$
$= 3z^2 + 5z + 2$

(b) $(5r - 3)(2r - 5)$
F O I L
$= 10r^2 - 25r - 6r + 15$
$= 10r^2 - 31r + 15$

(c) $(4p + 5q)(3p - 2q)$
F O I L
$= 12p^2 - 8pq + 15pq - 10q^2$
$= 12p^2 + 7pq - 10q^2$

5. Use $\boldsymbol{(x + y)(x - y) = x^2 - y^2}$.

(a) $(m + 5)(m - 5) = m^2 - 5^2$
$= m^2 - 25$

(b) $(x - 4y)(x + 4y) = x^2 - (4y)^2$
$= x^2 - 4^2y^2$
$= x^2 - 16y^2$

(c) $(7m - 2n)(7m + 2n) = (7m)^2 - (2n)^2$
$= 49m^2 - 4n^2$

(d) $4y^2(y + 7)(y - 7) = 4y^2(y^2 - 49)$
$= 4y^4 - 196y^2$

6. Use $\boldsymbol{(x + y)^2 = x^2 + 2xy + y^2}$
or $\boldsymbol{(x - y)^2 = x^2 - 2xy + y^2}$.

(a) $(a + 2)^2 = a^2 + 2 \cdot a \cdot 2 + 2^2$
$= a^2 + 4a + 4$

(b) $(2m-5)^2 = (2m)^2 - 2(2m)(5) + 5^2$
$= 4m^2 - 20m + 25$

(c) $(y+6z)^2 = y^2 + 2(y)(6z) + (6z)^2$
$= y^2 + 12yz + 36z^2$

(d) $(3k-2n)^2 = (3k)^2 - 2(3k)(2n) + (2n)^2$
$= 9k^2 - 12kn + 4n^2$

7. **(a)** $[(m-2n)-3]\cdot[(m-2n)+3]$
This is the product of a sum and a difference.
$= (m-2n)^2 - 3^2$
The first term is the square of a binomial.
$= m^2 - 2(m)(2n) + (2n)^2 - 3^2$
$= m^2 - 4mn + 4n^2 - 9$

(b) $[(k-5h)+2]^2$
This is the square of a binomial, with $(k-5h)$ as the first term.
$= (k-5h)^2 + 2(k-5h)(2) + 2^2$
$= k^2 - 2(k)(5h) + (5h)^2 + 4(k-5h) + 4$
$= k^2 - 10kh + 25h^2 + 4k - 20h + 4$

(c) $(p+2q)^3$
$= (p+2q)^2(p+2q)$
$= (p^2 + 4pq + 4q^2)(p+2q)$ *Square p + 2q.*
$= p^3 + 4p^2q + 4pq^2 + 2p^2q + 8pq^2 + 8q^3$
$= p^3 + 6p^2q + 12pq^2 + 8q^3$

(d) $(x+2)^4$
$= (x+2)^2(x+2)^2$
$= (x^2+4x+4)(x^2+4x+4)$
$= x^4 + 4x^3 + 4x^2 + 4x^3 + 16x^2 + 16x$
$+ 4x^2 + 16x + 16$
$= x^4 + 8x^3 + 24x^2 + 32x + 16$

8. $f(x) = 2x+7,\ g(x) = x^2 - 4$
$(fg)(x) = f(x)\cdot g(x)$
$= (2x+7)(x^2-4)$
$= 2x^3 - 8x + 7x^2 - 28$
$= 2x^3 + 7x^2 - 8x - 28$
$(fg)(2) = 2(2)^3 + 7(2)^2 - 8(2) - 28$
$= 16 + 28 - 16 - 28$
$= 0$

6.4 Section Exercises

1. $-8m^3(3m^2) = -8(3)m^{3+2} = -24m^5$

3. $3x(-2x+5) = 3x(-2x) + 3x(5) = -6x^2 + 15x$

5. $-q^3(2+3q) = -q^3(2) - q^3(3q) = -2q^3 - 3q^4$

7. $6k^2(3k^2+2k+1)$
$= 6k^2(3k^2) + 6k^2(2k) + 6k^2(1)$
$= 18k^4 + 12k^3 + 6k^2$

9. $(2m+3)(3m^2-4m-1)$
$= 2m(3m^2-4m-1) + 3(3m^2-4m-1)$
$= 2m(3m^2) + 2m(-4m) + 2m(-1)$
$+ 3(3m^2) + 3(-4m) + 3(-1)$
$= 6m^3 - 8m^2 - 2m + 9m^2 - 12m - 3$
$= 6m^3 - 8m^2 + 9m^2 - 2m - 12m - 3$
$= 6m^3 + m^2 - 14m - 3$

11. $4x^3(x-3)(x+2)$
$= 4x^3[(x-3)(x) + (x-3)(2)]$
$= 4x^3[x^2 - 3x + 2x - 6]$
$= 4x^3(x^2 - x - 6)$
$= 4x^3(x^2) + 4x^3(-x) + 4x^3(-6)$
$= 4x^5 - 4x^4 - 24x^3$

13. $(2y+3)(3y-4)$

$$\begin{array}{rrrl} & 2y & +\ 3 & \\ & 3y & -\ 4 & \\ \hline & -\ 8y & -\ 12 & \leftarrow -4(2y+3) \\ 6y^2 & +\ 9y & & \leftarrow 3y(2y+3) \\ \hline 6y^2 & +\ y & -\ 12 & \textit{Combine like terms.} \end{array}$$

15.

$$\begin{array}{rrr} & 5m & -\ 3n \\ & 5m & +\ 3n \\ \hline & 15mn & -\ 9n^2 \\ 25m^2 & -\ 15mn & \\ \hline 25m^2 & & -\ 9n^2 \end{array}$$

17.

$$\begin{array}{rrrr} & -b^2 & +\ 3b & +\ 3 \\ & & 2b & +\ 4 \\ \hline & -\ 4b^2 & +\ 12b & +\ 12 \\ -2b^3 & +\ 6b^2 & +\ 6b & \\ \hline -2b^3 & +\ 2b^2 & +\ 18b & +\ 12 \end{array}$$

19.

$$\begin{array}{rrrrr} & 2z^3 & -\ 5z^2 & +\ 8z & -\ 1 \\ & & & 4z & +\ 3 \\ \hline & 6z^3 & -\ 15z^2 & +\ 24z & -\ 3 \\ 8z^4 & -\ 20z^3 & +\ 32z^2 & -\ 4z & \\ \hline 8z^4 & -\ 14z^3 & +\ 17z^2 & +\ 20z & -\ 3 \end{array}$$

21.

$$\begin{array}{rrrrr} & & 2p^2 & +\ 3p & +\ 6 \\ & & 3p^2 & -\ 4p & -\ 1 \\ \hline & & -\ 2p^2 & -\ 3p & -\ 6 \\ & -\ 8p^3 & -\ 12p^2 & -\ 24p & \\ 6p^4 & +\ 9p^3 & +\ 18p^2 & & \\ \hline 6p^4 & +\ p^3 & +\ 4p^2 & -\ 27p & -\ 6 \end{array}$$

23. $(m+5)(m-8)$

F O I L

$= m^2 - 8m + 5m - 40$

$= m^2 - 3m - 40$

25. $(4k+3)(3k-2)$

F O I L

$= 12k^2 - 8k + 9k - 6$

$= 12k^2 + k - 6$

27. $(z-w)(3z+4w)$

F O I L

$= 3z^2 + 4zw - 3zw - 4w^2$

$= 3z^2 + zw - 4w^2$

29. $(6c-d)(2c+3d)$

F O I L

$= 12c^2 + 18cd - 2cd - 3d^2$

$= 12c^2 + 16cd - 3d^2$

31. $(0.2x+1.3)(0.5x-0.1)$

F O I L

$= 0.1x^2 - 0.02x + 0.65x - 0.13$

$= 0.1x^2 + 0.63x - 0.13$

33. $(3r+\frac{1}{4}y)(r-2y)$

F O I L

$= 3r^2 - 6ry + \frac{1}{4}ry - \frac{1}{2}y^2$

$= 3r^2 - \frac{23}{4}ry - \frac{1}{2}y^2$

35. Use the formula for the product of the sum and difference of two terms.

$$(2p-3)(2p+3) = (2p)^2 - (3)^2$$
$$= 4p^2 - 9$$

37. $(5m-1)(5m+1) = (5m)^2 - (1)^2$

$= 25m^2 - 1$

39. $(3a+2c)(3a-2c) = (3a)^2 - (2c)^2$

$= 9a^2 - 4c^2$

41. $(4x-\frac{2}{3})(4x+\frac{2}{3}) = (4x)^2 - (\frac{2}{3})^2$

$= 16x^2 - \frac{4}{9}$

43. $(4m+7n^2)(4m-7n^2) = (4m)^2 - (7n^2)^2$

$= 16m^2 - 49n^4$

45. $5y^3(y+2)(y-2) = 5y^3(y^2-4)$

$= 5y^5 - 20y^3$

47. Use the formula for the square of a binomial.

$$(y-5)^2 = y^2 - 2(y)(5) + 5^2$$
$$= y^2 - 10y + 25$$

49. $(2p+7)^2 = (2p)^2 + 2(2p)(7) + 7^2$

$= 4p^2 + 28p + 49$

51. $(4n-3m)^2 = (4n)^2 - 2(4n)(3m) + (3m)^2$

$= 16n^2 - 24nm + 9m^2$

53. $(k-\frac{5}{7}p)^2 = k^2 - 2(k)(\frac{5}{7}p) + (\frac{5}{7}p)^2$

$= k^2 - \frac{10}{7}kp + \frac{25}{49}p^2$

55. $[(5x+1)+6y]^2$

$= (5x+1)^2 + 2(5x+1)(6y) + (6y)^2$

Square of a binomial

$= (5x+1)^2 + 12y(5x+1) + 36y^2$

$= [(5x)^2 + 2(5x)(1) + 1^2]$
$+ 60xy + 12y + 36y^2$

Square of a binomial

$= 25x^2 + 10x + 1 + 60xy + 12y + 36y^2$

57. $[(2a+b)-3][(2a+b)+3]$

$= (2a+b)^2 - 3^2$

Product of the sum and difference of two terms

$= [(2a)^2 + 2(2a)(b) + b^2] - 9$

Square of a binomial

$= 4a^2 + 4ab + b^2 - 9$

59. $[(2h-k)+j][(2h-k)-j]$

$= (2h-k)^2 - j^2$

$= (2h)^2 - 2(2h)(k) + k^2 - j^2$

$= 4h^2 - 4hk + k^2 - j^2$

61. $(x+2)^3 = (x+2)^2(x+2)$

$= [x^2 + 2(x)(2) + 2^2](x+2)$

$= (x^2 + 4x + 4)(x+2)$

$$\begin{array}{r} x^2 + 4x + 4 \\ x + 2 \\ \hline 2x^2 + 8x + 8 \\ x^3 + 4x^2 + 4x \quad\quad \\ \hline x^3 + 6x^2 + 12x + 8 \end{array}$$

63. $(5r-s)^3$

$= (5r-s)^2(5r-s)$

$= (25r^2 - 10rs + s^2)(5r-s)$

$= 125r^3 - 50r^2s + 5rs^2 - 25r^2s + 10rs^2 - s^3$

$= 125r^3 - 75r^2s + 15rs^2 - s^3$

65. $(q-2)^4 = (q-2)^2(q-2)^2$

$= (q^2 - 4q + 4)(q^2 - 4q + 4)$

$$\begin{array}{r} q^2 - 4q + 4 \\ q^2 - 4q + 4 \\ \hline 4q^2 - 16q + 16 \\ -4q^3 + 16q^2 - 16q \quad\quad \\ q^4 - 4q^3 + 4q^2 \quad\quad\quad\quad \\ \hline q^4 - 8q^3 + 24q^2 - 32q + 16 \end{array}$$

67. The formula for the area of a triangle is $A = \frac{1}{2}bh$. Use $b = 3x + 2y$ and $h = 3x - 2y$.

$$\begin{aligned} A &= \tfrac{1}{2}(3x + 2y)(3x - 2y) \\ &= \tfrac{1}{2}(9x^2 - 4y^2) \\ &= \tfrac{9}{2}x^2 - 2y^2 \end{aligned}$$

69. The formula for the area of a parallelogram is $A = bh$. Use $b = 5x + 6$ and $h = 3x - 4$.

$$\begin{aligned} A &= (5x + 6)(3x - 4) \\ &= 15x^2 - 20x + 18x - 24 \\ &= 15x^2 - 2x - 24 \end{aligned}$$

71. The length of each side of the entire square is a. To find the length of each side of the blue square, subtract b, the length of a side of the green rectangle. The result is $a - b$.

72. The formula for the area A of a square with side s is $A = s^2$. Since $s = a - b$, the formula for the area of the blue square would be $A = (a - b)^2$.

73. The formula for the area of a rectangle is $A = LW$. The green rectangle has length $a - b$ and width b, so each green rectangle has an area of $(a - b)b$ or $ab - b^2$.

Since there are two green rectangles, the total area in green is $2(ab - b^2)$ or $2ab - 2b^2$.

74. The length of each side of the yellow square is b, so the yellow square has an area of b^2.

75. The area of the entire colored region is a^2, because each side of the entire colored region has length a.

76. Using the results from Exercises 71–75, the area of the blue square equals

$$a^2 - (2ab - 2b^2) - b^2 = a^2 - 2ab + b^2.$$

77. Both expressions for the area of the blue square must be equal to each other; that is, $(a - b)^2$ from Exercise 72 and $a^2 - 2ab + b^2$ from Exercise 76.

78. From Exercise 72, the area of the blue square is $(a - b)^2$. From Exercise 76, the area of the blue square is $a^2 - 2ab + b^2$. Since these expressions must be equal

$$(a - b)^2 = a^2 - 2ab + b^2.$$

This reinforces the special product for the square of a binomial difference.

79. $f(x) = 2x$, $g(x) = 5x - 1$

$$\begin{aligned} (fg)(x) &= f(x) \cdot g(x) \\ &= 2x(5x - 1) \\ &= 10x^2 - 2x \end{aligned}$$

81. $f(x) = x + 1$, $g(x) = 2x - 3$

$$\begin{aligned} (fg)(x) &= f(x) \cdot g(x) \\ &= (x + 1)(2x - 3) \\ &= 2x^2 - 3x + 2x - 3 \\ &= 2x^2 - x - 3 \end{aligned}$$

83. $f(x) = 2x - 3$, $g(x) = 4x^2 + 6x + 9$

$$\begin{aligned} (fg)(x) &= f(x) \cdot g(x) \\ &= (2x - 3)(4x^2 + 6x + 9) \end{aligned}$$

Multiply vertically.

$$\begin{array}{rrrr} & 4x^2 & +\ 6x & +\ 9 \\ & & 2x & -\ 3 \\ \hline & -\ 12x^2 & -\ 18x & -\ 27 \\ 8x^3 & +\ 12x^2 & +\ 18x & \\ \hline 8x^3 & & & -\ 27 \end{array}$$

85.
$$\begin{aligned} (fg)(2) &= f(2) \cdot g(2) \\ &= (2^2 - 9)[2(2)] \\ &= -5 \cdot 4 = -20 \end{aligned}$$

87.
$$\begin{aligned} (fh)(-1) &= f(-1) \cdot h(-1) \\ &= [(-1)^2 - 9] \cdot [(-1) - 3] \\ &= -8(-4) = 32 \end{aligned}$$

89.
$$\begin{aligned} (fg)(-2) &= f(-2) \cdot g(-2) \\ &= [(-2)^2 - 9] \cdot [2(-2)] \\ &= (4 - 9) \cdot (-4) \\ &= -5(-4) = 20 \end{aligned}$$

91. Substitute 2 for x and 3 for y.

$$(x + y)^3 = (2 + 3)^3 = 5^3 = 125$$

Again, substitute 2 for x and 3 for y.

$$x^3 + y^3 = 2^3 + 3^3 = 8 + 27 = 35$$

Since $125 \neq 35$, $(x + y)^3 \neq x^3 + y^3$.

The correct product is

$$\begin{aligned} (x + y)^3 &= (x + y)^2(x + y) \\ &= (x^2 + 2xy + y^2)(x + y) \\ &= x^3 + 2x^2y + xy^2 + x^2y + 2xy^2 + y^3 \\ &= x^3 + 3x^2y + 3xy^2 + y^3. \end{aligned}$$

6.5 Dividing Polynomials

6.5 Margin Exercises

1. (a) $\dfrac{12p + 30}{6} = \dfrac{12p}{6} + \dfrac{30}{6} = 2p + 5$

(b)
$$\begin{aligned} \frac{9y^3 - 4y^2 + 8y}{2y^2} &= \frac{9y^3}{2y^2} - \frac{4y^2}{2y^2} + \frac{8y}{2y^2} \\ &= \frac{9y}{2} - 2 + \frac{4}{y} \end{aligned}$$

(c) $$\frac{8a^2b^2 - 20ab^3}{4a^3b} = \frac{8a^2b^2}{4a^3b} - \frac{20ab^3}{4a^3b} = \frac{2b}{a} - \frac{5b^2}{a^2}$$

2. **(a)** $$\frac{2r^2 + r - 21}{r - 3}$$

$$\begin{array}{r|l}
 & \phantom{2r^2 +{}} 2r + 7 \\
r - 3 & \overline{2r^2 + r - 21} \\
 & \underline{2r^2 - 6r} \\
 & \phantom{2r^2 - {}} 7r - 21 \\
 & \phantom{2r^2 - {}} \underline{7r - 21} \\
 & \phantom{2r^2 - 7r - 2{}} 0
\end{array}$$

First step: $\dfrac{2r^2}{r} = 2r$

Check: $(r - 3)(2r + 7) = 2r^2 + r - 21$

Answer: $2r + 7$

(b) $$\frac{2k^2 + 17k + 30}{2k + 5}$$

$$\begin{array}{r|l}
 & \phantom{2k^2 +{}} k + 6 \\
2k + 5 & \overline{2k^2 + 17k + 30} \\
 & \underline{2k^2 + 5k} \\
 & \phantom{2k^2 +{}} 12k + 30 \\
 & \phantom{2k^2 +{}} \underline{12k + 30} \\
 & \phantom{2k^2 + 12k + 3{}} 0
\end{array}$$

First step: $\dfrac{2k^2}{2k} = k$

Check: $(2k + 5)(k + 6) = 2k^2 + 17k + 30$

Answer: $k + 6$

3. Add a term with 0 coefficient as a placeholder for the missing k^2-term.

$$\begin{array}{r|l}
 & 3k^2 + 6k + 21 \\
k - 2 & \overline{3k^3 + 0k^2 + 9k - 14} \\
 & \underline{3k^3 - 6k^2} \\
 & \phantom{3k^3 +{}} 6k^2 + 9k \\
 & \phantom{3k^3 +{}} \underline{6k^2 - 12k} \\
 & \phantom{3k^3 + 6k^2 +{}} 21k - 14 \\
 & \phantom{3k^3 + 6k^2 +{}} \underline{21k - 42} \\
 & \phantom{3k^3 + 6k^2 + 21k - {}} 28
\end{array}$$

Remainder

Check: $(k - 2)(3k^2 + 6k + 21) + 28$
$= (3k^3 + 9k - 42) + 28$
$= 3k^3 + 9k - 14$

Answer: $3k^2 + 6k + 21 + \dfrac{28}{k - 2}$

4. **(a)** $$\frac{3r^5 - 15r^4 - 2r^3 + 19r^2 - 7}{3r^2 - 2}$$

Each polynomial has a missing r-term.

$$\begin{array}{r|l}
 & r^3 - 5r^2 \phantom{{}+19r^2} + 3 \\
3r^2 + 0r - 2 & \overline{3r^5 - 15r^4 - 2r^3 + 19r^2 + 0r - 7} \\
 & \underline{3r^5 + 0r^4 - 2r^3} \\
 & - 15r^4 + 0r^3 + 19r^2 \\
 & \underline{-15r^4 + 0r^3 + 10r^2} \\
 & \phantom{3r^5 - 15r^4 + 0r^3 +{}} 9r^2 + 0r - 7 \\
 & \phantom{3r^5 - 15r^4 + 0r^3 +{}} \underline{9r^2 + 0r - 6} \\
 & \phantom{3r^5 - 15r^4 + 0r^3 + 9r^2 + 0r {}} - 1
\end{array}$$

Remainder

Check: $(3r^2 - 2)(r^3 - 5r^2 + 3) - 1$
$= (3r^5 - 15r^4 - 2r^3 + 19r^2 - 6) - 1$
$= 3r^5 - 15r^4 - 2r^3 + 19r^2 - 7$

Answer: $r^3 - 5r^2 + 3 + \dfrac{-1}{3r^2 - 2}$

(b) $$\frac{4x^4 - 7x^2 + x + 5}{2x^2 - x}$$

The numerator has a missing x^3-term.

$$\begin{array}{r|l}
 & \phantom{4x^4 + {}} 2x^2 + x - 3 \\
2x^2 - x & \overline{4x^4 + 0x^3 - 7x^2 + x + 5} \\
 & \underline{4x^4 - 2x^3} \\
 & \phantom{4x^4 +{}} 2x^3 - 7x^2 \\
 & \phantom{4x^4 +{}} \underline{2x^3 - x^2} \\
 & -6x^2 + x \\
 & \underline{-6x^2 + 3x} \\
 & - 2x + 5
\end{array}$$

Remainder

Check: $(2x^2 - x)(2x^2 + x - 3) - 2x + 5$
$= (4x^4 - 7x^2 + 3x) - 2x + 5$
$= 4x^4 - 7x^2 + x + 5$

Answer: $2x^2 + x - 3 + \dfrac{-2x + 5}{2x^2 - x}$

5.

$$\begin{array}{r|l}
 & p^2 + \tfrac{5}{2}p + 2 \\
2p + 2 & \overline{2p^3 + 7p^2 + 9p + 4} \\
 & \underline{2p^3 + 2p^2} \\
 & \phantom{2p^3 +{}} 5p^2 + 9p \\
 & \phantom{2p^3 +{}} \underline{5p^2 + 5p} \\
 & \phantom{2p^3 + 5p^2 +{}} 4p + 4 \\
 & \phantom{2p^3 + 5p^2 +{}} \underline{4p + 4} \\
 & \phantom{2p^3 + 5p^2 + 4p + {}} 0
\end{array}$$

Remainder

First step: $\dfrac{2p^3}{2p} = p^2$

Note: $\dfrac{5p^2}{2p} = \dfrac{5p}{2}$

Check: $(2p + 2)(p^2 + \frac{5}{2}p + 2)$
$= 2p^3 + 7p^2 + 9p + 4$

Answer: $p^2 + \frac{5}{2}p + 2$

6. $f(x) = 2x^2 + 17x + 30,\ g(x) = 2x + 5$

$$\left(\frac{f}{g}\right)(x) = \frac{f(x)}{g(x)} = \frac{2x^2 + 17x + 30}{2x + 5}$$

This quotient, found in Margin Exercise 2(b), with x replacing k, is $x + 6$. Thus, $\left(\frac{f}{g}\right)(x) = x + 6$, provided the denominator, $2x + 5$, is *not* equal to zero; that is, $x \neq -\frac{5}{2}$.

Using the above result, $\left(\frac{f}{g}\right)(x) = x + 6$, we have

$$\left(\frac{f}{g}\right)(-1) = (-1) + 6 = 5.$$

6.5 Section Exercises

1.
$$\frac{15x^3 - 10x^2 + 5}{5} = \frac{15x^3}{5} - \frac{10x^2}{5} + \frac{5}{5} = 3x^3 - 2x^2 + 1$$

3.
$$\frac{9y^2 + 12y - 15}{3y} = \frac{9y^2}{3y} + \frac{12y}{3y} - \frac{15}{3y} = 3y + 4 - \frac{5}{y}$$

5.
$$\frac{15m^3 + 25m^2 + 30m}{5m^2} = \frac{15m^3}{5m^2} + \frac{25m^2}{5m^2} + \frac{30m}{5m^2} = 3m + 5 + \frac{6}{m}$$

7.
$$\frac{14m^2n^2 - 21mn^3 + 28m^2n}{14m^2n} = \frac{14m^2n^2}{14m^2n} - \frac{21mn^3}{14m^2n} + \frac{28m^2n}{14m^2n} = n - \frac{3n^2}{2m} + 2$$

9. $\dfrac{y^2 + 3y - 18}{y + 6}$

$$\begin{array}{r|l} & y - 3 \\ y + 6 & y^2 + 3y - 18 \\ & \underline{y^2 + 6y} \\ & -3y - 18 \\ & \underline{-3y - 18} \\ & 0 \end{array}$$

Answer: $y - 3$

11. $\dfrac{3t^2 + 17t + 10}{3t + 2}$

$$\begin{array}{r|l} & t + 5 \\ 3t + 2 & 3t^2 + 17t + 10 \\ & \underline{3t^2 + 2t} \\ & 15t + 10 \\ & \underline{15t + 10} \\ & 0 \end{array}$$

Answer: $t + 5$

13. $\dfrac{p^2 + 2p + 20}{p + 6}$

$$\begin{array}{r|l} & p - 4 \\ p + 6 & p^2 + 2p + 20 \\ & \underline{p^2 + 6p} \\ & -4p + 20 \\ & \underline{-4p - 24} \\ & 44 \end{array}$$

Remainder

Answer: $p - 4 + \dfrac{44}{p + 6}$

15. $(2z^3 - 5z^2 + 6z - 15) \div (2z - 5)$

$$\begin{array}{r|l} & z^2 + 3 \\ 2z - 5 & 2z^3 - 5z^2 + 6z - 15 \\ & \underline{2z^3 - 5z^2} \\ & 6z - 15 \\ & \underline{6z - 15} \\ & 0 \end{array}$$

Answer: $z^2 + 3$

17. $(4x^3 + 9x^2 - 10x + 3) \div (4x + 1)$

$$\begin{array}{r|l} & x^2 + 2x - 3 \\ 4x + 1 & 4x^3 + 9x^2 - 10x + 3 \\ & \underline{4x^3 + x^2} \\ & 8x^2 - 10x \\ & \underline{8x^2 + 2x} \\ & -12x + 3 \\ & \underline{-12x - 3} \\ & 6 \end{array}$$

Remainder

Answer: $x^2 + 2x - 3 + \dfrac{6}{4x + 1}$

19. $\dfrac{14x + 6x^3 - 15 - 19x^2}{3x^2 - 2x + 4}$

Rewrite as $\dfrac{6x^3 - 19x^2 + 14x - 15}{3x^2 - 2x + 4}$.

$$\begin{array}{r|l} & 2x - 5 \\ 3x^2 - 2x + 4 & 6x^3 - 19x^2 + 14x - 15 \\ & \underline{6x^3 - 4x^2 + 8x} \\ & -15x^2 + 6x - 15 \\ & \underline{-15x^2 + 10x - 20} \\ & -4x + 5 \end{array}$$

Remainder

Answer: $2x - 5 + \dfrac{-4x + 5}{3x^2 - 2x + 4}$

21. $(3x^3 - x + 4) \div (x - 2)$

$$\begin{array}{r}
3x^2 + 6x + 11 \\
x - 2 \,\overline{)\, 3x^3 + 0x^2 - x + 4} \\
\underline{3x^3 - 6x^2} \\
6x^2 - x \\
\underline{6x^2 - 12x} \\
11x + 4 \\
\underline{11x - 22} \\
26
\end{array}$$

Answer: $3x^2 + 6x + 11 + \dfrac{26}{x - 2}$

23. $(2x^3 - 11x^2 + 28) \div (x - 5)$

$$\begin{array}{r}
2x^2 - x - 5 \\
x - 5 \,\overline{)\, 2x^3 - 11x^2 + 0x + 28} \\
\underline{2x^3 - 10x^2} \\
-x^2 + 0x \\
\underline{-x^2 + 5x} \\
-5x + 28 \\
\underline{-5x + 25} \\
3
\end{array}$$

Answer: $2x^2 - x - 5 + \dfrac{3}{x - 5}$

25. $\dfrac{4k^4 + 6k^3 + 3k - 1}{2k^2 + 1}$

$$\begin{array}{r}
2k^2 + 3k - 1 \\
2k^2 + 1 \,\overline{)\, 4k^4 + 6k^3 + 0k^2 + 3k - 1} \\
\underline{4k^4 + 2k^2} \\
6k^3 - 2k^2 + 3k \\
\underline{6k^3 + 3k} \\
-2k^2 - 1 \\
\underline{-2k^2 - 1} \\
0
\end{array}$$

Answer: $2k^2 + 3k - 1$

27. $(9z^4 - 13z^3 + 23z^2 - 10z + 8) \div (z^2 - z + 2)$

$$\begin{array}{r}
9z^2 - 4z + 1 \\
z^2 - z + 2 \,\overline{)\, 9z^4 - 13z^3 + 23z^2 - 10z + 8} \\
\underline{9z^4 - 9z^3 + 18z^2} \\
-4z^3 + 5z^2 - 10z \\
\underline{-4z^3 + 4z^2 - 8z} \\
z^2 - 2z + 8 \\
\underline{z^2 - z + 2} \\
-z + 6
\end{array}$$

Answer: $9z^2 - 4z + 1 + \dfrac{-z + 6}{z^2 - z + 2}$

29. $(2x^2 - \frac{7}{3}x - 1) \div (3x + 1)$

To start: $\dfrac{2x^2}{3x} = \dfrac{2}{3}x$

$$\begin{array}{r}
\frac{2}{3}x - 1 \\
3x + 1 \,\overline{)\, 2x^2 - \frac{7}{3}x - 1} \\
\underline{2x^2 + \frac{2}{3}x} \\
-3x - 1 \\
\underline{-3x - 1} \\
0
\end{array}$$

Answer: $\frac{2}{3}x - 1$

31. $(3a^2 - \frac{23}{4}a - 5) \div (4a + 3)$

$$\begin{array}{r}
\frac{3}{4}a - 2 \\
4a + 3 \,\overline{)\, 3a^2 - \frac{23}{4}a - 5} \\
\underline{3a^2 + \frac{9}{4}a} \\
-8a - 5 \\
\underline{-8a - 6} \\
1
\end{array}$$

Remainder

Answer: $\dfrac{3}{4}a - 2 + \dfrac{1}{4a + 3}$

33. $f(x) = 10x^2 - 2x,\ g(x) = 2x$

$$\begin{aligned}
\left(\frac{f}{g}\right)(x) = \frac{f(x)}{g(x)} &= \frac{10x^2 - 2x}{2x} \\
&= \frac{10x^2}{2x} - \frac{2x}{2x} \\
&= 5x - 1
\end{aligned}$$

The x-values that are not in the domain of the quotient function g are found by solving $g(x) = 0$.

$$\begin{aligned}
2x &= 0 \\
x &= 0
\end{aligned}$$

0 is not in the domain of f/g.

35. $f(x) = 2x^2 - x - 3,\ g(x) = x + 1$

$$\left(\frac{f}{g}\right)(x) = \frac{f(x)}{g(x)} = \frac{2x^2 - x - 3}{x + 1}$$

$$\begin{array}{r}
2x - 3 \\
x + 1 \,\overline{)\, 2x^2 - x - 3} \\
\underline{2x^2 + 2x} \\
-3x - 3 \\
\underline{-3x - 3} \\
0
\end{array}$$

Quotient: $2x - 3$

$$\begin{aligned}
g(x) &= 0 \\
x + 1 &= 0 \\
x &= -1
\end{aligned}$$

-1 is not in the domain of f/g.

37. $f(x) = 8x^3 - 27, g(x) = 2x - 3$

$$\left(\frac{f}{g}\right)(x) = \frac{f(x)}{g(x)} = \frac{8x^3 - 27}{2x - 3}$$

$$\begin{array}{r} 4x^2 + 6x + 9 \\ 2x - 3 \overline{)\,8x^3 + 0x^2 + 0x - 27} \\ \underline{8x^3 - 12x^2} \qquad\qquad \\ 12x^2 + 0x \qquad \\ \underline{12x^2 - 18x} \qquad \\ 18x - 27 \\ \underline{18x - 27} \\ 0 \end{array}$$

Quotient: $4x^2 + 6x + 9$

$$\begin{aligned} g(x) &= 0 \\ 2x - 3 &= 0 \\ 2x &= 3 \\ x &= \tfrac{3}{2} \end{aligned}$$

$\frac{3}{2}$ is not in the domain of f/g.

For Exercises 39–46, let $f(x) = x^2 - 9$, $g(x) = 2x$, and $h(x) = x - 3$.

39. $\left(\frac{f}{g}\right)(x) = \frac{f(x)}{g(x)} = \frac{x^2 - 9}{2x}$

We must exclude any values of x that make the denominator equal to zero, so $x \neq 0$.

41. $\left(\frac{f}{g}\right)(2) = \frac{f(2)}{g(2)} = \frac{2^2 - 9}{2(2)} = \frac{-5}{4} = -\frac{5}{4}$

43. $\left(\frac{h}{g}\right)(x) = \frac{h(x)}{g(x)} = \frac{x - 3}{2x}, x \neq 0$

45. $\left(\frac{h}{g}\right)(3) = \frac{h(3)}{g(3)} = \frac{(3) - 3}{2(3)} = \frac{0}{6} = 0$

47. The volume of a box is the product of the height, length, and width. Use the formula $V = LWH$.

$$\begin{aligned} V &= LWH \\ \frac{V}{LH} &= W \end{aligned}$$

Here, $L \cdot H = (p + 4)p = p^2 + 4p$, so

$$W = \frac{V}{LH} = \frac{2p^3 + 15p^2 + 28p}{p^2 + 4p}.$$

$$\begin{array}{r} 2p + 7 \\ p^2 + 4p \overline{)\,2p^3 + 15p^2 + 28p} \\ \underline{2p^3 + 8p^2} \qquad\quad \\ 7p^2 + 28p \\ \underline{7p^2 + 28p} \\ 0 \end{array}$$

The width is $(2p + 7)$ feet.

Chapter 6 Review Exercises

1. $4^3 = 4 \cdot 4 \cdot 4 = 64$

2. $\left(\frac{1}{3}\right)^4 = \frac{1}{3} \cdot \frac{1}{3} \cdot \frac{1}{3} \cdot \frac{1}{3} = \frac{1}{81}$

3. $(-5)^3 = (-5)(-5)(-5) = -125$

4.
$$\begin{aligned} \frac{2}{(-3)^{-2}} &= \frac{2}{\frac{1}{(-3)^2}} \\ &= 2 \cdot (-3)^2 \\ &= 2 \cdot (-3)(-3) \\ &= 18 \end{aligned}$$

5.
$$\begin{aligned} \left(\frac{2}{3}\right)^{-4} &= \left(\frac{3}{2}\right)^4 \\ &= \frac{3}{2} \cdot \frac{3}{2} \cdot \frac{3}{2} \cdot \frac{3}{2} \\ &= \frac{81}{16} \end{aligned}$$

6. $\left(\frac{5}{4}\right)^{-2} = \left(\frac{4}{5}\right)^2 = \frac{4}{5} \cdot \frac{4}{5} = \frac{16}{25}$

7. $5^{-1} + 6^{-1} = \frac{1}{5} + \frac{1}{6} = \frac{6}{30} + \frac{5}{30} = \frac{11}{30}$

8. $-3^0 + 3^0 = -1 + 1 = 0$

9.
$$\begin{aligned} (-3x^4y^3)(4x^{-2}y^5) &= -3(4)x^4x^{-2}y^3y^5 \\ &= -12x^{4+(-2)}y^{3+5} \\ &= -12x^2y^8 \end{aligned}$$

10.
$$\begin{aligned} \frac{6m^{-4}n^3}{-3mn^2} &= -2m^{-4-1}n^{3-2} \\ &= -2m^{-5}n^1 \\ &= \frac{-2n}{m^5} = -\frac{2n}{m^5} \end{aligned}$$

11.
$$\begin{aligned} \frac{(5p^{-2}q)(4p^5q^{-3})}{2p^{-5}q^5} &= \frac{20p^{-2+5}q^{1+(-3)}}{2p^{-5}q^5} \\ &= \frac{10p^3q^{-2}}{p^{-5}q^5} \\ &= 10p^{3-(-5)}q^{-2-5} \\ &= 10p^8q^{-7} \\ &= \frac{10p^8}{q^7} \end{aligned}$$

12.
$$\begin{aligned} \frac{x^{-2}y^{-4}}{x^{-4}y^{-2}} &= x^{-2-(-4)}y^{-4-(-2)} \\ &= x^{-2+4}y^{-4+2} \\ &= x^2y^{-2} \\ &= \frac{x^2}{y^2} \end{aligned}$$

13. $(3^{-4})^2 = 3^{(-4) \cdot 2} = 3^{-8} = \frac{1}{3^8}$

14. $(x^{-4})^{-2} = x^{-4(-2)} = x^8$

15. $(xy^{-3})^{-2} = x^{1(-2)}y^{(-3)(-2)}$
$= x^{-2}y^6$
$= \frac{1}{x^2} \cdot y^6 = \frac{y^6}{x^2}$

16. $(z^{-3})^3 z^{-6} = z^{-9}z^{-6}$
$= z^{-9+(-6)}$
$= z^{-15} = \frac{1}{z^{15}}$

17. $(5m^{-3})^2(m^4)^{-3} = 5^2m^{-6}m^{-12}$
$= 25m^{-6+(-12)}$
$= 25m^{-18} = \frac{25}{m^{18}}$

18. $\left(\frac{5z^{-3}}{z^{-1}}\right)\left(\frac{5}{z^2}\right)^{-1} = \frac{5z^{-3}}{z^{-1}} \cdot \frac{z^2}{5}$
$= \frac{z^{-3+2}}{z^{-1}}$
$= \frac{z^{-1}}{z^{-1}} = 1$

19. $\left(\frac{6m^{-4}}{m^{-9}}\right)^{-1}\left(\frac{m^{-2}}{16}\right) = \frac{6^{-1}m^4}{m^9} \cdot \frac{m^{-2}}{16}$
$= \frac{1}{6 \cdot 16}m^{4+(-2)-9}$
$= \frac{1}{96}m^{-7} = \frac{1}{96m^7}$

20. $\left(\frac{3r^5}{5r^{-3}}\right)^{-2}\left(\frac{9r^{-1}}{2r^{-5}}\right)^3$
$= \frac{3^{-2}r^{-10}}{5^{-2}r^6} \cdot \frac{9^3r^{-3}}{2^3r^{-15}}$
$= \frac{5^2}{3^2}r^{-10-6} \cdot \frac{9^3}{2^3}r^{-3-(-15)}$
$= \frac{25}{9}r^{-16} \cdot \frac{729}{8}r^{12}$
$= \frac{(25)(729)}{(9)(8)}r^{-16+12}$
$= \frac{2025}{8}r^{-4} = \frac{2025}{8r^4}$

21. $\left(\frac{3w^{-2}z^4}{-6wz^{-5}}\right)^{-2} = \left(\frac{w^{-2-1}z^{4-(-5)}}{-2}\right)^{-2}$
$= \left(\frac{w^{-3}z^9}{-2}\right)^{-2}$
$= \left(\frac{-2}{w^{-3}z^9}\right)^2$
$= \frac{(-2)^2}{(w^{-3})^2(z^9)^2}$
$= \frac{4}{w^{-6}z^{18}}$
$= \frac{4w^6}{z^{18}}$

22. 13,450 = 1‸3450.

Place a caret to the right of the first nonzero digit. Count 4 places.

Since the number 1.345 is to be made greater, the exponent on 10 is positive.

$$13{,}450 = 1.345 \times 10^4$$

23. 0.000 000 0765 = 0.0 0 0 0 0 0 0 7‸65

Count 8 places

Since the number 7.65 is to be made less, the exponent on 10 is negative.

$$0.000\,000\,0765 = 7.65 \times 10^{-8}$$

24. 0.138 = 0.1‸38

Count 1 place.

Since the number 1.38 is to be made less, the exponent on 10 is negative.

$$0.138 = 1.38 \times 10^{-1}$$

25. $299{,}400{,}000 = 2.994 \times 10^8$
$74{,}000 = 7.4 \times 10^4$
$100 = 1 \times 10^2$

26. $1.21 \times 10^6 = 1{,}210{,}000$

Move the decimal point 6 places to the right because the exponent is positive. Attach extra zeros.

27. $5.8 \times 10^{-3} = 0.0058$

Move the decimal point 3 places to the left because the exponent is negative.

28. $\frac{16 \times 10^4}{8 \times 10^8} = \frac{16}{8} \times 10^{4-8}$
$= 2 \times 10^{-4}$, or 0.0002

29. $\frac{6 \times 10^{-2}}{4 \times 10^{-5}} = \frac{6}{4} \times 10^{-2-(-5)}$
$= 1.5 \times 10^3$, or 1500

30. $\frac{0.000\,000\,0164}{0.0004} = \frac{1.64 \times 10^{-8}}{4 \times 10^{-4}}$
$= \frac{1.64}{4} \times 10^{-8-(-4)}$
$= 0.41 \times 10^{-4}$
$= 4.1 \times 10^{-5}$, or 0.000041

31. $\dfrac{0.0009 \times 12{,}000{,}000}{400{,}000}$
$= \dfrac{9 \times 10^{-4} \times 1.2 \times 10^{7}}{4 \times 10^{5}}$
$= \dfrac{9 \times 1.2}{4} \times \dfrac{10^{-4} \times 10^{7}}{10^{5}}$
$= \dfrac{10.8}{4} \times 10^{-4+7-5}$
$= 2.7 \times 10^{-2}$, or 0.027

32. The coefficient of $14p^5$ is 14.

33. The coefficient of $-z$ is -1.

34. The coefficient of $0.045x^4$ is 0.045.

35. The coefficient of $504p^3r^5$ is 504.

36. $9k + 11k^3 - 3k^2$

(a) In descending powers of k, the polynomial is

$$11k^3 - 3k^2 + 9k.$$

(b) The polynomial is a trinomial since it has three terms.

(c) The degree of the polynomial is 3 since the highest power of k is 3.

37. $14m^6 + 9m^7$

(a) In descending powers of m, the polynomial is

$$9m^7 + 14m^6.$$

(b) The polynomial is a binomial since it has two terms.

(c) The degree of the polynomial is 7 since the highest power of m is 7.

38. $-7q^5r^3$

(a) The polynomial is already written in descending powers.

(b) The polynomial is a monomial since it has just one term.

(c) The degree is $5 + 3 = 8$, the sum of the exponents of this term.

39. One example of a polynomial in the variable x that has degree 5, is lacking a third-degree term, and is written in descending powers of the variable is

$$x^5 + 2x^4 - x^2 + x + 2.$$

40. Add by columns.

$$\begin{array}{rcrcr} 3x^2 & - & 5x & + & 6 \\ -4x^2 & + & 2x & - & 5 \\ \hline -1x^2 & - & 3x & + & 1, \end{array} \quad \text{or} \quad -x^2 - 3x + 1$$

41. Subtract.

$$\begin{array}{rcrcrcr} -5y^3 & & & + & 8y & - & 3 \\ & & 4y^2 & + & 2y & + & 9 \\ \hline \end{array}$$

Change the signs in the second polynomial and add.

$$\begin{array}{rcrcrcr} -5y^3 & & & + & 8y & - & 3 \\ & - & 4y^2 & - & 2y & - & 9 \\ \hline -5y^3 & - & 4y^2 & + & 6y & - & 12 \end{array}$$

42. $(4a^3 - 9a + 15) - (-2a^3 + 4a^2 + 7a)$
$= 4a^3 - 9a + 15 + 2a^3 - 4a^2 - 7a$
$= 4a^3 + 2a^3 - 4a^2 - 9a - 7a + 15$
$= 6a^3 - 4a^2 - 16a + 15$

43. $(3y^2 + 2y - 1) + (5y^2 - 11y + 6)$
$= 3y^2 + 5y^2 + 2y - 11y - 1 + 6$
$= 8y^2 - 9y + 5$

44. To find the perimeter, add the measures of the three sides.

$(4x^2 + 2) + (6x^2 + 5x + 2) + (2x^2 + 3x + 1)$
$= 4x^2 + 6x^2 + 2x^2 + 5x + 3x + 2 + 2 + 1$
$= 12x^2 + 8x + 5$

The perimeter is $12x^2 + 8x + 5$.

45. $f(x) = -2x^2 + 5x + 7$

(a) $f(-2) = -2(-2)^2 + 5(-2) + 7$
$= -2(4) - 10 + 7$
$= -8 - 10 + 7 = -11$

(b) $f(3) = -2(3)^2 + 5(3) + 7$
$= -2(9) + 15 + 7$
$= -18 + 15 + 7 = 4$

46. **(a)** $(f + g)(x) = f(x) + g(x)$
$= (2x + 3) + (5x^2 - 3x + 2)$
$= 5x^2 + 2x - 3x + 3 + 2$
$= 5x^2 - x + 5$

(b) $(f - g)(x) = f(x) - g(x)$
$= (2x + 3) - (5x^2 - 3x + 2)$
$= 2x + 3 - 5x^2 + 3x - 2$
$= -5x^2 + 5x + 1$

(c) $(f + g)(-1)$
$= f(-1) + g(-1)$
$= [2(-1) + 3] + [5(-1)^2 - 3(-1) + 2]$
$= [1] + [10]$
$= 11$

Alternatively, use the result from part (a).

(d) $(f-g)(-1) = f(-1) - g(-1)$
$= 1 - 10$ *from part (c)*
$= -9$

Alternatively, use the result from part (b).

47. $f(x) =$
$-23.334x^3 + 645.75x^2 - 1861.4x + 95{,}072$

(a) $x = 1990 - 1990 = 0$

$$f(0) = -23.334(0)^3 + 645.75(0)^2 - 1861.4(0) + 95{,}072 = 95{,}072 \text{ twin births}$$

(b) $x = 2000 - 1990 = 10$

$$f(10) = -23.334(10)^3 + 645.75(10)^2 - 1861.4(10) + 95{,}072 = 117{,}699 \text{ twin births}$$

(c) $x = 2005 - 1990 = 15$

$$f(15) = -23.334(15)^3 + 645.75(15)^2 - 1861.4(15) + 95{,}072 = 133{,}692.5 \approx 133{,}693 \text{ twin births}$$

48. $y = f(x) = -2x + 5$

This is a linear function. The y-intercept is $(0, 5)$. To find another point, let $x = 2$.

$$y = -2(2) + 5 \quad \text{Let } x = 2.$$
$$= -4 + 5$$
$$= 1$$

Plot the points $(0, 5)$ and $(2, 1)$, and draw a line through them.

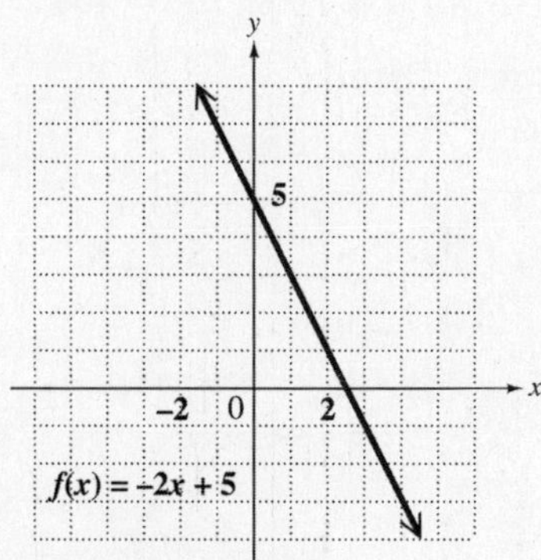

49.

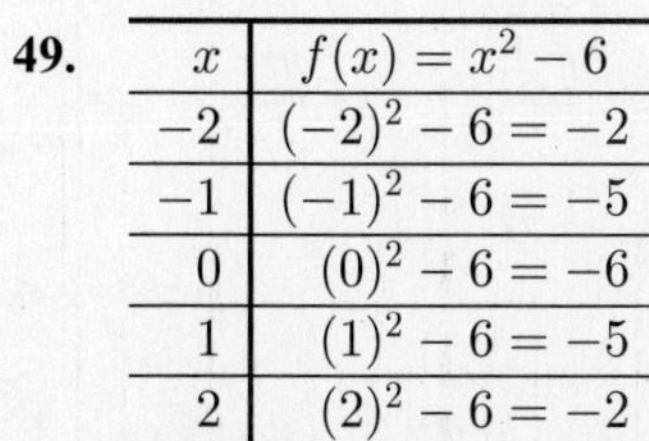

x	$f(x) = x^2 - 6$
-2	$(-2)^2 - 6 = -2$
-1	$(-1)^2 - 6 = -5$
0	$(0)^2 - 6 = -6$
1	$(1)^2 - 6 = -5$
2	$(2)^2 - 6 = -2$

Since the greatest exponent is 2, the graph of f is a parabola.

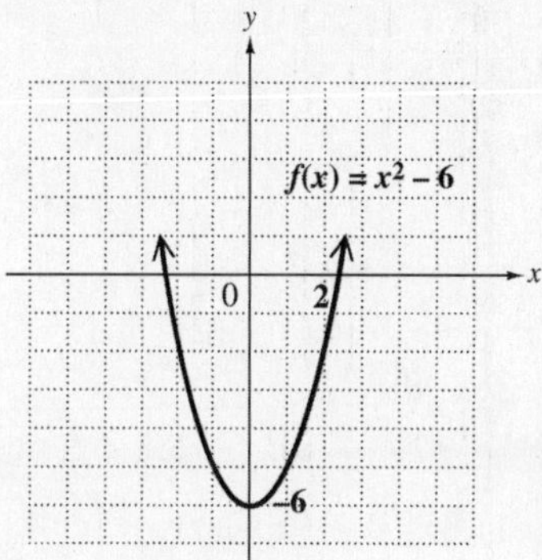

50.

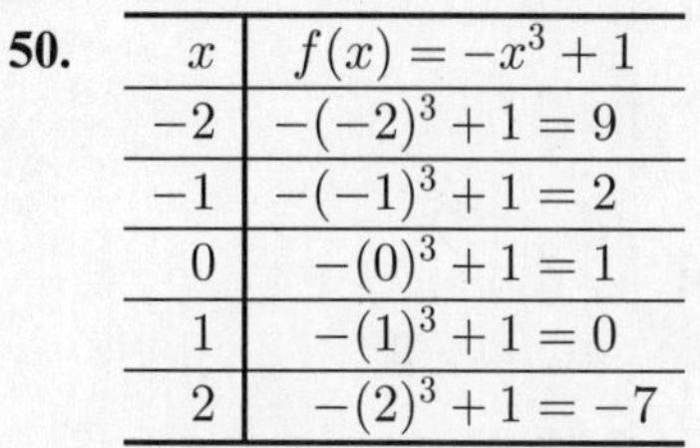

x	$f(x) = -x^3 + 1$
-2	$-(-2)^3 + 1 = 9$
-1	$-(-1)^3 + 1 = 2$
0	$-(0)^3 + 1 = 1$
1	$-(1)^3 + 1 = 0$
2	$-(2)^3 + 1 = -7$

The greatest exponent is 3, so the graph of f is s-shaped.

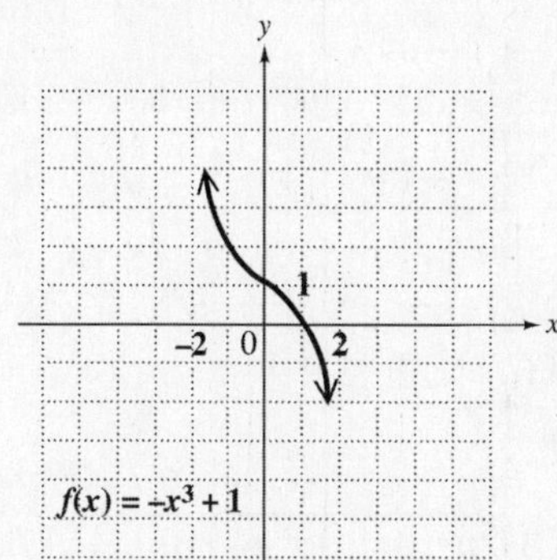

51. $-6k(2k^2 + 7) = -6k(2k^2) - 6k(7)$
$= -12k^3 - 42k$

52. $(7y - 8)(2y + 3)$
F O I L
$= 14y^2 + 21y - 16y - 24$
$= 14y^2 + 5y - 24$

53. $(3w - 2t)(2w - 3t)$
F O I L
$= 6w^2 - 9wt - 4wt + 6t^2$
$= 6w^2 - 13wt + 6t^2$

54. $(2p^2 + 6p)(5p^2 - 4)$
F O I L
$= 10p^4 - 8p^2 + 30p^3 - 24p$
$= 10p^4 + 30p^3 - 8p^2 - 24p$

55. $(3z^3 - 2z^2 + 4z - 1)(3z - 2)$
$= (3z^3 - 2z^2 + 4z - 1)(3z)$
$+ (3z^3 - 2z^2 + 4z - 1)(-2)$
$= 9z^4 - 6z^3 + 12z^2 - 3z$
$- 6z^3 + 4z^2 - 8z + 2$
$= 9z^4 - 6z^3 - 6z^3 + 12z^2 + 4z^2$
$- 3z - 8z + 2$
$= 9z^4 - 12z^3 + 16z^2 - 11z + 2$

56. $(6r^2 - 1)(6r^2 + 1) = (6r^2)^2 - 1^2$
$= 36r^4 - 1$

57. $(z + \frac{3}{5})(z - \frac{3}{5}) = z^2 - (\frac{3}{5})^2$
$= z^2 - \frac{9}{25}$

58. $(4m + 3)^2 = (4m)^2 + 2(4m)(3) + 3^2$
$= 16m^2 + 24m + 9$

59. $(2x + 5)^3 = (2x + 5)(2x + 5)^2$
$= (2x + 5)(4x^2 + 20x + 25)$
$= 8x^3 + 40x^2 + 50x + 20x^2 + 100x$
$+ 125$
$= 8x^3 + 60x^2 + 150x + 125$

60. $\dfrac{4y^3 - 12y^2 + 5y}{4y} = \dfrac{4y^3}{4y} - \dfrac{12y^2}{4y} + \dfrac{5y}{4y}$
$= y^2 - 3y + \frac{5}{4}$

61. $\dfrac{2p^3 + 9p^2 + 27}{2p - 3}$

$$
\begin{array}{r|l}
 & p^2 + 6p + 9 \\
\hline
2p - 3 & 2p^3 + 9p^2 + 0p + 27 \\
 & 2p^3 - 3p^2 \\
\hline
 & 12p^2 + 0p \\
 & 12p^2 - 18p \\
\hline
 & 18p + 27 \\
 & 18p - 27 \\
\hline
 & 54
\end{array}
$$

Remainder

Answer: $p^2 + 6p + 9 + \dfrac{54}{2p - 3}$

62. $\dfrac{5p^4 + 15p^3 - 33p^2 - 9p + 18}{5p^2 - 3}$

$$
\begin{array}{r|l}
 & p^2 + 3p - 6 \\
\hline
5p^2 - 3 & 5p^4 + 15p^3 - 33p^2 - 9p + 18 \\
 & 5p^4 - 3p^2 \\
\hline
 & 15p^3 - 30p^2 - 9p \\
 & 15p^3 - 9p \\
\hline
 & -30p^2 + 18 \\
 & -30p^2 + 18 \\
\hline
 & 0
\end{array}
$$

Answer: $p^2 + 3p - 6$

63. **[6.1]** **(a)** $4^{-2} = \dfrac{1}{4^2} = \dfrac{1}{16}$ **(A)**

(b) $-4^2 = -(4^2) = -16$ **(G)**

(c) $4^0 = 1$ **(C)**

(d) $(-4)^0 = 1$ **(C)**

(e) $(-4)^{-2} = \dfrac{1}{(-4)^2} = \dfrac{1}{16}$ **(A)**

(f) $-4^0 = -(4^0) = -1$ **(E)**

(g) $-4^0 + 4^0 = -1 + 1 = 0$ **(B)**

(h) $-4^0 - 4^0 = (-1) - 1 = -2$ **(H)**

(i) $4^{-2} + 4^{-1} = \dfrac{1}{4^2} + \dfrac{1}{4} = \dfrac{1}{16} + \dfrac{4}{16} = \dfrac{5}{16}$ **(F)**

64. **[6.1]** The density D is the population P divided by the area A.

$$D = \frac{P}{A}$$

We want to find the area, so solve the formula for A.

$$
\begin{aligned}
DA &= P \\
A &= \frac{P}{D} \\
&= \frac{4.86 \times 10^5}{487} \\
&= \frac{4.86 \times 10^5}{4.87 \times 10^2} \\
&\approx 0.998 \times 10^{5-2} \\
&= 0.998 \times 10^3 = 998
\end{aligned}
$$

The area is 998 mi^2.

65. **[6.4]** $(4x + 1)(2x - 3)$
F O I L
$= 8x^2 - 12x + 2x - 3$
$= 8x^2 - 10x - 3$

66. **[6.1]** $\dfrac{6^{-1}y^3(y^2)^{-2}}{6y^{-4}(y^{-1})} = \dfrac{y^3y^{-4}}{6^1 \cdot 6y^{-4}y^{-1}}$
$= \dfrac{y^{3+(-4)}}{36y^{-4+(-1)}} = \dfrac{y^{-1}}{36y^{-5}}$
$= \dfrac{y^{-1-(-5)}}{36} = \dfrac{y^4}{36}$

67. **[6.1]** $(y^6)^{-5}(2y^{-3})^{-4} = y^{-30}(2)^{-4}y^{12}$
$= \dfrac{y^{-30+12}}{2^4}$
$= \dfrac{y^{-18}}{2^4} = \dfrac{1}{16y^{18}}$

68. **[6.4]** $(2x-9)^2 = (2x)^2 - 2(2x)(9) + 9^2$
$= 4x^2 - 36x + 81$

69. **[6.5]** $\dfrac{20y^3x^3 + 15y^4x + 25yx^4}{10yx^2}$
$= \dfrac{20y^3x^3}{10yx^2} + \dfrac{15y^4x}{10yx^2} + \dfrac{25yx^4}{10yx^2}$
$= 2y^2x + \dfrac{3y^3}{2x} + \dfrac{5x^2}{2}$

70. **[6.4]** $7p^5(3p^4 + p^3 + 2p^2)$
$= 7p^5(3p^4) + 7p^5(p^3) + 7p^5(2p^2)$
$= 21p^9 + 7p^8 + 14p^7$

71. **[6.1]** $\dfrac{(-z^{-2})^3}{5(z^{-3})^{-1}} = \dfrac{(-1)^3z^{-2(3)}}{5z^{-3(-1)}}$
$= \dfrac{-z^{-6}}{5z^3}$
$= \dfrac{-z^{-6-3}}{5}$
$= \dfrac{-z^{-9}}{5} = -\dfrac{1}{5z^9}$

72. **[6.5]** $\dfrac{x^3 + 7x^2 + 7x - 12}{x+5}$

$$\begin{array}{r|l} & x^2 + 2x - 3 \\ x+5 & x^3 + 7x^2 + 7x - 12 \\ & \underline{x^3 + 5x^2} \\ & \quad 2x^2 + 7x \\ & \quad \underline{2x^2 + 10x} \\ & \qquad -3x - 12 \\ & \qquad \underline{-3x - 15} \\ & \qquad\qquad 3 \end{array}$$

Answer: $x^2 + 2x - 3 + \dfrac{3}{x+5}$

73. **[6.2]** $(-5 + 11w) + (6 + 5w) + (-15 - 8w^2)$
$= -5 + 6 - 15 + 11w + 5w - 8w^2$
$= -14 + 16w - 8w^2$

74. **[6.2]** $(2k-1) - (3k^2 - 2k + 6)$
$= 2k - 1 - 3k^2 + 2k - 6$
$= -3k^2 + 2k + 2k - 1 - 6$
$= -3k^2 + 4k - 7$

Chapter 6 Test

1. **(a)** $7^{-2} = \dfrac{1}{7^2} = \dfrac{1}{49}$ **(C)**

(b) $7^0 = 1$ **(A)**

(c) $-7^0 = -(1) = -1$ **(D)**

(d) $(-7)^0 = 1$ **(A)**

(e) $-7^2 = -49$ **(E)**

(f) $7^{-1} + 2^{-1} = \dfrac{1}{7} + \dfrac{1}{2}$
$= \dfrac{2}{14} + \dfrac{7}{14} = \dfrac{9}{14}$ **(F)**

(g) $(7+2)^{-1} = 9^{-1} = \dfrac{1}{9}$ **(B)**

(h) $\dfrac{7^{-1}}{2^{-1}} = \dfrac{2^1}{7^1} = \dfrac{2}{7}$ **(G)**

(i) $(-7)^{-2} = \dfrac{1}{(-7)^2} = \dfrac{1}{49}$ **(C)**

2. $(3x^{-2}y^3)^{-2}(4x^3y^{-4})$
$= 3^{-2}x^{-2(-2)}y^{3(-2)}4x^3y^{-4}$
$= 3^{-2}x^4y^{-6}4x^3y^{-4}$
$= \dfrac{4x^{4+3}y^{-6-4}}{3^2}$
$= \dfrac{4x^7y^{-10}}{9} = \dfrac{4x^7}{9y^{10}}$

3. $\dfrac{36r^{-4}(r^2)^{-3}}{6r^4} = \dfrac{36r^{-4}r^{2(-3)}}{6r^4}$
$= \dfrac{6r^{-4}r^{-6}}{r^4}$
$= \dfrac{6r^{-10}}{r^4}$
$= \dfrac{6}{r^4r^{10}} = \dfrac{6}{r^{14}}$

4. $\left(\dfrac{4p^2}{q^4}\right)^3\left(\dfrac{6p^8}{q^{-8}}\right)^{-2} = \dfrac{4^3p^6}{q^{12}} \cdot \dfrac{6^{-2}p^{-16}}{q^{16}}$
$= \dfrac{4^3p^{-10}}{6^2q^{28}}$
$= \dfrac{64}{36p^{10}q^{28}} = \dfrac{16}{9p^{10}q^{28}}$

5. $(-2x^4y^{-3})^0(-4x^{-3}y^{-8})^2$
$= 1(-4)^2x^{-6}y^{-16}$
$= \dfrac{16}{x^6y^{16}}$

6. $9.1 \times 10^{-7} = 0.000\,000\,91$

Move the decimal point 7 places to the left because the exponent is negative.

7. $\dfrac{2{,}500{,}000 \times 0.00003}{0.05 \times 5{,}000{,}000}$

$= \dfrac{2.5 \times 10^6 \times 3 \times 10^{-5}}{5 \times 10^{-2} \times 5 \times 10^6}$

$= \dfrac{7.5 \times 10^1}{25 \times 10^4}$

$= 0.3 \times 10^{1-4}$

$= 0.3 \times 10^{-3}$

$= 3 \times 10^{-4}$, or 0.0003

8. **(a)** $f(x) = -2x^2 + 5x - 6$

$f(4) = -2(4)^2 + 5(4) - 6$

$= -2 \cdot 16 + 20 - 6$

$= -32 + 20 - 6 = -18$

(b) $(f + g)(x) = f(x) + g(x)$

$= (-2x^2 + 5x - 6) + (7x - 3)$

$= -2x^2 + 12x - 9$

(c) $(f - g)(x) = f(x) - g(x)$

$= (-2x^2 + 5x - 6) - (7x - 3)$

$= -2x^2 + 5x - 6 - 7x + 3$

$= -2x^2 - 2x - 3$

(d) Using the answer in part (c), we have

$(f - g)(-2) = -2(-2)^2 - 2(-2) - 3$

$= -8 + 4 - 3 = -7.$

9.

x	$f(x) = -2x^2 + 3$
-2	$-2(-2)^2 + 3 = -5$
-1	$-2(-1)^2 + 3 = 1$
0	$-2(0)^2 + 3 = 3$
1	$-2(1)^2 + 3 = 1$
2	$-2(2)^2 + 3 = -5$

Since the greatest exponent is 2, the graph of f is a parabola.

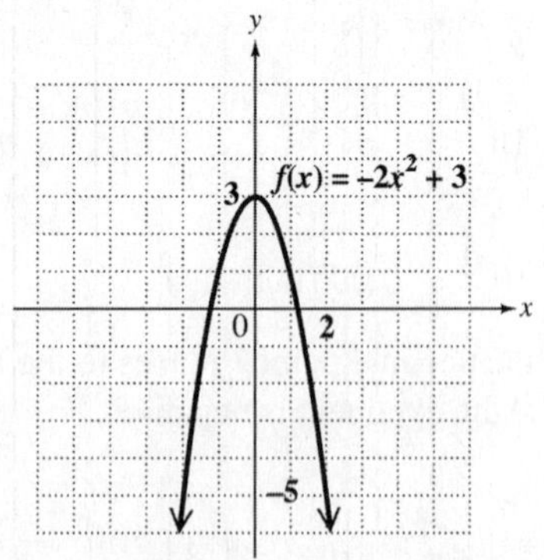

10. $f(x) = 0.139x^2 + 14.16x + 465.9$

$f(x)$ represents the number of medical doctors, in thousands, in the United States.

	Year	x	$f(x)$
(a)	1980	0	$465.900 \approx 466$
(b)	1995	15	$709.575 \approx 710$
(c)	2005	25	$906.775 \approx 907$

11. $(4x^3 - 3x^2 + 2x - 5)$

$- (3x^3 + 11x + 8) + (x^2 - x)$

$= 4x^3 - 3x^2 + 2x - 5 - 3x^3 - 11x$

$- 8 + x^2 - x$

$= x^3 - 2x^2 - 10x - 13$

12. $(5x - 3)(2x + 1)$

F O I L

$= 5x(2x) + 5x(1) + (-3)(2x) + (-3)(1)$

$= 10x^2 + 5x - 6x - 3$

$= 10x^2 - x - 3$

13. $(2m - 5)(3m^2 + 4m - 5)$

$= 2m(3m^2 + 4m - 5)$

$+ (-5)(3m^2 + 4m - 5)$

$= 6m^3 + 8m^2 - 10m$

$- 15m^2 - 20m + 25$

$= 6m^3 - 7m^2 - 30m + 25$

14. $(6x + y)(6x - y) = (6x)^2 - y^2$

$= 36x^2 - y^2$

15. $(3k + q)^2 = (3k)^2 + 2(3k)(q) + q^2$

$= 9k^2 + 6kq + q^2$

16. $[2y + (3z - x)][2y - (3z - x)]$

$= (2y)^2 - (3z - x)^2$

$= 4y^2 - (9z^2 - 6zx + x^2)$

$= 4y^2 - 9z^2 + 6zx - x^2$

17. $\dfrac{16p^3 - 32p^2 + 24p}{4p^2}$

$= \dfrac{16p^3}{4p^2} - \dfrac{32p^2}{4p^2} + \dfrac{24p}{4p^2}$

$= 4p - 8 + \dfrac{6}{p}$

18. $(x^3 + 3x^2 - 6) \div (x - 2)$

Insert $0x$ for the missing x-term.

$$\begin{array}{r|l} & x^2 + 5x + 10 \\ x - 2 & x^3 + 3x^2 + 0x - 6 \\ & \underline{x^3 - 2x^2} \\ & \quad 5x^2 + 0x \\ & \quad \underline{5x^2 - 10x} \\ & \qquad 10x - 6 \\ & \qquad \underline{10x - 20} \\ & \qquad\qquad 14 \end{array}$$

Answer: $x^2 + 5x + 10 + \dfrac{14}{x-2}$

19. **(a)** $(fg)(x) = f(x) \cdot g(x)$
$= (x^2 + 3x + 2)(x + 1)$
$= (x^2 + 3x + 2)(x)$
$\quad + (x^2 + 3x + 2)(1)$
$= x^3 + 3x^2 + 2x + x^2 + 3x + 2$
$= x^3 + 4x^2 + 5x + 2$

(b) $(fg)(-2) = f(-2) \cdot g(-2)$
$= [(-2)^2 + 3(-2) + 2] \cdot [(-2) + 1]$
$= [4 - 6 + 2] \cdot [-1]$
$= 0(-1) = 0$

Alternatively, we could have substituted -2 for x into our answer from part (a).

20. **(a)** $\left(\dfrac{f}{g}\right)(x) = \dfrac{f(x)}{g(x)} = \dfrac{x^2 + 3x + 2}{x + 1}$

$$\begin{array}{r|l} & x + 2 \\ x + 1 & x^2 + 3x + 2 \\ & \underline{x^2 + x} \\ & \quad 2x + 2 \\ & \quad \underline{2x + 2} \\ & \qquad\quad 0 \end{array}$$

Thus, $\left(\dfrac{f}{g}\right)(x) = x + 2$ if $x + 1 \neq 0$, that is, $x \neq -1$.

(b) Using our answer from part (a),

$$\left(\frac{f}{g}\right)(-2) = (-2) + 2 = 0.$$

Cumulative Review Exercises (Chapters 1–6)

1. 34 is a natural number, so it is also a whole number, an integer, a rational number, and a real number. **A, B, C, D, F**

2. 0 is a whole number, so it is also an integer, a rational number, and a real number. **B, C, D, F**

3. 2.16 is a rational number, so it is also a real number. **D, F**

4. $-\sqrt{36} = -6$ is an integer, so it is also a rational number and a real number. **C, D, F**

5. $\sqrt{13}$ is an irrational number, so it is also a real number. **E, F**

6. $-\frac{4}{5}$ is a rational number, so it is also a real number. **D, F**

7. $9 \cdot 4 - 16 \div 4 = (9 \cdot 4) - (16 \div 4) = 36 - 4 = 32$

8. $-|8 - 13| - |-4| + |-9| = -|-5| - 4 + 9$
$= -5 - 4 + 9$
$= -9 + 9 = 0$

9. $-5(8 - 2z) + 4(7 - z) = 7(8 + z) - 3$
$-40 + 10z + 28 - 4z = 56 + 7z - 3$ *Distributive property*
$6z - 12 = 7z + 53$ *Combine like terms*
$-65 = z$ *Subtract 6z; 53*

Thus, the solution set is $\{-65\}$.

10. $3(x + 2) - 5(x + 2) = -2x - 4$
$3x + 6 - 5x - 10 = -2x - 4$
$-2x - 4 = -2x - 4$

The last statement is true for all real numbers, so the solution set is $(-\infty, \infty)$.

11. $2(m + 5) - 3m + 1 > 5$
$2m + 10 - 3m + 1 > 5$
$-m + 11 > 5$
$-m > -6$
$m < 6$

Solution set: $(-\infty, 6)$

12. $|3x - 1| = 2$

$3x - 1 = 2$ or $3x - 1 = -2$
$3x = 3$ or $3x = -1$
$x = 1$ or $x = -\frac{1}{3}$

Solution set: $\{-\frac{1}{3}, 1\}$

13. $|3z + 1| \geq 7$

$3z + 1 \geq 7$ or $3z + 1 \leq -7$
$3z \geq 6$ or $3z \leq -8$
$z \geq 2$ or $z \leq -\frac{8}{3}$

Solution set: $(-\infty, -\frac{8}{3}] \cup [2, \infty)$

14. Personal computer: $\dfrac{480}{1500} = 32\%$

Pacemaker:

$$26\% \text{ of } 1500 = 0.26(1500) = 390$$

Wireless communication:

$$18\% \text{ of } 1500 = 0.18(1500) = 270$$

Television: $\dfrac{150}{1500} = 10\%$

15. The sum of the measures of the angles of any triangle is 180°, so

$$(x + 15) + (6x + 10) + (x - 5) = 180.$$

Solve this equation.

$$\begin{aligned} 8x + 20 &= 180 \\ 8x &= 160 \\ x &= 20 \end{aligned}$$

Substitute 20 for x to find the measures of the angles.

$$\begin{aligned} x - 5 &= 20 - 5 = 15 \\ x + 15 &= 20 + 15 = 35 \\ 6x + 10 &= 6(20) + 10 = 130 \end{aligned}$$

The measures of the angles of the triangle are 15°, 35°, and 130°.

16. Through $(-4, 5)$ and $(2, -3)$

Use the definition of slope with $x_1 = -4$, $y_1 = 5$, $x_2 = 2$, and $y_2 = -3$.

$$\begin{aligned} m = \frac{y_2 - y_1}{x_2 - x_1} &= \frac{-3 - 5}{2 - (-4)} \\ &= \frac{-8}{6} = -\frac{4}{3} \end{aligned}$$

Use the point-slope form of a line.

$$\begin{aligned} y - y_1 &= m(x - x_1) \\ y - 5 &= -\tfrac{4}{3}[x - (-4)] \\ 3y - 15 &= -4(x + 4) && \textit{Multiply by 3.} \\ 3y - 15 &= -4x - 16 \\ 4x + 3y &= -1 && \textit{Standard form} \end{aligned}$$

17. $-3x + 4y = 12$

If $y = 0$, $x = -4$, so the x-intercept is $(-4, 0)$.

If $x = 0$, $y = 3$, so the y-intercept is $(0, 3)$.

Draw a line through these intercepts. A third point may be used as a check.

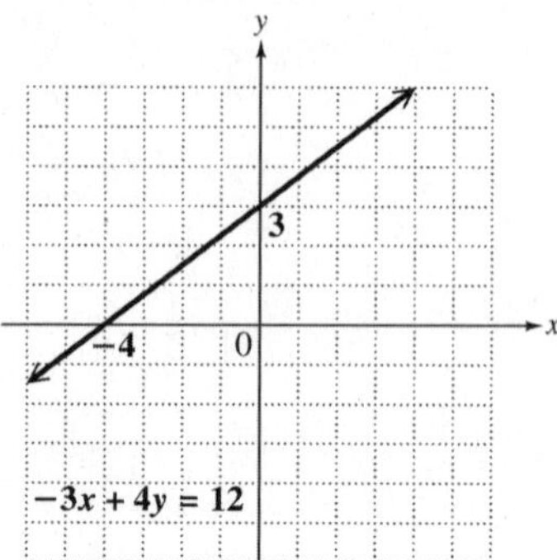

18. $y \leq 2x - 6$

Graph the boundary, $y = 2x - 6$, as a solid line through the intercepts $(3, 0)$ and $(0, -6)$. A third point such as $(1, -4)$ can be used as a check. Using $(0, 0)$ as a test point results in the false inequality $0 \leq -6$, so shade the region *not* containing the origin. This is the region below the line. The solid line shows that the boundary is part of the graph.

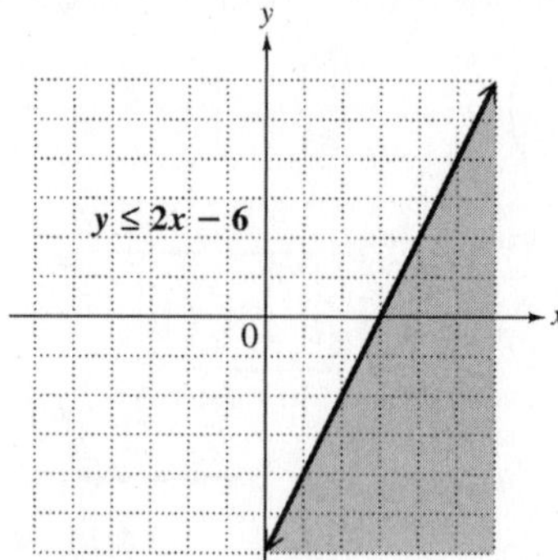

19. $3x + 2y < 0$

Graph the boundary, $3x + 2y = 0$, as a dashed line through $(0, 0)$, $(-2, 3)$, and $(2, -3)$. Choose a test point not on the line. Using $(1, 1)$ results in the false statement $5 < 0$, so shade the region *not* containing $(1, 1)$. This is the region below the line. The dashed line shows that the boundary is not part of the graph.

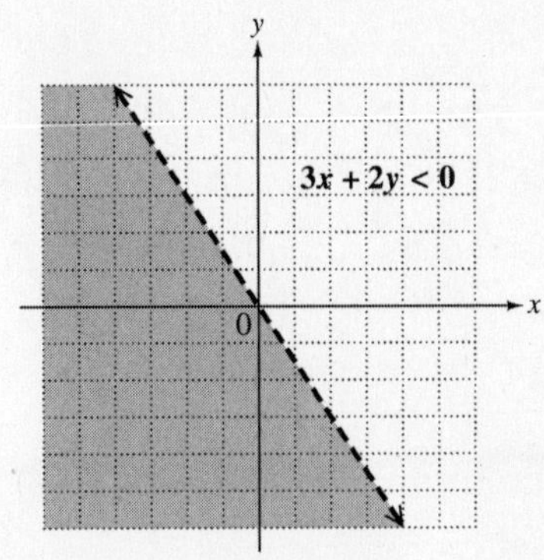

20. **(a)** $m = \dfrac{261{,}122 - 322{,}486}{5 - 0} = \dfrac{-61{,}364}{5}$
$= -12{,}272.8$

The average rate of change is $-12{,}272.8$ thousand pounds per year, that is, the number of pounds of shrimp caught in the United States decreased an average of $-12{,}272.8$ thousand pounds per year during the selected years.

(b) $y = mx + b$
$y = -12{,}272.8x + 322{,}486$

(c) For the year 2003, $x = 2003 - 2000 = 3$.

$$\begin{aligned} y &= -12{,}272.8(3) + 322{,}486 \\ &= 285{,}667.6 \approx 285{,}668 \end{aligned}$$

The model approximates about 285,668 thousand pounds of shrimp caught in 2003.

21. $\{(-4,-2), (-1,0), (2,0), (5,2)\}$

The domain is the set of first components, that is, $\{-4,-1,2,5\}$.
The range is the set of second components, that is, $\{-2,0,2\}$.
The relation is a function since each first component is paired with a unique second component.

22. $3x - 4y = 1$ (1)
$2x + 3y = 12$ (2)

To eliminate y, multiply equation (1) by 3 and equation (2) by 4. Then add the results.

$$\begin{array}{rcrl} 9x - 12y &=& 3 & 3 \times (1) \\ 8x + 12y &=& 48 & 4 \times (2) \\ \hline 17x &=& 51 & \\ x &=& 3 & \end{array}$$

Substitute $x = 3$ into equation (1).

$$\begin{aligned} 3x - 4y &= 1 \quad (1) \\ 3(3) - 4y &= 1 \\ 9 - 4y &= 1 \\ -4y &= -8 \\ y &= 2 \end{aligned}$$

Solution set: $\{(3,2)\}$

23. $3x - 2y = 4$ (1)
$-6x + 4y = 7$ (2)

Multiply equation (1) by 2 and add the result to equation (2).

$$\begin{array}{rcrl} 6x - 4y &=& 8 & 2 \times (1) \\ -6x + 4y &=& 7 & (2) \\ \hline 0 &=& 15 & \textit{False} \end{array}$$

Since a false statement results, the system is *inconsistent*. The solution set is $\emptyset$.

24. $x + 3y - 6z = 7$ (1)
$2x - y + z = 1$ (2)
$x + 2y + 2z = -1$ (3)

To eliminate x, multiply equation (1) by -2 and add the result to equation (2).

$$\begin{array}{rcrl} -2x - 6y + 12z &=& -14 & -2 \times (1) \\ 2x - y + z &=& 1 & (2) \\ \hline -7y + 13z &=& -13 & (4) \end{array}$$

To eliminate x again, multiply equation (3) by -2 and add the result to equation (2).

$$\begin{array}{rcrl} 2x - y + z &=& 1 & (2) \\ -2x - 4y - 4z &=& 2 & -2 \times (3) \\ \hline -5y - 3z &=& 3 & (5) \end{array}$$

Use equations (4) and (5) to eliminate z. Multiply equation (4) by 3 and add the result to 13 times equation (5).

$$\begin{array}{rcrl} -21y + 39z &=& -39 & 3 \times (4) \\ -65y - 39z &=& 39 & 13 \times (5) \\ \hline -86y &=& 0 & \\ y &=& 0 & \end{array}$$

From (5) with $y = 0$, $-3z = 3$, so $z = -1$.
From (3) with $y = 0$ and $z = -1$, $x - 2 = -1$, so $x = 1$.

Solution set: $\{(1,0,-1)\}$

25. The length L of the rectangular flag measured 12 feet more than its width W, so

$$L = W + 12. \quad (1)$$

The perimeter is 144 feet.

$$P = 2L + 2W \quad (2)$$

Substitute $W + 12$ for L into equation (2).

$$\begin{aligned} 144 &= 2(W + 12) + 2W \\ 144 &= 2W + 24 + 2W \\ 120 &= 4W \\ 30 &= W \end{aligned}$$

From (1), $L = 30 + 12 = 42$.

The length is 42 feet and the width is 30 feet.

26. $\left(\frac{2m^3n}{p^2}\right)^3 = \frac{2^3(m^3)^3n^3}{(p^2)^3} = \frac{8m^9n^3}{p^6}$

27. $\frac{x^{-6}y^3z^{-1}}{x^7y^{-4}z} = \frac{y^4y^3}{x^6x^7z^1z} = \frac{y^7}{x^{13}z^2}$

28. $(2m^{-2}n^3)^{-3} = 2^{-3}(m^{-2})^{-3}(n^3)^{-3}$

$= 2^{-3}m^{(-2)(-3)}n^{3(-3)}$

$= 2^{-3}m^6n^{-9}$

$= \frac{m^6}{2^3n^9} = \frac{m^6}{8n^9}$

29. $2(3x^2 - 8x + 1) - 4(x^2 - 3x - 9)$

$= 6x^2 - 16x + 2 - 4x^2 + 12x + 36$

Distributive property

$= (6x^2 - 4x^2) + (-16x + 12x) + (2 + 36)$

Combine like terms

$= 2x^2 - 4x + 38$

30. $(3x + 2y)(5x - y)$

F O I L

$= 3x(5x) + 3x(-y) + 2y(5x) + 2y(-y)$

$= 15x^2 - 3xy + 10xy - 2y^2$

$= 15x^2 + 7xy - 2y^2$

31. $(8m + 5n)(8m - 5n) = (8m)^2 - (5n)^2$

$= 64m^2 - 25n^2$

32. $\frac{m^3 - 3m^2 - 5m + 4}{m - 1}$

$$\begin{array}{r|l} & m^2 \quad -2m \quad -7 \\ \hline m-1 & m^3 - 3m^2 - 5m + 4 \\ & \underline{m^3 - m^2} \\ & \quad -2m^2 - 5m \\ & \quad \underline{-2m^2 + 2m} \\ & \qquad -7m + 4 \\ & \qquad \underline{-7m + 7} \\ & \qquad\qquad -3 \end{array}$$

Answer: $m^2 - 2m - 7 + \frac{-3}{m - 1}$

CHAPTER 7 FACTORING

7.1 Greatest Common Factors; Factoring by Grouping

7.1 Margin Exercises

1. **(a)** $7k+28=7\cdot k+7\cdot 4$ *GCF = 7*
$=7(k+4)$

Check: $7(k+4)$
$=7(k)+7(4)$ *Distributive property*
$=7k+28$ *Original polynomial*

(b) $32m+24=8\cdot 4m+8\cdot 3$ *GCF = 8*
$=8(4m+3)$

Check: $8(4m+3)$
$=8(4m)+8(3)$ *Dist. prop.*
$=32m+24$ *Original poly.*

(c) $8a-9$ *cannot be factored* because $8a$ and 9 do not have a common factor other than 1.

(d) $5z+5=5\cdot z+5\cdot 1$ *GCF = 5*
$=5(z+1)$

Check: $5(z+1)=5(z)+5(1)$
$=5z+5$

2. **(a)** $16y^4+8y^3$

The numerical part of the GCF is 8, the largest number that divides into both 16 and 8. For the variable parts, y^4 and y^3, use the least exponent that appears on y; here the least exponent is 3. The GCF is $8y^3$.

$$16y^4+8y^3=8y^3(2y)+8y^3(1)$$
$$=8y^3(2y+1)$$

(b) $14p^2-9p^3+6p^4$

The least exponent is 2. The GCF is p^2.

$$14p^2-9p^3+6p^4$$
$$=p^2(14)+p^2(-9p)+p^2(6p^2)$$
$$=p^2(14-9p+6p^2)$$

(c) $15z^2+45z^5-60z^6=15z^2(1+3z^3-4z^4)$

Check: $15z^2(1+3z^3-4z^4)$
$=15z^2(1)+15z^2(3z^3)+15z^2(-4z^4)$
$=15z^2+45z^5-60z^6$

(d) $4x^2z-2xz+8z^2=2z(2x^2-x+4z)$

Check: $2z(2x^2-x+4z)$
$=2z(2x^2)+2z(-x)+2z(4z)$
$=4x^2z-2xz+8z^2$

(e) $12y^5x^2+8y^3x^3$

The numerical part of the GCF is 4. The least exponent on x is 2, and the least exponent on y is 3. Therefore, the GCF is $4y^3x^2$.

$$12y^5x^2+8y^3x^3=4y^3x^2(3y^2+2x)$$

(f) $5m^4x^3+15m^5x^6-20m^4x^6$

The numerical part of the GCF is 5. The least exponent on m is 4, and the least exponent on x is 3. Therefore, the GCF is $5m^4x^3$.

$$5m^4x^3+15m^5x^6-20m^4x^6$$
$$=5m^4x^3(1+3mx^3-4x^3)$$

3. **(a)** $(a+2)(a-3)+(a+2)(a+6)$

The greatest common factor here is $(a+2)$.

$$=(a+2)[(a-3)+(a+6)]$$
$$=(a+2)(a-3+a+6)$$
$$=(a+2)(2a+3)$$

(b) $(y-1)(y+3)-(y-1)(y+4)$
$=(y-1)[(y+3)-(y+4)]$
$=(y-1)(y+3-y-4)$
$=(y-1)(-1)$, or $-y+1$

(c) $k^2(a+5b)+m^2(a+5b)^2$
$=(a+5b)[k^2+m^2(a+5b)]$ *GCF = a + 5b*
$=(a+5b)(k^2+m^2a+5m^2b)$

(d) $r^2(y+6)+r^2(y+3)$
$=r^2[(y+6)+(y+3)]$ *GCF = r^2*
$=r^2(y+6+y+3)$
$=r^2(2y+9)$

4. **(a)** $-k^2+3k$

Factor with GCF $=k$.

$$-k^2+3k=k(-k)+k(3)$$
$$=k(-k+3)$$

Factor with GCF $=-k$.

$$-k^2+3k=-k(k)+(-k)(-3)$$
$$=-k(k-3)$$

(b) $-6r^3 - 5r^2 + 14r$

Factor with GCF $= r$.

$$\begin{aligned}&-6r^3 - 5r^2 + 14r\\ &= r(-6r^2) + r(-5r) + r(14)\\ &= r(-6r^2 - 5r + 14)\end{aligned}$$

Factor with GCF $= -r$.

$$\begin{aligned}&-6r^3 - 5r^2 + 14r\\ &= -r(6r^2) + (-r)(5r) + (-r)(-14)\\ &= -r(6r^2 + 5r - 14)\end{aligned}$$

5. $$\begin{aligned}6p - 6q + rp - rq &= (6p - 6q) + (rp - rq)\\ &= 6(p - q) + r(p - q)\\ &= (p - q)(6 + r)\end{aligned}$$

6. $$\begin{aligned}xy - 2y - 4x + 8 &= (xy - 2y) - (4x - 8)\\ &= y(x - 2) - 4(x - 2)\\ &= (x - 2)(y - 4)\end{aligned}$$

7. $$\begin{aligned}2xy + 3y + 2x + 3 &= (2xy + 3y) + (2x + 3)\\ &= y(2x + 3) + 1(2x + 3)\\ &= (2x + 3)(y + 1)\end{aligned}$$

8. **(a)** $mn + 6 + 2n + 3m$

Rearrange the terms so that there is a common factor in the first two terms and a common factor in the last two terms.

$$= mn + 2n + 6 + 3m$$

Group the first two terms and the last two terms.

$$\begin{aligned}&= (mn + 2n) + (6 + 3m)\\ &= n(m + 2) + 3(2 + m)\\ &= n(m + 2) + 3(m + 2)\\ &= (m + 2)(n + 3)\end{aligned}$$

(b) $$\begin{aligned}&10x^2y^2 - 18 + 15y^2 - 12x^2\\ &= 10x^2y^2 + 15y^2 - 12x^2 - 18\\ &= (10x^2y^2 + 15y^2) + (-12x^2 - 18)\\ &= 5y^2(2x^2 + 3) - 6(2x^2 + 3)\\ &= (2x^2 + 3)(5y^2 - 6)\end{aligned}$$

9. **(a)** $$\begin{aligned}&12wy + 4wz - 24xy - 8xz\\ &= 4(3wy + wz - 6xy - 2xz)\\ &= 4[(3wy + wz) + (-6xy - 2xz)]\\ &= 4[w(3y + z) - 2x(3y + z)]\\ &= 4[(3y + z)(w - 2x)]\\ &= 4(3y + z)(w - 2x)\end{aligned}$$

(b) $$\begin{aligned}&6bxy + 3xyz + 6bxz + 3xz^2\\ &= 3x(2by + yz + 2bz + z^2)\\ &= 3x[(2by + yz) + (2bz + z^2)]\\ &= 3x[y(2b + z) + z(2b + z)]\\ &= 3x[(2b + z)(y + z)]\\ &= 3x(2b + z)(y + z)\end{aligned}$$

7.1 Section Exercises

1. $9m^3, 3m^2, 15m$

The numerical part of the GCF is 3. The least exponent on m is 1. Thus, the GCF is $3m$.

3. $16xy^3, 24x^2y^2, 8x^2y$

The numerical part of the GCF is 8. The least exponent on x is 1. The least exponent on y is 1. Thus, the GCF is $8xy$.

5. $6m(r + t)^2, 3p(r + t)^4$

The numerical part of the GCF is 3. The least exponent on $(r + t)$ is 2. Thus, the GCF is $3(r + t)^2$.

7. $6x^3y^4 - 12x^5y^2 + 24x^4y^8$

The numerical part of the GCF is 6. The least exponent that appears on x is 3, while 2 is the least exponent on y. Thus, $6x^3y^2$ is the GCF.

Since $6x^3y^2$ is one of the factors in $6x^3y^2(y^2 - 2x^2 + 4xy^6)$, choice **A** is the correct response.

9. $$\begin{aligned}10x - 30 &= 10 \cdot x - 10 \cdot 3 \quad \textit{GCF = 10}\\ &= 10(x - 3)\end{aligned}$$

11. $$\begin{aligned}8s + 16t &= 8(s) + 8(2t) \quad \textit{GCF = 8}\\ &= 8(s + 2t)\end{aligned}$$

13. $$\begin{aligned}6 + 12r &= 6(1) + 6(2r) \quad \textit{GCF = 6}\\ &= 6(1 + 2r)\end{aligned}$$

15. $$\begin{aligned}8k^3 + 24k &= 8k \cdot k^2 + 8k \cdot 3\\ &= 8k(k^2 + 3)\end{aligned}$$

17. $$\begin{aligned}3xy - 5xy^2 &= xy \cdot 3 - xy \cdot 5y\\ &= xy(3 - 5y)\end{aligned}$$

19. $$\begin{aligned}-4p^3q^4 - 2p^2q^5 &= -2p^2q^4 \cdot 2p - 2p^2q^4 \cdot q\\ &= -2p^2q^4(2p + q)\end{aligned}$$

21. $$\begin{aligned}&21x^5 + 35x^4 - 14x^3\\ &= 7x^3(3x^2) + 7x^3(5x) + 7x^3(-2)\\ &= 7x^3(3x^2 + 5x - 2)\end{aligned}$$

23. $$\begin{aligned}&36p^4 + 9p^2 - 27p^3\\ &= 9p^2(4p^2) + 9p^2(1) + 9p^2(-3p)\\ &= 9p^2(4p^2 + 1 - 3p)\end{aligned}$$

25. $$\begin{aligned}&15a^2c^3 - 25ac^2 + 5a^2c\\ &= 5ac(3ac^2) + 5ac(-5c) + 5ac(a)\\ &= 5ac(3ac^2 - 5c + a)\end{aligned}$$

27. $-27m^3p^5 + 5r^4s^3 - 8x^5z^4$ *cannot be factored* because there is no common factor (except ± 1).

29. $(m-4)(m+2)+(m-4)(m+3)$
The GCF is $(m-4)$.
$= (m-4)[(m+2)+(m+3)]$
$= (m-4)(m+2+m+3)$
$= (m-4)(2m+5)$

31. $(2z-1)(z+6)-(2z-1)(z-5)$
$= (2z-1)[(z+6)-(z-5)]$
$= (2z-1)[z+6-z+5]$
$= (2z-1)(11)$
$= 11(2z-1)$

33. $5(2-x)^3-(2-x)^4+4(2-x)^2$
$= (2-x)^2[5(2-x)-(2-x)^2+4]$
Factor out $(2-x)^2$.
$= (2-x)^2[10-5x-(4-4x+x^2)+4]$
$= (2-x)^2[10-5x-4+4x-x^2+4]$
Clear parentheses inside brackets.
$= (2-x)^2(10-x-x^2)$

35. $-r^3+3r^2+5r$

Factor out r.

$$-r^3+3r^2+5r = r(-r^2+3r+5)$$

Factor out $-r$.

$$-r^3+3r^2+5r = -r(r^2-3r-5)$$

37. $-12s^5+48s^4$

Factor out $12s^4$.

$$-12s^5+48s^4 = 12s^4(-s+4)$$

Factor out $-12s^4$.

$$-12s^5+48s^4 = -12s^4(s-4)$$

39. $-2x^5+6x^3+4x^2$

Factor out $2x^2$.

$$-2x^5+6x^3+4x^2 = 2x^2(-x^3+3x+2)$$

Factor out $-2x^2$.

$$-2x^5+6x^3+4x^2 = -2x^2(x^3-3x-2)$$

41. $mx+3qx+my+3qy$
$= (mx+3qx)+(my+3qy)$
$= x(m+3q)+y(m+3q)$
$= (m+3q)(x+y)$

43. $10m+2n+5mk+nk$
$= (10m+2n)+(5mk+nk)$
$= 2(5m+n)+k(5m+n)$
$= (5m+n)(2+k)$

45. $4-2q-6p+3pq$
$= (4-2q)+(-6p+3pq)$
$= 2(2-q)-3p(2-q)$
$= (2-q)(2-3p)$

47. $p^2-4zq+pq-4pz$
$= (p^2+pq)+(-4zq-4pz)$
$= p(p+q)-4z(q+p)$
$= (p+q)(p-4z)$

49. $7ab+35bc+a+5c$
$= (7ab+35bc)+(a+5c)$
$= 7b(a+5c)+1(a+5c)$
$= (a+5c)(7b+1)$

51. $m^3+4m^2-6m-24$
$= (m^3+4m^2)+(-6m-24)$
$= m^2(m+4)-6(m+4)$
$= (m+4)(m^2-6)$

53. $-3a^3-3ab^2+2a^2b+2b^3$
$= (-3a^3-3ab^2)+(2a^2b+2b^3)$
$= -3a(a^2+b^2)+2b(a^2+b^2)$
$= (a^2+b^2)(-3a+2b)$

55. $4+xy-2y-2x$
$= xy-2x-2y+4$
$= (xy-2x)+(-2y+4)$
$= x(y-2)-2(y-2)$
$= (y-2)(x-2)$

57. $8+9y^4-6y^3-12y$
$= 9y^4-6y^3-12y+8$
$= (9y^4-6y^3)+(-12y+8)$
$= 3y^3(3y-2)-4(3y-2)$
$= (3y-2)(3y^3-4)$

59. $2mx+6qx+2my+6qy$
$= 2(mx+3qx+my+3qy)$
$= 2[(mx+3qx)+(my+3qy)]$
$= 2[x(m+3q)+y(m+3q)]$
$= 2[(m+3q)(x+y)]$
$= 2(m+3q)(x+y)$

61. $2x^3y^2+x^2y^2-14xy^2-7y^2$
$= y^2(2x^3+x^2-14x-7)$
$= y^2[(2x^3+x^2)+(-14x-7)]$
$= y^2[x^2(2x+1)-7(2x+1)]$
$= y^2[(2x+1)(x^2-7)]$
$= y^2(2x+1)(x^2-7)$

63. $3m^{-5} + m^{-3}$

Factor out m^{-5} since -5 is the lesser exponent.

$$= m^{-5}(3) + m^{-5}(m^{-3-(-5)})$$
$$= m^{-5}(3 + m^2), \text{ or } \frac{3 + m^2}{m^5}$$

65. $3p^{-3} + 2p^{-2}$

Factor out p^{-3} since -3 is the lesser exponent.

$$= p^{-3}(3) + p^{-3}(2p^{-2-(-3)})$$
$$= p^{-3}(3 + 2p), \text{ or } \frac{3 + 2p}{p^3}$$

7.2 Factoring Trinomials

7.2 Margin Exercises

1. **(a)** $p^2 + 6p + 5$

Step 1 Find pairs of integers whose product is 5.	*Step 2* Write sums of those integers.
$5(1)$	$5 + 1 = 6 \leftarrow$
$-5(-1)$	$-5 + (-1) = -6$

The coefficient of the middle term is 6, so the required integers are 1 and 5. The factored form is

$$p^2 + 6p + 5 = (p + 1)(p + 5).$$

Check: $(p + 1)(p + 5) = p^2 + 6p + 5$

(b) $a^2 + 9a + 20$

Step 1 Find pairs of integers whose product is 20.	*Step 2* Write sums of those integers.
$20(1)$	$20 + 1 = 21$
$-20(-1)$	$-20 + (-1) = -21$
$10(2)$	$10 + 2 = 12$
$-10(-2)$	$-10 + (-2) = -12$
$5(4)$	$5 + 4 = 9 \leftarrow$
$-5(-4)$	$-5 + (-4) = -9$

The coefficient of the middle term is 9, so the required integers are 5 and 4. The factored form is

$$a^2 + 9a + 20 = (a + 5)(a + 4).$$

Check: $(a + 5)(a + 4) = a^2 + 9a + 20$

(c) $k^2 - k - 6$

Step 1 Find pairs of integers whose product is -6.	*Step 2* Write sums of those integers.
$-6(1)$	$-6 + 1 = -5$
$6(-1)$	$6 + (-1) = 5$
$-3(2)$	$-3 + 2 = -1 \leftarrow$
$3(-2)$	$3 + (-2) = 1$

The coefficient of the middle term is -1, so the required integers are -3 and 2. The factored form is

$$k^2 - k - 6 = (k - 3)(k + 2).$$

Check: $(k - 3)(k + 2) = k^2 - k - 6$

(d) $b^2 - 7b + 10$

Step 1 Find pairs of integers whose product is 10.	*Step 2* Write sums of those integers.
$10(1)$	$10 + 1 = 11$
$-10(-1)$	$-10 + (-1) = -11$
$5(2)$	$5 + 2 = 7$
$-5(-2)$	$-5 + (-2) = -7 \leftarrow$

The coefficient of the middle term is -7, so the required integers are -5 and -2. The factored form is

$$b^2 - 7b + 10 = (b - 5)(b - 2).$$

Check: $(b - 5)(b - 2) = b^2 - 7b + 10$

(e) $y^2 - 8y + 6$

Step 1 Find pairs of integers whose product is 6.	*Step 2* Write sums of those integers.
$6(1)$	$6 + 1 = 7$
$-6(-1)$	$-6 + (-1) = -7$
$3(2)$	$3 + 2 = 5$
$-3(-2)$	$-3 + (-2) = -5$

The coefficient of the middle term is -8, and none of the sums equal -8. Thus, $y^2 - 8y + 6$ cannot be factored with integer coefficients and is *prime*. (Note that we have assumed that the GCF has already been factored out.)

2. **(a)** $x^2 + 2nx - 8n^2$

Look for two expressions whose product is $-8n^2$ and whose sum is $2n$. The quantities $4n$ and $-2n$ have the necessary product and sum, so

$$x^2 + 2nx - 8n^2 = (x + 4n)(x - 2n).$$

Check:

$(x + 4n)(x - 2n)$

$= x^2 - 2nx + 4nx - 8n^2$ *FOIL*

$= x^2 + 2nx - 8n^2$ *Original polynomial*

(b) $x^2 - 7xz + 9z^2$

Look for two expressions whose product is $9z^2$ and whose sum is $-7z$. There are no such quantities. Therefore, the trinomial cannot be factored and is *prime*.

3. $5m^4 - 5m^3 - 100m^2$

Start by factoring out the GCF, $5m^2$.

$5m^4 - 5m^3 - 100m^2 = 5m^2(m^2 - m - 20)$

To factor $m^2 - m - 20$, look for two integers whose product is -20 and whose sum is -1. The necessary integers are -5 and 4. Remember to write the common factor $5m^2$ as part of the answer.

$$5m^4 - 5m^3 - 100m^2 = 5m^2(m - 5)(m + 4)$$

4. **(a)** $3y^2 - 11y - 4$

The product ac is $3(-4) = -12$. Look for two integers whose product is -12 and whose sum is -11. The necessary integers are -12 and 1. Write $-11y$ as $-12y + y$ and then factor by grouping.

$$\begin{aligned} &3y^2 - 11y - 4 \\ &= 3y^2 - 12y + y - 4 \\ &= (3y^2 - 12y) + (y - 4) \\ &= 3y(y - 4) + 1(y - 4) \\ &= (y - 4)(3y + 1) \end{aligned}$$

(b) $6k^2 - 19k + 10$

The product ac is $6(10) = 60$. Look for two integers whose product is 60 and whose sum is -19. The necessary integers are -15 and -4. Write $-19k$ as $-15k - 4k$ and then factor by grouping.

$$\begin{aligned} &6k^2 - 19k + 10 \\ &= 6k^2 - 15k - 4k + 10 \\ &= (6k^2 - 15k) + (-4k + 10) \\ &= 3k(2k - 5) - 2(2k - 5) \\ &= (2k - 5)(3k - 2) \end{aligned}$$

5. By trial and error, the following are factored.

(a) $10x^2 + 17x + 3 = (5x + 1)(2x + 3)$

(b) $16y^2 - 34y - 15 = (8y + 3)(2y - 5)$

(c) $8t^2 - 13t + 5 = (8t - 5)(t - 1)$

6. **(a)** $7p^2 + 15pq + 2q^2$

Try some possibilities using $7p$ and p along with $2q$ and q. Use FOIL to check.

$(7p + 2q)(p + q) = 7p^2 + 9pq + 2q^2$ *No*

$(7p + q)(p + 2q) = 7p^2 + 15pq + 2q^2$ *Yes*

The correct factoring is

$$7p^2 + 15pq + 2q^2 = (7p + q)(p + 2q).$$

(b) $6m^2 + 7mn - 5n^2$

Try some possibilities.

$(6m + n)(m - 5n) = 6m^2 - 29mn - 5n^2$ *No*

$(6m - 5n)(m + n) = 6m^2 + mn - 5n^2$ *No*

$(3m + n)(2m - 5n) = 6m^2 - 13mn - 5n^2$ *No*

$(3m + 5n)(2m - n) = 6m^2 + 7mn - 5n^2$ *Yes*

The correct factoring is

$$6m^2 + 7mn - 5n^2 = (3m + 5n)(2m - n).$$

(c) $12z^2 - 5zy - 2y^2$

Try some possibilities.

$(12z + y)(z - 2y) = 12z^2 - 23zy - 2y^2$ *No*

$(3z - 2y)(4z + y) = 12z^2 - 5zy - 2y^2$ *Yes*

The correct factoring is

$$12z^2 - 5zy - 2y^2 = (3z - 2y)(4z + y).$$

(d) $8m^2 + 18mx - 5x^2$

Try some possibilities.

$(8m + x)(m - 5x) = 8m^2 - 39mx - 5x^2$ *No*

$(8m + 5x)(m - x) = 8m^2 - 3mx - 5x^2$ *No*

$(4m + x)(2m - 5x) = 8m^2 - 18mx - 5x^2$ *No*

$(4m - x)(2m + 5x) = 8m^2 + 18mx - 5x^2$ *Yes*

The correct factoring is

$$8m^2 + 18mx - 5x^2 = (4m - x)(2m + 5x).$$

7. **(a)** $-6r^2 + 13r + 5$

First factor out -1, then proceed.

$$\begin{aligned} -6r^2 + 13r + 5 &= -1(6r^2 - 13r - 5) \\ &= -1(2r - 5)(3r + 1) \\ &= -(2r - 5)(3r + 1) \end{aligned}$$

(b) $-8x^2 + 10x - 3$

First factor out -1, then proceed.

$$\begin{aligned} -8x^2 + 10x - 3 &= -1(8x^2 - 10x + 3) \\ &= -1(4x - 3)(2x - 1) \\ &= -(4x - 3)(2x - 1) \end{aligned}$$

8. **(a)** $2m^3 - 4m^2 - 6m$

First, factor out the GCF, $2m$.

$$= 2m(m^2 - 2m - 3)$$

Look for two integers whose product is -3 and whose sum is -2. The integers are 1 and -3.

$$= 2m(m + 1)(m - 3)$$

(b) $12r^4 + 6r^3 - 90r^2$

First, factor out the GCF, $6r^2$.

$$= 6r^2(2r^2 + r - 15)$$

Look for two integers whose product is $2(-15) = -30$ and whose sum is 1. The integers are 6 and -5. Write r as $6r - 5r$.

$$\begin{aligned} &= 6r^2(2r^2 + 6r - 5r - 15) \\ &= 6r^2[(2r^2 + 6r) + (-5r - 15)] \\ &= 6r^2[(2r(r+3) - 5(r+3)] \\ &= 6r^2(r+3)(2r-5) \end{aligned}$$

(c) $30y^5 - 55y^4 - 50y^3$

First, factor out the GCF, $5y^3$.

$$= 5y^3(6y^2 - 11y - 10)$$

Look for two integers whose product is $6(-10) = -60$ and whose sum is -11. The integers are 4 and -15. Write $-11y$ as $4y - 15y$.

$$\begin{aligned} &= 5y^3(6y^2 + 4y - 15y - 10) \\ &= 5y^3[(6y^2 + 4y) + (-15y - 10)] \\ &= 5y^3[2y(3y+2) - 5(3y+2)] \\ &= 5y^3(3y+2)(2y-5) \end{aligned}$$

9. **(a)** $6(a-1)^2 + (a-1) - 2$

$$\begin{aligned} &= 6x^2 + x - 2 && \textit{Let } x = a - 1. \\ &= (2x-1)(3x+2) \end{aligned}$$

Now replace x with $a - 1$.

$$\begin{aligned} &6(a-1)^2 + (a-1) - 2 \\ &= [2(a-1) - 1][3(a-1) + 2] \\ &= (2a - 2 - 1)(3a - 3 + 2) \\ &= (2a-3)(3a-1) \end{aligned}$$

(b) $8(z+5)^2 - 2(z+5) - 3$

$$\begin{aligned} &= 8x^2 - 2x - 3 && \textit{Let } x = z + 5. \\ &= (2x+1)(4x-3) \end{aligned}$$

Now replace x with $z + 5$.

$$\begin{aligned} &8(z+5)^2 - 2(z+5) - 3 \\ &= [2(z+5) + 1][4(z+5) - 3] \\ &= (2z + 10 + 1)(4z + 20 - 3) \\ &= (2z+11)(4z+17) \end{aligned}$$

(c) $15(m-4)^2 - 11(m-4) + 2$

$$\begin{aligned} &= 15x^2 - 11x + 2 && \textit{Let } x = m - 4. \\ &= (5x-2)(3x-1) \end{aligned}$$

Now replace x with $m - 4$.

$$\begin{aligned} &15(m-4)^2 - 11(m-4) + 2 \\ &= [5(m-4) - 2][3(m-4) - 1] \\ &= (5m - 20 - 2)(3m - 12 - 1) \\ &= (5m-22)(3m-13) \end{aligned}$$

10. **(a)** $y^4 + y^2 - 6$

$$\begin{aligned} &= (y^2)^2 + y^2 - 6 \\ &= x^2 + x - 6 && \textit{Let } x = y^2. \\ &= (x-2)(x+3) && \textit{Factor.} \\ &= (y^2-2)(y^2+3) && \textit{Resubstitute.} \end{aligned}$$

(b) $2p^4 + 7p^2 - 15$

$$\begin{aligned} &= 2(p^2)^2 + 7p^2 - 15 \\ &= 2x^2 + 7x - 15 && \textit{Let } x = p^2. \\ &= (2x-3)(x+5) && \textit{Factor.} \\ &= (2p^2-3)(p^2+5) && \textit{Resubstitute.} \end{aligned}$$

(c) $6r^4 - 13r^2 + 5$

$$\begin{aligned} &= 6(r^2)^2 - 13r^2 + 5 \\ &= 6x^2 - 13x + 5 && \textit{Let } x = r^2. \\ &= (3x-5)(2x-1) && \textit{Factor.} \\ &= (3r^2-5)(2r^2-1) && \textit{Resubstitute.} \end{aligned}$$

7.2 Section Exercises

1. **D** is not valid.

$$(8x)(4x) = 32x^2 \neq 12x^2$$

3. **C** is the completely factored form.

$$\begin{aligned} 4(x+2)(x-3) &= 4(x^2 - x - 6) \\ &= 4x^2 - 4x - 24 \end{aligned}$$

5. To factor $y^2 + 7y - 30$, we need two integer factors whose sum is 7 (coefficient of the middle term) and whose product is -30 (the last term). Since $-3 + 10 = 7$ and $-3 \cdot 10 = -30$, we have

$$y^2 + 7y - 30 = (y-3)(y+10).$$

7. $p^2 - p - 56$

Look for two integers whose product is -56 and whose sum is -1. Since the sum is negative, we'll restrict the products in the following table to those that have negative sums.

Products	Sums
$-56 \cdot 1$	-55
$-28 \cdot 2$	-26
$-14 \cdot 4$	-10
$-8 \cdot 7$	-1 ←

Thus, $p^2 - p - 56 = (p-8)(p+7)$.

9. $m^2 - 11m + 60$

There are no integer factors of 60 that add up to -11, so this trinomial is *prime*.

11. $a^2 - 2ab - 35b^2$

Two integer factors whose product is -35 and whose sum is -2 are 5 and -7.

$$a^2 - 2ab - 35b^2 = (a+5b)(a-7b)$$

13. $y^2 - 3yq - 15q^2$

There are no integer factors of -15 that add up to -3, so this trinomial is *prime*.

15. $x^2 + 11xy + 18y^2$

Two integer factors whose product is 18 and whose sum is 11 are 9 and 2.

$$x^2 + 11xy + 18y^2 = (x + 9y)(x + 2y)$$

17. $-6m^2 - 13m + 15$

First factor -1 out of the trinomial to get a trinomial whose coefficient of the squared term is positive.

$$-6m^2 - 13m + 15 = -1(6m^2 + 13m - 15)$$

We'll try to factor $6m^2 + 13m - 15$.
Multiply the first and last coefficients to get $6(-15) = -90$.
Two integer factors whose product is -90 and whose sum is 13 are -5 and 18.
Rewrite the trinomial in a form that can be factored by grouping.

$$\begin{aligned} &6m^2 + 13m - 15 \\ &= 6m^2 + (-5m + 18m) - 15 \\ &= (6m^2 - 5m) + (18m - 15) \\ &= m(6m - 5) + 3(6m - 5) \\ &= (6m - 5)(m + 3) \end{aligned}$$

Thus, the final factored form is

$-1(6m - 5)(m + 3)$, or $-(6m - 5)(m + 3)$.

Note: These exercises can be worked using the alternative method of repeated combinations and FOIL, or the grouping method.

19. $10x^2 + 3x - 18$

Two integer factors whose product is $(10)(-18) = -180$ and whose sum is 3 are 15 and -12.
Rewrite the trinomial in a form that can be factored by grouping.

$$\begin{aligned} &10x^2 + 3x - 18 \\ &= 10x^2 + 15x - 12x - 18 \\ &= 5x(2x + 3) - 6(2x + 3) \\ &= (2x + 3)(5x - 6) \end{aligned}$$

21. $20k^2 + 47k + 24$

Two integer factors whose product is $(20)(24) = 480$ and whose sum is 47 are 15 and 32.
Rewrite the trinomial in a form that can be factored by grouping.

$$\begin{aligned} &= 20k^2 + 15k + 32k + 24 \\ &= 5k(4k + 3) + 8(4k + 3) \\ &= (4k + 3)(5k + 8) \end{aligned}$$

23. $15a^2 - 22ab + 8b^2$

Two integer factors whose product is $(15)(8) = 120$ and whose sum is -22 are -10 and -12.
Rewrite the trinomial in a form that can be factored by grouping.

$$\begin{aligned} &= 15a^2 - 10ab - 12ab + 8b^2 \\ &= 5a(3a - 2b) - 4b(3a - 2b) \\ &= (3a - 2b)(5a - 4b) \end{aligned}$$

25. $36m^2 - 60m + 25$

Use the alternative method and write $36m^2$ as $6m \cdot 6m$ and 25 as $5 \cdot 5$. Use these factors in the binomial factors to obtain

$$\begin{aligned} 36m^2 - 60m + 25 &= (6m - 5)(6m - 5) \\ &= (6m - 5)^2. \end{aligned}$$

27. $40x^2 + xy + 6y^2$

There are no integer factors of $(40)(6) = 240$ that add up to 1, so this trinomial is *prime*.

29. $6x^2z^2 + 5xz - 4$

Two integer factors whose product is $(6)(-4) = -24$ and whose sum is 5 are 8 and -3.
Rewrite the trinomial in a form that can be factored by grouping.

$$\begin{aligned} &= 6x^2z^2 + 8xz - 3xz - 4 \\ &= 2xz(3xz + 4) - 1(3xz + 4) \\ &= (3xz + 4)(2xz - 1) \end{aligned}$$

31. $24x^2 + 42x + 15$

Always factor out the GCF first.

$$= 3(8x^2 + 14x + 5)$$

Now factor $8x^2 + 14x + 5$ by the alternative method.

$$8x^2 + 14x + 5 = (4x + 5)(2x + 1)$$

The final factored form is

$$3(4x + 5)(2x + 1).$$

33. $-15a^2 - 70a + 120$

$$\begin{aligned} &= -5(3a^2 + 14a - 24) \\ &= -5(a + 6)(3a - 4) \end{aligned}$$

35. $11x^3 - 110x^2 + 264x$

$$\begin{aligned} &= 11x(x^2 - 10x + 24) \\ &= 11x(x - 6)(x - 4) \end{aligned}$$

37. $2x^3y^3 - 48x^2y^4 + 288xy^5$
$= 2xy^3(x^2 - 24xy + 144y^2)$
$= 2xy^3(x - 12y)(x - 12y),$
or $2xy^3(x - 12y)^2$

39. $10(k+1)^2 - 7(k+1) + 1$

Let $x = k + 1$ to obtain

$$10x^2 - 7x + 1 = (5x - 1)(2x - 1).$$

Replace x with $k + 1$.

$$\begin{aligned} &10(k+1)^2 - 7(k+1) + 1 \\ &= [5(k+1) - 1][2(k+1) - 1] \\ &= (5k + 5 - 1)(2k + 2 - 1) \\ &= (5k + 4)(2k + 1) \end{aligned}$$

41. $3(m+p)^2 - 7(m+p) - 20$

Let $x = m + p$ to obtain

$$3x^2 - 7x - 20 = (3x + 5)(x - 4).$$

Replace x with $m + p$.

$$\begin{aligned} &3(m+p)^2 - 7(m+p) - 20 \\ &= [3(m+p) + 5][(m+p) - 4] \\ &= (3m + 3p + 5)(m + p - 4) \end{aligned}$$

43. $a^2(a+b)^2 - ab(a+b) - 6b^2$

Let $x = a + b$ to obtain

$$a^2x^2 - abx - 6b^2 = (ax - 3b)(ax + 2b).$$

Replace x with $a + b$.

$$\begin{aligned} &a^2(a+b)^2 - ab(a+b) - 6b^2 \\ &= [a(a+b) - 3b][a(a+b) + 2b] \\ &= (a^2 + ab - 3b)(a^2 + ab + 2b) \end{aligned}$$

45. In $2x^4 - 9x^2 - 18$, let $y = x^2$ to obtain

$$2y^2 - 9y - 18 = (2y + 3)(y - 6).$$

Replace y with x^2.

$$2x^4 - 9x^2 - 18 = (2x^2 + 3)(x^2 - 6)$$

47. In $16x^4 + 16x^2 + 3$, let $m = x^2$ to obtain

$$16m^2 + 16m + 3 = (4m + 3)(4m + 1).$$

Replace m with x^2.

$$16x^4 + 16x^2 + 3 = (4x^2 + 3)(4x^2 + 1)$$

49. In $12p^6 - 32p^3r + 5r^2$, let $x = p^3$ to obtain

$$12x^2 - 32xr + 5r^2 = (6x - r)(2x - 5r).$$

Replace x with p^3.

$$12p^6 - 32p^3r + 5r^2 = (6p^3 - r)(2p^3 - 5r)$$

51. No, 2 is not a factor of 45 since 45 is an odd number.

52. The positive factors of 45 are 1 and 45, 3 and 15, and 5 and 9. No, 2 is not a factor of any of these factors.

53. No, 5 is not a factor of $10x^2 + 29x + 10$ since it is not a factor of 29.

54. $10x^2 + 29x + 10$

Look for two integers whose product is $10(10) = 100$ and whose sum is 29. The integers are 4 and 25. Write $29x$ as $4x + 25x$.

$$\begin{aligned} &10x^2 + 29x + 10 \\ &= 10x^2 + 4x + 25x + 10 \\ &= (10x^2 + 4x) + (25x + 10) \\ &= 2x(5x + 2) + 5(5x + 2) \\ &= (5x + 2)(2x + 5) \end{aligned}$$

Since 5 is not a factor of 2, 5 is not a factor of $5x + 2$ or $2x + 5$.

55. Since k is odd, 2 is not a factor of $2x^2 + kx + 8$, and because 2 is a factor of $2x + 4$, the binomial $2x + 4$ cannot be a factor.

56. $3y + 15$ cannot be a factor of $12y^2 - 11y - 15$ because 3 is a factor of $3y + 15$, but 3 is not a factor of $12y^2 - 11y - 15$.

7.3 Special Factoring

7.3 Margin Exercises

1. Use the difference of squares formula,

$$\boldsymbol{x^2 - y^2 = (x + y)(x - y).}$$

(a) $p^2 - 100 = p^2 - 10^2$
$= (p + 10)(p - 10)$

(b) $2x^2 - 18 = 2(x^2 - 9)$ *GCF = 2*
$= 2(x^2 - 3^2)$
$= 2(x + 3)(x - 3)$

(c) $9a^2 - 16b^2 = (3a)^2 - (4b)^2$
$= (3a + 4b)(3a - 4b)$

(d) $(m + 3)^2 - 49z^2$
$= (m + 3)^2 - (7z)^2$
$= (m + 3 + 7z)(m + 3 - 7z)$

(e) $y^4 - 16 = (y^2)^2 - 4^2$
$= (y^2 + 4)(y^2 - 4)$
$= (y^2 + 4)(y^2 - 2^2)$
$= (y^2 + 4)(y + 2)(y - 2)$

2. **(a)** $z^2 + 12z + 36 = z^2 + 12z + 6^2$

If this is a perfect square trinomial, the middle term must be

$$2(z)(6) = 12z.$$

Since this is true, $z^2 + 12z + 36$ is a perfect square trinomial.

(b) $2x^2 - 4x + 4$

This is not a perfect square trinomial since $2x^2$ is not a perfect square.

(c) $9a^2 + 12ab + 16b^2 = (3a)^2 + 12ab + (4b)^2$

If this is a perfect square trinomial, the middle term must be

$$2(3a)(4b) = 24ab.$$

Since the middle term is $12ab$, the trinomial is not a perfect square trinomial.

3. **(a)** $49z^2 - 14zk + k^2 = (7z)^2 - 14zk + k^2$
$= (7z - k)^2$

Check: $2(7z)(-k) = -14zk$, which is the middle term.

Thus, $49z^2 - 14zk + k^2 = (7z - k)^2$.

(b) $9a^2 + 48ab + 64b^2 = (3a)^2 + 48ab + (8b)^2$
$= (3a + 8b)^2$

Check: $2(3a)(8b) = 48ab$, which is the middle term.

Thus, $9a^2 + 48ab + 64b^2 = (3a + 8b)^2$.

(c) $(k + m)^2 - 12(k + m) + 36$
$= [(k + m) - 6]^2$, or $(k + m - 6)^2$

Check: $2(k + m)(-6) = -12(k + m)$, which is the middle term.

Thus, $(k + m)^2 - 12(k + m) + 36$
$= (k + m - 6)^2$.

(d) $x^2 - 2x + 1 - y^2$

Group the first three terms.

$= (x^2 - 2x + 1) - y^2$

Factor the perfect square trinomial.

$= (x - 1)^2 - y^2$

This is a difference of two squares.

$= [(x - 1) + y][(x - 1) - y]$
$= (x - 1 + y)(x - 1 - y)$

4. Use the difference of cubes formula,

$$x^3 - y^3 = (x - y)(x^2 + xy + y^2).$$

(a) $x^3 - 1000 = x^3 - 10^3$
$= (x - 10)(x^2 + 10x + 10^2)$
$= (x - 10)(x^2 + 10x + 100)$

(b) $8k^3 - y^3 = (2k)^3 - y^3$
$= (2k - y)[(2k)^2 + (2k)(y) + y^2]$
$= (2k - y)(4k^2 + 2ky + y^2)$

(c) $27a^3 - 64b^3$
$= (3a)^3 - (4b)^3$
$= (3a - 4b)[(3a)^2 + (3a)(4b) + (4b)^2]$
$= (3a - 4b)(9a^2 + 12ab + 16b^2)$

5. Use the sum of cubes formula,

$$x^3 + y^3 = (x + y)(x^2 - xy + y^2).$$

(a) $8p^3 + 125 = (2p)^3 + 5^3$
$= (2p + 5)[(2p)^2 - (2p)(5) + 5^2]$
$= (2p + 5)(4p^2 - 10p + 25)$

(b) $27m^3 + 125n^3$
$= (3m)^3 + (5n)^3$
$= (3m + 5n)[(3m)^2 - (3m)(5n) + (5n)^2]$
$= (3m + 5n)(9m^2 - 15mn + 25n^2)$

(c) $2x^3 + 2000$
$= 2(x^3 + 1000)$ *GCF = 2*
$= 2(x^3 + 10^3)$
$= 2(x + 10)(x^2 - 10x + 10^2)$
$= 2(x + 10)(x^2 - 10x + 100)$

(d) $(a - 4)^3 + b^3$
$= [(a - 4) + b][(a - 4)^2 - (a - 4)b + b^2]$
$= (a - 4 + b)(a^2 - 8a + 16 - ab + 4b + b^2)$

7.3 Section Exercises

1. **A.** Yes, 64 and m^2 are squares.

B. No, $2x^2$ is not a square.

C. No, $k^2 + 9$ is a *sum* of squares.

D. Yes, $4z^4$ and 49 are squares.

So, the binomials that are differences of squares are **A** and **D**.

3. **A.** Since $x^2 - 8x - 16$ has a negative third term, it is not a perfect square trinomial.

B. $4m^2 + 20m + 25 = (2m)^2 + 20m + 5^2$

$2(2m)(5) = 20m$, the middle term. Therefore, this trinomial is a perfect square.

C. $9z^4 + 30z^2 + 25 = (3z^2)^2 + 30z^2 + 5^2$

$2(3z^2)(5) = 30z^2$, the middle term. Therefore, this trinomial is a perfect square.

D. $25a^2 - 45a + 81 = (5a)^2 - 45a + (-9)^2$
$2(5a)(-9) = -90a$

This is not the middle term, so the trinomial is not a perfect square.

So, the perfect square trinomials are **B** and **C**.

5. $p^2 - 16 = p^2 - 4^2$
$= (p+4)(p-4)$

7. $25x^2 - 4 = (5x)^2 - 2^2$
$= (5x+2)(5x-2)$

9. $18a^2 - 98b^2 = 2(9a^2 - 49b^2)$
$= 2[(3a)^2 - (7b)^2]$
$= 2[(3a+7b)(3a-7b)]$
$= 2(3a+7b)(3a-7b)$

11. $64m^4 - 4y^4$
$= 4(16m^4 - y^4)$
$= 4[(4m^2)^2 - (y^2)^2]$

Factor the difference of squares.

$= 4(4m^2 + y^2)(4m^2 - y^2)$
$= 4(4m^2 + y^2)[(2m)^2 - y^2]$

Factor the difference of squares again.

$= 4(4m^2 + y^2)(2m + y)(2m - y)$

13. $(y+z)^2 - 81$
$= (y+z)^2 - 9^2$
$= [(y+z)+9][(y+z)-9]$
$= (y+z+9)(y+z-9)$

15. $16 - (x+3y)^2$
$= 4^2 - z^2$ *Let $z = (x+3y)$.*
$= (4+z)(4-z)$
Substitute $x + 3y$ for z.
$= [4+(x+3y)][4-(x+3y)]$
$= (4+x+3y)(4-x-3y)$

17. $p^4 - 256 = (p^2)^2 - 16^2$
$= (p^2+16)(p^2-16)$
$= (p^2+16)(p^2-4^2)$
$= (p^2+16)(p+4)(p-4)$

19. $k^2 - 6k + 9 = (k)^2 - 2(k)(3) + 3^2$
$= (k-3)^2$

21. $4z^2 + 4zw + w^2$
$= (2z)^2 + 2(2z)(w) + w^2$
$= (2z+w)^2$

23. $16m^2 - 8m + 1 - n^2$
Group the first three terms.
$= (16m^2 - 8m + 1) - n^2$
$= [(4m)^2 - 2(4m)(1) + 1^2] - n^2$
$= (4m-1)^2 - n^2$
$= [(4m-1)+n][(4m-1)-n]$
$= (4m-1+n)(4m-1-n)$

25. $4r^2 - 12r + 9 - s^2$
Group the first three terms.
$= (4r^2 - 12r + 9) - s^2$
$= [(2r)^2 - 2(2r)(3) + 3^2] - s^2$
$= (2r-3)^2 - s^2$
$= [(2r-3)+s][(2r-3)-s]$
$= (2r-3+s)(2r-3-s)$

27. $x^2 - y^2 + 2y - 1$
Group the last three terms.
$= x^2 - (y^2 - 2y + 1)$
$= x^2 - (y-1)^2$
$= [x+(y-1)][x-(y-1)]$
$= (x+y-1)(x-y+1)$

29. $98m^2 + 84mn + 18n^2$
$= 2(49m^2 + 42mn + 9n^2)$
$= 2[(7m)^2 + 2(7m)(3n) + (3n)^2]$
$= 2(7m+3n)^2$

31. $(p+q)^2 + 2(p+q) + 1$
$= x^2 + 2x + 1$ *Let $x = p + q$.*
$= (x+1)^2$
$= (p+q+1)^2$ *Resubstitute.*

33. $(a-b)^2 + 8(a-b) + 16$
$= (a-b)^2 + 2(a-b)(4) + 4^2$
$= [(a-b)+4]^2$
$= (a-b+4)^2$

35. $y^3 - 64 = y^3 - 4^3$
$= (y-4)(y^2 + y \cdot 4 + 4^2)$
$= (y-4)(y^2 + 4y + 16)$

37. $r^3 + 343 = r^3 + 7^3$
$= (r+7)(r^2 - r \cdot 7 + 7^2)$
$= (r+7)(r^2 - 7r + 49)$

39. $8x^3 - y^3$
$= (2x)^3 - y^3$
$= [2x - y]\,[(2x)^2 + (2x)(y) + y^2]$
$= (2x - y)(4x^2 + 2xy + y^2)$

41. $64g^3 + 27h^3$
$= (4g)^3 + (3h)^3$
$= (4g + 3h)[(4g)^2 - (4g)(3h) + (3h)^2]$
$= (4g + 3h)(16g^2 - 12gh + 9h^2)$

43. $24n^3 + 81p^3$
$= 3(8n^3 + 27p^3)$
$= 3[(2n)^3 + (3p)^3]$
$= 3[2n + 3p]\,[(2n)^2 - (2n)(3p) + (3p)^2]$
$= 3(2n + 3p)(4n^2 - 6np + 9p^2)$

45. $(y + z)^3 - 64$
$= (y + z)^3 - 4^3$
$= [(y + z) - 4]\,[(y + z)^2 + (y + z)(4) + 4^2]$
$= (y + z - 4)(y^2 + 2yz + z^2 + 4y + 4z + 16)$

47. $m^6 - 125 = (m^2)^3 - (5)^3$
$= (m^2 - 5)[(m^2)^2 + (m^2)(5) + 5^2]$
$= (m^2 - 5)(m^4 + 5m^2 + 25)$

49. $125y^6 + z^3$
$= (5y^2)^3 + z^3$
$= (5y^2 + z)[(5y^2)^2 - (5y^2)(z) + z^2]$
$= (5y^2 + z)(25y^4 - 5y^2z + z^2)$

50. $x^6 - y^6$
$= (x^3)^2 - (y^3)^2$
$= (x^3 + y^3)(x^3 - y^3)$
$= [(x + y)(x^2 - xy + y^2)]$
$\cdot [(x - y)(x^2 + xy + y^2)]$
$= (x + y)(x^2 - xy + y^2)(x - y)$
$\cdot (x^2 + xy + y^2)$

51. $x^6 - y^6 = (x - y)(x + y)$
$\cdot \underline{(x^2 + xy + y^2)(x^2 - xy + y^2)}$

52. $x^6 - y^6$
$= (x^2)^3 - (y^2)^3$
$= (x^2 - y^2)(x^4 + x^2y^2 + y^4)$
$= (x + y)(x - y)(x^4 + x^2y^2 + y^4)$

53. $x^6 - y^6 = (x - y)(x + y)$
$\cdot \underline{(x^4 + x^2y^2 + y^4)}$

54. The product written on the blank in Exercise 51 must equal the product written on the blank in Exercise 53. To verify this, multiply the two factors written in Exercise 51.

$(x^2 + xy + y^2)(x^2 - xy + y^2)$
$= x^2(x^2 - xy + y^2)$
$+ xy(x^2 - xy + y^2)$
$+ y^2(x^2 - xy + y^2)$
$= x^4 - x^3y + x^2y^2 + x^3y - x^2y^2$
$+ xy^3 + x^2y^2 - xy^3 + y^4$
$= x^4 + x^2y^2 + y^4$

They are equal.

55. Start by factoring as a difference of squares since doing so resulted in the complete factorization more directly.

7.4 A General Approach to Factoring

7.4 Margin Exercises

1. **(a)** $8x - 80$
$= 8(x - 10)$ *GCF = 8*

(b) $2x^3 + 10x^2 - 2x$
$= 2x(x^2 + 5x - 1)$ *GCF = 2x*

(c) $12m(p - q) - 7n(p - q)$
$= (p - q)(12m - 7n)$ *GCF = p − q*

2. **(a)** $36x^2 - y^2$
$= (6x)^2 - y^2$ *Difference of squares*
$= (6x + y)(6x - y)$

(b) $4t^2 + 1$ is a *sum* of squares without a GCF and cannot be factored. The binomial is prime.

(c) $125x^3 - 27y^3$
$= (5x)^3 - (3y)^3$ *Difference of cubes*
$= (5x - 3y)[(5x)^2 + 5x \cdot 3y + (3y)^2]$
$= (5x - 3y)(25x^2 + 15xy + 9y^2)$

(d) $x^3 + 343y^3$
$= x^3 + (7y)^3$ *Sum of cubes*
$= (x + 7y)[x^2 - x \cdot 7y + (7y)^2]$
$= (x + 7y)(x^2 - 7xy + 49y^2)$

3. **(a)** $16m^2 + 56m + 49$
$= (4m)^2 + 2(4m)(7) + 7^2$
Perfect square trinomial
$= (4m + 7)^2$

(b) $r^2 + 18r + 72$

Two integer factors whose product is 72 and whose sum is 18 are 6 and 12.

$= r^2 + 6r + 12r + 72$
$= r(r + 6) + 12(r + 6)$
$= (r + 6)(r + 12)$

(c) $8t^2 - 13t + 5$

Two integer factors whose product is $8(5) = 40$ and whose sum is -13 are -5 and -8.

$$= 8t^2 - 5t - 8t + 5$$
$$= t(8t - 5) - 1(8t - 5)$$
$$= (8t - 5)(t - 1)$$

(d) $6x^2 - 3x - 63 = 3(2x^2 - x - 21)$

Two integer factors whose product is $2(-21) = -42$ and whose sum is -1 are -7 and 6.

$$= 3(2x^2 - 7x + 6x - 21)$$
$$= 3[x(2x - 7) + 3(2x - 7)]$$
$$= 3[(2x - 7)(x + 3)]$$
$$= 3(2x - 7)(x + 3)$$

4. **(a)** $p^3 - 2pq^2 + p^2q - 2q^3$

$$= (p^3 - 2pq^2) + (p^2q - 2q^3)$$
$$= p(p^2 - 2q^2) + q(p^2 - 2q^2)$$
$$= (p^2 - 2q^2)(p + q)$$

(b) $9x^2 + 24x + 16 - y^2$

$$= (9x^2 + 24x + 16) - y^2$$
$$= (3x + 4)^2 - y^2$$
$$= [(3x + 4) + y][(3x + 4) - y]$$
$$= (3x + 4 + y)(3x + 4 - y)$$

(c) $64a^3 + 16a^2 + b^3 - b^2$

$$= (64a^3 + b^3) + (16a^2 - b^2)$$
$$= [(4a)^3 + b^3] + [(4a)^2 - b^2]$$
$$= \{(4a + b)[(4a)^2 - 4a \cdot b + b^2]\} + [(4a + b)(4a - b)]$$
$$= [(4a + b)(16a^2 - 4ab + b^2)] + [(4a + b)(4a - b)]$$
$$= (4a + b)(16a^2 - 4ab + b^2 + 4a - b)$$

7.4 Section Exercises

1. $100a^2 - 9b^2$

$$= (10a)^2 - (3b)^2 \quad \textit{Difference of squares}$$
$$= (10a + 3b)(10a - 3b)$$

3. $18p^5 - 24p^3 + 12p^6 = 6p^3(3p^2 - 4 + 2p^3)$

5. $x^2 + 2x - 35$

Look for two integers whose product is -35 and whose sum is 2. The integers are 7 and -5.

$$x^2 + 2x - 35 = (x + 7)(x - 5)$$

7. $225p^2 + 256$ is a *sum* of squares without a GCF and cannot be factored. The binomial is prime.

9. $6b^2 - 17b - 3$

Two integer factors whose product is $(6)(-3) = -18$ and whose sum is -17 are -18 and 1.

$$= 6b^2 - 18b + b - 3$$
$$= 6b(b - 3) + 1(b - 3)$$
$$= (b - 3)(6b + 1)$$

11. $18m^3n + 3m^2n^2 - 6mn^3$

$$= 3mn(6m^2 + mn - 2n^2)$$

Factor the trinomial.

$$= 3mn(3m + 2n)(2m - n)$$

13. $2p^2 + 11pq + 15q^2$

Look for two integers whose product is $2(15) = 30$ and whose sum is 11. The integers are 6 and 5. Write $11pq$ as $6pq + 5pq$ and factor by grouping.

$$= 2p^2 + 6pq + 5pq + 15q^2$$
$$= (2p^2 + 6pq) + (5pq + 15q^2)$$
$$= 2p(p + 3q) + 5q(p + 3q)$$
$$= (p + 3q)(2p + 5q)$$

15. $4k^2 + 28kr + 49r^2$

$$= (2k)^2 + 2(2k)(7r) + (7r)^2$$

Perfect square trinomial

$$= (2k + 7r)^2$$

17. $mn - 2n + 5m - 10$

$$= n(m - 2) + 5(m - 2)$$
$$= (m - 2)(n + 5)$$

19. $x^3 + 3x^2 - 9x - 27$

Factor by grouping.

$$= (x^3 + 3x^2) + (-9x - 27)$$
$$= x^2(x + 3) - 9(x + 3)$$
$$= (x + 3)(x^2 - 9)$$

Since $x^2 - 9$ is the difference of two squares, $x^2 - 3^2$, factor it as $(x + 3)(x - 3)$.

$$= (x + 3)(x + 3)(x - 3)$$
$$= (x + 3)^2(x - 3)$$

21. $9r^2 + 100 = (3r)^2 + 10^2$

There is no GCF and the sum of squares cannot be factored. This binomial is *prime*.

23. $6k^2 - k - 1$

Multiply $6(-1) = -6$. The integers -3 and 2 have the product -6 and the sum -1.

$$\begin{aligned} 6k^2 - k - 1 &= 6k^2 - 3k + 2k - 1 \\ &= (6k^2 - 3k) + (2k - 1) \\ &= 3k(2k - 1) + 1(2k - 1) \\ &= (2k - 1)(3k + 1) \end{aligned}$$

25. $x^4 - 625$

$= (x^2)^2 - 25^2$ *Difference of squares*

$= (x^2 + 25)(x^2 - 25)$

$= (x^2 + 25)(x + 5)(x - 5)$ *Difference of squares again*

27. $ab + 6b + ac + 6c$

$= (ab + 6b) + (ac + 6c)$ *Group.*

$= b(a + 6) + c(a + 6)$

$= (a + 6)(b + c)$

29. $4y^2 - 8y = 4y(y - 2)$

31. $14z^2 - 3zk - 2k^2$

Two integer factors whose product is $(14)(-2) = -28$ and whose sum is -3 are 4 and -7.

$$\begin{aligned} &= 14z^2 + 4zk - 7zk - 2k^2 \\ &= 2z(7z + 2k) - k(7z + 2k) \\ &= (7z + 2k)(2z - k) \end{aligned}$$

33. $256b^2 - 400c^2$

$= 16(16b^2 - 25c^2)$

$= 16[(4b)^2 - (5c)^2]$ *Difference of squares*

$= 16(4b + 5c)(4b - 5c)$

35. $1000z^3 + 512$

$= 8(125z^3 + 64)$

$= 8[(5z)^3 + 4^3]$ *Sum of cubes*

$= 8[5z + 4][(5z)^2 - (5z)(4) + 4^2]$

$= 8(5z + 4)(25z^2 - 20z + 16)$

37. $10r^2 + 23rs - 5s^2$

Two integer factors whose product is $(10)(-5) = -50$ and whose sum is 23 are 25 and -2.

$$\begin{aligned} &= 10r^2 + 25rs - 2rs - 5s^2 \\ &= 5r(2r + 5s) - s(2r + 5s) \\ &= (2r + 5s)(5r - s) \end{aligned}$$

39. $32x^2 + 16x^3 - 24x^5 = 8x^2(4 + 2x - 3x^3)$

41. $14x^2 - 25xq - 25q^2$

Two integer factors whose product is $(14)(-25) = -350$ and whose sum is -25 are -35 and 10.

$$\begin{aligned} &= 14x^2 - 35xq + 10xq - 25q^2 \\ &= 7x(2x - 5q) + 5q(2x - 5q) \\ &= (2x - 5q)(7x + 5q) \end{aligned}$$

43. $y^2 + 3y - 10 = (y + 5)(y - 2)$

45. $2a^3 + 6a^2 - 4a = 2a(a^2 + 3a - 2)$

Since -2 factors only as $1(-2)$ and $(-1)2$, and $1 + (-2) = -1 \neq 3$ and $-1 + 2 = 1 \neq 3$, we know that $a^2 + 3a - 2$ cannot be factored further.

47. $18p^2 + 53pr - 35r^2$

Two integer factors whose product is $(18)(-35) = -630$ and whose sum is 53 are -10 and 63.

$$\begin{aligned} &= 18p^2 - 10pr + 63pr - 35r^2 \\ &= 2p(9p - 5r) + 7r(9p - 5r) \\ &= (9p - 5r)(2p + 7r) \end{aligned}$$

49. $(x - 2y)^2 - 4$

$$\begin{aligned} &= (x - 2y)^2 - 2^2 \\ &= [(x - 2y) + 2][(x - 2y) - 2] \\ &= (x - 2y + 2)(x - 2y - 2) \end{aligned}$$

51. $(5r + 2s)^2 - 6(5r + 2s) + 9$

$= x^2 - 6x + 9$ *Let $x = 5r + 2s$.*

$= (x - 3)^2$

$= [(5r + 2s) - 3]^2$ *Resubstitute.*

$= (5r + 2s - 3)^2$

53. $z^4 - 9z^2 + 20$

$= x^2 - 9x + 20$ *Let $x = z^2$.*

$= (x - 5)(x - 4)$

$= (z^2 - 5)(z^2 - 4)$ *Resubstitute.*

$= (z^2 - 5)(z + 2)(z - 2)$

55. $4(p + 2) + m(p + 2) = (p + 2)(4 + m)$

57. $50p^2 - 162$

$$\begin{aligned} &= 2(25p^2 - 81) \\ &= 2[(5p)^2 - 9^2] \\ &= 2(5p + 9)(5p - 9) \end{aligned}$$

59. $16a^2 + 8ab + b^2$

$= (4a)^2 + 2(4a)(b) + (b)^2$

Perfect square trinomial

$= (4a + b)^2$

7.5 Solving Equations by Factoring

7.5 Margin Exercises

1. **(a)** $(3x+5)(x+1)=0$

$3x+5=0$ or $x+1=0$

$3x=-5$ or $x=-1$

$x=-\frac{5}{3}$

Check $x=-\frac{5}{3}$: $0(-\frac{2}{3})=0$ *True*

Check $x=-1$: $2(0)=0$ *True*

Solution set: $\{-\frac{5}{3},-1\}$

(b) $(3x+11)(5x-2)=0$

$3x+11=0$ or $5x-2=0$

$3x=-11$ or $5x=2$

$x=-\frac{11}{3}$ or $x=\frac{2}{5}$

Check $x=-\frac{11}{3}$: $0(-\frac{61}{3})=0$ *True*

Check $x=\frac{2}{5}$: $\frac{61}{5}(0)=0$ *True*

Solution set: $\{-\frac{11}{3},\frac{2}{5}\}$

2. **(a)** *Step 1 Standard form*

$3x^2-x=4$

$3x^2-x-4=0$

Step 2 Factor.

$(3x-4)(x+1)=0$

Step 3 Zero-factor property

$3x-4=0$ or $x+1=0$

Step 4 Solve each equation.

$3x=4$ or $x=-1$

$x=\frac{4}{3}$

Step 5

Check each solution in the original equation.

Check $x=\frac{4}{3}$: $\frac{16}{3}-\frac{4}{3}=4$ *True*

Check $x=-1$: $3-(-1)=4$ *True*

Solution set: $\{-1,\frac{4}{3}\}$

(b) $25x^2=-20x-4$

$25x^2+20x+4=0$ *Standard form*

$(5x+2)^2=0$ *Factor.*

$5x+2=0$ *Zero-factor property*

$5x=-2$

$x=-\frac{2}{5}$

Check $x=-\frac{2}{5}$: $4=8-4$ *True*

Solution set: $\{-\frac{2}{5}\}$

3. $x^2+12x=0$

$x(x+12)=0$

$x=0$ or $x+12=0$

$x=-12$

Check $x=0$: $0+0=0$ *True*

Check $x=-12$: $144-144=0$ *True*

Solution set: $\{-12,0\}$

4. $5x^2-80=0$

$5(x^2-16)=0$ *Factor out 5.*

$5(x+4)(x-4)=0$ *Difference of squares*

$x+4=0$ or $x-4=0$ *Zero-factor property*

$x=-4$ or $x=4$

Check $x=\pm 4$: $5(16)-80=0$ *True*

Solution set: $\{-4,4\}$

5. $(x+6)(x-2)=-8+x$

$x^2+4x-12=-8+x$

$x^2+3x-4=0$

$(x-1)(x+4)=0$

$x-1=0$ or $x+4=0$

$x=1$ or $x=-4$

Check $x=1$: $7(-1)=-7$ *True*

Check $x=-4$: $2(-6)=-12$ *True*

Solution set: $\{-4,1\}$

6. $3x^3+x^2=4x$

$3x^3+x^2-4x=0$

$x(3x^2+x-4)=0$

$x(x-1)(3x+4)=0$

$x=0$ or $x-1=0$ or $3x+4=0$

$x=1$ or $x=-\frac{4}{3}$

Check $x=0$: $0=0$ *True*

Check $x=1$: $4=4$ *True*

Check $x=-\frac{4}{3}$: $-\frac{64}{9}+\frac{16}{9}=-\frac{16}{3}$ *True*

Solution set: $\{-\frac{4}{3},0,1\}$

7. Let $x=$ the length of the deck. Then $\frac{1}{2}x-1=$ the width of the deck.

Use the formula $LW=A$, where $L=x$, $W=\frac{1}{2}x-1$, and $A=60$.

$x(\frac{1}{2}x-1)=60$

$\frac{1}{2}x^2-x=60$

$x^2-2x=120$ *Multiply by 2.*

$x^2-2x-120=0$

$(x-12)(x+10)=0$

$x-12=0$ or $x+10=0$

$x=12$ or $x=-10$

The deck cannot have a negative length, so reject $x=-10$. The length of the deck is 12 m and the width is $\frac{1}{2}(12)-1=5$ m.

8.
$$h(t) = -16t^2 + 128t$$
$$256 = -16t^2 + 128t$$
$$16t^2 - 128t + 256 = 0$$
$$t^2 - 8t + 16 = 0 \quad \textit{Divide by 16.}$$
$$(t-4)^2 = 0 \quad \textit{Factor.}$$
$$t - 4 = 0 \quad \textit{Zero-factor property}$$
$$t = 4$$

It will reach a height of 256 ft in 4 seconds.

7.5 Section Exercises

1. First rewrite the equation so that one side is 0. Factor the other side and set each factor equal to 0. The solutions of these linear equations are solutions of the quadratic equation.

In the exercises in this section, check all solutions to the equations by substituting them back in the original equations.

3. $(x+10)(x-5) = 0$

$x + 10 = 0$ or $x - 5 = 0$

$x = -10$ or $x = 5$

Check $x = -10$: $(0)(-15) = 0$ *True*

Check $x = 5$: $(15)(0) = 0$ *True*

Solution set: $\{-10, 5\}$

5. $(3k+8)(2k-5) = 0$

$3k + 8 = 0$ or $2k - 5 = 0$

$3k = -8$ or $2k = 5$

$k = -\frac{8}{3}$ or $k = \frac{5}{2}$

Solution set: $\{-\frac{8}{3}, \frac{5}{2}\}$

7. $m^2 - 3m - 10 = 0$

$(m+2)(m-5) = 0$

Use the zero-factor property.

$m + 2 = 0$ or $m - 5 = 0$

$m = -2$ or $m = 5$

Solution set: $\{-2, 5\}$

9. $z^2 + 9z + 18 = 0$

$(z+6)(z+3) = 0$

$z + 6 = 0$ or $z + 3 = 0$

$z = -6$ or $z = -3$

Solution set: $\{-6, -3\}$

11. $2x^2 = 7x + 4$

Get 0 on one side.

$2x^2 - 7x - 4 = 0$

$(2x+1)(x-4) = 0$

$2x + 1 = 0$ or $x - 4 = 0$

$2x = -1$ or $x = 4$

$x = -\frac{1}{2}$

Solution set: $\{-\frac{1}{2}, 4\}$

13. $15k^2 - 7k = 4$

$15k^2 - 7k - 4 = 0$

$(3k+1)(5k-4) = 0$

$3k + 1 = 0$ or $5k - 4 = 0$

$3k = -1$ or $5k = 4$

$k = -\frac{1}{3}$ or $k = \frac{4}{5}$

Solution set: $\{-\frac{1}{3}, \frac{4}{5}\}$

15. $16x^2 + 24x = -9$

$16x^2 + 24x + 9 = 0$

$(4x+3)^2 = 0$

$4x + 3 = 0$

$4x = -3$

$x = -\frac{3}{4}$

Solution set: $\{-\frac{3}{4}\}$

17. $2a^2 - 8a = 0$

$2a(a-4) = 0$

$2a = 0$ or $a - 4 = 0$

$a = 0$ $a = 4$

The solution set is $\{0, 4\}$.

19. $6m^2 - 36m = 0$

$6m(m-6) = 0$

$6m = 0$ or $m - 6 = 0$

$m = 0$ or $m = 6$

Solution set: $\{0, 6\}$

21. $4p^2 - 16 = 0$

$4(p^2 - 4) = 0$

$4(p+2)(p-2) = 0$

$p + 2 = 0$ or $p - 2 = 0$

$p = -2$ or $p = 2$

Solution set: $\{-2, 2\}$

23.
$$-3m^2 + 27 = 0$$
$$-3(m^2 - 9) = 0$$
$$-3(m+3)(m-3) = 0$$

Note that the leading -3 does not affect the solution set of the equation.

$m + 3 = 0$ or $m - 3 = 0$
$m = -3$ or $m = 3$

Solution set: $\{-3, 3\}$

25. $(x-3)(x+5) = -7$

Multiply the factors, and then add 7 on both sides of the equation to get 0 on the right.

$$x^2 + 5x - 3x - 15 = -7$$
$$x^2 + 2x - 8 = 0$$

Now factor the polynomial.

$$(x+4)(x-2) = 0$$

$x + 4 = 0$ or $x - 2 = 0$
$x = -4$ or $x = 2$

Solution set: $\{-4, 2\}$

27.
$$(2x+1)(x-3) = 6x + 3$$
$$2x^2 - 6x + x - 3 = 6x + 3$$
$$2x^2 - 5x - 3 = 6x + 3$$
$$2x^2 - 11x - 6 = 0$$
$$(2x+1)(x-6) = 0$$

$2x + 1 = 0$ or $x - 6 = 0$
$2x = -1$ or $x = 6$
$x = -\frac{1}{2}$

Solution set: $\{-\frac{1}{2}, 6\}$

29.
$$(5x+1)(x+3) = -2(5x+1)$$
$$(5x+1)(x+3) + 2(5x+1) = 0$$
$$(5x+1)[(x+3)+2] = 0$$
$$(5x+1)(x+5) = 0$$

$5x + 1 = 0$ or $x + 5 = 0$
$5x = -1$ or $x = -5$
$x = -\frac{1}{5}$

Solution set: $\{-5, -\frac{1}{5}\}$

31.
$$(x+3)(x-6) = (2x+2)(x-6)$$
$$x^2 - 3x - 18 = 2x^2 - 10x - 12$$
$$0 = x^2 - 7x + 6$$
$$0 = (x-1)(x-6)$$

$x - 1 = 0$ or $x - 6 = 0$
$x = 1$ or $x = 6$

Solution set: $\{1, 6\}$

33.
$$2x^3 - 9x^2 - 5x = 0$$
$$x(2x^2 - 9x - 5) = 0$$
$$x(2x+1)(x-5) = 0$$

$x = 0$ or $2x + 1 = 0$ or $x - 5 = 0$
$2x = -1$ or $x = 5$
$x = -\frac{1}{2}$

Solution set: $\{-\frac{1}{2}, 0, 5\}$

35.
$$x^3 - 2x^2 = 3x$$
$$x^3 - 2x^2 - 3x = 0$$
$$x(x^2 - 2x - 3) = 0$$
$$x(x-3)(x+1) = 0$$

$x = 0$ or $x - 3 = 0$ or $x + 1 = 0$
or $x = 3$ $x = -1$

The solution set is $\{-1, 0, 3\}$.

37.
$$9t^3 = 16t$$
$$9t^3 - 16t = 0$$
$$t(9t^2 - 16) = 0$$
$$t(3t+4)(3t-4) = 0$$

$t = 0$ or $3t + 4 = 0$ or $3t - 4 = 0$
$3t = -4$ or $3t = 4$
$t = -\frac{4}{3}$ or $t = \frac{4}{3}$

Solution set: $\{-\frac{4}{3}, 0, \frac{4}{3}\}$

39.
$$2r^3 + 5r^2 - 2r - 5 = 0$$

Factor by grouping.

$$(2r^3 - 2r) + (5r^2 - 5) = 0$$
$$2r(r^2 - 1) + 5(r^2 - 1) = 0$$
$$(r^2 - 1)(2r + 5) = 0$$
$$(r+1)(r-1)(2r+5) = 0$$

$r + 1 = 0$ or $r - 1 = 0$ or $2r + 5 = 0$
$r = -1$ or $r = 1$ or $r = -\frac{5}{2}$

Solution set: $\{-\frac{5}{2}, -1, 1\}$

41. By dividing each side by a variable expression, she "lost" the solution 0. The solution set is $\{-\frac{4}{3}, 0, \frac{4}{3}\}$.

43. Let $x =$ the width of the garden.
Then $x + 4 =$ the length of the garden.

The area of the rectangular-shaped garden is 320 ft^2, so use the formula $A = LW$ and substitute 320 for A, $x + 4$ for L, and x for W.

$$A = LW$$
$$320 = (x+4)x$$
$$320 = x^2 + 4x$$
$$0 = x^2 + 4x - 320$$
$$0 = (x-16)(x+20)$$

$$x-16=0 \quad \text{or} \quad x+20=0$$
$$x=16 \quad \text{or} \quad x=-20$$

A rectangle cannot have a width that is a negative measure, so reject -20 as a solution. The only possible solution is 16.

The width of the garden is 16 ft, and the length is $16+4=20$ ft.

45. Let $h =$ the height of the parallelogram. Then $h+7 =$ the base of the parallelogram.

The area is 60 ft^2, so use the formula $A = bh$ and substitute 60 for A, $h+7$ for b.

$$\begin{aligned} A &= bh \\ 60 &= (h+7)h \\ 60 &= h^2+7h \\ 0 &= h^2+7h-60 \\ 0 &= (h+12)(h-5) \end{aligned}$$

$$h+12=0 \quad \text{or} \quad h-5=0$$
$$h=-12 \quad \text{or} \quad h=5$$

A parallelogram cannot have a height that is negative, so reject -12 as a solution. The only possible solution is 5.
The height of the parallelogram is 5 ft and the base is $5+7=12$ ft.

47. Let $L =$ the length of the rectangular area and $W =$ the width.

Use the formula for perimeter, $P = 2L + 2W$, and solve for W in terms of L. The perimeter is 300 ft.

$$\begin{aligned} 300 &= 2L+2W \\ 300-2L &= 2W \\ 150-L &= W \end{aligned}$$

Now use the formula for area, $A = LW$, substitute 5000 for A, and solve for L.

$$\begin{aligned} 5000 &= L(150-L) \\ 5000 &= 150L-L^2 \\ L^2-150L+5000 &= 0 \\ (L-50)(L-100) &= 0 \end{aligned}$$

$$L-50=0 \quad \text{or} \quad L-100=0$$
$$L=50 \quad \text{or} \quad L=100$$

When $L=50$, $W=150-50=100$.
When $L=100$, $W=150-100=50$.
The dimensions should be 50 ft by 100 ft.

49. Let x and $x+1$ denote the two consecutive integers.

The sum of their squares is 61, so

$$\begin{aligned} x^2+(x+1)^2 &= 61. \\ x^2+x^2+2x+1 &= 61 \\ 2x^2+2x-60 &= 0 \quad \textit{Divide by 2.} \\ x^2+x-30 &= 0 \\ (x+6)(x-5) &= 0 \end{aligned}$$

$$x+6=0 \quad \text{or} \quad x-5=0$$
$$x=-6 \quad \text{or} \quad x=5$$

If $x=-6$, then $x+1=-5$.
If $x=5$, then $x+1=6$.

The two possible pairs of consecutive integers are -6 and -5 or 5 and 6.

51. Let $w =$ the width of the cardboard. Then $w+6 =$ the length of the cardboard.

If squares that measure 2 inches are cut from each corner of the cardboard, then the width becomes $w-4$ and the length becomes $(w+6)-4 = w+2$. Use the formula $V = LWH$ and substitute 110 for V, $w+2$ for L, $w-4$ for W, and 2 for H.

$$\begin{aligned} V &= LWH \\ 110 &= (w+2)(w-4)2 \\ 110 &= (w^2-2w-8)2 \\ 55 &= w^2-2w-8 \\ 0 &= w^2-2w-63 \\ 0 &= (w-9)(w+7) \end{aligned}$$

$$w-9=0 \quad \text{or} \quad w+7=0$$
$$w=9 \quad \text{or} \quad w=-7$$

A box cannot have a negative width, so reject -7 as a solution. The only possible solution is 9. The piece of cardboard has width 9 inches and length $9+6=15$ inches.

53. From Example 8,

$$\begin{aligned} h(t) &= -16t^2+128t. \\ 240 &= -16t^2+128t \quad \textit{Let } h(t) = 240. \\ 16t^2-128t+240 &= 0 \quad \textit{Standard form} \\ t^2-8t+15 &= 0 \quad \textit{Divide by 16.} \\ (t-3)(t-5) &= 0 \end{aligned}$$

$$t-3=0 \quad \text{or} \quad t-5=0$$
$$t=3 \quad \text{or} \quad t=5$$

continued

The height of the rocket will be 240 feet after 3 seconds (on the way up) and after 5 seconds (on the way down).

$$112 = -16t^2 + 128t \quad \textit{Let } h(t) = 112.$$
$$16t^2 - 128t + 112 = 0 \quad \textit{Standard form}$$
$$t^2 - 8t + 7 = 0 \quad \textit{Divide by 16.}$$
$$(t-1)(t-7) = 0$$

$$t - 1 = 0 \quad \text{or} \quad t - 7 = 0$$
$$t = 1 \quad \text{or} \quad t = 7$$

The height of the rocket will be 112 feet after 1 second (on the way up) and after 7 seconds (on the way down).

55. Use $f(t) = -16t^2 + 576$ with $f(t) = 0$.

$$0 = -16t^2 + 576$$
$$0 = t^2 - 36 \quad \textit{Divide by } -16.$$
$$0 = (t+6)(t-6)$$

$$t + 6 = 0 \quad \text{or} \quad t - 6 = 0$$
$$t = -6 \quad \text{or} \quad t = 6$$

Time cannot be negative, so reject -6 as a solution. The only possible solution is 6. The rock will hit the ground after 6 seconds.

Chapter 7 Review Exercises

1. $21y^2 + 35y = 7y(3y+5)$

2. $12q^2b + 8qb^2 - 20q^3b^2$
$= 4qb(3q + 2b - 5q^2b)$

3. $(x+3)(4x-1) - (x+3)(3x+2)$

The GCF is $(x+3)$.

$$= (x+3)[(4x-1) - (3x+2)]$$
$$= (x+3)(4x - 1 - 3x - 2)$$
$$= (x+3)(x-3)$$

4. $(z+1)(z-4) + (z+1)(2z+3)$
$= (z+1)[(z-4) + (2z+3)]$
$= (z+1)(3z-1)$

5. $4m + nq + mn + 4q$

Rearrange the terms.

$$= 4m + 4q + mn + nq$$
$$= 4(m+q) + n(m+q)$$
$$= (m+q)(4+n)$$

6. $x^2 + 5y + 5x + xy$
$= x^2 + xy + 5x + 5y$
$= x(x+y) + 5(x+y)$
$= (x+y)(x+5)$

7. $2m + 6 - am - 3a$
$= 2(m+3) - a(m+3)$
$= (m+3)(2-a)$

8. $2am - 2bm - ap + bp$
$= (2am - 2bm) + (-ap + bp)$
$= 2m(a-b) - p(a-b)$
$= (a-b)(2m-p)$

9. $3p^2 - p - 4$

Two integer factors whose product is $(3)(-4) = -12$ and whose sum is -1 are -4 and 3.

$$= 3p^2 - 4p + 3p - 4$$
$$= p(3p-4) + 1(3p-4)$$
$$= (3p-4)(p+1)$$

10. $12r^2 - 5r - 3$

Two integer factors whose product is $(12)(-3) = -36$ and whose sum is -5 are -9 and 4.

$$= 12r^2 - 9r + 4r - 3$$
$$= 3r(4r-3) + 1(4r-3)$$
$$= (4r-3)(3r+1)$$

11. $10m^2 + 37m + 30$

Two integer factors whose product is $(10)(30) = 300$ and whose sum is 37 are 12 and 25.

$$= 10m^2 + 12m + 25m + 30$$
$$= 2m(5m+6) + 5(5m+6)$$
$$= (5m+6)(2m+5)$$

12. $10k^2 - 11kh + 3h^2$

Two integer factors whose product is $(10)(3) = 30$ and whose sum is -11 are -6 and -5.

$$= 10k^2 - 6kh - 5kh + 3h^2$$
$$= 2k(5k-3h) - h(5k-3h)$$
$$= (5k-3h)(2k-h)$$

13. $9x^2 + 4xy - 2y^2$

There are no integers that have a product of $9(-2) = -18$ and a sum of 4. Therefore, the trinomial cannot be factored and is *prime*.

14. $24x - 2x^2 - 2x^3$
$= 2x(12 - x - x^2)$
$= 2x(4+x)(3-x)$

15. $2k^4 - 5k^2 - 3$
$= 2(k^2)^2 - 5k^2 - 3$
$= (2k^2+1)(k^2-3)$

16. $p^2(p+2)^2 + p(p+2)^2 - 6(p+2)^2$

Factor out $(p+2)^2$.

$= (p+2)^2(p^2+p-6)$
$= (p+2)^2(p+3)(p-2)$

17. $16x^2 - 25$
$= (4x)^2 - 5^2$ *Difference of squares*
$= (4x+5)(4x-5)$

18. $9t^2 - 49$
$= (3t)^2 - 7^2$ *Difference of squares*
$= (3t+7)(3t-7)$

19. $x^2 + 14x + 49$
$= x^2 + 2(x)(7) + 7^2$ *Perfect square tri.*
$= (x+7)^2$

20. $9k^2 - 12k + 4$
$= (3k)^2 - 2(3k)(2) + 2^2$ *Perfect square tri.*
$= (3k-2)^2$

21. $r^3 + 27$
$= r^3 + 3^3$ *Sum of cubes*
$= (r+3)(r^2 - 3r + 9)$

22. $125x^3 - 1$
$= (5x)^3 - 1^3$ *Difference of cubes*
$= (5x-1)(25x^2 + 5x + 1)$

23. $m^6 - 1$
$= (m^3)^2 - 1^2$ *Difference of squares*
$= (m^3+1)(m^3-1)$
$= (m^3+1^3)(m^3-1^3)$
Sum of cubes; difference of cubes
$= (m+1)(m^2-m+1)$
$\cdot (m-1)(m^2+m+1)$

24. $x^8 - 1$
$= (x^4)^2 - 1^2$ *Difference of squares*
$= (x^4+1)(x^4-1)$
$= (x^4+1)[(x^2)^2 - 1^2]$
Difference of squares again
$= (x^4+1)(x^2+1)(x^2-1)$
Difference of squares again
$= (x^4+1)(x^2+1)(x+1)(x-1)$

25. $x^2 + 6x + 9 - 25y^2$
$= (x^2+6x+9) - 25y^2$
$= [x^2 + 2(x)(3) + 3^2] - 25y^2$
Perfect square trinomial
$= (x+3)^2 - (5y)^2$ *Difference of squares*
$= [(x+3)+5y][(x+3)-5y]$
$= (x+3+5y)(x+3-5y)$

26. $(x+1)(5x+2) = 0$

$x+1=0$ or $5x+2=0$
$x=-1$ or $x=-\frac{2}{5}$

Solution set: $\{-1, -\frac{2}{5}\}$

27. $p^2 - 5p + 6 = 0$
$(p-2)(p-3) = 0$

$p-2=0$ or $p-3=0$
$p=2$ or $p=3$

Solution set: $\{2, 3\}$

28. $6z^2 = 5z + 50$
$6z^2 - 5z - 50 = 0$
$(3z-10)(2z+5) = 0$

$3z-10=0$ or $2z+5=0$
$3z=10$ or $2z=-5$
$z=\frac{10}{3}$ or $z=-\frac{5}{2}$

Solution set: $\{-\frac{5}{2}, \frac{10}{3}\}$

29. $6r^2 + 7r = 3$
$6r^2 + 7r - 3 = 0$
$(2r+3)(3r-1) = 0$

$2r+3=0$ or $3r-1=0$
$2r=-3$ or $3r=1$
$r=-\frac{3}{2}$ or $r=\frac{1}{3}$

Solution set: $\{-\frac{3}{2}, \frac{1}{3}\}$

30. $-4m^2 + 36 = 0$
$m^2 - 9 = 0$ *Divide by −4.*
$(m+3)(m-3) = 0$

$m+3=0$ or $m-3=0$
$m=-3$ or $m=3$

Solution set: $\{-3, 3\}$

31. $6x^2 + 9x = 0$
$3x(2x+3) = 0$

$3x=0$ or $2x+3=0$
$x=0$ or $2x=-3$
$x=-\frac{3}{2}$

Solution set: $\{-\frac{3}{2}, 0\}$

32. $(2x+1)(x-2) = -3$
$2x^2 - 3x - 2 = -3$
$2x^2 - 3x + 1 = 0$
$(2x-1)(x-1) = 0$

$2x-1=0$ or $x-1=0$
$2x=1$ or $x=1$
$x=\frac{1}{2}$

Solution set: $\{\frac{1}{2}, 1\}$

33.
$$x^2 - 8x + 16 = 0$$
$$(x-4)(x-4) = 0$$

$x - 4 = 0$ or $x - 4 = 0$
$x = 4$ or $x = 4$

Solution set: $\{4\}$

34.
$$2x^3 - x^2 - 28x = 0$$
$$x(2x^2 - x - 28) = 0$$
$$x(2x+7)(x-4) = 0$$

$x = 0$ or $2x + 7 = 0$ or $x - 4 = 0$
$2x = -7$ or $x = 4$
$x = -\frac{7}{2}$

Solution set: $\{-\frac{7}{2}, 0, 4\}$

35. Let x be the length of the shorter side. Then the length of the longer side will be $2x + 1$. The area is 10.5 ft^2. Use the formula for area of a triangle, $A = \frac{1}{2}bh$.

$$\tfrac{1}{2}(2x+1)(x) = 10.5$$
$$x(2x+1) = 21 \quad \textit{Multiply by 2.}$$
$$2x^2 + x = 21$$
$$2x^2 + x - 21 = 0$$
$$(2x+7)(x-3) = 0$$

$2x + 7 = 0$ or $x - 3 = 0$
$2x = -7$ or $x = 3$
$x = -\frac{7}{2}$

The side cannot have a negative length, so reject $x = -\frac{7}{2}$.
The length of the shorter side is 3 ft.

36. Let W be the width of the lot. Then $W + 20$ will be the length of the lot. The area is 2400 ft^2. Use the formula $LW = A$.

$$(W+20)W = 2400$$
$$W^2 + 20W = 2400$$
$$W^2 + 20W - 2400 = 0$$
$$(W-40)(W+60) = 0$$

$W - 40 = 0$ or $W + 60 = 0$
$W = 40$ or $W = -60$

The lot cannot have a negative width, so reject $W = -60$.
The width of the lot is 40 ft, and the length is $40 + 20 = 60$ ft.

37. The height is 0 when the rock returns to the ground.

$$f(t) = -16t^2 + 256t$$
$$0 = -16t^2 + 256t$$
$$0 = -16t(t-16)$$

$-16t = 0$ or $t - 16 = 0$
$t = 0$ or $t = 16$

The rock is on the ground when $t = 0$. It will return to the ground again after 16 seconds.

38.
$$f(t) = -16t^2 + 256t$$
$$240 = -16t^2 + 256t \quad \textit{Let } f(t) = 240.$$
$$0 = -16t^2 + 256t - 240$$
$$0 = -16(t^2 - 16t + 15)$$
$$0 = -16(t-15)(t-1)$$

$t - 15 = 0$ or $t - 1 = 0$
$t = 15$ or $t = 1$

The rock will be 240 ft above the ground after 1 second and again after 15 seconds.

39. The question in Exercise 38 has two answers because the rock will be 240 ft above the ground after 1 second on the way up and again after 15 seconds on the way back down.

40. **[7.2]** $30a + am - am^2$
$= a(30 + m - m^2)$
$= a(6 - m)(5 + m)$

41. **[7.3]** $8 - a^3$
$= 2^3 - a^3$
$= (2 - a)(2^2 + 2a + a^2)$
$= (2 - a)(4 + 2a + a^2)$

42. **[7.3]** $81k^2 - 16$
$= (9k)^2 - 4^2$
$= (9k + 4)(9k - 4)$

43. **[7.2]** $9x^2 + 13xy - 3y^2$

There are no integers that have a product of $9(-3) = -27$ and a sum of 13. Therefore, the trinomial cannot be factored and is *prime*.

44. **[7.1]** $15y^3 + 20y^2 = 5y^2(3y + 4)$

45. **[7.3]** $25z^2 - 30zm + 9m^2$
$= (5z)^2 - 2(5z)(3m) + (3m)^2$
$= (5z - 3m)^2$

46. **[7.5]** $5x^2 - 17x - 12 = 0$
$(5x + 3)(x - 4) = 0$

$5x + 3 = 0$ or $x - 4 = 0$
$5x = -3$ or $x = 4$
$x = -\frac{3}{5}$

Solution set: $\{-\frac{3}{5}, 4\}$

47. **[7.5]**
$$\begin{aligned} x^3 - x &= 0 \\ x(x^2-1) &= 0 \\ x(x+1)(x-1) &= 0 \end{aligned}$$

$$x = 0 \quad \text{or} \quad x+1=0 \quad \text{or} \quad x-1=0$$
$$x=-1 \quad \text{or} \quad x=1$$

Solution set: $\{-1, 0, 1\}$

48. **[7.5]**
$$\begin{aligned} 3m^2 - 9m &= 0 \\ 3m(m-3) &= 0 \end{aligned}$$

$$3m = 0 \quad \text{or} \quad m-3=0$$
$$m=0 \qquad\qquad m=3$$

Solution set: $\{0, 3\}$

49. **[7.5]** Let x be the width of the floor. Then $x+85$ will be the length of the floor. The area is 2750 ft^2. Use the formula $LW = A$.

$$\begin{aligned} (x+85)x &= 2750 \\ x^2+85x &= 2750 \\ x^2+85x-2750 &= 0 \\ (x+110)(x-25) &= 0 \end{aligned}$$

$$x+110=0 \quad \text{or} \quad x-25=0$$
$$x=-110 \quad \text{or} \quad x=25$$

The floor cannot have a negative width, so reject $x=-110$.
The width of the floor was 25 feet and the length was $25+85=110$ feet.

50. **[7.5]** Let x be the width of the frame. Then $x+2$ will be the length of the frame. The area is 48 in.^{2}. Use the formula $LW = A$.

$$\begin{aligned} (x+2)x &= 48 \\ x^2+2x &= 48 \\ x^2+2x-48 &= 0 \\ (x+8)(x-6) &= 0 \end{aligned}$$

$$x+8=0 \quad \text{or} \quad x-6=0$$
$$x=-8 \quad \text{or} \quad x=6$$

The frame cannot have a negative width, so reject $x=-8$.
The width of the frame is 6 inches.

Chapter 7 Test

1. $11z^2 - 44z = 11z(z-4)$

2. $10x^2y^5 - 5x^2y^3 - 25x^5y^3$
Factor out the GCF, $5x^2y^3$.
$= 5x^2y^3(2y^2 - 1 - 5x^3)$

3.
$$\begin{aligned} &3x + by + bx + 3y \\ &= 3x + 3y + bx + by \\ &= 3(x+y) + b(x+y) \\ &= (x+y)(3+b) \end{aligned}$$

4. $-2x^2 - x + 36 = -1(2x^2 + x - 36)$

Two integer factors whose product is $(2)(-36) = -72$ and whose sum is 1 are 9 and -8.

$$\begin{aligned} &2x^2 + x - 36 \\ &= 2x^2 + 9x - 8x - 36 \\ &= x(2x+9) - 4(2x+9) \\ &= (2x+9)(x-4) \end{aligned}$$

Thus, the final factored form is

$$-(2x+9)(x-4).$$

5. $6x^2 + 11x - 35$

Two integer factors whose product is $(6)(-35) = -210$ and whose sum is 11 are 21 and -10.

$$\begin{aligned} &6x^2 + 11x - 35 \\ &= 6x^2 + 21x - 10x - 35 \\ &= 3x(2x+7) - 5(2x+7) \\ &= (2x+7)(3x-5) \end{aligned}$$

6. $4p^2 + 3pq - q^2$

Two integer factors whose product is $(4)(-1) = -4$ and whose sum is 3 are 4 and -1.

$$\begin{aligned} &4p^2 + 3pq - q^2 \\ &= 4p^2 + 4pq - pq - q^2 \\ &= 4p(p+q) - q(p+q) \\ &= (p+q)(4p-q) \end{aligned}$$

7.
$$\begin{aligned} &16a^2 + 40ab + 25b^2 \\ &= (4a)^2 + 2(4a)(5b) + (5b)^2 \\ &= (4a+5b)^2 \end{aligned}$$

8.
$$\begin{aligned} &x^2 + 2x + 1 - 4z^2 \\ &= (x^2+2x+1) - 4z^2 \\ &= (x+1)^2 - (2z)^2 \\ &= [(x+1)+2z][(x+1)-2z] \\ &= (x+1+2z)(x+1-2z) \end{aligned}$$

9.
$$\begin{aligned} &a^3 + 2a^2 - ab^2 - 2b^2 \\ &= a^2(a+2) - b^2(a+2) \\ &= (a+2)(a^2-b^2) \\ &= (a+2)(a+b)(a-b) \end{aligned}$$

10.
$$\begin{aligned} &9k^2 - 121j^2 \\ &= (3k)^2 - (11j)^2 \quad \textit{Difference of squares} \\ &= (3k+11j)(3k-11j) \end{aligned}$$

11.
$$\begin{aligned} &y^3 - 216 \\ &= y^3 - 6^3 \quad \textit{Difference of cubes} \\ &= (y-6)(y^2+6y+6^2) \\ &= (y-6)(y^2+6y+36) \end{aligned}$$

12. $6k^4 - k^2 - 35 = 6(k^2)^2 - k^2 - 35$

Two integer factors whose product is $(6)(-35) = -210$ and whose sum is -1 are -15 and 14.

$$\begin{aligned} &6k^4 - k^2 - 35 \\ &= 6(k^2)^2 - 15k^2 + 14k^2 - 35 \\ &= 3k^2(2k^2 - 5) + 7(2k^2 - 5) \\ &= (2k^2 - 5)(3k^2 + 7) \end{aligned}$$

13. $$\begin{aligned} &27x^6 + 1 \\ &= (3x^2)^3 + (1)^3 \quad \textit{Sum of cubes} \\ &= (3x^2 + 1)[(3x^2)^2 - (3x^2)(1) + 1^2] \\ &= (3x^2 + 1)(9x^4 - 3x^2 + 1) \end{aligned}$$

14. $$\begin{aligned} &-x^2 + x + 30 \\ &= -1(x^2 - x - 30) \\ &= -1(x + 5)(x - 6) \\ &= -(x + 5)(x - 6) \end{aligned}$$

15. $$\begin{aligned} &(t^2 + 3)^2 + 4(t^2 + 3) - 5 \\ &= [(t^2 + 3) + 5][(t^2 + 3) - 1] \\ &= (t^2 + 3 + 5)(t^2 + 3 - 1) \\ &= (t^2 + 8)(t^2 + 2) \end{aligned}$$

16. It is not in factored form because there are two terms: $(x^2 + 2y)p$ and $3(x^2 + 2y)$. The common factor is $x^2 + 2y$, and the factored form is $(x^2 + 2y)(p + 3)$.

17. **A.** $$\begin{aligned} &(3 - x)(x + 4) \\ &= 3x + 12 - x^2 - 4x \\ &= -x^2 - x + 12 \end{aligned}$$

B. $$\begin{aligned} &-(x - 3)(x + 4) \\ &= -(x^2 + 4x - 3x - 12) \\ &= -(x^2 + x - 12) \\ &= -x^2 - x + 12 \end{aligned}$$

C. $$\begin{aligned} &(-x + 3)(x + 4) \\ &= -x^2 - 4x + 3x + 12 \\ &= -x^2 - x + 12 \end{aligned}$$

D. $$\begin{aligned} &(x - 3)(-x + 4) \\ &= -x^2 + 4x + 3x - 12 \\ &= -x^2 + 7x - 12 \end{aligned}$$

Therefore, only **D** is *not* a factored form of $-x^2 - x + 12$.

18. $$\begin{aligned} 3x^2 + 8x &= -4 \\ 3x^2 + 8x + 4 &= 0 \quad \textit{Add 4.} \\ (x + 2)(3x + 2) &= 0 \quad \textit{Factor.} \end{aligned}$$

$$\begin{aligned} x + 2 = 0 \quad &\text{or} \quad 3x + 2 = 0 \\ x = -2 \quad &\text{or} \quad 3x = -2 \\ & \quad\quad x = -\tfrac{2}{3} \end{aligned}$$

Solution set: $\{-2, -\frac{2}{3}\}$

19. $$\begin{aligned} 3x^2 - 5x &= 0 \\ x(3x - 5) &= 0 \end{aligned}$$

$$\begin{aligned} x = 0 \quad \text{or} \quad 3x - 5 &= 0 \\ 3x &= 5 \\ x &= \tfrac{5}{3} \end{aligned}$$

Solution set: $\{0, \frac{5}{3}\}$

20. $$\begin{aligned} 5m(m - 1) &= 2(1 - m) \\ 5m^2 - 5m &= 2 - 2m \\ 5m^2 - 3m - 2 &= 0 \\ (5m + 2)(m - 1) &= 0 \end{aligned}$$

$$\begin{aligned} 5m + 2 = 0 \quad &\text{or} \quad m - 1 = 0 \\ 5m = -2 \quad &\text{or} \quad m = 1 \\ m = -\tfrac{2}{5} \quad & \end{aligned}$$

Solution set: $\{-\frac{2}{5}, 1\}$

21. Using $A = LW$, substitute 40 for A, $x + 7$ for L, and $2x + 3$ for W.

$$\begin{aligned} A &= LW \\ 40 &= (x + 7)(2x + 3) \\ 40 &= 2x^2 + 3x + 14x + 21 \\ 40 &= 2x^2 + 17x + 21 \\ 0 &= 2x^2 + 17x - 19 \\ 0 &= (2x + 19)(x - 1) \end{aligned}$$

$$\begin{aligned} 2x + 19 = 0 \quad &\text{or} \quad x - 1 = 0 \\ 2x = -19 \quad &\text{or} \quad x = 1 \\ x = -\tfrac{19}{2} \quad & \end{aligned}$$

Length and width will be negative if $x = -\frac{19}{2}$, so reject it as a possible solution. Since $x = 1$, $x + 7 = 1 + 7 = 8$ and $2x + 3 = 2(1) + 3 = 5$.

The length is 8 inches, and the width is 5 inches.

22. Substitute 128 for $f(t)$ in the equation.

$$\begin{aligned} f(t) &= -16t^2 + 96t \\ 128 &= -16t^2 + 96t \\ 16t^2 - 96t + 128 &= 0 \\ 16(t^2 - 6t + 8) &= 0 \\ 16(t - 4)(t - 2) &= 0 \end{aligned}$$

$$\begin{aligned} t - 4 = 0 \quad &\text{or} \quad t - 2 = 0 \\ t = 4 \quad &\text{or} \quad t = 2 \end{aligned}$$

The ball is 128 ft high at 2 seconds (on the way up) and again at 4 seconds (on the way down).

Cumulative Review Exercises (Chapters 1–7)

1. $-2(m - 3) = -2(m) - 2(-3) = -2m + 6$

2. $-(-4m + 3) = -(-4m) - (3) = 4m - 3$

3. $3x^2 - 4x + 4 + 9x - x^2$
$= 3x^2 - x^2 - 4x + 9x + 4$
$= 2x^2 + 5x + 4$

For Exercises 4–7, let $p = -4, q = -2$, and $r = 5$.

4. $-3(2q - 3p) = -3[2(-2) - 3(-4)]$
$= -3(-4 + 12)$
$= -3(8) = -24$

5. $8r^2 + q^2 = 8(5)^2 + (-2)^2$
$= 8(25) + 4$
$= 200 + 4 = 204$

6. $\dfrac{\sqrt{r}}{-p + 2q} = \dfrac{\sqrt{5}}{-(-4) + 2(-2)}$
$= \dfrac{\sqrt{5}}{4 - 4} = \dfrac{\sqrt{5}}{0}$

This expression is *undefined* since the denominator is zero.

7. $\dfrac{rp + 6r^2}{p^2 + q - 1} = \dfrac{5(-4) + 6(5)^2}{(-4)^2 + (-2) - 1}$
$= \dfrac{-20 + 6(25)}{16 - 2 - 1}$
$= \dfrac{-20 + 150}{13}$
$= \dfrac{130}{13} = 10$

8. $2z - 5 + 3z = 4 - (z + 2)$
$5z - 5 = 4 - z - 2$
$5z - 5 = 2 - z$
$6z = 7$
$z = \frac{7}{6}$

Solution set: $\{\frac{7}{6}\}$

9. $\dfrac{3a - 1}{5} + \dfrac{a + 2}{2} = -\dfrac{3}{10}$
$2(3a - 1) + 5(a + 2) = -3$ *Multiply by 10.*
$6a - 2 + 5a + 10 = -3$
$11a + 8 = -3$
$11a = -11$
$a = -1$

Solution set: $\{-1\}$

10. $-\frac{4}{3}d \geq -5$
$-\frac{3}{4}(-\frac{4}{3}d) \leq -\frac{3}{4}(-5)$
$d \leq \frac{15}{4}$

Solution set: $(-\infty, \frac{15}{4}]$

11. $3 - 2(m + 3) < 4m$
$3 - 2m - 6 < 4m$
$-2m - 3 < 4m$
$-6m < 3$
$m > -\frac{3}{6}$ or $-\frac{1}{2}$

Solution set: $(-\frac{1}{2}, \infty)$

12. $2k + 4 < 10$ and $3k - 1 > 5$
$2k < 6$ and $3k > 6$
$k < 3$ and $k > 2$

The overlap of these inequalities is the set of all integers between 2 and 3.

Solution set: $(2, 3)$

13. $2k + 4 > 10$ or $3k - 1 < 5$
$2k > 6$ or $3k < 6$
$k > 3$ or $k < 2$

The solution set is the set of numbers that are either greater than 3 or less than 2.

Solution set: $(-\infty, 2) \cup (3, \infty)$

14. $|5x + 3| - 10 = 3$
$|5x + 3| = 13$

$5x + 3 = 13$ or $5x + 3 = -13$
$5x = 10$ or $5x = -16$
$x = 2$ or $x = -\frac{16}{5}$

Solution set: $\{-\frac{16}{5}, 2\}$

15. $|x + 2| < 9$
$-9 < x + 2 < 9$
$-11 < x < 7$

Solution set: $(-11, 7)$

16. $|2y - 5| \geq 9$

$2y - 5 \geq 9$ or $2y - 5 \leq -9$
$2y \geq 14$ or $2y \leq -4$
$y \geq 7$ or $y \leq -2$

Solution set: $(-\infty, -2] \cup [7, \infty)$

17. Solve $V = lwh$ for h.
$\dfrac{V}{lw} = \dfrac{lwh}{lw}$ *Divide by lw.*
$\dfrac{V}{lw} = h$

18. Let x be the time it takes for the planes to be 2100 mi apart. Use the formula $d = rt$ to complete the table.

Plane	r	t	d
Eastbound	550	x	$550x$
Westbound	500	x	$500x$

The total distance is 2100 mi.

$$\begin{aligned} 550x + 500x &= 2100 \\ 1050x &= 2100 \\ x &= 2 \end{aligned}$$

It will take 2 hours for the planes to be 2100 miles apart.

19. $4x + 2y = -8$

Draw a line through the x- and y-intercepts, $(-2, 0)$ and $(0, -4)$, respectively.

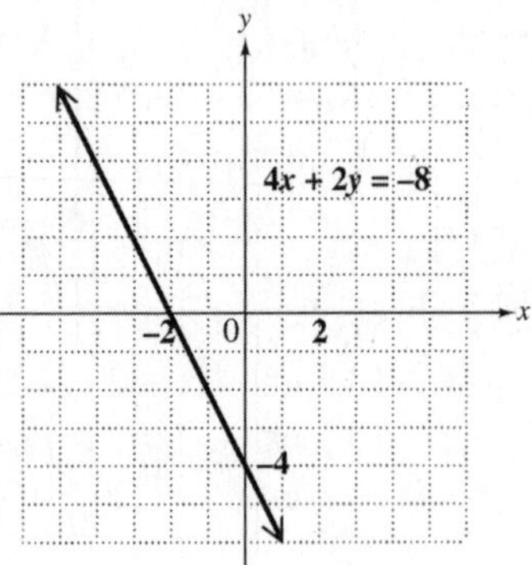

20. The slope m of the line through $(-4, 8)$ and $(-2, 6)$ is

$$m = \frac{6-8}{-2-(-4)} = \frac{-2}{2} = -1.$$

21. $y = -3$ is an equation of a horizontal line. Its slope is 0.

22.
$$\begin{aligned} f(x) &= 2x + 7 \\ f(-4) &= 2(-4) + 7 \\ &= -8 + 7 = -1 \end{aligned}$$

23. To find the x-intercept of the graph of $f(x) = 2x + 7$, let $f(x) = 0$ (which is the same as letting $y = 0$) and solve for x.

$$\begin{aligned} 0 &= 2x + 7 \\ -7 &= 2x \\ -\tfrac{7}{2} &= x \end{aligned}$$

The x-intercept is $(-\frac{7}{2}, 0)$.

24. $f(0) = 7$, so the y-intercept is $(0, 7)$.

25.
$$\begin{aligned} 3x - 2y &= -7 \quad (1) \\ 2x + 3y &= 17 \quad (2) \end{aligned}$$

To eliminate y, multiply (1) by 3 and (2) by 2, and then add the resulting equations.

$$\begin{array}{rcl l} 9x - 6y &=& -21 & 3 \times (1) \\ 4x + 6y &=& 34 & 2 \times (2) \\ \hline 13x &=& 13 & \\ x &=& 1 & \end{array}$$

Substitute 1 for x in (1).

$$\begin{aligned} 3x - 2y &= -7 \quad (1) \\ 3(1) - 2y &= -7 \\ -2y &= -10 \\ y &= 5 \end{aligned}$$

Solution set: $\{(1, 5)\}$

26.
$$\begin{aligned} 2x + 3y - 6z &= 5 \quad (1) \\ 8x - y + 3z &= 7 \quad (2) \\ 3x + 4y - 3z &= 7 \quad (3) \end{aligned}$$

To eliminate z, add (2) and (3).

$$\begin{array}{rcl l} 8x - y + 3z &=& 7 & (2) \\ 3x + 4y - 3z &=& 7 & (3) \\ \hline 11x + 3y &=& 14 & (4) \end{array}$$

To eliminate z again, multiply (2) by 2 and add the result to (1).

$$\begin{array}{rcl l} 2x + 3y - 6z &=& 5 & (1) \\ 16x - 2y + 6z &=& 14 & 2 \times (2) \\ \hline 18x + y &=& 19 & (5) \end{array}$$

To eliminate y, multiply (5) by -3 and add the result to (4).

$$\begin{array}{rcl l} 11x + 3y &=& 14 & (4) \\ -54x - 3y &=& -57 & -3 \times (5) \\ \hline -43x &=& -43 & \\ x &=& 1 & \end{array}$$

From (5), $18(1) + y = 19$, so $y = 1$.
To find z, let $x = 1$ and $y = 1$ in (1).

$$\begin{aligned} 2x + 3y - 6z &= 5 \quad (1) \\ 2(1) + 3(1) - 6z &= 5 \\ 5 - 6z &= 5 \\ -6z &= 0 \\ z &= 0 \end{aligned}$$

Solution set: $\{(1, 1, 0)\}$

27.
$$\begin{aligned} &(3x^2y^{-1})^{-2}(2x^{-3}y)^{-1} \\ &= 3^{-2}x^{-4}y^2 \cdot 2^{-1}x^3y^{-1} \\ &= 3^{-2}2^{-1}x^{-1}y \\ &= \frac{y}{3^2 \cdot 2x} = \frac{y}{18x} \end{aligned}$$

28. $\dfrac{5m^{-2}y^3}{3m^{-3}y^{-1}} = \dfrac{5}{3} \cdot \dfrac{m^{-2}}{m^{-3}} \cdot \dfrac{y^3}{y^{-1}}$
$= \dfrac{5}{3}m^{-2-(-3)}y^{3-(-1)}$
$= \dfrac{5}{3}m^1y^4$, or $\dfrac{5my^4}{3}$

29. $(3x^3 + 4x^2 - 7) - (2x^3 - 8x^2 + 3x)$
$= 3x^3 + 4x^2 - 7 - 2x^3 + 8x^2 - 3x$
$= x^3 + 12x^2 - 3x - 7$

30. $(7x + 3y)^2$
$= (7x)^2 + 2(7x)(3y) + (3y)^2$
$= 49x^2 + 42xy + 9y^2$

31. $(2p + 3)(5p^2 - 4p - 8)$
$= 10p^3 - 8p^2 - 16p + 15p^2 - 12p - 24$
$= 10p^3 + 7p^2 - 28p - 24$

32. $16w^2 + 50wz - 21z^2$

Two integer factors whose product is $(16)(-21) = -336$ and whose sum is 50 are 56 and -6.

$= 16w^2 + 56wz - 6wz - 21z^2$
$= 8w(2w + 7z) - 3z(2w + 7z)$
$= (2w + 7z)(8w - 3z)$

33. $4x^2 - 4x + 1 - y^2$
Group the first three terms.
$= (4x^2 - 4x + 1) - y^2$
$= (2x - 1)^2 - y^2$
$= [(2x - 1) + y][(2x - 1) - y]$
$= (2x - 1 + y)(2x - 1 - y)$

34. $4y^2 - 36y + 81$
$= (2y)^2 - 2(2y)(9) + 9^2$
$= (2y - 9)^2$

35. $100x^4 - 81$
$= (10x^2)^2 - 9^2$ *Difference of squares*
$= (10x^2 + 9)(10x^2 - 9)$

36. $8p^3 + 27$
$= (2p)^3 + 3^3$ *Sum of cubes*
$= (2p + 3)[(2p)^2 - (2p)(3) + 3^2]$
$= (2p + 3)(4p^2 - 6p + 9)$

37. $(p + 4)(2p + 3)(p - 1) = 0$

$p + 4 = 0$ or $2p + 3 = 0$ or $p - 1 = 0$
$p = -4$ or $p = -\frac{3}{2}$ or $p = 1$

Solution set: $\{-4, -\frac{3}{2}, 1\}$

38. $9q^2 = 6q - 1$
$9q^2 - 6q + 1 = 0$
$(3q - 1)^2 = 0$
$3q - 1 = 0$
$3q = 1$
$q = \frac{1}{3}$

Solution set: $\{\frac{1}{3}\}$

39. Let x be the length of the base. Then $x + 3$ will be the height. The area is 14 square feet. Use the formula $A = \frac{1}{2}bh$, and substitute 14 for A, x for b, and $x + 3$ for h.

$\frac{1}{2}bh = A$
$\frac{1}{2}(x)(x + 3) = 14$
$x(x + 3) = 28$ *Multiply by 2.*
$x^2 + 3x = 28$
$x^2 + 3x - 28 = 0$
$(x + 7)(x - 4) = 0$

$x + 7 = 0$ or $x - 4 = 0$
$x = -7$ or $x = 4$

The length cannot be negative, so reject -7 as a solution. The only possible solution is 4. The base is 4 feet long.

40. Let x be the distance between the longer sides. (This is actually the width.) Then $x + 2$ will be the length of the longer side. The area of the rectangle is 288 in^2. Use the formula $LW = A$. Substitute 288 for A, $x + 2$ for L, and x for W.

$(x + 2)x = 288$
$x^2 + 2x = 288$
$x^2 + 2x - 288 = 0$
$(x + 18)(x - 16) = 0$

$x + 18 = 0$ or $x - 16 = 0$
$x = -18$ or $x = 16$

The distance cannot be negative, so reject -18 as a solution. The only possible solution is 16. The distance between the longer sides is 16 inches, and the length of the longer sides is $16 + 2 = 18$ inches.

CHAPTER 8 RATIONAL EXPRESSIONS AND FUNCTIONS

8.1 Rational Expressions and Functions; Multiplying and Dividing

8.1 Margin Exercises

1. **(a)** $f(x) = \dfrac{x+4}{x-6}$

Set the denominator equal to zero, and solve the equation.

$$\begin{aligned} x - 6 &= 0 \\ x &= 0 \end{aligned}$$

The number 6 makes the function undefined, so it is not in the domain of f.

Domain: $\{x \mid x \neq 6\}$

(b) $f(x) = \dfrac{x+6}{x^2 - x - 6}$

Set the denominator equal to zero, and solve the equation.

$$\begin{aligned} x^2 - x - 6 &= 0 \\ (x-3)(x+2) &= 0 \\ x - 3 = 0 \quad \text{or} \quad x + 2 &= 0 \\ x = 3 \quad \text{or} \quad x &= -2 \end{aligned}$$

Both 3 and -2 make the function undefined.

Domain: $\{x \mid x \neq -2, 3\}$

(c) $f(x) = \dfrac{3+2x}{5}$

The denominator, 5, can never be 0, so there are *no* numbers that make the function undefined. The domain consists of all real numbers, $(-\infty, \infty)$.

(d) $f(x) = \dfrac{2}{x^2 + 1}$

Set the denominator equal to zero, and solve the equation.

$$\begin{aligned} x^2 + 1 &= 0 \\ x^2 &= -1 \end{aligned}$$

There is no real number whose square is -1, so there are *no* numbers that make the function undefined. The domain consists of all real numbers, $(-\infty, \infty)$.

2. **(a)**
$$\begin{aligned} \frac{y^2 + 2y - 3}{y^2 - 3y + 2} &= \frac{(y+3)(y-1)}{(y-2)(y-1)} && \textit{Factor.} \\ &= \frac{y+3}{y-2} \cdot 1 \\ &= \frac{y+3}{y-2} && \textit{Lowest terms} \end{aligned}$$

(b) $\dfrac{3y+9}{y^2-9} = \dfrac{3(y+3)}{(y-3)(y+3)} = \dfrac{3}{y-3}$

(c) $\dfrac{y+2}{y^2+4}$

The denominator cannot be factored, so this expression cannot be simplified further and is in lowest terms.

(d) $\dfrac{1+p^3}{1+p}$
$$\begin{aligned} &= \frac{(1+p)(1-p+p^2)}{1+p} && \textit{Factor the sum of cubes.} \\ &= 1 - p + p^2 && \textit{Lowest terms} \end{aligned}$$

(e) $\dfrac{3x + 3y + rx + ry}{5x + 5y - rx - ry}$
$$\begin{aligned} &= \frac{(3x+3y) + (rx+ry)}{(5x+5y) - (rx+ry)} && \textit{Group terms.} \\ &= \frac{3(x+y) + r(x+y)}{5(x+y) - r(x+y)} && \textit{Factor within groups.} \\ &= \frac{(x+y)(3+r)}{(x+y)(5-r)} && \textit{Factor by grouping.} \\ &= \frac{3+r}{5-r} && \textit{Lowest terms} \end{aligned}$$

3. **(a)** $\dfrac{y-2}{2-y} = \dfrac{y-2}{-1(y-2)} = \dfrac{1}{-1} = -1$

(b) $\dfrac{8-b}{8+b}$

Since this expression cannot be simplified further, it is in lowest terms.

(c)
$$\begin{aligned} \frac{p-2}{4-p^2} &= \frac{p-2}{(2-p)(2+p)} && \textit{Factor.} \\ &= \frac{p-2}{-1(p-2)(2+p)} \\ &= \frac{-1}{2+p} && \textit{Fundamental property} \end{aligned}$$

4. **(a)**
$$\begin{aligned} \frac{2r+4}{5r} \cdot \frac{3r}{5r+10} &= \frac{2(r+2)}{5r} \cdot \frac{3r}{5(r+2)} \\ &= \frac{2 \cdot 3}{5 \cdot 5} = \frac{6}{25} \end{aligned}$$

(b) $\frac{c^2+2c}{c^2-4} \cdot \frac{c^2-4c+4}{c^2-c}$

$= \frac{c(c+2)}{(c-2)(c+2)} \cdot \frac{(c-2)(c-2)}{c(c-1)}$

$= \frac{c-2}{c-1}$

(c) $\frac{m^2-16}{m+2} \cdot \frac{1}{m+4}$

$= \frac{(m-4)(m+4)}{m+2} \cdot \frac{1}{m+4}$

$= \frac{m-4}{m+2}$

(d) $\frac{x-3}{x^2+2x-15} \cdot \frac{x^2-25}{x^2+3x-40}$

$= \frac{x-3}{(x+5)(x-3)} \cdot \frac{(x-5)(x+5)}{(x+8)(x-5)}$

$= \frac{1}{x+8}$

5. To find the reciprocal, interchange the numerator and denominator of the rational expression.

	Expression	Reciprocal
(a)	$\frac{-3}{r}$	$\frac{r}{-3}$
(b)	$\frac{7}{y+8}$	$\frac{y+8}{7}$
(c)	$\frac{a^2+7a}{2a-1}$	$\frac{2a-1}{a^2+7a}$
(d)	$\frac{0}{-5}$	Undefined; there is no reciprocal for 0.

6. In each case, multiply by the reciprocal of the divisor.

(a) $\frac{16k^2}{5} \div \frac{3k}{10} = \frac{16k^2}{5} \cdot \frac{10}{3k}$

$= \frac{16k^2}{5} \cdot \frac{2 \cdot 5}{3k} = \frac{32k}{3}$

(b) $\frac{5p+2}{6} \div \frac{15p+6}{5} = \frac{5p+2}{6} \cdot \frac{5}{15p+6}$

$= \frac{5p+2}{6} \cdot \frac{5}{3(5p+2)}$

$= \frac{5}{18}$

(c) $\frac{y^2-2y-3}{y^2+4y+4} \div \frac{y^2-1}{y^2+y-2}$

$= \frac{y^2-2y-3}{y^2+4y+4} \cdot \frac{y^2+y-2}{y^2-1}$

$= \frac{(y-3)(y+1)}{(y+2)(y+2)} \cdot \frac{(y+2)(y-1)}{(y+1)(y-1)}$

$= \frac{y-3}{y+2}$

8.1 Section Exercises

1. $\frac{x-3}{x+4} = \frac{(-1)(x-3)}{(-1)(x+4)}$

$= \frac{-x+3}{-x-4}$, or $\frac{3-x}{-x-4}$ **(C)**

3. $\frac{x-3}{x-4} = \frac{(-1)(x-3)}{(-1)(x-4)}$

$= \frac{-x+3}{-x+4}$ **(D)**

5. $\frac{3-x}{x+4} = \frac{(-1)(3-x)}{(-1)(x+4)}$

$= \frac{-3+x}{-x-4}$, or $\frac{x-3}{-x-4}$ **(E)**

7. Replacing x with 2 makes the denominator 0 and the value of the expression undefined. To find the values excluded from the domain, set the denominator equal to 0 and solve the equation. All solutions of the equation are excluded from the domain.

9. $f(x) = \frac{x}{x-7}$

Set the denominator equal to zero, and solve the equation.

$$x - 7 = 0$$
$$x = 7$$

The number 7 makes the rational expression undefined, so 7 is not in the domain of the function. In set notation, the domain is $\{x \mid x \neq 7\}$.

11. $f(x) = \frac{6x-5}{7x+1}$

Set the denominator equal to zero, and solve the equation.

$$7x + 1 = 0$$
$$7x = -1$$
$$x = -\tfrac{1}{7}$$

The number $-\frac{1}{7}$ makes the rational expression undefined, so $-\frac{1}{7}$ is not in the domain of the function. In set notation, the domain is $\{x \mid x \neq -\frac{1}{7}\}$.

13. $f(x)=\dfrac{12x+3}{x}$

Set the denominator equal to zero and solve.

$$x=0$$

The number 0 makes the rational expression undefined, so 0 is not in the domain of the function. In set notation, the domain is $\{x\,|\,x\neq 0\}$.

15. $f(x)=\dfrac{3x+1}{2x^2+x-6}$

Set the denominator equal to zero and solve.

$$2x^2+x-6=0$$
$$(x+2)(2x-3)=0$$
$$x+2=0 \quad \text{or} \quad 2x-3=0$$
$$x=-2 \quad \text{or} \quad 2x=3$$
$$x=\tfrac{3}{2}$$

The numbers -2 and $\frac{3}{2}$ are not in the domain of the function. In set notation, the domain is $\{x\,|\,x\neq -2, \frac{3}{2}\}$.

17. $f(x)=\dfrac{x+2}{14}$

The denominator is never zero, so all numbers are in the domain of the function. In set notation, the domain is $(-\infty,\infty)$.

19. $f(x)=\dfrac{2x^2-3x+4}{3x^2+8}$

Set the denominator equal to zero and solve.

$$3x^2+8=0$$
$$3x^2=-8$$
$$x^2=-\tfrac{8}{3}$$

The square of any real number x is positive or zero, so this equation has no solution. There are no real numbers which make this rational expression undefined, so all numbers are in the domain of the function. In set notation, the domain is $(-\infty,\infty)$.

21. **(a)** $\dfrac{x^2+4x}{x+4}$

The two terms in the numerator are x^2 and $4x$. The two terms in the denominator are x and 4.

(b) First factor the numerator, getting $x(x+4)$, then divide the numerator and denominator by the common factor of $x+4$ to get $\frac{x}{1}$, or x.

$$\frac{x^2+4x}{x+4}=\frac{x(x+4)}{x+4}$$
$$=x\cdot 1=x$$

23. **A.** $\dfrac{3-x}{x-4}=\dfrac{-1(x-3)}{-1(4-x)}=\dfrac{x-3}{4-x}$

B. $\dfrac{x+3}{4+x}$ cannot be transformed to equal $\dfrac{x-3}{4-x}$.

C. $-\dfrac{3-x}{4-x}=\dfrac{-(3-x)}{4-x}=\dfrac{x-3}{4-x}$

D. $-\dfrac{x-3}{x-4}=\dfrac{x-3}{-(x-4)}=\dfrac{x-3}{4-x}$

Only the expression in **B** is *not* equivalent to $\dfrac{x-3}{4-x}$.

25. $\dfrac{x^2(x+1)}{x(x+1)}=\dfrac{x}{1}\cdot\dfrac{x(x+1)}{x(x+1)}=\dfrac{x}{1}\cdot 1=x$

27. $\dfrac{(x+4)(x-3)}{(x+5)(x+4)}=\dfrac{x-3}{x+5}\cdot\dfrac{x+4}{x+4}=\dfrac{x-3}{x+5}$

29. $\dfrac{4x(x+3)}{8x^2(x-3)}=\dfrac{(x+3)\cdot 4x}{2x(x-3)\cdot 4x}$

$$=\frac{x+3}{2x(x-3)}$$

31. $\dfrac{3x+7}{3}$ Since the numerator and denominator have no common factors, the expression is already in lowest terms.

33. $\dfrac{6m+18}{7m+21}=\dfrac{6(m+3)}{7(m+3)}=\dfrac{6}{7}$

35. $\dfrac{3z^2+z}{18z+6}=\dfrac{z(3z+1)}{6(3z+1)}=\dfrac{z}{6}$

37. $\dfrac{2t+6}{t^2-9}=\dfrac{2(t+3)}{(t-3)(t+3)}=\dfrac{2}{t-3}$

39. $\dfrac{x^2+2x-15}{x^2+6x+5}=\dfrac{(x+5)(x-3)}{(x+5)(x+1)}$

$$=\frac{x-3}{x+1}$$

41. $\dfrac{8x^2-10x-3}{8x^2-6x-9}=\dfrac{(4x+1)(2x-3)}{(4x+3)(2x-3)}$

$$=\frac{4x+1}{4x+3}$$

43. $\dfrac{a^3+b^3}{a+b}=\dfrac{(a+b)(a^2-ab+b^2)}{a+b}$

$$=a^2-ab+b^2$$

45. $\dfrac{2c^2+2cd-60d^2}{2c^2-12cd+10d^2}$

$$=\frac{2(c^2+cd-30d^2)}{2(c^2-6cd+5d^2)}$$
$$=\frac{2(c+6d)(c-5d)}{2(c-d)(c-5d)}$$
$$=\frac{c+6d}{c-d}$$

47. $\dfrac{ac-ad+bc-bd}{ac-ad-bc+bd}$
$=\dfrac{a(c-d)+b(c-d)}{a(c-d)-b(c-d)}$
$=\dfrac{(c-d)(a+b)}{(c-d)(a-b)}$ *Factor by grouping.*
$=\dfrac{a+b}{a-b}$

49. $\dfrac{7-b}{b-7}=\dfrac{-1(b-7)}{b-7}=-1$

In Exercises 51–56, there are other acceptable ways to express each answer.

51. $\dfrac{x^2-y^2}{y-x}=\dfrac{(x-y)(x+y)}{y-x}$
$=\dfrac{-1(y-x)(x+y)}{y-x}$
$=-(x+y)$

53. $\dfrac{(a-3)(x+y)}{(3-a)(x-y)}=\dfrac{(a-3)(x+y)}{-1(a-3)(x-y)}$
$=\dfrac{x+y}{-1(x-y)}$
$=-\dfrac{x+y}{x-y}$

55. $\dfrac{5k-10}{20-10k}=\dfrac{-5(2-k)}{10(2-k)}=\dfrac{-5}{10}=-\dfrac{1}{2}$

57. $\dfrac{a^2-b^2}{a^2+b^2}=\dfrac{(a+b)(a-b)}{a^2+b^2}$

The numerator and denominator have no common factors except 1, so the original expression is already in lowest terms.

59. $\dfrac{(x+2)(x+1)}{(x+3)(x-2)}\cdot\dfrac{(x+3)(x+4)}{(x+2)(x+1)}$
$=\dfrac{(x+1)(x+2)(x+3)}{(x+1)(x+2)(x+3)}\cdot\dfrac{x+4}{x-2}=\dfrac{x+4}{x-2}$

61. $\dfrac{(2x+3)(x-4)}{(x+8)(x-4)}\div\dfrac{(x-4)(x+2)}{(x-4)(x+8)}$
$=\dfrac{2x+3}{x+8}\div\dfrac{x+2}{x+8}$
$=\dfrac{2x+3}{x+8}\cdot\dfrac{x+8}{x+2}$
$=\dfrac{2x+3}{x+2}$

63. $\dfrac{7t+7}{-6}\div\dfrac{4t+4}{15}=\dfrac{7t+7}{-6}\cdot\dfrac{15}{4t+4}$
$=\dfrac{7(t+1)}{-2\cdot 3}\cdot\dfrac{3\cdot 5}{4(t+1)}$
$=-\dfrac{35}{8}$

65. $\dfrac{4x}{8x+4}\cdot\dfrac{14x+7}{6}=\dfrac{4x\cdot 7(2x+1)}{4(2x+1)\cdot 6}$
$=\dfrac{7x}{6}$

For Exercises 67–70, there are several other ways to express the answer.

67. $\dfrac{p^2-25}{4p}\cdot\dfrac{2}{5-p}=\dfrac{(p+5)(p-5)2}{2\cdot 2p(-1)(p-5)}$
$=\dfrac{p+5}{(2p)(-1)}=-\dfrac{p+5}{2p}$

69. $\dfrac{m^2-49}{m+1}\div\dfrac{7-m}{m}$
$=\dfrac{(m-7)(m+7)}{m+1}\cdot\dfrac{m}{7-m}$
$=\dfrac{(-1)(7-m)(m+7)}{m+1}\cdot\dfrac{m}{7-m}$
$=\dfrac{-m(m+7)}{m+1}$

71. $\dfrac{12x-10y}{3x+2y}\cdot\dfrac{6x+4y}{10y-12x}$
$=\dfrac{2(6x-5y)\cdot 2(3x+2y)}{(3x+2y)\cdot 2(5y-6x)}$
$=\dfrac{2(-1)(5y-6x)}{(5y-6x)}=-2$

73. $\dfrac{x^2-25}{x^2+x-20}\cdot\dfrac{x^2+7x+12}{x^2-2x-15}$
$=\dfrac{(x-5)(x+5)}{(x+5)(x-4)}\cdot\dfrac{(x+3)(x+4)}{(x-5)(x+3)}$
$=\dfrac{x+4}{x-4}$

75. $\dfrac{6x^2+5xy-6y^2}{12x^2-11xy+2y^2}\div\dfrac{4x^2-12xy+9y^2}{8x^2-14xy+3y^2}$
$=\dfrac{(3x-2y)(2x+3y)}{(3x-2y)(4x-y)}\cdot\dfrac{(4x-y)(2x-3y)}{(2x-3y)(2x-3y)}$
$=\dfrac{2x+3y}{2x-3y}$

77. $\dfrac{3k^2+17kp+10p^2}{6k^2+13kp-5p^2}\div\dfrac{6k^2+kp-2p^2}{6k^2-5kp+p^2}$
$=\dfrac{(3k+2p)(k+5p)}{(3k-p)(2k+5p)}\cdot\dfrac{(3k-p)(2k-p)}{(3k+2p)(2k-p)}$
$=\dfrac{k+5p}{2k+5p}$

79. $\left(\dfrac{6k^2-13k-5}{k^2+7k} \div \dfrac{2k-5}{k^3+6k^2-7k}\right) \cdot \dfrac{k^2-5k+6}{3k^2-8k-3}$

Factor k from the denominator of the divisor; multiply by the reciprocal.

$$= \left[\frac{6k^2-13k-5}{k^2+7k} \cdot \frac{k(k^2+6k-7)}{2k-5}\right] \cdot \frac{k^2-5k+6}{3k^2-8k-3}$$

$$= \left[\frac{(3k+1)(2k-5)}{k(k+7)} \cdot \frac{k(k+7)(k-1)}{2k-5}\right] \cdot \frac{(k-2)(k-3)}{(3k+1)(k-3)}$$

$$= (k-1)(k-2)$$

8.2 Adding and Subtracting Rational Expressions

8.2 Margin Exercises

1. **(a)** $\dfrac{3m}{8} + \dfrac{5n}{8} = \dfrac{3m+5n}{8}$

(b) $\dfrac{7}{3a} + \dfrac{10}{3a} = \dfrac{7+10}{3a} = \dfrac{17}{3a}$

(c) $\dfrac{2}{y^2} - \dfrac{5}{y^2} = \dfrac{2-5}{y^2} = \dfrac{-3}{y^2} = -\dfrac{3}{y^2}$

(d) $\dfrac{a}{a+b} + \dfrac{b}{a+b} = \dfrac{a+b}{a+b} = 1$

(e) $\dfrac{2y-1}{y^2+y-2} - \dfrac{y}{y^2+y-2} = \dfrac{(2y-1)-y}{y^2+y-2}$

$$= \frac{y-1}{(y+2)(y-1)}$$

$$= \frac{1}{y+2}$$

2. **(a)** $5k^3s, 10ks^4$

Factor the denominators.

$$5k^3s = 5 \cdot k^3 \cdot s$$
$$10ks^4 = 2 \cdot 5 \cdot k \cdot s^4$$

The least common denominator (LCD) is the product of all the different factors, with each factor raised to the greatest power in any denominator.

$$\text{LCD} = 2 \cdot 5 \cdot k^3 \cdot s^4 = 10k^3s^4$$

(b) $3-x, 9-x^2$

Factor the denominators.

$$3-x = 3-x$$
$$9-x^2 = (3+x)(3-x)$$

The least common denominator (LCD) is the product of all the different factors, with each factor raised to the greatest power in any denominator.

$\text{LCD} = (3+x)(3-x)$

(c) $z, z+6$

Each denominator is already factored.

$\text{LCD} = z(z+6)$

(d) $2y^2-3y-2, 2y^2+3y+1$

Factor the denominators.

$$2y^2-3y-2 = (2y+1)(y-2)$$
$$2y^2+3y+1 = (2y+1)(y+1)$$

$\text{LCD} = (y-2)(2y+1)(y+1)$

(e) $x^2-2x+1, x^2-4x+3, 4x-4$

Factor the denominators.

$$x^2-2x+1 = (x-1)(x-1)$$
$$x^2-4x+3 = (x-1)(x-3)$$
$$4x-4 = 4(x-1)$$

$\text{LCD} = 4(x-1)^2(x-3)$

3. **(a)** $\dfrac{6}{7} + \dfrac{1}{5}$

The LCD of 7 and 5 is 35. Multiply $\frac{6}{7}$ by $\frac{5}{5}$ and $\frac{1}{5}$ by $\frac{7}{7}$ so that each fraction has denominator 35. Then add the numerators.

$$\frac{6}{7} + \frac{1}{5} = \frac{6 \cdot 5}{7 \cdot 5} + \frac{1 \cdot 7}{5 \cdot 7} = \frac{30}{35} + \frac{7}{35} = \frac{30+7}{35} = \frac{37}{35}$$

(b) $\dfrac{8}{3k} - \dfrac{2}{9k}$

The LCD for $3k$ and $9k$ is $9k$. To write $\dfrac{8}{3k}$ with a denominator of $9k$, multiply by $\frac{3}{3}$.

$$\frac{8}{3k} - \frac{2}{9k} = \frac{8 \cdot 3}{3k \cdot 3} - \frac{2}{9k} = \frac{24}{9k} - \frac{2}{9k} = \frac{24-2}{9k} = \frac{22}{9k}$$

(c) $\frac{2}{y} - \frac{1}{y+4}$

The LCD is $y(y+4)$. Rewrite each rational expression with this denominator.

$$\begin{aligned}\frac{2}{y} - \frac{1}{y+4} &= \frac{2(y+4)}{y(y+4)} - \frac{(1)y}{(y+4)y} \\ &= \frac{2(y+4)}{y(y+4)} - \frac{y}{y(y+4)} \\ &= \frac{2y+8-y}{y(y+4)} \\ &= \frac{y+8}{y(y+4)}\end{aligned}$$

4. (a) $\frac{5x+7}{2x+7} - \frac{-x-14}{2x+7}$

The denominator is already the same for both rational expressions, so write them as a single expression. Be careful to apply the subtraction sign to both terms of the second expression.

$$\begin{aligned}&= \frac{5x+7-(-x-14)}{2x+7} \\ &= \frac{5x+7+x+14}{2x+7} \\ &= \frac{6x+21}{2x+7} \\ &= \frac{3(2x+7)}{2x+7} \\ &= 3\end{aligned}$$

(b) $\frac{2}{r-2} - \frac{r}{r-1}$

The LCD is $(r-2)(r-1)$.

$$\begin{aligned}&= \frac{2(r-1)}{(r-2)(r-1)} - \frac{r(r-2)}{(r-1)(r-2)} \\ &= \frac{2r-2-(r^2-2r)}{(r-2)(r-1)} \\ &= \frac{2r-2-r^2+2r}{(r-2)(r-1)} \\ &= \frac{-r^2+4r-2}{(r-2)(r-1)}\end{aligned}$$

5. (a) $\frac{8}{x-4} + \frac{2}{4-x}$

To get a common denominator of $x-4$, multiply both the numerator and denominator of the second expression by -1.

$$\begin{aligned}&= \frac{8}{x-4} + \frac{2(-1)}{(4-x)(-1)} \\ &= \frac{8}{x-4} + \frac{-2}{x-4} \\ &= \frac{8+(-2)}{x-4} \\ &= \frac{6}{x-4}\end{aligned}$$

An equivalent answer is

$$\frac{-1(6)}{-1(x-4)} = \frac{-6}{4-x}.$$

(b) $\frac{9}{2x-9} - \frac{4}{9-2x}$

$$\begin{aligned}&= \frac{9}{2x-9} - \frac{4(-1)}{(9-2x)(-1)} \\ &= \frac{9}{2x-9} - \frac{-4}{2x-9} \\ &= \frac{9-(-4)}{2x-9} \\ &= \frac{13}{2x-9}\end{aligned}$$

An equivalent answer is

$$\frac{-1(13)}{-1(2x-9)} = \frac{-13}{9-2x}.$$

6. $\frac{4}{x-5} + \frac{-2}{x} - \frac{10}{x^2-5x}$

$$= \frac{4}{x-5} + \frac{-2}{x} - \frac{10}{x(x-5)}$$

Factor denominators.

$$= \frac{4x}{(x-5)x} + \frac{-2(x-5)}{x(x-5)} - \frac{10}{x(x-5)}$$

LCD = x(x – 5)

$$\begin{aligned}&= \frac{4x+(-2)(x-5)-10}{x(x-5)} \\ &= \frac{4x-2x+10-10}{x(x-5)} \\ &= \frac{2x}{x(x-5)} \\ &= \frac{2}{x-5}\end{aligned}$$

7. $\frac{-a}{a^2+3a-4} - \frac{4a}{a^2+7a+12}$

$= \frac{-a}{(a+4)(a-1)} - \frac{4a}{(a+3)(a+4)}$

The LCD is $(a+4)(a-1)(a+3)$.

$= \frac{-a(a+3)}{(a+4)(a-1)(a+3)}$

$- \frac{4a(a-1)}{(a+3)(a+4)(a-1)}$

$= \frac{-a(a+3)-4a(a-1)}{(a+4)(a-1)(a+3)}$

$= \frac{-a^2-3a-4a^2+4a}{(a+4)(a-1)(a+3)}$

$= \frac{-5a^2+a}{(a+4)(a-1)(a+3)}$

8. $\frac{4}{p^2-6p+9} + \frac{1}{p^2+2p-15}$

$= \frac{4}{(p-3)(p-3)} + \frac{1}{(p+5)(p-3)}$

The LCD is $(p-3)^2(p+5)$.

$= \frac{4(p+5)}{(p-3)^2(p+5)} + \frac{1(p-3)}{(p+5)(p-3)^2}$

$= \frac{4(p+5)+1(p-3)}{(p-3)^2(p+5)}$

$= \frac{4p+20+p-3}{(p-3)^2(p+5)}$

$= \frac{5p+17}{(p-3)^2(p+5)}$

8.2 Section Exercises

1. To add or subtract rational expressions that have a common denominator, first add or subtract the numerators. Then place the result over the common denominator. Write the answer in lowest terms.

3. $\frac{7}{t} + \frac{2}{t} = \frac{7+2}{t} = \frac{9}{t}$

5. $\frac{11}{5x} - \frac{1}{5x} = \frac{11-1}{5x} = \frac{10}{5x} = \frac{2}{x}$

7. $\frac{5x+4}{6x+5} + \frac{x+1}{6x+5}$

$= \frac{5x+4+x+1}{6x+5}$

$= \frac{6x+5}{6x+5} = 1$

9. $\frac{x^2}{x+5} - \frac{25}{x+5} = \frac{x^2-25}{x+5}$

$= \frac{(x+5)(x-5)}{x+5}$

$= x-5$

11. $\frac{4}{p^2+7p+12} + \frac{p}{p^2+7p+12}$

$= \frac{4+p}{p^2+7p+12}$

$= \frac{p+4}{(p+3)(p+4)}$

$= \frac{1}{p+3}$

13. $\frac{a^3}{a^2+ab+b^2} - \frac{b^3}{a^2+ab+b^2}$

$= \frac{a^3-b^3}{a^2+ab+b^2}$

$= \frac{(a-b)(a^2+ab+b^2)}{a^2+ab+b^2}$

$= a-b$

15. $18x^2y^3$, $24x^4y^5$

Factor each denominator.

$$18x^2y^3 = 2\cdot 3\cdot 3\cdot x^2\cdot y^3 = 2\cdot 3^2\cdot x^2\cdot y^3$$

$$24x^4y^5 = 2\cdot 2\cdot 2\cdot 3\cdot x^4\cdot y^5 = 2^3\cdot 3\cdot x^4\cdot y^5$$

The least common denominator (LCD) is the product of all the different factors, with each factor raised to the greatest power in any denominator.

$$\text{LCD} = 2^3\cdot 3^2\cdot x^4\cdot y^5 = 8\cdot 9\cdot x^4\cdot y^5 = 72x^4y^5$$

The LCD is $72x^4y^5$.

17. $z-2$, z

Both $z-2$ and z have only 1 and themselves for factors.

$\text{LCD} = z(z-2)$

19. $2y+8$, $y+4$

Factor each denominator.

$$2y+8 = 2(y+4)$$

The second denominator, $y+4$, is already factored. The LCD is

$$2(y+4).$$

21. x^2-81, $x^2+18x+81$

Factor each denominator.

$$x^2-81 = (x+9)(x-9)$$
$$x^2+18x+81 = (x+9)(x+9)$$

$\text{LCD} = (x+9)^2(x-9)$

23. $m+n,\ m-n,\ m^2-n^2$

Both $m+n$ and $m-n$ have only 1 and themselves for factors, while $m^2-n^2=(m+n)(m-n)$.

$\text{LCD}=(m+n)(m-n)$

25. $x^2-3x-4,\ x+x^2$

Factor each denominator.

$$x^2-3x-4=(x-4)(x+1)$$
$$x+x^2=x(1+x)=x(x+1)$$

The LCD is $x(x-4)(x+1)$.

27. $2t^2+7t-15,\ t^2+3t-10$

Factor each denominator.

$$2t^2+7t-15=(2t-3)(t+5)$$
$$t^2+3t-10=(t+5)(t-2)$$

The LCD is $(2t-3)(t+5)(t-2)$.

29. $2y+6,\ y^2-9,\ y$

Factor each denominator.

$$2y+6=2(y+3)$$
$$y^2-9=(y+3)(y-3)$$

Remember the factor y from the third denominator. The LCD is

$$2y(y+3)(y-3).$$

31. Yes, they could both be correct because the expressions are equivalent. Multiplying $\frac{3}{5-y}$ by 1 in the form $\frac{-1}{-1}$ gives $\frac{-3}{y-5}$.

33. $\frac{8}{t}+\frac{7}{3t}$ LCD = 3t

$$=\frac{8\cdot 3}{t\cdot 3}+\frac{7}{3t}$$
$$=\frac{24+7}{3t}$$
$$=\frac{31}{3t}$$

35. $\frac{5}{12x^2y}-\frac{11}{6xy}$ LCD = $12x^2y$

$$=\frac{5}{12x^2y}-\frac{11\cdot 2x}{6xy\cdot 2x}$$
$$=\frac{5}{12x^2y}-\frac{22x}{12x^2y}$$
$$=\frac{5-22x}{12x^2y}$$

37. $\frac{1}{x-1}-\frac{1}{x}$ LCD = $x(x-1)$

$$=\frac{1\cdot x}{(x-1)x}-\frac{1\cdot(x-1)}{x(x-1)}$$
$$=\frac{x-(x-1)}{x(x-1)}$$
$$=\frac{x-x+1}{x(x-1)}$$
$$=\frac{1}{x(x-1)}$$

39. $\frac{3a}{a+1}+\frac{2a}{a-3}$ LCD = $(a+1)(a-3)$

$$=\frac{3a(a-3)}{(a+1)(a-3)}+\frac{2a(a+1)}{(a-3)(a+1)}$$
$$=\frac{3a(a-3)+2a(a+1)}{(a+1)(a-3)}$$
$$=\frac{3a^2-9a+2a^2+2a}{(a+1)(a-3)}$$
$$=\frac{5a^2-7a}{(a+1)(a-3)}$$

41. $\frac{17y+3}{9y+7}-\frac{-10y-18}{9y+7}$

$$=\frac{17y+3-(-10y-18)}{9y+7}$$
$$=\frac{17y+3+10y+18}{9y+7}$$
$$=\frac{27y+21}{9y+7}$$
$$=\frac{3(9y+7)}{9y+7}$$
$$=3$$

43. $\frac{2}{4-x}+\frac{5}{x-4}$

To get a common denominator of $x-4$, multiply both the numerator and denominator of the first expression by -1.

$$=\frac{(2)(-1)}{(4-x)(-1)}+\frac{5}{x-4}$$
$$=\frac{-2}{x-4}+\frac{5}{x-4}$$
$$=\frac{-2+5}{x-4}$$
$$=\frac{3}{x-4}$$

If you chose $4-x$ for the LCD, then you should have obtained the equivalent answer, $\frac{-3}{4-x}$.

45. $\dfrac{w}{w-z}-\dfrac{z}{z-w}$

$w-z$ and $z-w$ are opposites, so factor out -1 from $z-w$ to get a common denominator.

$$=\frac{w}{w-z}-\frac{z}{-1(w-z)}$$
$$=\frac{w}{w-z}+\frac{z}{w-z}$$
$$=\frac{w+z}{w-z},\text{ or }\frac{-w-z}{z-w}$$

47. $\dfrac{5}{12+4x}-\dfrac{7}{9+3x}$

$$=\frac{5}{4(3+x)}-\frac{7}{3(3+x)}$$

LCD = 3 • 4(3 + x)

$$=\frac{5\cdot 3}{4(3+x)\cdot 3}-\frac{7\cdot 4}{3(3+x)\cdot 4}$$
$$=\frac{15-28}{12(3+x)}$$
$$=\frac{-13}{12(3+x)}$$

49. $\dfrac{4x}{x-1}-\dfrac{2}{x+1}-\dfrac{4}{x^2-1}$

$$=\frac{4x(x+1)}{(x-1)(x+1)}-\frac{2(x-1)}{(x+1)(x-1)}$$
$$-\frac{4}{(x+1)(x-1)}$$

The LCD is $(x+1)(x-1)$.

$$=\frac{4x(x+1)-2(x-1)-4}{(x+1)(x-1)}$$
$$=\frac{4x^2+4x-2x+2-4}{(x-1)(x+1)}$$
$$=\frac{4x^2+2x-2}{(x-1)(x+1)}$$
$$=\frac{2(2x^2+x-1)}{(x-1)(x+1)}$$
$$=\frac{2(2x-1)(x+1)}{(x-1)(x+1)}$$
$$=\frac{2(2x-1)}{x-1}$$

51. $\dfrac{15}{y^2+3y}+\dfrac{2}{y}+\dfrac{5}{y+3}$

$$=\frac{15}{y(y+3)}+\frac{2(y+3)}{y(y+3)}+\frac{5y}{(y+3)y}$$

The LCD is $y(y+3)$.

$$=\frac{15+2(y+3)+5y}{y(y+3)}$$
$$=\frac{15+2y+6+5y}{y(y+3)}$$
$$=\frac{7y+21}{y(y+3)}$$
$$=\frac{7(y+3)}{y(y+3)}=\frac{7}{y}$$

53. $\dfrac{5}{x-2}+\dfrac{1}{x}+\dfrac{2}{x^2-2x}$

$$=\frac{5x}{(x-2)x}+\frac{1(x-2)}{x(x-2)}+\frac{2}{x(x-2)}$$

The LCD is $x(x-2)$.

$$=\frac{5x+x-2+2}{x(x-2)}$$
$$=\frac{6x}{x(x-2)}=\frac{6}{x-2}$$

55. $\dfrac{3x}{x+1}+\dfrac{4}{x-1}-\dfrac{6}{x^2-1}$

$$=\frac{3x(x-1)}{(x+1)(x-1)}+\frac{4(x+1)}{(x-1)(x+1)}$$
$$-\frac{6}{(x+1)(x-1)}$$

The LCD is $(x+1)(x-1)$.

$$=\frac{3x(x-1)+4(x+1)-6}{(x+1)(x-1)}$$
$$=\frac{3x^2-3x+4x+4-6}{(x+1)(x-1)}$$
$$=\frac{3x^2+x-2}{(x+1)(x-1)}$$
$$=\frac{(3x-2)(x+1)}{(x+1)(x-1)}$$
$$=\frac{3x-2}{x-1}$$

57. $\dfrac{4}{x+1}+\dfrac{1}{x^2-x+1}-\dfrac{12}{x^3+1}$

$=\dfrac{4(x^2-x+1)}{(x+1)(x^2-x+1)}$
$+\dfrac{1\cdot(x+1)}{(x^2-x+1)(x+1)}$
$-\dfrac{12}{(x+1)(x^2-x+1)}$

The LCD is $(x+1)(x^2-x+1)$.

$=\dfrac{4(x^2-x+1)+(x+1)-12}{(x+1)(x^2-x+1)}$

$=\dfrac{4x^2-4x+4+x+1-12}{(x+1)(x^2-x+1)}$

$=\dfrac{4x^2-3x-7}{(x+1)(x^2-x+1)}$

$=\dfrac{(4x-7)(x+1)}{(x+1)(x^2-x+1)}$

$=\dfrac{4x-7}{x^2-x+1}$

59. $\dfrac{2x+4}{x+3}+\dfrac{3}{x}-\dfrac{6}{x^2+3x}$

$=\dfrac{(2x+4)x}{(x+3)x}+\dfrac{3(x+3)}{x(x+3)}-\dfrac{6}{x(x+3)}$

The LCD is $x(x+3)$.

$=\dfrac{(2x+4)x+3(x+3)-6}{x(x+3)}$

$=\dfrac{2x^2+4x+3x+9-6}{x(x+3)}$

$=\dfrac{2x^2+7x+3}{x(x+3)}$

$=\dfrac{(2x+1)(x+3)}{x(x+3)}=\dfrac{2x+1}{x}$

61. $\dfrac{3}{x^2-5x+6}-\dfrac{2}{x^2-4x+4}$

$=\dfrac{3}{(x-2)(x-3)}-\dfrac{2}{(x-2)(x-2)}$

The LCD is $(x-2)^2(x-3)$.

$=\dfrac{3(x-2)}{(x-2)(x-3)(x-2)}$
$-\dfrac{2(x-3)}{(x-2)(x-2)(x-3)}$

$=\dfrac{3x-6-2x+6}{(x-2)^2(x-3)}$

$=\dfrac{x}{(x-2)^2(x-3)}$

63. $\dfrac{3}{x^2+4x+4}+\dfrac{7}{x^2+5x+6}$

$=\dfrac{3}{(x+2)^2}+\dfrac{7}{(x+2)(x+3)}$

The LCD is $(x+2)^2(x+3)$.

$=\dfrac{3(x+3)}{(x+2)^2(x+3)}+\dfrac{7(x+2)}{(x+2)^2(x+3)}$

$=\dfrac{3x+9+7x+14}{(x+2)^2(x+3)}$

$=\dfrac{10x+23}{(x+2)^2(x+3)}$

65. $\dfrac{5x}{x^2+xy-2y^2}-\dfrac{3x}{x^2+5xy-6y^2}$

Factor each denominator.

$$x^2+xy-2y^2=(x+2y)(x-y)$$
$$x^2+5xy-6y^2=(x+6y)(x-y)$$

The LCD is $(x+2y)(x-y)(x+6y)$.

$\dfrac{5x}{(x+2y)(x-y)}-\dfrac{3x}{(x+6y)(x-y)}$

$=\dfrac{(5x)(x+6y)}{(x+2y)(x-y)(x+6y)}$
$-\dfrac{(3x)(x+2y)}{(x+6y)(x-y)(x+2y)}$

$=\dfrac{(5x)(x+6y)-(3x)(x+2y)}{(x+6y)(x-y)(x+2y)}$

$=\dfrac{5x^2+30xy-(3x^2+6xy)}{(x+2y)(x-y)(x+6y)}$

$=\dfrac{2x^2+24xy}{(x+2y)(x-y)(x+6y)}$

$=\dfrac{2x(x+12y)}{(x+2y)(x-y)(x+6y)}$

67. $c(x)=\dfrac{1010}{49(101-x)}-\dfrac{10}{49}$

$=\dfrac{1010}{49(101-x)}-\dfrac{10(101-x)}{49(101-x)}$

$=\dfrac{1010-1010+10x}{49(101-x)}$

$=\dfrac{10x}{49(101-x)}$

69. $\dfrac{3}{7}+\dfrac{5}{9}-\dfrac{6}{63}$

The LCD is $7(9)=63$.

$=\dfrac{3\cdot9}{7\cdot9}+\dfrac{5\cdot7}{9\cdot7}-\dfrac{6}{63}$

$=\dfrac{27+35-6}{63}$

$=\dfrac{56}{63}=\dfrac{8}{9}$

70. From Example 6,

$$\frac{3}{x-2}+\frac{5}{x}-\frac{6}{x^2-2x}.$$

Substitute 9 for x.

$$\frac{3}{9-2}+\frac{5}{9}-\frac{6}{9^2-2(9)}=\frac{3}{7}+\frac{5}{9}-\frac{6}{63}$$

The problems in Exercises 69 and 70 are the same.

71. From Exercise 69,

$$\frac{3}{7}+\frac{5}{9}-\frac{6}{63}=\frac{8}{9}.$$

From Example 6, the answer is $\frac{8}{x}$. If we substitute 9 for x, the answer becomes $\frac{8}{9}$. The answers agree.

72. Answers will vary. For example, suppose the last name is Sosa so that $x = 4$. The problem in Example 6 becomes

$$\frac{3}{4-2}+\frac{5}{4}-\frac{6}{4^2-2(4)}=\frac{3}{2}+\frac{5}{4}-\frac{6}{8}.$$

The predicted answer is

$$\frac{8}{x}, \text{ or } \frac{8}{4}=2.$$

Perform the operations to verify our prediction.

$$\begin{aligned}\frac{3}{2}+\frac{5}{4}-\frac{6}{8}&=\frac{3\cdot 4}{2\cdot 4}+\frac{5\cdot 2}{4\cdot 2}-\frac{6}{8}\\&=\frac{12+10-6}{8}\\&=\frac{16}{8}=2\end{aligned}$$

The prediction is correct.

73. If $x = 2$, then the problem from Example 6,

$$\frac{3}{x-2}+\frac{5}{x}-\frac{6}{x^2-2x},$$

becomes

$$\frac{3}{2-2}+\frac{5}{2}-\frac{6}{2^2-2(2)}=\frac{3}{0}+\frac{5}{2}-\frac{6}{0}.$$

Thus, if $x = 2$, then

$$\frac{3}{x-2} \text{ and } \frac{6}{x^2-2x}$$

are undefined.

74. If $x = 0$, then

$$\frac{5}{x}=\frac{5}{0} \text{ and } \frac{6}{x^2-2x}=\frac{6}{0},$$

which are undefined. Therefore, 0 is not allowed as a value of x.

8.3 Complex Fractions

8.3 Margin Exercises

1. (a) $$\frac{\frac{a+2}{5a}}{\frac{a-3}{7a}}$$

Both the numerator and denominator are already simplified.

$$\begin{aligned}&=\frac{a+2}{5a}\div\frac{a-3}{7a}\\&=\frac{a+2}{5a}\cdot\frac{7a}{a-3} \quad \textit{Multiply by the reciprocal of the denominator.}\\&=\frac{7(a+2)}{5(a-3)}\end{aligned}$$

(b) $$\frac{2+\frac{1}{k}}{2-\frac{1}{k}}$$

$$\begin{aligned}&=\frac{\frac{2k}{k}+\frac{1}{k}}{\frac{2k}{k}-\frac{1}{k}} \quad \textit{Simplify the numerator and denominator.}\\&=\frac{\frac{2k+1}{k}}{\frac{2k-1}{k}}\\&=\frac{2k+1}{k}\cdot\frac{k}{2k-1} \quad \textit{Multiply by the reciprocal of the denominator.}\\&=\frac{2k+1}{2k-1}\end{aligned}$$

(c) $$\frac{\frac{r^2-4}{4}}{1+\frac{2}{r}}$$

$$\begin{aligned}&=\frac{\frac{r^2-4}{4}}{\frac{r+2}{r}} \quad \textit{Simplify the denominator.}\\&=\frac{r^2-4}{4}\cdot\frac{r}{r+2} \quad \textit{Multiply by the reciprocal of the denominator.}\\&=\frac{(r-2)(r+2)}{4}\cdot\frac{r}{r+2}\\&=\frac{r(r-2)}{4}\end{aligned}$$

2. **(a)** $\dfrac{\frac{5}{y}+6}{\frac{8}{3y}-1}$

The LCD of $\frac{5}{y}$ and $\frac{8}{3y}$ is $3y$. Multiply the numerator and denominator by $3y$.

$$= \frac{\left(\frac{5}{y}+6\right)\cdot 3y}{\left(\frac{8}{3y}-1\right)\cdot 3y}$$

$$= \frac{\frac{5}{y}\cdot 3y + 6\cdot 3y}{\frac{8}{3y}\cdot 3y - 1\cdot 3y} \quad \textit{Distributive property}$$

$$= \frac{15+18y}{8-3y} \quad \textit{Simplify.}$$

(b) $\dfrac{\frac{1}{y}+\frac{1}{y-1}}{\frac{1}{y}-\frac{2}{y-1}}$

Multiply the numerator and denominator by $y(y-1)$, the LCD of all the fractions.

$$= \frac{\left(\frac{1}{y}+\frac{1}{y-1}\right)\cdot y(y-1)}{\left(\frac{1}{y}-\frac{2}{y-1}\right)\cdot y(y-1)}$$

$$= \frac{\frac{1}{y}\cdot y(y-1) + \frac{1}{y-1}\cdot y(y-1)}{\frac{1}{y}\cdot y(y-1) - \frac{2}{y-1}\cdot y(y-1)}$$

$$= \frac{(y-1)+y}{(y-1)-2y}$$

$$= \frac{2y-1}{-y-1}$$

An equivalent answer is

$$\frac{(-1)(2y-1)}{(-1)(-y-1)} = \frac{1-2y}{y+1}.$$

3. **(a) Method 1**

$$\frac{\frac{5}{y+2}}{\frac{-3}{y^2-4}} = \frac{\frac{5}{y+2}}{\frac{-3}{(y+2)(y-2)}}$$

$$= \frac{5}{y+2} \div \frac{-3}{(y+2)(y-2)}$$

$$= \frac{5}{y+2}\cdot\frac{(y+2)(y-2)}{-3}$$

$$= \frac{5(y-2)}{-3}$$

Method 2

$$\frac{\frac{5}{y+2}}{\frac{-3}{y^2-4}} = \frac{\frac{5}{y+2}}{\frac{-3}{(y+2)(y-2)}}$$

The LCD of the numerator and denominator is $(y+2)(y-2)$.

$$= \frac{\frac{5}{y+2}\cdot (y+2)(y-2)}{\frac{-3}{(y+2)(y-2)}\cdot (y+2)(y-2)}$$

$$= \frac{5(y-2)}{-3}$$

(b) Method 1

$$\frac{\frac{1}{a}-\frac{1}{b}}{\frac{1}{a^2}-\frac{1}{b^2}} = \frac{\frac{b}{ab}-\frac{a}{ab}}{\frac{b^2}{a^2b^2}-\frac{a^2}{a^2b^2}} \quad \begin{matrix} LCD = ab \\ LCD = a^2b^2 \end{matrix}$$

$$= \frac{\frac{b-a}{ab}}{\frac{b^2-a^2}{a^2b^2}}$$

$$= \frac{\frac{b-a}{ab}}{\frac{(b+a)(b-a)}{a^2b^2}}$$

$$= \frac{b-a}{ab} \div \frac{(b+a)(b-a)}{a^2b^2}$$

$$= \frac{b-a}{ab}\cdot\frac{a^2b^2}{(b+a)(b-a)}$$

$$= \frac{ab}{b+a}$$

Method 2

$$\frac{\frac{1}{a}-\frac{1}{b}}{\frac{1}{a^2}-\frac{1}{b^2}}$$

The LCD of the numerator and the denominator is a^2b^2.

$$= \frac{\left(\frac{1}{a}-\frac{1}{b}\right)\cdot a^2b^2}{\left(\frac{1}{a^2}-\frac{1}{b^2}\right)\cdot a^2b^2}$$

$$= \frac{ab^2-a^2b}{b^2-a^2}$$

$$= \frac{ab(b-a)}{(b+a)(b-a)}$$

$$= \frac{ab}{b+a}$$

4. **(a)** $\dfrac{r^{-2}-s^{-1}}{4r^{-1}+s^{-2}} = \dfrac{\dfrac{1}{r^2}-\dfrac{1}{s}}{\dfrac{4}{r}+\dfrac{1}{s^2}}$ $\quad LCD = r^2s^2$

$$= \frac{r^2s^2\left(\frac{1}{r^2}-\frac{1}{s}\right)}{r^2s^2\left(\frac{4}{r}+\frac{1}{s^2}\right)}$$

$$= \frac{r^2s^2\cdot\frac{1}{r^2}-r^2s^2\cdot\frac{1}{s}}{r^2s^2\cdot\frac{4}{r}+r^2s^2\cdot\frac{1}{s^2}}$$

$$= \frac{s^2-r^2s}{4rs^2+r^2}$$

(b) $\dfrac{b^{-4}}{b^{-5}+2} = \dfrac{\dfrac{1}{b^4}}{\dfrac{1}{b^5}+2}$ $\quad LCD = b^5$

$$= \frac{b^5\left(\frac{1}{b^4}\right)}{b^5\left(\frac{1}{b^5}+2\right)}$$

$$= \frac{b^5\cdot\frac{1}{b^4}}{b^5\cdot\frac{1}{b^5}+b^5\cdot 2}$$

$$= \frac{b}{1+2b^5}$$

8.3 Section Exercises

1. Answers will vary. Begin by simplifying the numerator to a single fraction. Then simplify the denominator to a single fraction. Write as a division problem, and multiply by the reciprocal of the denominator. Simplify the result, if possible.

3. $\dfrac{\dfrac{12}{x-1}}{\dfrac{6}{x}} = \dfrac{12}{x-1} \div \dfrac{6}{x}$

Multiply by the reciprocal of the denominator.

$$= \frac{12}{x-1}\cdot\frac{x}{6}$$

$$= \frac{2x}{x-1}$$

5. $\dfrac{\dfrac{k+1}{2k}}{\dfrac{3k-1}{4k}} = \dfrac{k+1}{2k}\cdot\dfrac{4k}{3k-1}$

$$= \frac{4k(k+1)}{2k(3k-1)}$$

$$= \frac{2(k+1)}{3k-1}$$

7. $\dfrac{\dfrac{4z^2x^4}{9}}{\dfrac{12x^2z^5}{15}} = \dfrac{\dfrac{4z^2x^4}{9}}{\dfrac{4x^2z^5}{5}}$

$$= \frac{4z^2x^4}{9} \div \frac{4x^2z^5}{5}$$

$$= \frac{4z^2x^4}{9}\cdot\frac{5}{4x^2z^5}$$

$$= \frac{5z^2x^4}{9x^2z^5} = \frac{5x^2}{9z^3}$$

9. $\dfrac{\dfrac{1}{x}+1}{-\dfrac{1}{x}+1}$

Multiply the numerator and denominator by x, the LCD of all the fractions.

$$= \frac{x\left(\frac{1}{x}+1\right)}{x\left(-\frac{1}{x}+1\right)}$$

$$= \frac{x\cdot\frac{1}{x}+x\cdot 1}{x\left(-\frac{1}{x}\right)+x\cdot 1}$$

$$= \frac{1+x}{-1+x}$$

11. $\dfrac{\dfrac{3}{x}+\dfrac{3}{y}}{\dfrac{3}{x}-\dfrac{3}{y}}$

Multiply the numerator and denominator by xy, the LCD of all the fractions.

$$= \frac{\left(\frac{3}{x}+\frac{3}{y}\right)xy}{\left(\frac{3}{x}-\frac{3}{y}\right)xy}$$

$$= \frac{\frac{3}{x}\cdot xy+\frac{3}{y}\cdot xy}{\frac{3}{x}\cdot xy-\frac{3}{y}\cdot xy}$$

$$= \frac{3y+3x}{3y-3x}$$

$$= \frac{3(y+x)}{3(y-x)}$$

$$= \frac{y+x}{y-x}$$

13. $$\frac{\frac{8x-24y}{10}}{\frac{x-3y}{5x}} = \frac{8x-24y}{10} \cdot \frac{5x}{x-3y}$$
$$= \frac{8(x-3y)5x}{10(x-3y)}$$
$$= \frac{40x}{10} = 4x$$

15. $$\frac{\frac{x^2-16y^2}{xy}}{\frac{1}{y}-\frac{4}{x}}$$

Multiply the numerator and denominator by xy, the LCD of all the fractions.

$$= \frac{\left(\frac{x^2-16y^2}{xy}\right)xy}{\left(\frac{1}{y}-\frac{4}{x}\right)xy}$$
$$= \frac{x^2-16y^2}{\frac{1}{y}\cdot xy - \frac{4}{x}\cdot xy}$$
$$= \frac{x^2-16y^2}{x-4y}$$
$$= \frac{(x+4y)(x-4y)}{x-4y}$$
$$= x+4y$$

17. $$\frac{y-\frac{y-3}{3}}{\frac{4}{9}+\frac{2}{3y}}$$

Multiply the numerator and denominator by $9y$, the LCD of all the fractions.

$$= \frac{9y\left(y-\frac{y-3}{3}\right)}{9y\left(\frac{4}{9}+\frac{2}{3y}\right)}$$
$$= \frac{9y^2-3y(y-3)}{4y+6}$$
$$= \frac{9y^2-3y^2+9y}{4y+6}$$
$$= \frac{6y^2+9y}{4y+6}$$
$$= \frac{3y(2y+3)}{2(2y+3)} = \frac{3y}{2}$$

19. $$\frac{\frac{x+2}{x}+\frac{1}{x+2}}{\frac{5}{x}+\frac{x}{x+2}}$$

Multiply the numerator and denominator by $x(x+2)$, the LCD of all the fractions.

$$= \frac{x(x+2)\left(\frac{x+2}{x}+\frac{1}{x+2}\right)}{x(x+2)\left(\frac{5}{x}+\frac{x}{x+2}\right)}$$
$$= \frac{x(x+2)\left(\frac{x+2}{x}\right)+x(x+2)\left(\frac{1}{x+2}\right)}{x(x+2)\left(\frac{5}{x}\right)+x(x+2)\left(\frac{x}{x+2}\right)}$$
$$= \frac{(x+2)(x+2)+x}{5(x+2)+x^2}$$
$$= \frac{x^2+4x+4+x}{5x+10+x^2}$$
$$= \frac{x^2+5x+4}{x^2+5x+10}$$

21. To add the fractions in the numerator, use the LCD, $m(m-1)$.

$$\frac{4}{m}+\frac{m+2}{m-1} = \frac{4(m-1)}{m(m-1)}+\frac{m(m+2)}{m(m-1)}$$
$$= \frac{4m-4+m^2+2m}{m(m-1)}$$
$$= \frac{m^2+6m-4}{m(m-1)}$$

22. To subtract the fractions in the denominator, use the same LCD, $m(m-1)$.

$$\frac{m+2}{m}-\frac{2}{m-1}$$
$$= \frac{(m+2)(m-1)}{m(m-1)}-\frac{m\cdot 2}{m(m-1)}$$
$$= \frac{m^2+m-2-2m}{m(m-1)}$$
$$= \frac{m^2-m-2}{m(m-1)}$$

23.

Exercise 21 answer ↓ Exercise 22 answer ↓

$$\frac{m^2+6m-4}{m(m-1)} \div \frac{m^2-m-2}{m(m-1)}$$

Multiply by the reciprocal.

$$= \frac{m^2+6m-4}{m(m-1)} \cdot \frac{m(m-1)}{m^2-m-2}$$
$$= \frac{m^2+6m-4}{m^2-m-2}$$

24. The LCD of all the denominators in the complex fraction is $m(m-1)$.

25. $$\frac{\left(\frac{4}{m}+\frac{m+2}{m-1}\right)\cdot m(m-1)}{\left(\frac{m+2}{m}-\frac{2}{m-1}\right)\cdot m(m-1)}$$
$$=\frac{4(m-1)+m(m+2)}{(m+2)(m-1)-2m}$$
$$=\frac{4m-4+m^2+2m}{m^2+m-2-2m}$$
$$=\frac{m^2+6m-4}{m^2-m-2}$$

26. Answers will vary. Because of the complicated nature of the numerator and denominator of the complex fraction, using Method 1 takes much longer to simplify the complex fraction. Method 2 is a simpler, more direct means of simplifying and is most likely the preferred method.

27. $$\frac{1}{x^{-2}+y^{-2}}=\frac{1}{\frac{1}{x^2}+\frac{1}{y^2}}$$
$$=\frac{x^2y^2(1)}{x^2y^2\left(\frac{1}{x^2}+\frac{1}{y^2}\right)} \qquad LCD = x^2y^2$$
$$=\frac{x^2y^2}{y^2+x^2}$$

29. $$\frac{x^{-2}+y^{-2}}{x^{-1}+y^{-1}}$$
$$=\frac{\frac{1}{x^2}+\frac{1}{y^2}}{\frac{1}{x}+\frac{1}{y}}$$

Multiply the numerator and denominator by x^2y^2, the LCD of all the fractions.

$$=\frac{x^2y^2\left(\frac{1}{x^2}+\frac{1}{y^2}\right)}{x^2y^2\left(\frac{1}{x}+\frac{1}{y}\right)}$$
$$=\frac{x^2y^2\cdot\frac{1}{x^2}+x^2y^2\cdot\frac{1}{y^2}}{x^2y^2\cdot\frac{1}{x}+x^2y^2\cdot\frac{1}{y}}$$
$$=\frac{y^2+x^2}{xy^2+x^2y}, \text{ or } \frac{y^2+x^2}{xy(y+x)}$$

31. $$\frac{x^{-1}+2y^{-1}}{2y+4x}=\frac{\frac{1}{x}+\frac{2}{y}}{2y+4x}$$

Multiply the numerator and denominator by xy, the LCD of all the fractions.

$$=\frac{xy\left(\frac{1}{x}+\frac{2}{y}\right)}{xy(2y+4x)}$$
$$=\frac{y+2x}{2xy(y+2x)}$$
$$=\frac{1}{2xy}$$

8.4 Equations with Rational Expressions and Graphs

8.4 Margin Exercises

1. (a) $\frac{3}{x}+\frac{1}{2}=\frac{5}{6x}$

Find the domain of each term.

For $\frac{3}{x}$, $x\neq 0$, so the domain is $\{x\,|\,x\neq 0\}$.

The domain of $\frac{1}{2}$ is $(-\infty,\infty)$.

For $\frac{5}{6x}$, $6x\neq 0$, so the domain is $\{x\,|\,x\neq 0\}$.

The intersection of these three domains is $\{x\,|\,x\neq 0\}$.

(b) $\frac{4}{x-5}-\frac{2}{x+5}=\frac{1}{x^2-25}$

Find the domain of each term.

For $\frac{4}{x-5}$, $x-5\neq 0$, so the domain is $\{x\,|\,x\neq 5\}$.

For $\frac{2}{x+5}$, $x+5\neq 0$, so the domain is $\{x\,|\,x\neq -5\}$.

For $\frac{1}{x^2-25}$, $x^2-25=(x+5)(x-5)\neq 0$, so the domain is $\{x\,|\,x\neq \pm 5\}$.

The intersection of these three domains is $\{x\,|\,x\neq \pm 5\}$.

2. $-\frac{3}{20}+\frac{2}{x}=\frac{5}{4x}$

The domain is $\{x \mid x \neq 0\}$.

Multiply each side by the LCD, $20x$.

$$20x\left(-\frac{3}{20}+\frac{2}{x}\right)=20x\left(\frac{5}{4x}\right)$$
$$20x\left(-\frac{3}{20}\right)+20x\left(\frac{2}{x}\right)=20x\left(\frac{5}{4x}\right)$$
$$-3x+40=25$$
$$-3x=-15$$
$$x=5$$

Check $x=5$: $-\frac{3}{20}+\frac{8}{20}=\frac{5}{20}$ *True*

Solution set: $\{5\}$

3. **(a)** $\frac{3}{x+1}=\frac{1}{x-1}-\frac{2}{x^2-1}$

The domain is $\{x \mid x \neq \pm 1\}$.

Multiply each side by the LCD, $(x+1)(x-1)=x^2-1$.

$$(x+1)(x-1)\left(\frac{3}{x+1}\right)=(x+1)(x-1)\left(\frac{1}{x-1}-\frac{2}{x^2-1}\right)$$
$$(x+1)(x-1)\left(\frac{3}{x+1}\right)=(x+1)(x-1)\left(\frac{1}{x-1}\right)-(x+1)(x-1)\left(\frac{2}{x^2-1}\right)$$
$$3(x-1)=x+1-2$$
$$3x-3=x-1$$
$$2x=2$$
$$x=1$$

Since 1 is not in the domain, the solution set is $\emptyset$.

(b) $\frac{1}{x-3}+\frac{1}{x+3}=\frac{6}{x^2-9}$

The domain is $\{x \mid x \neq \pm 3\}$.

Multiply each side by the LCD, $(x+3)(x-3)=x^2-9$.

$$(x+3)(x-3)\left(\frac{1}{x-3}+\frac{1}{x+3}\right)=(x+3)(x-3)\left(\frac{6}{x^2-9}\right)$$
$$(x+3)(x-3)\left(\frac{1}{x-3}\right)+(x+3)(x-3)\left(\frac{1}{x+3}\right)=(x+3)(x-3)\left(\frac{6}{x^2-9}\right)$$
$$x+3+x-3=6$$
$$2x=6$$
$$x=3$$

Since 3 is not in the domain, the solution set is $\emptyset$.

4. $\frac{4}{x^2+x-6}-\frac{1}{x^2-4}=\frac{2}{x^2+5x+6}$

Factor the denominators and determine the domain.

$$x^2+x-6=(x+3)(x-2),\text{ so } x\neq -3, 2.$$
$$x^2-4=(x+2)(x-2),\text{ so } x\neq \pm 2.$$
$$x^2+5x+6=(x+3)(x+2),\text{ so } x\neq -3, -2.$$

The domain is $\{x \mid x \neq -3, \pm 2\}$.

Multiply each side by the LCD, $(x+3)(x+2)(x-2)$.

$$(x+3)(x+2)(x-2)\left[\frac{4}{(x+3)(x-2)}-\frac{1}{(x+2)(x-2)}\right]=(x+3)(x+2)(x-2)\frac{2}{(x+3)(x+2)}$$
$$4(x+2)-(x+3)=2(x-2)$$
$$4x+8-x-3=2x-4$$
$$3x+5=2x-4$$
$$x=-9$$

The solution set is $\{-9\}$.

5. $\frac{1}{x+4}+\frac{x}{x-4}=\frac{-8}{x^2-16}$

The domain is $\{x \mid x \neq \pm 4\}$.

Multiply each side by the LCD, $(x+4)(x-4)=x^2-16$.

$$(x+4)(x-4)\left(\frac{1}{x+4}+\frac{x}{x-4}\right)$$
$$=(x+4)(x-4)\left(\frac{-8}{x^2-16}\right)$$
$$(x+4)(x-4)\left(\frac{1}{x+4}\right)$$
$$+(x+4)(x-4)\left(\frac{x}{x-4}\right)$$
$$=(x+4)(x-4)\left(\frac{-8}{x^2-16}\right)$$
$$x-4+x(x+4)=-8$$
$$x-4+x^2+4x=-8$$
$$x^2+5x+4=0$$
$$(x+4)(x+1)=0$$

$$x+4=0 \quad \text{or} \quad x+1=0$$
$$x=-4 \quad \text{or} \quad x=-1$$

Because -4 is not in the domain of the variable, it is not a solution.

Check $x=-1$: $\frac{1}{3}+\frac{1}{5}=\frac{8}{15}$ *True*
Solution set: $\{-1\}$

6. **(a)** $f(x)=-\frac{1}{x}$

x	-10	-1	$-\frac{1}{10}$	$\frac{1}{10}$	1	10
y	$\frac{1}{10}$	1	10	-10	-1	$-\frac{1}{10}$

Since $x \neq 0$, the equation of the vertical asymptote is $x=0$. As the values of x get larger, the values of y get closer to 0, so $y=0$ is the equation of the horizontal asymptote.

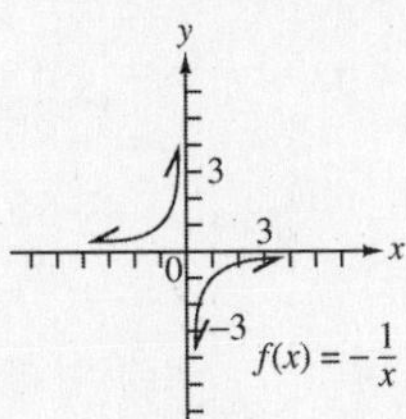

(b) $f(x)=\frac{2}{x+3}$

x	-13	-4	-3.1	-2.9	-2	7
y	$-\frac{1}{5}$	-2	-20	20	2	$\frac{1}{5}$

Set the denominator equal to zero and solve for x.

$$x+3=0$$
$$x=-3$$

Since $x \neq -3$, the equation of the vertical asymptote is $x=-3$. As the values of x get larger, the values of y get closer to 0, so $y=0$ is the equation of the horizontal asymptote.

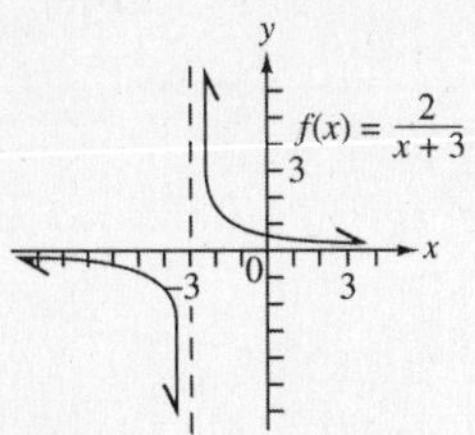

8.4 Section Exercises

1. **(a)** $\frac{1}{x+1}-\frac{1}{x-2}=0$

Set each denominator equal to 0 and solve.

$$x+1=0 \quad \text{or} \quad x-2=0$$
$$x=-1 \quad \text{or} \quad x=2$$

Solutions of -1 and 2 would be rejected since these values would make a denominator of the original equation equal to 0.

(b) The domain is $\{x \mid x \neq -1, 2\}$.

3. **(a)** $\frac{5}{3x+5}-\frac{1}{x}=\frac{1}{2x+3}$

Set each denominator equal to 0 and solve.

$$3x+5=0 \quad \text{or} \quad x=0 \quad \text{or} \quad 2x+3=0$$
$$3x=-5 \qquad\qquad 2x=-3$$
$$x=-\tfrac{5}{3} \qquad\qquad x=-\tfrac{3}{2}$$

So $-\frac{5}{3}$, 0, and $-\frac{3}{2}$ would have to be rejected as proposed solutions.

(b) The domain is $\{x \mid x \neq -\frac{5}{3}, 0, -\frac{3}{2}\}$.

5. **(a)** $\frac{1}{3x}+\frac{1}{2x}=\frac{x}{3}$

Only 0 would make any of the denominators equal to 0, so 0 would have to be rejected as a proposed solution.

(b) The domain is $\{x \mid x \neq 0\}$.

7. **(a)** $\frac{3x+1}{x-4}=\frac{6x+5}{2x-7}$

Set each denominator equal to 0 and solve.

$$x-4=0 \quad \text{or} \quad 2x-7=0$$
$$x=4 \quad \text{or} \quad 2x=7$$
$$x=\tfrac{7}{2}$$

So 4 and $\frac{7}{2}$ would have to be rejected as proposed solutions.

(b) The domain is $\{x \mid x \neq 4, \frac{7}{2}\}$.

9. **(a)** $$\frac{2}{x^2-x}+\frac{1}{x+3}=\frac{4}{x-2}$$

$x^2-x=x(x-1)$ is 0 if $x=0$ or $x=1$.
$x+3$ is 0 if $x=-3$.
$x-2$ is 0 if $x=2$.

So 0, 1, -3, and 2 would have to be rejected as proposed solutions.

(b) The domain is $\{x\,|\,x\neq 0,1,-3,2\}$.

In Exercises 11–37, check each proposed solution in the original equation.

11. $$\frac{-5}{2x}+\frac{3}{4x}=\frac{-7}{4}$$

Multiply by the LCD, $4x$. $(x\neq 0)$

$$4x\left(\frac{-5}{2x}+\frac{3}{4x}\right)=4x\left(\frac{-7}{4}\right)$$
$$2(-5)+1(3)=x(-7)$$
$$-10+3=-7x$$
$$-7=-7x$$
$$1=x$$

Check $x=1$: $-\frac{5}{2}+\frac{3}{4}=-\frac{7}{4}$ *True*
Solution set: $\{1\}$

13. $$x-\frac{24}{x}=-2$$

Multiply by the LCD, x. $(x\neq 0)$

$$x\left(x-\frac{24}{x}\right)=-2\cdot x$$
$$x^2-24=-2x$$
$$x^2+2x-24=0$$
$$(x+6)(x-4)=0$$

$x+6=0$ or $x-4=0$
$x=-6$ or $x=4$

Check $x=-6$: $-6+4=-2$ *True*
Check $x=4$: $4-6=-2$ *True*
Solution set: $\{-6,4\}$

15. $$\frac{x-4}{x+6}=\frac{2x+3}{2x-1}$$

Multiply by the LCD, $(x+6)(2x-1)$.
Note that $x\neq -6$ and $x\neq \frac{1}{2}$.

$$(x+6)(2x-1)\left(\frac{x-4}{x+6}\right)=(x+6)(2x-1)\left(\frac{2x+3}{2x-1}\right)$$
$$(2x-1)(x-4)=(x+6)(2x+3)$$
$$2x^2-9x+4=2x^2+15x+18$$
$$-24x=14$$
$$x=\frac{14}{-24}=-\frac{7}{12}$$

A calculator check is suggested.

```
-7/12→X:(X-4)/(X
+6)▸Frac
                -11/13
(2X+3)/(2X-1)▸Fr
ac
                -11/13
```

Check $x=-\frac{7}{12}$: $-\frac{11}{13}=-\frac{11}{13}$ *True*
Solution set: $\{-\frac{7}{12}\}$

17. $$\frac{3x+1}{x-4}=\frac{6x+5}{2x-7}$$

Multiply by the LCD, $(x-4)(2x-7)$.
Note that $x\neq 4$ and $x\neq \frac{7}{2}$.

$$(x-4)(2x-7)\left(\frac{3x+1}{x-4}\right)=(x-4)(2x-7)\left(\frac{6x+5}{2x-7}\right)$$
$$(2x-7)(3x+1)=(x-4)(6x+5)$$
$$6x^2-19x-7=6x^2-19x-20$$
$$-7=-20 \quad \textit{False}$$

The false statement indicates that the original equation has no solution.

Solution set: $\emptyset$

19. $$\frac{1}{y-1}+\frac{5}{12}=\frac{-2}{3y-3}$$
$$\frac{1}{y-1}+\frac{5}{12}=\frac{-2}{3(y-1)}$$

Multiply by the LCD, $12(y-1)$. $(y\neq 1)$

$$12(y-1)\left(\frac{1}{y-1}+\frac{5}{12}\right)=12(y-1)\left(\frac{-2}{3(y-1)}\right)$$
$$12+5(y-1)=-8$$
$$12+5y-5=-8$$
$$5y+7=-8$$
$$5y=-15$$
$$y=-3$$

Check $y=-3$: $-\frac{1}{4}+\frac{5}{12}=\frac{1}{6}$ *True*
Solution set: $\{-3\}$

21. $$\frac{-2}{3t-6}-\frac{1}{36}=\frac{-3}{4t-8}$$
$$\frac{-2}{3(t-2)}-\frac{1}{36}=\frac{-3}{4(t-2)}$$

Multiply by the LCD, $36(t-2)$. $(t\neq 2)$

$$36(t-2)\left(\frac{-2}{3(t-2)}-\frac{1}{36}\right)=36(t-2)\left(\frac{-3}{4(t-2)}\right)$$
$$12(-2)-1(t-2)=9(-3)$$
$$-24-t+2=-27$$
$$-t-22=-27$$
$$-t=-5$$
$$t=5$$

Check $t=5$: $-\frac{2}{9}-\frac{1}{36}=\frac{-3}{12}$ *True*
Solution set: $\{5\}$

23.
$$\frac{3}{k+2}-\frac{2}{k^2-4}=\frac{1}{k-2}$$
$$\frac{3}{k+2}-\frac{2}{(k+2)(k-2)}=\frac{1}{k-2}$$
Multiply by the LCD, $(k+2)(k-2)$. $(k \neq -2, 2)$
$$(k+2)(k-2)\left(\frac{3}{k+2}-\frac{2}{(k+2)(k-2)}\right) = (k+2)(k-2)\left(\frac{1}{k-2}\right)$$
$$\begin{aligned} 3(k-2)-2 &= k+2 \\ 3k-6-2 &= k+2 \\ 3k-8 &= k+2 \\ 2k &= 10 \\ k &= 5 \end{aligned}$$
Check $k=5$: $\frac{3}{7}-\frac{2}{21}=\frac{1}{3}$ *True*

Solution set: $\{5\}$

25.
$$\frac{1}{y+2}+\frac{3}{y+7}=\frac{5}{y^2+9y+14}$$
$$\frac{1}{y+2}+\frac{3}{y+7}=\frac{5}{(y+2)(y+7)}$$
Multiply by the LCD, $(y+2)(y+7)$. $(y \neq -2, -7)$
$$(y+2)(y+7)\left(\frac{1}{y+2}+\frac{3}{y+7}\right) = (y+2)(y+7)\left(\frac{5}{(y+2)(y+7)}\right)$$
$$\begin{aligned} (y+7)+3(y+2) &= 5 \\ y+7+3y+6 &= 5 \\ 4y+13 &= 5 \\ 4y &= -8 \\ y &= -2 \end{aligned}$$
But y cannot equal -2 because that would make the denominator $y+2$ equal to 0. Since division by 0 is undefined, the equation has no solution.

Solution set: $\emptyset$

27.
$$\frac{9}{x}+\frac{4}{6x-3}=\frac{2}{6x-3}$$
Multiply by the LCD, $x(6x-3)$. $(x \neq 0, \frac{1}{2})$
$$x(6x-3)\left(\frac{9}{x}+\frac{4}{6x-3}\right) = x(6x-3)\left(\frac{2}{6x-3}\right)$$
$$\begin{aligned} 9(6x-3)+4x &= 2x \\ 54x-27+4x &= 2x \\ 56x &= 27 \\ x &= \tfrac{27}{56} \end{aligned}$$
Check $x=\frac{27}{56}$: $\frac{56}{3}+(-\frac{112}{3})=-\frac{56}{3}$ *True*

Solution set: $\{\frac{27}{56}\}$

29.
$$\frac{6}{w+3}+\frac{-7}{w-5}=\frac{-48}{w^2-2w-15}$$
$$\frac{6}{w+3}+\frac{-7}{w-5}=\frac{-48}{(w+3)(w-5)}$$
Multiply by the LCD, $(w+3)(w-5)$. $(w \neq -3, 5)$
$$(w+3)(w-5)\left(\frac{6}{w+3}+\frac{-7}{w-5}\right) = (w+3)(w-5)\left[\frac{-48}{(w+3)(w-5)}\right]$$
$$\begin{aligned} 6(w-5)-7(w+3) &= -48 \\ 6w-30-7w-21 &= -48 \\ -w-51 &= -48 \\ -w &= 3 \\ w &= -3 \end{aligned}$$
But w cannot equal -3 because that would make the denominator $w+3$ equal to 0. Since division by 0 is undefined, the equation has no solution.

Solution set: $\emptyset$

31.
$$\frac{x}{x-3}+\frac{4}{x+3}=\frac{18}{x^2-9}$$
$$\frac{x}{x-3}+\frac{4}{x+3}=\frac{18}{(x-3)(x+3)}$$
Multiply by the LCD, $(x-3)(x+3)$. $(x \neq 3, -3)$
$$(x-3)(x+3)\left(\frac{x}{x-3}+\frac{4}{x+3}\right) = (x-3)(x+3)\left(\frac{18}{(x-3)(x+3)}\right)$$
$$\begin{aligned} x(x+3)+4(x-3) &= 18 \\ x^2+3x+4x-12 &= 18 \\ x^2+7x-30 &= 0 \\ (x-3)(x+10) &= 0 \end{aligned}$$
$$x-3=0 \quad \text{or} \quad x+10=0$$
$$x=3 \quad \text{or} \quad x=-10$$
But $x \neq 3$ since a denominator of 0 results. The only solution to check is -10.

Check $x=-10$: $\frac{10}{13}+(-\frac{4}{7})=\frac{18}{91}$ *True*

Solution set: $\{-10\}$

33.
$$\frac{6}{x-4}+\frac{5}{x}=\frac{-20}{x^2-4x}$$
$$\frac{6}{x-4}+\frac{5}{x}=\frac{-20}{x(x-4)}$$
Multiply by the LCD, $x(x-4)$. $(x \neq 0, 4)$
$$x(x-4)\left(\frac{6}{x-4}+\frac{5}{x}\right)=x(x-4)\left(\frac{-20}{x(x-4)}\right)$$
$$6x+5(x-4)=-20$$
$$6x+5x-20=-20$$
$$11x=0$$
$$x=0$$

Since 0 cannot appear in the denominator, this equation has no solution.

Solution set: $\emptyset$

35.
$$\frac{2}{4x+7}+\frac{x}{3}=\frac{6}{12x+21}$$
$$\frac{2}{4x+7}+\frac{x}{3}=\frac{6}{3(4x+7)}$$
Multiply by the LCD, $3(4x+7)$. $(x \neq -\frac{7}{4})$
$$2(3)+x(4x+7)=6$$
$$6+4x^2+7x=6$$
$$4x^2+7x=0$$
$$x(4x+7)=0$$
$$x=0 \quad \text{or} \quad 4x+7=0$$
$$4x=-7$$
$$x=-\tfrac{7}{4}$$

But x cannot equal $-\frac{7}{4}$, so we only need to check 0.

Check $x=0$: $\frac{2}{7}+0=\frac{6}{21}$ *True*
Solution set: $\{0\}$

37.
$$\frac{4x-7}{4x^2-9}=\frac{-2x^2+5x-4}{4x^2-9}+\frac{x+1}{2x+3}$$
$$\frac{4x-7}{(2x+3)(2x-3)}=\frac{-2x^2+5x-4}{(2x+3)(2x-3)}+\frac{x+1}{2x+3}$$
Multiply by the LCD, $(2x+3)(2x-3)$.
$(x \neq -\frac{3}{2}, \frac{3}{2})$
$$4x-7=-2x^2+5x-4+(2x-3)(x+1)$$
$$4x-7=-2x^2+5x-4+2x^2-x-3$$
$$4x-7=4x-7 \quad \textit{True}$$

This equation is true for every real number value of x, but we have already determined that $x \neq -\frac{3}{2}$ or $x \neq \frac{3}{2}$. So every real number except $-\frac{3}{2}$ and $\frac{3}{2}$ is a solution.

Solution set: $\{x \mid x \neq -\frac{3}{2}, \frac{3}{2}\}$

39. $f(x)=\dfrac{2}{x}$ is not defined when $x=0$, so an equation of the vertical asymptote is $x=0$. As the values of x get larger, the values of y get closer to 0, so $y=0$ is the equation of the horizontal asymptote.

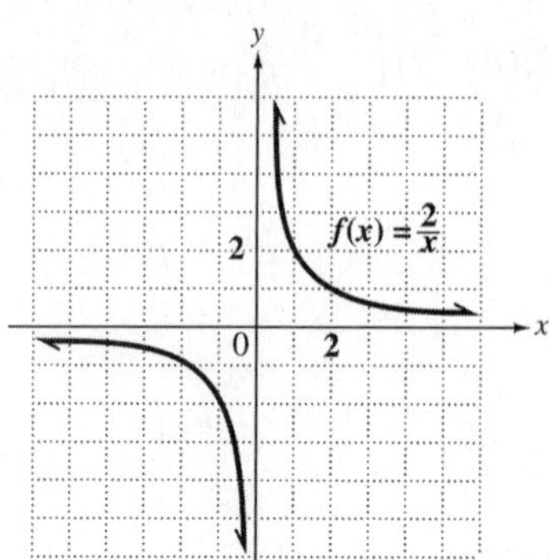

41. $f(x)=\dfrac{1}{x-2}$ is not defined when $x=2$, so an equation of the vertical asymptote is $x=2$. As the values of x get larger, the values of y get closer to 0, so $y=0$ is the equation of the horizontal asymptote.

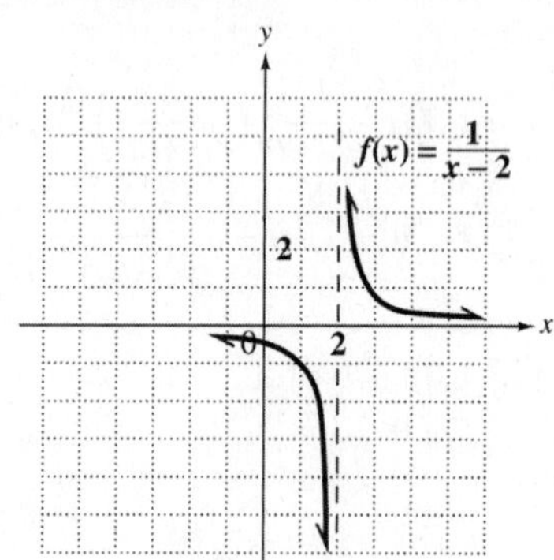

43. $w(x)=\dfrac{x^2}{2(1-x)}$

(a) $w(0.1)=\dfrac{(0.1)^2}{2(1-0.1)}$
$$=\frac{0.01}{2(0.9)}\approx 0.006$$

To the nearest tenth, $w(0.1)$ is 0.

(b) $w(0.8)=\dfrac{(0.8)^2}{2(1-0.8)}$
$$=\frac{0.64}{2(0.2)}=1.6$$

(c) $w(0.9)=\dfrac{(0.9)^2}{2(1-0.9)}$
$$=\frac{0.81}{2(0.1)}=4.05\approx 4.1$$

(d) Based on the answers in (a), (b), and (c), we see that as the traffic intensity increases, the waiting time also increases.

45. Substituting -1 for x gives a true statement, $\frac{4}{3} = \frac{4}{3}$. Substituting -2 for x leads to 0 in the first and third denominators.

46. If $A = 2$, and $B = 1$, then

$$C = -2AB = -2(2)(1) = -4.$$

The equation determined by A, B, and C is

$$\frac{2}{x+1} + \frac{x}{x-1} = \frac{-4}{x^2-1}.$$
$$\frac{2}{x+1} + \frac{x}{x-1} = \frac{-4}{(x+1)(x-1)}$$

Multiply by the LCD, $(x+1)(x-1)$. $(x \neq \pm 1)$

$$(x+1)(x-1)\left(\frac{2}{x+1} + \frac{x}{x-1}\right)$$
$$= (x+1)(x-1)\left(\frac{-4}{(x+1)(x-1)}\right)$$
$$2(x-1) + x(x+1) = -4$$
$$2x - 2 + x^2 + x = -4$$
$$x^2 + 3x + 2 = 0$$
$$(x+2)(x+1) = 0$$

$x + 2 = 0$ or $x + 1 = 0$

$x = -2$ or $x = -1$

But $x \neq -1$, so it must be rejected.

Check $x = -2$: $-2 + \frac{2}{3} = -\frac{4}{3}$ *True*

Solution set: $\{-2\}$

47. If $A = 4$ and $B = -3$, then

$$C = -2AB = -2(4)(-3) = 24.$$

The equation determined by A, B, and C is

$$\frac{4}{x-3} + \frac{x}{x+3} = \frac{24}{x^2-9}.$$
$$\frac{4}{x-3} + \frac{x}{x+3} = \frac{24}{(x+3)(x-3)}$$

Multiply by the LCD, $(x+3)(x-3)$. $(x \neq \pm 3)$

$$(x+3)(x-3)\left(\frac{4}{x-3} + \frac{x}{x+3}\right)$$
$$= (x+3)(x-3)\left(\frac{24}{(x+3)(x-3)}\right)$$
$$4(x+3) + x(x-3) = 24$$
$$4x + 12 + x^2 - 3x = 24$$
$$x^2 + x - 12 = 0$$
$$(x+4)(x-3) = 0$$

$x + 4 = 0$ or $x - 3 = 0$

$x = -4$ or $x = 3$

But $x \neq 3$, so it must be rejected.

Check $x = -4$: $-\frac{4}{7} + 4 = \frac{24}{7}$ *True*

Solution set: $\{-4\}$

48. Answers will vary. However, in every case, $-B$ will be the rejected solution, and $\{-A\}$ will be the solution set.

Summary Exercises on Rational Expressions and Equations

1. $\frac{x}{2} - \frac{x}{4} = 5$

There is an equals sign, so this is an *equation.*

$$4\left(\frac{x}{2} - \frac{x}{4}\right) = 4(5) \quad \textit{Multiply by 4.}$$
$$2x - x = 20$$
$$x = 20$$

Check $x = 20$: $10 - 5 = 5$ *True*

Solution set: $\{20\}$

3. $\frac{6}{7x} - \frac{4}{x}$

No equals sign appears so this is an *expression.*

$$= \frac{6}{7x} - \frac{4 \cdot 7}{x \cdot 7} \quad \textit{LCD} = 7x$$
$$= \frac{6-28}{7x} = \frac{-22}{7x} \text{ or } -\frac{22}{7x}$$

5. $\frac{5}{7t} = \frac{52}{7} - \frac{3}{t}$

There is an equals sign, so this is an *equation.* Multiply each side by the LCD, $7t$. $(t \neq 0)$

$$7t\left(\frac{5}{7t}\right) = 7t\left(\frac{52}{7} - \frac{3}{t}\right)$$
$$5 = 52t - 3(7)$$
$$5 = 52t - 21$$
$$26 = 52t$$
$$t = \frac{26}{52} = \frac{1}{2}$$

Check $t = \frac{1}{2}$: $\frac{10}{7} = \frac{52}{7} - 6$ *True*

Solution set: $\{\frac{1}{2}\}$

7. $\frac{7}{6x} + \frac{5}{8x}$

No equals sign appears so this is an *expression.*

$$= \frac{7}{3 \cdot 2x} + \frac{5}{4 \cdot 2x}$$
$$= \frac{4(7)}{4 \cdot 3 \cdot 2x} + \frac{3(5)}{3 \cdot 4 \cdot 2x} \quad \textit{LCD} = 24x$$
$$= \frac{28+15}{3 \cdot 4 \cdot 2x} = \frac{43}{24x}$$

9. $\dfrac{\dfrac{6}{x+1}-\dfrac{1}{x}}{\dfrac{2}{x}-\dfrac{4}{x+1}}$

No equals sign appears so this is an *expression.* Multiply the numerator and denominator by the LCD of all the fractions, $x(x+1)$.

$$=\frac{6(x)-1(x+1)}{2(x+1)-4(x)}$$
$$=\frac{6x-x-1}{2x+2-4x}$$
$$=\frac{5x-1}{-2x+2}, \text{ or } \frac{5x-1}{-2(x-1)}$$

11. $\dfrac{x}{x+y}+\dfrac{2y}{x-y}$

No equals sign appears so this is an *expression.*

$$=\frac{x(x-y)}{(x+y)(x-y)}+\frac{2y(x+y)}{(x+y)(x-y)}$$
LCD $=(x+y)(x-y)$
$$=\frac{x^2-xy+2xy+2y^2}{(x+y)(x-y)}$$
$$=\frac{x^2+xy+2y^2}{(x+y)(x-y)}$$

13. $\dfrac{x-2}{9}\cdot\dfrac{5}{8-4x}$

No equals sign appears so this is an *expression.*

$$=\frac{x-2}{9}\cdot\frac{5}{-4(x-2)}$$
$$=\frac{5}{-36}=-\frac{5}{36}$$

15. $\dfrac{b^2+b-6}{b^2+2b-8}\cdot\dfrac{b^2+8b+16}{3b+12}$

No equals sign appears so this is an *expression.*

$$=\frac{(b+3)(b-2)}{(b+4)(b-2)}\cdot\frac{(b+4)(b+4)}{3(b+4)}$$
$$=\frac{b+3}{3}$$

17. $\dfrac{5}{x^2-2x}-\dfrac{3}{x^2-4}$

No equals sign appears so this is an *expression.*

$$=\frac{5}{x(x-2)}-\frac{3}{(x+2)(x-2)}$$
LCD $=x(x-2)(x+2)$
$$=\frac{5(x+2)}{x(x-2)(x+2)}-\frac{3x}{x(x-2)(x+2)}$$
$$=\frac{5x+10-3x}{x(x-2)(x+2)}$$
$$=\frac{2x+10}{x(x-2)(x+2)}$$

19. $\dfrac{\dfrac{5}{x}-\dfrac{3}{y}}{\dfrac{9x^2-25y^2}{x^2y}}$

No equals sign appears so this is an *expression.* Multiply the numerator and denominator by the LCD of all the fractions, x^2y.

$$=\frac{x^2y\left(\frac{5}{x}-\frac{3}{y}\right)}{x^2y\left(\frac{9x^2-25y^2}{x^2y}\right)}$$
$$=\frac{5xy-3x^2}{9x^2-25y^2}$$
$$=\frac{-x(3x-5y)}{(3x+5y)(3x-5y)}$$
$$=\frac{-x}{(3x+5y)}$$

21. $\dfrac{4y^2-13y+3}{2y^2-9y+9}\div\dfrac{4y^2+11y-3}{6y^2-5y-6}$

No equals sign appears so this is an *expression.*

$$=\frac{(4y-1)(y-3)}{(2y-3)(y-3)}\cdot\frac{(2y-3)(3y+2)}{(4y-1)(y+3)}$$
$$=\frac{3y+2}{y+3}$$

23. $\dfrac{3r}{r-2}=1+\dfrac{6}{r-2}$

There is an equals sign, so this is an *equation.* Multiply by the LCD, $r-2$. $(r\neq 2)$

$$3r=r-2+6$$
$$2r=4$$
$$r=2$$

But $r\neq 2$.

Solution set: $\emptyset$

25. $\dfrac{-1}{3-x}-\dfrac{2}{x-3}$

No equals sign appears so this is an *expression.*

$$=\frac{-1}{-(x-3)}-\frac{2}{x-3}$$
$$=\frac{1}{x-3}-\frac{2}{x-3}$$
$$=\frac{-1}{x-3}, \text{ or } \frac{-1}{-(3-x)}=\frac{1}{3-x}$$

27. $$\frac{2}{y+1}-\frac{3}{y^2-y-2}=\frac{3}{y-2}$$
$$\frac{2}{y+1}-\frac{3}{(y-2)(y+1)}=\frac{3}{y-2}$$
There is an equals sign, so this is an *equation.*
Multiply by the LCD, $(y-2)(y+1)$.
$(y \neq -1, 2)$
$$\begin{aligned}2(y-2)-3&=3(y+1)\\2y-4-3&=3y+3\\2y-7&=3y+3\\-10&=y\end{aligned}$$

Check $y=-10$: $-\frac{2}{9}-\frac{1}{36}=-\frac{1}{4}$ *True*
Solution set: $\{-10\}$

29. $$\frac{3}{y-3}-\frac{3}{y^2-5y+6}=\frac{2}{y-2}$$
$$\frac{3}{y-3}-\frac{3}{(y-3)(y-2)}=\frac{2}{y-2}$$
There is an equals sign, so this is an *equation.*
Multiply by the LCD, $(y-3)(y-2)$. $(y \neq 2, 3)$
$$\begin{aligned}3(y-2)-3&=2(y-3)\\3y-6-3&=2y-6\\3y-9&=2y-6\\y&=3\end{aligned}$$

But y cannot equal 3 because that would make the denominator $y-3$ equal to 0. Since division by 0 is undefined, the equation has no solution.

Solution set: $\emptyset$

8.5 Applications of Rational Expressions

8.5 Margin Exercises

1. (a) $$\frac{1}{f}=\frac{1}{p}+\frac{1}{q}$$
$$\frac{1}{15}=\frac{1}{p}+\frac{1}{25} \quad \textit{Let } f=15,\ q=25.$$

Multiply by the LCD, $75p$.

$$75p\cdot\frac{1}{15}=75p\left(\frac{1}{p}+\frac{1}{25}\right)$$
$$75p\cdot\frac{1}{15}=75p\left(\frac{1}{p}\right)+75p\left(\frac{1}{25}\right)$$
$$\begin{aligned}5p&=75+3p\\2p&=75\\p&=\tfrac{75}{2}\end{aligned}$$

The distance from the object to the lens is $\frac{75}{2}$ cm.

(b) $$\frac{1}{f}=\frac{1}{p}+\frac{1}{q}$$
$$\frac{1}{f}=\frac{1}{6}+\frac{1}{9} \quad \textit{Let } p=6,\ q=9.$$

Multiply by the LCD, $18f$.

$$18f\cdot\frac{1}{f}=18f\left(\frac{1}{6}+\frac{1}{9}\right)$$
$$18f\cdot\frac{1}{f}=18f\left(\frac{1}{6}\right)+18f\left(\frac{1}{9}\right)$$
$$\begin{aligned}18&=3f+2f\\18&=5f\\\tfrac{18}{5}&=f\end{aligned}$$

The focal length of the lens is $\frac{18}{5}$ cm.

(c) $$\frac{1}{f}=\frac{1}{p}+\frac{1}{q}$$
$$\frac{1}{12}=\frac{1}{16}+\frac{1}{q} \quad \textit{Let } f=12,\ p=16.$$

Multiply by the LCD, $48q$.

$$48q\cdot\frac{1}{12}=48q\left(\frac{1}{16}+\frac{1}{q}\right)$$
$$48q\cdot\frac{1}{12}=48q\left(\frac{1}{16}\right)+48q\left(\frac{1}{q}\right)$$
$$\begin{aligned}4q&=3q+48\\q&=48\end{aligned}$$

The distance from the lens to the image is 48 cm.

2. Solve $\frac{3}{p}+\frac{3}{q}=\frac{5}{r}$ for q.

Multiply by the LCD, pqr.

$$pqr\left(\frac{3}{p}+\frac{3}{q}\right)=pqr\left(\frac{5}{r}\right)$$
$$pqr\left(\frac{3}{p}\right)+pqr\left(\frac{3}{q}\right)=pqr\left(\frac{5}{r}\right)$$
$$3qr+3pr=5pq$$

Get the q terms on one side.

$$\begin{aligned}3qr-5pq&=-3pr\\q(3r-5p)&=-3pr \quad \textit{Factor out } q.\\q&=\frac{-3pr}{3r-5p}, \text{ or } \frac{3pr}{5p-3r}\end{aligned}$$

3. Solve $A=\frac{Rr}{R+r}$ for R.

Multiply by the LCD, $R+r$.

$$\begin{aligned}A(R+r)&=\frac{Rr}{R+r}(R+r)\\A(R+r)&=Rr\\AR+Ar&=Rr\end{aligned}$$

Get the R terms on one side.

$$\begin{aligned}AR-Rr&=-Ar\\R(A-r)&=-Ar \quad \textit{Factor out } R.\\R&=\frac{-Ar}{A-r}, \text{ or } \frac{Ar}{r-A}\end{aligned}$$

4. *Step 2*
Let $x =$ the number (in millions) who had no health insurance.

Step 3
Set up a proportion.

$$\frac{11.7}{100} = \frac{x}{73{,}740{,}000}$$

Step 4
To solve the equation, multiply by the LCD.

$$73{,}740{,}000\left(\frac{11.7}{100}\right) = 73{,}740{,}000\left(\frac{x}{73{,}740{,}000}\right)$$
$$737{,}400(11.7) = x$$
$$x = 8{,}627{,}580$$

Step 5
There were 8,627,580 children under 18 years of age in the United States with no health insurance in 2006.

Step 6
The ratio of 8,627,580 to 73,740,000 equals $\frac{11.7}{100}$.

5. *Step 2*
Let $x =$ the additional number of gallons of gasoline needed.

Step 3
She knows that she can drive 495 miles with 15 gallons of gasoline. She wants to drive 600 miles using $(6 + x)$ gallons of gasoline. Set up a proportion.

$$\frac{495}{15} = \frac{600}{6+x}$$
$$\frac{33}{1} = \frac{600}{6+x} \quad \textit{Reduce.}$$

Step 4
Find the cross products and solve for x.

$$33(6+x) = 1(600)$$
$$198 + 33x = 600 \quad \textit{Distributive property}$$
$$33x = 402 \quad \textit{Subtract 198.}$$
$$x = \tfrac{402}{33} \quad \textit{Divide by 33.}$$
$$x \approx 12.2 \quad \textit{Approximate.}$$

Step 5
She will need about 12.2 more gallons of gasoline.

Step 6
Check The 6 gallons plus the 12.2 gallons equals 18.2 gallons. We'll check the rates (miles/gallon). Note that we could also use gallons/mile.

$$\tfrac{495}{15} = 33 \text{ mpg} \qquad \tfrac{600}{18.2} \approx 33 \text{ mpg}$$

The rates are approximately equal, so the solution is correct. Note that we could have used our exact value for x to get the exact rate

$$\frac{600}{6 + (402/33)} = 33.$$

6. **(a)** Let $x =$ the speed of the plane in still air.

Use $d = rt$, or $t = \frac{d}{r}$, to complete the table.

	d	r	t
Against Wind	100	$x - 20$	$\frac{100}{x-20}$
With Wind	120	$x + 20$	$\frac{120}{x+20}$

(b) Because the time against the wind equals the time with the wind,

$$\frac{100}{x-20} = \frac{120}{x+20}.$$

Multiply by the LCD, $(x-20)(x+20)$.

$$\frac{100}{x-20}(x-20)(x+20)$$
$$= \frac{120}{x+20}(x-20)(x+20)$$
$$100(x+20) = 120(x-20)$$
$$100x + 2000 = 120x - 2400$$
$$4400 = 20x$$
$$220 = x$$

The speed of the plane is 220 mph in still air.

7. Let $x =$ the distance at reduced speed.

Use $d = rt$, or $t = \frac{d}{r}$, to complete the table.

	d	r	t
Normal Speed	$300 - x$	55	$\frac{300-x}{55}$
Reduced Speed	x	15	$\frac{x}{15}$

Now write an equation.

Time on the freeway	plus	time at reduced speed	equals 6 hr.
↓	↓	↓	↓
$\frac{300-x}{55}$	$+$	$\frac{x}{15}$	$= 6$

Multiply by the LCD, 165.

$$165\left(\frac{300-x}{55} + \frac{x}{15}\right) = 165(6)$$
$$3(300-x) + 11x = 990$$
$$900 - 3x + 11x = 990$$
$$8x = 90$$
$$x = \tfrac{90}{8}, \text{ or } 11\tfrac{1}{4}$$

She drove $11\frac{1}{4}$ mi at reduced speed.

8. **(a)** Let $x =$ the time it will take them working together.

Make a table.

	Rate	Time Working Together	Fractional Part of the Job Done
Stan	$\frac{1}{45}$	x	$\frac{1}{45}x$
Deb	$\frac{1}{30}$	x	$\frac{1}{30}x$

Part done by Stan plus Part done by Deb equals 1 whole job.

$$\frac{x}{45} + \frac{x}{30} = 1$$

Multiply by the LCD, 90.

$$\begin{aligned} 90\left(\frac{1}{45}x + \frac{1}{30}x\right) &= 90 \cdot 1 \\ 2x + 3x &= 90 \\ 5x &= 90 \\ x &= 18 \end{aligned}$$

It will take them 18 min working together.

(b) Let $x =$ the time for Deb to do the job alone.

Make a table.

	Rate	Time Working Together	Fractional Part of the Job Done
Stan	$\frac{1}{35}$	15	$\frac{15}{35}$
Deb	$\frac{1}{x}$	15	$\frac{15}{x}$

Since together they complete one job, the sum of the fractional parts should be 1.

$$\begin{aligned} \frac{15}{35} + \frac{15}{x} &= 1 \\ \frac{3}{7} + \frac{15}{x} &= 1 \end{aligned}$$

Multiply by the LCD, $7x$.

$$\begin{aligned} 7x\left(\frac{3}{7} + \frac{15}{x}\right) &= 7x \cdot 1 \\ 3x + 105 &= 7x \\ 105 &= 4x \\ x &= \tfrac{105}{4}, \text{ or } 26\tfrac{1}{4} \end{aligned}$$

Deb can do the dishes alone in $26\frac{1}{4}$ min.

8.5 Section Exercises

1. **A.** $b = \dfrac{p}{r}$ is the same as $p = br$.

B. $r = \dfrac{b}{p}$ is the same as $b = pr$.

C. $b = \dfrac{r}{p}$ is the same as $r = bp$.

D. $p = \dfrac{r}{b}$ is the same as $r = bp$.

Choice **A** is correct.

3. **A.** $a = mF$ is the same as $m = \dfrac{a}{F}$.

B. $F = \dfrac{m}{a}$ is the same as $m = Fa$.

C. $F = \dfrac{a}{m}$ is the same as $Fm = a$, which is the same as $m = \dfrac{a}{F}$.

D. $F = ma$ is the same as $m = \dfrac{F}{a}$.

Choice **D** is correct.

5. $$\frac{PV}{T} = \frac{pv}{t}$$

Suppose that $T = 300$, $t = 350$, $V = 9$, $P = 50$, $v = 8$. Find p.

$$\begin{aligned} \frac{50 \cdot 9}{300} &= \frac{p \cdot 8}{350} \\ p &= \frac{50 \cdot 9 \cdot 350}{300 \cdot 8} \\ &= 65.625 \end{aligned}$$

7. $$c = \frac{100b}{L}$$

Let $c = 80$, $b = 5$. Find L.

$$\begin{aligned} 80 &= \frac{100 \cdot 5}{L} \\ 80L &= 500 && \textit{Multiply by } L. \\ L &= \tfrac{500}{80} = \tfrac{25}{4} \end{aligned}$$

9. Solve $F = \dfrac{GMm}{d^2}$ for G.

$$\begin{aligned} Fd^2 &= GMm && \textit{Multiply by } d^2. \\ \frac{Fd^2}{Mm} &= G && \textit{Divide by } Mm. \end{aligned}$$

11. Solve $\dfrac{1}{a} = \dfrac{1}{b} + \dfrac{1}{c}$ for a.

Multiply each side by the LCD, abc.

$$\begin{aligned} abc\left(\frac{1}{a}\right) &= abc\left(\frac{1}{b} + \frac{1}{c}\right) \\ bc &= ac + ab \\ bc &= a(c + b) && \textit{Factor out } a. \\ \frac{bc}{c + b} &= a && \textit{Divide by } c + b. \end{aligned}$$

13. Solve $\dfrac{PV}{T} = \dfrac{pv}{t}$ for v.

$$\begin{aligned} PVt &= pvT && \textit{Multiply by } Tt. \\ \frac{PVt}{pT} &= v && \textit{Divide by } pT. \end{aligned}$$

15. Solve $I = \dfrac{nE}{R+nr}$ for r.

$$I(R+nr) = nE$$
$$IR + Inr = nE$$
$$Inr = nE - IR$$
$$r = \frac{nE - IR}{In}$$

17. Solve $A = \frac{1}{2}h(b+B)$ for b.

$$2A = h(b+B) \qquad \textit{Multiply by 2.}$$
$$\frac{2A}{h} = b + B \qquad \textit{Divide by } h.$$
$$\frac{2A}{h} - B = b, \text{ or } b = \frac{2A - Bh}{h}$$

19. $\dfrac{E}{e} = \dfrac{R+r}{r}$ for r

$$Er = e(R+r) \qquad \textit{Multiply by er.}$$
$$Er = eR + er$$
$$Er - er = eR \qquad \textit{Subtract er.}$$
$$r(E-e) = eR \qquad \textit{Factor out r.}$$
$$r = \frac{eR}{E-e} \qquad \textit{Divide by } E-e.$$

21. To solve the equation $m = \dfrac{ab}{a-b}$ for a, the first step is to multiply each side of the equation by the LCD, $a-b$.

23. Let $x =$ the number of girls in the class.
Write and solve a proportion.

$$\frac{3}{4} = \frac{x}{28}$$
$$28\left(\frac{3}{4}\right) = 28\left(\frac{x}{28}\right) \qquad \textit{Multiply by the LCD, 28.}$$
$$21 = x$$

There are 21 girls and $28 - 21 = 7$ boys in the class.

25.
$$\text{Marin's rate} = \frac{1 \text{ job}}{\text{time to complete 1 job}} = \frac{1 \text{ job}}{2 \text{ hours}} = \tfrac{1}{2} \text{ job per hour}$$

27. Let $x =$ the distance between Chicago and El Paso on the map (in inches).

Set up a proportion with one ratio involving map distances and the other involving actual distances.

$$\frac{x \text{ inches}}{4.125 \text{ inches}} = \frac{1606 \text{ miles}}{1238 \text{ miles}}$$
$$1238x = 4.125(1606)$$
$$1238x = 6624.75$$
$$x \approx 5.351$$

The distance on the map between Chicago and El Paso would be about 5.4 inches.

29. Let $x =$ the distance between Madrid and Rio de Janeiro on the map (in inches).

Set up a proportion with one ratio involving map distances and the other involving actual distances.

$$\frac{x \text{ inches}}{8.5 \text{ inches}} = \frac{5045 \text{ miles}}{5619 \text{ miles}}$$
$$5619x = 8.5(5045)$$
$$5619x = 42{,}882.5$$
$$x \approx 7.632$$

The distance on the map between Madrid and Rio de Janeiro would be about 7.6 inches.

31. Let $x =$ the number of games the Red Sox would win.
Write a proportion.

$$\frac{x}{162} = \frac{31}{50}$$
$$50x = 162(31)$$
$$50x = 5022$$
$$x = 100.44 \approx 100$$

The Red Sox would have won 100 games if the team would have continued to win the same fraction of games.

33. Let $x =$ the number of fish in the lake.
Write and solve a proportion.

$$\frac{\text{total in lake}}{\text{tagged in lake}} = \frac{\text{total in sample}}{\text{tagged in sample}}$$
$$\frac{x}{500} = \frac{400}{8}$$
$$\frac{x}{500} = 50$$
$$x = 500(50) = 25{,}000$$

There are approximately 25,000 fish in the lake.

35. *Step 2*
Let $x =$ the additional number of gallons of gasoline needed.

Step 3
He knows that he can drive 156 miles with 5 gallons of gasoline. He wants to drive 300 miles using $(3+x)$ gallons of gasoline. Set up a proportion.

$$\frac{156}{5} = \frac{300}{3+x}$$

Step 4
Find the cross products and solve for x.

$$156(3+x) = 5(300)$$
$$468 + 156x = 1500$$
$$156x = 1032$$
$$x = \tfrac{1032}{156}$$
$$x \approx 6.6$$

Step 5
He will need about 6.6 more gallons of gasoline.

Step 6
Check The 3 gallons plus the 6.6 gallons equals 9.6 gallons. We'll check the rates (miles/gallon). Note that we could also use gallons/mile.

$$\tfrac{156}{5} = 31.2 \text{ mpg} \qquad \tfrac{300}{9.6} = 31.25 \text{ mpg}$$

The rates are approximately equal, so the solution is correct. Note that we could have used our exact value for x to get the exact rate

$$\frac{300}{3 + (1032/156)} = 31.2.$$

37. Let x represent the amount to administer in milliliters.

$$\frac{100 \text{ mg}}{2 \text{ mL}} = \frac{120 \text{ mg}}{x \text{ mL}}$$
$$2x\left(\frac{100}{2}\right) = 2x\left(\frac{120}{x}\right) \quad \textit{Multiply by the LCD, 2x.}$$
$$100x = 240$$
$$x = 2.4$$

The correct dose is 2.4 mL.

39. Since $\frac{2}{3} = \frac{4}{6} = \frac{6}{9}$, use the proportion

$$\frac{2}{3} = \frac{2x+1}{2x+5}.$$
$$2(2x+5) = 3(2x+1)$$
$$4x + 10 = 6x + 3$$
$$7 = 2x$$
$$\tfrac{7}{2} = x$$

Since $x = \frac{7}{2}$,

$$AC = 2x + 1 = 2(\tfrac{7}{2}) + 1 = 8$$
$$\text{and } DF = 2x + 5 = 2(\tfrac{7}{2}) + 5 = 12.$$

41. *Step 2*
Let x represent the speed of the current of the river. The boat goes 12 mph, so the downstream speed is $12 + x$ and the upstream speed is $12 - x$.

Use $t = \dfrac{d}{r}$ and make a table.

	Distance	Rate	Time
Downstream	10	$12+x$	$\dfrac{10}{12+x}$
Upstream	6	$12-x$	$\dfrac{6}{12-x}$

Step 3
Because the time upstream equals the time downstream,

$$\frac{6}{12-x} = \frac{10}{12+x}$$

Step 4
Multiply by the LCD, $(12-x)(12+x)$.

$$(12-x)(12+x)\left(\frac{6}{12-x}\right) = (12-x)(12+x)\left(\frac{10}{12+x}\right)$$
$$6(12+x) = 10(12-x)$$
$$72 + 6x = 120 - 10x$$
$$16x = 48$$
$$x = 3$$

Step 5
The speed of the current of the river is 3 mph.

Step 6
Check The rate downstream is $12 + 3 = 15$ mph, so she can go 10 miles in $\frac{10}{15} = \frac{2}{3}$ hour. The rate upstream is $12 - 3 = 9$ mph, so she can go 6 miles in $\frac{6}{9} = \frac{2}{3}$ hour. The times are the same, as required.

43. *Step 2*
Find the distance from Montpelier to Columbia. Let x represent that distance.

Complete the table.

	d	r	t
Actual Trip	x	51	$\dfrac{x}{51}$
Alternative Trip	x	60	$\dfrac{x}{60}$

Step 3
At 60 mph, his time at 51 mph would be decreased 3 hr.

$$\frac{x}{60} = \frac{x}{51} - 3$$

Step 4
Multiply by the LCD, 1020.

$$17x = 20x - 3060$$
$$3060 = 3x$$
$$1020 = x$$

Step 5
The distance from Montpelier to Columbia is 1020 miles.

Step 6
Check 1020 miles at 51 mph takes $\frac{1020}{51}$ or 20 hours; 1020 miles at 60 mph takes $\frac{1020}{60}$ or 17 hours; $17 = 20 - 3$ as required.

45. *Step 2*
Let $x =$ the distance from San Francisco to the secret rendezvous.

Make a table.

	d	r	t
First Trip	x	200	$\frac{x}{200}$
Return Trip	x	300	$\frac{x}{300}$

Step 3

Time there	plus	time back	equals	4 hr.
$\frac{x}{200}$	$+$	$\frac{x}{300}$	$=$	4

Step 4
Multiply by the LCD, 600.

$$600\left(\frac{x}{200} + \frac{x}{300}\right) = 600(4)$$
$$3x + 2x = 2400$$
$$5x = 2400$$
$$x = 480$$

Step 5
The distance is 480 miles.

Step 6
Check 480 miles at 200 mph takes $\frac{480}{200}$ or 2.4 hours; 480 miles at 300 mph takes $\frac{480}{300}$ or 1.6 hours. The total time is 4 hours, as required.

47. *Step 2*
Let $x =$ the distance on the first part of the trip.

Make a table.

	d	r	t
First Part	x	60	$\frac{x}{60}$
Second Part	$x + 10$	50	$\frac{x+10}{50}$

Step 3
From the problem, the equation is stated in words: (Note that 30 min $= \frac{1}{2}$ hr.) Time for the second part $=$ time for the first part $+ \frac{1}{2}$. Use the times given in the table to write the equation.

$$\frac{x+10}{50} = \frac{x}{60} + \frac{1}{2}$$

Step 4
Multiply by the LCD, 300.

$$300\left(\frac{x+10}{50}\right) = 300\left(\frac{x}{60} + \frac{1}{2}\right)$$
$$6(x + 10) = 5x + 150$$
$$6x + 60 = 5x + 150$$
$$x = 90$$

Step 5
The distance for both parts of the trip is given by

$$x + (x + 10) = 90 + (90 + 10) = 190.$$

The distance is 190 miles.

Step 6
Check 90 miles at 60 mph takes $\frac{90}{60}$ or $1\frac{1}{2}$ hours; 100 miles at 50 mph takes 2 hours. The second part of the trip takes $\frac{1}{2}$ hour more than the first part, as required.

49. Let $x =$ the time it would take them working together.

Complete the table.

Worker	Rate	Time Working Together	Fractional Part of the Job Done
Butch	$\frac{1}{15}$	x	$\frac{1}{15}x$
Peggy	$\frac{1}{12}$	x	$\frac{1}{12}x$

Part done by Butch	$+$	part done by Peggy	$=$	1 whole job.
$\frac{1}{15}x$	$+$	$\frac{1}{12}x$	$=$	1

Multiply by the LCD, 60.

$$60\left(\tfrac{1}{15}x + \tfrac{1}{12}x\right) = 60 \cdot 1$$
$$4x + 5x = 60$$
$$9x = 60$$
$$x = \tfrac{60}{9} = \tfrac{20}{3} \text{ or } 6\tfrac{2}{3}$$

Together they could do the job in $\frac{20}{3}$ or $6\frac{2}{3}$ minutes.

51. Let $x =$ the time it would take Kuba working alone.

Worker	Rate	Time Working Together	Fractional Part of the Job Done
Jerry	$\frac{1}{20}$	12	$\frac{1}{20}(12) = \frac{3}{5}$
Kuba	$\frac{1}{x}$	12	$\frac{1}{x}(12) = \frac{12}{x}$

Part done by Jerry + part done by Kuba = 1 whole job.

$$\frac{3}{5} + \frac{12}{x} = 1$$

Multiply by the LCD, $5x$.

$$3x + 60 = 5x$$
$$60 = 2x$$
$$30 = x$$

It would take Kuba 30 hours to do the job alone.

53. Let $x =$ the time it will take to fill the vat if both pipes are open.

	Rate	Time to Fill the Vat	Fractional Part of the Job Done
Inlet Pipe	$\frac{1}{10}$	x	$\frac{1}{10}x$
Outlet Pipe	$-\frac{1}{20}$	x	$-\frac{1}{20}x$

Notice that the rate of the outlet pipe is negative because it will empty the vat, not fill it.

Part done with the inlet pipe open + Part done with the outlet pipe open = 1 whole job.

$$\frac{1}{10}x + \left(-\frac{1}{20}x\right) = 1$$

Multiply by the LCD, 20.

$$2x - x = 20$$
$$x = 20$$

It will take 20 hours to fill the vat.

55. Let $x =$ the time from Mimi's arrival home to the time the place is a shambles.

	Rate	Time to Mess up House	Fractional Part of the Job Done
Hortense and Mort	$-\frac{1}{7}$	x	$-\frac{1}{7}x$
Mimi	$\frac{1}{2}$	x	$\frac{1}{2}x$

Notice that Hortense and Mort's rate is negative since they are opposing the messing up by cleaning the house.

Part done by Hortense and Mort + Part done by Mimi = 1 whole job of messing up.

$$-\frac{1}{7}x + \frac{1}{2}x = 1$$

Multiply by the LCD, 14.

$$-2x + 7x = 14$$
$$5x = 14$$
$$x = \frac{14}{5} \text{ or } 2\tfrac{4}{5}$$

It would take $\frac{14}{5}$ or $2\frac{4}{5}$ hours after Mimi got home for the house to be a shambles.

8.6 Variation

8.6 Margin Exercises

1. **(a)** Let E represent her earnings for d days. E varies directly as d, so

$$E = kd,$$

where k represents Ginny's daily wage. Let $d = 17$ and $E = 1334.50$.

$$1334.50 = 17k$$
$$78.50 = k$$

Her daily wage is \$78.50. Thus,

$$E = 78.50d.$$

(b) Let d represent the distance traveled in h hours. d varies directly as h, so

$$d = kh,$$

where k represents the speed. Let $d = 100$ and $h = 2$.

$$100 = k \cdot 2$$
$$50 = k$$

Then $d = 50h$.

2. Let c represent the cost of using h kilowatt hours. Use $c = kh$ with $c = 52$ and $h = 800$ to find k.

$$\begin{aligned} c &= kh \\ 52 &= k(800) \\ \tfrac{52}{800} &= k \\ \tfrac{13}{200} &= k \end{aligned}$$

So $c = \frac{13}{200}h$.

Let $h = 1000$. Find c.

$$c = \tfrac{13}{200}(1000) = 65$$

1000 kilowatt-hours cost \$65.

3. **(a)** Let A represent the area of a circle and r its radius.

A varies directly as r^2, so

$$A = kr^2,$$

for some constant k. Since $A = 28.278$ when $r = 3$, substitute these values in the equation and solve for k.

$$\begin{aligned} A &= kr^2 \\ 28.278 &= k(3)^2 \\ 28.278 &= 9k \\ 3.142 &= k \end{aligned}$$

So $A = 3.142r^2$.

(b) Let $r = 4.1$. Find A.

$$\begin{aligned} A &= 3.142r^2 \\ A &= 3.142(4.1)^2 \\ A &= 52.817 \end{aligned}$$

(to the nearest thousandth)

The area is 52.817 in.^{2}.

4. **(a)** Let V represent the volume, and P the pressure. V varies inversely as P, so

$$V = \frac{k}{P},$$

for some constant k. Since $V = 10$ when $P = 6$, find k.

$$\begin{aligned} V &= \frac{k}{P} \\ 10 &= \frac{k}{6} \\ 60 &= k \end{aligned}$$

So $V = \dfrac{60}{P}$.

(b) Let $P = 12$. Find V.

$$V = \frac{60}{P} = \frac{60}{12} = 5$$

The volume is 5 cm^3.

5. Let V represent the volume, W the width, and L the length. The volume is proportional to the width and length of the box, so

$$V = kWL.$$

Let $W = 2$, $L = 4$, and $V = 12$. Find k.

$$\begin{aligned} 12 &= k(2)(4) \\ 12 &= 8k \\ k &= \tfrac{12}{8} = \tfrac{3}{2} \end{aligned}$$

So $V = \frac{3}{2}WL$.

Now let $L = 5$ and $W = 3$. Find V.

$$\begin{aligned} V &= \tfrac{3}{2}(3)(5) \\ &= \tfrac{45}{2} \quad \text{or} \quad 22.5 \end{aligned}$$

The volume is 22.5 ft^3.

6. Let L represent the maximum load, d the diameter of the cross section, and h the height.

L varies directly as d^4 and inversely as h^2, so

$$L = \frac{kd^4}{h^2},$$

for some constant k. Let $L = 8$ when $h = 9$ and $d = 1$. Find k.

$$\begin{aligned} L &= \frac{kd^4}{h^2} \\ 8 &= \frac{k(1)^4}{9^2} \\ 648 &= k \end{aligned}$$

So $L = \dfrac{648d^4}{h^2}$.

Now let $h = 12$ and $d = \frac{2}{3}$.

$$\begin{aligned} L &= \frac{648(\frac{2}{3})^4}{12^2} \\ &= \frac{648(\frac{16}{81})}{144} = \frac{128}{144} = \frac{8}{9} \end{aligned}$$

The column can support $\frac{8}{9}$ metric ton.

8.6 Section Exercises

1. The equation $y = \dfrac{3}{x}$ represents *inverse* variation. y varies inversely as x because x is in the denominator.

3. The equation $y = 10x^2$ represents *direct* variation. The number 10 is the constant of variation, and y varies directly as the square of x.

5. The equation $y = 3xz^4$ represents *joint* variation. y varies directly as x and z^4.

7. The equation $y = \frac{4x}{wz}$ represents *combined* variation because it is a combination of direct and inverse variation.

9. "x varies directly as y" means

$$x = ky$$

for some constant k.
Substitute $x = 9$ and $y = 3$ in the equation and solve for k.

$$\begin{aligned} x &= ky \\ 9 &= k(3) \\ k &= \tfrac{9}{3} = 3 \end{aligned}$$

So $x = 3y$.
To find x when $y = 12$, substitute 12 for y in the equation.

$$x = 3y = 3(12) = 36$$

11. "z varies inversely as w" means

$$z = \frac{k}{w}$$

for some constant k. Since $z = 10$ when $w = 0.5$, substitute these values in the equation and solve for k.

$$\begin{aligned} z &= \frac{k}{w} \\ 10 &= \frac{k}{0.5} \\ k &= 10(0.5) = 5 \end{aligned}$$

So $z = \frac{5}{w}$.
To find z when $w = 8$, substitute 8 for w in the equation.

$$z = \frac{5}{w} = \frac{5}{8} \text{ or } 0.625$$

13. "p varies jointly as q and r^2" means

$$p = kqr^2$$

for some constant k. Given that $p = 200$ when $q = 2$ and $r = 3$, solve for k.

$$\begin{aligned} p &= kqr^2 \\ 200 &= k(2)(3)^2 \\ 200 &= 18k \\ k &= \tfrac{200}{18} = \tfrac{100}{9} \end{aligned}$$

So $p = \frac{100}{9}qr^2$. Now let $q = 5$ and $r = 2$.

$$\begin{aligned} p &= \tfrac{100}{9}qr^2 \\ p &= \tfrac{100}{9}(5)(2)^2 \\ &= \tfrac{100}{9}(20) \\ &= \tfrac{2000}{9} \text{ or } 222\tfrac{2}{9} \end{aligned}$$

15. For $k > 0$, if y varies directly as x (then $y = kx$), when x increases, y increases, and when x decreases, y decreases.

17. If y varies inversely as x, x is in the denominator; however, if y varies directly as x, x is in the numerator. Also, for $k > 0$, with inverse variation, as x increases, y decreases. With direct variation, y increases as x increases.

19. Let $x =$ the number of gallons he bought and let $C =$ the cost.
C varies directly as x, so

$$C = kx,$$

where k represents the cost per gallon.
Since $C = 36.79$ when $x = 8$,

$$\begin{aligned} 36.79 &= k(8) \\ k &= \frac{36.79}{8} = 4.59875 \approx 4.599. \end{aligned}$$

The price per gallon is $\$4.59\frac{9}{10}$.

21. Let $V =$ the volume of the can and let $h =$ the height of the can.
V varies directly as h, so

$$V = kh.$$

Since $V = 300$ when $h = 10.62$,

$$\begin{aligned} 300 &= k(10.62) \\ k &= \frac{300}{10.62} \approx 28.25. \end{aligned}$$

So $V = 28.25h$. Now let $h = 15.92$.

$$\begin{aligned} V &= 28.25h \\ V &= 28.25(15.92) = 449.74 \end{aligned}$$

The volume is about 450 cm^3.

23. Let $d =$ the distance and $t =$ the time.
d varies directly as the square of t, so $d = kt^2$.
Let $d = -576$ and $t = 6$. (You could also use $d = 576$, but the negative sign indicates the direction of the body.)

$$\begin{aligned} -576 &= k(6)^2 \\ -576 &= 36k \\ -16 &= k \end{aligned}$$

So $d = -16t^2$. Now let $t = 4$.

$$d = -16(4)^2 = -256$$

The object fell 256 ft in the first 4 seconds.

25. Let $C =$ the current and $R =$ the resistance. C varies inversely as R, so

$$C = \frac{k}{R}$$

for some constant k. Since $C = 20$ when $R = 5$, substitute these values in the equation and solve for k.

$$C = \frac{k}{R}$$
$$20 = \frac{k}{5}$$
$$k = 20(5) = 100$$

So $C = \dfrac{100}{R}$. Now let $R = 7.5$.

$$C = \frac{100}{7.5} = \frac{1000}{75} = \frac{40}{3}, \text{ or } 13\frac{1}{3}$$

The current is $13\frac{1}{3}$ amperes.

27. Let $I =$ the illumination (amount of light) produced by a light source and $d =$ the distance from the source. I varies inversely as d^2, so

$$I = \frac{k}{d^2}$$

for some constant k. Since $I = 768$ when $d = 1$, substitute these values in the equation and solve for k.

$$I = \frac{k}{d^2}$$
$$768 = \frac{k}{1^2}$$
$$768 = k$$

So $I = \dfrac{768}{d^2}$. Now let $d = 6$.

$$I = \frac{768}{d^2}$$
$$I = \frac{768}{6^2} = \frac{768}{36} = \frac{64}{3}, \text{ or } 21\frac{1}{3}$$

The illumination produced by the light source is $21\frac{1}{3}$ foot-candles.

29. Let $I =$ the simple interest, P the principal, and t the time. Since I varies jointly as the principal and time, there is a constant k such that $I = kPt$. Find k by replacing I with 280, P with 2000, and t with 4.

$$I = kPt$$
$$280 = k(2000)(4)$$
$$k = \tfrac{280}{8000} = 0.035$$

So $I = 0.035Pt$. Now let $t = 6$.

$$I = 0.035(2000)(6)$$
$$= 420$$

The interest would be \$420.

31. Let $F =$ the force, $w =$ the weight of the car, $s =$ the speed, and $r =$ the radius. The force varies inversely as the radius and jointly as the weight and the square of the speed, so

$$F = \frac{kws^2}{r}.$$

Let $F = 242$, $w = 2000$, $r = 500$, and $s = 30$.

$$242 = \frac{k(2000)(30)^2}{500}$$
$$k = \frac{242(500)}{2000(900)} = \frac{121}{1800}$$

So $F = \dfrac{121ws^2}{1800r}$.

Let $r = 750$, $s = 50$, and $w = 2000$.

$$F = \frac{121(2000)(50)^2}{1800(750)} \approx 448.1$$

Approximately 448.1 lb of force would be needed.

33. Let $N =$ the number of long distance calls,
$p_1 =$ the population of City 1,
$p_2 =$ the population of City 2,
and $d =$ the distance between them.

$$N = \frac{kp_1p_2}{d}$$

Let $N = 80{,}000$, $p_1 = 70{,}000$, $p_2 = 100{,}000$, and $d = 400$.

$$80{,}000 = \frac{k(70{,}000)(100{,}000)}{400}$$
$$80{,}000 = 17{,}500{,}000k$$
$$k = \frac{80{,}000}{17{,}500{,}000} = \frac{4}{875}$$
$$N = \frac{4}{875}\left(\frac{p_1p_2}{d}\right)$$

Let $p_1 = 50{,}000$, $p_2 = 75{,}000$, and $d = 250$.

$$N = \frac{4}{875}\left(\frac{50{,}000 \cdot 75{,}000}{250}\right)$$
$$= \frac{480{,}000}{7}, \text{ or } 68{,}571\frac{3}{7}$$

Rounded to the nearest hundred, there are approximately 68,600 calls.

35. The weight W of a bass varies jointly as its girth G and the square of its length L, so

$$W = kGL^2$$

for some constant k. Substitute 22.7 for W, 21 for G, and 36 for L.

$$\begin{aligned} 22.7 &= k(21)(36)^2 \\ k &= \frac{22.7}{27{,}216} \approx 0.000834 \end{aligned}$$

So $W = 0.000834GL^2$. Now let $G = 18$ and $L = 28$.

$$\begin{aligned} W &= 0.000834GL^2 \\ &= 0.000834(18)(28)^2 \\ &\approx 11.8 \end{aligned}$$

The bass would weigh about 11.8 pounds.

37. The ordered pairs in the form of (gallons, price) are $(0, 0)$ and $(1, 4.45)$.

38. Let $(x_1, y_1) = (0, 0)$ and $(x_2, y_2) = (1, 4.45)$. Then

$$m = \frac{\text{change in } y}{\text{change in } x} = \frac{4.45 - 0}{1 - 0} = 4.45.$$

The slope is 4.45.

39. Since $m = 4.45$ and $b = 0$, the equation is

$$\begin{aligned} y &= 4.45x + 0 \\ \text{or} \quad y &= 4.45x. \end{aligned}$$

40. If $f(x) = ax + b$, then $a = 4.45$ and $b = 0$.

41. The value of a, 4.45, is the price in dollars per gallon of gasoline. It is also the slope of the line.

42. Since $f(x) = 4.45x$, it fits the form for a direct variation, that is, $y = kx$. The value of a, 4.45, is the constant of variation $(k = a)$.

Chapter 8 Review Exercises

1. **(a)** $f(x) = \dfrac{-7}{3x + 18}$

Set the denominator equal to zero and solve.

$$\begin{aligned} 3x + 18 &= 0 \\ 3x &= -18 \\ x &= -6 \end{aligned}$$

The number -6 makes the expression undefined, so it is excluded from the domain.

(b) The domain is $\{x \mid x \neq -6\}$.

2. **(a)** $f(x) = \dfrac{5x + 17}{x^2 - 7x + 10}$

Set the denominator equal to zero and solve.

$$\begin{aligned} x^2 - 7x + 10 &= 0 \\ (x - 5)(x - 2) &= 0 \end{aligned}$$

$$\begin{aligned} x - 5 = 0 \quad &\text{or} \quad x - 2 = 0 \\ x = 5 \quad &\text{or} \quad x = 2 \end{aligned}$$

The numbers 2 and 5 make the expression undefined, so they are excluded from the domain.

(b) The domain is $\{x \mid x \neq 2, 5\}$.

3. **(a)** $f(x) = \dfrac{9}{x^2 - 18x + 81}$

Set the denominator equal to zero and solve.

$$\begin{aligned} x^2 - 18x + 81 &= 0 \\ (x - 9)^2 &= 0 \\ x - 9 &= 0 \\ x &= 9 \end{aligned}$$

The number 9 makes the expression undefined, so it is excluded from the domain.

(b) The domain is $\{x \mid x \neq 9\}$.

4. $\dfrac{12x^2 + 6x}{24x + 12} = \dfrac{6x(2x + 1)}{12(2x + 1)} = \dfrac{x}{2}$

5. $$\begin{aligned} \frac{25m^2 - n^2}{25m^2 - 10mn + n^2} &= \frac{(5m + n)(5m - n)}{(5m - n)(5m - n)} \\ &= \frac{5m + n}{5m - n} \end{aligned}$$

6. $$\begin{aligned} \frac{r - 2}{4 - r^2} &= \frac{r - 2}{(2 + r)(2 - r)} \\ &= \frac{(-1)(2 - r)}{(2 + r)(2 - r)} \\ &= \frac{-1}{2 + r} \end{aligned}$$

7. The reciprocal of a rational expression is another rational expression such that the two rational expressions have a product of 1.

8. $$\begin{aligned} &\frac{(2y + 3)^2}{5y} \cdot \frac{15y^3}{4y^2 - 9} \\ &= \frac{15y^3(2y + 3)^2}{5y(2y + 3)(2y - 3)} \\ &= \frac{3y^2(2y + 3)}{2y - 3} \end{aligned}$$

9. $\dfrac{w^2-16}{w} \cdot \dfrac{3}{4-w}$

$= \dfrac{(w-4)(w+4)}{w} \cdot \dfrac{3}{4-w}$

$= \dfrac{(-1)(4-w)(w+4)}{w} \cdot \dfrac{3}{4-w}$

$= \dfrac{-3(w+4)}{w}$

10. $\dfrac{z^2-z-6}{z-6} \div \dfrac{z^2+2z-15}{z^2-6z}$

$= \dfrac{z^2-z-6}{z-6} \cdot \dfrac{z^2-6z}{z^2+2z-15}$

$= \dfrac{(z-3)(z+2)}{z-6} \cdot \dfrac{z(z-6)}{(z-3)(z+5)}$

$= \dfrac{z(z+2)}{z+5}$

11. $\dfrac{m^3-n^3}{m^2-n^2} \div \dfrac{m^2+mn+n^2}{m+n}$

Multiply by the reciprocal.

$= \dfrac{m^3-n^3}{m^2-n^2} \cdot \dfrac{m+n}{m^2+mn+n^2}$

$= \dfrac{(m-n)(m^2+mn+n^2)}{(m-n)(m+n)} \cdot \dfrac{m+n}{m^2+mn+n^2}$

$= 1$

12. $32b^3, 24b^5$

Factor each denominator.

$$32b^3 = 2 \cdot 2 \cdot 2 \cdot 2 \cdot 2 \cdot b^3 = 2^5 \cdot b^3$$
$$24b^5 = 2 \cdot 2 \cdot 2 \cdot 3 \cdot b^5 = 2^3 \cdot 3 \cdot b^5$$

$\text{LCD} = 2^5 \cdot 3 \cdot b^5 = 96b^5$

13. $9r^2,\ 3r+1$

Factor each denominator.

$$9r^2 = 3^2 \cdot r^2$$

The second denominator is already in factored form. The LCD is

$$3^2 \cdot r^2 \cdot (3r+1), \quad \text{or} \quad 9r^2(3r+1).$$

14. $6x^2+13x-5,\ 9x^2+9x-4$

Factor each denominator.

$$6x^2+13x-5 = (3x-1)(2x+5)$$
$$9x^2+9x-4 = (3x-1)(3x+4)$$

The LCD is $(3x-1)(2x+5)(3x+4)$.

15. $\dfrac{8}{z} - \dfrac{3}{2z^2}$

$= \dfrac{8 \cdot 2z}{z \cdot 2z} - \dfrac{3}{2z^2} \qquad LCD = 2z^2$

$= \dfrac{16z}{2z^2} - \dfrac{3}{2z^2}$

$= \dfrac{16z-3}{2z^2}$

16. $\dfrac{5y+13}{y+1} - \dfrac{1-7y}{y+1}$

$= \dfrac{5y+13-(1-7y)}{y+1}$

$= \dfrac{5y+13-1+7y}{y+1}$

$= \dfrac{12y+12}{y+1}$

$= \dfrac{12(y+1)}{y+1} = 12$

17. $\dfrac{6}{5a+10} + \dfrac{7}{6a+12}$

$= \dfrac{6}{5(a+2)} + \dfrac{7}{6(a+2)}$

The LCD is $30(a+2)$.

$= \dfrac{6 \cdot 6}{5(a+2) \cdot 6} + \dfrac{7 \cdot 5}{6(a+2) \cdot 5}$

$= \dfrac{36}{30(a+2)} + \dfrac{35}{30(a+2)}$

$= \dfrac{36+35}{30(a+2)} = \dfrac{71}{30(a+2)}$

18. $\dfrac{3r}{10r^2-3rs-s^2} + \dfrac{2r}{2r^2+rs-s^2}$

$= \dfrac{3r}{(5r+s)(2r-s)} + \dfrac{2r}{(2r-s)(r+s)}$

The LCD is $(5r+s)(2r-s)(r+s)$.

$= \dfrac{3r(r+s)}{(5r+s)(2r-s)(r+s)}$

$\quad + \dfrac{2r(5r+s)}{(2r-s)(r+s)(5r+s)}$

$= \dfrac{3r^2+3rs+10r^2+2rs}{(5r+s)(2r-s)(r+s)}$

$= \dfrac{13r^2+5rs}{(5r+s)(2r-s)(r+s)}$

19. $\dfrac{\frac{3}{t}+2}{\frac{4}{t}-7}$

Multiply the numerator and denominator by the LCD of all the fractions, t.

$$= \frac{t\left(\frac{3}{t}+2\right)}{t\left(\frac{4}{t}-7\right)} = \frac{3+2t}{4-7t}$$

20. $\dfrac{\frac{2}{m-3n}}{\frac{1}{3n-m}} = \dfrac{2}{m-3n} \div \dfrac{1}{3n-m}$

$$= \frac{2}{m-3n} \cdot \frac{3n-m}{1}$$
$$= \frac{2}{m-3n} \cdot \frac{-1(m-3n)}{1}$$
$$= \frac{2 \cdot (-1)}{1} = -2$$

21. $\dfrac{\frac{3}{p}-\frac{2}{q}}{\frac{9q^2-4p^2}{qp}}$

Multiply the numerator and denominator by the LCD of all the fractions, qp.

$$= \frac{qp\left(\frac{3}{p}-\frac{2}{q}\right)}{qp\left(\frac{9q^2-4p^2}{qp}\right)}$$
$$= \frac{3q-2p}{9q^2-4p^2}$$
$$= \frac{3q-2p}{(3q+2p)(3q-2p)}$$
$$= \frac{1}{3q+2p}$$

22. $\dfrac{x^{-2}-y^{-2}}{x^{-1}-y^{-1}} = \dfrac{\frac{1}{x^2}-\frac{1}{y^2}}{\frac{1}{x}-\frac{1}{y}}$

Multiply the numerator and denominator by the LCD of all the fractions, x^2y^2.

$$= \frac{x^2y^2\left(\frac{1}{x^2}-\frac{1}{y^2}\right)}{x^2y^2\left(\frac{1}{x}-\frac{1}{y}\right)}$$
$$= \frac{y^2-x^2}{xy^2-x^2y}$$
$$= \frac{(y+x)(y-x)}{xy(y-x)}$$
$$= \frac{y+x}{xy}$$

23. $\dfrac{1}{t+4}+\dfrac{1}{2}=\dfrac{3}{2t+8}$

$$\frac{1}{t+4}+\frac{1}{2}=\frac{3}{2(t+4)}$$

Multiply by the LCD, $2(t+4)$. $(t \neq -4)$

$$2(t+4)\left(\frac{1}{t+4}+\frac{1}{2}\right) = 2(t+4)\left(\frac{3}{2(t+4)}\right)$$
$$2+(t+4)=3$$
$$t+6=3$$
$$t=-3$$

Check $t=-3$: $1+\frac{1}{2}=\frac{3}{2}$ *True*

Solution set: $\{-3\}$

24. $\dfrac{-5m}{m+1}+\dfrac{m}{3m+3}=\dfrac{56}{6m+6}$

$$\frac{-5m}{m+1}+\frac{m}{3(m+1)}=\frac{56}{6(m+1)}$$
$$\frac{-5m}{m+1}+\frac{m}{3(m+1)}=\frac{28}{3(m+1)}$$

Multiply by the LCD, $3(m+1)$. $(m \neq -1)$

$$3(m+1)\left(\frac{-5m}{m+1}+\frac{m}{3(m+1)}\right) = 3(m+1)\left(\frac{28}{3(m+1)}\right)$$
$$-15m+m=28$$
$$-14m=28$$
$$m=-2$$

Check $m=-2$: $-10+\frac{2}{3}=-\frac{56}{6}$ *True*

Solution set: $\{-2\}$

25. $$\frac{2}{k-1}-\frac{4k+1}{k^2-1}=\frac{-1}{k+1}$$

$$\frac{2}{k-1}-\frac{4k+1}{(k+1)(k-1)}=\frac{-1}{k+1}$$

Multiply by the LCD, $(k+1)(k-1)$. $(k \neq \pm 1)$

$$(k+1)(k-1)\left(\frac{2}{k-1}-\frac{4k+1}{(k+1)(k-1)}\right)=(k+1)(k-1)\left(\frac{-1}{k+1}\right)$$

$$2(k+1)-(4k+1)=-1(k-1)$$
$$2k+2-4k-1=-k+1$$
$$-2k+1=-k+1$$
$$0=k$$

Check $k=0$: $-2+1=-1$ *True*
Solution set: $\{0\}$

26. $$\frac{5}{x+2}+\frac{3}{x+3}=\frac{x}{x^2+5x+6}$$

$$\frac{5}{x+2}+\frac{3}{x+3}=\frac{x}{(x+2)(x+3)}$$

Multiply by the LCD, $(x+2)(x+3)$. $(x \neq -3, -2)$

$$(x+2)(x+3)\left(\frac{5}{x+2}+\frac{3}{x+3}\right)=(x+2)(x+3)\left(\frac{x}{(x+2)(x+3)}\right)$$

$$5(x+3)+3(x+2)=x$$
$$5x+15+3x+6=x$$
$$8x+21=x$$
$$7x=-21$$
$$x=-3$$

Substituting -3 in the original equation results in division by 0, so -3 is not a solution.

Solution set: $\emptyset$

27. Although her algebra was correct, 3 is not a solution because it is not in the domain of the variable in the equation. Thus, $\emptyset$ is correct.

28. In simplifying the expression, we are combining terms to get a single fraction with a denominator of $6x$. In solving the equation, we are finding a value for x that makes the equation true.

29. The graph in choice **C** has a vertical asymptote. Its equation is $x=0$. The equation of its horizontal asymptote is $y=0$.

30. $$\frac{1}{A}=\frac{1}{B}+\frac{1}{C}$$

Let $B=30$ and $C=10$. Find A.

$$\frac{1}{A}=\frac{1}{30}+\frac{1}{10}$$

To solve for A, multiply each side by the LCD, $30A$.

$$30A\left(\frac{1}{A}\right)=30A\left(\frac{1}{30}+\frac{1}{10}\right)$$
$$30=A+3A$$
$$30=4A$$
$$A=\frac{30}{4}=\frac{15}{2}$$

31. Solve $V=\frac{1}{3}\pi r^2 h$ for h.

$3V=\pi r^2 h$ *Multiply by 3.*

$\frac{3V}{\pi r^2}=h$ *Divide by πr^2.*

32. Solve $\mu=\frac{Mv}{M+m}$ for M.

$\mu(M+m)=Mv$ *Multiply by $M+m$.*

$$\mu M+\mu m=Mv$$
$$\mu m=Mv-\mu M$$
$$m\mu=M(v-\mu)$$
$$M=\frac{m\mu}{v-\mu}$$

33. Let $x=$ the speed of the boat in still water.

Use $d=rt$, or $t=\frac{d}{r}$, to make a table.

	Distance	**Rate**	**Time**
Downstream	40	$x+4$	$\frac{40}{x+4}$
Upstream	24	$x-4$	$\frac{24}{x-4}$

Because the times are equal,

$$\frac{40}{x+4}=\frac{24}{x-4}.$$

Multiply by the LCD, $(x+4)(x-4)$. $(x \neq -4, 4)$

$$(x+4)(x-4)\left(\frac{40}{x+4}\right)=(x+4)(x-4)\left(\frac{24}{x-4}\right)$$
$$40(x-4)=24(x+4)$$
$$40x-160=24x+96$$
$$16x=256$$
$$x=16$$

The speed of the boat in still water is 16 km/hr.

34. Let $x =$ the time it takes to fill the sink with both taps open.

Make a table.

	Rate	Time Working Together	Fractional Part of the Job Done
Cold	$\frac{1}{8}$	x	$\frac{x}{8}$
Hot	$\frac{1}{12}$	x	$\frac{x}{12}$

Part done by cold	plus	part done by hot	equals	1 whole job.
$\frac{x}{8}$	$+$	$\frac{x}{12}$	$=$	1

Multiply by the LCD, 24.

$$24\left(\frac{x}{8}+\frac{x}{12}\right) = 24\cdot 1$$
$$3x + 2x = 24$$
$$5x = 24$$
$$x = \tfrac{24}{5}, \quad \text{or} \quad 4\tfrac{4}{5}$$

The sink will be filled in $\frac{24}{5}$ or $4\frac{4}{5}$ minutes.

35. If y varies inversely as x, then $y = \dfrac{k}{x}$, for some constant k. This form fits choice **C**.

36. "m varies inversely as p^2" means

$$m = \frac{k}{p^2}$$

for some constant k. Since $m = 20$ when $p = 2$, substitute these values in the equation and solve for k.

$$m = \frac{k}{p^2}$$
$$20 = \frac{k}{2^2}$$
$$k = 20(4) = 80$$

So $m = \dfrac{80}{p^2}$. Now let $p = 5$.

$$m = \frac{80}{p^2} = \frac{80}{5^2} = \frac{16}{5}$$

37. Let $v =$ the viewing distance and $e =$ the amount of enlargement.
v varies directly as e, so

$$v = ke$$

for some constant k. Since $v = 250$ when $e = 5$, substitute these values in the equation and solve for k.

$$v = ke$$
$$250 = k(5)$$
$$50 = k$$

So $v = 50e$. Now let $e = 8.6$.

$$v = 50(8.6) = 430$$

It should be viewed from 430 mm.

38. The volume V of a rectangular box of a given height is proportional to its width W and length L, so

$$V = kWL$$

for some constant k. Substitute 4 for W, 8 for L, and 64 for V.

$$64 = k(4)(8)$$
$$k = \tfrac{64}{32} = 2$$

So $V = 2WL$. Now let $W = 3$ and $L = 6$.

$$V = 2(3)(6) = 36$$

The volume is 36 ft^3.

39. **[8.1]** $$\frac{x+2y}{x^2-4y^2} = \frac{x+2y}{(x+2y)(x-2y)} = \frac{1}{x-2y}$$

40. **[8.1]** $$\frac{x^2+2x-15}{x^2-x-6} = \frac{(x+5)(x-3)}{(x-3)(x+2)} = \frac{x+5}{x+2}$$

41. **[8.2]** $$\frac{2}{m} + \frac{5}{3m^2}$$

The LCD is $3m^2$.

$$= \frac{2\cdot 3m}{m\cdot 3m} + \frac{5}{3m^2}$$
$$= \frac{6m}{3m^2} + \frac{5}{3m^2} = \frac{6m+5}{3m^2}$$

42. **[8.1]** $$\frac{k^2-6k+9}{1-216k^3}\cdot\frac{6k^2+17k-3}{9-k^2}$$

Factor $1-216k^3$ as the difference of cubes, $1^3-(6k)^3$.

$$= \frac{(k-3)(k-3)}{(1-6k)(1+6k+36k^2)} \cdot \frac{(6k-1)(k+3)}{(3-k)(3+k)}$$
$$= \frac{(k-3)(k-3)}{(-1)(6k-1)(1+6k+36k^2)} \cdot \frac{(6k-1)(k+3)}{(-1)(k-3)(k+3)}$$
$$= \frac{k-3}{1+6k+36k^2} \quad \text{or} \quad \frac{k-3}{36k^2+6k+1}$$

43. **[8.3]** $\dfrac{\dfrac{-3}{x}+\dfrac{x}{2}}{1+\dfrac{x+1}{x}}$

Multiply the numerator and denominator by the LCD of all the fractions, $2x$.

$$=\frac{2x\left(\dfrac{-3}{x}+\dfrac{x}{2}\right)}{2x\left(1+\dfrac{x+1}{x}\right)}$$

$$=\frac{-6+x^2}{2x+2(x+1)}$$

$$=\frac{x^2-6}{2x+2x+2}$$

$$=\frac{x^2-6}{4x+2}=\frac{x^2-6}{2(2x+1)}$$

44. **[8.1]** $\dfrac{9x^2+46x+5}{3x^2-2x-1}\div\dfrac{x^2+11x+30}{x^3+5x^2-6x}$

Multiply by the reciprocal.

$$=\frac{9x^2+46x+5}{3x^2-2x-1}\cdot\frac{x(x^2+5x-6)}{x^2+11x+30}$$

$$=\frac{(9x+1)(x+5)}{(3x+1)(x-1)}\cdot\frac{x(x+6)(x-1)}{(x+6)(x+5)}$$

$$=\frac{x(9x+1)}{3x+1}$$

45. **[8.3]** $\dfrac{\dfrac{3}{x}-5}{6+\dfrac{1}{x}}$

Multiply the numerator and denominator by the LCD of all the fractions, x.

$$=\frac{x\left(\dfrac{3}{x}-5\right)}{x\left(6+\dfrac{1}{x}\right)}$$

$$=\frac{3-5x}{6x+1}$$

46. **[8.2]** $\dfrac{9}{3-x}-\dfrac{2}{x-3}$

$$=\frac{9}{3-x}-\frac{2(-1)}{(x-3)(-1)}$$

$$=\frac{9}{3-x}-\frac{-2}{3-x}$$

$$=\frac{9-(-2)}{3-x}$$

$$=\frac{11}{3-x},\ \text{or}\ \frac{-11}{x-3}$$

47. **[8.1]** $\dfrac{4y+16}{30}\div\dfrac{2y+8}{5}$

Multiply by the reciprocal.

$$=\frac{4y+16}{30}\cdot\frac{5}{2y+8}$$

$$=\frac{4(y+4)}{30}\cdot\frac{5}{2(y+4)}$$

$$=\frac{4\cdot 5}{2\cdot 30}=\frac{2}{6}=\frac{1}{3}$$

48. **[8.3]** $\dfrac{t^{-2}+s^{-2}}{t^{-1}-s^{-1}}=\dfrac{\dfrac{1}{t^2}+\dfrac{1}{s^2}}{\dfrac{1}{t}-\dfrac{1}{s}}$

Multiply the numerator and denominator by the LCD of all the fractions, t^2s^2.

$$=\frac{t^2s^2\left(\dfrac{1}{t^2}+\dfrac{1}{s^2}\right)}{t^2s^2\left(\dfrac{1}{t}-\dfrac{1}{s}\right)}$$

$$=\frac{s^2+t^2}{ts^2-t^2s}$$

$$=\frac{s^2+t^2}{st(s-t)}$$

49. **[8.2]**

$$\frac{4a}{a^2-ab-2b^2}-\frac{6b-a}{a^2+4ab+3b^2}$$

$$=\frac{4a}{(a-2b)(a+b)}-\frac{6b-a}{(a+3b)(a+b)}$$

The LCD is $(a+3b)(a-2b)(a+b)$.

$$=\frac{4a(a+3b)}{(a-2b)(a+b)(a+3b)}-\frac{(6b-a)(a-2b)}{(a+3b)(a+b)(a-2b)}$$

$$=\frac{4a(a+3b)-(6b-a)(a-2b)}{(a+3b)(a+b)(a-2b)}$$

$$=\frac{4a^2+12ab-(6ab-12b^2-a^2+2ab)}{(a+3b)(a+b)(a-2b)}$$

$$=\frac{4a^2+12ab-6ab+12b^2+a^2-2ab}{(a+3b)(a+b)(a-2b)}$$

$$=\frac{5a^2+4ab+12b^2}{(a+3b)(a+b)(a-2b)}$$

50. **[8.2]** $\dfrac{a}{b}+\dfrac{b}{c}+\dfrac{c}{d}$

The LCD is bcd.

$$=\frac{a\cdot cd}{b\cdot cd}+\frac{b\cdot bd}{c\cdot bd}+\frac{c\cdot bc}{d\cdot bc}$$

$$=\frac{acd+b^2d+bc^2}{bcd}$$

51. **[8.4]**

$$\frac{x+3}{x^2-5x+4}-\frac{1}{x}=\frac{2}{x^2-4x}$$
$$\frac{x+3}{(x-4)(x-1)}-\frac{1}{x}=\frac{2}{x(x-4)}$$

Multiply by the LCD, $x(x-4)(x-1)$. $(x \neq 0, 1, 4)$

$$x(x-4)(x-1)\left(\frac{x+3}{(x-4)(x-1)}-\frac{1}{x}\right)$$
$$= x(x-4)(x-1)\cdot\left(\frac{2}{x(x-4)}\right)$$
$$x(x+3)-(x-4)(x-1)=2(x-1)$$
$$x^2+3x-(x^2-5x+4)=2x-2$$
$$x^2+3x-x^2+5x-4=2x-2$$
$$8x-4=2x-2$$
$$6x=2$$
$$x=\tfrac{1}{3}$$

Check $x=\frac{1}{3}$: $\frac{15}{11}-3=-\frac{18}{11}$ *True*

Solution set: $\{\frac{1}{3}\}$

52. **[8.4]** Solve $A=\dfrac{Rr}{R+r}$ for r.

$$A(R+r)=Rr$$
$$AR+Ar=Rr$$
$$AR=Rr-Ar$$
$$AR=(R-A)r$$
$$\frac{AR}{R-A}=r, \quad \text{or} \quad r=\frac{-AR}{A-R}$$

53. **[8.4]**

$$1-\frac{5}{r}=\frac{-4}{r^2}$$

Multiply by the LCD, r^2. $(r \neq 0)$

$$r^2\left(1-\frac{5}{r}\right)=r^2\left(\frac{-4}{r^2}\right)$$
$$r^2-5r=-4$$
$$r^2-5r+4=0$$
$$(r-4)(r-1)=0$$
$$r-4=0 \quad \text{or} \quad r-1=0$$
$$r=4 \quad \text{or} \quad r=1$$

Check $r=1$: $1-5=-4$ *True*

Check $r=4$: $1-\frac{5}{4}=-\frac{1}{4}$ *True*

Solution set: $\{1, 4\}$

54. **[8.4]**

$$\frac{3x}{x-4}+\frac{2}{x}=\frac{48}{x^2-4x}$$
$$\frac{3x}{x-4}+\frac{2}{x}=\frac{48}{x(x-4)}$$

Multiply by the LCD, $x(x-4)$. $(x \neq 0, 4)$

$$x(x-4)\left(\frac{3x}{x-4}+\frac{2}{x}\right)=x(x-4)\left(\frac{48}{x(x-4)}\right)$$
$$3x^2+2(x-4)=48$$
$$3x^2+2x-8=48$$
$$3x^2+2x-56=0$$
$$(3x+14)(x-4)=0$$
$$3x+14=0 \quad \text{or} \quad x-4=0$$
$$3x=-14 \quad \text{or} \quad x=4$$
$$x=-\tfrac{14}{3}$$

The number 4 is not allowed as a solution because substituting it in the original equation results in division by 0.

Check $x=-\frac{14}{3}$: $\frac{21}{13}-\frac{3}{7}=\frac{108}{91}$ *True*

Solution set: $\{-\frac{14}{3}\}$

55. **[8.5]** Let $x=$ the time to do the job working together.

Make a table.

Worker	Rate	Time Working Together	Fractional Part of the Job Done
Anna	$\frac{1}{9}$	x	$\frac{x}{9}$
Matthew	$\frac{1}{6}$	x	$\frac{x}{6}$

Part done by Anna	plus	part done by Matthew	equals	1 whole job.
$\frac{x}{9}$	$+$	$\frac{x}{6}$	$=$	1

Multiply by the LCD, 36.

$$36\left(\frac{x}{9}+\frac{x}{6}\right)=36\cdot 1$$
$$4x+6x=36$$
$$10x=36$$
$$x=\tfrac{36}{10}$$
$$x=\tfrac{18}{5} \quad \text{or} \quad 3\tfrac{3}{5}$$

Working together, they can do the job in $\frac{18}{5}$ or $3\frac{3}{5}$ hours.

56. **[8.5]** *Step 2*

Let $x=$ the distance between her apartment and campus. Make a table using the information in the problem and the formula $t=\dfrac{d}{r}$.

continued

	d	r	t
Biking	x	12	$\frac{x}{12}$
Walking	x	3	$\frac{x}{3}$

Step 3
Her biking time is 36 minutes $= \frac{3}{5}$ hour less than her walking time.

$$\frac{x}{12} = \frac{x}{3} - \frac{3}{5}$$

Step 4
Multiply by the LCD, 60.

$$5x = 20x - 36$$
$$36 = 15x$$
$$2.4 = x$$

Step 5
It is 2.4 miles from her apartment to campus.

Step 6
Check Her biking time is $\frac{2.4}{12} = 0.2$ hour. Her walking time is $\frac{2.4}{3} = 0.8$ hour. The difference is $0.8 - 0.2 = 0.6$, or $\frac{3}{5}$ hour, as required.

57. **[8.6]** The frequency f of a vibrating guitar string varies inversely as its length L, so

$$f = \frac{k}{L}$$

for some constant k. Substitute 0.65 for L and 4.3 for f.

$$4.3 = \frac{k}{6.5}$$
$$k = 4.3(0.65) = 2.795$$

So $f = \dfrac{2.795}{L}$. Now let $L = 0.5$.

$$f = \frac{2.795}{0.5} = 5.59$$

The frequency would be 5.59 vibrations per second.

58. **[8.6]** The area A of a triangle varies jointly as the lengths of the base b and height h, so

$$A = kbh.$$

When $b = 10$ and $h = 4$, $A = 20$.

$$20 = k(10)(4)$$
$$k = \tfrac{20}{40} = \tfrac{1}{2}$$

Thus, $A = \frac{1}{2}bh$. When $b = 3$ and $h = 8$,

$$A = \tfrac{1}{2}(3)(8) = 12.$$

The area of the triangle is 12 square feet.

Chapter 8 Test

1. $f(x) = \dfrac{x+3}{3x^2 + 2x - 8}$

Set the denominator equal to zero and solve.

$$3x^2 + 2x - 8 = 0$$
$$(3x - 4)(x + 2) = 0$$
$$3x - 4 = 0 \quad \text{or} \quad x + 2 = 0$$
$$3x = 4 \quad \text{or} \quad x = -2$$
$$x = \tfrac{4}{3}$$

The numbers -2 and $\frac{4}{3}$ make the rational expression undefined and are excluded from the domain of f, which can be written in set notation as $\{x \mid x \neq -2, \frac{4}{3}\}$.

2. $$\frac{6x^2 - 13x - 5}{9x^3 - x} = \frac{(3x+1)(2x-5)}{x(9x^2 - 1)} = \frac{(3x+1)(2x-5)}{x(3x+1)(3x-1)} = \frac{2x-5}{x(3x-1)}$$

3. $$\frac{(x+3)^2}{4} \cdot \frac{6}{2x+6} = \frac{(x+3)^2}{4} \cdot \frac{2 \cdot 3}{2(x+3)} = \frac{3(x+3)}{4}$$

4. $$\frac{y^2 - 16}{y^2 - 25} \cdot \frac{y^2 + 2y - 15}{y^2 - 7y + 12}$$
$$= \frac{(y+4)(y-4)}{(y+5)(y-5)} \cdot \frac{(y+5)(y-3)}{(y-4)(y-3)}$$
$$= \frac{y+4}{y-5}$$

5. $$\frac{x^2 - 9}{x^3 + 3x^2} \div \frac{x^2 + x - 12}{x^3 + 9x^2 + 20x}$$

Multiply by the reciprocal.

$$= \frac{x^2 - 9}{x^3 + 3x^2} \cdot \frac{x(x^2 + 9x + 20)}{x^2 + x - 12}$$
$$= \frac{(x+3)(x-3)}{x^2(x+3)} \cdot \frac{x(x+5)(x+4)}{(x+4)(x-3)}$$
$$= \frac{x+5}{x}$$

6. $t^2 + t - 6$, $t^2 + 3t$, t^2

Factor each denominator.

$$t^2 + t - 6 = (t+3)(t-2)$$
$$t^2 + 3t = t(t+3)$$

The third denominator is already in factored form. The LCD is

$$t^2(t+3)(t-2).$$

7. $\dfrac{7}{6t^2} - \dfrac{1}{3t}$

$= \dfrac{7}{6t^2} - \dfrac{1 \cdot 2t}{3t \cdot 2t}$ *The LCD is $6t^2$.*

$= \dfrac{7}{6t^2} - \dfrac{2t}{6t^2}$

$= \dfrac{7 - 2t}{6t^2}$

8. $\dfrac{9}{x-7} + \dfrac{4}{x+7}$

The LCD is $(x-7)(x+7)$.

$= \dfrac{9(x+7)}{(x-7)(x+7)} + \dfrac{4(x-7)}{(x+7)(x-7)}$

$= \dfrac{9(x+7) + 4(x-7)}{(x-7)(x+7)}$

$= \dfrac{9x + 63 + 4x - 28}{(x-7)(x+7)}$

$= \dfrac{13x + 35}{(x-7)(x+7)}$

9. $\dfrac{6}{x+4} + \dfrac{1}{x+2} - \dfrac{3x}{x^2 + 6x + 8}$

$= \dfrac{6}{x+4} + \dfrac{1}{x+2} - \dfrac{3x}{(x+4)(x+2)}$

The LCD is $(x+4)(x+2)$.

$= \dfrac{6(x+2)}{(x+4)(x+2)} + \dfrac{1(x+4)}{(x+2)(x+4)}$

$- \dfrac{3x}{(x+4)(x+2)}$

$= \dfrac{6(x+2) + x + 4 - 3x}{(x+4)(x+2)}$

$= \dfrac{6x + 12 + x + 4 - 3x}{(x+4)(x+2)}$

$= \dfrac{4x + 16}{(x+4)(x+2)}$

$= \dfrac{4(x+4)}{(x+4)(x+2)} = \dfrac{4}{x+2}$

10. $\dfrac{\dfrac{12}{r+4}}{\dfrac{11}{6r+24}} = \dfrac{12}{r+4} \div \dfrac{11}{6r+24}$

Multiply by the reciprocal.

$= \dfrac{12}{r+4} \cdot \dfrac{6r+24}{11}$

$= \dfrac{12}{r+4} \cdot \dfrac{6(r+4)}{11} = \dfrac{72}{11}$

11. $\dfrac{\dfrac{1}{a} - \dfrac{1}{b}}{\dfrac{a}{b} - \dfrac{b}{a}}$

Multiply the numerator and the denominator by the LCD of all the fractions, ab.

$= \dfrac{ab\left(\dfrac{1}{a} - \dfrac{1}{b}\right)}{ab\left(\dfrac{a}{b} - \dfrac{b}{a}\right)} = \dfrac{b-a}{a^2 - b^2}$

$= \dfrac{b-a}{(a-b)(a+b)} = \dfrac{(-1)(a-b)}{(a-b)(a+b)}$

$= \dfrac{-1}{a+b}$, or $-\dfrac{1}{a+b}$

12. $\dfrac{2x^{-2} + y^{-2}}{x^{-1} - y^{-1}} = \dfrac{\dfrac{2}{x^2} + \dfrac{1}{y^2}}{\dfrac{1}{x} - \dfrac{1}{y}}$

Multiply the numerator and denominator by the LCD of all the fractions, x^2y^2.

$= \dfrac{x^2y^2\left(\dfrac{2}{x^2} + \dfrac{1}{y^2}\right)}{x^2y^2\left(\dfrac{1}{x} - \dfrac{1}{y}\right)}$

$= \dfrac{2y^2 + x^2}{xy^2 - x^2y}$

$= \dfrac{2y^2 + x^2}{xy(y-x)}$

13. (a) $\dfrac{2x}{3} + \dfrac{x}{4} - \dfrac{11}{2}$ No equals sign appears so this is an *expression.*

$= \dfrac{2x \cdot 4}{3 \cdot 4} + \dfrac{x \cdot 3}{4 \cdot 3} - \dfrac{11 \cdot 6}{2 \cdot 6}$ *The LCD is 12.*

$= \dfrac{8x}{12} + \dfrac{3x}{12} - \dfrac{66}{12}$

$= \dfrac{8x + 3x - 66}{12}$

$= \dfrac{11x - 66}{12}$

$= \dfrac{11(x-6)}{12}$

(b) $\frac{2x}{3}+\frac{x}{4}=\frac{11}{2}$

There is an equals sign, so this is an *equation.* Multiply by the LCD, 12.

$$12\left(\frac{2x}{3}+\frac{x}{4}\right)=12\left(\frac{11}{2}\right)$$
$$8x+3x=66$$
$$11x=66$$
$$x=6$$

Check $x=6$: $4+\frac{3}{2}=\frac{11}{2}$ *True*

Solution set: $\{6\}$

14. $\frac{1}{x}-\frac{4}{3x}=\frac{1}{x-2}$

Multiply by the LCD, $3x(x-2)$. $(x\neq 0,2)$

$$3x(x-2)\left(\frac{1}{x}-\frac{4}{3x}\right)=3x(x-2)\left(\frac{1}{x-2}\right)$$
$$3(x-2)-4(x-2)=3x$$
$$3x-6-4x+8=3x$$
$$-x+2=3x$$
$$-4x=-2$$
$$x=\tfrac{1}{2}$$

Check $x=\frac{1}{2}$: $2-\frac{8}{3}=-\frac{2}{3}$ *True*

Solution set: $\{\frac{1}{2}\}$

15. $\frac{y}{y+2}-\frac{1}{y-2}=\frac{8}{y^2-4}$

$$\frac{y}{y+2}-\frac{1}{y-2}=\frac{8}{(y+2)(y-2)}$$

Multiply by the LCD, $(y+2)(y-2)$. $(y\neq -2,2)$

$$(y+2)(y-2)\left(\frac{y}{y+2}-\frac{1}{y-2}\right)$$
$$=(y+2)(y-2)\left(\frac{8}{(y+2)(y-2)}\right)$$
$$y(y-2)-1(y+2)=8$$
$$y^2-2y-y-2=8$$
$$y^2-3y-10=0$$
$$(y-5)(y+2)=0$$

$y-5=0$ or $y+2=0$

$y=5$ or $y=-2$

The number -2 is not allowed as a solution because substituting it in the original equation results in division by 0.

Check $y=5$: $\frac{5}{7}-\frac{1}{3}=\frac{8}{21}$ *True*

Solution set: $\{5\}$

16. $f(x)=\frac{-2}{x+1}$ is not defined when $x=-1$, so an equation of the vertical asymptote is $x=-1$. As the values of x get larger, the values of y get closer to 0, so $y=0$ is the equation of the horizontal asymptote.

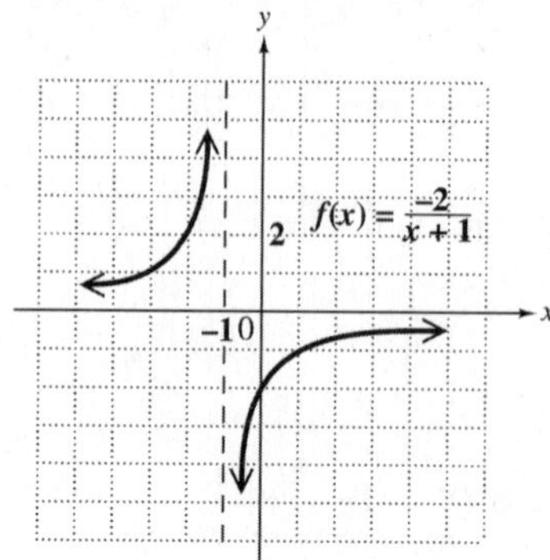

17. Let $x=$ the time to do the job working together. Make a table.

Worker	Rate	Time Working Together	Fractional Part of the Job Done
Wayne	$\frac{1}{9}$	x	$\frac{x}{9}$
Sandra	$\frac{1}{5}$	x	$\frac{x}{5}$

Part done by Wayne	plus	part done by Sandra	equals	1 whole job.
$\frac{x}{9}$	$+$	$\frac{x}{5}$	$=$	1

Multiply by the LCD, 45.

$$45\left(\frac{x}{9}+\frac{x}{5}\right)=45\cdot 1$$
$$5x+9x=45$$
$$14x=45$$
$$x=\tfrac{45}{14},\text{ or } 3\tfrac{3}{14}$$

Working together, they can do the job in $\frac{45}{14}$ or $3\frac{3}{14}$ hours.

18. Let $x =$ the speed of the boat in still water. Make a table. Use $d = rt$, or $t = \frac{d}{r}$.

	d	r	t
Downstream	36	$x+3$	$\frac{36}{x+3}$
Upstream	24	$x-3$	$\frac{24}{x-3}$

Because the times are equal,

$$\frac{36}{x+3} = \frac{24}{x-3}.$$

Multiply by the LCD, $(x+3)(x-3)$. $(x \neq -3, 3)$

$$(x+3)(x-3)\left(\frac{36}{x+3}\right) = (x+3)(x-3)\left(\frac{24}{x-3}\right)$$
$$36(x-3) = 24(x+3)$$
$$36x - 108 = 24x + 72$$
$$12x = 180$$
$$x = 15$$

The speed of the boat in still water is 15 mph.

19. Let $x =$ the number of fish in Lake Linda. Write a proportion.

$$\frac{x}{600} = \frac{800}{10}$$
$$600\left(\frac{x}{600}\right) = 600\left(\frac{800}{10}\right) \quad \textit{Multiply by the LCD, 600.}$$
$$x = 60 \cdot 800$$
$$x = 48{,}000$$

There are about 48,000 fish in the lake.

20. $g(x) = \dfrac{5x}{2+x}$

(a) $$3 = \frac{5x}{2+x} \quad \textit{Let } g(x) = 3.$$
$$3(2+x) = 5x$$
$$6 + 3x = 5x$$
$$6 = 2x$$
$$x = 3 \text{ units}$$

3 units of food will produce a growth rate of 3 units of growth per unit of food.

(b) If no food is available, then $x = 0$.

$$g(0) = \frac{5(0)}{2+0} = \frac{0}{2} = 0$$

The growth rate is 0.

21. The current I is inversely proportional to the resistance R, so

$$I = \frac{k}{R}$$

for some constant k. Let $I = 80$ and $R = 30$. Find k.

$$80 = \frac{k}{30}$$
$$k = 80(30) = 2400$$

So $I = \dfrac{2400}{R}$. Now let $R = 12$.

$$I = \frac{2400}{12} = 200$$

The current is 200 amps.

22. The force F of the wind blowing on a vertical surface varies jointly as the area A of the surface and the square of the velocity V, so

$$F = kAV^2$$

for some constant k. Let $F = 50$, $A = 500$, and $V = 40$. Find k.

$$50 = k(500)(40)^2$$
$$k = \frac{50}{500(1600)} = \frac{1}{16{,}000}$$

So $F = \frac{1}{16{,}000}AV^2$. Now let $A = 2$ and $V = 80$.

$$F = \frac{1}{16{,}000}(2)(80)^2 = 0.8$$

The force of the wind is 0.8 lb.

Cumulative Review Exercises (Chapters 1–8)

1. $$7(2x+3) - 4(2x+1) = 2(x+1)$$
$$14x + 21 - 8x - 4 = 2x + 2$$
$$6x + 17 = 2x + 2$$
$$4x = -15$$
$$x = -\tfrac{15}{4}$$

Solution set: $\{-\frac{15}{4}\}$

2. $$|6x - 8| - 4 = 0$$
$$|6x - 8| = 4$$

$$6x - 8 = 4 \quad \text{or} \quad 6x - 8 = -4$$
$$6x = 12 \quad \text{or} \quad 6x = 4$$
$$x = 2 \quad \text{or} \quad x = \tfrac{4}{6} = \tfrac{2}{3}$$

Solution set: $\{\frac{2}{3}, 2\}$

3. $$\frac{2}{3}x + \frac{5}{12}x \le 20$$

$$12(\tfrac{2}{3}x + \tfrac{5}{12}x) \le 12(20)$$ *Multiply by the LCD, 12.*

$$8x + 5x \le 240$$
$$13x \le 240$$
$$x \le \tfrac{240}{13}$$

Solution set: $(-\infty, \frac{240}{13}]$

4. Let $x =$ the amount of money invested at 4% and $2x =$ the amount of money invested at 3%.

Use $I = prt$ with the time, t, equal to 1 yr.

$$0.04x + 0.03(2x) = 400$$
$$0.04x + 0.06x = 400$$
$$0.10x = 400$$

Multiply by 10 to clear the decimal.

$$x = 4000$$

Since $x = 4000$, $2x = 2(4000) = 8000$.

He invested \$4000 at 4% and \$8000 at 3%.

5. Let $h =$ the height of the triangle.
Use the formula $A = \frac{1}{2}bh$. Here, $A = 42$ and $b = 14$.

$$A = \tfrac{1}{2}bh$$
$$42 = \tfrac{1}{2}(14)h$$
$$42 = 7h$$
$$6 = h$$

The height is 6 meters.

6. **(a)** Through $(-5, 8)$ and $(-1, 2)$

Let $(x_1, y_1) = (-5, 8)$ and $(x_2, y_2) = (-1, 2)$. Then,

$$m = \frac{y_2 - y_1}{x_2 - x_1} = \frac{2 - 8}{-1 - (-5)} = \frac{-6}{4} = -\frac{3}{2}.$$

The slope is $-\frac{3}{2}$.

(b) Perpendicular to $4x + 3y = 12$, through $(5, 2)$

Solve for y to write the equation in slope-intercept form and find the slope.

$$4x + 3y = 12$$
$$3y = -4x + 12$$
$$y = -\tfrac{4}{3}x + 4$$

The slope is $-\frac{4}{3}$. Perpendicular lines have slopes that are negative reciprocals of each other. The negative reciprocal of $-\frac{4}{3}$ is $\frac{3}{4}$. The slope of a line perpendicular to the given line is $\frac{3}{4}$.

7. **(a)** Use $(x_1, y_1) = (-5, 8)$ and $m = -\frac{3}{2}$ in the point-slope form.

$$y - y_1 = m(x - x_1)$$
$$y - 8 = -\tfrac{3}{2}[x - (-5)]$$
$$y - 8 = -\tfrac{3}{2}(x + 5)$$
$$y - 8 = -\tfrac{3}{2}x - \tfrac{15}{2}$$
$$y = -\tfrac{3}{2}x + \tfrac{1}{2}$$

(b) Use $(x_1, y_1) = (5, 2)$ and $m = \frac{3}{4}$ in the point-slope form.

$$y - y_1 = m(x - x_1)$$
$$y - 2 = \tfrac{3}{4}(x - 5)$$
$$y - 2 = \tfrac{3}{4}x - \tfrac{15}{4}$$
$$y = \tfrac{3}{4}x - \tfrac{7}{4}$$

8. $-4x + 2y = 8$

To find the x-intercept, let $y = 0$.

$$-4x + 2(0) = 8$$
$$-4x = 8$$
$$x = -2$$

The x-intercept is $(-2, 0)$.
To find the y-intercept, let $x = 0$.

$$-4(0) + 2y = 8$$
$$2y = 8$$
$$y = 4$$

The y-intercept is $(0, 4)$. Plot the intercepts, and draw the line through them.

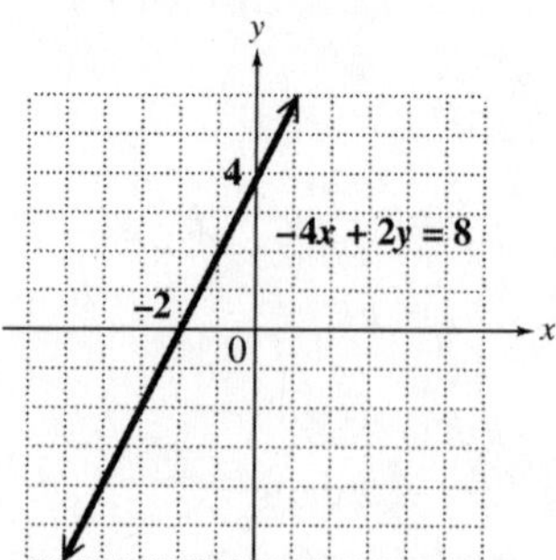

9. $2x + 5y > 10$

Graph the line $2x + 5y = 10$ by drawing a dashed line (since the inequality involves $>$) through the intercepts $(5, 0)$ and $(0, 2)$.
Test a point not on this line, such as $(0, 0)$.

$$2x + 5y > 10$$
$$2(0) + 5(0) \overset{?}{>} 10$$
$$0 > 10 \quad \textit{False}$$

Shade the side of the line not containing $(0, 0)$.

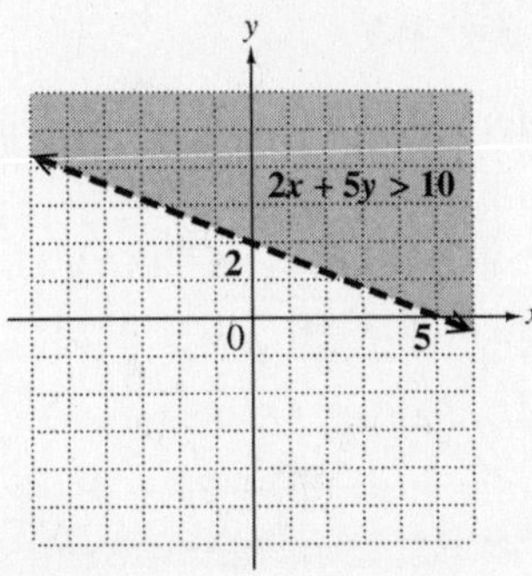

10. $x - y \geq 3$ and $3x + 4y \leq 12$

Graph the solid line $x - y = 3$ through $(3, 0)$ and $(0, -3)$. The inequality $x - y \geq 3$ can be written as $y \leq x - 3$, so shade the region below the boundary line.
Graph the solid line $3x + 4y = 12$ through $(4, 0)$ and $(0, 3)$. The inequality $3x + 4y \leq 12$ can be written as $y \leq -\frac{3}{4}x + 3$, so shade the region below the boundary line.
The required graph is the common shaded area as well as the portions of the lines that bound it.

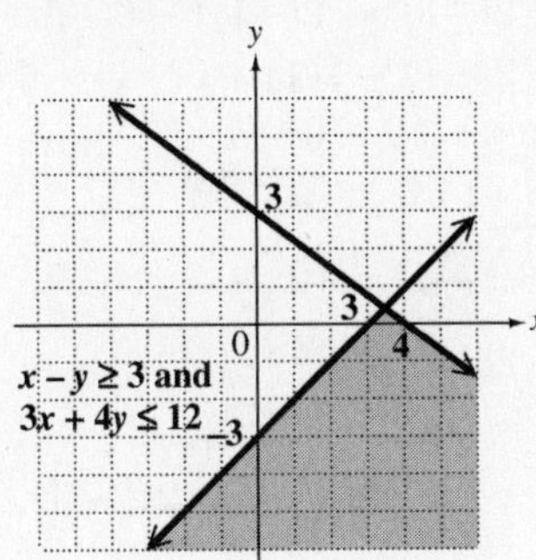

11. **(a)** Solve the equation for y.

$$\begin{aligned} 5x - 3y &= 8 \\ 5x - 8 &= 3y \\ \frac{5x - 8}{3} &= y \end{aligned}$$

So $f(x) = \dfrac{5x - 8}{3}$, or $f(x) = \dfrac{5}{3}x - \dfrac{8}{3}$.

(b) $f(1) = \dfrac{5(1) - 8}{3} = \dfrac{-3}{3} = -1$

12.
$$\begin{aligned} f(x) &= 3x + 6 \\ f(x + 3) &= 3(x + 3) + 6 \\ &= 3x + 9 + 6 \\ &= 3x + 15 \end{aligned}$$

13.
$$\begin{aligned} 4x - y &= -7 \quad (1) \\ 5x + 2y &= 1 \quad (2) \end{aligned}$$

To eliminate y, multiply equation (1) by 2 and add the result to (2).

$$\begin{array}{rcrl} 8x - 2y &=& -14 & 2 \times (1) \\ 5x + 2y &=& 1 & (2) \\ \hline 13x &=& -13 & \\ x &=& -1 & \end{array}$$

To find y, substitute -1 for x in (1).

$$\begin{aligned} 4x - y &= -7 \quad (1) \\ 4(-1) - y &= -7 \\ -4 - y &= -7 \\ -y &= -3 \\ y &= 3 \end{aligned}$$

Solution set: $\{(-1, 3)\}$

14.
$$\begin{aligned} x + y - 2z &= -1 \quad (1) \\ 2x - y + z &= -6 \quad (2) \\ 3x + 2y - 3z &= -3 \quad (3) \end{aligned}$$

Add (1) and (2) to eliminate y.

$$\begin{array}{rcrl} x + y - 2z &=& -1 & (1) \\ 2x - y + z &=& -6 & (2) \\ \hline 3x - z &=& -7 & (4) \end{array}$$

Multiply (2) by 2 and add it to (3) to eliminate y.

$$\begin{array}{rcrl} 4x - 2y + 2z &=& -12 & 2 \times (2) \\ 3x + 2y - 3z &=& -3 & (3) \\ \hline 7x - z &=& -15 & (5) \end{array}$$

Multiply (4) by -1 and add to (5) to eliminate z.

$$\begin{array}{rcrl} -3x + z &=& 7 & -1 \times (4) \\ 7x - z &=& -15 & (5) \\ \hline 4x &=& -8 & \\ x &=& -2 & \end{array}$$

Substitute -2 for x into (4) and solve for z.

$$\begin{aligned} 3(-2) - z &= -7 \\ -6 - z &= -7 \\ -z &= -1 \\ z &= 1 \end{aligned}$$

Substitute -2 for x and 1 for z into (1) and solve for y.

$$\begin{aligned} -2 + y - 2(1) &= -1 \\ -2 + y - 2 &= -1 \\ y - 4 &= -1 \\ y &= 3 \end{aligned}$$

Solution set: $\{(-2, 3, 1)\}$

15.
$$\begin{aligned} x + 2y + z &= 5 \quad (1) \\ x - y + z &= 3 \quad (2) \\ 2x + 4y + 2z &= 11 \quad (3) \end{aligned}$$

Multiply (1) by -1 and add to (2).

$$\begin{array}{rcrcrcrl} -x & - & 2y & - & z & = & -5 & \quad -1 \times (1) \\ x & - & y & + & z & = & 3 & \quad (2) \\ \hline & - & 3y & & & = & -2 & \\ & & & & y & = & \frac{2}{3} & \end{array}$$

Substitute $\frac{2}{3}$ for y in (1) and (3), then add the resulting equations.

$$\begin{aligned} x + 2(\tfrac{2}{3}) + z &= 5 \quad \text{Let } y = \tfrac{2}{3} \text{ in (1).} \\ x + \tfrac{4}{3} + z &= 5 \\ x + z &= \tfrac{11}{3} \quad (4) \end{aligned}$$

$$\begin{aligned} 2x + 4(\tfrac{2}{3}) + 2z &= 11 \quad \text{Let } y = \tfrac{2}{3} \text{ in (3).} \\ 2x + \tfrac{8}{3} + 2z &= 11 \\ 2x + 2z &= \tfrac{25}{3} \quad (5) \end{aligned}$$

Multiply (4) by -2 and add to (5).

$$\begin{array}{rcrcrl} -2x & - & 2z & = & -\frac{22}{3} & \quad -2 \times (4) \\ 2x & + & 2z & = & \frac{25}{3} & \quad (5) \\ \hline & & 0 & = & 1 & \quad \textit{False} \end{array}$$

Since this statement is false, the solution is $\emptyset$.

16.
$$\begin{aligned} &(3y^2 - 2y + 6) - (-y^2 + 5y + 12) \\ &= 3y^2 - 2y + 6 + y^2 - 5y - 12 \\ &= 4y^2 - 7y - 6 \end{aligned}$$

17. $(3x^3 + 13x^2 - 17x - 7) \div (3x + 1)$

$$\begin{array}{r|l} & x^2 + 4x - 7 \\ \hline 3x + 1 & 3x^3 + 13x^2 - 17x - 7 \\ & 3x^3 + x^2 \\ & \quad 12x^2 - 17x \\ & \quad 12x^2 + 4x \\ & \qquad -21x - 7 \\ & \qquad -21x - 7 \\ & \qquad\qquad 0 \end{array}$$

Answer: $x^2 + 4x - 7$

18.
$$\begin{aligned} (4f + 3)(3f - 1) &= 12f^2 - 4f + 9f - 3 \\ &= 12f^2 + 5f - 3 \end{aligned}$$

19. $(7t^3 + 8)(7t^3 - 8)$

This is the product of the sum and difference of two terms.

$$\begin{aligned} (7t^3 + 8)(7t^3 - 8) &= (7t^3)^2 - 8^2 \\ &= 49t^6 - 64 \end{aligned}$$

20. $(\frac{1}{4}x + 5)^2$

Use the formula for the square of a binomial, $(a + b)^2 = a^2 + 2ab + b^2$.

$$\begin{aligned} (\tfrac{1}{4}x + 5)^2 &= (\tfrac{1}{4}x)^2 + 2(\tfrac{1}{4}x)(5) + 5^2 \\ &= \tfrac{1}{16}x^2 + \tfrac{5}{2}x + 25 \end{aligned}$$

21. **(a)**
$$\begin{aligned} (f + g)(x) &= f(x) + g(x) \\ &= (x^2 + 2x - 3) \\ &\quad + (2x^3 - 3x^2 + 4x - 1) \\ &= 2x^3 - 2x^2 + 6x - 4 \end{aligned}$$

(b)
$$\begin{aligned} (g - f)(x) &= g(x) - f(x) \\ &= (2x^3 - 3x^2 + 4x - 1) \\ &\quad - (x^2 + 2x - 3) \\ &= 2x^3 - 3x^2 + 4x - 1 \\ &\quad - x^2 - 2x + 3 \\ &= 2x^3 - 4x^2 + 2x + 2 \end{aligned}$$

(c) Use part (a).

$$\begin{aligned} &(f + g)(-1) \\ &= 2(-1)^3 - 2(-1)^2 + 6(-1) - 4 \\ &= 2(-1) - 2(1) + 6(-1) - 4 \\ &= -2 - 2 - 6 - 4 \\ &= -14 \end{aligned}$$

22. $2x^2 - 13x - 45 = (2x + 5)(x - 9)$

23.
$$\begin{aligned} 100t^4 - 25 &= 25(4t^4 - 1) \\ &= 25[(2t^2)^2 - 1^2] \\ &= 25(2t^2 + 1)(2t^2 - 1) \end{aligned}$$

24. Use the sum of cubes formula,

$$x^3 + y^3 = (x + y)(x^2 - xy + y^2).$$

$$\begin{aligned} 8p^3 + 125 &= (2p)^3 + 5^3 \\ &= (2p + 5)[(2p)^2 - (2p)(5) + 5^2] \\ &= (2p + 5)(4p^2 - 10p + 25) \end{aligned}$$

25.
$$\begin{aligned} \frac{2a^2}{a + b} \cdot \frac{a - b}{4a} &= \frac{2a^2(a - b)}{4a(a + b)} \\ &= \frac{a(a - b)}{2(a + b)} \end{aligned}$$

26.
$$\begin{aligned} \frac{x + 4}{x - 2} + \frac{2x - 10}{x - 2} &= \frac{x + 4 + 2x - 10}{x - 2} \\ &= \frac{3x - 6}{x - 2} \\ &= \frac{3(x - 2)}{x - 2} = 3 \end{aligned}$$

27. $\dfrac{2x}{2x-1}+\dfrac{4}{2x+1}+\dfrac{8}{4x^2-1}$

$=\dfrac{2x}{2x-1}+\dfrac{4}{2x+1}+\dfrac{8}{(2x+1)(2x-1)}$

The LCD is $(2x+1)(2x-1)$.

$=\dfrac{2x(2x+1)}{(2x-1)(2x+1)}+\dfrac{4(2x-1)}{(2x+1)(2x-1)}$

$+\dfrac{8}{(2x+1)(2x-1)}$

$=\dfrac{2x(2x+1)+4(2x-1)+8}{(2x+1)(2x-1)}$

$=\dfrac{4x^2+2x+8x-4+8}{(2x+1)(2x-1)}$

$=\dfrac{4x^2+10x+4}{(2x+1)(2x-1)}$

$=\dfrac{2(2x^2+5x+2)}{(2x+1)(2x-1)}$

$=\dfrac{2(2x+1)(x+2)}{(2x+1)(2x-1)}$

$=\dfrac{2(x+2)}{2x-1}$

28.
$$3x^2+4x=7$$
$$3x^2+4x-7=0$$
$$(3x+7)(x-1)=0$$

$3x+7=0$ or $x-1=0$

$3x=-7$ or $x=1$

$x=-\frac{7}{3}$

Solution set: $\{-\frac{7}{3},1\}$

29.
$$\frac{-3x}{x+1}+\frac{4x+1}{x}=\frac{-3}{x^2+x}$$
$$\frac{-3x}{x+1}+\frac{4x+1}{x}=\frac{-3}{x(x+1)}$$

Multiply by the LCD, $x(x+1)$. $(x\neq-1,0)$

$$x(x+1)\left(\frac{-3x}{x+1}+\frac{4x+1}{x}\right)=x(x+1)\left(\frac{-3}{x(x+1)}\right)$$
$$x(-3x)+(x+1)(4x+1)=-3$$
$$-3x^2+4x^2+x+4x+1=-3$$
$$x^2+5x+4=0$$
$$(x+4)(x+1)=0$$

$x+4=0$ or $x+1=0$

$x=-4$ or $x=-1$

The number -1 is not allowed as a solution because substituting it in the original equation results in division by 0.

Check $x=-4$: $-4+\frac{15}{4}=-\frac{1}{4}$ *True*

Solution set: $\{-4\}$

30. Solve $\dfrac{1}{f}=\dfrac{1}{p}+\dfrac{1}{q}$ for q.

$fpq\left(\dfrac{1}{f}\right)=fpq\left(\dfrac{1}{p}+\dfrac{1}{q}\right)$ *Multiply by the LCD, fpq.*

$pq=fq+fp$

$pq-fq=fp$ *Get all terms with q on one side.*

$q(p-f)=fp$

$q=\dfrac{fp}{p-f}$,

or $q=\dfrac{-fp}{f-p}$

CHAPTER 9 ROOTS, RADICALS, AND ROOT FUNCTIONS

9.1 Radical Expressions and Graphs

9.1 Margin Exercises

1. **(a)** $\sqrt[3]{27} = 3$, because $3^3 = 27$.

   (b) $\sqrt[3]{1000} = 10$, because $10^3 = 1000$.

   (c) $\sqrt[4]{256} = 4$, because $4^4 = 256$.

   (d) $\sqrt[5]{243} = 3$, because $3^5 = 243$.

   (e) $\sqrt[4]{\frac{16}{81}} = \frac{2}{3}$, because $(\frac{2}{3})^4 = \frac{16}{81}$.

   (f) $\sqrt[3]{0.064} = 0.4$, because $(0.4)^3 = 0.064$.

2. **(a)** $\sqrt{36} = 6$
   There are two square roots, 6 and -6, and we want the principal square root, which is 6.

   (b) $\sqrt{36} = 6$, because $6^2 = 36$, so $-\sqrt{36} = -6$. The negative sign outside the radical makes the answer negative.

   (c) $\sqrt[4]{16} = 2$, because $2^4 = 16$.

   (d) $-\sqrt[4]{16} = -(\sqrt[4]{16}) = -(2) = -2$

   (e) $\sqrt[4]{-16}$ is *not a real number*, since the index, 4, is even and the radicand, -16, is negative.

   (f) $\sqrt[5]{1024} = 4$, because $4^5 = 1024$.

   (g) $\sqrt[5]{-1024} = -4$, because $(-4)^5 = -1024$.

3. **(a)** $f(x) = \sqrt{x} + 2$

   For the radicand to be nonnegative, we must have $x \geq 0$. Therefore, the domain is $[0, \infty)$. The function values are at least 2 since $\sqrt{x} \geq 0$ and we add 2 to $\sqrt{x}$, so the range is $[2, \infty)$.

x	$f(x) = \sqrt{x} + 2$
0	$\sqrt{0} + 2 = 2$
1	$\sqrt{1} + 2 = 3$
4	$\sqrt{4} + 2 = 4$
9	$\sqrt{9} + 2 = 5$

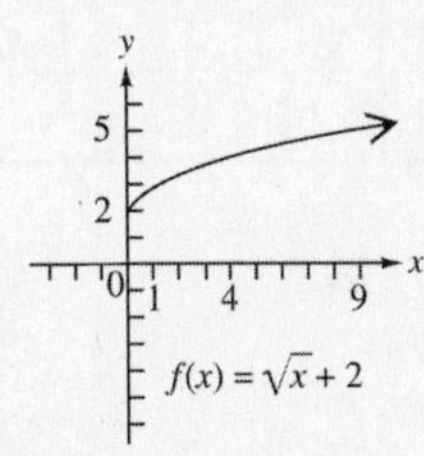

   (b) $f(x) = \sqrt[3]{x - 1}$

   Since we can take the cube root of any real number, the domain is $(-\infty, \infty)$.
   The result of a cube root can be any real number, so the range is $(-\infty, \infty)$.

x	$f(x) = \sqrt[3]{x - 1}$
-7	$\sqrt[3]{-7 - 1} = -2$
0	$\sqrt[3]{0 - 1} = -1$
1	$\sqrt[3]{1 - 1} = 0$
2	$\sqrt[3]{2 - 1} = 1$
9	$\sqrt[3]{9 - 1} = 2$

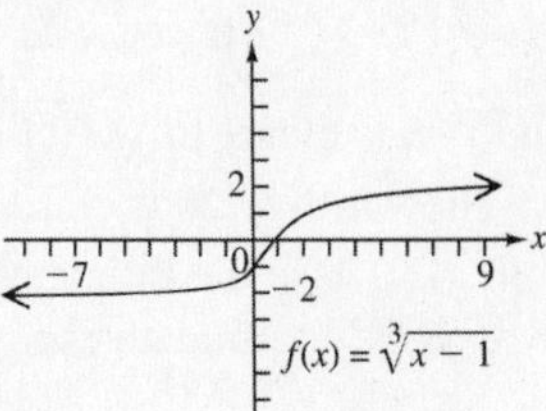

4. **(a)** $\sqrt{15^2} = |15| = 15$

   (b) $\sqrt{(-12)^2} = |-12| = 12$

   (c) $\sqrt{r^2} = |r|$

   (d) $\sqrt{(-r)^2} = |-r| = |r|$

5. **(a)** $\sqrt[4]{(-5)^4} = |-5| = 5$ (n is even)

   (b) $\sqrt[5]{(-7)^5} = -7$ (n is odd)

   (c) $-\sqrt[6]{(-3)^6} = -|-3| = -3$ (n is even)

   (d) $-\sqrt[4]{m^8} = -\sqrt[4]{(m^2)^4} = -|m^2| = -m^2$ (n is even) No absolute value bars are needed here because m^2 is nonnegative for any real number value of m.

   (e) $\sqrt[3]{x^{24}} = \sqrt[3]{(x^8)^3} = x^8$

   (f) $\sqrt[6]{y^{18}} = \sqrt[6]{(y^3)^6} = |y^3|$
   Since y^3 could be negative, we need the absolute value bars.

6. Use a calculator, and round to three decimal places.

   (a) $\sqrt{17} \approx 4.123$

   (b) $-\sqrt{362} \approx -19.026$

   (c) $\sqrt[3]{9482} \approx 21.166$

   (d) $\sqrt[4]{6825} \approx 9.089$

7. $f = \dfrac{1}{2\pi\sqrt{LC}}$
$= \dfrac{1}{2\pi\sqrt{(6 \times 10^{-5})(4 \times 10^{-9})}}$
$\approx 324{,}874$

The resonant frequency is about 325,000 cycles per second.

9.1 Section Exercises

1. $-\sqrt{16} = -(4) = -4$ **(E)**

3. $\sqrt[3]{-27} = -3$ because $(-3)^3 = -27$. **(D)**

5. $\sqrt[4]{16} = 2$ because $2^4 = 16$. **(C)**

7. $\sqrt{123.5} \approx \sqrt{121} = 11$ because $11^2 = 121$. **(C)**

9. The length $\sqrt{98}$ is closer to $\sqrt{100} = 10$ than to $\sqrt{81} = 9$. The width $\sqrt{26}$ is closer to $\sqrt{25} = 5$ than to $\sqrt{36} = 6$. Use the estimates $L = 10$ and $W = 5$ in $A = LW$ to find an estimate of the area.

$$A \approx 10 \cdot 5 = 50$$

Choice **C** is the best estimate.

11. **(a)** If $a > 0$, then $-a < 0$ and $\sqrt{-a}$ is not a real number. Thus, the given expression, $-\sqrt{-a}$, is *not a real number.*

(b) If $a < 0$, then $-a > 0$, so $\sqrt{-a}$ is a positive number and $-\sqrt{-a}$ is a *negative* number.

(c) If $a = 0$, then $-a = 0$, so $\sqrt{-a} = \sqrt{0} = 0$.

13. $\sqrt{81} = 9$, because $9^2 = 81$, so $-\sqrt{81} = -9$.

15. $\sqrt[3]{216} = 6$, because $6^3 = 216$.

17. $\sqrt[3]{-64} = -4$, because $(-4)^3 = -64$.

19. $\sqrt[3]{512} = 8$, because $8^3 = 512$, so $-\sqrt[3]{512} = -8$.

21. $\sqrt[4]{1296} = 6$, because $6^4 = 1296$.

23. $\sqrt[4]{16} = 2$, because $2^4 = 16$, so $-\sqrt[4]{16} = -2$.

25. $\sqrt[4]{-625}$ is not a real number since no real number to the fourth power equals -625. Any real number raised to the fourth power is 0 or positive.

27. $\sqrt[6]{729} = 3$, because $3^6 = 729$.

29. $\sqrt[6]{-64}$ is not a real number, because the index, 6, is even and the radicand, -64, is negative.

31. $\sqrt{\frac{64}{81}} = \frac{8}{9}$, because $(\frac{8}{9})^2 = \frac{64}{81}$.

33. $\sqrt{0.49} = 0.7$, because $(0.7)^2 = 0.49$.

35. $\sqrt[3]{\frac{64}{27}} = \frac{4}{3}$, because $(\frac{4}{3})^3 = \frac{64}{27}$.

37. $\sqrt[6]{\frac{1}{64}} = \frac{1}{2}$, because $(\frac{1}{2})^6 = \frac{1}{64}$, so $-\sqrt[6]{\frac{1}{64}} = -\frac{1}{2}$.

39. $\sqrt[3]{0.001} = 0.1$, because $(0.1)^3 = 0.001$.

41. $f(x) = \sqrt{x+3}$
For the radicand to be nonnegative, we must have

$$x + 3 \geq 0 \quad \text{or} \quad x \geq -3.$$

Thus, the domain is $[-3, \infty)$.
The function values are positive or zero (the result of the radical), so the range is $[0, \infty)$.

x	$f(x) = \sqrt{x+3}$
-3	$\sqrt{-3+3} = 0$
-2	$\sqrt{-2+3} = 1$
1	$\sqrt{1+3} = 2$

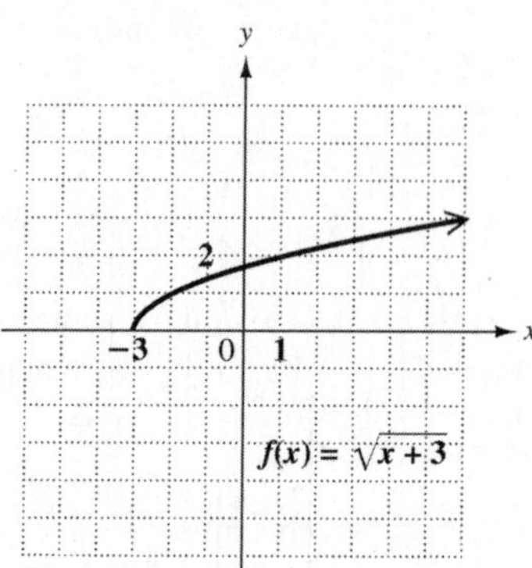

43. $f(x) = \sqrt{x} - 2$
For the radicand to be nonnegative, we must have

$$x \geq 0.$$

Note that the "-2" does not affect the domain, which is $[0, \infty)$.
The result of the radical is positive or zero, but the function values are 2 less than those values, so the range is $[-2, \infty)$.

x	$f(x) = \sqrt{x} - 2$
0	$\sqrt{0} - 2 = -2$
1	$\sqrt{1} - 2 = -1$
4	$\sqrt{4} - 2 = 0$

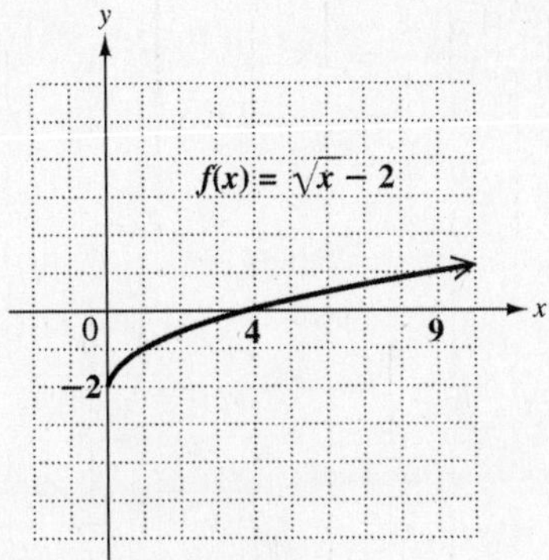

45. $f(x) = \sqrt[3]{x} - 3$
Since we can take the cube root of any real number, the domain is $(-\infty, \infty)$.
The result of a cube root can be any real number, so the range is $(-\infty, \infty)$. (The "-3" does not affect the range.)

x	$f(x) = \sqrt[3]{x} - 3$
-8	$\sqrt[3]{-8} - 3 = -5$
-1	$\sqrt[3]{-1} - 3 = -4$
0	$\sqrt[3]{0} - 3 = -3$
1	$\sqrt[3]{1} - 3 = -2$
8	$\sqrt[3]{8} - 3 = -1$

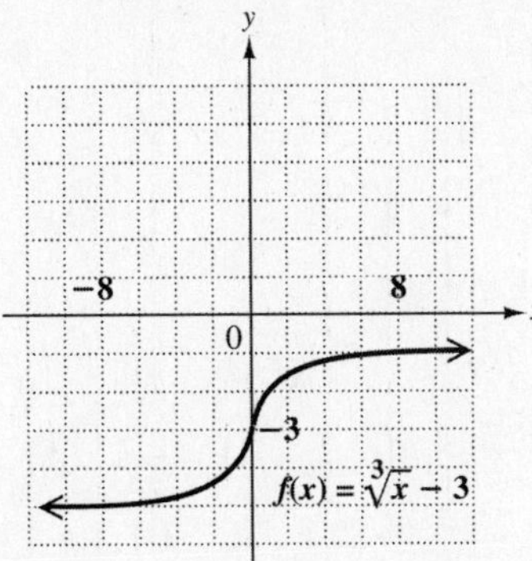

47. $\sqrt{12^2} = |12| = 12$

49. $\sqrt{(-10)^2} = |-10| = -(-10) = 10$

51. Since 6 is an even positive integer, $\sqrt[6]{a^6} = |a|$, so
$$\sqrt[6]{(-2)^6} = |-2| = 2.$$

53. Since 5 is odd, $\sqrt[5]{a^5} = a$, so $\sqrt[5]{(-9)^5} = -9$.

55. $\sqrt[6]{(-5)^6} = |-5| = 5$, so $-\sqrt[6]{(-5)^6} = -5$.

57. Since the index is even, $\sqrt{x^2} = |x|$.

59. $\sqrt{(-z)^2} = |-z| = |z|$

61. Since the index is odd, $\sqrt[3]{x^3} = x$.

63. $\sqrt[3]{x^{15}} = \sqrt[3]{(x^5)^3} = x^5$ (3 is odd)

65. $\sqrt[6]{x^{30}} = \sqrt[6]{(x^5)^6} = |x^5|$, or $|x|^5$ (6 is even)

In Exercises 67–78, use a calculator and round to three decimal places

67. $\sqrt{9483} \approx 97.381$

69. $\sqrt{284.361} \approx 16.863$

71. $-\sqrt{82} \approx -9.055$

73. $\sqrt[3]{423} \approx 7.507$

75. $\sqrt[4]{100} \approx 3.162$

77. $\sqrt[5]{23.8} \approx 1.885$

79. $$f = \frac{1}{2\pi\sqrt{LC}} = \frac{1}{2\pi\sqrt{(7.237 \times 10^{-5})(2.5 \times 10^{-10})}} \approx 1{,}183{,}235$$
The resonant frequency is about 1,183,000 cycles per second.

81. Since $H = 44 + 6 = 50$ ft, substitute 50 for H in the formula.
$$D = \sqrt{2H} = \sqrt{2 \cdot 50} = \sqrt{100} = 10$$
She will be able to see about 10 miles.

83. Let $a = 850$, $b = 925$, and $c = 1300$.
First find the semiperimeter s.
$$s = \tfrac{1}{2}(a + b + c) = \tfrac{1}{2}(850 + 925 + 1300) = \tfrac{3075}{2} = 1537.5$$
Now find the area A using Heron's formula.
$$A = \sqrt{s(s-a)(s-b)(s-c)} = \sqrt{1537.5(687.5)(612.5)(237.5)} \approx 392{,}128.8$$
The area of the Bermuda Triangle is about 392,000 square miles.

85. $$I = \sqrt{\frac{2P}{L}} = \sqrt{\frac{2(120)}{80}} \quad \textit{Let } P = 120, L = 80.$$
$$= \sqrt{3} \approx 1.732 \text{ amps}$$

9.2 Rational Exponents

9.2 Margin Exercises

1. (a) $8^{1/3} = \sqrt[3]{8} = 2$

(b) $9^{1/2} = \sqrt{9} = 3$

(c) $-81^{1/4} = -\sqrt[4]{81} = -3$

(d) $(-81)^{1/4} = \sqrt[4]{-81}$ is *not a real number* because the radicand, -81, is negative and the index, 4, is even.

(e) $(-64)^{1/3} = \sqrt[3]{-64} = -4 \quad ((-4)^3 = -64)$

(f) $\left(\frac{1}{32}\right)^{1/5} = \sqrt[5]{\frac{1}{32}} = \frac{1}{2}$

2. **(a)** $25^{3/2} = (25^{1/2})^3 = (\sqrt{25})^3 = 5^3 = 125$

(b) $27^{2/3} = (27^{1/3})^2 = (\sqrt[3]{27})^2 = 3^2 = 9$

(c) $-16^{3/2} = -(16^{1/2})^3 = -(\sqrt{16})^3$
$= -4^3 = -64$

(d) $(-64)^{2/3} = [(-64)^{1/3}]^2 = (\sqrt[3]{-64})^2$
$= (-4)^2 = 16$

(e) $(-36)^{3/2} = [(-36)^{1/2}]^3$ is *not a real number*, since $(-36)^{1/2}$, or $\sqrt{-36}$, is not a real number.

3. **(a)** $36^{-3/2} = \frac{1}{36^{3/2}} = \frac{1}{(36^{1/2})^3} = \frac{1}{(\sqrt{36})^3}$
$= \frac{1}{6^3} = \frac{1}{216}$

(b) $32^{-4/5} = \frac{1}{32^{4/5}} = \frac{1}{(32^{1/5})^4} = \frac{1}{(\sqrt[5]{32})^4}$
$= \frac{1}{2^4} = \frac{1}{16}$

(c) $\left(\frac{4}{9}\right)^{-5/2} = \left(\frac{9}{4}\right)^{5/2} = \left(\sqrt{\frac{9}{4}}\right)^5$
$= \left(\frac{3}{2}\right)^5 = \frac{243}{32}$

4. **(a)** $19^{1/2} = (\sqrt[2]{19})^1 = \sqrt{19}$

(b) $5^{2/3} = (\sqrt[3]{5})^2$

(c) $4k^{3/5} = 4(k^{3/5}) = 4(\sqrt[5]{k})^3$

(d) $5x^{3/5} - (2x)^{3/5} = 5(\sqrt[5]{x})^3 - (\sqrt[5]{2x})^3$

(e) $x^{-5/7} = \frac{1}{x^{5/7}} = \frac{1}{(\sqrt[7]{x})^5}$

(f) $(m^3 + n^3)^{1/3} = \sqrt[3]{m^3 + n^3}$

5. **(a)** $\sqrt{37} = 37^{1/2}$

(b) $\sqrt[4]{9^8} = 9^{8/4} = 9^2$, or 81

(c) $\sqrt[4]{t^4} = t^{4/4} = t^1 = t$ (t is positive)

6. **(a)** $11^{3/4} \cdot 11^{5/4} = 11^{3/4+5/4} = 11^{8/4}$
$= 11^2$, or 121

(b) $\frac{7^{3/4}}{7^{7/4}} = 7^{3/4-7/4} = 7^{-4/4} = 7^{-1} = \frac{1}{7}$

(c) $\frac{9^{2/3}(x^{1/3})^4}{9^{-1/3}} = 9^{2/3-(-1/3)}x^{4/3}$
$= 9^{2/3+1/3}x^{4/3}$
$= 9^1x^{4/3}$
$= 9x^{4/3}$

(d) $\left(\frac{a^3b^{-4}}{a^{-2}b^{1/5}}\right)^{-1/2} = \frac{(a^3)^{-1/2}(b^{-4})^{-1/2}}{(a^{-2})^{-1/2}(b^{1/5})^{-1/2}}$
$= \frac{a^{-3/2}b^2}{a^1b^{-1/10}}$
$= a^{-3/2-1}b^{2-(-1/10)}$
$= a^{-5/2}b^{21/10}$
$= \frac{b^{21/10}}{a^{5/2}}$

(e) $a^{2/3}(a^{7/3} + a^{1/3}) = a^{2/3} \cdot a^{7/3} + a^{2/3} \cdot a^{1/3}$
$= a^{2/3+7/3} + a^{2/3+1/3}$
$= a^{9/3} + a^{3/3}$
$= a^3 + a$

7. **(a)** $\sqrt[5]{m^3} \cdot \sqrt{m} = m^{3/5} \cdot m^{1/2}$
$= m^{3/5+1/2}$
$= m^{6/10+5/10}$
$= m^{11/10}$

(b) $\frac{\sqrt[3]{p^5}}{\sqrt{p^3}} = \frac{p^{5/3}}{p^{3/2}} = p^{5/3-3/2} = p^{10/6-9/6} = p^{1/6}$

(c) $\sqrt[4]{\sqrt[3]{x}} = \sqrt[4]{x^{1/3}} = (x^{1/3})^{1/4}$
$= x^{(1/3)(1/4)} = x^{1/12}$

9.2 Section Exercises

1. $2^{1/2} = \sqrt[2]{2^1} = \sqrt{2}$ **(C)**

3. $-16^{1/2} = -\sqrt{16} = -(4) = -4$ **(A)**

5. $(-32)^{1/5} = \sqrt[5]{-32} = -2$ **(H)**

7. $4^{3/2} = (4^{1/2})^3 = 2^3 = 8$ **(B)**

9. $-6^{2/4} = -(6^{1/2}) = -\sqrt{6}$ **(D)**

11. $169^{1/2} = \sqrt{169} = 13$

13. $729^{1/3} = \sqrt[3]{729} = 9$

15. $16^{1/4} = \sqrt[4]{16} = 2$

17. $\left(\frac{64}{81}\right)^{1/2} = \sqrt{\frac{64}{81}} = \frac{8}{9}$

19. $(-27)^{1/3} = \sqrt[3]{-27} = -3$

21. $(-144)^{1/2} = \sqrt{-144}$ is not a real number because no real number squared equals -144.

23. $100^{3/2} = (100^{1/2})^3 = 10^3 = 1000$

25. $81^{3/4} = (81^{1/4})^3 = \left(\sqrt[4]{81}\right)^3 = 3^3 = 27$

27. $-16^{5/2} = -(16^{1/2})^5 = -(4)^5 = -1024$

29. $(-8)^{4/3} = \left[(-8)^{1/3}\right]^4 = \left(\sqrt[3]{-8}\right)^4$
$= (-2)^4 = 16$

31. $32^{-3/5} = \frac{1}{32^{3/5}} = \frac{1}{(32^{1/5})^3} = \frac{1}{\left(\sqrt[5]{32}\right)^3}$
$= \frac{1}{2^3} = \frac{1}{8}$

33. $64^{-3/2} = \frac{1}{64^{3/2}} = \frac{1}{(64^{1/2})^3} = \frac{1}{8^3} = \frac{1}{512}$

35. $\left(\frac{125}{27}\right)^{-2/3} = \left(\frac{27}{125}\right)^{2/3}$
$= \left(\left[\left(\frac{3}{5}\right)^3\right]^{1/3}\right)^2$
$= \left(\frac{3}{5}\right)^2 = \frac{9}{25}$

37. $12^{1/2} = \sqrt{12}$

39. $8^{3/4} = \left(8^{1/4}\right)^3 = \left(\sqrt[4]{8}\right)^3$

41. $(9q)^{5/8} - (2x)^{2/3}$
$= \left[(9q)^{1/8}\right]^5 - \left[(2x)^{1/3}\right]^2$
$= \left(\sqrt[8]{9q}\right)^5 - \left(\sqrt[3]{2x}\right)^2$

43. $(2m)^{-3/2} = \frac{1}{(2m)^{3/2}}$
$= \frac{1}{[(2m)^{1/2}]^3}$
$= \frac{1}{\left(\sqrt{2m}\right)^3}$

45. $(2y + x)^{2/3} = \left[(2y + x)^{1/3}\right]^2$
$= \left(\sqrt[3]{2y + x}\right)^2$

47. $(3m^4 + 2k^2)^{-2/3} = \frac{1}{(3m^4 + 2k^2)^{2/3}}$
$= \frac{1}{\left[(3m^4 + 2k^2)^{1/3}\right]^2}$
$= \frac{1}{\left(\sqrt[3]{3m^4 + 2k^2}\right)^2}$

49. We are to show that, in general,

$$\sqrt{a^2 + b^2} \neq a + b.$$

When $a = 3$ and $b = 4$,

$$\sqrt{a^2 + b^2} = \sqrt{3^2 + 4^2} = \sqrt{9 + 16} = \sqrt{25} = 5,$$

but

$$a + b = 3 + 4 = 7.$$

Since $5 \neq 7$, $\sqrt{a^2 + b^2} \neq a + b$.

51. $\sqrt{2^{12}} = (2^{12})^{1/2} = 2^{12/2} = 2^6 = 64$

53. $\sqrt[3]{4^9} = 4^{9/3} = 4^3 = 64$

55. $\sqrt{x^{20}} = x^{20/2} = x^{10}$

57. $\sqrt[3]{x} \cdot \sqrt{x} = x^{1/3} \cdot x^{1/2}$
$= x^{1/3 + 1/2} = x^{2/6 + 3/6}$
$= x^{5/6} = \sqrt[6]{x^5}$

59. $\frac{\sqrt[3]{t^4}}{\sqrt[5]{t^4}} = \frac{t^{4/3}}{t^{4/5}} = \frac{t^{20/15}}{t^{12/15}}$
$= t^{20/15 - 12/15} = t^{8/15} = \sqrt[15]{t^8}$

61. $3^{1/2} \cdot 3^{3/2} = 3^{1/2 + 3/2} = 3^{4/2} = 3^2 = 9$

63. $\frac{64^{5/3}}{64^{4/3}} = 64^{5/3 - 4/3} = 64^{1/3} = \sqrt[3]{64} = 4$

65. $y^{7/3} \cdot y^{-4/3} = y^{7/3 + (-4/3)}$
$= y^{3/3} = y^1 = y$

67. $x^{2/3} \cdot x^{-1/4} = x^{2/3 + (-1/4)} = x^{8/12 - 3/12} = x^{5/12}$

69. $\frac{k^{1/3}}{k^{2/3} \cdot k^{-1}} = \frac{k^{1/3}}{k^{-1/3}} = k^{1/3 - (-1/3)} = k^{2/3}$

71. $\frac{\left(x^{1/4}y^{2/5}\right)^{20}}{x^2} = \frac{x^{(1/4)\cdot 20}y^{(2/5)\cdot 20}}{x^2}$
$= \frac{x^5y^8}{x^2}$
$= x^3y^8$

73. $\frac{\left(x^{2/3}\right)^2}{\left(x^2\right)^{7/3}} = \frac{x^{4/3}}{x^{14/3}} = x^{4/3 - 14/3}$
$= x^{-10/3} = \frac{1}{x^{10/3}}$

75. $\frac{m^{3/4}n^{-1/4}}{(m^2n)^{1/2}} = \frac{m^{3/4}n^{-1/4}}{m^1n^{1/2}}$
$= m^{3/4 - 1}n^{-1/4 - 1/2}$
$= m^{3/4 - 4/4}n^{-1/4 - 2/4}$
$= m^{-1/4}n^{-3/4}$
$= \frac{1}{m^{1/4}n^{3/4}}$

77. $\frac{p^{1/5}p^{7/10}p^{1/2}}{(p^3)^{-1/5}} = \frac{p^{2/10 + 7/10 + 5/10}}{p^{-3/5}}$
$= \frac{p^{14/10}}{p^{-6/10}}$
$= p^{14/10 - (-6/10)}$
$= p^{20/10} = p^2$

79. $\left(\frac{b^{-3/2}}{c^{-5/3}}\right)^2 \left(b^{-1/4}c^{-1/3}\right)^{-1}$

$= \left(\frac{c^{5/3}}{b^{3/2}}\right)^2 \left(b^{1/4}c^{1/3}\right)$

$= \frac{c^{10/3}}{b^3}\left(b^{1/4}c^{1/3}\right)$

$= \frac{c^{10/3}b^{1/4}c^{1/3}}{b^3}$

$= c^{10/3+1/3}b^{1/4-3}$

$= c^{11/3}b^{-11/4}$

$= \frac{c^{11/3}}{b^{11/4}}$

81. $\left(\frac{p^{-1/4}q^{-3/2}}{3^{-1}p^{-2}q^{-2/3}}\right)^{-2}$

$= \frac{p^{1/2}q^3}{3^2p^4q^{4/3}}$

$= \frac{p^{1/2-4}q^{3-4/3}}{9}$

$= \frac{p^{-7/2}q^{5/3}}{9}$

$= \frac{q^{5/3}}{9p^{7/2}}$

83. $p^{2/3}\left(p^{1/3}+2p^{4/3}\right)$

$= p^{2/3}p^{1/3} + p^{2/3}\left(2p^{4/3}\right)$

$= p^{2/3+1/3} + 2p^{2/3+4/3}$

$= p^{3/3} + 2p^{6/3}$

$= p^1 + 2p^2$

$= p + 2p^2$

85. $k^{1/4}\left(k^{3/2} - k^{1/2}\right)$

$= k^{1/4+3/2} - k^{1/4+1/2}$

$= k^{1/4+6/4} - k^{1/4+2/4}$

$= k^{7/4} - k^{3/4}$

87. $6a^{7/4}\left(a^{-7/4}+3a^{-3/4}\right)$

$= 6a^{7/4+(-7/4)} + 18a^{7/4+(-3/4)}$

$= 6a^0 + 18a^{4/4}$

$= 6(1) + 18a^1 = 6 + 18a$

89. $5m^{-2/3}\left(m^{2/3}+m^{-7/3}\right)$

$= 5m^{-2/3}\cdot m^{2/3} + 5m^{-2/3}\cdot m^{-7/3}$

$= 5m^{-2/3+2/3} + 5m^{-2/3+(-7/3)}$

$= 5m^0 + 5m^{-9/3}$

$= 5\cdot 1 + 5m^{-3}$

$= 5 + \frac{5}{m^3}$

91. $\sqrt[6]{y^5}\cdot\sqrt[3]{y^2} = y^{5/6}\cdot y^{2/3}$

$= y^{5/6+2/3}$

$= y^{5/6+4/6}$

$= y^{9/6} = y^{3/2}$

93. $\frac{\sqrt[3]{k^5}}{\sqrt[3]{k^7}} = \frac{\left(k^5\right)^{1/3}}{\left(k^7\right)^{1/3}}$

$= \frac{k^{5(1/3)}}{k^{7(1/3)}}$

$= \frac{k^{5/3}}{k^{7/3}}$

$= k^{5/3-7/3}$

$= k^{-2/3} = \frac{1}{k^{2/3}}$

95. $\sqrt[3]{xz}\cdot\sqrt{z} = (xz)^{1/3}\cdot z^{1/2}$

$= x^{1/3}z^{1/3}z^{1/2}$

$= x^{1/3}z^{1/3+1/2}$

$= x^{1/3}z^{5/6}$

97. $\sqrt[3]{\sqrt{k}} = \sqrt[3]{k^{1/2}} = \left(k^{1/2}\right)^{1/3} = k^{1/6}$

99. $\sqrt[3]{\sqrt[5]{\sqrt{y}}} = \sqrt[3]{\sqrt[5]{y^{1/2}}}$

$= \sqrt[3]{\left(y^{1/2}\right)^{1/5}}$

$= \left(y^{1/10}\right)^{1/3}$

$= y^{1/30}$

101. $\sqrt[3]{x^{5/9}} = \left(x^{5/9}\right)^{1/3} = x^{(5/9)(1/3)} = x^{5/27}$

103. Use $h(T) = (1860.867T)^{1/3}$ with $T = 200$.

$$h(200) = [1860.867(200)]^{1/3}$$
$$\approx 71.93 \approx 72 \text{ inches}$$

The height is 6.0 feet (to the nearest tenth of a foot).

105. As in the previous exercise, for $T = 10°\text{F}$ and $V = 30$ mph, $W \approx -12.3°\text{F}$. The table gives $-12°\text{F}$.

9.3 Simplifying Radical Expressions

9.3 Margin Exercises

1. **(a)** $\sqrt{5}\cdot\sqrt{13} = \sqrt{5\cdot 13} = \sqrt{65}$

(b) $\sqrt{10y}\cdot\sqrt{3k} = \sqrt{10y\cdot 3k} = \sqrt{30yk}$

2. **(a)** $\sqrt[3]{2}\cdot\sqrt[3]{7} = \sqrt[3]{2\cdot 7} = \sqrt[3]{14}$

(b) $\sqrt[6]{8r^2}\cdot\sqrt[6]{2r^3} = \sqrt[6]{8r^2\cdot 2r^3} = \sqrt[6]{16r^5}$

(c) $\sqrt[5]{9y^2x}\cdot\sqrt[5]{8xy^2} = \sqrt[5]{9y^2x\cdot 8xy^2} = \sqrt[5]{72y^4x^2}$

(d) $\sqrt{7} \cdot \sqrt[3]{5}$ cannot be simplified using the product rule for radicals, because the indexes (2 and 3) are different.

3. **(a)** $\sqrt{\frac{100}{81}} = \frac{\sqrt{100}}{\sqrt{81}} = \frac{10}{9}$

(b) $\sqrt{\frac{11}{25}} = \frac{\sqrt{11}}{\sqrt{25}} = \frac{\sqrt{11}}{5}$

(c) $\sqrt[3]{-\frac{125}{216}} = \sqrt[3]{\frac{-125}{216}} = \frac{\sqrt[3]{-125}}{\sqrt[3]{216}}$

$$= \frac{-5}{6} = -\frac{5}{6}$$

(d) $\sqrt{\frac{y^8}{16}} = \frac{\sqrt{y^8}}{\sqrt{16}} = \frac{y^4}{4}$

(e) $-\sqrt[3]{\frac{x^2}{r^{12}}} = -\frac{\sqrt[3]{x^2}}{\sqrt[3]{r^{12}}} = -\frac{\sqrt[3]{x^2}}{r^4}$

4. **(a)** $\sqrt{32} = \sqrt{16 \cdot 2} = \sqrt{16} \cdot \sqrt{2} = 4\sqrt{2}$

(b) $\sqrt{45} = \sqrt{9 \cdot 5} = \sqrt{9} \cdot \sqrt{5} = 3\sqrt{5}$

(c) $\sqrt{300} = \sqrt{100 \cdot 3} = \sqrt{100} \cdot \sqrt{3} = 10\sqrt{3}$

(d) $\sqrt{35}$

No perfect square (other than 1) divides into 35, so $\sqrt{35}$ *cannot be simplified further.*

(e) $-\sqrt[3]{54} = -\sqrt[3]{27 \cdot 2} = -\sqrt[3]{27} \cdot \sqrt[3]{2} = -3\sqrt[3]{2}$

(f) $\sqrt[4]{243} = \sqrt[4]{81 \cdot 3} = \sqrt[4]{81} \cdot \sqrt[4]{3} = 3\sqrt[4]{3}$

5. **(a)** $\sqrt{25p^7} = \sqrt{25p^6 \cdot p}$

$$= \sqrt{25p^6} \cdot \sqrt{p} = 5p^3\sqrt{p}$$

(b) $\sqrt{72y^3x} = \sqrt{36y^2 \cdot 2yx}$

$$= \sqrt{36y^2} \cdot \sqrt{2yx}$$
$$= 6y\sqrt{2yx}$$

(c) $\sqrt[3]{-27y^7x^5z^6} = \sqrt[3]{-27y^6x^3z^6 \cdot yx^2}$

$$= \sqrt[3]{-27y^6x^3z^6} \cdot \sqrt[3]{yx^2}$$
$$= -3y^2xz^2\sqrt[3]{yx^2}$$

(d) $-\sqrt[4]{32a^5b^7} = -\sqrt[4]{16a^4b^4 \cdot 2ab^3}$

$$= -\sqrt[4]{16a^4b^4} \cdot \sqrt[4]{2ab^3}$$
$$= -2ab\sqrt[4]{2ab^3}$$

6. **(a)** $\sqrt[12]{2^3} = 2^{3/12} = 2^{1/4} = \sqrt[4]{2}$

(b) $\sqrt[6]{t^2} = t^{2/6} = t^{1/3} = \sqrt[3]{t}$

7. $\sqrt{5} \cdot \sqrt[3]{4}$

The least common index of 2 and 3 is 6. Write each radical as a sixth root.

$$\sqrt{5} = 5^{1/2} = 5^{3/6} = \sqrt[6]{5^3} = \sqrt[6]{125}$$
$$\sqrt[3]{4} = 4^{1/3} = 4^{2/6} = \sqrt[6]{4^2} = \sqrt[6]{16}$$

Therefore,

$$\sqrt{5} \cdot \sqrt[3]{4} = \sqrt[6]{125} \cdot \sqrt[6]{16} = \sqrt[6]{2000}.$$

8. **(a)** Substitute 14 for a and 8 for b in the Pythagorean formula to find c.

$$c^2 = a^2 + b^2$$
$$c^2 = 14^2 + 8^2$$
$$c^2 = 260$$
$$c = \sqrt{260}$$
$$c = \sqrt{4 \cdot 65}$$
$$c = \sqrt{4} \cdot \sqrt{65}$$
$$c = 2\sqrt{65}$$

The length of the hypotenuse is $2\sqrt{65}$.

(b) Substitute 4 for a and 6 for c to find b.

$$a^2 + b^2 = c^2$$
$$b^2 = c^2 - a^2$$
$$b^2 = 6^2 - 4^2$$
$$b^2 = 20$$
$$b = \sqrt{20}$$
$$b = \sqrt{4 \cdot 5}$$
$$b = \sqrt{4} \cdot \sqrt{5}$$
$$b = 2\sqrt{5}$$

The length of the unknown side is $2\sqrt{5}$.

9. Use the distance formula,

$$d = \sqrt{(x_2 - x_1)^2 + (y_2 - y_1)^2}.$$

(a) $(x_1, y_1) = (2, -1)$ and $(x_2, y_2) = (5, 3)$

$$d = \sqrt{(5-2)^2 + [3 - (-1)]^2}$$
$$= \sqrt{3^2 + 4^2}$$
$$= \sqrt{9 + 16}$$
$$= \sqrt{25}$$
$$= 5$$

(b) $(x_1, y_1) = (-3, 2)$ and $(x_2, y_2) = (0, -4)$

$$d = \sqrt{[0 - (-3)]^2 + (-4 - 2)^2}$$
$$= \sqrt{3^2 + (-6)^2}$$
$$= \sqrt{9 + 36}$$
$$= \sqrt{45}$$

Note that $\sqrt{45}$ can be written as $\sqrt{9} \cdot \sqrt{5} = 3\sqrt{5}$.

9.3 Section Exercises

1. Does $2\sqrt{12} = \sqrt{48}$?

$2\sqrt{12} = 2\sqrt{4 \cdot 3}$
$= 2\sqrt{4} \cdot \sqrt{3} = 2 \cdot 2 \cdot \sqrt{3} = 4\sqrt{3}$
$\sqrt{48} = \sqrt{16 \cdot 3} = \sqrt{16} \cdot \sqrt{3} = 4\sqrt{3}$

The calculator approximation for each expression is 6.92820323. The statement is true.

3. Does $3\sqrt{8} = 2\sqrt{18}$?

$3\sqrt{8} = 3\sqrt{4 \cdot 2} = 3\sqrt{4}\sqrt{2} = 3 \cdot 2\sqrt{2} = 6\sqrt{2}$
$2\sqrt{18} = 2\sqrt{9 \cdot 2} = 2\sqrt{9}\sqrt{2} = 2 \cdot 3\sqrt{2} = 6\sqrt{2}$

The calculator approximation for each expression is 8.485281374. The statement is true.

5. $\sqrt[3]{x} \cdot \sqrt[3]{x}$ is not equal to x because there are only two factors of $\sqrt[3]{x}$, and we would need three factors of $\sqrt[3]{x}$ to equal x.

$$\sqrt[3]{x} \cdot \sqrt[3]{x} = \left(\sqrt[3]{x}\right)^2 \quad \text{or} \quad \sqrt[3]{x^2}$$

7. **A.** $0.5 = \frac{1}{2}$ so $\sqrt{0.5} = \sqrt{\frac{1}{2}}$

B. $\frac{2}{4} = \frac{1}{2}$ so $\sqrt{\frac{2}{4}} = \sqrt{\frac{1}{2}}$

C. $\frac{3}{6} = \frac{1}{2}$ so $\sqrt{\frac{3}{6}} = \sqrt{\frac{1}{2}}$

D. $\frac{\sqrt{4}}{\sqrt{16}} = \sqrt{\frac{4}{16}} = \sqrt{\frac{1}{4}} \neq \sqrt{\frac{1}{2}}$

Choice **D** is not equal to $\sqrt{\frac{1}{2}}$.

9. $\sqrt{5} \cdot \sqrt{6} = \sqrt{5 \cdot 6} = \sqrt{30}$

11. $\sqrt{14} \cdot \sqrt{x} = \sqrt{14 \cdot x} = \sqrt{14x}$

13. $\sqrt{14} \cdot \sqrt{3pqr} = \sqrt{14 \cdot 3pqr} = \sqrt{42pqr}$

15. $\sqrt[3]{7x} \cdot \sqrt[3]{2y} = \sqrt[3]{7x \cdot 2y} = \sqrt[3]{14xy}$

17. $\sqrt[4]{11} \cdot \sqrt[4]{3} = \sqrt[4]{11 \cdot 3} = \sqrt[4]{33}$

19. $\sqrt[4]{2x} \cdot \sqrt[4]{3y^2} = \sqrt[4]{2x \cdot 3y^2} = \sqrt[4]{6xy^2}$

21. $\sqrt[3]{7} \cdot \sqrt[4]{3}$ cannot be multiplied using the product rule, because the indexes (3 and 4) are different.

23. $\sqrt{\frac{64}{121}} = \frac{\sqrt{64}}{\sqrt{121}} = \frac{8}{11}$

25. $\sqrt{\frac{3}{25}} = \frac{\sqrt{3}}{\sqrt{25}} = \frac{\sqrt{3}}{5}$

27. $\sqrt{\frac{x}{25}} = \frac{\sqrt{x}}{\sqrt{25}} = \frac{\sqrt{x}}{5}$

29. $\sqrt{\frac{p^6}{81}} = \frac{\sqrt{p^6}}{\sqrt{81}} = \frac{\sqrt{(p^3)^2}}{9} = \frac{p^3}{9}$

31. $\sqrt[3]{-\frac{27}{64}} = \sqrt[3]{\frac{-27}{64}} = \frac{\sqrt[3]{-27}}{\sqrt[3]{64}} = \frac{-3}{4} = -\frac{3}{4}$

33. $\sqrt[3]{\frac{r^2}{8}} = \frac{\sqrt[3]{r^2}}{\sqrt[3]{8}} = \frac{\sqrt[3]{r^2}}{2}$

35. $-\sqrt[4]{\frac{81}{x^4}} = -\frac{\sqrt[4]{3^4}}{\sqrt[4]{x^4}} = -\frac{3}{x}$

37. $\sqrt[5]{\frac{1}{x^{15}}} = \frac{\sqrt[5]{1}}{\sqrt[5]{x^{15}}} = \frac{\sqrt[5]{1^5}}{\sqrt[5]{(x^3)^5}} = \frac{1}{x^3}$

39. $\sqrt{12} = \sqrt{4 \cdot 3} = \sqrt{4} \cdot \sqrt{3} = 2\sqrt{3}$

41. $\sqrt{288} = \sqrt{144 \cdot 2} = \sqrt{144} \cdot \sqrt{2} = 12\sqrt{2}$

43. $-\sqrt{32} = -\sqrt{16 \cdot 2} = -\sqrt{16} \cdot \sqrt{2} = -4\sqrt{2}$

45. $-\sqrt{28} = -\sqrt{4 \cdot 7} = -\sqrt{4} \cdot \sqrt{7} = -2\sqrt{7}$

47. $\sqrt{30}$ cannot be simplified further.

49. $\sqrt[3]{128} = \sqrt[3]{64 \cdot 2} = \sqrt[3]{64} \cdot \sqrt[3]{2} = 4\sqrt[3]{2}$

51. $\sqrt[3]{-16} = \sqrt[3]{-8 \cdot 2} = \sqrt[3]{-8} \cdot \sqrt[3]{2} = -2\sqrt[3]{2}$

53. $\sqrt[3]{40} = \sqrt[3]{8 \cdot 5} = \sqrt[3]{8} \cdot \sqrt[3]{5} = 2\sqrt[3]{5}$

55. $-\sqrt[4]{512} = -\sqrt[4]{256 \cdot 2} = -\sqrt[4]{4^4} \cdot \sqrt[4]{2} = -4\sqrt[4]{2}$

57. $\sqrt[5]{64} = \sqrt[5]{32 \cdot 2} = \sqrt[5]{2^5} \cdot \sqrt[5]{2} = 2\sqrt[5]{2}$

59. His reasoning was incorrect. The radicand 14 must be written as a product of two factors (not a sum of two terms) where one of the two factors is a perfect cube.

61. $\sqrt{72k^2} = \sqrt{36k^2 \cdot 2} = \sqrt{36k^2} \cdot \sqrt{2} = 6k\sqrt{2}$

63. $\sqrt{144x^3y^9} = \sqrt{144x^2y^8 \cdot xy}$
$= \sqrt{(12xy^4)^2} \cdot \sqrt{xy}$
$= 12xy^4\sqrt{xy}$

65. $\sqrt{121x^6} = \sqrt{(11x^3)^2} = 11x^3$

67. $-\sqrt[3]{27t^{12}} = -\sqrt[3]{(3t^4)^3} = -3t^4$

69. $-\sqrt{100m^8z^4} = -\sqrt{(10m^4z^2)^2} = -10m^4z^2$

71. $-\sqrt[3]{-125a^6b^9c^{12}} = -\sqrt[3]{(-5a^2b^3c^4)^3}$
$= -(-5a^2b^3c^4)$
$= 5a^2b^3c^4$

73. $\sqrt[4]{\frac{1}{16}r^8t^{20}} = \sqrt[4]{\left(\frac{1}{2}r^2t^5\right)^4} = \frac{1}{2}r^2t^5$

75. $$\begin{aligned}\sqrt{50x^3} &= \sqrt{25x^2 \cdot 2x}\\ &= \sqrt{(5x)^2} \cdot \sqrt{2x}\\ &= 5x\sqrt{2x}\end{aligned}$$

77. $$\begin{aligned}-\sqrt{500r^{11}} &= -\sqrt{100r^{10} \cdot 5r}\\ &= -\sqrt{(10r^5)^2} \cdot \sqrt{5r}\\ &= -10r^5\sqrt{5r}\end{aligned}$$

79. $$\begin{aligned}\sqrt{13x^7y^8} &= \sqrt{(x^6y^8)(13x)}\\ &= \sqrt{x^6y^8} \cdot \sqrt{13x}\\ &= x^3y^4\sqrt{13x}\end{aligned}$$

81. $\sqrt[3]{8z^6w^9} = \sqrt[3]{(2z^2w^3)^3} = 2z^2w^3$

83. $$\begin{aligned}\sqrt[3]{-16z^5t^7} &= \sqrt[3]{(-2^3z^3t^6)(2z^2t)}\\ &= -2zt^2\sqrt[3]{2z^2t}\end{aligned}$$

85. $\sqrt[4]{81x^{12}y^{16}} = \sqrt[4]{(3x^3y^4)^4} = 3x^3y^4$

87. $$\begin{aligned}-\sqrt[4]{162r^{15}s^{10}} &= -\sqrt[4]{81r^{12}s^8(2r^3s^2)}\\ &= -\sqrt[4]{81r^{12}s^8} \cdot \sqrt[4]{2r^3s^2}\\ &= -3r^3s^2\sqrt[4]{2r^3s^2}\end{aligned}$$

89. $\sqrt{\frac{y^{11}}{36}} = \frac{\sqrt{y^{11}}}{\sqrt{36}} = \frac{\sqrt{y^{10} \cdot y}}{6} = \frac{y^5\sqrt{y}}{6}$

91. $\sqrt[3]{\frac{x^{16}}{27}} = \frac{\sqrt[3]{x^{15} \cdot x}}{\sqrt[3]{27}} = \frac{x^5\sqrt[3]{x}}{3}$

93. $$\begin{aligned}\sqrt[4]{48^2} &= 48^{2/4} = 48^{1/2}\\ &= \sqrt{48} = \sqrt{16 \cdot 3}\\ &= \sqrt{16} \cdot \sqrt{3} = 4\sqrt{3}\end{aligned}$$

95. $\sqrt[4]{25} = 25^{1/4} = (5^2)^{1/4} = 5^{2/4} = 5^{1/2} = \sqrt{5}$

97. $$\begin{aligned}\sqrt[10]{x^{25}} &= x^{25/10} = x^{5/2} = \sqrt{x^5}\\ &= \sqrt{x^4 \cdot x} = x^2\sqrt{x}\end{aligned}$$

99. $\sqrt[3]{4} \cdot \sqrt{3}$

The least common index of 3 and 2 is 6. Write each radical as a sixth root.

$$\begin{aligned}\sqrt[3]{4} &= 4^{1/3} = 4^{2/6} = \sqrt[6]{4^2} = \sqrt[6]{16}\\ \sqrt{3} &= 3^{1/2} = 3^{3/6} = \sqrt[6]{3^3} = \sqrt[6]{27}\end{aligned}$$

Therefore,

$$\begin{aligned}\sqrt[3]{4} \cdot \sqrt{3} &= \sqrt[6]{16} \cdot \sqrt[6]{27}\\ &= \sqrt[6]{16 \cdot 27} = \sqrt[6]{432}.\end{aligned}$$

101. $\sqrt[4]{3} \cdot \sqrt[3]{4}$

The least common index of 4 and 3 is 12. Write each radical as a twelfth root.

$$\begin{aligned}\sqrt[4]{3} &= 3^{1/4} = 3^{3/12} = \sqrt[12]{3^3} = \sqrt[12]{27}\\ \sqrt[3]{4} &= 4^{1/3} = 4^{4/12} = \sqrt[12]{4^4} = \sqrt[12]{256}\end{aligned}$$

Therefore,

$$\begin{aligned}\sqrt[4]{3} \cdot \sqrt[3]{4} &= \sqrt[12]{27} \cdot \sqrt[12]{256}\\ &= \sqrt[12]{27 \cdot 256} = \sqrt[12]{6912}.\end{aligned}$$

103. $\sqrt{x} \cdot \sqrt[3]{x}$

$$\begin{aligned}\sqrt{x} &= x^{1/2} = x^{3/6} = \sqrt[6]{x^3}\\ \sqrt[3]{x} &= x^{1/3} = x^{2/6} = \sqrt[6]{x^2}\end{aligned}$$

So $\sqrt{x} \cdot \sqrt[3]{x} = \sqrt[6]{x^3} \cdot \sqrt[6]{x^2}$

$$= \sqrt[6]{x^3 \cdot x^2} = \sqrt[6]{x^5}.$$

105. Substitute 3 for a and 4 for b in the Pythagorean formula to find c.

$$\begin{aligned}c^2 &= a^2 + b^2\\ c &= \sqrt{a^2 + b^2} = \sqrt{3^2 + 4^2}\\ &= \sqrt{9 + 16} = \sqrt{25} = 5\end{aligned}$$

The length of the hypotenuse is 5.

107. Substitute 4 for a and 12 for c in the Pythagorean formula to find b.

$$\begin{aligned}a^2 + b^2 &= c^2\\ b &= \sqrt{c^2 - a^2} = \sqrt{12^2 - 4^2}\\ &= \sqrt{144 - 16} = \sqrt{128}\\ &= \sqrt{64}\sqrt{2} = 8\sqrt{2}\end{aligned}$$

The length of the unknown leg is $8\sqrt{2}$.

109. Substitute 5 for b and 9 for c in the Pythagorean formula to find a.

$$\begin{aligned}a^2 + b^2 &= c^2\\ a &= \sqrt{c^2 - b^2} = \sqrt{9^2 - 5^2}\\ &= \sqrt{81 - 25} = \sqrt{56}\\ &= \sqrt{4}\sqrt{14} = 2\sqrt{14}\end{aligned}$$

The length of the unknown leg is $2\sqrt{14}$.

In Exercises 111–122, use the distance formula,

$$d = \sqrt{(x_2 - x_1)^2 + (y_2 - y_1)^2}.$$

111. $(6, 13)$ and $(1, 1)$

$$\begin{aligned}d &= \sqrt{(1-6)^2 + (1-13)^2}\\ &= \sqrt{(-5)^2 + (-12)^2}\\ &= \sqrt{25 + 144} = \sqrt{169} = 13\end{aligned}$$

113. $(-6, 5)$ and $(3, -4)$

$$\begin{aligned} d &= \sqrt{[3-(-6)]^2 + (-4-5)^2} \\ &= \sqrt{9^2 + (-9)^2} = \sqrt{81+81} \\ &= \sqrt{162} = \sqrt{81} \cdot \sqrt{2} = 9\sqrt{2} \end{aligned}$$

115. $(-8, 2)$ and $(-4, 1)$

$$\begin{aligned} d &= \sqrt{[-4-(-8)]^2 + (1-2)^2} \\ &= \sqrt{4^2 + (-1)^2} \\ &= \sqrt{16+1} = \sqrt{17} \end{aligned}$$

117. $(4.7, 2.3)$ and $(1.7, -1.7)$

$$\begin{aligned} d &= \sqrt{(1.7-4.7)^2 + (-1.7-2.3)^2} \\ &= \sqrt{(-3)^2 + (-4)^2} \\ &= \sqrt{9+16} = \sqrt{25} = 5 \end{aligned}$$

119. $(\sqrt{2}, \sqrt{6})$ and $(-2\sqrt{2}, 4\sqrt{6})$

$$\begin{aligned} d &= \sqrt{\left(-2\sqrt{2}-\sqrt{2}\right)^2 + \left(4\sqrt{6}-\sqrt{6}\right)^2} \\ &= \sqrt{\left(-3\sqrt{2}\right)^2 + \left(3\sqrt{6}\right)^2} \\ &= \sqrt{9 \cdot 2 + 9 \cdot 6} = \sqrt{18+54} \\ &= \sqrt{72} = \sqrt{36} \cdot \sqrt{2} = 6\sqrt{2} \end{aligned}$$

121. $(x+y, y)$ and $(x-y, x)$

$$\begin{aligned} d &= \sqrt{[(x-y)-(x+y)]^2 + (x-y)^2} \\ &= \sqrt{(-2y)^2 + (x-y)^2} \\ &= \sqrt{4y^2 + x^2 - 2xy + y^2} \\ &= \sqrt{5y^2 - 2xy + x^2} \end{aligned}$$

123. Substitute 21.7 for a and 16 for b in the Pythagorean formula to find c.

$$\begin{aligned} c^2 &= a^2 + b^2 \\ &= (21.7)^2 + 16^2 = 726.89 \\ c &= \sqrt{726.89} \approx 26.96 \end{aligned}$$

To the nearest tenth of an inch, the diagonal of the screen is 27.0 inches.

125. Substitute 282 for E, 100 for R, 264 for L, and 120π for ω in the formula to find I.

$$\begin{aligned} I &= \frac{E}{\sqrt{R^2 + \omega^2 L^2}} \\ &= \frac{282}{\sqrt{100^2 + (120\pi)^2 (264)^2}} \\ &\approx \frac{282}{99{,}526} \approx 0.003 \end{aligned}$$

127. $d = 1.224\sqrt{h}$
$= 1.224\sqrt{156} \approx 15.3$ miles

9.4 Adding and Subtracting Radical Expressions

9.4 Margin Exercises

1. **(a)** $3\sqrt{5} + 7\sqrt{5} = (3+7)\sqrt{5} = 10\sqrt{5}$

(b) $2\sqrt{11} - \sqrt{11} + 3\sqrt{44}$

$$\begin{aligned} &= 2\sqrt{11} - 1\sqrt{11} + 3\sqrt{4} \cdot \sqrt{11} \\ &= 2\sqrt{11} - 1\sqrt{11} + 3 \cdot 2 \cdot \sqrt{11} \\ &= (2-1+6)\sqrt{11} \\ &= 7\sqrt{11} \end{aligned}$$

(c) $5\sqrt{12y} + 6\sqrt{75y},\ y \ge 0$

$$\begin{aligned} &= 5\sqrt{4} \cdot \sqrt{3y} + 6\sqrt{25} \cdot \sqrt{3y} \\ &= 5 \cdot 2\sqrt{3y} + 6 \cdot 5\sqrt{3y} \\ &= 10\sqrt{3y} + 30\sqrt{3y} \\ &= (10+30)\sqrt{3y} \\ &= 40\sqrt{3y} \end{aligned}$$

(d) $3\sqrt{8} - 6\sqrt{50} + 2\sqrt{200}$

$$\begin{aligned} &= 3\sqrt{4} \cdot \sqrt{2} - 6\sqrt{25} \cdot \sqrt{2} + 2\sqrt{100} \cdot \sqrt{2} \\ &= 3 \cdot 2\sqrt{2} - 6 \cdot 5\sqrt{2} + 2 \cdot 10\sqrt{2} \\ &= 6\sqrt{2} - 30\sqrt{2} + 20\sqrt{2} \\ &= (6-30+20)\sqrt{2} \\ &= -4\sqrt{2} \end{aligned}$$

(e) $9\sqrt{5} - 4\sqrt{10}$

Here the radicals differ and are already simplified, so this expression *cannot be simplified further.*

2. **(a)** $7\sqrt[3]{81} + 3\sqrt[3]{24}$

$$\begin{aligned} &= 7\sqrt[3]{27 \cdot 3} + 3\sqrt[3]{8 \cdot 3} \\ &= 7\sqrt[3]{27} \cdot \sqrt[3]{3} + 3\sqrt[3]{8} \cdot \sqrt[3]{3} \\ &= 7 \cdot 3\sqrt[3]{3} + 3 \cdot 2\sqrt[3]{3} \\ &= 21\sqrt[3]{3} + 6\sqrt[3]{3} \\ &= 27\sqrt[3]{3} \end{aligned}$$

(b) $-2\sqrt[4]{32} - 7\sqrt[4]{162}$

$$\begin{aligned} &= -2\sqrt[4]{16 \cdot 2} - 7\sqrt[4]{81 \cdot 2} \\ &= -2 \cdot 2\sqrt[4]{2} - 7 \cdot 3\sqrt[4]{2} \\ &= -4\sqrt[4]{2} - 21\sqrt[4]{2} \\ &= -25\sqrt[4]{2} \end{aligned}$$

(c) $\sqrt[3]{p^4q^7} - \sqrt[3]{64pq}$

$$\begin{aligned} &= \sqrt[3]{p^3q^6 \cdot pq} - \sqrt[3]{64 \cdot pq} \\ &= pq^2\sqrt[3]{pq} - 4\sqrt[3]{pq} \\ &= (pq^2 - 4)\sqrt[3]{pq} \end{aligned}$$

3. **(a)** $2\sqrt{\frac{8}{9}}-2\frac{\sqrt{27}}{\sqrt{108}}$

$=2\frac{\sqrt{4\cdot 2}}{\sqrt{9}}-2\frac{\sqrt{9\cdot 3}}{\sqrt{36\cdot 3}}$

$=2\left(\frac{2\sqrt{2}}{3}\right)-2\left(\frac{3\sqrt{3}}{6\sqrt{3}}\right)$

$=\frac{4\sqrt{2}}{3}-\frac{3}{3}$

$=\frac{4\sqrt{2}-3}{3}$

(b) $\sqrt{\frac{80}{y^4}}+\sqrt{\frac{81}{y^{10}}}$

$=\frac{\sqrt{16\cdot 5}}{\sqrt{y^4}}+\frac{\sqrt{81}}{\sqrt{y^{10}}}$

$=\frac{4\sqrt{5}}{y^2}+\frac{9}{y^5}$

$=\frac{4y^3\sqrt{5}}{y^5}+\frac{9}{y^5}$

$=\frac{4y^3\sqrt{5}+9}{y^5}$

9.4 Section Exercises

1. Only choice **B** has like radical factors, so it can be simplified without first simplifying the individual radical expressions.

$$3\sqrt{6}+9\sqrt{6}=12\sqrt{6}$$

3. $\sqrt{64}+\sqrt[3]{125}+\sqrt[4]{16}$

$=\sqrt{8^2}+\sqrt[3]{5^3}+\sqrt[4]{2^4}$

$=8+5+2=15$

This sum can be found easily since each radicand has a whole number power corresponding to the index of the radical; that is, each radical expression simplifies to a whole number.

5. Simplify each radical and subtract.

$$\sqrt{36}-\sqrt{100}=6-10=-4$$

7. $-2\sqrt{48}+3\sqrt{75}$

$=-2\sqrt{16\cdot 3}+3\sqrt{25\cdot 3}$

$=-2\cdot 4\sqrt{3}+3\cdot 5\sqrt{3}$

$=-8\sqrt{3}+15\sqrt{3}$

$=7\sqrt{3}$

9. $\sqrt[3]{16}+4\sqrt[3]{54}$

$=\sqrt[3]{8\cdot 2}+4\sqrt[3]{27\cdot 2}$

$=\sqrt[3]{8}\sqrt[3]{2}+4\sqrt[3]{27}\sqrt[3]{2}$

$=2\sqrt[3]{2}+4\cdot 3\sqrt[3]{2}$

$=2\sqrt[3]{2}+12\sqrt[3]{2}$

$=14\sqrt[3]{2}$

11. $\sqrt[4]{32}+3\sqrt[4]{2}$

$=\sqrt[4]{16\cdot 2}+3\sqrt[4]{2}$

$=\sqrt[4]{16}\sqrt[4]{2}+3\sqrt[4]{2}$

$=2\sqrt[4]{2}+3\sqrt[4]{2}$

$=5\sqrt[4]{2}$

13. $6\sqrt{18}-\sqrt{32}+2\sqrt{50}$

$=6\sqrt{9\cdot 2}-\sqrt{16\cdot 2}+2\sqrt{25\cdot 2}$

$=6\cdot 3\sqrt{2}-4\sqrt{2}+2\cdot 5\sqrt{2}$

$=18\sqrt{2}-4\sqrt{2}+10\sqrt{2}$

$=24\sqrt{2}$

15. $5\sqrt{6}+2\sqrt{10}$

The radicals differ and are already simplified, so the expression cannot be simplified further.

17. $2\sqrt{5}+3\sqrt{20}+4\sqrt{45}$

$=2\sqrt{5}+3\sqrt{4\cdot 5}+4\sqrt{9\cdot 5}$

$=2\sqrt{5}+3\cdot 2\sqrt{5}+4\cdot 3\sqrt{5}$

$=2\sqrt{5}+6\sqrt{5}+12\sqrt{5}$

$=20\sqrt{5}$

19. $8\sqrt{2x}-\sqrt{8x}+\sqrt{72x}$

$=8\sqrt{2x}-\sqrt{4\cdot 2x}+\sqrt{36\cdot 2x}$

$=8\sqrt{2x}-2\sqrt{2x}+6\sqrt{2x}$

$=12\sqrt{2x}$

21. $3\sqrt{72m^2}-5\sqrt{32m^2}$

$=3\sqrt{36m^2\cdot 2}-5\sqrt{16m^2\cdot 2}$

$=3\cdot 6m\sqrt{2}-5\cdot 4m\sqrt{2}$

$=18m\sqrt{2}-20m\sqrt{2}$

$=-2m\sqrt{2}$

23. $-\sqrt[3]{54}+2\sqrt[3]{16}$

$=-\sqrt[3]{27\cdot 2}+2\sqrt[3]{8\cdot 2}$

$=-3\sqrt[3]{2}+2\cdot 2\sqrt[3]{2}$

$=-3\sqrt[3]{2}+4\sqrt[3]{2}$

$=\sqrt[3]{2}$

25. $2\sqrt[3]{27x} - 2\sqrt[3]{8x}$
$= 2\sqrt[3]{27 \cdot x} - 2\sqrt[3]{8 \cdot x}$
$= 2 \cdot 3\sqrt[3]{x} - 2 \cdot 2\sqrt[3]{x}$
$= 6\sqrt[3]{x} - 4\sqrt[3]{x}$
$= 2\sqrt[3]{x}$

27. $\sqrt[3]{x^2y} - \sqrt[3]{8x^2y}$
$= \sqrt[3]{x^2y} - \sqrt[3]{8}\sqrt[3]{x^2y}$
$= 1\sqrt[3]{x^2y} - 2\sqrt[3]{x^2y}$
$= (1-2)\sqrt[3]{x^2y}$
$= -\sqrt[3]{x^2y}$

29. $3x\sqrt[3]{xy^2} - 2\sqrt[3]{8x^4y^2}$
$= 3x\sqrt[3]{xy^2} - 2\sqrt[3]{8x^3} \cdot \sqrt[3]{xy^2}$
$= 3x\sqrt[3]{xy^2} - 2 \cdot 2x \cdot \sqrt[3]{xy^2}$
$= (3x - 4x)\sqrt[3]{xy^2}$
$= -x\sqrt[3]{xy^2}$

31. $5\sqrt[4]{32} + 3\sqrt[4]{162}$
$= 5\sqrt[4]{16 \cdot 2} + 3\sqrt[4]{81 \cdot 2}$
$= 5 \cdot 2\sqrt[4]{2} + 3 \cdot 3\sqrt[4]{2}$
$= 10\sqrt[4]{2} + 9\sqrt[4]{2}$
$= 19\sqrt[4]{2}$

33. $3\sqrt[4]{x^5y} - 2x\sqrt[4]{xy}$
$= 3\sqrt[4]{x^4 \cdot xy} - 2x\sqrt[4]{xy}$
$= 3x\sqrt[4]{xy} - 2x\sqrt[4]{xy}$
$= (3x - 2x)\sqrt[4]{xy}$
$= x\sqrt[4]{xy}$

35. $2\sqrt[4]{32a^3} + 5\sqrt[4]{2a^3}$
$= 2\sqrt[4]{16} \cdot \sqrt[4]{2a^3} + 5\sqrt[4]{2a^3}$
$= 2 \cdot 2 \cdot \sqrt[4]{2a^3} + 5\sqrt[4]{2a^3}$
$= (4+5)\sqrt[4]{2a^3}$
$= 9\sqrt[4]{2a^3}$

37. $\frac{2\sqrt{5}}{3} + \frac{\sqrt{5}}{6}$
$= \frac{4\sqrt{5}}{6} + \frac{1\sqrt{5}}{6}$
$= \frac{4\sqrt{5} + 1\sqrt{5}}{6}$
$= \frac{5\sqrt{5}}{6}$

39. $\sqrt{\frac{8}{9}} + \sqrt{\frac{18}{36}}$
$= \frac{\sqrt{8}}{\sqrt{9}} + \frac{\sqrt{18}}{\sqrt{36}}$
$= \frac{\sqrt{4}\sqrt{2}}{3} + \frac{\sqrt{9}\sqrt{2}}{6}$
$= \frac{2\sqrt{2}}{3} + \frac{3\sqrt{2}}{6}$
$= \frac{4\sqrt{2}}{6} + \frac{3\sqrt{2}}{6}$
$= \frac{4\sqrt{2} + 3\sqrt{2}}{6} = \frac{7\sqrt{2}}{6}$

41. $\frac{\sqrt{32}}{3} + \frac{2\sqrt{2}}{3} - \frac{\sqrt{2}}{\sqrt{9}}$
$= \frac{\sqrt{16}\sqrt{2}}{3} + \frac{2\sqrt{2}}{3} - \frac{\sqrt{2}}{3}$
$= \frac{4\sqrt{2} + 2\sqrt{2} - \sqrt{2}}{3} = \frac{5\sqrt{2}}{3}$

43. $3\sqrt{\frac{50}{9}} + 8\frac{\sqrt{2}}{\sqrt{8}}$
$= 3\frac{\sqrt{50}}{\sqrt{9}} + 8\frac{\sqrt{2}}{2\sqrt{2}}$
$= 3 \cdot \frac{5\sqrt{2}}{3} + 8 \cdot \frac{1}{2}$
$= 5\sqrt{2} + 4$

45. $\sqrt{\frac{25}{x^8}} - \sqrt{\frac{9}{x^6}}$
$= \frac{\sqrt{25}}{\sqrt{x^8}} - \frac{\sqrt{9}}{\sqrt{x^6}}$
$= \frac{5}{x^4} - \frac{3}{x^3}$
$= \frac{5}{x^4} - \frac{3 \cdot x}{x^3 \cdot x}$ $LCD = x^4$
$= \frac{5 - 3x}{x^4}$

47. $3\sqrt[3]{\frac{m^5}{27}} - 2m\sqrt[3]{\frac{m^2}{64}}$
$= \frac{3\sqrt[3]{m^5}}{\sqrt[3]{27}} - \frac{2m\sqrt[3]{m^2}}{\sqrt[3]{64}}$
$= \frac{3\sqrt[3]{m^3}\sqrt[3]{m^2}}{3} - \frac{2m\sqrt[3]{m^2}}{4}$
$= \frac{m\sqrt[3]{m^2}}{1} - \frac{m\sqrt[3]{m^2}}{2}$
$= \frac{2m\sqrt[3]{m^2} - m\sqrt[3]{m^2}}{2} = \frac{m\sqrt[3]{m^2}}{2}$

49. $3\sqrt[3]{\frac{2}{x^6}} - 4\sqrt[3]{\frac{5}{x^9}}$

$= 3\frac{\sqrt[3]{2}}{\sqrt[3]{x^6}} - 4\frac{\sqrt[3]{5}}{\sqrt[3]{x^9}}$

$= 3\frac{\sqrt[3]{2}}{x^2} - 4\frac{\sqrt[3]{5}}{x^3}$

$= \frac{3 \cdot x \cdot \sqrt[3]{2}}{x^2 \cdot x} - \frac{4\sqrt[3]{5}}{x^3} \quad LCD = x^3$

$= \frac{3x\sqrt[3]{2} - 4\sqrt[3]{5}}{x^3}$

51. The perimeter, P, of a triangle is the sum of the measures of the sides.

$P = 3\sqrt{20} + 2\sqrt{45} + \sqrt{75}$

$= 3\sqrt{4 \cdot 5} + 2\sqrt{9 \cdot 5} + \sqrt{25 \cdot 3}$

$= 3 \cdot 2\sqrt{5} + 2 \cdot 3\sqrt{5} + 5\sqrt{3}$

$= 6\sqrt{5} + 6\sqrt{5} + 5\sqrt{3}$

$= 12\sqrt{5} + 5\sqrt{3}$

The perimeter is $(12\sqrt{5} + 5\sqrt{3})$ inches.

53. To find the perimeter, add the lengths of the sides.

$4\sqrt{18} + \sqrt{108} + 2\sqrt{72} + 3\sqrt{12}$

$= 4\sqrt{9}\sqrt{2} + \sqrt{36}\sqrt{3} + 2\sqrt{36}\sqrt{2} + 3\sqrt{4}\sqrt{3}$

$= 4 \cdot 3\sqrt{2} + 6\sqrt{3} + 2 \cdot 6\sqrt{2} + 3 \cdot 2\sqrt{3}$

$= 12\sqrt{2} + 6\sqrt{3} + 12\sqrt{2} + 6\sqrt{3}$

$= 24\sqrt{2} + 12\sqrt{3}$

The perimeter is $(24\sqrt{2} + 12\sqrt{3})$ inches.

9.5 Multiplying and Dividing Radical Expressions

9.5 Margin Exercises

1. Use the FOIL method and multiply as two binomials.

(a) $(2 + \sqrt{3})(1 + \sqrt{5})$

F O I L

$= 2 \cdot 1 + 2\sqrt{5} + \sqrt{3} \cdot 1 + \sqrt{3} \cdot \sqrt{5}$

$= 2 + 2\sqrt{5} + \sqrt{3} + \sqrt{15}$

(b) $(4 + \sqrt{3})(4 - \sqrt{3})$

$= 4 \cdot 4 - 4\sqrt{3} + 4\sqrt{3} - \sqrt{3} \cdot \sqrt{3}$

$= 16 - 3$

$= 13$

(c) $(\sqrt{13} - 2)^2$

$= (\sqrt{13} - 2)(\sqrt{13} - 2)$

$= \sqrt{13} \cdot \sqrt{13} - 2 \cdot \sqrt{13} - 2 \cdot \sqrt{13} + 2 \cdot 2$

$= 13 - 4\sqrt{13} + 4$

$= 17 - 4\sqrt{13}$

(d) $(4 + \sqrt[3]{7})(4 - \sqrt[3]{7})$

$= 4 \cdot 4 - 4\sqrt[3]{7} + 4\sqrt[3]{7} - \sqrt[3]{7} \cdot \sqrt[3]{7}$

$= 16 - \sqrt[3]{49}$

(e) $(\sqrt{p} + \sqrt{s})(\sqrt{p} - \sqrt{s})$

$= (\sqrt{p})^2 - (\sqrt{s})^2$

$= p - s, \quad p \geq 0 \text{ and } s \geq 0$

2. **(a)** $\frac{8}{\sqrt{3}} = \frac{8 \cdot \sqrt{3}}{\sqrt{3} \cdot \sqrt{3}} = \frac{8\sqrt{3}}{3}$

(b) $\frac{5\sqrt{6}}{\sqrt{5}} = \frac{5\sqrt{6} \cdot \sqrt{5}}{\sqrt{5} \cdot \sqrt{5}} = \frac{5\sqrt{30}}{5} = \sqrt{30}$

(c) $\frac{3}{\sqrt{48}} = \frac{3}{\sqrt{16 \cdot 3}} = \frac{3}{4\sqrt{3}}$

$= \frac{3 \cdot \sqrt{3}}{4\sqrt{3} \cdot \sqrt{3}} = \frac{3\sqrt{3}}{4 \cdot 3}$

$= \frac{\sqrt{3}}{4}$

(d) $\frac{-16}{\sqrt{32}} = -\frac{16}{\sqrt{16 \cdot 2}} = -\frac{16}{4\sqrt{2}}$

$= -\frac{4}{\sqrt{2}} = -\frac{4 \cdot \sqrt{2}}{\sqrt{2} \cdot \sqrt{2}}$

$= -\frac{4\sqrt{2}}{2} = -2\sqrt{2}$

3. **(a)** $\sqrt{\frac{8}{45}} = \frac{\sqrt{8}}{\sqrt{45}} = \frac{\sqrt{4 \cdot 2}}{\sqrt{9 \cdot 5}} = \frac{2\sqrt{2}}{3\sqrt{5}}$

$= \frac{2\sqrt{2} \cdot \sqrt{5}}{3\sqrt{5} \cdot \sqrt{5}} = \frac{2\sqrt{10}}{3 \cdot 5} = \frac{2\sqrt{10}}{15}$

(b) $\sqrt{\frac{72}{y}} = \frac{\sqrt{72} \cdot \sqrt{y}}{\sqrt{y} \cdot \sqrt{y}}$

$= \frac{\sqrt{72y}}{y} = \frac{\sqrt{36 \cdot 2y}}{y} = \frac{6\sqrt{2y}}{y}$

(c) $\sqrt{\dfrac{200k^6}{y^7}} = \dfrac{\sqrt{100k^6 \cdot 2}}{\sqrt{y^6 \cdot y}}$

$= \dfrac{10k^3\sqrt{2}}{y^3\sqrt{y}}$

$= \dfrac{10k^3\sqrt{2} \cdot \sqrt{y}}{y^3\sqrt{y} \cdot \sqrt{y}}$

$= \dfrac{10k^3\sqrt{2y}}{y^3 \cdot y}$

$= \dfrac{10k^3\sqrt{2y}}{y^4}$

4. (a) $\sqrt[3]{\dfrac{15}{32}} = \dfrac{\sqrt[3]{15}}{\sqrt[3]{32}} = \dfrac{\sqrt[3]{15}}{\sqrt[3]{8} \cdot \sqrt[3]{4}} = \dfrac{\sqrt[3]{15}}{2\sqrt[3]{4}}$

$= \dfrac{\sqrt[3]{15} \cdot \sqrt[3]{2}}{2\sqrt[3]{4} \cdot \sqrt[3]{2}} = \dfrac{\sqrt[3]{30}}{2\sqrt[3]{8}}$

$= \dfrac{\sqrt[3]{30}}{2 \cdot 2} = \dfrac{\sqrt[3]{30}}{4}$

(b) $\sqrt[3]{\dfrac{m^{12}}{n}} = \dfrac{\sqrt[3]{m^{12}}}{\sqrt[3]{n}}$

$= \dfrac{m^4 \cdot \sqrt[3]{n^2}}{\sqrt[3]{n} \cdot \sqrt[3]{n^2}}$

$= \dfrac{m^4\sqrt[3]{n^2}}{n} \quad (n \neq 0)$

(c) $\sqrt[4]{\dfrac{6y}{w^2}} = \dfrac{\sqrt[4]{6y}}{\sqrt[4]{w^2}} \cdot \dfrac{\sqrt[4]{w^2}}{\sqrt[4]{w^2}}$

$= \dfrac{\sqrt[4]{6yw^2}}{\sqrt[4]{w^4}}$

$= \dfrac{\sqrt[4]{6yw^2}}{w} \quad (y \geq 0, w \neq 0)$

5. (a) $\dfrac{-4}{\sqrt{5}+2}$

Multiply both the numerator and denominator by the conjugate of the denominator, $\sqrt{5} - 2$.

$= \dfrac{-4(\sqrt{5}-2)}{(\sqrt{5}+2)(\sqrt{5}-2)}$

$= \dfrac{-4(\sqrt{5}-2)}{5-4}$

$= -4(\sqrt{5}-2)$

(b) $\dfrac{15}{\sqrt{7}+\sqrt{2}}$

Multiply both the numerator and denominator by the conjugate of the denominator, $\sqrt{7} - \sqrt{2}$.

$= \dfrac{15(\sqrt{7}-\sqrt{2})}{(\sqrt{7}+\sqrt{2})(\sqrt{7}-\sqrt{2})}$

$= \dfrac{15(\sqrt{7}-\sqrt{2})}{7-2}$

$= \dfrac{15(\sqrt{7}-\sqrt{2})}{5}$

$= 3(\sqrt{7}-\sqrt{2})$

(c) $\dfrac{\sqrt{3}+\sqrt{5}}{\sqrt{2}-\sqrt{7}}$

Multiply both the numerator and denominator by the conjugate of the denominator, $\sqrt{2} + \sqrt{7}$.

$= \dfrac{(\sqrt{3}+\sqrt{5})(\sqrt{2}+\sqrt{7})}{(\sqrt{2}-\sqrt{7})(\sqrt{2}+\sqrt{7})}$

$= \dfrac{\sqrt{6}+\sqrt{21}+\sqrt{10}+\sqrt{35}}{2-7}$

$= \dfrac{-(\sqrt{6}+\sqrt{21}+\sqrt{10}+\sqrt{35})}{5}$

(d) $\dfrac{2}{\sqrt{k}+\sqrt{z}}$

Multiply both the numerator and denominator by the conjugate of the denominator, $\sqrt{k} - \sqrt{z}$.

$= \dfrac{2(\sqrt{k}-\sqrt{z})}{(\sqrt{k}+\sqrt{z})(\sqrt{k}-\sqrt{z})}$

$= \dfrac{2(\sqrt{k}-\sqrt{z})}{k-z} \quad (k \neq z,\ k > 0, z > 0)$

6. (a) $\dfrac{24 - 36\sqrt{7}}{16} = \dfrac{4(6-9\sqrt{7})}{16} = \dfrac{6-9\sqrt{7}}{4}$

(b) $\dfrac{2x+\sqrt{32x^2}}{6x} = \dfrac{2x+\sqrt{16x^2 \cdot 2}}{6x}$

$= \dfrac{2x+4x\sqrt{2}}{6x} \quad (x > 0)$

$= \dfrac{2x(1+2\sqrt{2})}{2x \cdot 3}$

$= \dfrac{1+2\sqrt{2}}{3}$

9.5 Section Exercises

1. $(x+\sqrt{y})(x-\sqrt{y})$
$= x^2 - (\sqrt{y})^2$
$= x^2 - y$ **(E)**

3. $(\sqrt{x}+\sqrt{y})(\sqrt{x}-\sqrt{y})$
$= (\sqrt{x})^2 - (\sqrt{y})^2$
$= x - y$ **(A)**

5. $(\sqrt{x}-\sqrt{y})^2$
$= (\sqrt{x})^2 - 2\sqrt{x}\sqrt{y} + (\sqrt{y})^2$
$= x - 2\sqrt{xy} + y$ **(D)**

7. $\sqrt{3}(\sqrt{12}-4) = \sqrt{3}\cdot\sqrt{12} - \sqrt{3}\cdot 4$
$= \sqrt{3\cdot 12} - 4\sqrt{3}$
$= \sqrt{36} - 4\sqrt{3}$
$= 6 - 4\sqrt{3}$

9. $\sqrt{2}(\sqrt{18}-\sqrt{3}) = \sqrt{2}\cdot\sqrt{18} - \sqrt{2}\cdot\sqrt{3}$
$= \sqrt{36} - \sqrt{6}$
$= 6 - \sqrt{6}$

11. $(\sqrt{6}+2)(\sqrt{6}-2)$
$= \sqrt{6}\cdot\sqrt{6} - 2\sqrt{6} + 2\sqrt{6} - 2\cdot 2$
$= 6 - 4$
$= 2$

13. $(\sqrt{12}-\sqrt{3})(\sqrt{12}+\sqrt{3})$
$= (\sqrt{12})^2 - (\sqrt{3})^2$
$= 12 - 3$
$= 9$

15. $(\sqrt{3}+2)(\sqrt{6}-5)$
$= \sqrt{3}\cdot\sqrt{6} - 5\sqrt{3} + 2\sqrt{6} - 10$
$= \sqrt{18} - 5\sqrt{3} + 2\sqrt{6} - 10$
$= \sqrt{9\cdot 2} - 5\sqrt{3} + 2\sqrt{6} - 10$
$= 3\sqrt{2} - 5\sqrt{3} + 2\sqrt{6} - 10$

17. $(\sqrt{3x}+2)(\sqrt{3x}-2) = (\sqrt{3x})^2 - 2^2$
$= 3x - 4$

19. $(2\sqrt{x}+\sqrt{y})(2\sqrt{x}-\sqrt{y})$
$= (2\sqrt{x})^2 - (\sqrt{y})^2$
$= 2^2(\sqrt{x})^2 - y$
$= 4x - y$

21. $(4\sqrt{x}+3)^2 = (4\sqrt{x})^2 + 2(4\sqrt{x})(3) + 3^2$
$= 16x + 24\sqrt{x} + 9$

23. $(9-\sqrt[3]{2})(9+\sqrt[3]{2}) = 9^2 - (\sqrt[3]{2})^2$
$= 81 - \sqrt[3]{4}$

25. Because 6 and $4\sqrt{3}$ are not like terms, they cannot be combined.

27. $\dfrac{7}{\sqrt{7}} = \dfrac{7\cdot\sqrt{7}}{\sqrt{7}\cdot\sqrt{7}} = \dfrac{7\sqrt{7}}{7} = \sqrt{7}$

29. $\dfrac{15}{\sqrt{3}} = \dfrac{15\cdot\sqrt{3}}{\sqrt{3}\cdot\sqrt{3}} = \dfrac{15\sqrt{3}}{3} = 5\sqrt{3}$

31. $\dfrac{\sqrt{3}}{\sqrt{2}} = \dfrac{\sqrt{3}\cdot\sqrt{2}}{\sqrt{2}\cdot\sqrt{2}} = \dfrac{\sqrt{6}}{2}$

33. $\dfrac{9\sqrt{3}}{\sqrt{5}} = \dfrac{9\sqrt{3}\cdot\sqrt{5}}{\sqrt{5}\cdot\sqrt{5}} = \dfrac{9\sqrt{15}}{5}$

35. $\dfrac{-6}{\sqrt{18}} = \dfrac{-6}{\sqrt{9\cdot 2}} = \dfrac{-6}{3\sqrt{2}} = \dfrac{-2}{\sqrt{2}} = \dfrac{-2\cdot\sqrt{2}}{\sqrt{2}\cdot\sqrt{2}}$
$= \dfrac{-2\sqrt{2}}{2} = -\sqrt{2}$

37. $\dfrac{-8\sqrt{3}}{\sqrt{k}} = \dfrac{-8\sqrt{3}\cdot\sqrt{k}}{\sqrt{k}\cdot\sqrt{k}} = \dfrac{-8\sqrt{3k}}{k}$

39. $\dfrac{6\sqrt{3y}}{\sqrt{y^3}} = \dfrac{6\sqrt{3y}\cdot\sqrt{y}}{\sqrt{y^3}\cdot\sqrt{y}}$
$= \dfrac{6\sqrt{3y^2}}{\sqrt{y^4}}$
$= \dfrac{6y\sqrt{3}}{y^2}$
$= \dfrac{6\sqrt{3}}{y}$

41. Multiply the numerator and the denominator by $\sqrt[3]{2}$:

$$\frac{1}{\sqrt[3]{2}} = \frac{1\cdot\sqrt[3]{2}}{\sqrt[3]{2}\cdot\sqrt[3]{2}} = \frac{\sqrt[3]{2}}{\sqrt[3]{4}}.$$

The denominator is not yet rationalized. Multiply both the numerator and the denominator by $\sqrt[3]{4}$:

$$\frac{1}{\sqrt[3]{2}} = \frac{1\cdot\sqrt[3]{4}}{\sqrt[3]{2}\cdot\sqrt[3]{4}} = \frac{\sqrt[3]{4}}{\sqrt[3]{8}} = \frac{\sqrt[3]{4}}{2}.$$

To rationalize a cube root, three factors of the quantity under the radical sign are needed. We must multiply by $\sqrt[3]{2^2}$ or $\sqrt[3]{4}$ to rationalize $\sqrt[3]{2}$.

43. $\sqrt{\dfrac{7}{2}} = \dfrac{\sqrt{7}}{\sqrt{2}} = \dfrac{\sqrt{7}\cdot\sqrt{2}}{\sqrt{2}\cdot\sqrt{2}} = \dfrac{\sqrt{14}}{2}$

45. $-\sqrt{\dfrac{7}{50}} = -\dfrac{\sqrt{7}}{\sqrt{25\cdot 2}} = -\dfrac{\sqrt{7}}{5\sqrt{2}}$
$= -\dfrac{\sqrt{7}\cdot\sqrt{2}}{5\sqrt{2}\cdot\sqrt{2}} = -\dfrac{\sqrt{14}}{5\cdot 2} = -\dfrac{\sqrt{14}}{10}$

47. $\sqrt{\dfrac{24}{x}} = \dfrac{\sqrt{24}}{\sqrt{x}} = \dfrac{\sqrt{4\cdot 6}}{\sqrt{x}} = \dfrac{2\sqrt{6}}{\sqrt{x}}$
$= \dfrac{2\sqrt{6}\cdot\sqrt{x}}{\sqrt{x}\cdot\sqrt{x}} = \dfrac{2\sqrt{6x}}{x}$

49. $-\sqrt{\frac{98r^3}{s}} = -\frac{\sqrt{98r^3}}{\sqrt{s}} = -\frac{\sqrt{49r^2 \cdot 2r}}{\sqrt{s}}$

$= -\frac{7r\sqrt{2r}}{\sqrt{s}} = -\frac{7r\sqrt{2r} \cdot \sqrt{s}}{\sqrt{s} \cdot \sqrt{s}}$

$= -\frac{7r\sqrt{2rs}}{s}$

51. $\sqrt{\frac{288x^7}{y^9}} = \frac{\sqrt{288x^7}}{\sqrt{y^9}} = \frac{\sqrt{144x^6 \cdot 2x}}{\sqrt{y^8 \cdot y}}$

$= \frac{12x^3\sqrt{2x}}{y^4\sqrt{y}} = \frac{12x^3\sqrt{2x} \cdot \sqrt{y}}{y^4\sqrt{y} \cdot \sqrt{y}}$

$= \frac{12x^3\sqrt{2xy}}{y^4 \cdot y} = \frac{12x^3\sqrt{2xy}}{y^5}$

53. $\sqrt[3]{\frac{2}{3}} = \frac{\sqrt[3]{2} \cdot \sqrt[3]{9}}{\sqrt[3]{3} \cdot \sqrt[3]{9}} = \frac{\sqrt[3]{18}}{\sqrt[3]{27}} = \frac{\sqrt[3]{18}}{3}$

55. $\sqrt[3]{\frac{4}{9}} = \frac{\sqrt[3]{4}}{\sqrt[3]{9}} = \frac{\sqrt[3]{4}}{\sqrt[3]{3^2}} = \frac{\sqrt[3]{4} \cdot \sqrt[3]{3}}{\sqrt[3]{3^2} \cdot \sqrt[3]{3}}$

$= \frac{\sqrt[3]{12}}{\sqrt[3]{3^3}} = \frac{\sqrt[3]{12}}{3}$

57. $-\sqrt[3]{\frac{2p}{r^2}} = -\frac{\sqrt[3]{2p}}{\sqrt[3]{r^2}} = -\frac{\sqrt[3]{2p} \cdot \sqrt[3]{r}}{\sqrt[3]{r^2} \cdot \sqrt[3]{r}}$

$= -\frac{\sqrt[3]{2pr}}{\sqrt[3]{r^3}} = -\frac{\sqrt[3]{2pr}}{r}$

59. $\sqrt[4]{\frac{16}{x}} = \frac{\sqrt[4]{16}}{\sqrt[4]{x}} = \frac{2}{\sqrt[4]{x}} = \frac{2 \cdot \sqrt[4]{x^3}}{\sqrt[4]{x} \cdot \sqrt[4]{x^3}}$

$= \frac{2\sqrt[4]{x^3}}{\sqrt[4]{x^4}} = \frac{2\sqrt[4]{x^3}}{x}$

61. $\sqrt[4]{\frac{2y}{z}} = \frac{\sqrt[4]{2y}}{\sqrt[4]{z}} \cdot \frac{\sqrt[4]{z^3}}{\sqrt[4]{z^3}}$

$= \frac{\sqrt[4]{2yz^3}}{z}$

63. $\frac{2}{4+\sqrt{3}}$

Multiply both the numerator and denominator by the conjugate of the denominator, $4-\sqrt{3}$.

$= \frac{2(4-\sqrt{3})}{(4+\sqrt{3})(4-\sqrt{3})}$

$= \frac{2(4-\sqrt{3})}{16-3}$

$= \frac{2(4-\sqrt{3})}{13}$

65. $\frac{6}{\sqrt{5}+\sqrt{3}}$

Multiply both the numerator and denominator by the conjugate of the denominator, $\sqrt{5}-\sqrt{3}$.

$= \frac{6(\sqrt{5}-\sqrt{3})}{(\sqrt{5}+\sqrt{3})(\sqrt{5}-\sqrt{3})}$

$= \frac{6(\sqrt{5}-\sqrt{3})}{(\sqrt{5})^2-(\sqrt{3})^2}$

$= \frac{6(\sqrt{5}-\sqrt{3})}{5-3}$

$= \frac{6(\sqrt{5}-\sqrt{3})}{2}$

$= 3(\sqrt{5}-\sqrt{3})$

67. $\frac{-4}{\sqrt{3}-\sqrt{7}}$

Multiply both the numerator and denominator by the conjugate of the denominator, $\sqrt{3}+\sqrt{7}$.

$= \frac{-4(\sqrt{3}+\sqrt{7})}{(\sqrt{3}-\sqrt{7})(\sqrt{3}+\sqrt{7})}$

$= \frac{-4(\sqrt{3}+\sqrt{7})}{3-7}$

$= \frac{-4(\sqrt{3}+\sqrt{7})}{-4}$

$= \sqrt{3}+\sqrt{7}$

69. $\frac{1-\sqrt{2}}{\sqrt{7}+\sqrt{6}}$

Multiply both the numerator and denominator by the conjugate of the denominator, $\sqrt{7}-\sqrt{6}$.

$= \frac{(1-\sqrt{2})(\sqrt{7}-\sqrt{6})}{(\sqrt{7}+\sqrt{6})(\sqrt{7}-\sqrt{6})}$

$= \frac{\sqrt{7}-\sqrt{6}-\sqrt{14}+\sqrt{12}}{7-6}$

$= \frac{\sqrt{7}-\sqrt{6}-\sqrt{14}+\sqrt{4 \cdot 3}}{1}$

$= \sqrt{7}-\sqrt{6}-\sqrt{14}+2\sqrt{3}$

71. $\dfrac{\sqrt{2}-\sqrt{3}}{\sqrt{6}-\sqrt{5}}$

Multiply both the numerator and denominator by the conjugate of the denominator, $\sqrt{6}+\sqrt{5}$.

$$=\frac{(\sqrt{2}-\sqrt{3})(\sqrt{6}+\sqrt{5})}{(\sqrt{6}-\sqrt{5})(\sqrt{6}+\sqrt{5})}$$
$$=\frac{\sqrt{12}+\sqrt{10}-\sqrt{18}-\sqrt{15}}{(\sqrt{6})^2-(\sqrt{5})^2}$$
$$=\frac{\sqrt{4}\cdot\sqrt{3}+\sqrt{10}-\sqrt{9}\cdot\sqrt{2}-\sqrt{15}}{6-5}$$
$$=2\sqrt{3}+\sqrt{10}-3\sqrt{2}-\sqrt{15}$$

73. $\dfrac{4}{\sqrt{x}-2\sqrt{y}}$

Multiply both the numerator and denominator by the conjugate of the denominator, $\sqrt{x}+2\sqrt{y}$.

$$=\frac{4(\sqrt{x}+2\sqrt{y})}{(\sqrt{x}-2\sqrt{y})(\sqrt{x}+2\sqrt{y})}$$
$$=\frac{4(\sqrt{x}+2\sqrt{y})}{x-4y}$$

75. $\dfrac{\sqrt{x}-\sqrt{y}}{\sqrt{2x}+\sqrt{3y}}$

Multiply both the numerator and denominator by the conjugate of the denominator, $\sqrt{2x}-\sqrt{3y}$.

$$=\frac{(\sqrt{x}-\sqrt{y})(\sqrt{2x}-\sqrt{3y})}{(\sqrt{2x}+\sqrt{3y})(\sqrt{2x}-\sqrt{3y})}$$
$$=\frac{\sqrt{2x^2}-\sqrt{3xy}-\sqrt{2xy}+\sqrt{3y^2}}{2x-3y}$$
$$=\frac{x\sqrt{2}-\sqrt{3xy}-\sqrt{2xy}+y\sqrt{3}}{2x-3y}$$

77. $\dfrac{25+10\sqrt{6}}{20}=\dfrac{5(5+2\sqrt{6})}{5\cdot 4}$

$$=\frac{5+2\sqrt{6}}{4}$$

79. $\dfrac{16+4\sqrt{8}}{12}=\dfrac{16+4(2\sqrt{2})}{12}$

$$=\frac{4(4+2\sqrt{2})}{4\cdot 3}=\frac{4+2\sqrt{2}}{3}$$

81. $\dfrac{6x+\sqrt{24x^3}}{3x}$

$$=\frac{6x+\sqrt{4x^2\cdot 6x}}{3x}$$
$$=\frac{6x+2x\sqrt{6x}}{3x}$$
$$=\frac{x(6+2\sqrt{6x})}{3x}=\frac{6+2\sqrt{6x}}{3}$$

83. $\dfrac{8\sqrt{5}-1}{6}=\dfrac{(8\sqrt{5}-1)(8\sqrt{5}+1)}{6(8\sqrt{5}+1)}$

$$=\frac{(8\sqrt{5})^2-1^2}{6(8\sqrt{5}+1)}$$
$$=\frac{64(5)-1}{6(8\sqrt{5}+1)}$$
$$=\frac{319}{6(8\sqrt{5}+1)}$$

84. $\dfrac{3\sqrt{a}+\sqrt{b}}{\sqrt{b}-\sqrt{a}}=\dfrac{(3\sqrt{a}+\sqrt{b})(3\sqrt{a}-\sqrt{b})}{(\sqrt{b}-\sqrt{a})(3\sqrt{a}-\sqrt{b})}$

$$=\frac{(3\sqrt{a})^2-(\sqrt{b})^2}{(\sqrt{b}-\sqrt{a})(3\sqrt{a}-\sqrt{b})}$$
$$=\frac{9a-b}{(\sqrt{b}-\sqrt{a})(3\sqrt{a}-\sqrt{b})}$$

85. $\dfrac{3\sqrt{a}+\sqrt{b}}{\sqrt{b}-\sqrt{a}}=\dfrac{(3\sqrt{a}+\sqrt{b})(\sqrt{b}+\sqrt{a})}{(\sqrt{b}-\sqrt{a})(\sqrt{b}+\sqrt{a})}$

$$=\frac{(3\sqrt{a}+\sqrt{b})(\sqrt{b}+\sqrt{a})}{(\sqrt{b})^2-(\sqrt{a})^2}$$
$$=\frac{(3\sqrt{a}+\sqrt{b})(\sqrt{b}+\sqrt{a})}{b-a}$$

86. In Exercise 84, we multiplied the numerator and denominator by the conjugate of the numerator, while in Exercise 85 we multiplied by the conjugate of the denominator.

Summary Exercises on Operations with Radicals and Rational Exponents

1. $6\sqrt{10}-12\sqrt{10}=(6-12)\sqrt{10}$

$$=-6\sqrt{10}$$

3. $(1-\sqrt{3})(2+\sqrt{6})$

F O I L

$$=2+\sqrt{6}-2\sqrt{3}-\sqrt{18}$$
$$=2+\sqrt{6}-2\sqrt{3}-\sqrt{9}\cdot\sqrt{2}$$
$$=2+\sqrt{6}-2\sqrt{3}-3\sqrt{2}$$

5. $$\begin{aligned}&(3\sqrt{5}+2\sqrt{7})^2\\ &= (3\sqrt{5})^2+2(3\sqrt{5})(2\sqrt{7})+(2\sqrt{7})^2\\ &= 9\cdot 5+12\sqrt{35}+4\cdot 7\\ &= 45+12\sqrt{35}+28\\ &= 73+12\sqrt{35}\end{aligned}$$

7. $$\begin{aligned}\frac{8}{\sqrt{7}+\sqrt{5}} &= \frac{8(\sqrt{7}-\sqrt{5})}{(\sqrt{7}+\sqrt{5})(\sqrt{7}-\sqrt{5})}\\ &= \frac{8(\sqrt{7}-\sqrt{5})}{7-5}\\ &= \frac{8(\sqrt{7}-\sqrt{5})}{2}\\ &= 4(\sqrt{7}-\sqrt{5})\end{aligned}$$

9. $$\begin{aligned}\frac{1-\sqrt{2}}{1+\sqrt{2}} &= \frac{(1-\sqrt{2})(1-\sqrt{2})}{(1+\sqrt{2})(1-\sqrt{2})}\\ &= \frac{(1-\sqrt{2})^2}{1^2-(\sqrt{2})^2}\\ &= \frac{1-2\sqrt{2}+2}{1-2}\\ &= \frac{3-2\sqrt{2}}{-1}\\ &= -3+2\sqrt{2}\end{aligned}$$

11. $$\begin{aligned}(\sqrt{5}+7)(\sqrt{5}-7) &= (\sqrt{5})^2-7^2\\ &= 5-49\\ &= -44\end{aligned}$$

13. $$\begin{aligned}\sqrt[3]{8a^3b^5c^9} &= \sqrt[3]{8a^3b^3c^9}\cdot\sqrt[3]{b^2}\\ &= 2abc^3\sqrt[3]{b^2}\end{aligned}$$

15. $$\begin{aligned}\frac{3}{\sqrt{5}+2} &= \frac{3(\sqrt{5}-2)}{(\sqrt{5}+2)(\sqrt{5}-2)}\\ &= \frac{3(\sqrt{5}-2)}{5-4}\\ &= 3(\sqrt{5}-2)\end{aligned}$$

17. $$\begin{aligned}\frac{16\sqrt{3}}{5\sqrt{12}} &= \frac{16\sqrt{3}}{5\cdot\sqrt{4}\cdot\sqrt{3}}\\ &= \frac{16}{5\cdot 2}\\ &= \frac{8}{5}\end{aligned}$$

19. $$\begin{aligned}\frac{-10}{\sqrt[3]{10}} &= \frac{-10}{\sqrt[3]{10}}\cdot\frac{\sqrt[3]{100}}{\sqrt[3]{100}}\\ &= \frac{-10\sqrt[3]{100}}{\sqrt[3]{1000}}\\ &= \frac{-10\sqrt[3]{100}}{10} = -\sqrt[3]{100}\end{aligned}$$

21. $$\begin{aligned}\sqrt{12x}-\sqrt{75x} &= \sqrt{4}\cdot\sqrt{3x}-\sqrt{25}\cdot\sqrt{3x}\\ &= 2\sqrt{3x}-5\sqrt{3x}\\ &= -3\sqrt{3x}\end{aligned}$$

23. $$(\sqrt{74}-\sqrt{73})(\sqrt{74}+\sqrt{73}) = 74-73 = 1$$

25. $$\begin{aligned}&-t^2\sqrt[4]{t}+3\sqrt[4]{t^9}-t\sqrt[4]{t^5}\\ &= -t^2\sqrt[4]{t}+3\sqrt[4]{t^8}\cdot\sqrt[4]{t}-t\cdot\sqrt[4]{t^4}\cdot\sqrt[4]{t}\\ &= -t^2\sqrt[4]{t}+3t^2\sqrt[4]{t}-t\cdot t\cdot\sqrt[4]{t}\\ &= -t^2\sqrt[4]{t}+3t^2\sqrt[4]{t}-t^2\sqrt[4]{t}\\ &= (-t^2+3t^2-t^2)\sqrt[4]{t} = t^2\sqrt[4]{t}\end{aligned}$$

27. $$\begin{aligned}\frac{6}{\sqrt[4]{3}} &= \frac{6}{\sqrt[4]{3}}\cdot\frac{\sqrt[4]{3^3}}{\sqrt[4]{3^3}}\\ &= \frac{6\sqrt[4]{27}}{3} = 2\sqrt[4]{27}\end{aligned}$$

29. $$\begin{aligned}\sqrt[3]{\frac{x^2y}{x^{-3}y^4}} &= \sqrt[3]{x^{2-(-3)}y^{1-4}}\\ &= \sqrt[3]{x^5y^{-3}}\\ &= \frac{\sqrt[3]{x^5}}{\sqrt[3]{y^3}}\\ &= \frac{\sqrt[3]{x^3\cdot x^2}}{y} = \frac{x\sqrt[3]{x^2}}{y}\end{aligned}$$

31. $$\begin{aligned}\frac{x^{-2/3}y^{4/5}}{x^{-5/3}y^{-2/5}} &= x^{-2/3-(-5/3)}y^{4/5-(-2/5)}\\ &= x^{3/3}y^{6/5}\\ &= xy^{6/5}\end{aligned}$$

33. $$\begin{aligned}(125x^3)^{-2/3} &= \frac{1}{(125x^3)^{2/3}}\\ &= \frac{1}{(\sqrt[3]{125x^3})^2}\\ &= \frac{1}{(5x)^2}\\ &= \frac{1}{25x^2}\end{aligned}$$

35. $$\begin{aligned}\frac{4^{1/2}+3^{1/2}}{4^{1/2}-3^{1/2}} &= \frac{2+\sqrt{3}}{2-\sqrt{3}}\\ &= \frac{(2+\sqrt{3})(2+\sqrt{3})}{(2-\sqrt{3})(2+\sqrt{3})}\\ &= \frac{2^2+2(2)(\sqrt{3})+(\sqrt{3})^2}{2^2-(\sqrt{3})^2}\\ &= \frac{4+4\sqrt{3}+3}{4-3}\\ &= 7+4\sqrt{3},\quad\text{or}\quad 7+4\cdot 3^{1/2}\end{aligned}$$

9.6 Solving Equations with Radicals

9.6 Margin Exercises

1. **(a)**

$$\sqrt{r} = 3$$
$$\left(\sqrt{r}\right)^2 = 3^2 \quad \textit{Square both sides.}$$
$$r = 9$$

To check, substitute the proposed solution in the *original* equation.

$$\sqrt{r} = 3$$
$$\sqrt{9} \stackrel{?}{=} 3 \quad \textit{Let } r = 9.$$
$$3 = 3 \quad \textit{True}$$

Solution set: $\{9\}$

(b)

$$\sqrt{5x+1} = 4$$
$$\left(\sqrt{5x+1}\right)^2 = 4^2 \quad \textit{Square both sides.}$$
$$5x + 1 = 16$$
$$5x = 15$$
$$x = 3$$

Check $x = 3$ in the original equation.

$$\sqrt{5x+1} = 4$$
$$\sqrt{5 \cdot 3 + 1} \stackrel{?}{=} 4 \quad \textit{Let } x = 3.$$
$$\sqrt{16} \stackrel{?}{=} 4$$
$$4 = 4 \quad \textit{True}$$

Solution set: $\{3\}$

2. **(a)** $\sqrt{5x+3} + 2 = 0$

Step 1 Isolate the radical on one side.

$$\sqrt{5x+3} = -2$$

The equation has no solution, because the square root of a real number must be nonnegative.

Solution set: $\emptyset$

(b) $\sqrt{x-9} - 3 = 0$

Step 1 Isolate the radical on one side.

$$\sqrt{x-9} = 3$$

Step 2 Square both sides.

$$\left(\sqrt{x-9}\right)^2 = 3^2$$

Step 3 Solve the resulting equation.

$$x - 9 = 9$$
$$x = 18$$

Step 4 ***Check*** 18 in the original equation.

$$\sqrt{x-9} - 3 = 0$$
$$\sqrt{18-9} - 3 \stackrel{?}{=} 0$$
$$\sqrt{9} - 3 \stackrel{?}{=} 0$$
$$3 - 3 \stackrel{?}{=} 0$$
$$0 = 0 \quad \textit{True}$$

The proposed solution, 18, checks.

Solution set: $\{18\}$

3. **(a)**

$$\sqrt{3x-5} = x - 1$$
$$\left(\sqrt{3x-5}\right)^2 = (x-1)^2 \quad \textit{Square.}$$
$$3x - 5 = x^2 - 2x + 1$$
$$0 = x^2 - 5x + 6$$
$$0 = (x-3)(x-2)$$
$$x - 3 = 0 \quad \text{or} \quad x - 2 = 0$$
$$x = 3 \quad \text{or} \quad x = 2$$

Check $x = 3$: $\sqrt{4} = 2$ *True*

Check $x = 2$: $\sqrt{1} = 1$ *True*

Solution set: $\{2, 3\}$

(b)

$$x + 1 = \sqrt{-2x-2}$$
$$(x+1)^2 = \left(\sqrt{-2x-2}\right)^2 \quad \textit{Square.}$$
$$x^2 + 2x + 1 = -2x - 2$$
$$x^2 + 4x + 3 = 0$$
$$(x+3)(x+1) = 0$$
$$x + 3 = 0 \quad \text{or} \quad x + 1 = 0$$
$$x = -3 \quad \text{or} \quad x = -1$$

Check $x = -3$: $-2 = \sqrt{4}$ *False*

Check $x = -1$: $0 = \sqrt{0}$ *True*

Solution set: $\{-1\}$

4.

$$\sqrt{4x^2 + 2x - 3} = 2x + 7$$
$$\left(\sqrt{4x^2 + 2x - 3}\right)^2 = (2x+7)^2 \quad \textit{Square.}$$
$$4x^2 + 2x - 3 = 4x^2 + 28x + 49$$
$$-52 = 26x$$
$$-2 = x$$

Check $x = -2$: $\sqrt{9} = 3$ *True*

Solution set: $\{-2\}$

5. **(a)** ***Check*** $x = 15$:

$$\sqrt{5x+6} + \sqrt{3x+4} = 2$$
$$\sqrt{81} + \sqrt{49} \stackrel{?}{=} 2 \quad \textit{Let } x = 15.$$
$$9 + 7 \stackrel{?}{=} 2$$
$$16 = 2 \quad \textit{False}$$

Thus, 15 is not a solution.

(b) $\sqrt{2x+3}+\sqrt{x+1}=1$

Get one radical on each side of the equals sign.

$$\sqrt{2x+3}=1-\sqrt{x+1}$$

Square both sides.

$$(\sqrt{2x+3})^2=(1-\sqrt{x+1})^2$$
$$2x+3=1-2\sqrt{x+1}+(x+1)$$

Isolate the remaining radical.

$$2x+3=-2\sqrt{x+1}+x+2$$
$$x+1=-2\sqrt{x+1}$$

Square both sides again.

$$(x+1)^2=(-2\sqrt{x+1})^2$$
$$x^2+2x+1=(-2)^2(\sqrt{x+1})^2$$
$$x^2+2x+1=4(x+1)$$
$$x^2+2x+1=4x+4$$
$$x^2-2x-3=0$$
$$(x-3)(x+1)=0$$

$$x-3=0 \quad \text{or} \quad x+1=0$$
$$x=3 \quad \text{or} \quad x=-1$$

Check $x=3$: $\sqrt{9}+\sqrt{4}=1$ *False*

Check $x=-1$: $\sqrt{1}+\sqrt{0}=1$ *True*

Solution set: $\{-1\}$

6. **(a)** $\sqrt[3]{2x+7}=\sqrt[3]{3x-2}$

$$(\sqrt[3]{2x+7})^3=(\sqrt[3]{3x-2})^3 \quad \textit{Cube.}$$
$$2x+7=3x-2$$
$$9=x$$

Check $x=9$: $\sqrt[3]{25}=\sqrt[3]{25}$ *True*

Solution set: $\{9\}$

(b) $\sqrt[4]{2x+5}+1=0$

$$\sqrt[4]{2x+5}=-1 \quad \textit{Isolate.}$$

Since no real number has a principal fourth root that is negative, the equation has no solution.

Solution set: $\emptyset$

9.6 Section Exercises

1. $\sqrt{3x+18}=x$

(a) ***Check*** $x=6$.

$$\sqrt{3(6)+18}\stackrel{?}{=}6$$
$$\sqrt{18+18}\stackrel{?}{=}6$$
$$\sqrt{36}\stackrel{?}{=}6$$
$$6=6 \quad \textit{True}$$

The number 6 is a solution.

(b) ***Check*** $x=-3$.

$\sqrt{3(-3)+18}=-3$ is a false statement since the principal square root of a number is nonnegative. The number -3 is not a solution.

3. $\sqrt{x+2}=\sqrt{9x-2}-2\sqrt{x-1}$

(a) ***Check*** $x=2$.

$$\sqrt{2+2}\stackrel{?}{=}\sqrt{9(2)-2}-2\sqrt{2-1}$$
$$\sqrt{4}\stackrel{?}{=}\sqrt{16}-2\sqrt{1}$$
$$2\stackrel{?}{=}4-2$$
$$2=2 \quad \textit{True}$$

The number 2 is a solution.

(b) ***Check*** $x=7$.

$$\sqrt{7+2}\stackrel{?}{=}\sqrt{9(7)-2}-2\sqrt{7-1}$$
$$\sqrt{9}\stackrel{?}{=}\sqrt{61}-2\sqrt{6}$$
$$3=\sqrt{61}-2\sqrt{6} \quad \textit{False}$$

The number 7 is not a solution.

5. $\sqrt{9}=3$, not -3. There is no solution of $\sqrt{x}=-3$ since the value of a principal square root cannot equal a negative number.

7. $\sqrt{x-2}=3$

$$(\sqrt{x-2})^2=3^2 \quad \textit{Square both sides.}$$
$$x-2=9$$
$$x=11$$

Check the proposed solution, 11.

Check $x=11$: $\sqrt{9}=3$ *True*

Solution set: $\{11\}$

9. $\sqrt{6x-1}=1$

$$(\sqrt{6x-1})^2=1^2 \quad \textit{Square both sides.}$$
$$6x-1=1$$
$$6x=2$$
$$x=\tfrac{2}{6}=\tfrac{1}{3}$$

Check the proposed solution, $\frac{1}{3}$.

Check $x=\frac{1}{3}$: $\sqrt{1}=1$ *True*

Solution set: $\{\frac{1}{3}\}$

11. $\sqrt{4x+3}+1=0$

$$\sqrt{4x+3}=-1 \quad \textit{Isolate the radical.}$$

This equation has no solution because $\sqrt{4x+3}$ cannot be negative.

Solution set: $\emptyset$

13. $\sqrt{3k+1}-4=0$

$\sqrt{3k+1}=4$ *Isolate the radical.*

$(\sqrt{3k+1})^2=4^2$ *Square both sides.*

$3k+1=16$

$3k=15$

$k=5$

Check $k=5$: $\sqrt{16}-4=0$ *True*
Solution set: $\{5\}$

15. $4-\sqrt{x-2}=0$

$4=\sqrt{x-2}$ *Isolate.*

$4^2=(\sqrt{x-2})^2$ *Square.*

$16=x-2$

$18=x$

Check $x=18$: $4-\sqrt{16}=0$ *True*
Solution set: $\{18\}$

17. $\sqrt{9a-4}=\sqrt{8a+1}$

$(\sqrt{9a-4})^2=(\sqrt{8a+1})^2$ *Square.*

$9a-4=8a+1$

$a=5$

Check $a=5$: $\sqrt{41}=\sqrt{41}$ *True*
Solution set: $\{5\}$

19. $2\sqrt{x}=\sqrt{3x+4}$

$(2\sqrt{x})^2=(\sqrt{3x+4})^2$ *Square.*

$4x=3x+4$

$x=4$

Check $x=4$: $4=\sqrt{16}$ *True*
Solution set: $\{4\}$

21. $3\sqrt{z-1}=2\sqrt{2z+2}$

$(3\sqrt{z-1})^2=(2\sqrt{2z+2})^2$ *Square.*

$9(z-1)=4(2z+2)$

$9z-9=8z+8$

$z=17$

Check $z=17$: $3(4)=2(6)$ *True*
Solution set: $\{17\}$

23. $k=\sqrt{k^2+4k-20}$

$k^2=(\sqrt{k^2+4k-20})^2$ *Square.*

$k^2=k^2+4k-20$

$20=4k$

$5=k$

Check $k=5$: $5=\sqrt{25}$ *True*
Solution set: $\{5\}$

25. $x=\sqrt{x^2+3x+9}$

$x^2=(\sqrt{x^2+3x+9})^2$ *Square.*

$x^2=x^2+3x+9$

$-3x=9$

$x=-3$

Substituting -3 for x makes the left side of the original equation negative, but the right side is nonnegative, so the solution set is $\emptyset$.

27. $\sqrt{9-x}=x+3$

$(\sqrt{9-x})^2=(x+3)^2$ *Square.*

$9-x=x^2+6x+9$

$0=x^2+7x$

$0=x(x+7)$

$x=0$ or $x+7=0$

$x=-7$

Check $x=-7$: $\sqrt{16}=-4$ *False*

Check $x=0$: $\sqrt{9}=3$ *True*

Solution set: $\{0\}$

29. $\sqrt{k^2+2k+9}=k+3$

$(\sqrt{k^2+2k+9})^2=(k+3)^2$ *Square.*

$k^2+2k+9=k^2+6k+9$

$0=4k$

$0=k$

Check $k=0$: $\sqrt{9}=3$ *True*
Solution set: $\{0\}$

31. $\sqrt{r^2+9r+3}=-r$

$(\sqrt{r^2+9r+3})^2=(-r)^2$ *Square.*

$r^2+9r+3=r^2$

$9r=-3$

$r=-\frac{1}{3}$

Check $r=-\frac{1}{3}$: $\sqrt{\frac{1}{9}}=\frac{1}{3}$ *True*
Solution set: $\{-\frac{1}{3}\}$

33. $\sqrt{z^2+12z-4}+4-z=0$

$\sqrt{z^2+12z-4}=z-4$ *Isolate.*

$(\sqrt{z^2+12z-4})^2=(z-4)^2$ *Square.*

$z^2+12z-4=z^2-8z+16$

$20z=20$

$z=1$

Substituting 1 for z makes the left side of the original equation positive, but the right side is zero, so the solution set is $\emptyset$.

35. $\sqrt{3x+4} = 8 - x$

$(8-x)^2$ equals $64 - 16x + x^2$, *not* $64 + x^2$. We cannot just square each term. The correct first step and solution follows.

$$3x + 4 = 64 - 16x + x^2$$
$$0 = x^2 - 19x + 60$$
$$0 = (x-4)(x-15)$$
$$x - 4 = 0 \quad \text{or} \quad x - 15 = 0$$
$$x = 4 \quad \text{or} \quad x = 15$$

Check $x = 4$: $\sqrt{16} = 8 - 4$ *True*
Check $x = 15$: $\sqrt{49} = 8 - 15$ *False*
Solution set: $\{4\}$

37. $\sqrt[3]{2x+5} = \sqrt[3]{6x+1}$

$$\left(\sqrt[3]{2x+5}\right)^3 = \left(\sqrt[3]{6x+1}\right)^3 \quad \textit{Cube both sides.}$$
$$2x + 5 = 6x + 1$$
$$4 = 4x$$
$$1 = x$$

Check $x = 1$: $\sqrt[3]{7} = \sqrt[3]{7}$ *True*
Solution set: $\{1\}$

39. $\sqrt[3]{a^2+5a+1} = \sqrt[3]{a^2+4a}$

$$\left(\sqrt[3]{a^2+5a+1}\right)^3 = \left(\sqrt[3]{a^2+4a}\right)^3 \quad \textit{Cube.}$$
$$a^2 + 5a + 1 = a^2 + 4a$$
$$a = -1$$

Check $a = -1$: $\sqrt[3]{-3} = \sqrt[3]{-3}$ *True*
Solution set: $\{-1\}$

41. $\sqrt[3]{2m-1} = \sqrt[3]{m+13}$

$$\left(\sqrt[3]{2m-1}\right)^3 = \left(\sqrt[3]{m+13}\right)^3 \quad \textit{Cube.}$$
$$2m - 1 = m + 13$$
$$m = 14$$

Check $m = 14$: $\sqrt[3]{27} = \sqrt[3]{27}$ *True*
Solution set: $\{14\}$

43. $\sqrt[4]{a+8} = \sqrt[4]{2a}$

Raise each side to the fourth power.

$$\left(\sqrt[4]{a+8}\right)^4 = \left(\sqrt[4]{2a}\right)^4$$
$$a + 8 = 2a$$
$$8 = a$$

Check $a = 8$: $\sqrt[4]{16} = \sqrt[4]{16}$ *True*
Solution set: $\{8\}$

45. $\sqrt[3]{x-8} + 2 = 0$

$$\sqrt[3]{x-8} = -2 \quad \textit{Isolate.}$$
$$\left(\sqrt[3]{x-8}\right)^3 = (-2)^3 \quad \textit{Cube.}$$
$$x - 8 = -8$$
$$x = 0$$

Check $x = 0$: $\sqrt[3]{-8} + 2 = 0$ *True*
Solution set: $\{0\}$

47. $\sqrt[4]{2k-5} + 4 = 0$

$$\sqrt[4]{2k-5} = -4 \quad \textit{Isolate.}$$

This equation has no solution because $\sqrt[4]{2k-5}$ cannot be negative.

Solution set: $\emptyset$

49. $\sqrt{k+2} - \sqrt{k-3} = 1$

Get one radical on each side of the equals sign.

$$\sqrt{k+2} = 1 + \sqrt{k-3}$$
$$\left(\sqrt{k+2}\right)^2 = \left(1 + \sqrt{k-3}\right)^2 \quad \textit{Square.}$$
$$k + 2 = 1 + 2\sqrt{k-3} + k - 3$$
$$4 = 2\sqrt{k-3} \quad \textit{Isolate.}$$
$$2 = \sqrt{k-3} \quad \textit{Divide by 2.}$$
$$2^2 = \left(\sqrt{k-3}\right)^2 \quad \textit{Square again.}$$
$$4 = k - 3$$
$$7 = k$$

Check $k = 7$: $\sqrt{9} - \sqrt{4} = 1$ *True*
Solution set: $\{7\}$

51. $\sqrt{2r+11} - \sqrt{5r+1} = -1$

Get one radical on each side of the equals sign.

$$\sqrt{2r+11} = -1 + \sqrt{5r+1}$$
$$\left(\sqrt{2r+11}\right)^2 = \left(-1 + \sqrt{5r+1}\right)^2 \quad \textit{Square.}$$
$$2r + 11 = 1 - 2\sqrt{5r+1} + 5r + 1$$
$$2\sqrt{5r+1} = 3r - 9$$
$$\left(2\sqrt{5r+1}\right)^2 = (3r-9)^2 \quad \textit{Square.}$$
$$4(5r+1) = 9r^2 - 54r + 81$$
$$20r + 4 = 9r^2 - 54r + 81$$
$$0 = 9r^2 - 74r + 77$$
$$0 = (9r - 11)(r - 7)$$
$$9r - 11 = 0 \quad \text{or} \quad r - 7 = 0$$
$$r = \tfrac{11}{9} \quad \text{or} \quad r = 7$$

Check $r = \frac{11}{9}$: $\frac{11}{3} - \frac{8}{3} = -1$ *False*
Check $r = 7$: $5 - 6 = -1$ *True*
Solution set: $\{7\}$

53. $\sqrt{3p+4} - \sqrt{2p-4} = 2$

Get one radical on each side of the equals sign.

$$\sqrt{3p+4} = 2 + \sqrt{2p-4}$$
$$\left(\sqrt{3p+4}\right)^2 = \left(2+\sqrt{2p-4}\right)^2 \quad \textit{Square.}$$
$$3p+4 = 4 + 4\sqrt{2p-4} + 2p - 4$$
$$p+4 = 4\sqrt{2p-4} \quad \textit{Isolate.}$$
$$(p+4)^2 = \left(4\sqrt{2p-4}\right)^2 \quad \textit{Square.}$$
$$p^2 + 8p + 16 = 16(2p-4)$$
$$p^2 + 8p + 16 = 32p - 64$$
$$p^2 - 24p + 80 = 0$$
$$(p-4)(p-20) = 0$$

$$p-4=0 \quad \text{or} \quad p-20=0$$
$$p=4 \quad \text{or} \quad p=20$$

Check $p=4$: $\sqrt{16}-\sqrt{4}=2$ *True*

Check $p=20$: $\sqrt{64}-\sqrt{36}=2$ *True*

Solution set: $\{4, 20\}$

55. $\sqrt{3-3p} - 3 = \sqrt{3p+2}$

Square both sides.

$$\left(\sqrt{3-3p}-3\right)^2 = \left(\sqrt{3p+2}\right)^2$$
$$3 - 3p - 6\sqrt{3-3p} + 9 = 3p + 2$$

Isolate the remaining radical.

$$-6\sqrt{3-3p} = 6p - 10$$
$$-3\sqrt{3-3p} = 3p - 5$$

Square both sides again.

$$\left(-3\sqrt{3-3p}\right)^2 = (3p-5)^2$$
$$9(3-3p) = 9p^2 - 30p + 25$$
$$27 - 27p = 9p^2 - 30p + 25$$
$$0 = 9p^2 - 3p - 2$$
$$0 = (3p+1)(3p-2)$$

$$3p+1=0 \quad \text{or} \quad 3p-2=0$$
$$p=-\tfrac{1}{3} \quad \text{or} \quad p=\tfrac{2}{3}$$

Check $p=-\frac{1}{3}$: $\sqrt{4}-3=\sqrt{1}$ *False*

Check $p=\frac{2}{3}$: $\sqrt{1}-3=\sqrt{4}$ *False*

Solution set: $\emptyset$

57. $\sqrt{2\sqrt{x+11}} = \sqrt{4x+2}$

$$2\sqrt{x+11} = 4x+2 \quad \textit{Square.}$$
$$\left(2\sqrt{x+11}\right)^2 = (4x+2)^2 \quad \textit{Square again.}$$
$$4(x+11) = 16x^2 + 16x + 4$$
$$4x + 44 = 16x^2 + 16x + 4$$
$$0 = 16x^2 + 12x - 40$$
$$0 = 4x^2 + 3x - 10$$
$$0 = (x+2)(4x-5)$$

$$x+2=0 \quad \text{or} \quad 4x-5=0$$
$$x=-2 \quad \text{or} \quad x=\tfrac{5}{4}$$

Check $x=-2$: $\sqrt{6}=\sqrt{-6}$ *False*

Check $x=\frac{5}{4}$: $\sqrt{7}=\sqrt{7}$ *True*

Solution set: $\{\frac{5}{4}\}$

59. $(2x-9)^{1/2} = 2 + (x-8)^{1/2}$

$$\sqrt{2x-9} = 2 + \sqrt{x-8}$$
$$\left(\sqrt{2x-9}\right)^2 = \left(2+\sqrt{x-8}\right)^2$$
$$2x - 9 = 4 + 4\sqrt{x-8} + x - 8$$
$$x - 5 = 4\sqrt{x-8}$$
$$(x-5)^2 = \left(4\sqrt{x-8}\right)^2$$
$$x^2 - 10x + 25 = 16(x-8)$$
$$x^2 - 10x + 25 = 16x - 128$$
$$x^2 - 26x + 153 = 0$$
$$(x-9)(x-17) = 0$$

$$x=9 \quad \text{or} \quad x=17$$

Check $x=9$: $9^{1/2} \stackrel{?}{=} 2 + 1^{1/2}$

$3 = 2 + 1$ *True*

Check $x=17$: $25^{1/2} \stackrel{?}{=} 2 + 9^{1/2}$

$5 = 2 + 3$ *True*

Solution set: $\{9, 17\}$

61. $(2w-1)^{2/3} - w^{1/3} = 0$

$$\sqrt[3]{(2w-1)^2} = \sqrt[3]{w}$$
$$\left[\sqrt[3]{(2w-1)^2}\right]^3 = \left(\sqrt[3]{w}\right)^3$$
$$(2w-1)^2 = w$$
$$4w^2 - 4w + 1 = w$$
$$4w^2 - 5w + 1 = 0$$
$$(4w-1)(w-1) = 0$$

$$4w-1=0 \quad \text{or} \quad w-1=0$$
$$w=\tfrac{1}{4} \quad \text{or} \quad w=1$$

Check $w=\frac{1}{4}$: $(-\frac{1}{2})^{2/3} - (\frac{1}{4})^{1/3} \stackrel{?}{=} 0$

$(\frac{1}{4})^{1/3} - (\frac{1}{4})^{1/3} = 0$ *True*

Check $w=1$: $1^{2/3} - 1^{1/3} = 0$ *True*

Solution set: $\{\frac{1}{4}, 1\}$

63. Solve $V = \sqrt{\dfrac{2K}{m}}$ for K.

$$(V)^2 = \left(\sqrt{\frac{2K}{m}}\right)^2 \quad \textit{Square.}$$
$$V^2 = \frac{2K}{m}$$
$$\frac{V^2 m}{2} = K \quad \textit{Multiply by } \frac{m}{2}.$$

65. Solve $Z = \sqrt{\frac{L}{C}}$ for C.

$$(Z)^2 = \left(\sqrt{\frac{L}{C}}\right)^2 \quad \textit{Square.}$$
$$Z^2 = \frac{L}{C}$$
$$CZ^2 = L \quad \textit{Multiply by } C.$$
$$C = \frac{L}{Z^2} \quad \textit{Divide by } Z^2.$$

67. Solve $f = \frac{1}{2\pi\sqrt{LC}}$ for L.

$$2\pi f\sqrt{LC} = 1$$
$$\left(2\pi f\sqrt{LC}\right)^2 = 1^2$$
$$4\pi^2 f^2 LC = 1$$
$$L = \frac{1}{4\pi^2 f^2 C}$$

69. Solve $N = \frac{1}{2\pi}\sqrt{\frac{a}{r}}$ for r.

$$2\pi N = \sqrt{\frac{a}{r}}$$
$$(2\pi N)^2 = \left(\sqrt{\frac{a}{r}}\right)^2 \quad \textit{Square.}$$
$$4\pi^2 N^2 = \frac{a}{r}$$
$$4\pi^2 N^2 r = a \quad \textit{Multiply by } r.$$
$$r = \frac{a}{4\pi^2 N^2} \quad \textit{Divide by } 4\pi^2 N^2.$$

9.7 Complex Numbers

9.7 Margin Exercises

1. **(a)** $\sqrt{-16} = i\sqrt{16} = 4i$

(b) $-\sqrt{-81} = -i\sqrt{81} = -9i$

(c) $\sqrt{-7} = i\sqrt{7}$

(d) $\sqrt{-32} = i\sqrt{32} = i\sqrt{16}\cdot\sqrt{2} = 4i\sqrt{2}$

2. **(a)** $\sqrt{-7}\cdot\sqrt{-5} = i\sqrt{7}\cdot i\sqrt{5}$
$$= i^2\sqrt{7\cdot 5}$$
$$= -1\sqrt{35}$$
$$= -\sqrt{35}$$

(b) $\sqrt{-5}\cdot\sqrt{-10} = i\sqrt{5}\cdot i\sqrt{10}$
$$= i^2\sqrt{50}$$
$$= i^2\sqrt{25\cdot 2}$$
$$= -1(5\sqrt{2})$$
$$= -5\sqrt{2}$$

(c) $\sqrt{-15}\cdot\sqrt{2} = i\sqrt{15}\cdot\sqrt{2} = i\sqrt{30}$

3. **(a)** $\frac{\sqrt{-32}}{\sqrt{-2}} = \frac{i\sqrt{32}}{i\sqrt{2}} = \sqrt{\frac{32}{2}} = \sqrt{16} = 4$

(b) $\frac{\sqrt{-27}}{\sqrt{-3}} = \frac{i\sqrt{27}}{i\sqrt{3}} = \sqrt{\frac{27}{3}} = \sqrt{9} = 3$

(c) $\frac{\sqrt{-40}}{\sqrt{10}} = \frac{i\sqrt{40}}{\sqrt{10}} = i\sqrt{\frac{40}{10}} = i\sqrt{4} = 2i$

4. To add complex numbers, add the real parts and add the imaginary parts.

(a) $(4+6i) + (-3+5i)$
$$= [4+(-3)] + (6+5)i$$
$$= 1 + 11i$$

(b) $(-1+8i) + (9-3i)$
$$= (-1+9) + (8-3)i$$
$$= 8 + 5i$$

5. To subtract complex numbers, subtract the real parts and subtract the imaginary parts.

(a) $(7+3i) - (4+2i)$
$$= (7-4) + (3-2)i$$
$$= 3 + i$$

(b) $(-6-i) - (-5-4i)$
$$= (-6+5) + (-1+4)i$$
$$= -1 + 3i$$

(c) $8 - (3-2i) = (8-3) + 2i$
$$= 5 + 2i$$

6. **(a)** $6i(4+3i) = 6i(4) + 6i(3i)$
$$= 24i + 18i^2$$
$$= 24i + 18(-1)$$
$$= -18 + 24i$$

(b) $(6-4i)(2+4i)$

F O I L
$$= 6(2) + 6(4i) + 2(-4i) + 4i(-4i)$$
$$= 12 + 24i - 8i - 16i^2$$
$$= 12 + 16i - 16(-1)$$
$$= 12 + 16i + 16$$
$$= 28 + 16i$$

(c) $(3-2i)(3+2i)$

F O I L
$$= 3(3) + 3(2i) + (-2i)(3) + (-2i)(2i)$$
$$= 9 + 6i - 6i - 4i^2$$
$$= 9 - 4(-1)$$
$$= 9 + 4$$
$$= 13$$

7. In each case, multiply the numerator and denominator by the conjugate of the denominator.

(a) $\frac{2+i}{3-i} = \frac{(2+i)(3+i)}{(3-i)(3+i)}$

$= \frac{6+2i+3i+i^2}{3^2-i^2}$

$= \frac{6+2i+3i-1}{9-(-1)}$

$= \frac{5+5i}{10}$

$= \frac{5(1+i)}{10}$

$= \frac{1+i}{2}$

$= \frac{1}{2}+\frac{1}{2}i$

(b) $\frac{8-4i}{1-i} = \frac{(8-4i)(1+i)}{(1-i)(1+i)}$

$= \frac{8+8i-4i-4i^2}{1^2+1^2}$

$= \frac{8+4i+4}{1+1}$ *$i^2 = -1$*

$= \frac{12+4i}{2}$

$= \frac{2(6+2i)}{2}$

$= 6+2i$

(c) $\frac{5}{3-2i} = \frac{5(3+2i)}{(3-2i)(3+2i)}$

$= \frac{5(3+2i)}{3^2+2^2}$

$= \frac{5(3+2i)}{9+4}$

$= \frac{5(3+2i)}{13}$

$= \frac{15+10i}{13}$

$= \frac{15}{13}+\frac{10}{13}i$

(d) $\frac{5-i}{i}$

The conjugate of i is $-i$.

$= \frac{(5-i)(-i)}{i(-i)}$

$= \frac{-5i+i^2}{-i^2}$

$= \frac{-5i+(-1)}{-(-1)}$

$= \frac{-5i-1}{1}$

$= -1-5i$

8. **(a)** $i^{21} = i^{20} \cdot i = (i^4)^5 \cdot i = 1^5 \cdot i = i$

(b) $i^{36} = (i^4)^9 = 1^9 = 1$

(c) $i^{50} = i^{48} \cdot i^2 = (i^4)^{12} \cdot i^2 = 1^{12} \cdot (-1) = -1$

(d) $i^{-9} = \frac{1}{i^9} = \frac{1}{i^8 \cdot i} = \frac{1}{(i^4)^2 \cdot i} = \frac{1}{1^2 \cdot i} = \frac{1}{i}$

Now multiply by $-i$, the conjugate of i.

$\frac{1}{i} = \frac{1(-i)}{i(-i)} = \frac{-i}{-i^2} = \frac{-i}{-(-1)} = \frac{-i}{1} = -i$

9.7 Section Exercises

1. $\sqrt{-1} = i$

3. $\frac{1}{i} = \frac{1}{i} \cdot \frac{-i}{-i}$ *$-i$ is the conjugate of i*

$= \frac{-i}{-i^2} = \frac{i}{i^2} = \frac{i}{-1} = -i$

5. Since any real number a can be written as $a + bi$, where $b = 0$, every real number is also a complex number.

7. $\sqrt{-169} = i\sqrt{169} = 13i$

9. $-\sqrt{-144} = -i\sqrt{144} = -12i$

11. $\sqrt{-5} = i\sqrt{5}$

13. $\sqrt{-48} = i\sqrt{48} = i\sqrt{16 \cdot 3} = 4i\sqrt{3}$

15. $\sqrt{-15} \cdot \sqrt{-15} = i\sqrt{15} \cdot i\sqrt{15}$

$= i^2\left(\sqrt{15}\right)^2$

$= -1(15) = -15$

17. $\sqrt{-3} \cdot \sqrt{-19} = i\sqrt{3} \cdot i\sqrt{19}$

$= i^2\sqrt{3 \cdot 19}$

$= -1\sqrt{57} = -\sqrt{57}$

19. $\sqrt{-4} \cdot \sqrt{-25} = i\sqrt{4} \cdot i\sqrt{25}$

$= 2i \cdot 5i$

$= 10i^2 = 10(-1) = -10$

21. $\sqrt{-3} \cdot \sqrt{11} = i\sqrt{3} \cdot \sqrt{11}$

$= i\sqrt{33}$

23. $\frac{\sqrt{-300}}{\sqrt{-100}} = \frac{i\sqrt{300}}{i\sqrt{100}} = \sqrt{\frac{300}{100}} = \sqrt{3}$

25. $\frac{\sqrt{-75}}{\sqrt{3}} = \frac{i\sqrt{75}}{\sqrt{3}} = i\sqrt{\frac{75}{3}} = i\sqrt{25} = 5i$

27. $(3+2i)+(-4+5i)$

$= [(3+(-4)]+(2+5)i$

$= -1+7i$

29. $(5-i)+(-5+i)$
$=(5-5)+(-1+1)i$
$=0$

31. $(4+i)-(-3-2i)$
$=[(4-(-3)]+[(1-(-2)]i$
$=7+3i$

33. $(-3-4i)-(-1-4i)$
$=[-3-(-1)]+[-4-(-4)]i$
$=-2$

35. $(-4+11i)+(-2-4i)+(7+6i)$
$=(-4-2+7)+(11-4+6)i$
$=1+13i$

37. $[(7+3i)-(4-2i)]+(3+i)$
Work inside the brackets first.
$=[(7-4)+(3+2)i]+(3+i)$
$=(3+5i)+(3+i)$
$=(3+3)+(5+1)i$
$=6+6i$

39. If $a-c=b$, then $b+c=a$.
So, $(4+2i)-(3+i)=1+i$ implies that

$$(1+i)+(3+i)=\underline{4+2i}.$$

41. $(3i)(27i)=81i^2=81(-1)=-81$

43. $(-8i)(-2i)=16i^2=16(-1)=-16$

45. $5i(-6+2i)=(5i)(-6)+(5i)(2i)$
$=-30i+10i^2$
$=-30i+10(-1)$
$=-10-30i$

47. $(4+3i)(1-2i)$
F O I L
$=(4)(1)+4(-2i)+(3i)(1)+(3i)(-2i)$
$=4-8i+3i-6i^2$
$=4-5i-6(-1)$
$=4-5i+6=10-5i$

49. $(4+5i)^2$
$=4^2+2(4)(5i)+(5i)^2$ *Square binomial.*
$=16+40i+25i^2$
$=16+40i+25(-1)$
$=16+40i-25$
$=-9+40i$

51. $(12+3i)(12-3i)$
$=12^2-(3i)^2$
$=144-9i^2$
$=144-9(-1)=144+9=153$

53. **(a)** The conjugate of $a+bi$ is $a-bi$.

(b) $(a+bi)(a-bi)$
$=a^2-abi+abi-b^2i^2$
$=a^2-b^2(-1)$
$=\underline{a^2}+\underline{b^2}$,
which is always a real number.

55. $\dfrac{2}{1-i}$

Multiply the numerator and the denominator by the conjugate of the denominator, $1+i$.

$$=\frac{2(1+i)}{(1-i)(1+i)}$$
$$=\frac{2(1+i)}{1^2-i^2}$$
$$=\frac{2(1+i)}{1-(-1)}$$
$$=\frac{2(1+i)}{2}=1+i$$

57. $\dfrac{-7+4i}{3+2i}$

Multiply the numerator and the denominator by the conjugate of the denominator, $3-2i$.

$$=\frac{(-7+4i)(3-2i)}{(3+2i)(3-2i)}$$

In the denominator, we make use of the fact that $(a+bi)(a-bi)=a^2+b^2$.

$$=\frac{-21+14i+12i+8}{3^2+2^2}$$
$$=\frac{-13+26i}{13}$$
$$=\frac{13(-1+2i)}{13}=-1+2i$$

59. $\dfrac{8i}{2+2i}$

$$=\frac{2\cdot 4i}{2(1+i)}=\frac{4i}{1+i}\quad \textit{Write in lowest terms.}$$

Multiply the numerator and the denominator by the conjugate of the denominator, $1-i$.

$$=\frac{4i(1-i)}{(1+i)(1-i)}$$
$$=\frac{4(i-i^2)}{1^2+1^2}$$
$$=\frac{4(i+1)}{2}=2(i+1)=2+2i$$

61. $\dfrac{2-3i}{2+3i}$

Multiply the numerator and the denominator by the conjugate of the denominator, $2-3i$.

$$=\frac{(2-3i)(2-3i)}{(2+3i)(2-3i)}$$
$$=\frac{2^2-2(2)(3i)+(3i)^2}{2^2+3^2}$$
$$=\frac{4-12i+9i^2}{4+9}$$
$$=\frac{4-12i-9}{13}$$
$$=\frac{-5-12i}{13}=-\frac{5}{13}-\frac{12}{13}i$$

63. **(a)** $(x+2)+(3x-1)$
$=(1+3)x+(2-1)$
$=4x+1$

(b) $(1+2i)+(3-i)$
$=(1+3)+(2-1)i$
$=4+i$

64. **(a)** $(x+2)-(3x-1)$
$=(x-3x)+[2-(-1)]$
$=(1-3)x+(2+1)$
$=-2x+3$

(b) $(1+2i)-(3-i)$
$=(1-3)+[2-(-1)]i$
$=-2+3i$

65. **(a)** $(x+2)(3x-1)$
$=3x^2-x+6x-2$
$=3x^2+5x-2$

(b) $(1+2i)(3-i)$
$=3-i+6i-2i^2$
$=3+5i-2(-1)$
$=3+5i+2$
$=5+5i$

66. **(a)**
$$\frac{\sqrt{3}-1}{1+\sqrt{2}}=\frac{(\sqrt{3}-1)(1-\sqrt{2})}{(1+\sqrt{2})(1-\sqrt{2})}$$
$$=\frac{\sqrt{3}-\sqrt{6}-1+\sqrt{2}}{1^2-(\sqrt{2})^2}$$
$$=\frac{\sqrt{3}-\sqrt{6}-1+\sqrt{2}}{1-2}$$
$$=\frac{\sqrt{3}-\sqrt{6}-1+\sqrt{2}}{-1}$$
$$=-(\sqrt{3}-\sqrt{6}-1+\sqrt{2})$$
$$=-\sqrt{3}+\sqrt{6}+1-\sqrt{2}$$

(b)
$$\frac{3-i}{1+2i}=\frac{(3-i)(1-2i)}{(1+2i)(1-2i)}$$
$$=\frac{3-6i-i+2i^2}{1-(2i)^2}$$
$$=\frac{3-7i+2(-1)}{1-4(-1)}$$
$$=\frac{3-7i-2}{1+4}$$
$$=\frac{1-7i}{5}$$
$$=\tfrac{1}{5}-\tfrac{7}{5}i$$

67. In parts (a) and (b) of Exercises 63 and 64, real and imaginary parts are added, just like coefficients of similar terms in the binomials, and the answers correspond. In Exercise 65, introducing $i^2=-1$ when a product is found leads to answers that do not correspond.

68. In parts (a) and (b) of Exercises 63 and 64, real and imaginary parts are added, just like coefficients of similar terms in the binomials, and the answers correspond. In Exercise 66, introducing $i^2=-1$ when performing the division leads to answers that do not correspond.

69. The reciprocal of $5-4i$ is $\dfrac{1}{5-4i}$.

$$\frac{1}{5-4i}=\frac{1\cdot(5+4i)}{(5-4i)(5+4i)}$$
$$=\frac{5+4i}{5^2+4^2}=\frac{5+4i}{25+16}$$
$$=\frac{5+4i}{41}=\frac{5}{41}+\frac{4}{41}i$$

71. $i^{18}=i^{16}\cdot i^2$
$=(i^4)^4\cdot i^2$
$=1^4\cdot(-1)=1\cdot(-1)=-1$

73. $i^{89}=i^{88}\cdot i$
$=(i^4)^{22}\cdot i$
$=1^{22}\cdot i=1\cdot i=i$

75. $i^{96}=(i^4)^{24}=1^{24}=1$

77. $i^{-5}=\dfrac{1}{i^5}=\dfrac{1}{i^4\cdot i}=\dfrac{1}{1\cdot i}=\dfrac{1}{i}$

From Exercise 3, $\dfrac{1}{i}=-i$.

79. Since $i^{20}=(i^4)^5=1^5=1$, the student multiplied by 1, which is justified by the identity property for multiplication.

81. $I = \dfrac{E}{R + (X_L - X_c)i}$

Substitute $2 + 3i$ for E, 5 for R, 4 for X_L, and 3 for X_c.

$$I = \frac{2+3i}{5+(4-3)i} = \frac{2+3i}{5+i}$$
$$= \frac{(2+3i)(5-i)}{(5+i)(5-i)} = \frac{10-2i+15i-3i^2}{5^2+1^2}$$
$$= \frac{10+3+13i}{25+1} = \frac{13+13i}{26}$$
$$= \frac{13(1+i)}{13 \cdot 2} = \frac{1+i}{2} = \frac{1}{2} + \frac{1}{2}i$$

83. To check that $1 + 5i$ is a solution of the equation, substitute $1 + 5i$ for x.

$$x^2 - 2x + 26 = 0$$
$$(1+5i)^2 - 2(1+5i) + 26 \stackrel{?}{=} 0$$
$$(1 + 10i + 25i^2) - 2 - 10i + 26 \stackrel{?}{=} 0$$
$$1 + 10i - 25 - 2 - 10i + 26 \stackrel{?}{=} 0$$
$$(1 - 25 - 2 + 26) + (10 - 10)i \stackrel{?}{=} 0$$
$$0 = 0 \quad \textit{True}$$

Thus, $1 + 5i$ is a solution of

$$x^2 - 2x + 26 = 0.$$

Chapter 9 Review Exercises

1. $\sqrt{1764} = 42$, because $42^2 = 1764$.

2. $\sqrt{289} = 17$, because $17^2 = 289$, so $-\sqrt{289} = -17$.

3. $-\sqrt{-841}$ is *not a real number*, since the square of a real number is nonnegative.

4. $\sqrt[3]{216} = 6$, because $6^3 = 216$.

5. $\sqrt[5]{-32} = -2$, because $(-2)^5 = -32$.

6. $\sqrt{x^2} = |x|$

7. $\sqrt[3]{x^3} = x$

8. $\sqrt[4]{x^{20}} = \sqrt[4]{(x^5)^4} = |x^5|$, or $x^4|x|$.

9. $f(x) = \sqrt{x-1}$

For the radicand to be nonnegative, we must have

$$x - 1 \geq 0 \quad \text{or} \quad x \geq 1.$$

Thus, the domain is $[1, \infty)$.
The function values are positive or zero (the result of the radical), so the range is $[0, \infty)$.

x	$f(x) = \sqrt{x-1}$
1	$\sqrt{1-1} = 0$
2	$\sqrt{2-1} = 1$
5	$\sqrt{5-1} = 2$

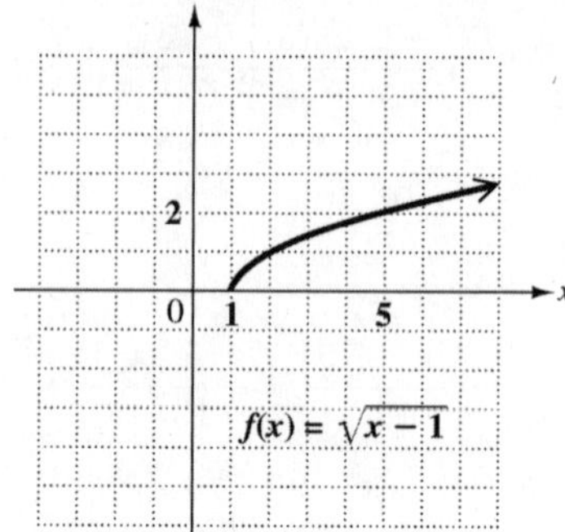

10. $f(x) = \sqrt[3]{x} + 4$

Since we can take the cube root of any real number, the domain is $(-\infty, \infty)$.
The result of a cube root can be any real number, so the range is $(-\infty, \infty)$. (The "+4" does not affect that range.)

x	$f(x) = \sqrt[3]{x} + 4$
-8	$\sqrt[3]{-8} + 4 = 2$
-1	$\sqrt[3]{-1} + 4 = 3$
0	$\sqrt[3]{0} + 4 = 4$
1	$\sqrt[3]{1} + 4 = 5$
8	$\sqrt[3]{8} + 4 = 6$

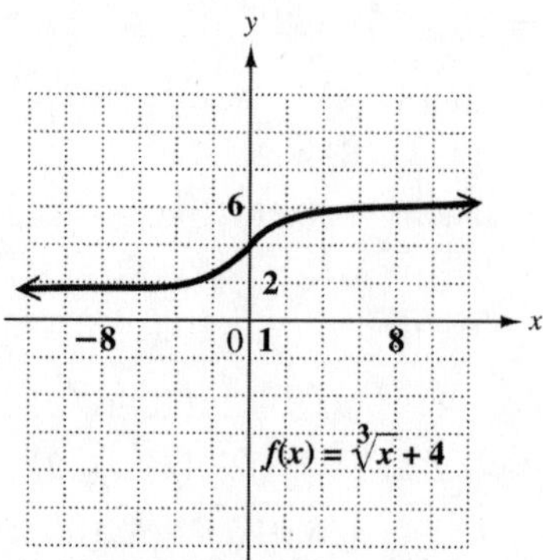

11. $\sqrt[n]{a}$ is not a real number if n is even and a is negative.

12. If a is negative and n is even, then $a^{1/n}$ *is not a real number*. An example is $(-4)^{1/2}$, not a real number.

13. $\sqrt{40} \approx 6.325$

14. $\sqrt{77} \approx 8.775$

15. $\sqrt{310} \approx 17.607$

16. Substitute 3 for L and 32 for g.

$$t = 2\pi\sqrt{\frac{L}{g}} = 2\pi\sqrt{\frac{3}{32}} \approx 1.9$$

The time to complete one swing is approximately 1.9 seconds.

17. Let $a = 11$, $b = 13$, and $c = 20$.
First find the semiperimeter s.

$$\begin{aligned} s &= \tfrac{1}{2}(a+b+c) \\ &= \tfrac{1}{2}(11+13+20) \\ &= \tfrac{44}{2} = 22 \end{aligned}$$

Now find the area A using Heron's formula.

$$\begin{aligned} A &= \sqrt{s(s-a)(s-b)(s-c)} \\ &= \sqrt{22(22-11)(22-13)(22-20)} \\ &= \sqrt{22(11)(9)(2)} \\ &= 66 \end{aligned}$$

The area of the triangle is 66 in.^{2}.

18. $49^{1/2} = \sqrt{49} = 7$

19. $-8^{1/3} = -\sqrt[3]{8} = -2$

20. $(-16)^{1/4} = \sqrt[4]{-16}$ is *not a real number* because the radicand, -16, is negative and the index, 4, is even.

21. The expression with fractional exponents, $a^{m/n}$, is equivalent to the radical expression, $\sqrt[n]{a^m}$. The denominator of the exponent is the index of the radical. For example, $8^{2/3} = (8^{1/3})^2 = 2^2 = 4$, and $\sqrt[3]{8^2} = \sqrt[3]{64} = 4$.

22. $16^{5/4} = (16^{1/4})^5 = 2^5 = 32$

23. $-8^{2/3} = -(8^{1/3})^2 = -2^2 = -4$

24.
$$\begin{aligned} -\left(\frac{36}{25}\right)^{3/2} &= -\left(\sqrt{\frac{36}{25}}\right)^3 \\ &= -\left(\frac{6}{5}\right)^3 = -\frac{216}{125} \end{aligned}$$

25.
$$\begin{aligned} \left(-\frac{1}{8}\right)^{-5/3} &= (-8)^{5/3} \\ &= \left[(-8)^{1/3}\right]^5 \\ &= (-2)^5 = -32 \end{aligned}$$

26.
$$\begin{aligned} \left(\frac{81}{10{,}000}\right)^{-3/4} &= \left(\frac{10{,}000}{81}\right)^{3/4} \\ &= \left(\sqrt[4]{\frac{10{,}000}{81}}\right)^3 \\ &= \left(\frac{10}{3}\right)^3 = \frac{1000}{27} \end{aligned}$$

27. $7^{1/3} \cdot 7^{5/3} = 7^{1/3+5/3} = 7^{6/3} = 7^2 = 49$

28. $\dfrac{96^{2/3}}{96^{-1/3}} = 96^{2/3-(-1/3)} = 96^1 = 96$

29.
$$\begin{aligned} \frac{k^{2/3}k^{-1/2}k^{3/4}}{2(k^2)^{-1/4}} &= \frac{k^{2/3-1/2+3/4}}{2k^{-2/4}} \\ &= \frac{k^{8/12-6/12+9/12}}{2k^{-6/12}} \\ &= \frac{k^{11/12}}{2k^{-6/12}} \\ &= \frac{k^{17/12}}{2} \end{aligned}$$

30. $2^{4/5} = (2^4)^{1/5} = \sqrt[5]{2^4}$, or $\sqrt[5]{16}$

31. $\sqrt{3^{18}} = 3^{18/2} = 3^9$

32.
$$\begin{aligned} \sqrt{7^9} &= 7^{9/2} \\ &= 7^{8/2+1/2} \\ &= 7^{4+1/2} = 7^4 \cdot 7^{1/2} = 7^4\sqrt{7} \end{aligned}$$

33.
$$\begin{aligned} \sqrt[3]{m^5} \cdot \sqrt[3]{m^8} &= (m^5)^{1/3}(m^8)^{1/3} \\ &= m^{5/3}m^{8/3} \\ &= m^{5/3+8/3} \\ &= m^{13/3} \\ &= m^{12/3}m^{1/3} \\ &= m^4\sqrt[3]{m} \end{aligned}$$

34.
$$\begin{aligned} \sqrt[4]{k^2} \cdot \sqrt[4]{k^7} &= k^{2/4} \cdot k^{7/4} \\ &= k^{2/4+7/4} \\ &= k^{9/4} \\ &= k^{2+1/4} \\ &= k^2 \cdot k^{1/4} \\ &= k^2\sqrt[4]{k} \end{aligned}$$

35.
$$\begin{aligned} \sqrt[3]{\sqrt{m}} = \left(\sqrt{m}\right)^{1/3} &= (m^{1/2})^{1/3} \\ &= m^{(1/2)(1/3)} \\ &= m^{1/6} = \sqrt[6]{m} \end{aligned}$$

36.
$$\begin{aligned} \sqrt[4]{16y^5} &= (16y^5)^{1/4} \\ &= 16^{1/4}y^{5/4} \\ &= 2y^{4/4}y^{1/4} \\ &= 2y\sqrt[4]{y} \end{aligned}$$

37. $\sqrt[5]{y} \cdot \sqrt[3]{y} = y^{1/5} \cdot y^{1/3}$
$= y^{1/5 + 1/3}$
$= y^{3/15 + 5/15}$
$= y^{8/15}$
$= \sqrt[15]{y^8}$

38. $\dfrac{\sqrt[3]{y^2}}{\sqrt[4]{y}} = \dfrac{y^{2/3}}{y^{1/4}}$
$= y^{2/3 - 1/4}$
$= y^{8/12 - 3/12}$
$= y^{5/12} = \sqrt[12]{y^5}$

39. $\sqrt{6} \cdot \sqrt{11} = \sqrt{6 \cdot 11} = \sqrt{66}$

40. $\sqrt{5} \cdot \sqrt{r} = \sqrt{5 \cdot r} = \sqrt{5r}$

41. $\sqrt[3]{6} \cdot \sqrt[3]{5} = \sqrt[3]{6 \cdot 5} = \sqrt[3]{30}$

42. $\sqrt[4]{7} \cdot \sqrt[4]{3} = \sqrt[4]{7 \cdot 3} = \sqrt[4]{21}$

43. $\sqrt{20} = \sqrt{4 \cdot 5} = \sqrt{4}\sqrt{5} = 2\sqrt{5}$

44. $-\sqrt{125} = -\sqrt{25 \cdot 5} = -5\sqrt{5}$

45. $\sqrt[3]{-108x^4y} = \sqrt[3]{-1 \cdot 27 \cdot x^3 \cdot 4xy}$
$= -3x\sqrt[3]{4xy}$

46. $\sqrt[3]{64p^4q^6} = \sqrt[3]{64p^3q^6 \cdot p}$
$= 4pq^2\sqrt[3]{p}$

47. $\sqrt{\dfrac{49}{81}} = \dfrac{\sqrt{49}}{\sqrt{81}} = \dfrac{7}{9}$

48. $\sqrt{\dfrac{y^3}{144}} = \dfrac{\sqrt{y^3}}{\sqrt{144}} = \dfrac{\sqrt{y^2 \cdot y}}{12} = \dfrac{y\sqrt{y}}{12}$

49. $\sqrt[3]{\dfrac{m^{15}}{27}} = \dfrac{\sqrt[3]{m^{15}}}{\sqrt[3]{27}} = \dfrac{\sqrt[3]{(m^5)^3}}{\sqrt[3]{3^3}} = \dfrac{m^5}{3}$

50. $\sqrt[3]{\dfrac{r^2}{8}} = \dfrac{\sqrt[3]{r^2}}{\sqrt[3]{8}} = \dfrac{\sqrt[3]{r^2}}{2}$

51. $\dfrac{\sqrt[3]{2^4}}{\sqrt[4]{32}} = \dfrac{\sqrt[3]{2^3}\sqrt[3]{2}}{\sqrt[4]{16}\sqrt[4]{2}} = \dfrac{2 \cdot 2^{1/3}}{2 \cdot 2^{1/4}}$
$= 2^{1/3 - 1/4}$
$= 2^{4/12 - 3/12}$
$= 2^{1/12}$
$= \sqrt[12]{2}$

52. $\dfrac{\sqrt{x}}{\sqrt[5]{x}} = \dfrac{x^{1/2}}{x^{1/5}} = x^{1/2 - 1/5}$
$= x^{5/10 - 2/10} = x^{3/10} = \sqrt[10]{x^3}$

53. $(x_1, y_1) = (2, 7)$ and $(x_2, y_2) = (-1, -4)$
$d = \sqrt{(x_2 - x_1)^2 + (y_2 - y_1)^2}$
$= \sqrt{(-1 - 2)^2 + (-4 - 7)^2}$
$= \sqrt{(-3)^2 + (-11)^2}$
$= \sqrt{9 + 121}$
$= \sqrt{130}$

54. $(x_1, y_1) = (-3, -5)$ and $(x_2, y_2) = (4, -3)$
$d = \sqrt{(x_2 - x_1)^2 + (y_2 - y_1)^2}$
$= \sqrt{[4 - (-3)]^2 + [-3 - (-5)]^2}$
$= \sqrt{7^2 + 2^2}$
$= \sqrt{49 + 4}$
$= \sqrt{53}$

55. $2\sqrt{8} - 3\sqrt{50} = 2\sqrt{4 \cdot 2} - 3\sqrt{25 \cdot 2}$
$= 2 \cdot 2\sqrt{2} - 3 \cdot 5\sqrt{2}$
$= 4\sqrt{2} - 15\sqrt{2} = -11\sqrt{2}$

56. $8\sqrt{80} - 3\sqrt{45} = 8\sqrt{16 \cdot 5} - 3\sqrt{9 \cdot 5}$
$= 8 \cdot 4\sqrt{5} - 3 \cdot 3\sqrt{5}$
$= 32\sqrt{5} - 9\sqrt{5} = 23\sqrt{5}$

57. $-\sqrt{27y} + 2\sqrt{75y} = -\sqrt{9 \cdot 3y} + 2\sqrt{25 \cdot 3y}$
$= -3\sqrt{3y} + 2 \cdot 5\sqrt{3y}$
$= -3\sqrt{3y} + 10\sqrt{3y}$
$= 7\sqrt{3y}$

58. $2\sqrt{54m^3} + 5\sqrt{96m^3}$
$= 2\sqrt{9m^2 \cdot 6m} + 5\sqrt{16m^2 \cdot 6m}$
$= 2 \cdot 3m\sqrt{6m} + 5 \cdot 4m\sqrt{6m}$
$= 6m\sqrt{6m} + 20m\sqrt{6m} = 26m\sqrt{6m}$

59. $3\sqrt[3]{54} + 5\sqrt[3]{16} = 3\sqrt[3]{27 \cdot 2} + 5\sqrt[3]{8 \cdot 2}$
$= 3 \cdot 3\sqrt[3]{2} + 5 \cdot 2\sqrt[3]{2}$
$= 9\sqrt[3]{2} + 10\sqrt[3]{2} = 19\sqrt[3]{2}$

60. $-6\sqrt[4]{32} + \sqrt[4]{512} = -6\sqrt[4]{16 \cdot 2} + \sqrt[4]{256 \cdot 2}$
$= -6 \cdot 2\sqrt[4]{2} + 4\sqrt[4]{2}$
$= -12\sqrt[4]{2} + 4\sqrt[4]{2} = -8\sqrt[4]{2}$

61. $(\sqrt{3} + 1)(\sqrt{3} - 2) = 3 - 2\sqrt{3} + \sqrt{3} - 2$
$= 1 - \sqrt{3}$

62. $(\sqrt{7} + \sqrt{5})(\sqrt{7} - \sqrt{5}) = (\sqrt{7})^2 - (\sqrt{5})^2$
$= 7 - 5 = 2$

63. $(3\sqrt{2} + 1)(2\sqrt{2} - 3)$
$= 6 \cdot 2 - 9\sqrt{2} + 2\sqrt{2} - 3$
$= 12 - 7\sqrt{2} - 3 = 9 - 7\sqrt{2}$

64. $(\sqrt{11}+3\sqrt{5})(\sqrt{11}+5\sqrt{5})$
$= 11+5\sqrt{55}+3\sqrt{55}+15\cdot 5$
$= 11+8\sqrt{55}+75$
$= 86+8\sqrt{55}$

65. $(\sqrt{13}-\sqrt{2})^2$
$= (\sqrt{13})^2 - 2\cdot\sqrt{13}\cdot\sqrt{2}+(\sqrt{2})^2$
$= 13-2\sqrt{26}+2 = 15-2\sqrt{26}$

66. $(\sqrt{5}-\sqrt{7})^2$
$= (\sqrt{5})^2-2\cdot\sqrt{5}\cdot\sqrt{7}+(\sqrt{7})^2$
$= 5-2\sqrt{35}+7$
$= 12-2\sqrt{35}$

67. $\frac{-6\sqrt{3}}{\sqrt{2}} = \frac{-6\sqrt{3}\cdot\sqrt{2}}{\sqrt{2}\cdot\sqrt{2}} = \frac{-6\sqrt{6}}{2} = -3\sqrt{6}$

68. $\frac{3\sqrt{7p}}{\sqrt{y}} = \frac{3\sqrt{7p}\cdot\sqrt{y}}{\sqrt{y}\cdot\sqrt{y}} = \frac{3\sqrt{7py}}{y}$

69. $-\sqrt[3]{\frac{9}{25}} = -\frac{\sqrt[3]{9}}{\sqrt[3]{5^2}} = -\frac{\sqrt[3]{9}\cdot\sqrt[3]{5}}{\sqrt[3]{5^2}\cdot\sqrt[3]{5}}$
$= -\frac{\sqrt[3]{45}}{\sqrt[3]{5^3}} = -\frac{\sqrt[3]{45}}{5}$

70. $\sqrt[3]{\frac{108m^3}{n^5}} = \frac{\sqrt[3]{108m^3}}{\sqrt[3]{n^5}} = \frac{\sqrt[3]{27m^3\cdot 4}}{\sqrt[3]{n^3\cdot n^2}}$
$= \frac{3m\sqrt[3]{4}}{n\sqrt[3]{n^2}} = \frac{3m\sqrt[3]{4}\cdot\sqrt[3]{n}}{n\sqrt[3]{n^2}\cdot\sqrt[3]{n}}$
$= \frac{3m\sqrt[3]{4n}}{n\cdot n} = \frac{3m\sqrt[3]{4n}}{n^2}$

71. $\frac{1}{\sqrt{2}+\sqrt{7}}$

Multiply the numerator and denominator by the conjugate of the denominator, $\sqrt{2}-\sqrt{7}$.

$= \frac{1(\sqrt{2}-\sqrt{7})}{(\sqrt{2}+\sqrt{7})(\sqrt{2}-\sqrt{7})}$
$= \frac{\sqrt{2}-\sqrt{7}}{2-7}$
$= \frac{\sqrt{2}-\sqrt{7}}{-5}$

72. $\frac{-5}{\sqrt{6}-\sqrt{3}}$

Multiply the numerator and denominator by the conjugate of the denominator, $\sqrt{6}+\sqrt{3}$.

$= \frac{-5(\sqrt{6}+\sqrt{3})}{(\sqrt{6}-\sqrt{3})(\sqrt{6}+\sqrt{3})}$
$= \frac{-5(\sqrt{6}+\sqrt{3})}{6-3}$
$= \frac{-5(\sqrt{6}+\sqrt{3})}{3}$

73. $\sqrt{8x+9} = 5$
$(\sqrt{8x+9})^2 = 5^2$ *Square.*
$8x+9 = 25$
$8x = 16$
$x = 2$

Check $x=2$: $\sqrt{25}=5$ *True*
Solution set: $\{2\}$

74. $\sqrt{2z-3}-3=0$
$\sqrt{2z-3}=3$ *Isolate.*
$(\sqrt{2z-3})^2 = 3^2$ *Square.*
$2z-3=9$
$2z=12$
$z=6$

Check $z=6$: $\sqrt{9}-3=0$ *True*
Solution set: $\{6\}$

75. $\sqrt{3m+1}=-1$

This equation has no solution because $\sqrt{3m+1}$ cannot be negative.

Solution set: $\emptyset$

76. $\sqrt{7z+1}=z+1$
$(\sqrt{7z+1})^2=(z+1)^2$ *Square.*
$7z+1=z^2+2z+1$
$0=z^2-5z$
$0=z(z-5)$

$z=0$ or $z=5$

Check $z=0$: $\sqrt{1}=1$ *True*
Check $z=5$: $\sqrt{36}=6$ *True*
Solution set: $\{0,5\}$

77. $3\sqrt{m}=\sqrt{10m-9}$
$(3\sqrt{m})^2=(\sqrt{10m-9})^2$ *Square.*
$9m=10m-9$
$9=m$

Check $m=9$: $3\sqrt{9}=\sqrt{81}$ *True*
Solution set: $\{9\}$

78. $\sqrt{p^2+3p+7} = p+2$

$\left(\sqrt{p^2+3p+7}\right)^2 = (p+2)^2$ *Square.*

$p^2+3p+7 = p^2+4p+4$

$3 = p$

Check $p=3$: $\sqrt{25}=5$ *True*

Solution set: $\{3\}$

79. $\sqrt{x+2} - \sqrt{x-3} = 1$

Get one radical on each side of the equals sign.

$\sqrt{x+2} = 1 + \sqrt{x-3}$

$(\sqrt{x+2})^2 = (1+\sqrt{x-3})^2$ *Square.*

$x+2 = 1 + 2\sqrt{x-3} + x - 3$

$4 = 2\sqrt{x-3}$

$2 = \sqrt{x-3}$

$2^2 = (\sqrt{x-3})^2$ *Square.*

$4 = x - 3$

$7 = x$

Check $x=7$: $\sqrt{9}-\sqrt{4}=1$ *True*

Solution set: $\{7\}$

80. $\sqrt[3]{5m-1} = \sqrt[3]{3m-2}$

$(\sqrt[3]{5m-1})^3 = (\sqrt[3]{3m-2})^3$ *Cube both sides.*

$5m-1 = 3m-2$

$2m = -1$

$m = -\frac{1}{2}$

Check $m=-\frac{1}{2}$: $\sqrt[3]{-\frac{7}{2}} = \sqrt[3]{-\frac{7}{2}}$ *True*

Solution set: $\{-\frac{1}{2}\}$

81. $\sqrt[4]{x+6} = \sqrt[4]{2x}$

Raise both sides to the fourth power.

$(\sqrt[4]{x+6})^4 = (\sqrt[4]{2x})^4$

$x+6 = 2x$

$6 = x$

Check $x=6$: $\sqrt[4]{12} = \sqrt[4]{12}$ *True*

Solution set: $\{6\}$

82. $\sqrt{-25} = i\sqrt{25} = 5i$

83. $\sqrt{-200} = i\sqrt{100\cdot 2} = 10i\sqrt{2}$

84. $\sqrt{-160} = i\sqrt{16\cdot 10} = 4i\sqrt{10}$

85. $(-2+5i)+(-8-7i)$

$= [-2+(-8)]+[5+(-7)]i$

$= -10-2i$

86. $(5+4i)-(-9-3i)$

$= [(5-(-9)]+[(4-(-3)]i$

$= 14+7i$

87. $\sqrt{-5}\cdot\sqrt{-7}$

$= i\sqrt{5}\cdot i\sqrt{7}$

$= i^2\sqrt{35}$

$= -1\sqrt{35} = -\sqrt{35}$

88. $\sqrt{-25}\cdot\sqrt{-81}$

$= i\sqrt{25}\cdot i\sqrt{81}$

$= 5i\cdot 9i$

$= 45i^2 = 45(-1) = -45$

89. $\dfrac{\sqrt{-72}}{\sqrt{-8}} = \dfrac{i\sqrt{72}}{i\sqrt{8}} = \sqrt{\dfrac{72}{8}} = \sqrt{9} = 3$

90. $(2+3i)(1-i) = 2-2i+3i-3i^2$

$= 2+i-3(-1)$

$= 2+i+3$

$= 5+i$

91. $(6-2i)^2 = 6^2 - 2\cdot 6\cdot 2i + (2i)^2$

$= 36-24i+4i^2$

$= 36-24i+4(-1)$

$= 36-24i-4$

$= 32-24i$

92. $\dfrac{3-i}{2+i}$

Multiply the numerator and denominator by the conjugate of the denominator, $2-i$.

$= \dfrac{(3-i)(2-i)}{(2+i)(2-i)}$

$= \dfrac{6-3i-2i+i^2}{4-i^2}$

$= \dfrac{6-5i-1}{4-(-1)}$

$= \dfrac{5-5i}{5} = \dfrac{5(1-i)}{5} = 1-i$

93. $\dfrac{5+14i}{2+3i}$

Multiply the numerator and denominator by the conjugate of the denominator, $2-3i$.

$= \dfrac{(5+14i)(2-3i)}{(2+3i)(2-3i)}$

$= \dfrac{10-15i+28i-42i^2}{2^2+3^2}$

$= \dfrac{10+13i-42(-1)}{4+9}$

$= \dfrac{52+13i}{13} = \dfrac{13(4+i)}{13} = 4+i$

94. $i^{11} = i^8\cdot i^3$

$= (i^4)^2\cdot i^2\cdot i$

$= 1^2\cdot(-1)\cdot i = -i$

95. $i^{52} = (i^4)^{13} = 1^{13} = 1$

96. $i^{-13} = i^{-13} \cdot i^{16}$, since $i^{16} = (i^4)^4 = 1^4 = 1$. So, $i^{-13} = i^{-13} \cdot i^{16} = i^3 = i^2 \cdot i = -i$.

97. **[9.1]** $-\sqrt{169a^2b^4} = -\sqrt{(13ab^2)^2} = -13ab^2$

98. **[9.2]** $1000^{-2/3} = \dfrac{1}{1000^{2/3}} = \dfrac{1}{(1000^{1/3})^2}$

$$= \frac{1}{10^2} = \frac{1}{100}$$

99. **[9.2]** $\dfrac{y^{-1/3} \cdot y^{5/6}}{y} = \dfrac{y^{-2/6+5/6}}{y}$

$$= \frac{y^{3/6}}{y} = \frac{y^{1/2}}{y}$$

$$= y^{1/2-1} = y^{-1/2} = \frac{1}{y^{1/2}}$$

100. **[9.2]** $\dfrac{z^{-1/4}x^{1/2}}{z^{1/2}x^{-1/4}} = z^{-1/4-1/2} \cdot x^{1/2-(-1/4)}$

$$= z^{-1/4-2/4} \cdot x^{2/4+1/4}$$

$$= z^{-3/4} \cdot x^{3/4}$$

$$= \frac{x^{3/4}}{z^{3/4}}$$

101. **[9.3]** $\sqrt[4]{k^{24}} = k^{24/4} = k^6$

102. **[9.3]** $\sqrt[3]{54z^9t^8} = \sqrt[3]{27z^9t^6 \cdot 2t^2} = 3z^3t^2\sqrt[3]{2t^2}$

103. **[9.4]** $-5\sqrt{18} + 12\sqrt{72}$

$$= -5\sqrt{9 \cdot 2} + 12\sqrt{36 \cdot 2}$$

$$= -5 \cdot 3\sqrt{2} + 12 \cdot 6\sqrt{2}$$

$$= -15\sqrt{2} + 72\sqrt{2} = 57\sqrt{2}$$

104. **[9.4]** $8\sqrt[3]{x^3y^2} - 2x\sqrt[3]{y^2} = 8x\sqrt[3]{y^2} - 2x\sqrt[3]{y^2}$

$$= 6x\sqrt[3]{y^2}$$

105. **[9.5]** $(\sqrt{5} - \sqrt{3})(\sqrt{7} + \sqrt{3})$

$$= \sqrt{35} + \sqrt{15} - \sqrt{21} - 3$$

106. **[9.5]** $\dfrac{-1}{\sqrt{12}} = \dfrac{-1}{\sqrt{4 \cdot 3}} = \dfrac{-1}{2\sqrt{3}}$

$$= \frac{-1 \cdot \sqrt{3}}{2\sqrt{3} \cdot \sqrt{3}}$$

$$= \frac{-\sqrt{3}}{2 \cdot 3} = \frac{-\sqrt{3}}{6}$$

107. **[9.5]** $\sqrt[3]{\dfrac{12}{25}} = \dfrac{\sqrt[3]{12}}{\sqrt[3]{25}} = \dfrac{\sqrt[3]{12}}{\sqrt[3]{5^2}}$

$$= \frac{\sqrt[3]{12} \cdot \sqrt[3]{5}}{\sqrt[3]{5^2} \cdot \sqrt[3]{5}}$$

$$= \frac{\sqrt[3]{60}}{\sqrt[3]{5^3}} = \frac{\sqrt[3]{60}}{5}$$

108. **[9.5]** $\dfrac{2\sqrt{z}}{\sqrt{z} - 2}$

Multiply the numerator and denominator by the conjugate of the denominator, $\sqrt{z} + 2$.

$$= \frac{2\sqrt{z}(\sqrt{z} + 2)}{(\sqrt{z} - 2)(\sqrt{z} + 2)}$$

$$= \frac{2\sqrt{z}(\sqrt{z} + 2)}{z - 4}$$

109. **[9.7]** $\sqrt{-49} = i\sqrt{49} = 7i$

110. **[9.7]** $(4 - 9i) + (-1 + 2i)$

$$= (4 - 1) + (-9 + 2)i$$

$$= 3 - 7i$$

111. **[9.7]** $\dfrac{\sqrt{50}}{\sqrt{-2}} = \dfrac{\sqrt{25 \cdot 2}}{i\sqrt{2}} = \dfrac{5\sqrt{2}}{i\sqrt{2}} = \dfrac{5}{i}$

The conjugate of i is $-i$.

$$\frac{5(-i)}{i(-i)} = \frac{-5i}{-i^2} = \frac{-5i}{-(-1)} = -5i$$

112. **[9.6]**

$$\sqrt{x + 4} = x - 2$$

$$\left(\sqrt{x + 4}\right)^2 = (x - 2)^2 \quad \textit{Square.}$$

$$x + 4 = x^2 - 4x + 4$$

$$0 = x^2 - 5x$$

$$0 = x(x - 5)$$

$$x = 0 \quad \text{or} \quad x = 5$$

Check $x = 0$: $\sqrt{4} = -2$ *False*
Check $x = 5$: $\sqrt{9} = 3$ *True*
Solution set: $\{5\}$

113. **[9.6]**

$$\sqrt{6 + 2x} - 1 = \sqrt{7 - 2x}$$

$$\left(\sqrt{6 + 2x} - 1\right)^2 = \left(\sqrt{7 - 2x}\right)^2 \quad \textit{Square.}$$

$$6 + 2x - 2\sqrt{6 + 2x} + 1 = 7 - 2x$$

$$4x = 2\sqrt{6 + 2x}$$

$$2x = \sqrt{6 + 2x}$$

$$(2x)^2 = \left(\sqrt{6 + 2x}\right)^2 \quad \textit{Square.}$$

$$4x^2 = 6 + 2x$$

$$4x^2 - 2x - 6 = 0$$

$$2x^2 - x - 3 = 0$$

$$(2x - 3)(x + 1) = 0$$

$$2x - 3 = 0 \quad \text{or} \quad x + 1 = 0$$

$$x = \tfrac{3}{2} \quad \text{or} \quad x = -1$$

Check $x = \frac{3}{2}$: $\sqrt{9} - 1 = \sqrt{4}$ *True*
Check $x = -1$: $\sqrt{4} - 1 = \sqrt{9}$ *False*
Solution set: $\{\frac{3}{2}\}$

114. [9.6] Substitute 12 for L and 9 for W.

$$L = \sqrt{H^2 + W^2}$$
$$12 = \sqrt{H^2 + 9^2}$$
$$12^2 = \left(\sqrt{H^2 + 81}\right)^2$$
$$144 = H^2 + 81$$
$$63 = H^2$$
$$\sqrt{63} = H$$

To the nearest tenth of a foot, the height is approximately 7.9 feet.

115. [9.4] Add the measures of the sides.

$$P = a + b + c$$
$$P = 2\sqrt{27} + \sqrt{108} + \sqrt{50}$$
$$= 2\sqrt{9 \cdot 3} + \sqrt{36 \cdot 3} + \sqrt{25 \cdot 2}$$
$$= 2 \cdot 3\sqrt{3} + 6\sqrt{3} + 5\sqrt{2}$$
$$= 6\sqrt{3} + 6\sqrt{3} + 5\sqrt{2}$$
$$= 12\sqrt{3} + 5\sqrt{2}$$

The perimeter is $(12\sqrt{3} + 5\sqrt{2})$ feet.

Chapter 9 Test

1. $\sqrt{841} = 29$, because $29^2 = 841$, so $-\sqrt{841} = -29$.

2. $\sqrt[3]{-512} = -8$ because $(-8)^3 = -512$.

3. $125^{1/3} = \sqrt[3]{125} = 5$, because $5^3 = 125$.

4. $\sqrt{146.25} \approx \sqrt{144} = 12$, because $12^2 = 144$, so choice **C** is the best estimate.

5. $\sqrt{478} \approx 21.8632 \approx 21.863$

6. $\sqrt[3]{-832} \approx -9.4053 \approx -9.405$

7. $f(x) = \sqrt{x + 6}$

For the radicand to be nonnegative, we must have

$$x + 6 \geq 0 \quad \text{or} \quad x \geq -6.$$

Thus, the domain is $[-6, \infty)$.
The function values are positive or zero (the result of the radical), so the range is $[0, \infty)$.

x	$f(x) = \sqrt{x+6}$
-6	$\sqrt{-6+6} = 0$
-5	$\sqrt{-5+6} = 1$
-2	$\sqrt{-2+6} = 2$

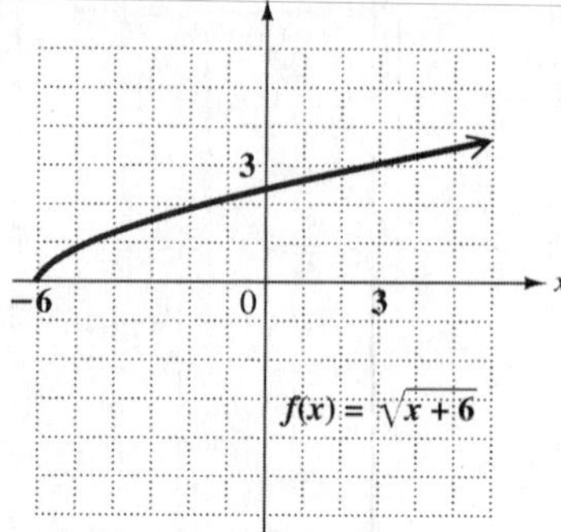

8. $$(-64)^{-4/3} = \frac{1}{(-64)^{4/3}} = \frac{1}{\left(\sqrt[3]{-64}\right)^4} = \frac{1}{(-4)^4} = \frac{1}{256}$$

9. $$\frac{3^{2/5}x^{-1/4}y^{2/5}}{3^{-8/5}x^{7/4}y^{1/10}}$$
$$= 3^{2/5-(-8/5)}x^{-1/4-7/4}y^{2/5-1/10}$$
$$= 3^{10/5}x^{-8/4}y^{4/10-1/10}$$
$$= 3^2x^{-2}y^{3/10} = \frac{9y^{3/10}}{x^2}$$

10. $$\sqrt{54x^5y^6} = \sqrt{9x^4y^6 \cdot 6x}$$
$$= \sqrt{9x^4y^6} \cdot \sqrt{6x}$$
$$= 3x^2y^3\sqrt{6x}$$

11. $$\sqrt[4]{32a^7b^{13}} = \sqrt[4]{16a^4b^{12} \cdot 2a^3b}$$
$$= \sqrt[4]{16a^4b^{12}} \cdot \sqrt[4]{2a^3b}$$
$$= 2ab^3\sqrt[4]{2a^3b}$$

12. $$\sqrt{2} \cdot \sqrt[3]{5} = 2^{1/2} \cdot 5^{1/3} = 2^{3/6} \cdot 5^{2/6}$$
$$= (2^3 \cdot 5^2)^{1/6} = \sqrt[6]{2^3 \cdot 5^2}$$
$$= \sqrt[6]{8 \cdot 25} = \sqrt[6]{200}$$

13. $$3\sqrt{20} - 5\sqrt{80} + 4\sqrt{500}$$
$$= 3\sqrt{4 \cdot 5} - 5\sqrt{16 \cdot 5} + 4\sqrt{100 \cdot 5}$$
$$= 3 \cdot 2\sqrt{5} - 5 \cdot 4\sqrt{5} + 4 \cdot 10\sqrt{5}$$
$$= 6\sqrt{5} - 20\sqrt{5} + 40\sqrt{5} = 26\sqrt{5}$$

14. $$(7\sqrt{5} + 4)(2\sqrt{5} - 1)$$
$$= 14 \cdot 5 - 7\sqrt{5} + 8\sqrt{5} - 4 \quad \textit{FOIL}$$
$$= 70 + \sqrt{5} - 4 = 66 + \sqrt{5}$$

15. $(\sqrt{3}-2\sqrt{5})^2$
$= (\sqrt{3})^2 - 2(\sqrt{3})(2\sqrt{5}) + (2\sqrt{5})^2$
$= 3 - 4\sqrt{15} + 4 \cdot 5$
$= 3 - 4\sqrt{15} + 20$
$= 23 - 4\sqrt{15}$

16. $\dfrac{-5}{\sqrt{40}} = \dfrac{-5}{\sqrt{4 \cdot 10}} = \dfrac{-5}{2\sqrt{10}}$
$= \dfrac{-5 \cdot \sqrt{10}}{2\sqrt{10} \cdot \sqrt{10}} = \dfrac{-5\sqrt{10}}{2 \cdot 10}$
$= \dfrac{-5\sqrt{10}}{20} = \dfrac{-\sqrt{10}}{4}$

17. $\dfrac{2}{\sqrt[3]{5}} = \dfrac{2 \cdot \sqrt[3]{5^2}}{\sqrt[3]{5} \cdot \sqrt[3]{5^2}} = \dfrac{2\sqrt[3]{25}}{5}$

18. $\dfrac{-4}{\sqrt{7}+\sqrt{5}}$

Multiply the numerator and the denominator by the conjugate of the denominator, $\sqrt{7}-\sqrt{5}$.

$= \dfrac{-4(\sqrt{7}-\sqrt{5})}{(\sqrt{7}+\sqrt{5})(\sqrt{7}-\sqrt{5})}$
$= \dfrac{-4(\sqrt{7}-\sqrt{5})}{7-5}$
$= \dfrac{-4(\sqrt{7}-\sqrt{5})}{2} = -2(\sqrt{7}-\sqrt{5})$

19. $\dfrac{6+\sqrt{24}}{2}$
$= \dfrac{6+\sqrt{4 \cdot 6}}{2}$
$= \dfrac{6+2\sqrt{6}}{2}$
$= \dfrac{2(3+\sqrt{6})}{2} = 3+\sqrt{6}$

20. The distance d between $(x_1, y_1) = (-3, 8)$ and $(x_2, y_2) = (2, 7)$ is

$d = \sqrt{(x_2-x_1)^2 + (y_2-y_1)^2}$
$= \sqrt{[2-(-3)]^2 + (7-8)^2}$
$= \sqrt{5^2 + (-1)^2}$
$= \sqrt{26}.$

21. $a^2 + b^2 = c^2$
$12^2 + b^2 = 17^2$ *Let a = 12, c = 17.*
$144 + b^2 = 289$
$b^2 = 145$
$b = \sqrt{145}$

22. $\sqrt[3]{5x} = \sqrt[3]{2x-3}$
$(\sqrt[3]{5x})^3 = (\sqrt[3]{2x-3})^3$ *Cube.*
$5x = 2x - 3$
$3x = -3$
$x = -1$

Check $x = -1$: $\sqrt[3]{-5} = \sqrt[3]{-5}$ *True*
Solution set: $\{-1\}$

23. $\sqrt{7-x} + 5 = x$
$\sqrt{7-x} = x - 5$
$(\sqrt{7-x})^2 = (x-5)^2$ *Square.*
$7 - x = x^2 - 10x + 25$
$0 = x^2 - 9x + 18$
$0 = (x-3)(x-6)$

$x - 3 = 0$ or $x - 6 = 0$
$x = 3$ or $x = 6$

Check $x = 3$: $\sqrt{4} + 5 = 3$ *False*
Check $x = 6$: $\sqrt{1} + 5 = 6$ *True*
Solution set: $\{6\}$

24. $\sqrt{x+4} - \sqrt{1-x} = -1$
$\sqrt{x+4} = -1 + \sqrt{1-x}$
$(\sqrt{x+4})^2 = (-1+\sqrt{1-x})^2$
$x + 4 = 1 - 2\sqrt{1-x} + 1 - x$
$2x + 2 = -2\sqrt{1-x}$
$x + 1 = -\sqrt{1-x}$
$(x+1)^2 = (-\sqrt{1-x})^2$
$x^2 + 2x + 1 = 1 - x$
$x^2 + 3x = 0$
$x(x+3) = 0$

$x = 0$ or $x + 3 = 0$
$x = -3$

Check $x = -3$: $1 - 2 = -1$ *True*
Check $x = 0$: $2 - 1 = -1$ *False*
Solution set: $\{-3\}$

25. $(-2+5i) - (3+6i) - 7i$
$= (-2-3) + (5-6-7)i$
$= -5 - 8i$

26. $(-4+2i)(3-i)$
F O I L
$= -4(3) + (-4)(-i) + 2i(3) + 2i(-i)$
$= -12 + 4i + 6i - 2i^2$
$= -12 + 10i + 2$
$= -10 + 10i$

27. $\dfrac{7+i}{1-i}$

Multiply the numerator and the denominator by the conjugate of the denominator, $1+i$.

$$= \frac{(7+i)(1+i)}{(1-i)(1+i)}$$
$$= \frac{7+7i+i+i^2}{1-i^2}$$
$$= \frac{7+8i-1}{1-(-1)}$$
$$= \frac{6+8i}{2} = \frac{2(3+4i)}{2} = 3+4i$$

28. $i^{35} = i^{32} \cdot i^3$
$= (i^4)^8 \cdot i^3$
$= 1^8 \cdot i^2 \cdot i$
$= 1(-1) \cdot i = -i$

Cumulative Review Exercises (Chapters 1–9)

1. $7-(4+3t)+2t = -6(t-2)-5$
$7-4-3t+2t = -6t+12-5$
$3-t = -6t+7$
$5t = 4$
$t = \frac{4}{5}$

Check $t=\frac{4}{5}$: $\frac{35}{5} - \frac{32}{5} + \frac{8}{5} = \frac{36}{5} - \frac{25}{5}$ *True*
Solution set: $\{\frac{4}{5}\}$

2. $\frac{1}{3}x + \frac{1}{4}(x+8) = x+7$
Multiply by the LCD, 12.
$12\left[\frac{1}{3}x + \frac{1}{4}(x+8)\right] = 12(x+7)$
$4x+3(x+8) = 12x+84$
$4x+3x+24 = 12x+84$
$7x+24 = 12x+84$
$-5x = 60$
$x = -12$

Check $x=-12$: $-4-1=-5$ *True*
Solution set: $\{-12\}$

3. $|6x-9| = |-4x+2|$

$6x-9 = -4x+2$ or $6x-9 = -(-4x+2)$
$10x = 11$ or $6x-9 = 4x-2$
$2x = 7$
$x = \frac{11}{10}$ or $x = \frac{7}{2}$

Solution set: $\{\frac{11}{10}, \frac{7}{2}\}$

4. $-5-3(x-2) < 11-2(x+2)$
$-5-3x+6 < 11-2x-4$
$1-3x < 7-2x$
$-x < 6$
Multiply by -1; reverse the inequality.
$x > -6$

Solution set: $(-6, \infty)$

5. $1+4x > 5$ and $-2x > -6$
$4x > 4$
$x > 1$ and $x < 3$

Solution set: $(1, 3)$

6. $-2 < 1-3x < 7$
$-3 < -3x < 6$
$1 > x > -2$ or $-2 < x < 1$

Solution set: $(-2, 1)$

7. To write an equation of the line through $(-4, 6)$ and $(7, -6)$, first find the slope.

$$m = \frac{-6-6}{7-(-4)} = \frac{-12}{11} = -\frac{12}{11}$$

Now use the point-slope form with $(x_1, y_1) = (-4, 6)$ and $m = -\frac{12}{11}$.

$y - y_1 = m(x-x_1)$
$y - 6 = -\frac{12}{11}[x-(-4)]$
$11y - 66 = -12(x+4)$ *Multiply by 11.*
$11y - 66 = -12x - 48$
$12x + 11y = 18$

8. $2x+3y = 8$ and $6y = 4x+16$

Find the slope of each line by writing each equation in slope-intercept form.

$2x+3y = 8$
$3y = -2x+8$
$y = -\frac{2}{3}x + \frac{8}{3}$

The slope of the first line is $-\frac{2}{3}$.

$6y = 4x+16$
$y = \frac{4}{6}x + \frac{16}{6}$
$y = \frac{2}{3}x + \frac{8}{3}$

The slope of this line is $\frac{2}{3}$.

The slopes are not the same, so the lines are not parallel. The slopes are not negative reciprocals of one another, so the lines are not perpendicular. Therefore, the answer is **C**.

9. $f(x) = -3x + 6$ or $y = -3x + 6$

(a) From the equation, the y-intercept is $(0, 6)$.

(b) Let $y = 0$ to find the x-intercept.

$$\begin{aligned} 0 &= -3x + 6 \\ 3x &= 6 \\ x &= 2 \end{aligned}$$

The x-intercept is $(2, 0)$.

10. To graph the inequality $-2x + y < -6$, graph the line $-2x + y = -6$, which has intercepts $(3, 0)$ and $(0, -6)$, as a dashed line since the inequality involves " $<$ ". Test $(0, 0)$, which yields $0 < -6$, a false statement. Shade the region that does not include $(0, 0)$.

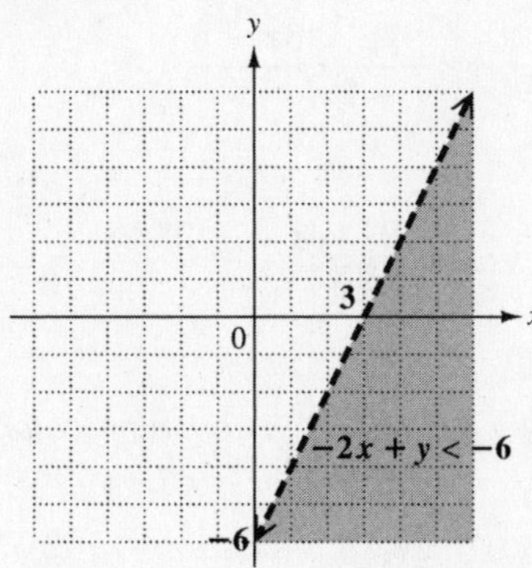

11. The two angles have the same measure, so

$$\begin{aligned} 10x - 70 &= 7x - 25 \\ 3x &= 45 \\ x &= 15. \end{aligned}$$

Since $x = 15$, $10x - 70 = 10(15) - 70$
$= 150 - 70 = 80.$

Each angle measures 80°.

12. $3x - y = 23$ (1)
$2x + 3y = 8$ (2)

To eliminate y, multiply equation (1) by 3 and add the result to equation (2).

$$\begin{aligned} 9x - 3y &= 69 \quad 3 \times (1) \\ 2x + 3y &= 8 \quad (2) \\ \hline 11x &= 77 \\ x &= 7 \end{aligned}$$

Substitute 7 for x in (2).

$$\begin{aligned} 2(7) + 3y &= 8 \\ 14 + 3y &= 8 \\ 3y &= -6 \\ y &= -2 \end{aligned}$$

Solution set: $\{(7, -2)\}$

13. $5x + 2y = 7$ (1)
$10x + 4y = 12$ (2)

Multiplying equation (1) by -2 and adding the result to equation (2) yields the false statement $0 = -2$. The system is inconsistent.

Solution set: $\emptyset$

14. $2x + y - z = 5$ (1)
$6x + 3y - 3z = 15$ (2)
$4x + 2y - 2z = 10$ (3)

Multiplying equation (1) by -2 and adding the result to equation (3) yields the true statement $0 = 0$. Equations (1) and (3) are dependent. Since the two equations that are left include *three* variables, it is not possible to find a unique solution. Therefore, the system has an *infinite number of solutions.*

Solution set: $\{(x, y, z) \mid 2x + y - z = 5\}$

15. Let $x =$ the number of 2-ounce letters and
$y =$ the number of 3-ounce letters.

$5x + 3y = 5.76$ (1)
$3x + 5y = 6.24$ (2)

To eliminate x, multiply (1) by -3 and (2) by 5 and add the results.

$$\begin{aligned} -15x - 9y &= -17.28 \quad -3 \times (1) \\ 15x + 25y &= 31.20 \quad 5 \times (2) \\ \hline 16y &= 13.92 \\ y &= 0.87 \end{aligned}$$

Substitute $y = 0.87$ in (1).

$$\begin{aligned} 5x + 3y &= 5.76 \quad (1) \\ 5x + 3(0.87) &= 5.76 \\ 5x + 2.61 &= 5.76 \\ 5x &= 3.15 \\ x &= 0.63 \end{aligned}$$

The 2006 postage rate for a 2-ounce letter was \$0.63 and for a 3-ounce letter, \$0.87.

16. $(3k^3 - 5k^2 + 8k - 2) - (4k^3 + 11k + 7)$
$+ (2k^2 - 5k)$
$= 3k^3 - 4k^3 - 5k^2 + 2k^2$
$+ 8k - 11k - 5k - 2 - 7$
$= -k^3 - 3k^2 - 8k - 9$

17. $(8x - 7)(x + 3)$
$= 8x^2 + 24x - 7x - 21$
$= 8x^2 + 17x - 21$

18. $\dfrac{8z^3-16z^2+24z}{8z^2}=\dfrac{8z^3}{8z^2}-\dfrac{16z^2}{8z^2}+\dfrac{24z}{8z^2}$

$=z-2+\dfrac{3}{z}$

19. $\dfrac{6y^4-3y^3+5y^2+6y-9}{2y+1}$

$$\begin{array}{r|l} & 3y^3 - 3y^2 + 4y + 1 \\ 2y+1 & 6y^4 - 3y^3 + 5y^2 + 6y - 9 \\ & \underline{6y^4 + 3y^3} \\ & -6y^3 + 5y^2 \\ & \underline{-6y^3 - 3y^2} \\ & 8y^2 + 6y \\ & \underline{8y^2 + 4y} \\ & 2y - 9 \\ & \underline{2y + 1} \\ & -10 \end{array}$$

The answer is

$$3y^3-3y^2+4y+1+\frac{-10}{2y+1}.$$

20. $2p^2-5pq+3q^2$

Two integer factors whose product is $(2)(3)=6$ and whose sum is -5 are -3 and -2. Rewrite the trinomial in a form that can be factored by grouping.

$=2p^2-3pq-2pq+3q^2$
$=p(2p-3q)-q(2p-3q)$
$=(2p-3q)(p-q)$

21. In $18k^4+9k^2-20$, let $y=k^2$ to obtain

$18y^2+9y-20=(3y+4)(6y-5).$

Replace y with k^2.

$18k^4+9k^2-20=(3k^2+4)(6k^2-5)$

22. x^3+512

$=x^3+8^3$ *Factor; sum of cubes*
$=(x+8)(x^2-8x+64)$

23. $\dfrac{y^2+y-12}{y^3+9y^2+20y}\div\dfrac{y^2-9}{y^3+3y^2}$

$=\dfrac{y^2+y-12}{y(y^2+9y+20)}\cdot\dfrac{y^3+3y^2}{y^2-9}$

$=\dfrac{(y+4)(y-3)}{y(y+4)(y+5)}\cdot\dfrac{y^2(y+3)}{(y+3)(y-3)}$

$=\dfrac{y}{y+5}$

24. $\dfrac{1}{x+y}+\dfrac{3}{x-y}$ *The LCD is $(x+y)(x-y)$.*

$=\dfrac{1(x-y)}{(x+y)(x-y)}+\dfrac{3(x+y)}{(x-y)(x+y)}$

$=\dfrac{(x-y)+3(x+y)}{(x+y)(x-y)}$

$=\dfrac{x-y+3x+3y}{(x+y)(x-y)}$

$=\dfrac{4x+2y}{(x+y)(x-y)}$

25. $\dfrac{\frac{-6}{x-2}}{\frac{8}{3x-6}}=\dfrac{-6}{x-2}\div\dfrac{8}{3x-6}$

$=\dfrac{-6}{x-2}\cdot\dfrac{3x-6}{8}$

$=\dfrac{-6}{x-2}\cdot\dfrac{3(x-2)}{8}$

$=\dfrac{-2\cdot 3\cdot 3}{2\cdot 4}=-\dfrac{9}{4}$

26. $\dfrac{\frac{1}{a}-\frac{1}{b}}{\frac{a}{b}-\frac{b}{a}}$ *The LCD in both numerator and denominator is ab.*

$=\dfrac{\frac{b-a}{ab}}{\frac{a^2-b^2}{ab}}$

$=\dfrac{b-a}{ab}\div\dfrac{a^2-b^2}{ab}$

$=\dfrac{b-a}{ab}\cdot\dfrac{ab}{a^2-b^2}$

$=\dfrac{b-a}{a^2-b^2}$

$=\dfrac{-(a-b)}{(a-b)(a+b)}=\dfrac{-1}{a+b}$, or $-\dfrac{1}{a+b}$

27. $2x^2+11x+15=0$

$(x+3)(2x+5)=0$

$x+3=0$ or $2x+5=0$
$x=-3$ or $x=-\frac{5}{2}$

Solution set: $\{-3,-\frac{5}{2}\}$

28.
$$\begin{aligned} 5t(t-1) &= 2(1-t) \\ 5t^2 - 5t &= 2 - 2t \\ 5t^2 - 3t - 2 &= 0 \\ (5t+2)(t-1) &= 0 \end{aligned}$$

$$5t + 2 = 0 \quad \text{or} \quad t - 1 = 0$$
$$t = -\tfrac{2}{5} \quad \text{or} \quad t = 1$$

Solution set: $\{-\frac{2}{5}, 1\}$

29.
$$27^{-5/3} = \frac{1}{27^{5/3}} = \frac{1}{(27^{1/3})^5} = \frac{1}{3^5} = \frac{1}{243}$$

30.
$$\frac{x^{-2/3}}{x^{-3/4}} = x^{-2/3-(-3/4)} = x^{-8/12+9/12} = x^{1/12}$$

31.
$$\begin{aligned} &8\sqrt{20} + 3\sqrt{80} - 2\sqrt{500} \\ &= 8\sqrt{4 \cdot 5} + 3\sqrt{16 \cdot 5} - 2\sqrt{100 \cdot 5} \\ &= 8 \cdot 2\sqrt{5} + 3 \cdot 4\sqrt{5} - 2 \cdot 10\sqrt{5} \\ &= 16\sqrt{5} + 12\sqrt{5} - 20\sqrt{5} \\ &= 8\sqrt{5} \end{aligned}$$

32.
$$\frac{-9}{\sqrt{80}} = \frac{-9}{\sqrt{16 \cdot 5}} = \frac{-9}{4\sqrt{5}} = \frac{-9}{4\sqrt{5}} \cdot \frac{\sqrt{5}}{\sqrt{5}} = \frac{-9\sqrt{5}}{20}$$

33.
$$\frac{4}{\sqrt{6}-\sqrt{5}} = \frac{4}{\sqrt{6}-\sqrt{5}} \cdot \frac{\sqrt{6}+\sqrt{5}}{\sqrt{6}+\sqrt{5}} = \frac{4(\sqrt{6}+\sqrt{5})}{6-5} = 4(\sqrt{6}+\sqrt{5})$$

34.
$$\frac{12}{\sqrt[3]{2}} = \frac{12}{\sqrt[3]{2}} \cdot \frac{\sqrt[3]{4}}{\sqrt[3]{4}} = \frac{12\sqrt[3]{4}}{\sqrt[3]{8}} = \frac{12\sqrt[3]{4}}{2} = 6\sqrt[3]{4}$$

35. The distance d between $(x_1, y_1) = (-4, 4)$ and $(x_2, y_2) = (-2, 9)$ is

$$\begin{aligned} d &= \sqrt{(x_2 - x_1)^2 + (y_2 - y_1)^2} \\ &= \sqrt{[-2-(-4)]^2 + (9-4)^2} \\ &= \sqrt{2^2 + 5^2} \\ &= \sqrt{4 + 25} \\ &= \sqrt{29}. \end{aligned}$$

36.
$$\begin{aligned} \sqrt{8x-4} - \sqrt{7x+2} &= 0 \\ \sqrt{8x-4} &= \sqrt{7x+2} \\ 8x - 4 &= 7x + 2 \quad \textit{Square.} \\ x &= 6 \end{aligned}$$

Check $x = 6$: $\sqrt{44} - \sqrt{44} = 0$ *True*

Solution set: $\{6\}$

37. Let $x =$ the speed of the boat in still water. Make a table. Use $d = rt$, or $t = \frac{d}{r}$.

Direction	Distance	Rate	Time
Downstream	36	$x+3$	$\frac{36}{x+3}$
Upstream	24	$x-3$	$\frac{24}{x-3}$

Both times are the same.

$$\frac{36}{x+3} = \frac{24}{x-3}$$

Multiply each side by the LCD, $(x-3)(x+3)$.

$$\begin{aligned} (x-3)(x+3)\left(\frac{36}{x+3}\right) &= (x-3)(x+3)\left(\frac{24}{x-3}\right) \\ 36(x-3) &= 24(x+3) \\ 36x - 108 &= 24x + 72 \\ 12x &= 180 \\ x &= 15 \end{aligned}$$

The speed of the boat in still water is 15 mph.

38. Let $x =$ the amount of pure alcohol.

Solution	Strength	Amount	Pure Alcohol
#1	100%	x	$1 \cdot x = x$
#2	18%	40	$0.18(40) = 7.2$
Mixture	22%	$x + 40$	$0.22(x+40)$

The last column gives the equation.

$$\begin{aligned} x + 7.2 &= 0.22(x+40) \\ x + 7.2 &= 0.22x + 8.8 \\ 0.78x &= 1.6 \\ x &= \frac{1.6}{0.78} = \frac{160}{78} = \frac{80}{39}, \text{ or } 2\frac{2}{39} \end{aligned}$$

The required amount is $\frac{80}{39}$ or $2\frac{2}{39}$ L.

39. Let $x =$ the number of dimes.
Then $29 - x =$ the number of quarters.

The total value of the coins is \$4.70, so

$$0.10x + 0.25(29 - x) = 4.70.$$

Multiply by 100 to clear the decimals.

$$\begin{aligned} 10x + 25(29 - x) &= 470 \\ 10x + 725 - 25x &= 470 \\ -15x &= -255 \\ x &= 17 \end{aligned}$$

Since $x = 17$, $29 - x = 29 - 17 = 12$.

There are 17 dimes and 12 quarters.

40. Let $x =$ Chuck's speed.
Then $x + 4 =$ Brenda's speed.

Make a table. Use $d = rt$, or $t = \frac{d}{r}$.

	Distance	Rate	Time
Chuck	24	x	$\frac{24}{x}$
Brenda	48	$x + 4$	$\frac{48}{x+4}$

The times are the same.

$$\frac{24}{x} = \frac{48}{x+4}$$

Multiply each side by the LCD, $x(x+4)$.

$$x(x+4)\left(\frac{24}{x}\right) = x(x+4)\left(\frac{48}{x+4}\right)$$
$$24(x+4) = 48x$$
$$24x + 96 = 48x$$
$$-24x = -96$$
$$x = 4$$

Chuck's speed is 4 mph and Brenda's speed is $x + 4 = 4 + 4 = 8$ mph.

CHAPTER 10 QUADRATIC EQUATIONS, INEQUALITIES, AND FUNCTIONS

10.1 The Square Root Property and Completing the Square

10.1 Margin Exercises

1. **(a)** **B** and **C** are quadratic equations because they are equations with a squared term and no terms of higher degree.

   (b) Equation **B**, $x^2 - 8x + 16 = 0$, is in standard form, $ax^2 + bx + c = 0$.

2. **(a)** $x^2 + 3x + 2 = 0$
   $(x+1)(x+2) = 0$

   $x + 1 = 0$ or $x + 2 = 0$
   $x = -1$ or $x = -2$

   Solution set: $\{-2, -1\}$

   (b) $3m^2 = 3 - 8m$
   $3m^2 + 8m - 3 = 0$
   $(m+3)(3m-1) = 0$

   $m + 3 = 0$ or $3m - 1 = 0$
   $m = -3$ or $m = \frac{1}{3}$

   Solution set: $\{-3, \frac{1}{3}\}$

3. **(a)** $m^2 = 64 \qquad (64 = 8^2)$
   By the square root property,

   $m = 8$ or $m = -8$.

   Solution set: $\{-8, 8\}$

   (b) $p^2 = 7$
   By the square root property,

   $p = \sqrt{7}$ or $p = -\sqrt{7}$.

   Solution set: $\{-\sqrt{7}, \sqrt{7}\}$

   (c) $3x^2 - 54 = 0$
   $3x^2 = 54$
   $x^2 = 18$
   By the square root property,

   $x = \sqrt{18}$ or $x = -\sqrt{18}$.
   $x = 3\sqrt{2}$ or $x = -3\sqrt{2}$

   Solution set: $\{3\sqrt{2}, -3\sqrt{2}\}$

4. $d = 16t^2$
   $4 = 16t^2 \qquad$ *Let d = 4.*
   $\frac{1}{4} = t^2$
   By the square root property,

   $t = \frac{1}{2}$ or $t = -\frac{1}{2}$.

   Since time cannot be negative, we discard the negative solution. Therefore, 0.5 second elapses between the dropping of the coin and the shot.

5. **(a)** $(x-3)^2 = 25$
   By the square root property,

   $x - 3 = 5$ or $x - 3 = -5$
   $x = 8$ or $x = -2$.

   Solution set: $\{-2, 8\}$

   (b) $(3k+1)^2 = 2$

   $3k + 1 = \sqrt{2}$ or $3k + 1 = -\sqrt{2}$
   $3k = -1 + \sqrt{2}$ or $3k = -1 - \sqrt{2}$
   $k = \dfrac{-1+\sqrt{2}}{3}$ or $k = \dfrac{-1-\sqrt{2}}{3}$

   Solution set: $\left\{\dfrac{-1+\sqrt{2}}{3}, \dfrac{-1-\sqrt{2}}{3}\right\}$

   (c) $(2r+3)^2 = 8$

   $2r + 3 = \sqrt{8}$ or $2r + 3 = -\sqrt{8}$
   $2r + 3 = 2\sqrt{2}$ or $2r + 3 = -2\sqrt{2}$
   $2r = -3 + 2\sqrt{2}$ or $2r = -3 - 2\sqrt{2}$
   $r = \dfrac{-3+2\sqrt{2}}{2}$ or $r = \dfrac{-3-2\sqrt{2}}{2}$

   Solution set: $\left\{\dfrac{-3+2\sqrt{2}}{2}, \dfrac{-3-2\sqrt{2}}{2}\right\}$

6. **(a)** $x^2 + 4x + \underline{4}$ since $\left[\frac{1}{2}(4)\right]^2 = 2^2 = 4$

   (b) $t^2 - 2t + \underline{1}$ since $\left[\frac{1}{2}(-2)\right]^2 = (-1)^2 = 1$

   (c) $m^2 + 5m + \underline{\frac{25}{4}}$ since $\left[\frac{1}{2}(5)\right]^2 = (\frac{5}{2})^2 = \frac{25}{4}$

   (d) $x^2 - \frac{2}{3}x + \underline{\frac{1}{9}}$ since $\left[\frac{1}{2}(-\frac{2}{3})\right]^2 = (-\frac{1}{3})^2 = \frac{1}{9}$

7. $n^2 + 6n + 4 = 0$

$$n^2 + 6n = -4$$

Complete the square. Take half of 6, the coefficient of n, and square the result.

$$\left[\tfrac{1}{2}(6)\right]^2 = 3^2 = 9$$

Add 9 to each side of the equation.

$$n^2 + 6n + 9 = -4 + 9$$
$$(n+3)^2 = 5$$

Use the square root property.

$$n + 3 = \sqrt{5} \quad \text{or} \quad n + 3 = -\sqrt{5}$$
$$n = -3 + \sqrt{5} \quad \text{or} \quad n = -3 - \sqrt{5}$$

Solution set: $\{-3+\sqrt{5}, -3-\sqrt{5}\}$

8. **(a)** $x^2 + 2x - 10 = 0$

$$x^2 + 2x = 10$$

Complete the square.

$$\left[\tfrac{1}{2}(2)\right]^2 = 1^2 = 1$$

Add 1 to each side.

$$x^2 + 2x + 1 = 10 + 1$$
$$(x+1)^2 = 11$$

Use the square root property.

$$x + 1 = \sqrt{11} \quad \text{or} \quad x + 1 = -\sqrt{11}$$
$$x = -1 + \sqrt{11} \quad \text{or} \quad x = -1 - \sqrt{11}$$

Solution set: $\{-1+\sqrt{11}, -1-\sqrt{11}\}$

(b) $r^2 + 3r - 1 = 0$

$$r^2 + 3r = 1$$

Complete the square.

$$\left[\tfrac{1}{2}(3)\right]^2 = (\tfrac{3}{2})^2 = \tfrac{9}{4}$$

Add $\frac{9}{4}$ to each side.

$$r^2 + 3r + \tfrac{9}{4} = 1 + \tfrac{9}{4}$$
$$(r + \tfrac{3}{2})^2 = \tfrac{13}{4}$$

Use the square root property.

$$r + \frac{3}{2} = \sqrt{\frac{13}{4}} \quad \text{or} \quad r + \frac{3}{2} = -\sqrt{\frac{13}{4}}$$
$$r + \frac{3}{2} = \frac{\sqrt{13}}{2} \quad \text{or} \quad r + \frac{3}{2} = -\frac{\sqrt{13}}{2}$$
$$r = \frac{-3+\sqrt{13}}{2} \quad \text{or} \quad r = \frac{-3-\sqrt{13}}{2}$$

Solution set: $\left\{\frac{-3+\sqrt{13}}{2}, \frac{-3-\sqrt{13}}{2}\right\}$

9. **(a)** $2r^2 - 4r + 1 = 0$

$$2r^2 - 4r = -1$$
$$r^2 - 2r = -\tfrac{1}{2}$$

Complete the square.

$$\left[\tfrac{1}{2}(-2)\right]^2 = (-1)^2 = 1$$

Add 1 to each side.

$$r^2 - 2r + 1 = -\tfrac{1}{2} + 1$$
$$(r-1)^2 = \tfrac{1}{2}$$

Use the square root property.

$$r - 1 = \sqrt{\frac{1}{2}} \quad \text{or} \quad r - 1 = -\sqrt{\frac{1}{2}}$$
$$r - 1 = \frac{\sqrt{2}}{2} \quad \text{or} \quad r - 1 = -\frac{\sqrt{2}}{2}$$
$$r = 1 + \frac{\sqrt{2}}{2} \quad \text{or} \quad r = 1 - \frac{\sqrt{2}}{2}$$
$$r = \frac{2+\sqrt{2}}{2} \quad \text{or} \quad r = \frac{2-\sqrt{2}}{2}$$

Solution set: $\left\{\frac{2+\sqrt{2}}{2}, \frac{2-\sqrt{2}}{2}\right\}$

(b) $3z^2 - 6z - 2 = 0$

$$3z^2 - 6z = 2$$
$$z^2 - 2z = \tfrac{2}{3}$$

Complete the square.

$$\left[\tfrac{1}{2}(-2)\right]^2 = (-1)^2 = 1$$

Add 1 to each side.

$$z^2 - 2z + 1 = \tfrac{2}{3} + 1$$
$$(z-1)^2 = \tfrac{5}{3}$$

Use the square root property.

$$z - 1 = \sqrt{\frac{5}{3}} \quad \text{or} \quad z - 1 = -\sqrt{\frac{5}{3}}$$
$$z - 1 = \frac{\sqrt{15}}{3} \quad \text{or} \quad z - 1 = -\frac{\sqrt{15}}{3}$$
$$z = 1 + \frac{\sqrt{15}}{3} \quad \text{or} \quad z = 1 - \frac{\sqrt{15}}{3}$$
$$z = \frac{3+\sqrt{15}}{3} \quad \text{or} \quad z = \frac{3-\sqrt{15}}{3}$$

Solution set: $\left\{\frac{3+\sqrt{15}}{3}, \frac{3-\sqrt{15}}{3}\right\}$

(c) $8x^2 - 4x - 2 = 0$

$$8x^2 - 4x = 2$$
$$x^2 - \tfrac{1}{2}x = \tfrac{1}{4}$$

Complete the square.

$$\left[\frac{1}{2}\left(-\frac{1}{2}\right)\right]^2 = \left(-\frac{1}{4}\right)^2 = \frac{1}{16}$$

Add $\frac{1}{16}$ to each side.

$$x^2 - \frac{1}{2}x + \frac{1}{16} = \frac{1}{4} + \frac{1}{16}$$
$$\left(x - \frac{1}{4}\right)^2 = \frac{5}{16}$$

$$x - \frac{1}{4} = \sqrt{\frac{5}{16}} \quad \text{or} \quad x - \frac{1}{4} = -\sqrt{\frac{5}{16}}$$
$$x - \frac{1}{4} = \frac{\sqrt{5}}{4} \quad \text{or} \quad x - \frac{1}{4} = -\frac{\sqrt{5}}{4}$$
$$x = \frac{1 + \sqrt{5}}{4} \quad \text{or} \quad x = \frac{1 - \sqrt{5}}{4}$$

Solution set: $\left\{\frac{1+\sqrt{5}}{4}, \frac{1-\sqrt{5}}{4}\right\}$

10. **(a)** $x^2 = -17$

$$x = \sqrt{-17} \quad \text{or} \quad x = -\sqrt{-17}$$
$$x = i\sqrt{17} \quad \text{or} \quad x = -i\sqrt{17}$$

Solution set: $\{i\sqrt{17}, -i\sqrt{17}\}$

(b) $(k+5)^2 = -100$

$$k + 5 = \sqrt{-100} \quad \text{or} \quad k + 5 = -\sqrt{-100}$$
$$k + 5 = 10i \quad \text{or} \quad k + 5 = -10i$$
$$k = -5 + 10i \quad \text{or} \quad k = -5 - 10i$$

Solution set: $\{-5 + 10i, -5 - 10i\}$

(c) $5t^2 - 15t + 12 = 0$

$$5t^2 - 15t = -12$$
$$t^2 - 3t = -\frac{12}{5}$$

Complete the square.

$$\left[\frac{1}{2}(-3)\right]^2 = \left(-\frac{3}{2}\right)^2 = \frac{9}{4}$$

Add $\frac{9}{4}$ to each side.

$$t^2 - 3t + \frac{9}{4} = -\frac{12}{5} + \frac{9}{4}$$
$$\left(t - \frac{3}{2}\right)^2 = -\frac{3}{20}$$

$$t - \frac{3}{2} = \sqrt{-\frac{3}{20}} \quad \text{or} \quad t - \frac{3}{2} = -\sqrt{-\frac{3}{20}}$$
$$t - \frac{3}{2} = \frac{i\sqrt{3}}{\sqrt{20}} \cdot \frac{\sqrt{5}}{\sqrt{5}} \quad \text{or} \quad t - \frac{3}{2} = \frac{-i\sqrt{3}}{\sqrt{20}} \cdot \frac{\sqrt{5}}{\sqrt{5}}$$
$$t - \frac{3}{2} = \frac{i\sqrt{15}}{10} \quad \text{or} \quad t - \frac{3}{2} = \frac{-i\sqrt{15}}{10}$$
$$t = \frac{3}{2} + \frac{\sqrt{15}}{10}i \quad \text{or} \quad t = \frac{3}{2} - \frac{\sqrt{15}}{10}i$$

Solution set: $\left\{\frac{3}{2} + \frac{\sqrt{15}}{10}i, \frac{3}{2} - \frac{\sqrt{15}}{10}i\right\}$

10.1 Section Exercises

1. By the square root property, if $x^2 = 16$, then

$$x = +\sqrt{16} \quad \text{or} \quad x = -\sqrt{16}.$$

Thus, the equation is also true for $x = -4$.

3. **(a)** A quadratic equation in standard form has a second-degree polynomial in decreasing powers equal to 0.

(b) The zero-factor property states that if a product equals 0, then at least one of the factors equals 0.

(c) The square root property states that if the square of a quantity equals a number, then the quantity equals the positive or negative square root of the number.

5. $x^2 = 81$

$$x = 9 \quad \text{or} \quad x = -9$$

Solution set: $\{9, -9\}$

7. $t^2 = 17$

$$t = \sqrt{17} \quad \text{or} \quad t = -\sqrt{17}$$

Solution set: $\{\sqrt{17}, -\sqrt{17}\}$

9. $m^2 = 32$

$$m = \sqrt{32} \quad \text{or} \quad m = -\sqrt{32}$$
$$m = 4\sqrt{2} \quad \text{or} \quad m = -4\sqrt{2}$$

Solution set: $\{4\sqrt{2}, -4\sqrt{2}\}$

11. $t^2 - 20 = 0$

$$t^2 = 20$$
$$t = \sqrt{20} \quad \text{or} \quad t = -\sqrt{20}$$
$$t = 2\sqrt{5} \quad \text{or} \quad t = -2\sqrt{5}$$

Solution set: $\{2\sqrt{5}, -2\sqrt{5}\}$

13. $3n^2 - 72 = 0$

$$3n^2 = 72$$
$$n = 24$$
$$n = \sqrt{24} \quad \text{or} \quad n = -\sqrt{24}$$
$$n = 2\sqrt{6} \quad \text{or} \quad n = -2\sqrt{6}$$

Solution set: $\{2\sqrt{6}, -2\sqrt{6}\}$

15. $(x+2)^2 = 25$

$$x + 2 = \sqrt{25} \quad \text{or} \quad x + 2 = -\sqrt{25}$$
$$x + 2 = 5 \quad \text{or} \quad x + 2 = -5$$
$$x = 3 \quad \text{or} \quad x = -7$$

Solution set: $\{-7, 3\}$

17. $(x-4)^2 = 3$

$x-4=\sqrt{3}$ or $x-4=-\sqrt{3}$

$x=4+\sqrt{3}$ or $x=4-\sqrt{3}$

Solution set: $\{4+\sqrt{3}, 4-\sqrt{3}\}$

19. $(t+5)^2 = 48$

$t+5=\sqrt{48}$ or $t+5=-\sqrt{48}$

$t+5=4\sqrt{3}$ or $t+5=-4\sqrt{3}$

$t=-5+4\sqrt{3}$ or $t=-5-4\sqrt{3}$

Solution set: $\{-5+4\sqrt{3}, -5-4\sqrt{3}\}$

21. $(3k-1)^2 = 7$

$3k-1=\sqrt{7}$ or $3k-1=-\sqrt{7}$

$3k=1+\sqrt{7}$ or $3k=1-\sqrt{7}$

$k=\frac{1+\sqrt{7}}{3}$ or $k=\frac{1-\sqrt{7}}{3}$

Solution set: $\left\{\frac{1+\sqrt{7}}{3}, \frac{1-\sqrt{7}}{3}\right\}$

23. $(4p+1)^2 = 24$

$4p+1=\sqrt{24}$ or $4p+1=-\sqrt{24}$

$4p+1=2\sqrt{6}$ or $4p+1=-2\sqrt{6}$

$4p=-1+2\sqrt{6}$ or $4p=-1-2\sqrt{6}$

$p=\frac{-1+2\sqrt{6}}{4}$ or $p=\frac{-1-2\sqrt{6}}{4}$

Solution set: $\left\{\frac{-1+2\sqrt{6}}{4}, \frac{-1-2\sqrt{6}}{4}\right\}$

25. $d=16t^2$

$500=16t^2$ *Let d = 500.*

$t^2=\frac{500}{16}=31.25$

$t=\sqrt{31.25}$ or $t=-\sqrt{31.25}$

$t\approx 5.6$ or $t\approx -5.6$

Since time cannot be negative, -5.6 is discarded. Therefore, it takes the object about 5.6 seconds to fall 500 feet.

27. To solve $(2x+1)^2=5$, we use the square root property and find that $2x+1=\sqrt{5}$ or $2x+1=-\sqrt{5}$. Then we solve for x in both equations.
To solve $x^2+4x=12$, we find that

$$x^2+4x+4=12+4$$
$$(x+2)^2=16$$

by completing the square.

29. **(a)** We need to add the square of half the coefficient of x to get a perfect square trinomial.

$\frac{1}{2}(6)=3$ and $3^2=\underline{9}$

Add 9 to x^2+6x to get a perfect square trinomial.

(b) $\frac{1}{2}(14)=7$ and $7^2=\underline{49}$

(c) $\frac{1}{2}(-12)=-6$ and $(-6)^2=\underline{36}$

(d) $\frac{1}{2}(3)=\frac{3}{2}$ and $(\frac{3}{2})^2=\underline{\frac{9}{4}}$

(e) $\frac{1}{2}(-9)=-\frac{9}{2}$ and $(-\frac{9}{2})^2=\underline{\frac{81}{4}}$

(f) $\frac{1}{2}(-\frac{1}{2})=-\frac{1}{4}$ and $(-\frac{1}{4})^2=\underline{\frac{1}{16}}$

31. $x^2+4x-2=0$

$x^2+4x=2$

$\left[\frac{1}{2}(4)\right]^2=2^2=4$

33. $x^2+10x+18=0$

$x^2+10x=-18$

$\left[\frac{1}{2}(10)\right]^2=5^2=25$

35. $3w^2-w-24=0$

$w^2-\frac{1}{3}w-8=0$ *Divide by 3.*

$w^2-\frac{1}{3}w=8$

$\left[\frac{1}{2}(-\frac{1}{3})\right]^2=(-\frac{1}{6})^2=\frac{1}{36}$

37. $x^2-2x-24=0$

Get the variable terms alone on the left side.

$$x^2-2x=24$$

Complete the square by taking half of -2, the coefficient of x, and squaring the result.

$$\left[\frac{1}{2}(-2)\right]^2=(-1)^2=1$$

Add 1 to each side.

$$x^2-2x+1=24+1$$

Factor the left side.

$$(x-1)^2=25$$

Use the square root property.

$x-1=\sqrt{25}$ or $x-1=-\sqrt{25}$

$x-1=5$ or $x-1=-5$

$x=6$ or $x=-4$

Solution set: $\{-4, 6\}$

39. $x^2 + 4x - 2 = 0$

$x^2 + 4x = 2$

$x^2 + 4x + 4 = 2 + 4 \qquad \left[\frac{1}{2}(4)\right]^2 = 4$

$(x+2)^2 = 6$

$x + 2 = \sqrt{6}$ or $x + 2 = -\sqrt{6}$

$x = -2 + \sqrt{6}$ or $x = -2 - \sqrt{6}$

Solution set: $\{-2 + \sqrt{6}, -2 - \sqrt{6}\}$

41. $x^2 + 10x + 18 = 0$

$x^2 + 10x = -18$

$x^2 + 10x + 25 = -18 + 25 \qquad \left[\frac{1}{2}(10)\right]^2 = 25$

$(x+5)^2 = 7$

$x + 5 = \sqrt{7}$ or $x + 5 = -\sqrt{7}$

$x = -5 + \sqrt{7}$ or $x = -5 - \sqrt{7}$

Solution set: $\{-5 + \sqrt{7}, -5 - \sqrt{7}\}$

43. $3w^2 - w = 24$

$w^2 - \frac{1}{3}w = 8$ *Divide by 3.*

By Exercise 35, add $\frac{1}{36}$ to each side.

$w^2 - \frac{1}{3}w + \frac{1}{36} = 8 + \frac{1}{36}$

$(w - \frac{1}{6})^2 = \frac{288}{36} + \frac{1}{36}$

$(w - \frac{1}{6})^2 = \frac{289}{36}$

$w - \frac{1}{6} = \sqrt{\frac{289}{36}}$ or $w - \frac{1}{6} = -\sqrt{\frac{289}{36}}$

$w = \frac{1}{6} + \frac{\sqrt{289}}{\sqrt{36}}$ or $w = \frac{1}{6} - \frac{\sqrt{289}}{\sqrt{36}}$

$w = \frac{1}{6} + \frac{17}{6}$ or $w = \frac{1}{6} - \frac{17}{6}$

$w = \frac{18}{6}$ or $w = -\frac{16}{6}$

$w = 3$ or $w = -\frac{8}{3}$

Solution set: $\{-\frac{8}{3}, 3\}$

45. $2k^2 + 5k - 2 = 0$

$2k^2 + 5k = 2$

$k^2 + \frac{5}{2}k = 1$ *Divide by 2.*

Complete the square.

$\left[\frac{1}{2}(\frac{5}{2})\right]^2 = (\frac{5}{4})^2 = \frac{25}{16}$

Add $\frac{25}{16}$ to each side.

$k^2 + \frac{5}{2}k + \frac{25}{16} = 1 + \frac{25}{16}$

$(k + \frac{5}{4})^2 = \frac{41}{16}$

$k + \frac{5}{4} = \sqrt{\frac{41}{16}}$ or $k + \frac{5}{4} = -\sqrt{\frac{41}{16}}$

$k = -\frac{5}{4} + \frac{\sqrt{41}}{4}$ or $k = -\frac{5}{4} - \frac{\sqrt{41}}{4}$

$k = \frac{-5 + \sqrt{41}}{4}$ or $k = \frac{-5 - \sqrt{41}}{4}$

Solution set: $\left\{\frac{-5 + \sqrt{41}}{4}, \frac{-5 - \sqrt{41}}{4}\right\}$

47. $5x^2 - 10x + 2 = 0$

$5x^2 - 10x = -2$

$x^2 - 2x = -\frac{2}{5}$ *Divide by 5.*

Complete the square.

$\left[\frac{1}{2}(-2)\right]^2 = (-1)^2 = 1$

Add 1 to each side.

$x^2 - 2x + 1 = -\frac{2}{5} + 1$

$(x-1)^2 = \frac{3}{5}$

$x - 1 = \sqrt{\frac{3}{5}}$ or $x - 1 = -\sqrt{\frac{3}{5}}$

$x - 1 = \frac{\sqrt{3}}{\sqrt{5}} \cdot \frac{\sqrt{5}}{\sqrt{5}}$ or $x - 1 = -\frac{\sqrt{3}}{\sqrt{5}} \cdot \frac{\sqrt{5}}{\sqrt{5}}$

$x = \frac{5}{5} + \frac{\sqrt{15}}{5}$ or $x = \frac{5}{5} - \frac{\sqrt{15}}{5}$

$x = \frac{5 + \sqrt{15}}{5}$ or $x = \frac{5 - \sqrt{15}}{5}$

Solution set: $\left\{\frac{5 + \sqrt{15}}{5}, \frac{5 - \sqrt{15}}{5}\right\}$

49. $9x^2 - 24x = -13$

$x^2 - \frac{24}{9}x = \frac{-13}{9}$ *Divide by 9.*

$x^2 - \frac{8}{3}x = \frac{-13}{9}$

Complete the square.

$\left[\frac{1}{2}(-\frac{8}{3})\right]^2 = (-\frac{4}{3})^2 = \frac{16}{9}$

Add $\frac{16}{9}$ to each side.

$x^2 - \frac{8}{3}x + \frac{16}{9} = \frac{-13}{9} + \frac{16}{9}$

$(x - \frac{4}{3})^2 = \frac{3}{9}$

$x - \frac{4}{3} = \sqrt{\frac{3}{9}}$ or $x - \frac{4}{3} = -\sqrt{\frac{3}{9}}$

$x = \frac{4}{3} + \frac{\sqrt{3}}{3}$ or $x = \frac{4}{3} - \frac{\sqrt{3}}{3}$

$x = \frac{4 + \sqrt{3}}{3}$ or $x = \frac{4 - \sqrt{3}}{3}$

Solution set: $\left\{\frac{4 + \sqrt{3}}{3}, \frac{4 - \sqrt{3}}{3}\right\}$

51. $z^2 - \frac{4}{3}z = -\frac{1}{9}$

Complete the square.

$\left[\frac{1}{2}(-\frac{4}{3})\right]^2 = (-\frac{2}{3})^2 = \frac{4}{9}$

Add $\frac{4}{9}$ to each side.

$z^2 - \frac{4}{3}z + \frac{4}{9} = -\frac{1}{9} + \frac{4}{9}$

$(z - \frac{2}{3})^2 = \frac{3}{9}$

$z - \dfrac{2}{3} = \sqrt{\dfrac{3}{9}}$ or $z - \dfrac{2}{3} = -\sqrt{\dfrac{3}{9}}$

$z = \dfrac{2}{3} + \dfrac{\sqrt{3}}{3}$ or $z = \dfrac{2}{3} - \dfrac{\sqrt{3}}{3}$

$z = \dfrac{2+\sqrt{3}}{3}$ or $z = \dfrac{2-\sqrt{3}}{3}$

Solution set: $\left\{\dfrac{2+\sqrt{3}}{3}, \dfrac{2-\sqrt{3}}{3}\right\}$

53. $0.1x^2 - 0.2x - 0.1 = 0$

Multiply each side by 10 to clear the decimals.

$x^2 - 2x - 1 = 0$

$x^2 - 2x = 1$

Complete the square.

$\left[\frac{1}{2}(-2)\right]^2 = (-1)^2 = 1$

Add 1 to each side.

$x^2 - 2x + 1 = 1 + 1$

$(x - 1)^2 = 2$

$x - 1 = \sqrt{2}$ or $x - 1 = -\sqrt{2}$

$x = 1 + \sqrt{2}$ or $x = 1 - \sqrt{2}$

Solution set: $\{1 + \sqrt{2}, 1 - \sqrt{2}\}$

55. $x^2 = -12$

$x = \sqrt{-12}$ or $x = -\sqrt{-12}$

$x = i\sqrt{12}$ or $x = -i\sqrt{12}$

$x = 2i\sqrt{3}$ or $x = -2i\sqrt{3}$

Solution set: $\{2i\sqrt{3}, -2i\sqrt{3}\}$

57. $(r - 5)^2 = -3$

$r - 5 = \sqrt{-3}$ or $r - 5 = -\sqrt{-3}$

$r = 5 + i\sqrt{3}$ or $r = 5 - i\sqrt{3}$

Solution set: $\{5 + i\sqrt{3}, 5 - i\sqrt{3}\}$

59. $(6k - 1)^2 = -8$

$6k - 1 = \sqrt{-8}$ or $6k - 1 = -\sqrt{-8}$

$6k - 1 = i\sqrt{8}$ or $6k - 1 = -i\sqrt{8}$

$6k - 1 = 2i\sqrt{2}$ or $6k - 1 = -2i\sqrt{2}$

$6k = 1 + 2i\sqrt{2}$ or $6k = 1 - 2i\sqrt{2}$

$k = \dfrac{1 + 2i\sqrt{2}}{6}$ or $k = \dfrac{1 - 2i\sqrt{2}}{6}$

$k = \dfrac{1}{6} + \dfrac{\sqrt{2}}{3}i$ or $k = \dfrac{1}{6} - \dfrac{\sqrt{2}}{3}i$

Solution set: $\left\{\dfrac{1}{6} + \dfrac{\sqrt{2}}{3}i, \dfrac{1}{6} - \dfrac{\sqrt{2}}{3}i\right\}$

61. $m^2 + 4m + 13 = 0$

$m^2 + 4m = -13$

$m^2 + 4m + 4 = -13 + 4$ $\quad\left[\frac{1}{2}(4)\right]^2 = 2^2 = 4$

$(m + 2)^2 = -9$

$m + 2 = \sqrt{-9}$ or $m + 2 = -\sqrt{-9}$

$m = -2 + 3i$ or $m = -2 - 3i$

Solution set: $\{-2 + 3i, -2 - 3i\}$

63. $3r^2 + 4r + 4 = 0$

$3r^2 + 4r = -4$

$r^2 + \frac{4}{3}r = \frac{-4}{3}$ *Divide by 3.*

Complete the square.

$\left[\frac{1}{2}(\frac{4}{3})\right]^2 = (\frac{2}{3})^2 = \frac{4}{9}$

Add $\frac{4}{9}$ to each side.

$r^2 + \frac{4}{3}r + \frac{4}{9} = \frac{-4}{3} + \frac{4}{9}$

$(r + \frac{2}{3})^2 = \frac{-8}{9}$

$r + \dfrac{2}{3} = \dfrac{\sqrt{-8}}{\sqrt{9}}$ or $r + \dfrac{2}{3} = -\dfrac{\sqrt{-8}}{\sqrt{9}}$

$r = -\dfrac{2}{3} + \dfrac{2i\sqrt{2}}{3}$ or $r = -\dfrac{2}{3} - \dfrac{2i\sqrt{2}}{3}$

$r = -\dfrac{2}{3} + \dfrac{2\sqrt{2}}{3}i$ or $r = -\dfrac{2}{3} - \dfrac{2\sqrt{2}}{3}i$

Solution set: $\left\{-\dfrac{2}{3} + \dfrac{2\sqrt{2}}{3}i, -\dfrac{2}{3} - \dfrac{2\sqrt{2}}{3}i\right\}$

65. $-m^2 - 6m - 12 = 0$

Multiply each side by -1.

$$m^2 + 6m + 12 = 0$$
$$m^2 + 6m = -12$$

Complete the square.

$$\left[\tfrac{1}{2}(6)\right]^2 = 3^2 = 9$$

Add 9 to each side.

$$m^2 + 6m + 9 = -12 + 9$$
$$(m+3)^2 = -3$$

$$m + 3 = \sqrt{-3} \quad \text{or} \quad m + 3 = -\sqrt{-3}$$
$$m = -3 + i\sqrt{3} \quad \text{or} \quad m = -3 - i\sqrt{3}$$

Solution set: $\{-3 + i\sqrt{3}, -3 - i\sqrt{3}\}$

67. The area of the original square is $x \cdot x$, or x^2.

68. Each rectangular strip has length x and width 1, so each strip has an area of $x \cdot 1$, or x.

69. From Exercise 68, the area of a rectangular strip is x. The area of 6 rectangular strips is $6x$.

70. These are 1 by 1 squares, so each has an area of $1 \cdot 1$, or 1.

71. There are 9 small squares, each with area 1 (from Exercise 70), so the total area is $9 \cdot 1$, or 9.

72. The area of the larger square is $(x+3)^2$. Using the results from Exercises 67–71,

$$(x+3)^2 = x^2 + 6x + 9.$$

10.2 The Quadratic Formula

10.2 Margin Exercises

1. **(a)** $-3x^2 + 9x - 4 = 0$

Here $a = -3$, $b = 9$, and $c = -4$.

(b) $3x^2 = 6x + 2$

$$3x^2 - 6x - 2 = 0$$

Here $a = 3$, $b = -6$, and $c = -2$.

2. $4x^2 - 11x - 3 = 0$

Here $a = 4$, $b = -11$, and $c = -3$.

$$x = \frac{-b \pm \sqrt{b^2 - 4ac}}{2a}$$
$$x = \frac{-(-11) \pm \sqrt{(-11)^2 - 4(4)(-3)}}{2(4)}$$
$$x = \frac{11 \pm \sqrt{121 + 48}}{8}$$
$$x = \frac{11 \pm \sqrt{169}}{8}$$
$$x = \frac{11 \pm 13}{8}$$
$$x = \frac{11+13}{8} = \frac{24}{8} = 3$$
$$\text{or} \quad x = \frac{11-13}{8} = \frac{-2}{8} = -\frac{1}{4}$$

Solution set: $\{-\frac{1}{4}, 3\}$

3. **(a)** $6x^2 + 4x - 1 = 0$

Here $a = 6$, $b = 4$, and $c = -1$.

$$x = \frac{-b \pm \sqrt{b^2 - 4ac}}{2a}$$
$$x = \frac{-4 \pm \sqrt{4^2 - 4(6)(-1)}}{2(6)}$$
$$= \frac{-4 \pm \sqrt{16 + 24}}{12}$$
$$= \frac{-4 \pm \sqrt{40}}{12}$$
$$= \frac{-4 \pm 2\sqrt{10}}{12}$$
$$= \frac{2(-2 \pm \sqrt{10})}{12}$$
$$= \frac{-2 \pm \sqrt{10}}{6}$$

Solution set: $\left\{\frac{-2 + \sqrt{10}}{6}, \frac{-2 - \sqrt{10}}{6}\right\}$

(b)

$$\begin{aligned} 2x^2 + 19 &= 14x \\ 2x^2 - 14x + 19 &= 0 \end{aligned}$$

Here $a = 2,\ b = -14,$ and $c = 19.$

$$\begin{aligned} x &= \frac{-b \pm \sqrt{b^2 - 4ac}}{2a} \\ x &= \frac{-(-14) \pm \sqrt{(-14)^2 - 4(2)(19)}}{2(2)} \\ &= \frac{14 \pm \sqrt{196 - 152}}{4} \\ &= \frac{14 \pm \sqrt{44}}{4} \\ &= \frac{14 \pm 2\sqrt{11}}{4} \\ &= \frac{2(7 \pm \sqrt{11})}{4} \\ &= \frac{7 \pm \sqrt{11}}{2} \end{aligned}$$

Solution set: $\left\{\frac{7+\sqrt{11}}{2}, \frac{7-\sqrt{11}}{2}\right\}$

4. **(a)** $x^2 + x + 1 = 0$

Here $a = 1,\ b = 1,$ and $c = 1.$

$$\begin{aligned} x &= \frac{-b \pm \sqrt{b^2 - 4ac}}{2a} \\ x &= \frac{-1 \pm \sqrt{1^2 - 4(1)(1)}}{2(1)} \\ &= \frac{-1 \pm \sqrt{1 - 4}}{2} \\ &= \frac{-1 \pm \sqrt{-3}}{2} \\ &= \frac{-1}{2} \pm \frac{i\sqrt{3}}{2} \end{aligned}$$

Solution set: $\left\{-\frac{1}{2} + \frac{\sqrt{3}}{2}i, -\frac{1}{2} - \frac{\sqrt{3}}{2}i\right\}$

(b)

$$\begin{aligned} (x+2)(x-6) &= -17 \\ x^2 + 2x - 6x - 12 &= -17 \\ x^2 - 4x - 12 &= -17 \\ x^2 - 4x + 5 &= 0 \end{aligned}$$

Here $a = 1,\ b = -4,$ and $c = 5.$

$$\begin{aligned} x &= \frac{-b \pm \sqrt{b^2 - 4ac}}{2a} \\ x &= \frac{-(-4) \pm \sqrt{(-4)^2 - 4(1)(5)}}{2(1)} \\ &= \frac{4 \pm \sqrt{16 - 20}}{2} \\ &= \frac{4 \pm \sqrt{-4}}{2} \\ &= \frac{4 \pm 2i}{2} \\ &= \frac{2(2 \pm i)}{2} \\ &= 2 \pm i \end{aligned}$$

Solution set: $\{2 + i, 2 - i\}$

5. **(a)**

$$\begin{aligned} 2x^2 + 3x &= 4 \\ 2x^2 + 3x - 4 &= 0 \end{aligned}$$

Here $a = 2,\ b = 3,$ and $c = -4,$ so the discriminant is

$$\begin{aligned} b^2 - 4ac &= 3^2 - 4(2)(-4) \\ &= 9 + 32 \\ &= 41. \end{aligned}$$

The discriminant is positive but not a perfect square and a, b, and c are integers, so there are two different irrational solutions.

(b) $2x^2 + 3x + 4 = 0$

Here $a = 2,\ b = 3,$ and $c = 4,$ so the discriminant is

$$\begin{aligned} b^2 - 4ac &= 3^2 - 4(2)(4) \\ &= 9 - 32 \\ &= -23. \end{aligned}$$

The discriminant is negative and a, b, and c are integers, so there are two different nonreal complex solutions.

(c) $x^2 + 20x + 100 = 0$

Here $a = 1,\ b = 20,$ and $c = 100,$ so the discriminant is

$$\begin{aligned} b^2 - 4ac &= 20^2 - 4(1)(100) \\ &= 400 - 400 \\ &= 0. \end{aligned}$$

The discriminant is zero, so there is one rational solution.

(d)

$$\begin{aligned} 15x^2 + 11x &= 14 \\ 15x^2 + 11x - 14 &= 0 \end{aligned}$$

Here $a = 15,\ b = 11,$ and $c = -14,$ so the discriminant is

$$\begin{aligned} b^2 - 4ac &= 11^2 - 4(15)(-14) \\ &= 121 + 840 \\ &= 961 \quad \text{or} \quad 31^2. \end{aligned}$$

The discriminant is a perfect square and a, b, and c are integers, so there are two different rational solutions.

(e) If the discriminant is a perfect square (including 0), then the equation can be solved by factoring, so the answer is (c) and (d).

10.2 Section Exercises

1. If the expression $\dfrac{-b \pm \sqrt{b^2 - 4ac}}{2a}$ were written as two terms, we would have $-\dfrac{b}{2a} \pm \dfrac{\sqrt{b^2 - 4ac}}{2a}$.

Therefore,

$$\frac{-b \pm \sqrt{b^2 - 4ac}}{2a} \neq -b \pm \frac{\sqrt{b^2 - 4ac}}{2a}.$$

The student was incorrect, since the fraction bar should extend under the term $-b$.

3. $x^2 - 8x + 15 = 0$

Here $a = 1$, $b = -8$, and $c = 15$.

$$x = \frac{-b \pm \sqrt{b^2 - 4ac}}{2a}$$

$$x = \frac{-(-8) \pm \sqrt{(-8)^2 - 4(1)(15)}}{2(1)}$$

$$= \frac{8 \pm \sqrt{64 - 60}}{2}$$

$$= \frac{8 \pm \sqrt{4}}{2} = \frac{8 \pm 2}{2}$$

$$x = \frac{8+2}{2} = \frac{10}{2} = 5$$

$$\text{or} \quad x = \frac{8-2}{2} = \frac{6}{2} = 3$$

Solution set: $\{3, 5\}$

5. $2x^2 + 4x + 1 = 0$

Here $a = 2$, $b = 4$, and $c = 1$.

$$x = \frac{-b \pm \sqrt{b^2 - 4ac}}{2a}$$

$$x = \frac{-4 \pm \sqrt{4^2 - 4(2)(1)}}{2(2)}$$

$$= \frac{-4 \pm \sqrt{16 - 8}}{4}$$

$$= \frac{-4 \pm \sqrt{8}}{4} = \frac{-4 \pm 2\sqrt{2}}{4}$$

$$= \frac{2\left(-2 \pm \sqrt{2}\right)}{2 \cdot 2} = \frac{-2 \pm \sqrt{2}}{2}$$

Solution set: $\left\{\dfrac{-2 + \sqrt{2}}{2}, \dfrac{-2 - \sqrt{2}}{2}\right\}$

7. $2x^2 - 2x = 1$

$2x^2 - 2x - 1 = 0$

Here $a = 2$, $b = -2$, and $c = -1$.

$$x = \frac{-b \pm \sqrt{b^2 - 4ac}}{2a}$$

$$x = \frac{-(-2) \pm \sqrt{(-2)^2 - 4(2)(-1)}}{2(2)}$$

$$= \frac{2 \pm \sqrt{4 + 8}}{4} = \frac{2 \pm \sqrt{12}}{4}$$

$$= \frac{2 \pm 2\sqrt{3}}{4} = \frac{2\left(1 \pm \sqrt{3}\right)}{2 \cdot 2}$$

$$= \frac{1 \pm \sqrt{3}}{2}$$

Solution set: $\left\{\dfrac{1 + \sqrt{3}}{2}, \dfrac{1 - \sqrt{3}}{2}\right\}$

9. $x^2 + 18 = 10x$

$x^2 - 10x + 18 = 0$

Here $a = 1$, $b = -10$, and $c = 18$.

$$x = \frac{-b \pm \sqrt{b^2 - 4ac}}{2a}$$

$$x = \frac{-(-10) \pm \sqrt{(-10)^2 - 4(1)(18)}}{2(1)}$$

$$= \frac{10 \pm \sqrt{100 - 72}}{2} = \frac{10 \pm \sqrt{28}}{2}$$

$$= \frac{10 \pm 2\sqrt{7}}{2} = \frac{2\left(5 \pm \sqrt{7}\right)}{2}$$

$$= 5 \pm \sqrt{7}$$

Solution set: $\{5 + \sqrt{7}, 5 - \sqrt{7}\}$

11. $4k^2 + 4k - 1 = 0$

Here $a = 4$, $b = 4$, and $c = -1$.

$$k = \frac{-b \pm \sqrt{b^2 - 4ac}}{2a}$$

$$k = \frac{-4 \pm \sqrt{4^2 - 4(4)(-1)}}{2(4)}$$

$$= \frac{-4 \pm \sqrt{16 + 16}}{8} = \frac{-4 \pm \sqrt{32}}{8}$$

$$= \frac{-4 \pm 4\sqrt{2}}{8} = \frac{4(-1 \pm \sqrt{2})}{2 \cdot 4}$$

$$= \frac{-1 \pm \sqrt{2}}{2}$$

Solution set: $\left\{\dfrac{-1 + \sqrt{2}}{2}, \dfrac{-1 - \sqrt{2}}{2}\right\}$

13. $2 - 2x = 3x^2$

$0 = 3x^2 + 2x - 2$

Here $a = 3$, $b = 2$, and $c = -2$.

$$x = \frac{-b \pm \sqrt{b^2 - 4ac}}{2a}$$

$$x = \frac{-2 \pm \sqrt{2^2 - 4(3)(-2)}}{2(3)}$$

$$= \frac{-2 \pm \sqrt{4 + 24}}{6} = \frac{-2 \pm \sqrt{28}}{6}$$

$$= \frac{-2 \pm 2\sqrt{7}}{6} = \frac{2(-1 \pm \sqrt{7})}{2 \cdot 3}$$

$$= \frac{-1 \pm \sqrt{7}}{3}$$

Solution set: $\left\{\frac{-1 + \sqrt{7}}{3}, \frac{-1 - \sqrt{7}}{3}\right\}$

15. $\frac{x^2}{4} - \frac{x}{2} = 1$

$\frac{x^2}{4} - \frac{x}{2} - 1 = 0$

$x^2 - 2x - 4 = 0$ *Multiply by 4.*

Here $a = 1$, $b = -2$, and $c = -4$.

$$x = \frac{-b \pm \sqrt{b^2 - 4ac}}{2a}$$

$$x = \frac{-(-2) \pm \sqrt{(-2)^2 - 4(1)(-4)}}{2(1)}$$

$$= \frac{2 \pm \sqrt{4 + 16}}{2} = \frac{2 \pm \sqrt{20}}{2}$$

$$= \frac{2 \pm 2\sqrt{5}}{2} = 1 \pm \sqrt{5}$$

Solution set: $\{1 + \sqrt{5}, 1 - \sqrt{5}\}$

17. $-2t(t + 2) = -3$

$-2t^2 - 4t = -3$

$-2t^2 - 4t + 3 = 0$

Here $a = -2$, $b = -4$, and $c = 3$.

$$t = \frac{-b \pm \sqrt{b^2 - 4ac}}{2a}$$

$$t = \frac{-(-4) \pm \sqrt{(-4)^2 - 4(-2)(3)}}{2(-2)}$$

$$= \frac{4 \pm \sqrt{16 + 24}}{-4} = \frac{4 \pm \sqrt{40}}{-4}$$

$$= \frac{4 \pm 2\sqrt{10}}{-4} = \frac{2(2 \pm \sqrt{10})}{-2 \cdot 2}$$

$$= \frac{2 \pm \sqrt{10}}{-2} \cdot \frac{-1}{-1} = \frac{-2 \mp \sqrt{10}}{2}$$

$$= \frac{-2 \pm \sqrt{10}}{2}$$

Solution set: $\left\{\frac{-2 + \sqrt{10}}{2}, \frac{-2 - \sqrt{10}}{2}\right\}$

19. $(r - 3)(r + 5) = 2$

$r^2 + 2r - 15 = 2$

$r^2 + 2r - 17 = 0$

Here $a = 1$, $b = 2$, and $c = -17$.

$$r = \frac{-b \pm \sqrt{b^2 - 4ac}}{2a}$$

$$r = \frac{-2 \pm \sqrt{2^2 - 4(1)(-17)}}{2(1)}$$

$$= \frac{-2 \pm \sqrt{4 + 68}}{2} = \frac{-2 \pm \sqrt{72}}{2}$$

$$= \frac{-2 \pm 6\sqrt{2}}{2} = \frac{2(-1 \pm 3\sqrt{2})}{2}$$

$$= -1 \pm 3\sqrt{2}$$

Solution set: $\{-1 + 3\sqrt{2}, -1 - 3\sqrt{2}\}$

21. $x^2 - 3x + 17 = 0$

Here $a = 1$, $b = -3$, and $c = 17$.

$$x = \frac{-b \pm \sqrt{b^2 - 4ac}}{2a}$$

$$x = \frac{-(-3) \pm \sqrt{(-3)^2 - 4(1)(17)}}{2(1)}$$

$$= \frac{3 \pm \sqrt{9 - 68}}{2}$$

$$= \frac{3 \pm \sqrt{-59}}{2}$$

$$= \frac{3}{2} \pm \frac{i\sqrt{59}}{2}$$

Solution set: $\left\{\frac{3}{2} + \frac{\sqrt{59}}{2}i, \frac{3}{2} - \frac{\sqrt{59}}{2}i\right\}$

23. $r^2 - 6r + 14 = 0$

Here $a = 1$, $b = -6$, and $c = 14$.

$$r = \frac{-b \pm \sqrt{b^2 - 4ac}}{2a}$$

$$r = \frac{-(-6) \pm \sqrt{(-6)^2 - 4(1)(14)}}{2(1)}$$

$$= \frac{6 \pm \sqrt{36 - 56}}{2}$$

$$= \frac{6 \pm \sqrt{-20}}{2} = \frac{6 \pm 2i\sqrt{5}}{2}$$

$$= \frac{2(3 \pm i\sqrt{5})}{2} = 3 \pm i\sqrt{5}$$

Solution set: $\{3 + i\sqrt{5}, 3 - i\sqrt{5}\}$

25. $4x^2 - 4x = -7$

$4x^2 - 4x + 7 = 0$

Here $a = 4$, $b = -4$, and $c = 7$.

$$x = \frac{-b \pm \sqrt{b^2 - 4ac}}{2a}$$
$$x = \frac{-(-4) \pm \sqrt{(-4)^2 - 4(4)(7)}}{2(4)}$$
$$= \frac{4 \pm \sqrt{16 - 112}}{8} = \frac{4 \pm \sqrt{-96}}{8}$$
$$= \frac{4 \pm 4i\sqrt{6}}{8} = \frac{4(1 \pm i\sqrt{6})}{2 \cdot 4}$$
$$= \frac{1 \pm i\sqrt{6}}{2} = \frac{1}{2} \pm \frac{i\sqrt{6}}{2}$$

Solution set: $\left\{\frac{1}{2} + \frac{\sqrt{6}}{2}i, \frac{1}{2} - \frac{\sqrt{6}}{2}i\right\}$

27. $x(3x + 4) = -2$

$3x^2 + 4x = -2$

$3x^2 + 4x + 2 = 0$

Here $a = 3$, $b = 4$, and $c = 2$.

$$x = \frac{-b \pm \sqrt{b^2 - 4ac}}{2a}$$
$$x = \frac{-4 \pm \sqrt{4^2 - 4(3)(2)}}{2(3)}$$
$$= \frac{-4 \pm \sqrt{16 - 24}}{6} = \frac{-4 \pm \sqrt{-8}}{6}$$
$$= \frac{-4 \pm 2i\sqrt{2}}{6} = \frac{2(-2 \pm i\sqrt{2})}{2 \cdot 3}$$
$$= \frac{-2 \pm i\sqrt{2}}{3} = \frac{-2}{3} \pm \frac{i\sqrt{2}}{3}$$

Solution set: $\left\{-\frac{2}{3} + \frac{\sqrt{2}}{3}i, -\frac{2}{3} - \frac{\sqrt{2}}{3}i\right\}$

29. $25x^2 + 70x + 49 = 0$

Here $a = 25$, $b = 70$, and $c = 49$, so the discriminant is

$$b^2 - 4ac = 70^2 - 4(25)(49) = 4900 - 4900 = 0.$$

Since the discriminant is 0, the quantity under the radical in the quadratic formula is 0, and there is only one rational solution. The answer is **B**.

31. $x^2 + 4x + 2 = 0$

Here $a = 1$, $b = 4$, and $c = 2$, so the discriminant is

$$b^2 - 4ac = 4^2 - 4(1)(2) = 16 - 8 = 8.$$

Since the discriminant is positive, but not a perfect square, there are two distinct irrational number solutions. The answer is **C**.

33. $3x^2 = 5x + 2$

$3x^2 - 5x - 2 = 0$

Here $a = 3$, $b = -5$, and $c = -2$, so the discriminant is

$$b^2 - 4ac = (-5)^2 - 4(3)(-2) = 25 + 24 = 49.$$

Since the discriminant is a perfect square, there are two distinct rational solutions. The answer is **A**.

35. $3m^2 - 10m + 15 = 0$

Here $a = 3$, $b = -10$, and $c = 15$, so the discriminant is

$$b^2 - 4ac = (-10)^2 - 4(3)(15) = 100 - 180 = -80.$$

Since the discriminant is negative, there are two distinct nonreal complex number solutions. The answer is **D**.

37. The equations in Exercises 29, 30, 33, and 34 can be solved by factoring because the discriminant is a perfect square.

10.3 Equations Quadratic in Form

10.3 Margin Exercises

1. **(a)** $\frac{5}{m} + \frac{12}{m^2} = 2$

Clear fractions by multiplying each term by the LCD, m^2.

$$m^2\left(\frac{5}{m}\right) + m^2\left(\frac{12}{m^2}\right) = m^2(2)$$
$$5m + 12 = 2m^2$$
$$-2m^2 + 5m + 12 = 0$$
$$2m^2 - 5m - 12 = 0$$
$$(2m + 3)(m - 4) = 0$$
$$2m + 3 = 0 \quad \text{or} \quad m - 4 = 0$$
$$m = -\tfrac{3}{2} \quad \text{or} \quad m = 4$$

Check $m = -\frac{3}{2}$: $-\frac{10}{3} + \frac{16}{3} = 2$ *True*

Check $m = 4$: $\frac{5}{4} + \frac{3}{4} = 2$ *True*

Solution set: $\{-\frac{3}{2}, 4\}$

(b) $\frac{2}{x}+\frac{1}{x-2}=\frac{5}{3}$

Multiply each side by the LCD, $3x(x-2)$.

$$3x(x-2)\left(\frac{2}{x}+\frac{1}{x-2}\right)=3x(x-2)\left(\frac{5}{3}\right)$$
$$6(x-2)+3x=5x(x-2)$$
$$6x-12+3x=5x^2-10x$$
$$-5x^2+19x-12=0$$
$$5x^2-19x+12=0$$
$$(5x-4)(x-3)=0$$
$$5x-4=0 \quad\text{or}\quad x-3=0$$
$$x=\tfrac{4}{5} \quad\text{or}\quad x=3$$

Check $x=\frac{4}{5}$: $\frac{5}{2}-\frac{5}{6}=\frac{5}{3}$ *True*

Check $x=3$: $\frac{2}{3}+1=\frac{5}{3}$ *True*

Solution set: $\{\frac{4}{5},3\}$

(c) $\frac{4}{m-1}+9=-\frac{7}{m}$

Multiply by the LCD, $m(m-1)$.

$$m(m-1)\left(\frac{4}{m-1}+9\right)=m(m-1)\left(-\frac{7}{m}\right)$$
$$4m+9m(m-1)=-7(m-1)$$
$$4m+9m^2-9m=-7m+7$$
$$9m^2+2m-7=0$$
$$(9m-7)(m+1)=0$$
$$9m-7=0 \quad\text{or}\quad m+1=0$$
$$m=\tfrac{7}{9} \quad\text{or}\quad m=-1$$

Check $m=\frac{7}{9}$: $-18+9=-9$ *True*

Check $m=-1$: $-2+9=7$ *True*

Solution set: $\{-1,\frac{7}{9}\}$

2. **(a)** Let $x=$ the speed of the boat in still water.

Since the speed of the current is 5 mph, the speed upriver (*against* the current) is $x-5$ and the speed downriver (*with* the current) is $x+5$.

Kerrie went 15 mi each way. Complete the table. Use $d=rt$, or $t=\frac{d}{r}$.

	d	r	t
Up	15	$x-5$	$\frac{15}{x-5}$
Down	15	$x+5$	$\frac{15}{x+5}$

(b) To solve, use the fact that she can make the entire trip in 4 hr; that is, the time going upriver added to the time going downriver is 4.

$$\frac{15}{x-5}+\frac{15}{x+5}=4$$

Multiply each term by the LCD, $(x-5)(x+5)$.

$$\frac{15}{x-5}(x-5)(x+5)+\frac{15}{x+5}(x-5)(x+5)$$
$$=4(x-5)(x+5)$$
$$15(x+5)+15(x-5)=4(x^2-25)$$
$$15x+75+15x-75=4x^2-100$$
$$30x=4x^2-100$$
$$0=4x^2-30x-100$$
$$0=2x^2-15x-50$$
$$0=(2x+5)(x-10)$$
$$2x+5=0 \quad\text{or}\quad x-10=0$$
$$x=-\tfrac{5}{2} \quad\text{or}\quad x=10$$

The speed must be nonnegative, so $-\frac{5}{2}$ is not a solution. The speed of the boat is 10 mph.

(c) Let $x=$ the speed Ken can row.

Make a table. Use $t=\frac{d}{r}$.

	d	r	t
Up	5	$x-3$	$\frac{5}{x-3}$
Down	5	$x+3$	$\frac{5}{x+3}$

The time going upriver added to the time going downriver is $1\frac{3}{4}$ or $\frac{7}{4}$ hr.

$$\frac{5}{x-3}+\frac{5}{x+3}=\frac{7}{4}$$

Multiply each term by the LCD, $4(x-3)(x+3)$.

$$4(x-3)(x+3)\frac{5}{x-3}+4(x-3)(x+3)\frac{5}{x+3}$$
$$=4(x-3)(x+3)(\tfrac{7}{4})$$
$$20(x+3)+20(x-3)=7(x-3)(x+3)$$
$$20x+60+20x-60=7(x^2-9)$$
$$40x=7x^2-63$$
$$0=7x^2-40x-63$$
$$0=(7x+9)(x-7)$$
$$7x+9=0 \quad\text{or}\quad x-7=0$$
$$x=-\tfrac{9}{7} \quad\text{or}\quad x=7$$

Speed can't be negative, so Ken rows at the speed of 7 mph.

3. **(a)** Let $x =$ Jaime's time alone. Then $x - 2 =$ Carlos' time alone.

Make a table.

	Rate	Time Working Together	Fractional Part of the Job Done
Carlos	$\frac{1}{x-2}$	2	$\frac{2}{x-2}$
Jaime	$\frac{1}{x}$	2	$\frac{2}{x}$

Since together they complete 1 whole job,

$$\frac{2}{x-2} + \frac{2}{x} = 1.$$

Multiply each side by the LCD, $x(x-2)$.

$$x(x-2)\left(\frac{2}{x-2} + \frac{2}{x}\right) = x(x-2)(1)$$

$$2x + 2(x-2) = x(x-2)$$

$$2x + 2x - 4 = x^2 - 2x$$

$$0 = x^2 - 6x + 4$$

Use the quadratic formula.

Here $a = 1$, $b = -6$, and $c = 4$.

$$x = \frac{-b \pm \sqrt{b^2 - 4ac}}{2a}$$

$$x = \frac{-(-6) \pm \sqrt{(-6)^2 - 4(1)(4)}}{2(1)}$$

$$= \frac{6 \pm \sqrt{36 - 16}}{2}$$

$$= \frac{6 \pm \sqrt{20}}{2} = \frac{6 \pm 2\sqrt{5}}{2}$$

$$= \frac{2(3 \pm \sqrt{5})}{2} = 3 \pm \sqrt{5}$$

$x = 3 + \sqrt{5} \approx 5.2$ or $x = 3 - \sqrt{5} \approx 0.8$

Jaime's time cannot be 0.8, since Carlos' time would then be $0.8 - 2$ or -1.2. So, Jaime's time alone is about 5.2 hr, and Carlos' time alone is about 3.2 hr.

(b) Let $x =$ the slow chef's time alone. Then $x - 2 =$ the fast chef's time alone.

Make a table.

	Rate	Time Working Together	Fractional Part of the Job Done
Slow	$\frac{1}{x}$	5	$\frac{5}{x}$
Fast	$\frac{1}{x-2}$	5	$\frac{5}{x-2}$

Since together they complete 1 whole job,

$$\frac{5}{x} + \frac{5}{x-2} = 1.$$

Multiply each term by the LCD, $x(x-2)$.

$$x(x-2)\left(\frac{5}{x}\right) + x(x-2)\left(\frac{5}{x-2}\right)$$

$$= x(x-2)(1)$$

$$5(x-2) + 5x = x(x-2)$$

$$5x - 10 + 5x = x^2 - 2x$$

$$0 = x^2 - 12x + 10$$

Here $a = 1$, $b = -12$, and $c = 10$.

$$x = \frac{-b \pm \sqrt{b^2 - 4ac}}{2a}$$

$$x = \frac{-(-12) \pm \sqrt{(-12)^2 - 4(1)(10)}}{2(1)}$$

$$= \frac{12 \pm \sqrt{144 - 40}}{2}$$

$$= \frac{12 \pm \sqrt{104}}{2} = \frac{12 \pm 2\sqrt{26}}{2}$$

$$= \frac{2(6 \pm \sqrt{26})}{2} = 6 \pm \sqrt{26}$$

$x = 6 + \sqrt{26} \approx 11.1$ or $x = 6 - \sqrt{26} \approx 0.9$

The slow chef's time cannot be 0.9 since the fast chef's time would then be $0.9 - 2$ or -1.1. So the slow chef's time working alone is about 11.1 hr and the fast chef's time working alone is about $11.1 - 2 = 9.1$ hr.

4. **(a)**

$$x = \sqrt{7x - 10}$$

$$x^2 = \left(\sqrt{7x - 10}\right)^2 \quad \textit{Square.}$$

$$x^2 = 7x - 10$$

$$x^2 - 7x + 10 = 0$$

$$(x-2)(x-5) = 0$$

$x - 2 = 0$ or $x - 5 = 0$

$x = 2$ or $x = 5$

Check $x = 2$: $2 = \sqrt{4}$ *True*

Check $x = 5$: $5 = \sqrt{25}$ *True*

Solution set: $\{2, 5\}$

(b)

$$2x = \sqrt{x} + 1$$

$$2x - 1 = \sqrt{x} \quad \textit{Isolate.}$$

$$(2x-1)^2 = \left(\sqrt{x}\right)^2 \quad \textit{Square.}$$

$$4x^2 - 4x + 1 = x$$

$$4x^2 - 5x + 1 = 0$$

$$(4x-1)(x-1) = 0$$

continued

$$4x - 1 = 0 \quad \text{or} \quad x - 1 = 0$$
$$x = \tfrac{1}{4} \quad \text{or} \quad x = 1$$

Check $x = \frac{1}{4}$: $\frac{1}{2} = \frac{1}{2} + 1$ *False*

Check $x = 1$: $2 = \sqrt{1} + 1$ *True*

Solution set: $\{1\}$

5. **(a)** $2x^4 + 5x^2 - 12 = 0$

Since $x^4 = (x^2)^2$, define $u = x^2$, and the original equation becomes $2u^2 + 5u - 12 = 0$.

(b) $2(x+5)^2 - 7(x+5) + 6 = 0$

Because this equation involves both $(x+5)^2$ and $(x+5)$, define $u = x + 5$, and the original equation becomes $2u^2 - 7u + 6 = 0$.

(c) $x^{4/3} - 8x^{2/3} + 16 = 0$

Since $x^{4/3} = (x^{2/3})^2$, define $u = x^{2/3}$, and the original equation becomes $u^2 - 8u + 16 = 0$.

6. **(a)** $m^4 - 10m^2 + 9 = 0$

Let $y = m^2$, so $y^2 = (m^2)^2 = m^4$.

$$y^2 - 10y + 9 = 0$$
$$(y-9)(y-1) = 0$$

$$y - 9 = 0 \quad \text{or} \quad y - 1 = 0$$
$$y = 9 \quad \text{or} \quad y = 1$$

To find m, substitute m^2 for y.

$$m^2 = 9 \quad \text{or} \quad m^2 = 1$$
$$m = \pm 3 \quad \text{or} \quad m = \pm 1$$

Check $m = \pm 3$: $81 - 90 + 9 = 0$ *True*

Check $m = \pm 1$: $1 - 10 + 9 = 0$ *True*

Solution set: $\{-3, -1, 1, 3\}$

(b) $9k^4 - 37k^2 + 4 = 0$

Let $y = k^2$, so $y^2 = (k^2)^2 = k^4$.

$$9y^2 - 37y + 4 = 0$$
$$(y-4)(9y-1) = 0$$

$$y - 4 = 0 \quad \text{or} \quad 9y - 1 = 0$$
$$y = 4 \quad \text{or} \quad y = \tfrac{1}{9}$$

To find k, substitute k^2 for y.

$$k^2 = 4 \quad \text{or} \quad k^2 = \tfrac{1}{9}$$
$$k = \pm 2 \quad \text{or} \quad k = \pm \tfrac{1}{3}$$

Check $k = \pm 2$: $144 - 148 + 4 = 0$ *True*

Check $k = \pm \frac{1}{3}$: $\frac{1}{9} - \frac{37}{9} + 4 = 0$ *True*

Solution set: $\{-2, -\frac{1}{3}, \frac{1}{3}, 2\}$

(c) $x^4 - 4x^2 = -2$

$$x^4 - 4x^2 + 2 = 0$$

Let $y = x^2$, so $(x^2)^2 = x^4$.

$$y^2 - 4y + 2 = 0$$

Use the quadratic formula with $a = 1$, $b = -4$, and $c = 2$.

$$y = \frac{-b \pm \sqrt{b^2 - 4ac}}{2a}$$
$$y = \frac{-(-4) \pm \sqrt{(-4)^2 - 4(1)(2)}}{2(1)}$$
$$= \frac{4 \pm \sqrt{16 - 8}}{2}$$
$$= \frac{4 \pm \sqrt{8}}{2} = \frac{4 \pm 2\sqrt{2}}{2}$$
$$= \frac{2(2 \pm \sqrt{2})}{2} = 2 \pm \sqrt{2}$$

To find x, substitute x^2 for y.

$$x^2 = 2 \pm \sqrt{2}$$
$$x = \pm\sqrt{2 \pm \sqrt{2}}$$

Check $x = \pm\sqrt{2 + \sqrt{2}}$:

$$(2 + \sqrt{2})^2 - 4(2 + \sqrt{2}) \stackrel{?}{=} -2$$
$$4 + 4\sqrt{2} + 2 - 8 - 4\sqrt{2} \stackrel{?}{=} -2$$
$$-2 = -2 \quad \textit{True}$$

Check $x = \pm\sqrt{2 - \sqrt{2}}$:

$$(2 - \sqrt{2})^2 - 4(2 - \sqrt{2}) \stackrel{?}{=} -2$$
$$4 - 4\sqrt{2} + 2 - 8 + 4\sqrt{2} \stackrel{?}{=} -2$$
$$-2 = -2 \quad \textit{True}$$

The solution set contains four numbers:

$$\left\{\sqrt{2+\sqrt{2}}, -\sqrt{2+\sqrt{2}}, \sqrt{2-\sqrt{2}}, -\sqrt{2-\sqrt{2}}\right\}$$

7. **(a)** $5(r+3)^2 + 9(r+3) = 2$

Let $x = r + 3$. The equation becomes:

$$5x^2 + 9x = 2$$
$$5x^2 + 9x - 2 = 0$$
$$(5x - 1)(x + 2) = 0$$

$$5x - 1 = 0 \quad \text{or} \quad x + 2 = 0$$
$$x = \tfrac{1}{5} \quad \text{or} \quad x = -2$$

To find r, substitute $r + 3$ for x.

$$r + 3 = \tfrac{1}{5} \quad \text{or} \quad r + 3 = -2$$
$$r = -\tfrac{14}{5} \quad \text{or} \quad r = -5$$

Check $r = -\frac{14}{5}$: $\frac{1}{5} + \frac{9}{5} = 2$ *True*

Check $r = -5$: $20 - 18 = 2$ *True*

Solution set: $\{-5, -\frac{14}{5}\}$

(b) $4m^{2/3} = 3m^{1/3} + 1$

Let $x = m^{1/3}$, so $x^2 = (m^{1/3})^2 = m^{2/3}$.

$$4x^2 = 3x + 1$$
$$4x^2 - 3x - 1 = 0$$
$$(4x+1)(x-1) = 0$$

$4x + 1 = 0$ or $x - 1 = 0$

$x = -\frac{1}{4}$ or $x = 1$

To find m, substitute $m^{1/3}$ for x.

$m^{1/3} = -\frac{1}{4}$ or $m^{1/3} = 1$

Cube each side of each equation.

$(m^{1/3})^3 = (-\frac{1}{4})^3$ or $(m^{1/3})^3 = 1^3$

$m = -\frac{1}{64}$ or $m = 1$

Check $m = -\frac{1}{64}$: $\frac{1}{4} = -\frac{3}{4} + 1$ *True*

Check $m = 1$: $4 = 3 + 1$ *True*

Solution set: $\{-\frac{1}{64}, 1\}$

10.3 Section Exercises

1. $\frac{14}{x} = x - 5$

This is a rational equation, so multiply both sides by the LCD, x.

3. $(r^2 + r)^2 - 8(r^2 + r) + 12 = 0$

This is quadratic in form, so substitute a variable for $r^2 + r$.

5. The proposed solution -1 does not check. The solution set is $\{4\}$.

7. $1 - \frac{3}{x} - \frac{28}{x^2} = 0$

Multiply by the LCD, x^2.

$$x^2(1) - x^2\left(\frac{3}{x}\right) - x^2\left(\frac{28}{x^2}\right) = x^2 \cdot 0$$
$$x^2 - 3x - 28 = 0$$
$$(x+4)(x-7) = 0$$

$x + 4 = 0$ or $x - 7 = 0$

$x = -4$ or $x = 7$

Check $x = -4$: $1 + \frac{3}{4} - \frac{7}{4} = 0$ *True*

Check $x = 7$: $1 - \frac{3}{7} - \frac{4}{7} = 0$ *True*

Solution set: $\{-4, 7\}$

9. $3 - \frac{1}{t} = \frac{2}{t^2}$

Multiply each term by the LCD, t^2.

$$t^2(3) - t^2\left(\frac{1}{t}\right) = t^2\left(\frac{2}{t^2}\right)$$
$$3t^2 - t = 2$$
$$3t^2 - t - 2 = 0$$
$$(3t+2)(t-1) = 0$$

$3t + 2 = 0$ or $t - 1 = 0$

$t = -\frac{2}{3}$ or $t = 1$

Check $t = -\frac{2}{3}$: $3 + \frac{3}{2} = \frac{9}{2}$ *True*

Check $t = 1$: $3 - 1 = 2$ *True*

Solution set: $\{-\frac{2}{3}, 1\}$

11. $\frac{1}{x} + \frac{2}{x+2} = \frac{17}{35}$

Multiply by the LCD, $35x(x+2)$.

$$35x(x+2)\left(\frac{1}{x}\right) + 35x(x+2)\left(\frac{2}{x+2}\right) = 35x(x+2)(\tfrac{17}{35})$$
$$35(x+2) + 35x(2) = 17x(x+2)$$
$$35x + 70 + 70x = 17x^2 + 34x$$
$$70 + 105x = 17x^2 + 34x$$
$$0 = 17x^2 - 71x - 70$$
$$0 = (17x+14)(x-5)$$

$17x + 14 = 0$ or $x - 5 = 0$

$x = -\frac{14}{17}$ or $x = 5$

Check $x = -\frac{14}{17}$: $-\frac{17}{14} + \frac{17}{10} = \frac{17}{35}$ *True*

Check $x = 5$: $\frac{1}{5} + \frac{2}{7} = \frac{17}{35}$ *True*

Solution set: $\{-\frac{14}{17}, 5\}$

13. $\frac{2}{x+1} + \frac{3}{x+2} = \frac{7}{2}$

Multiply by the LCD, $2(x+1)(x+2)$.

$$2(x+1)(x+2)\left(\frac{2}{x+1} + \frac{3}{x+2}\right) = 2(x+1)(x+2)(\tfrac{7}{2})$$
$$2(x+2)(2) + 2(x+1)(3) = (x+1)(x+2)(7)$$
$$4x + 8 + 6x + 6 = (x^2 + 3x + 2)(7)$$
$$10x + 14 = 7x^2 + 21x + 14$$
$$7x^2 + 11x = 0$$
$$x(7x + 11) = 0$$

$x = 0$ or $7x + 11 = 0$

$x = -\frac{11}{7}$

Check $x = -\frac{11}{7}$: $-\frac{7}{2} + 7 = \frac{7}{2}$ *True*

Check $x = 0$: $2 + \frac{3}{2} = \frac{7}{2}$ *True*

Solution set: $\{-\frac{11}{7}, 0\}$

15. $\frac{3}{2x} - \frac{1}{2(x+2)} = 1$

Multiply by the LCD, $2x(x+2)$.

$$2x(x+2)\left(\frac{3}{2x} - \frac{1}{2(x+2)}\right) = 2x(x+2) \cdot 1$$
$$3(x+2) - x(1) = 2x(x+2)$$
$$3x + 6 - x = 2x^2 + 4x$$
$$0 = 2x^2 + 2x - 6$$
$$0 = x^2 + x - 3$$

Use $a = 1$, $b = 1$, $c = -3$ in the quadratic formula.

$$x = \frac{-b \pm \sqrt{b^2 - 4ac}}{2a}$$
$$x = \frac{-1 \pm \sqrt{1^2 - 4(1)(-3)}}{2(1)}$$
$$= \frac{-1 \pm \sqrt{1+12}}{2} = \frac{-1 \pm \sqrt{13}}{2}$$

Use a calculator to check both proposed solutions. Both solutions check.

Solution set: $\left\{\frac{-1+\sqrt{13}}{2}, \frac{-1-\sqrt{13}}{2}\right\}$

17. $\frac{6}{p} = 2 + \frac{p}{p+1}$

Multiply each term by the LCD, $p(p+1)$.

$$6(p+1) = 2p(p+1) + p \cdot p$$
$$6p + 6 = 2p^2 + 2p + p^2$$
$$0 = 3p^2 - 4p - 6$$

Use $a = 3$, $b = -4$, and $c = -6$ in the quadratic formula.

$$p = \frac{-b \pm \sqrt{b^2 - 4ac}}{2a}$$
$$p = \frac{-(-4) \pm \sqrt{(-4)^2 - 4(3)(-6)}}{2(3)}$$
$$= \frac{4 \pm \sqrt{16+72}}{2(3)} = \frac{4 \pm \sqrt{88}}{2(3)}$$
$$= \frac{4 \pm 2\sqrt{22}}{2(3)} = \frac{2 \pm \sqrt{22}}{3}$$

Use a calculator to check both proposed solutions. Both solutions check.

Solution set: $\left\{\frac{2+\sqrt{22}}{3}, \frac{2-\sqrt{22}}{3}\right\}$

19. Rate in still water: 20 mph
Rate of current: t mph

(a) When the boat travels upstream, the current works *against* the rate of the boat in still water, so the rate is $(20 - t)$ mph.

(b) When the boat travels downstream, the current works *with* the rate of the boat in still water, so the rate is $(20 + t)$ mph.

21. Let x = rate of the boat in still water.
With the speed of the current at 15 mph, then

$$x - 15 = \text{rate going upstream and}$$
$$x + 15 = \text{rate going downstream.}$$

Complete a table using the information in the problem, the rates given above, and the formula $d = rt$, or $t = d/r$.

	d	r	t
Upstream	4	$x - 15$	$\frac{4}{x-15}$
Downstream	16	$x + 15$	$\frac{16}{x+15}$

The time, 48 min, is written as $\frac{48}{60} = \frac{4}{5}$ hr. The time upstream plus the time downstream equals $\frac{4}{5}$. So, from the table, the equation is written as

$$\frac{4}{x-15} + \frac{16}{x+15} = \frac{4}{5}.$$

Multiply by the LCD, $5(x-15)(x+15)$.

$$5(x-15)(x+15)\left(\frac{4}{x-15} + \frac{16}{x+15}\right) = 5(x-15)(x+15) \cdot \tfrac{4}{5}$$
$$20(x+15) + 80(x-15) = 4(x-15)(x+15)$$
$$20x + 300 + 80x - 1200 = 4(x^2 - 225)$$
$$100x - 900 = 4x^2 - 900$$
$$0 = 4x^2 - 100x$$
$$0 = 4x(x - 25)$$

$$4x = 0 \quad \text{or} \quad x - 25 = 0$$
$$x = 0 \quad \text{or} \quad x = 25$$

Reject $x = 0$ mph as a possible boat speed. Yoshiaki's boat had a top speed of 25 mph.

23. Let x = Harry's average speed.
Then $x - 20$ = Yoshi's average speed.

	d	r	t
Harry	300	x	$\frac{300}{x}$
Yoshi	300	$x - 20$	$\frac{300}{x-20}$

It takes Harry $1\frac{1}{4}$ or $\frac{5}{4}$ hours less time than Yoshi.

$$\frac{300}{x} = \frac{300}{x-20} - \frac{5}{4}$$

Multiply by the LCD, $4x(x-20)$.

$$4x(x-20)\left(\frac{300}{x}\right) = 4x(x-20)\left(\frac{300}{x-20} - \frac{5}{4}\right)$$
$$1200(x-20) = 4x(300) - x(x-20) \cdot 5$$
$$1200x - 24{,}000 = 1200x - 5x^2 + 100x$$
$$5x^2 - 100x - 24{,}000 = 0$$
$$x^2 - 20x - 4800 = 0$$
$$(x-80)(x+60) = 0$$

$x - 80 = 0$ or $x + 60 = 0$
$x = 80$ or $x = -60$

Reject $x = -60$. Harry's average speed is 80 km/hr.

25. Let x be the time in hours required for the faster person to cut the lawn. Then the slower person requires $x+1$ hours. Complete the chart.

	Rate	Time Working Together	Fractional Part of the Job Done
Faster Worker	$\frac{1}{x}$	2	$\frac{2}{x}$
Slower Worker	$\frac{1}{x+1}$	2	$\frac{2}{x+1}$

Part done by faster person + Part done by slower person = one whole job.

$$\frac{2}{x} + \frac{2}{x+1} = 1$$

Multiply each side by the LCD, $x(x+1)$.

$$x(x+1)\left(\frac{2}{x} + \frac{2}{x+1}\right) = x(x+1) \cdot 1$$
$$2(x+1) + 2x = x^2 + x$$
$$2x + 2 + 2x = x^2 + x$$
$$0 = x^2 - 3x - 2$$

Solve for x using the quadratic formula with $a = 1$, $b = -3$, and $c = -2$.

$$x = \frac{-(-3) \pm \sqrt{(-3)^2 - 4(1)(-2)}}{2(1)}$$
$$= \frac{3 \pm \sqrt{9+8}}{2} = \frac{3 \pm \sqrt{17}}{2}$$

$x = \frac{3+\sqrt{17}}{2}$ or $x = \frac{3-\sqrt{17}}{2}$
$x \approx 3.6$ or $x \approx -0.6$

Discard -0.6 as a solution since time cannot be negative.
It would take the faster person approximately 3.6 hours.

27. Let $x =$ the number of minutes it takes for the cold water tap alone to fill the washer.
$x + 9 =$ the number of minutes it takes for the hot water tap alone to fill the washer.

Working together, both taps can fill the washer in 6 minutes. Complete a chart using the above information.

Tap	Rate	Time	Fractional Part of Washer Filled
Cold	$\frac{1}{x}$	6	$\frac{6}{x}$
Hot	$\frac{1}{x+9}$	6	$\frac{6}{x+9}$

Since together the hot and cold taps fill one washer, the sum of their fractional parts is 1; that is,

$$\frac{6}{x} + \frac{6}{x+9} = 1.$$

Multiply by the LCD, $x(x+9)$.

$$x(x+9)\left(\frac{6}{x} + \frac{6}{x+9}\right) = x(x+9) \cdot 1$$
$$6(x+9) + 6x = x(x+9)$$
$$6x + 54 + 6x = x^2 + 9x$$
$$0 = x^2 - 3x - 54$$
$$0 = (x-9)(x+6)$$

$x - 9 = 0$ or $x + 6 = 0$
$x = 9$ or $x = -6$

Reject -6 as a possible time. The cold water tap can fill the washer in 9 minutes.

29.
$$z = \sqrt{5z - 4}$$
$$(z)^2 = \left(\sqrt{5z-4}\right)^2$$
$$z^2 = 5z - 4$$
$$z^2 - 5z + 4 = 0$$
$$(z-1)(z-4) = 0$$

$z - 1 = 0$ or $z - 4 = 0$
$z = 1$ or $z = 4$

Check $z = 1$: $1 = \sqrt{1}$ *True*
Check $z = 4$: $4 = \sqrt{16}$ *True*

Solution set: $\{1, 4\}$

31.

$$2x = \sqrt{11x+3}$$
$$(2x)^2 = \left(\sqrt{11x+3}\right)^2$$
$$4x^2 = 11x + 3$$
$$4x^2 - 11x - 3 = 0$$
$$(4x+1)(x-3) = 0$$
$$4x + 1 = 0 \quad \text{or} \quad x - 3 = 0$$
$$x = -\tfrac{1}{4} \quad \text{or} \quad x = 3$$

Check $x = -\frac{1}{4}$: $-\frac{1}{2} = \sqrt{\frac{1}{4}}$ *False*

Check $x = 3$: $6 = \sqrt{36}$ *True*

Solution set: $\{3\}$

33.

$$3x = \sqrt{16 - 10x}$$
$$(3x)^2 = \left(\sqrt{16-10x}\right)^2$$
$$9x^2 = 16 - 10x$$
$$9x^2 + 10x - 16 = 0$$
$$(9x-8)(x+2) = 0$$
$$9x - 8 = 0 \quad \text{or} \quad x + 2 = 0$$
$$x = \tfrac{8}{9} \quad \text{or} \quad x = -2$$

Check $x = \frac{8}{9}$: $\frac{8}{3} = \sqrt{\frac{64}{9}}$ *True*

Check $x = -2$: $-6 = \sqrt{36}$ *False*

Solution set: $\{\frac{8}{9}\}$

35.

$$p - 2\sqrt{p} = 8$$
$$p - 8 = 2\sqrt{p}$$
$$(p-8)^2 = \left(2\sqrt{p}\right)^2$$
$$p^2 - 16p + 64 = 4p$$
$$p^2 - 20p + 64 = 0$$
$$(p-4)(p-16) = 0$$
$$p - 4 = 0 \quad \text{or} \quad p - 16 = 0$$
$$p = 4 \quad \text{or} \quad p = 16$$

Check $p = 4$: $4 - 4 = 8$ *False*

Check $p = 16$: $16 - 8 = 8$ *True*

Solution set: $\{16\}$

37.

$$m = \sqrt{\frac{6 - 13m}{5}}$$
$$m^2 = \frac{6 - 13m}{5}$$
$$5m^2 = 6 - 13m$$
$$5m^2 + 13m - 6 = 0$$
$$(5m-2)(m+3) = 0$$
$$5m - 2 = 0 \quad \text{or} \quad m + 3 = 0$$
$$m = \tfrac{2}{5} \quad \text{or} \quad m = -3$$

Check $m = \frac{2}{5}$: $\frac{2}{5} = \sqrt{\frac{4}{25}}$ *True*

Check $m = -3$: $-3 = \sqrt{9}$ *False*

Solution set: $\{\frac{2}{5}\}$

39. $t^4 - 18t^2 + 81 = 0$

Let $u = t^2$, so $u^2 = t^4$.

$$u^2 - 18u + 81 = 0$$
$$(u-9)^2 = 0$$
$$u - 9 = 0$$
$$u = 9$$

To find t, substitute t^2 for u.

$$t^2 = 9$$
$$t = 3 \quad \text{or} \quad t = -3$$

Check $t = \pm 3$: $81 - 162 + 81 = 0$ *True*

Solution set: $\{-3, 3\}$

41. $4k^4 - 13k^2 + 9 = 0$

Let $u = k^2$, so $u^2 = k^4$.

$$4u^2 - 13u + 9 = 0$$
$$(4u-9)(u-1) = 0$$
$$4u - 9 = 0 \quad \text{or} \quad u - 1 = 0$$
$$u = \tfrac{9}{4} \quad \text{or} \quad u = 1.$$

To find k, substitute k^2 for u.

$$k^2 = \tfrac{9}{4} \quad \text{or} \quad k^2 = 1$$
$$k = \pm\tfrac{3}{2} \quad \text{or} \quad k = \pm 1$$

Check $k = \pm\frac{3}{2}$: $\frac{81}{4} - \frac{117}{4} + 9 = 0$ *True*

Check $k = \pm 1$: $4 - 13 + 9 = 0$ *True*

Solution set: $\{-\frac{3}{2}, -1, 1, \frac{3}{2}\}$

43.

$$x^4 + 48 = 16x^2$$
$$x^4 - 16x^2 + 48 = 0$$

Let $u = x^2$, so $u^2 = x^4$.

$$u^2 - 16u + 48 = 0$$
$$(u-4)(u-12) = 0$$
$$u - 4 = 0 \quad \text{or} \quad u - 12 = 0$$
$$u = 4 \quad \text{or} \quad u = 12$$

To find x substitute x^2 for u.

$$x^2 = 4 \quad \text{or} \quad x^2 = 12$$
$$x = \pm\sqrt{4} \quad \text{or} \quad x = \pm\sqrt{12}$$
$$x = \pm 2 \quad \text{or} \quad x = \pm 2\sqrt{3}$$

Check $x = \pm 2$: $16 + 48 = 64$ *True*

Check $x = \pm 2\sqrt{3}$: $144 + 48 = 192$ *True*

Solution set: $\{-2\sqrt{3}, -2, 2, 2\sqrt{3}\}$

45. $2x^4 - 9x^2 = -2$

$$2x^4 - 9x^2 + 2 = 0$$

Let $u = x^2$, so $u^2 = x^4$.

$$2u^2 - 9u + 2 = 0$$

Use $a = 2$, $b = -9$, and $c = 2$ in the quadratic formula.

$$u = \frac{-b \pm \sqrt{b^2 - 4ac}}{2a}$$

$$u = \frac{-(-9) \pm \sqrt{(-9)^2 - 4(2)(2)}}{2(2)}$$

$$= \frac{9 \pm \sqrt{81 - 16}}{4}$$

$$= \frac{9 \pm \sqrt{65}}{4}$$

To find x, substitute x^2 for u.

$$x^2 = \frac{9 \pm \sqrt{65}}{4}$$

$$x = \pm\sqrt{\frac{9 \pm \sqrt{65}}{4}}$$

$$= \pm\frac{\sqrt{9 \pm \sqrt{65}}}{2}$$

Note: the last expression represents four numbers. All four proposed solutions check.

Solution set: $\left\{\frac{\sqrt{9+\sqrt{65}}}{2}, -\frac{\sqrt{9+\sqrt{65}}}{2}, \frac{\sqrt{9-\sqrt{65}}}{2}, -\frac{\sqrt{9-\sqrt{65}}}{2}\right\}$

47. $(x+3)^2 + 5(x+3) + 6 = 0$

Let $u = x + 3$, so $u^2 = (x+3)^2$.

$$u^2 + 5u + 6 = 0$$

$$(u+3)(u+2) = 0$$

$u + 3 = 0$ or $u + 2 = 0$

$u = -3$ or $u = -2$

To find x, substitute $x + 3$ for u.

$x + 3 = -3$ or $x + 3 = -2$

$x = -6$ or $x = -5$

Check $x = -6$: $9 - 15 + 6 = 0$ *True*

Check $x = -5$: $4 - 10 + 6 = 0$ *True*

Solution set: $\{-6, -5\}$

49. $(t+5)^2 + 6 = 7(t+5)$

Let $u = t + 5$, so $u^2 = (t+5)^2$.

$$u^2 + 6 = 7u$$

$$u^2 - 7u + 6 = 0$$

$$(u-6)(u-1) = 0$$

$u - 6 = 0$ or $u - 1 = 0$

$u = 6$ or $u = 1$

To find t, substitute $t + 5$ for u.

$t + 5 = 6$ or $t + 5 = 1$

$t = 1$ or $t = -4$

Check $t = 1$: $36 + 6 = 42$ *True*

Check $t = -4$: $1 + 6 = 7$ *True*

Solution set: $\{-4, 1\}$

51. $2 + \frac{5}{3k-1} = \frac{-2}{(3k-1)^2}$

Let $u = 3k - 1$, so $u^2 = (3k-1)^2$.

$$2 + \frac{5}{u} = -\frac{2}{u^2}$$

Multiply by the LCD, u^2.

$$u^2\left(2 + \frac{5}{u}\right) = u^2\left(-\frac{2}{u^2}\right)$$

$$2u^2 + 5u = -2$$

$$2u^2 + 5u + 2 = 0$$

$$(2u+1)(u+2) = 0$$

$2u + 1 = 0$ or $u + 2 = 0$

$u = -\frac{1}{2}$ or $u = -2$

To find k, substitute $3k - 1$ for u.

$3k - 1 = -\frac{1}{2}$ or $3k - 1 = -2$

$3k = \frac{1}{2}$ or $3k = -1$

$k = \frac{1}{6}$ or $k = -\frac{1}{3}$

Check $k = \frac{1}{6}$: $2 - 10 = -8$ *True*

Check $k = -\frac{1}{3}$: $2 - \frac{5}{2} = -\frac{1}{2}$ *True*

Solution set: $\{-\frac{1}{3}, \frac{1}{6}\}$

53. $x^{2/3} + x^{1/3} - 2 = 0$

Let $u = x^{1/3}$, so $u^2 = x^{2/3}$.

$$u^2 + u - 2 = 0$$

$$(u+2)(u-1) = 0$$

$u + 2 = 0$ or $u - 1 = 0$

$u = -2$ or $u = 1$

To find x, substitute $x^{1/3}$ for u.

$x^{1/3} = -2$ or $x^{1/3} = 1$

Cube both sides of each equation.

$(x^{1/3})^3 = (-2)^3$ or $(x^{1/3})^3 = 1^3$

$x = -8$ or $x = 1$

Check $x = -8$: $4 - 2 - 2 = 0$ *True*

Check $x = 1$: $1 + 1 - 2 = 0$ *True*

Solution set: $\{-8, 1\}$

55. $r^{2/3} + r^{1/3} - 12 = 0$

Let $u = r^{1/3}$, so $u^2 = r^{2/3}$.

$$u^2 + u - 12 = 0$$
$$(u + 4)(u - 3) = 0$$

$$u + 4 = 0 \quad \text{or} \quad u - 3 = 0$$
$$u = -4 \quad \text{or} \quad u = 3$$

To find r, substitute $r^{1/3}$ for u.

$$r^{1/3} = -4 \quad \text{or} \quad r^{1/3} = 3$$
$$(r^{1/3})^3 = (-4)^3 \quad \text{or} \quad (r^{1/3})^3 = 3^3$$
$$r = -64 \quad \text{or} \quad r = 27$$

Check $r = -64$: $16 - 4 - 12 = 0$ *True*
Check $r = 27$: $9 + 3 - 12 = 0$ *True*

Solution set: $\{-64, 27\}$

57. $2\left(1 + \sqrt{r}\right)^2 = 13\left(1 + \sqrt{r}\right) - 6$

Let $u = 1 + \sqrt{r}$.

$$2u^2 = 13u - 6$$
$$2u^2 - 13u + 6 = 0$$
$$(2u - 1)(u - 6) = 0$$

$$2u - 1 = 0 \quad \text{or} \quad u - 6 = 0$$
$$u = \tfrac{1}{2} \quad \text{or} \quad u = 6$$

To find r, substitute $1 + \sqrt{r}$ for u.

$$1 + \sqrt{r} = \tfrac{1}{2} \quad \text{or} \quad 1 + \sqrt{r} = 6$$
$$\sqrt{r} = -\tfrac{1}{2} \quad \text{or} \quad \sqrt{r} = 5$$

Not possible, since $\sqrt{r} \geq 0$. $\qquad r = 25$

Check $r = 25$: $72 = 78 - 6$ *True*
Solution set: $\{25\}$

For Exercises 59–64, use the equation

$$\frac{x^2}{(x-3)^2} + \frac{3x}{x-3} - 4 = 0.$$

59. Substituting 3 for x would cause both denominators to equal 0, and division by 0 is undefined.

60.
$$(x-3)^2\left(\frac{x^2}{(x-3)^2}\right) + (x-3)^2\left(\frac{3x}{x-3}\right) - (x-3)^2 \cdot 4 = (x-3)^2 \cdot 0$$
$$x^2 + 3x(x-3) - 4(x^2 - 6x + 9) = 0$$
$$x^2 + 3x^2 - 9x - 4x^2 + 24x - 36 = 0$$
$$15x - 36 = 0$$
$$15x = 36$$
$$x = \tfrac{36}{15} = \tfrac{12}{5}$$

Check $x = \frac{12}{5}$: $16 + (-12) - 4 = 0$ *True*
The solution is $\frac{12}{5}$.

61.
$$\frac{x^2}{(x-3)^2} + \frac{3x}{x-3} - 4 = 0$$
$$\left(\frac{x}{x-3}\right)^2 + 3\left(\frac{x}{x-3}\right) - 4 = 0$$

62. If a fraction is equal to 1, then the numerator must be equal to the denominator. But in this case the numerator can never equal the denominator, since the denominator is 3 less than the numerator.

63. From Exercise 61,

$$\left(\frac{x}{x-3}\right)^2 + 3\left(\frac{x}{x-3}\right) - 4 = 0.$$

Let $t = \frac{x}{x-3}$, so $t^2 = \left(\frac{x}{x-3}\right)^2$.

$$t^2 + 3t - 4 = 0$$
$$(t - 1)(t + 4) = 0$$

$$t - 1 = 0 \quad \text{or} \quad t + 4 = 0$$
$$t = 1 \quad \text{or} \quad t = -4$$

To find x, substitute $\frac{x}{x-3}$ for t.

$$\frac{x}{x-3} = 1 \quad \text{or} \quad \frac{x}{x-3} = -4$$

The equation $\frac{x}{x-3} = 1$ has no solution since there is no value of x for which $x = x - 3$. (See Exercise 62.) Therefore, $t = 1$ is impossible.

$$\frac{x}{x-3} = -4$$
$$x = -4(x - 3)$$
$$x = -4x + 12$$
$$5x = 12$$
$$x = \tfrac{12}{5}$$

$x = \frac{12}{5}$ was already checked in Exercise 60.
Solution set: $\{\frac{12}{5}\}$

64. $x^2(x-3)^{-2} + 3x(x-3)^{-1} - 4 = 0$

Let $s = (x-3)^{-1}$, so $s^2 = (x-3)^{-2}$.

$$x^2s^2 + 3xs - 4 = 0$$
$$(xs + 4)(xs - 1) = 0$$

$$xs + 4 = 0 \quad \text{or} \quad xs - 1 = 0$$
$$xs = -4 \quad \text{or} \quad xs = 1$$
$$s = -\frac{4}{x} \quad \text{or} \quad s = \frac{1}{x}$$

To find x, substitute $\frac{1}{x-3}$ for s.

$$\frac{1}{x-3} = -\frac{4}{x} \quad \text{or} \quad \frac{1}{x-3} = \frac{1}{x}$$

$x = -4(x-3)$ Since $x \neq x-3$,

$x = -4x + 12$ this equation has

$5x = 12$ no solutions.

$x = \frac{12}{5}$ Thus, $s = \frac{1}{x}$ is impossible.

$x = \frac{12}{5}$ was already checked in Exercise 60.

Solution set: $\{\frac{12}{5}\}$

Summary Exercises on Solving Quadratic Equations

1. $(2x+3)^2 = 4$

Since the equation has the form $(ax+b)^2 = c$, use the *square root property*.

3. $z^2 + 5z - 8 = 0$

The discriminant is

$$b^2 - 4ac = 5^2 - 4(1)(-8) = 25 + 32 = 57.$$

Since the discriminant is not a perfect square, use the *quadratic formula*.

5. $3m^2 = 2 - 5m$

$3m^2 + 5m - 2 = 0$

The discriminant is

$$b^2 - 4ac = 5^2 - 4(3)(-2) = 25 + 24 = 49.$$

Since the discriminant is a perfect square, use *factoring*.

7. $p^2 = 47$

$$p = \sqrt{47} \quad \text{or} \quad p = -\sqrt{47}$$

Solution set: $\{\sqrt{47}, -\sqrt{47}\}$

9. $n^2 + 8n + 6 = 0$

$n^2 + 8n = -6$

$n^2 + 8n + 16 = -6 + 16 \qquad \left[\frac{1}{2}(8)\right]^2 = 16$

$(n+4)^2 = 10$

$n + 4 = \sqrt{10}$ or $n + 4 = -\sqrt{10}$

$n = -4 + \sqrt{10}$ or $n = -4 - \sqrt{10}$

Solution set: $\{-4 + \sqrt{10}, -4 - \sqrt{10}\}$

11. $\frac{9}{m} + \frac{5}{m^2} = 2$

Multiply by the LCD, m^2.

$9m + 5 = 2m^2$

$0 = 2m^2 - 9m - 5$

$0 = (2m+1)(m-5)$

$2m + 1 = 0$ or $m - 5 = 0$

$m = -\frac{1}{2}$ or $m = 5$

Solution set: $\{-\frac{1}{2}, 5\}$

13. $3x^2 - 9x + 4 = 0$

Use $a = 3$, $b = -9$, and $c = 4$ in the quadratic formula.

$$x = \frac{-b \pm \sqrt{b^2 - 4ac}}{2a}$$

$$x = \frac{-(-9) \pm \sqrt{(-9)^2 - 4(3)(4)}}{2(3)} = \frac{9 \pm \sqrt{81 - 48}}{6} = \frac{9 \pm \sqrt{33}}{6}$$

Solution set: $\left\{\frac{9+\sqrt{33}}{6}, \frac{9-\sqrt{33}}{6}\right\}$

15. $x\sqrt{2} = \sqrt{5x-2}$

$(x\sqrt{2})^2 = (\sqrt{5x-2})^2$

$x^2 \cdot 2 = 5x - 2$

$2x^2 - 5x + 2 = 0$

$(2x-1)(x-2) = 0$

$2x - 1 = 0$ or $x - 2 = 0$

$x = \frac{1}{2}$ $\qquad x = 2$

Check $x = \frac{1}{2}$: $\frac{1}{2}\sqrt{2} = \sqrt{\frac{1}{2}}$ *True*

Check $x = 2$: $2\sqrt{2} = \sqrt{8}$ *True*

Solution set: $\{\frac{1}{2}, 2\}$

17. $(2k+5)^2 = 12$

$2k + 5 = \sqrt{12}$ or $2k + 5 = -\sqrt{12}$

$2k = -5 + 2\sqrt{3}$ or $2k = -5 - 2\sqrt{3}$

$k = \frac{-5 + 2\sqrt{3}}{2}$ or $k = \frac{-5 - 2\sqrt{3}}{2}$

Solution set: $\left\{\frac{-5+2\sqrt{3}}{2}, \frac{-5-2\sqrt{3}}{2}\right\}$

19.
$$t^4 + 14 = 9t^2$$
$$t^4 - 9t^2 + 14 = 0$$
$$(t^2 - 2)(t^2 - 7) = 0$$
$$t^2 - 2 = 0 \quad \text{or} \quad t^2 - 7 = 0$$
$$t^2 = 2 \qquad t^2 = 7$$
$$t = \pm\sqrt{2} \quad \text{or} \quad t = \pm\sqrt{7}$$

Solution set: $\{-\sqrt{7}, -\sqrt{2}, \sqrt{2}, \sqrt{7}\}$

21. $z^2 + z + 2 = 0$

Use $a = 1$, $b = 1$, and $c = 2$ in the quadratic formula.

$$z = \frac{-b \pm \sqrt{b^2 - 4ac}}{2a}$$
$$z = \frac{-1 \pm \sqrt{1^2 - 4(1)(2)}}{2(1)}$$
$$= \frac{-1 \pm \sqrt{1 - 8}}{2} = \frac{-1 \pm \sqrt{-7}}{2}$$
$$= \frac{-1 \pm i\sqrt{7}}{2} = \frac{-1}{2} \pm \frac{i\sqrt{7}}{2}$$

Solution set: $\left\{-\frac{1}{2} + \frac{\sqrt{7}}{2}i, -\frac{1}{2} - \frac{\sqrt{7}}{2}i\right\}$

23.
$$4t^2 - 12t + 9 = 0$$
$$(2t - 3)(2t - 3) = 0$$
$$(2t - 3)^2 = 0$$
$$2t - 3 = 0$$
$$t = \tfrac{3}{2}$$

Solution set: $\{\frac{3}{2}\}$

25. $r^2 - 72 = 0$
$$r^2 = 72$$
$$r = \pm\sqrt{72} = \pm 6\sqrt{2}$$

Solution set: $\{6\sqrt{2}, -6\sqrt{2}\}$

27.
$$x^2 - 5x - 36 = 0$$
$$(x + 4)(x - 9) = 0$$
$$x + 4 = 0 \quad \text{or} \quad x - 9 = 0$$
$$x = -4 \quad \text{or} \quad x = 9$$

Solution set: $\{-4, 9\}$

29.
$$3p^2 = 6p - 4$$
$$3p^2 - 6p + 4 = 0$$

Use $a = 3$, $b = -6$, and $c = 4$ in the quadratic formula.

$$p = \frac{-b \pm \sqrt{b^2 - 4ac}}{2a}$$
$$p = \frac{-(-6) \pm \sqrt{(-6)^2 - 4(3)(4)}}{2(3)}$$
$$= \frac{6 \pm \sqrt{36 - 48}}{2(3)} = \frac{6 \pm \sqrt{-12}}{2(3)}$$
$$= \frac{6 \pm 2i\sqrt{3}}{2(3)} = \frac{3 \pm i\sqrt{3}}{3}$$
$$= \frac{3}{3} \pm \frac{i\sqrt{3}}{3} = 1 \pm \frac{\sqrt{3}}{3}i$$

Solution set: $\left\{1 + \frac{\sqrt{3}}{3}i, 1 - \frac{\sqrt{3}}{3}i\right\}$

31. $2(3k - 1)^2 + 5(3k - 1) = -2$

Let $u = 3k - 1$.
$$2u^2 + 5u + 2 = 0$$
$$(2u + 1)(u + 2) = 0$$
$$2u + 1 = 0 \quad \text{or} \quad u + 2 = 0$$
$$u = -\tfrac{1}{2} \qquad u = -2$$
$$3k - 1 = -\tfrac{1}{2} \quad \text{or} \quad 3k - 1 = -2$$
$$3k = \tfrac{1}{2} \qquad 3k = -1$$
$$k = \tfrac{1}{6} \qquad k = -\tfrac{1}{3}$$

Solution set: $\{-\frac{1}{3}, \frac{1}{6}\}$

33. $x - \sqrt{15 - 2x} = 0$
$$x = \sqrt{15 - 2x} \qquad \textit{Isolate.}$$
$$x^2 = \left(\sqrt{15 - 2x}\right)^2 \qquad \textit{Square.}$$
$$x^2 = 15 - 2x$$
$$x^2 + 2x - 15 = 0$$
$$(x + 5)(x - 3) = 0$$
$$x + 5 = 0 \quad \text{or} \quad x - 3 = 0$$
$$x = -5 \quad \text{or} \quad x = 3$$

Check $x = -5$: $\quad -5 - \sqrt{25} = 0 \quad$ *False*

Check $x = 3$: $\quad 3 - \sqrt{9} = 0 \quad$ *True*

Solution set: $\{3\}$

35. $4k^4 + 5k^2 + 1 = 0$

Let $x = k^2$, so $x^2 = k^4$.
$$4x^2 + 5x + 1 = 0$$
$$(4x + 1)(x + 1) = 0$$
$$4x + 1 = 0 \quad \text{or} \quad x + 1 = 0$$
$$x = -\tfrac{1}{4} \quad \text{or} \quad x = -1$$
$$k^2 = -\tfrac{1}{4} \quad \text{or} \quad k^2 = -1$$
$$k = \pm\sqrt{-\tfrac{1}{4}} \quad \text{or} \quad k = \pm\sqrt{-1}$$
$$k = \pm\tfrac{1}{2}i \quad \text{or} \quad k = \pm i$$

Check $k = \pm\frac{1}{2}i$: $\frac{1}{4} - \frac{5}{4} + 1 = 0$ *True*

Check $k = \pm i$: $4 - 5 + 1 = 0$ *True*

Solution set: $\{-i, i, -\frac{1}{2}i, \frac{1}{2}i\}$

10.4 Formulas and Further Applications

10.4 Margin Exercises

1. **(a)** Solve $A = \pi r^2$ for r.

$\frac{A}{\pi} = r^2$ *Divide by π.*

$r = \pm\sqrt{\frac{A}{\pi}}$ *Square root property*

$r = \frac{\pm\sqrt{A}}{\sqrt{\pi}} \cdot \frac{\sqrt{\pi}}{\sqrt{\pi}}$ *Rationalize the denominator.*

$r = \frac{\pm\sqrt{A\pi}}{\pi}$

(b) Solve $s = 30\sqrt{\frac{a}{p}}$ for a.

$s^2 = 900 \cdot \frac{a}{p}$ *Square both sides.*

$ps^2 = 900a$ *Multiply by p.*

$\frac{ps^2}{900} = a$ *Divide by 900.*

2. Solve $2t^2 - 5t + k = 0$ for t.

Use $a = 2$, $b = -5$, and $c = k$ in the quadratic formula.

$$t = \frac{-b \pm \sqrt{b^2 - 4ac}}{2a}$$

$$t = \frac{-(-5) \pm \sqrt{(-5)^2 - 4(2)k}}{2(2)}$$

$$t = \frac{5 \pm \sqrt{25 - 8k}}{4}$$

The solutions are

$$t = \frac{5 + \sqrt{25 - 8k}}{4} \text{ and } t = \frac{5 - \sqrt{25 - 8k}}{4}.$$

3. *Step 2*
Let $x =$ the distance to the top of the ladder from the ground. Then $x - 7 =$ the distance to the bottom of the ladder from the house.

Step 3
The wall of the house is perpendicular to the ground, so this is a right triangle. Use the Pythagorean formula.

$$a^2 + b^2 = c^2$$
$$x^2 + (x - 7)^2 = 13^2$$

Step 4

$$x^2 + x^2 - 14x + 49 = 169$$
$$2x^2 - 14x - 120 = 0$$
$$x^2 - 7x - 60 = 0$$
$$(x - 12)(x + 5) = 0$$

$x - 12 = 0$ or $x + 5 = 0$

$x = 12$ or $x = -5$

Step 5
Since x represents a length, it must be positive, so reject -5. Since $x = 12$, $x - 7 = 12 - 7 = 5$. The bottom of the ladder is 5 ft from the house.

Step 6
Since $5^2 + 12^2 = 13^2$ and 5 ft is 7 ft less than 12 ft, the distances are correct.

4. *Step 2*
Let $x =$ the width of the grass strip.

Step 3
The width of the large rectangle is $20 + 2x$, and the length is $40 + 2x$.

Area of rectangle − area of pool = area of grass.

$$(20 + 2x)(40 + 2x) - 20(40) = 700$$

Step 4

$$800 + 120x + 4x^2 - 800 = 700$$
$$4x^2 + 120x - 700 = 0$$
$$x^2 + 30x - 175 = 0$$
$$(x + 35)(x - 5) = 0$$

$x + 35 = 0$ or $x - 5 = 0$

$x = -35$ or $x = 5$

Step 5
The width cannot be -35, so the grass strip should be 5 ft wide.

Step 6
If $x = 5$, then the area of the large rectangle is

$$(40 + 2 \cdot 5)(20 + 2 \cdot 5) = 50 \cdot 30 = 1500 \text{ ft}^2.$$

The area of the pool is $40 \cdot 20 = 800 \text{ ft}^2$. So, the area of the grass strip is $1500 - 800 = 700 \text{ ft}^2$, as required. The answer is correct.

5.

$$s(t) = -16t^2 + 64t$$
$$32 = -16t^2 + 64t \quad \textit{Let } s(t) = 32.$$
$$16t^2 - 64t + 32 = 0$$
$$t^2 - 4t + 2 = 0 \quad \textit{Divide by 16.}$$

Use $a = 1$, $b = -4$, and $c = 2$ in the quadratic formula.

$$t = \frac{-b \pm \sqrt{b^2 - 4ac}}{2a}$$
$$t = \frac{-(-4) \pm \sqrt{(-4)^2 - 4(1)(2)}}{2(1)}$$
$$= \frac{4 \pm \sqrt{16 - 8}}{2} = \frac{4 \pm \sqrt{8}}{2}$$
$$= \frac{4 \pm 2\sqrt{2}}{2} = 2 \pm \sqrt{2}$$
$$t = 2 + \sqrt{2} \approx 3.4 \quad \text{or} \quad t = 2 - \sqrt{2} \approx 0.6$$

The ball will be at a height of 32 ft at about 0.6 sec and 3.4 sec.

6.

$$x = \frac{-14.8 \pm \sqrt{14.8^2 - 4(-0.065)(-301)}}{2(-0.065)}$$
$$= \frac{-14.8 \pm \sqrt{140.78}}{-0.13}$$
$$x \approx 22.58 \approx 23 \quad \text{or} \quad x \approx 205.12 \approx 205$$

The solution $x \approx 23$ is valid for this problem. The solution $x \approx 205$ is rejected since it corresponds to a year far beyond the period covered by the model.

10.4 Section Exercises

1. The first step in solving a formula like $gw^2 = 2r$ for w is to divide both sides by g. This isolates w^2 and allows us to apply the square root property.

3. Since the triangle is a right triangle, use the Pythagorean formula with legs m and n and hypotenuse p.

$$m^2 + n^2 = p^2$$
$$m^2 = p^2 - n^2$$
$$m = \sqrt{p^2 - n^2}$$

Only the positive square root is given since m represents the side of a triangle.

5. Solve $d = kt^2$ for t.

$$kt^2 = d$$
$$t^2 = \frac{d}{k} \quad \textit{Divide by k.}$$
$$t = \pm\sqrt{\frac{d}{k}} \quad \textit{Use square root property.}$$
$$= \frac{\pm\sqrt{d}}{\sqrt{k}} \cdot \frac{\sqrt{k}}{\sqrt{k}} \quad \textit{Rationalize denominator.}$$
$$t = \frac{\pm\sqrt{dk}}{k} \quad \textit{Simplify}$$

7. Solve $I = \dfrac{ks}{d^2}$ for d.

$$Id^2 = ks \quad \textit{Multiply by } d^2.$$
$$d^2 = \frac{ks}{I} \quad \textit{Divide by I.}$$
$$d = \pm\sqrt{\frac{ks}{I}} \quad \textit{Use square root property.}$$
$$= \pm\frac{\sqrt{ks}}{\sqrt{I}} \cdot \frac{\sqrt{I}}{\sqrt{I}} \quad \textit{Rationalize denominator.}$$
$$d = \frac{\pm\sqrt{ksI}}{I} \quad \textit{Simplify}$$

9. Solve $F = \dfrac{kA}{v^2}$ for v.

$$v^2F = kA \quad \textit{Multiply by } v^2.$$
$$v^2 = \frac{kA}{F} \quad \textit{Divide by F.}$$
$$v = \pm\sqrt{\frac{kA}{F}} \quad \textit{Use square root property.}$$
$$= \frac{\pm\sqrt{kA}}{\sqrt{F}} \cdot \frac{\sqrt{F}}{\sqrt{F}} \quad \textit{Rationalize denominator.}$$
$$v = \frac{\pm\sqrt{kAF}}{F} \quad \textit{Simplify}$$

11. Solve $V = \dfrac{1}{3}\pi r^2 h$ for r.

$$3V = \pi r^2 h \quad \textit{Multiply by 3.}$$
$$\frac{3V}{\pi h} = r^2 \quad \textit{Divide by } \pi h.$$
$$r = \pm\sqrt{\frac{3V}{\pi h}} \quad \textit{Use square root property.}$$
$$= \frac{\pm\sqrt{3V} \cdot \sqrt{\pi h}}{\sqrt{\pi h} \cdot \sqrt{\pi h}} \quad \textit{Rationalize denominator.}$$
$$r = \frac{\pm\sqrt{3\pi V h}}{\pi h} \quad \textit{Simplify}$$

13. Solve $At^2 + Bt = -C$ for t.

$$At^2 + Bt + C = 0$$

Use the quadratic formula.

$$t = \frac{-B \pm \sqrt{B^2 - 4AC}}{2A}$$

15. Solve $D = \sqrt{kh}$ for h.

$D^2 = kh$ *Square both sides.*

$\frac{D^2}{k} = h$ *Divide by k.*

17. Solve $p = \sqrt{\frac{k\ell}{g}}$ for ℓ.

$p^2 = \frac{k\ell}{g}$ *Square both sides.*

$p^2 g = k\ell$ *Multiply by g.*

$\frac{p^2 g}{k} = \ell$ *Divide by k.*

19. Apply the Pythagorean formula.

$$(5m)^2 = (2m)^2 + (2m+3)^2$$
$$25m^2 = 4m^2 + 4m^2 + 12m + 9$$
$$17m^2 - 12m - 9 = 0$$

Here $a = 17$, $b = -12$, and $c = -9$.

$$m = \frac{-(-12) \pm \sqrt{(-12)^2 - 4(17)(-9)}}{2(17)}$$
$$= \frac{12 \pm \sqrt{144 + 612}}{34} = \frac{12 \pm \sqrt{756}}{34}$$
$$m = \frac{12 + \sqrt{756}}{34} \approx 1.16 \text{ or}$$
$$m = \frac{12 - \sqrt{756}}{34} \approx -0.46$$

Reject the negative solution.
If $m = 1.16$, then

$$5m = 5(1.16) = 5.80,$$
$$2m = 2(1.16) = 2.32, \text{ and}$$
$$2m + 3 = 2(1.16) + 3 = 5.32.$$

The lengths of the sides of the triangle are approximately 2.3, 5.3, and 5.8.

21. Let $x =$ the distance traveled by the eastbound ship. Then $x + 70 =$ the distance traveled by the southbound ship.

Since the ships are traveling at right angles to one another, the distance d between them can be found using the Pythagorean formula.

$$c^2 = a^2 + b^2$$
$$d^2 = x^2 + (x+70)^2$$

Let $d = 170$, and solve for x.

$$170^2 = x^2 + (x+70)^2$$
$$28{,}900 = x^2 + x^2 + 140x + 4900$$
$$0 = 2x^2 + 140x - 24{,}000$$
$$0 = x^2 + 70x - 12{,}000$$
$$0 = (x+150)(x-80)$$

$$x + 150 = 0 \quad \text{or} \quad x - 80 = 0$$
$$x = -150 \quad \text{or} \quad x = 80$$

Distance cannot be negative, so reject -150. If $x = 80$, then $x + 70 = 150$. The eastbound ship traveled 80 miles, and the southbound ship traveled 150 miles.

23. Let $x =$ the width of the uncovered strip of flooring.

From the problem,
(length of the rug) • (width of the rug) = 234.

The rug is centered in the room a distance x from the walls (width of the strip x), so the length of the rug

= length of the room − 2 • (width of the strip)
$= 20 - 2x$,

and the width of the rug

= width of the room − 2 • (width of the strip)
$= 15 - 2x$.

The equation

(length of rug) • (width of rug) = 234

becomes

$$(20 - 2x)(15 - 2x) = 234.$$
$$300 - 70x + 4x^2 = 234$$
$$4x^2 - 70x + 66 = 0$$

Divide by 2.

$$2x^2 - 35x + 33 = 0$$
$$(2x - 33)(x - 1) = 0$$

$$2x - 33 = 0 \quad \text{or} \quad x - 1 = 0$$
$$x = \tfrac{33}{2} \quad \text{or} \quad x = 1$$

Reject $\frac{33}{2} = 16\frac{1}{2}$ since $16\frac{1}{2}$ is wider than the room itself. The width of the uncovered strip is 1 foot.

25. Let x be the width of the sheet metal. Then the length is $2x - 4$.

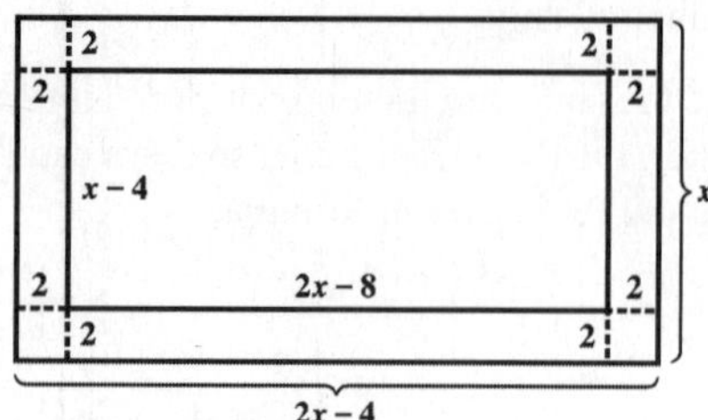

By cutting out 2-inch squares from each corner we get a rectangle with width $x - 4$ and length $(2x - 4) - 4 = 2x - 8$. The uncovered box then has height 2 inches, length $(2x - 8)$ inches, and width $(x - 4)$ inches.
Use the formula $V = LWH$ or $V = HLW$.

$$256 = 2(2x - 8)(x - 4)$$
$$256 = 4(x - 4)(x - 4) \quad \textit{Factor out 2.}$$
$$64 = (x - 4)^2 \quad \textit{Divide by 4.}$$

Use the square root property.

$$\pm 8 = x - 4$$

$$x - 4 = 8 \quad \text{or} \quad x - 4 = -8$$
$$x = 12 \quad \text{or} \quad x = -4$$

Since x represents width, discard the negative solution.
The width is 12 inches, and the length is $2(12) - 4 = 20$ inches.

27. $s(t) = -16t^2 + 128t$

$$213 = -16t^2 + 128t \quad \textit{Let } s(t) = 213.$$
$$0 = -16t^2 + 128t - 213$$
$$0 = 16t^2 - 128t + 213 \quad \textit{Divide by } -1.$$

Here $a = 16$, $b = -128$, and $c = 213$.

$$t = \frac{-b \pm \sqrt{b^2 - 4ac}}{2a}$$
$$t = \frac{-(-128) \pm \sqrt{(-128)^2 - 4(16)(213)}}{2(16)}$$
$$= \frac{128 \pm \sqrt{16{,}384 - 13{,}632}}{32}$$
$$= \frac{128 \pm \sqrt{2752}}{32}$$
$$t = \frac{128 + \sqrt{2752}}{32} \quad \text{or} \quad t = \frac{128 - \sqrt{2752}}{32}$$
$$t \approx 2.4 \quad \text{or} \quad t \approx 5.6$$

The ball will be 213 feet from the ground after 2.4 seconds and again after 5.6 seconds.

29. $D(t) = 13t^2 - 100t$

$$180 = 13t^2 - 100t \quad \textit{Let } D(t) = 180.$$
$$0 = 13t^2 - 100t - 180$$

Here $a = 13$, $b = -100$, and $c = -180$.

$$t = \frac{-b \pm \sqrt{b^2 - 4ac}}{2a}$$
$$t = \frac{-(-100) \pm \sqrt{(-100)^2 - 4(13)(-180)}}{2(13)}$$
$$= \frac{100 \pm \sqrt{10{,}000 + 9360}}{26}$$
$$= \frac{100 \pm \sqrt{19{,}360}}{26}$$
$$t = \frac{100 + \sqrt{19{,}360}}{26} \quad \text{or} \quad t = \frac{100 - \sqrt{19{,}360}}{26}$$
$$t \approx 9.2 \quad \text{or} \quad t \approx -1.5$$

Discard the negative solution. The car will skid 180 feet in approximately 9.2 seconds.

31. $s(t) = -16t^2 + 160t$

$$400 = -16t^2 + 160t \quad \textit{Let } s(t) = 400.$$
$$0 = -16t^2 + 160t - 400$$
$$0 = t^2 - 10t + 25 \quad \textit{Divide by } -16.$$
$$0 = (t - 5)(t - 5)$$
$$0 = (t - 5)^2$$
$$0 = t - 5$$
$$5 = t$$

The ball reaches a height of 400 feet after 5 seconds. This is its maximum height since this is the only time it reaches 400 feet.

33. Supply and demand are equal when

$$3p - 200 = \frac{3200}{p}.$$

Solve for p.

$$3p^2 - 200p = 3200$$
$$3p^2 - 200p - 3200 = 0$$

Use the quadratic formula with $a = 3$, $b = -200$, and $c = -3200$.

$$p = \frac{-(-200) \pm \sqrt{(-200)^2 - 4(3)(-3200)}}{2(3)}$$
$$= \frac{200 \pm \sqrt{40{,}000 + 38{,}400}}{6}$$
$$= \frac{200 \pm \sqrt{78{,}400}}{6} = \frac{200 \pm 280}{6}$$
$$p = \tfrac{480}{6} = 80 \quad \text{or} \quad p = \tfrac{-80}{6} = -\tfrac{40}{3}$$

Discard the negative solution. The supply and demand are equal when the price is 80 cents or \$0.80.

35. **(a)** From the graph, the number of miles traveled in 2000 appears to be 2750 billion (to the nearest ten billion).

(b) For 2000, $x = 2000 - 1994 = 6$.

$$\begin{aligned} f(x) &= -1.705x^2 + 75.93x + 2351 \\ f(6) &= -1.705(6)^2 + 75.93(6) + 2351 \\ &= 2745.2 \end{aligned}$$

To the nearest ten billion, the model gives 2750 billion, the same as the estimate in part (a).

37. Use $f(x) = -1.705x^2 + 75.93x + 2351$ with $f(x) = 2800$.

$$\begin{aligned} 2800 &= -1.705x^2 + 75.93x + 2351 \\ 0 &= -1.705x^2 + 75.93x - 449 \end{aligned}$$

Here $a = -1.705$, $b = 75.93$, and $c = -449$.

$$x = \frac{-b \pm \sqrt{b^2 - 4ac}}{2a}$$

$$\begin{aligned} x &= \frac{-75.93 \pm \sqrt{(75.93)^2 - 4(-1.705)(-449)}}{2(-1.705)} \\ &= \frac{-75.93 \pm \sqrt{2703.1849}}{-3.41} \\ &\approx 7.02 \text{ or } 37.51 \end{aligned}$$

The model indicates that the number of miles traveled was 2800 billion in the year $1994 + 7 = 2001$. The other value represents a future year. The graph indicates that vehicle-miles reached 2800 billion in 2001.

39. Let F denote the Froude number. Solve

$$F = \frac{v^2}{g\ell}$$

for v.

$$\begin{aligned} v^2 &= Fg\ell \\ v &= \pm\sqrt{Fg\ell} \end{aligned}$$

v is positive, so

$$v = \sqrt{Fg\ell}.$$

For the rhinoceros, $\ell = 1.2$ and $F = 2.57$.

$$v = \sqrt{(2.57)(9.8)(1.2)} \approx 5.5$$

or 5.5 meters per second.

41. Write a proportion.

$$\frac{x-4}{3x-19} = \frac{4}{x-3}$$

Multiply by the LCD, $(3x - 19)(x - 3)$.

$$\begin{aligned} (3x-19)(x-3)\left(\frac{x-4}{3x-19}\right) &= (3x-19)(x-3)\left(\frac{4}{x-3}\right) \\ (x-3)(x-4) &= (3x-19)4 \\ x^2 - 7x + 12 &= 12x - 76 \\ x^2 - 19x + 88 &= 0 \\ (x-8)(x-11) &= 0 \end{aligned}$$

$$\begin{aligned} x - 8 = 0 \quad &\text{or} \quad x - 11 = 0 \\ x = 8 \quad &\text{or} \quad x = 11 \end{aligned}$$

If $x = 8$, then

$$3x - 19 = 3(8) - 19 = 5.$$

If $x = 11$, then

$$3x - 19 = 3(11) - 19 = 14.$$

Thus, $AC = 5$ or $AC = 14$.

10.5 Graphs of Quadratic Functions

10.5 Margin Exercises

1. **(a)**

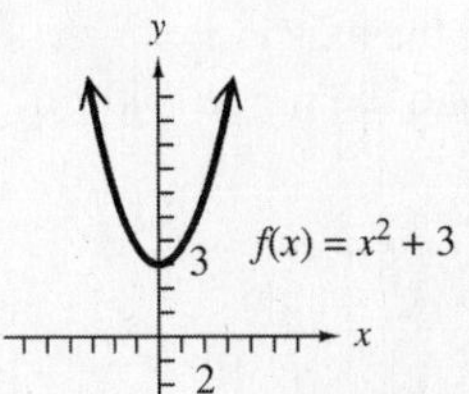

The graph of $f(x) = x^2 + 3$ has the same shape as the graph of $f(x) = x^2$. Here $k = 3$, so the graph is shifted up 3 units and has vertex $(0, 3)$.

Since x can be any real number, the domain is $(-\infty, \infty)$. The value of y is always greater than or equal to 3, so the range is $[3, \infty)$.

(b)

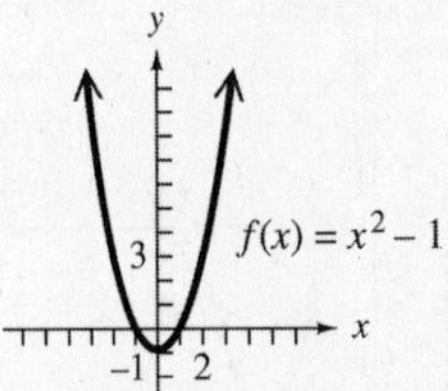

The graph of $f(x) = x^2 - 1$ has the same shape as the graph of $f(x) = x^2$. Here $k = -1$, so the graph is shifted 1 unit down and has vertex $(0, -1)$.

Since x can be any real number, the domain is $(-\infty, \infty)$. The value of y is always greater than or equal to -1, so the range is $[-1, \infty)$.

2. **(a)**

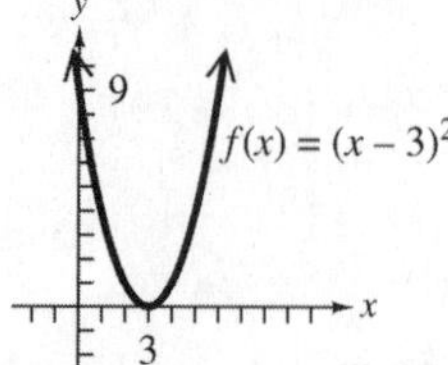

The parabola $f(x) = (x - 3)^2$ has the same shape as $f(x) = x^2$ but is shifted 3 units to the right, since 3 would cause $x - 3$ to equal 0. The vertex is $(3, 0)$ and the axis is $x = 3$. The domain is $(-\infty, \infty)$ and the range is $[0, \infty)$.

(b)

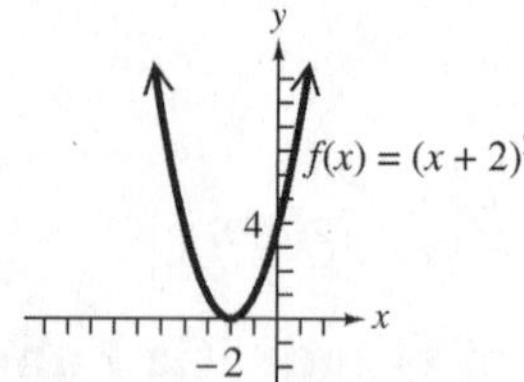

The parabola $f(x) = (x + 2)^2$ has the same shape as $f(x) = x^2$ but is shifted 2 units to the left, since -2 would cause $x + 2$ to equal 0. The vertex is $(-2, 0)$ and the axis is $x = -2$. The domain is $(-\infty, \infty)$ and the range is $[0, \infty)$.

3. **(a)**

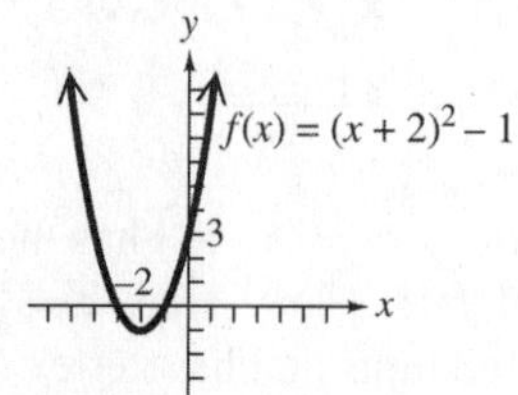

$f(x) = (x + 2)^2 - 1$ has the same shape as $f(x) = x^2$ but is shifted 2 units to the left (since $x + 2 = 0$ if $x = -2$) and 1 unit down (because of the -1). The vertex is $(-2, -1)$ and the axis is $x = -2$. The domain is $(-\infty, \infty)$ and the range is $[-1, \infty)$.

(b)

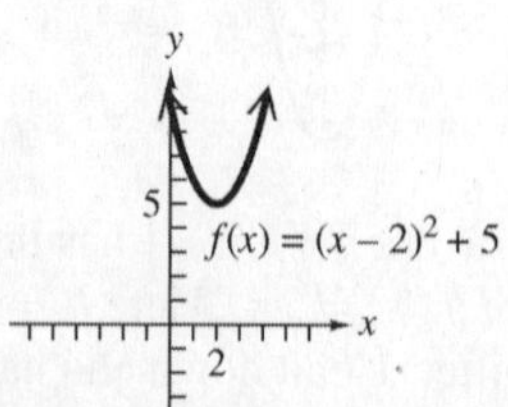

$f(x) = (x - 2)^2 + 5$ has the same shape as $f(x) = x^2$ but is shifted 2 units to the right (since $x - 2 = 0$ if $x = 2$) and 5 units up (because of the $+5$). The vertex is $(2, 5)$ and the axis is $x = 2$. The domain is $(-\infty, \infty)$ and the range is $[5, \infty)$.

4. An equation of the form $f(x) = a(x - h)^2 + k$ has a parabola for its graph. If $a > 0$, the parabola opens up; if $a < 0$, the parabola opens down.

(a) $f(x) = -\frac{2}{3}x^2$

Here, $a = -\frac{2}{3} < 0$, so the parabola opens *down*.

(b) $f(x) = \frac{3}{4}x^2 + 1$

Here, $a = \frac{3}{4} > 0$, so the parabola opens *up*.

(c) $f(x) = -2x^2 - 3$

Here, $a = -2 < 0$, so the parabola opens *down*.

(d) $f(x) = 3x^2 + 2$

Here, $a = 3 > 0$, so the parabola opens *up*.

5. For $f(x) = a(x - h)^2 + k$; if $0 < |a| < 1$, then the graph is wider than the graph of $f(x) = x^2$; if $|a| > 1$, then the graph is narrower.

(a) $f(x) = -\frac{2}{3}x^2$

Here, $|a| = \left|-\frac{2}{3}\right| = \frac{2}{3} < 1$, so the graph is *wider*.

(b) $f(x) = \frac{3}{4}x^2 + 1$

Here, $|a| = \left|\frac{3}{4}\right| = \frac{3}{4} < 1$, so the parabola is *wider*.

(c) $f(x) = -2x^2 - 3$

Here, $|a| = |-2| = 2 > 1$, so the parabola is *narrower*.

(d) $f(x) = 3x^2 + 2$

Here, $|a| = |3| = 3 > 1$, so the parabola is *narrower*.

6. $f(x) = \frac{1}{2}(x - 2)^2 + 1$

This equation has a graph like $f(x) = x^2$ but is shifted 2 units to the right and 1 unit up. It has vertex $(2, 1)$. Since $a = \frac{1}{2} > 0$, the parabola opens up. Since $|a| = \left|\frac{1}{2}\right| = \frac{1}{2} < 1$, the parabola is wider than the graph of $f(x) = x^2$. If $x = 0$, then $f(x) = 3$, so $(0, 3)$ is on the graph. By symmetry about the axis $x = 2$, the point $(4, 3)$ is also on the graph.

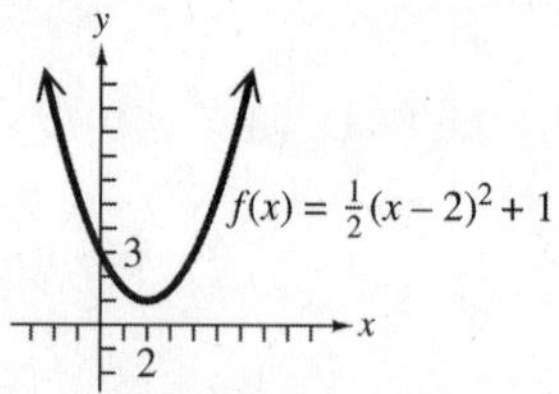

7. **(a)** The points appear to lie on a line, so a *linear* function would be a more appropriate model. The line would rise, so it would have a *positive* slope.

(b) The points appear to lie on a parabola, so a *quadratic* function would be a more appropriate model. The parabola would open up, so a would be *positive.*

8. Use the points with

$$y = ax^2 + bx + c.$$

$$\begin{aligned}
&\text{Use } (1, 5939)\text{:} && a(1)^2 + b(1) + c = 5939 \\
&\text{Use } (6, 7471)\text{:} && a(6)^2 + b(6) + c = 7471 \\
&\text{Use } (10, 6694)\text{:} && a(10)^2 + b(10) + c = 6694
\end{aligned}$$

Simplifying the system gives us:

$$\begin{aligned}
a + b + c &= 5939 \quad (1) \\
36a + 6b + c &= 7471 \quad (2) \\
100a + 10b + c &= 6694 \quad (3)
\end{aligned}$$

To eliminate c, multiply equation (1) by -1 and add the result to equation (2).

$$\begin{aligned}
-a - b - c &= -5939 \quad -1 \times (1) \\
36a + 6b + c &= 7471 \quad (2) \\
\hline
35a + 5b &= 1532 \quad (4)
\end{aligned}$$

To eliminate c again, multiply equation (2) by -1 and add the result to equation (3).

$$\begin{aligned}
-36a - 6b - c &= -7471 \quad -1 \times (2) \\
100a + 10b + c &= 6694 \quad (3) \\
\hline
64a + 4b &= -777 \quad (5)
\end{aligned}$$

To eliminate b, multiply equation (4) by -4 and add the result to 5 times equation (5).

$$\begin{aligned}
-140a - 20b &= -6128 \quad -4 \times (4) \\
320a + 20b &= -3885 \quad 5 \times (5) \\
\hline
180a &= -10{,}013 \\
a &= \tfrac{-10{,}013}{180} \approx -55.63
\end{aligned}$$

To find b, substitute -55.63 for a in (4).

$$\begin{aligned}
35a + 5b &= 1532 \quad (4) \\
35(-55.63) + 5b &= 1532 \\
-1947.05 + 5b &= 1532 \\
5b &= 3479.05 \\
b &= 695.81
\end{aligned}$$

To find c, use (1).

$$\begin{aligned}
c &= 5939 - a - b \\
&= 5939 - (-55.63) - 695.81 = 5298.82
\end{aligned}$$

So using these three points gives us the quadratic model

$$y = -55.63x^2 + 695.81x + 5298.82.$$

10.5 Section Exercises

1. A parabola with equation $f(x) = a(x - h)^2 + k$ has vertex $V(h, k)$. We'll identify the vertex for each quadratic function.

(a) $f(x) = (x + 2)^2 - 1$

$V(-2, -1)$, choice **B**

(b) $f(x) = (x + 2)^2 + 1$

$V(-2, 1)$, choice **C**

(c) $f(x) = (x - 2)^2 - 1$

$V(2, -1)$, choice **A**

(d) $f(x) = (x - 2)^2 + 1$

$V(2, 1)$, choice **D**

For Exercises 3–10, we write $f(x)$ in the form $f(x) = a(x - h)^2 + k$ and then list the vertex (h, k).

3. $f(x) = -3x^2 = -3(x - 0)^2 + 0$
The vertex (h, k) is $(0, 0)$.

5. $f(x) = x^2 + 4 = 1(x - 0)^2 + 4$
The vertex (h, k) is $(0, 4)$.

7. $f(x) = (x - 1)^2 = 1(x - 1)^2 + 0$
The vertex (h, k) is $(1, 0)$.

9. $f(x) = (x + 3)^2 - 4 = 1[x - (-3)]^2 - 4$
The vertex (h, k) is $(-3, -4)$.

11. In Exercise 9, the parabola is shifted 3 units to the left and 4 units down. The parabola in Exercise 10 is shifted 5 units to the right and 8 units down.

13. $f(x) = -\frac{2}{5}x^2$

Since $a = -\frac{2}{5} < 0$, the graph opens down. Since $|a| = \left|-\frac{2}{5}\right| = \frac{2}{5} < 1$, the graph is wider than the graph of $f(x) = x^2$.

15. $f(x) = 3x^2 + 1$

Since $a = 3 > 0$, the graph opens up. Since $|a| = |3| = 3 > 1$, the graph is narrower than the graph of $f(x) = x^2$.

17. Consider $f(x) = a(x - h)^2 + k$.

(a) If $h > 0$ and $k > 0$ in $f(x) = a(x - h)^2 + k$, the shift is to the right and upward, so the vertex is in quadrant I.

(b) If $h > 0$ and $k < 0$, the shift is to the right and downward, so the vertex is in quadrant IV.

(c) If $h < 0$ and $k > 0$, the shift is to the left and upward, so the vertex is in quadrant II.

(d) If $h < 0$ and $k < 0$, the shift is to the left and downward, so the vertex is in quadrant III.

19. $f(x) = -2x^2$ written in the form
$f(x) = a(x-h)^2 + k$ is
$f(x) = -2(x-0)^2 + 0.$

Here, $h = 0$ and $k = 0$, so the vertex (h, k) is $(0, 0)$. Since $a = -2 < 0$, the graph opens down. Since $|a| = |-2| = 2 > 1$, the graph is narrower than the graph of $f(x) = x^2$. By evaluating the function with $x = 2$ and $x = -2$, we see that the points $(2, -8)$ and $(-2, -8)$ are on the graph.

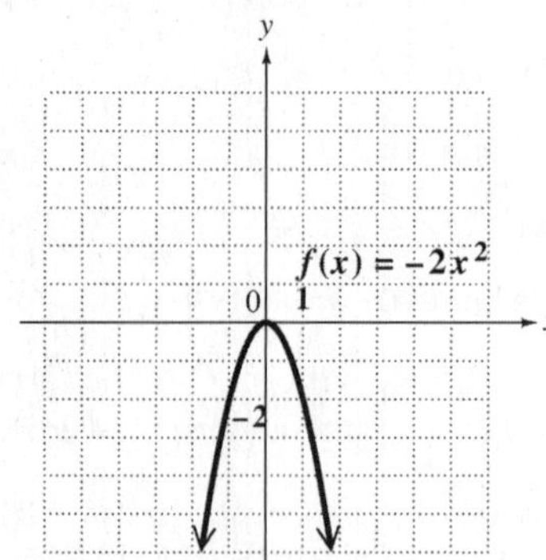

21. $f(x) = x^2 - 1$ written in the form
$f(x) = a(x-h)^2 + k$ is
$f(x) = 1(x-0)^2 + (-1).$

Here, $h = 0$ and $k = -1$, so the vertex is $(0, -1)$. The graph opens up and has the same shape as $f(x) = x^2$ because $a = 1$. Two other points on the graph are $(-2, 3)$ and $(2, 3)$.

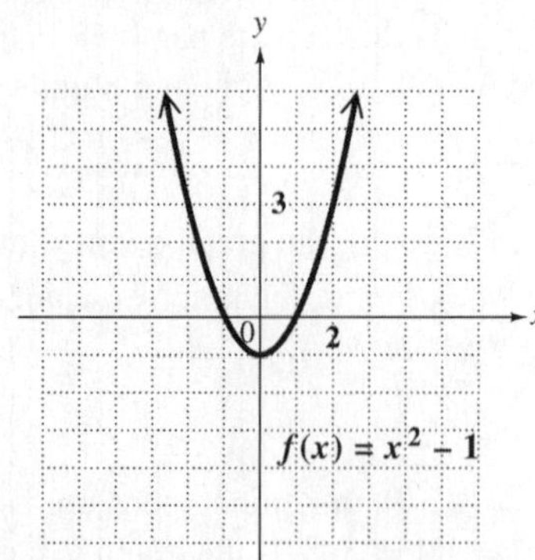

23. $f(x) = -x^2 + 2$ written in the form
$f(x) = a(x-h)^2 + k$ is
$f(x) = -1(x-0)^2 + 2.$

Here, $h = 0$ and $k = 2$, so the vertex (h, k) is $(0, 2)$. Since $a = -1 < 0$, the graph opens down. Since $|a| = |-1| = 1$, the graph has the same shape as $f(x) = x^2$. The points $(2, -2)$ and $(-2, -2)$ are on the graph.

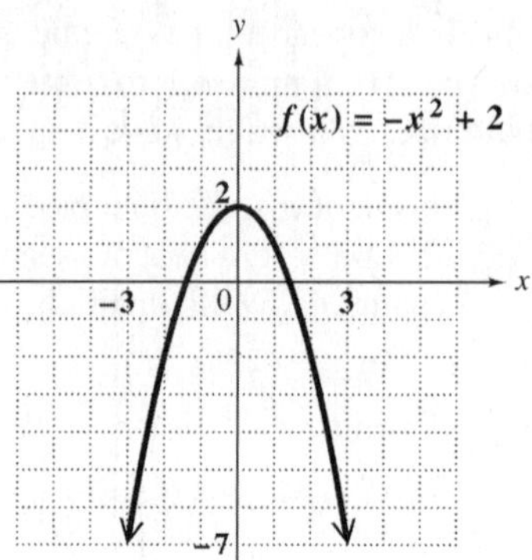

25. $f(x) = \frac{1}{2}(x-4)^2$ written in the form
$f(x) = a(x-h)^2 + k$ is
$f(x) = \frac{1}{2}(x-4)^2 + 0.$

Here, $h = 4$ and $k = 0$, so the vertex (h, k) is $(4, 0)$ and the axis is $x = 4$. The graph opens up since a is positive and is wider than the graph of $f(x) = x^2$ because $|a| = \frac{1}{2} < 1$. Two other points on the graph are $(2, 2)$ and $(6, 2)$. We can substitute any value for x, so the domain is $(-\infty, \infty)$. The range is $[0, \infty)$ since the smallest y-value is 0.

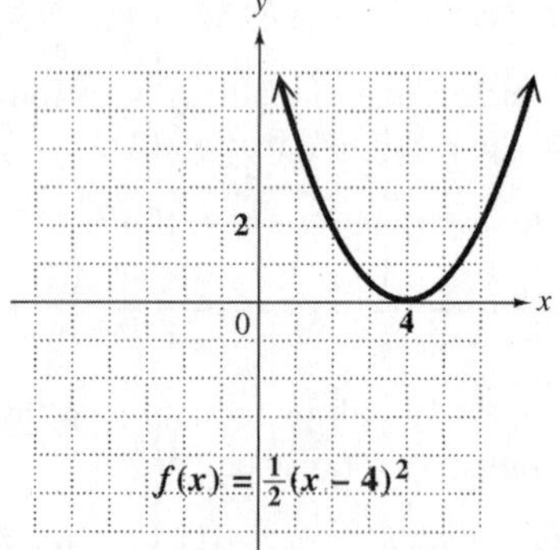

27. $f(x) = (x+2)^2 - 1$ written in the form
$f(x) = a(x-h)^2 + k$ is
$f(x) = 1[x - (-2)]^2 + (-1).$

Since $h = -2$ and $k = -1$, the vertex (h, k) is $(-2, -1)$ and the axis is $x = -2$. Here, $a = 1$, so the graph opens up and has the same shape as $f(x) = x^2$. The points $(-1, 0)$ and $(-3, 0)$ are on the graph. The domain is $(-\infty, \infty)$. The range is $[-1, \infty)$ since the smallest y-value is -1.

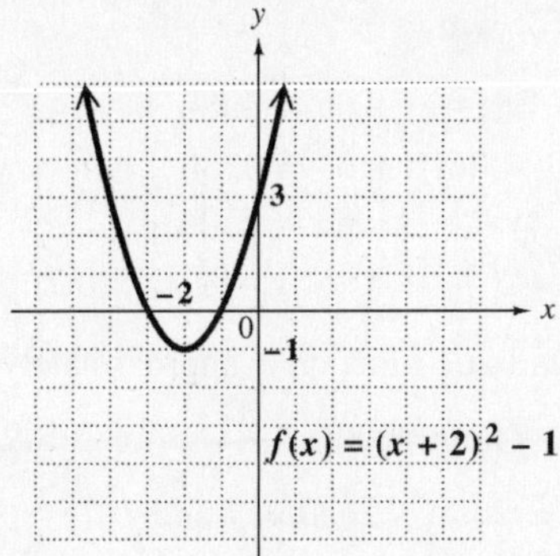

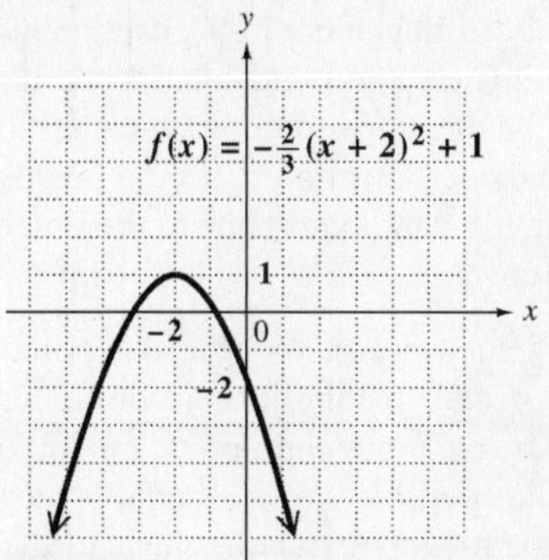

29. $f(x) = -2(x+3)^2 + 4$ written in the form $f(x) = a(x-h)^2 + k$ is $f(x) = -2[x - (-3)]^2 + 4$.

Since $h = -3$ and $k = 4$, the vertex (h, k) is $(-3, 4)$ and the axis is $x = -3$. Here, $a = -2 < 0$, so the graph opens down. Also, $|a| = 2 > 1$, so the graph is narrower than the graph of $f(x) = x^2$. The points $(-5, -4)$ and $(-1, -4)$ are on the graph.
We can substitute any value for x, so the domain is $(-\infty, \infty)$. The value of y is less than or equal to 4, so the range is $(-\infty, 4]$.

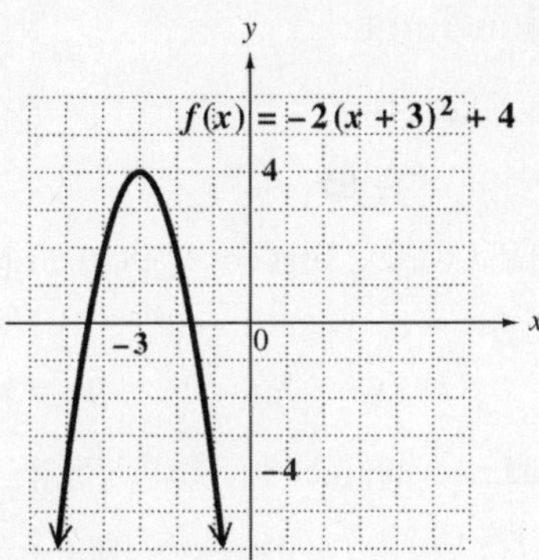

31. $f(x) = -\frac{2}{3}(x+2)^2 + 1$

Because $a = -\frac{2}{3}$, the graph opens down and is wider than the graph of $f(x) = x^2$. Because $h = -2$ and $k = 1$, the graph is shifted 2 units to the left and 1 unit upward. The vertex is at $(-2, 1)$ and the axis is $x = -2$. Two other points on the graph are $(-4, -\frac{5}{3})$ and $(0, -\frac{5}{3})$.
We can substitute any value for x, so the domain is $(-\infty, \infty)$. The value of y is less than or equal to 1, so the range is $(-\infty, 1]$.

33. The graph of $F(x) = x^2 + 6$ would be shifted 6 units upward from the graph of $f(x) = x^2$.

34. To graph $G(x) = x + 6$, plot the intercepts $(-6, 0)$ and $(0, 6)$, and draw the line through them.

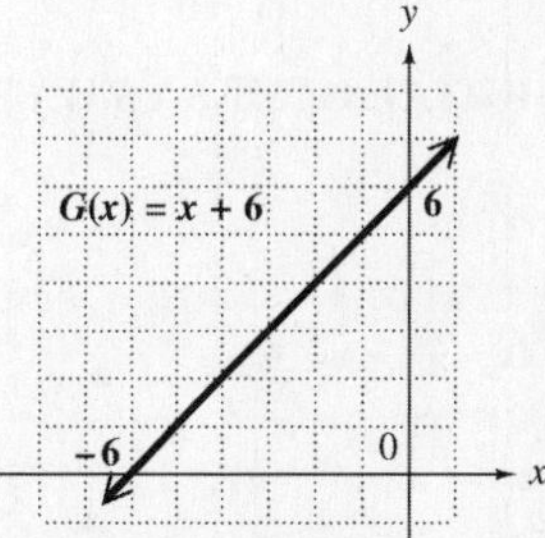

35. When considering the graph of $G(x) = x + 6$, the y-intercept is 6. The graph of $g(x) = x$ has y-intercept 0. Therefore, the graph of $G(x) = x + 6$ is shifted 6 units upward compared to the graph of $g(x) = x$.

36. The graph of $F(x) = (x-6)^2$ is shifted 6 units to the right compared to the graph of $f(x) = x^2$.

37. To graph $G(x) = x - 6$, plot the intercepts $(6, 0)$ and $(0, -6)$, and draw the line through them.

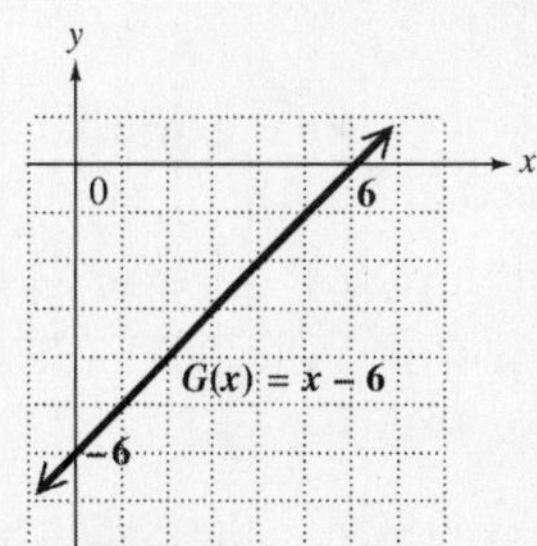

38. When considering the graph of $G(x) = x - 6$, its x-intercept is 6 as compared to the graph of $g(x) = x$ with x-intercept 0. The graph of $G(x) = x - 6$ is shifted 6 units to the right compared to the graph of $g(x) = x$.

39. Since the arrangement of the data points is approximately parabolic, a quadratic function would be the more appropriate model for the data set. The coefficient of x^2 should be positive, since the roughly parabolic shape of the graphed data set opens upward.

41. Since the arrangement of the data points is approximately parabolic, a quadratic function would be the more appropriate model for the data set. The coefficient of x^2 should be negative, since the roughly parabolic shape of the graphed data set opens downward.

43. Since the points lie approximately along a straight line, a linear function would be the more appropriate model for the data set. As the x-values of the points increase, the y-values increase, so the slope should be positive.

45. **(a)**

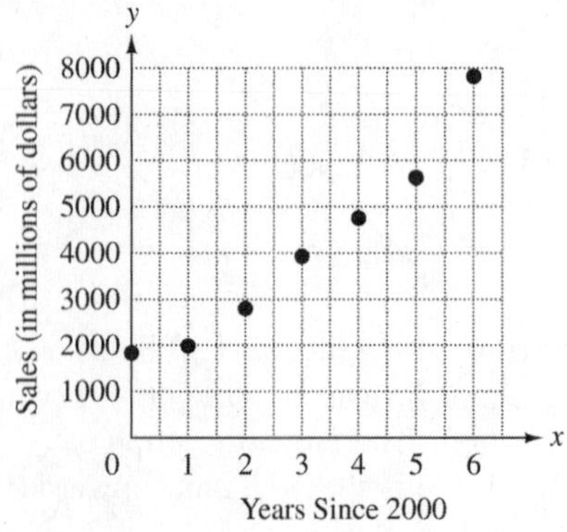

(b) Since the arrangement of the data points is approximately parabolic, a quadratic function would be the more appropriate model for the data set. The coefficient of x^2 should be positive, since the roughly parabolic shape of the graphed data set opens upward.

(c) Use $ax^2 + bx + c = y$ with $(0, 1825)$, $(3, 3921)$, and $(6, 7805)$.

$$\begin{aligned} 0a + 0b + c &= 1825 \quad (1) \\ 9a + 3b + c &= 3921 \quad (2) \\ 36a + 6b + c &= 7805 \quad (3) \end{aligned}$$

From (1), $c = 1825$, so the system becomes

$$\begin{aligned} 9a + 3b &= 2096 \quad (4) \\ 36a + 6b &= 5980 \quad (5) \end{aligned}$$

Now eliminate b.

$$\begin{aligned} -18a - 6b &= -4192 \qquad -2 \times (4) \\ 36a + 6b &= 5980 \qquad (5) \\ \hline 18a &= 1788 \\ a &= \tfrac{1788}{18} = 99\tfrac{1}{3} \approx 99.3 \end{aligned}$$

From (5) with $a = 99\frac{1}{3} = \frac{298}{3}$,

$$\begin{aligned} 36(\tfrac{298}{3}) + 6b &= 5980 \\ 3576 + 6b &= 5980 \\ 6b &= 2404 \\ b &= \tfrac{2404}{6} \approx 400.7 \end{aligned}$$

The quadratic function is approximately

$$y = f(x) = 99.3x^2 + 400.7x + 1825.$$

(d) $x = 2007 - 2000 = 7$ and $f(7) \approx 9496$. The sales of digital cameras in the United States in 2007 were about \$9496 million.

(e) No. The number of digital cameras sold in 2007 is far below the number approximated by the model. Rather than continuing to increase, sales of digital cameras fell in 2007.

47. **(a)** $y = f(x) = -0.3x^2 + 3.8x + 24.5$
$x = 2005 - 1990 = 15$ and $f(15) = 14$.
$x = 2007 - 1990 = 17$ and $f(17) = 2.4$.

(b) The approximations using the model are far too low.

10.6 More about Parabolas and Their Applications

10.6 Margin Exercises

1. **(a)** $f(x) = x^2 - 6x + 7$

To find the vertex, first complete the square on $x^2 - 6x$.

$$\left[\tfrac{1}{2}(-6)\right]^2 = (-3)^2 = 9$$

Now add and subtract 9 on the right.

$$\begin{aligned} f(x) &= (x^2 - 6x + 9 - 9) + 7 \\ &= (x^2 - 6x + 9) - 9 + 7 \\ f(x) &= (x - 3)^2 - 2 \end{aligned}$$

The vertex of this parabola is $(3, -2)$.

(b) $f(x) = x^2 + 4x - 9$

To find the vertex, first complete the square on $x^2 + 4x$.

$$\left[\tfrac{1}{2}(4)\right]^2 = 2^2 = 4$$

Now add and subtract 4 on the right.

$$\begin{aligned} f(x) &= (x^2 + 4x + 4 - 4) - 9 \\ &= (x^2 + 4x + 4) - 4 - 9 \\ f(x) &= (x + 2)^2 - 13 \end{aligned}$$

The vertex of this parabola is $(-2, -13)$.

2. **(a)** $f(x) = 2x^2 - 4x + 1$

Factor out 2 from the first two terms.

$$f(x) = 2(x^2 - 2x) + 1$$

Complete the square on $x^2 - 2x$.

$$\left[\tfrac{1}{2}(-2)\right]^2 = (-1)^2 = 1$$

Add and subtract 1.

$$\begin{aligned} f(x) &= 2(x^2 - 2x + 1 - 1) + 1 \\ &= 2(x^2 - 2x + 1) + 2(-1) + 1 \\ &= 2(x^2 - 2x + 1) - 2 + 1 \\ f(x) &= 2(x - 1)^2 - 1 \end{aligned}$$

The vertex of this parabola is $(1, -1)$.

(b) $f(x) = -\dfrac{1}{2}x^2 + 2x - 3$

Factor out $-\frac{1}{2}$ from the first two terms.

$$f(x) = -\tfrac{1}{2}(x^2 - 4x) - 3$$

Complete the square on $x^2 - 4x$.

$$\left[\tfrac{1}{2}(-4)\right]^2 = (-2)^2 = 4$$

Add and subtract 4.

$$\begin{aligned} f(x) &= -\tfrac{1}{2}(x^2 - 4x + 4 - 4) - 3 \\ &= -\tfrac{1}{2}(x^2 - 4x + 4) - \tfrac{1}{2}(-4) - 3 \\ &= -\tfrac{1}{2}(x^2 - 4x + 4) + 2 - 3 \\ f(x) &= -\tfrac{1}{2}(x - 2)^2 - 1 \end{aligned}$$

The vertex of this parabola is $(2, -1)$.

3. **(a)** For $f(x) = -2x^2 + 3x - 1$, $a = -2$, $b = 3$, and $c = -1$.

The vertex is $\left(\dfrac{-b}{2a}, f\left(\dfrac{-b}{2a}\right)\right)$, so the x-coordinate is

$$\frac{-b}{2a} = \frac{-3}{2(-2)} = \frac{-3}{-4} = \frac{3}{4},$$

and the y-coordinate is

$$\begin{aligned} f(\tfrac{3}{4}) &= -2(\tfrac{3}{4})^2 + 3(\tfrac{3}{4}) - 1 \\ &= -2(\tfrac{9}{16}) + \tfrac{9}{4} - 1 \\ &= -\tfrac{9}{8} + \tfrac{18}{8} - \tfrac{8}{8} \\ &= \tfrac{1}{8}. \end{aligned}$$

The vertex is $(\frac{3}{4}, \frac{1}{8})$.

(b) For $f(x) = 4x^2 - x + 5$, $a = 4$, $b = -1$, and $c = 5$.

The vertex is $\left(\dfrac{-b}{2a}, f\left(\dfrac{-b}{2a}\right)\right)$, so the x-coordinate is

$$\frac{-b}{2a} = \frac{-(-1)}{2(4)} = \frac{1}{8},$$

and the y-coordinate is

$$\begin{aligned} f(\tfrac{1}{8}) &= 4(\tfrac{1}{8})^2 - \tfrac{1}{8} + 5 \\ &= 4(\tfrac{1}{64}) - \tfrac{1}{8} + 5 \\ &= \tfrac{1}{16} - \tfrac{2}{16} + \tfrac{80}{16} \\ &= \tfrac{79}{16}. \end{aligned}$$

The vertex is $(\frac{1}{8}, \frac{79}{16})$.

4. $f(x) = x^2 - 6x + 5$

Step 1
The graph opens up because $a = 1 > 0$.

Step 2
To find the vertex, complete the square on $x^2 - 6x$.

$$\left[\tfrac{1}{2}(-6)\right]^2 = (-3)^2 = 9$$

Add and subtract 9.

$$\begin{aligned} f(x) &= x^2 - 6x + 9 - 9 + 5 \\ f(x) &= (x - 3)^2 - 4 \end{aligned}$$

The vertex is at $(3, -4)$.

Step 3
Find any x-intercepts. Let $f(x) = 0$.

$$\begin{aligned} f(x) &= x^2 - 6x + 5 \\ 0 &= x^2 - 6x + 5 \\ 0 &= (x - 5)(x - 1) \end{aligned}$$

$$x - 5 = 0 \quad \text{or} \quad x - 1 = 0$$
$$x = 5 \quad \text{or} \quad x = 1$$

The x-intercepts are $(5, 0)$ and $(1, 0)$.

Find the y-intercept. Let $x = 0$.

$$f(0) = 0^2 - 6(0) + 5 = 5$$

The y-intercept is $(0, 5)$.

Step 4
The axis is $x = 3$. By symmetry, another point on the graph is $(6, 5)$. Since x can be any real number, the domain is $(-\infty, \infty)$. Since the lowest point of the parabola is $(3, -4)$, the range is $[-4, \infty)$.

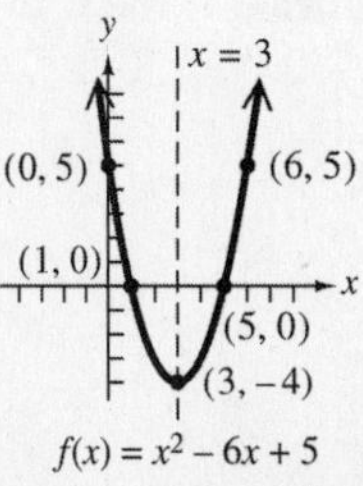

5. **(a)** $f(x) = 4x^2 - 20x + 25$

Here $a = 4$, $b = -20$, and $c = 25$. The discriminant is

$$\begin{aligned} b^2 - 4ac &= (-20)^2 - 4(4)(25) \\ &= 400 - 400 \\ &= 0. \end{aligned}$$

Since the discriminant is 0, the graph has only one x-intercept.

(b) $f(x) = 2x^2 + 3x + 5$

Here $a = 2$, $b = 3$, and $c = 5$. The discriminant is

$$\begin{aligned} b^2 - 4ac &= 3^2 - 4(2)(5) \\ &= 9 - 40 \\ &= -31. \end{aligned}$$

Since the discriminant is negative, the graph has no x-intercepts.

(c) $f(x) = -3x^2 - x + 2$

Here $a = -3$, $b = -1$, and $c = 2$. The discriminant is

$$\begin{aligned} b^2 - 4ac &= (-1)^2 - 4(-3)(2) \\ &= 1 + 24 \\ &= 25. \end{aligned}$$

Since the discriminant is positive, the graph has two x-intercepts.

6. In Example 6, replace 120 with 100, so

$$A(x) = -2x^2 + 100x.$$

Here $a = -2$, $b = 100$, and $c = 0$, so the x-coordinate of the vertex is

$$h = \frac{-b}{2a} = \frac{-100}{2(-2)} = 25,$$

and the y-coordinate of the vertex, $f(h)$, is

$$\begin{aligned} f(25) &= -2(25)^2 + 100(25) \\ &= -2(625) + 2500 \\ &= -1250 + 2500 \\ &= 1250. \end{aligned}$$

The graph is a parabola that opens down with vertex at $(25, 1250)$. The vertex of the graph shows that the maximum area will be 1250 ft^2. This area will occur if the width, x, is 25 ft and the length is $100 - 2x$, or 50 ft.

7. For $s(t) = -16t^2 + 208t$, $a = -16$, $b = 208$, and $c = 0$.

The vertex is $\left(\frac{-b}{2a}, f\left(\frac{-b}{2a}\right)\right)$, so the x-coordinate of the vertex is

$$\frac{-b}{2a} = \frac{-208}{2(-16)} = \frac{13}{2} = 6.5,$$

and the y-coordinate of the vertex is

$$\begin{aligned} f(\tfrac{13}{2}) &= -16(\tfrac{13}{2})^2 + 208(\tfrac{13}{2}) \\ &= -16(\tfrac{169}{4}) + 1352 \\ &= -676 + 1352 \\ &= 676. \end{aligned}$$

Therefore, the toy rocket reaches a maximum height of 676 ft in 6.5 sec.

8. $x = (y + 1)^2 - 4$

Since the roles of x and y are reversed, this graph has its vertex at $(-4, -1)$. It opens to the right, the positive x-direction, and has the same shape as $y = x^2$. The axis is $y = -1$, the domain is $[-4, \infty)$, and the range is $(-\infty, \infty)$. The points $(0, 1)$ and $(0, -3)$ are on the graph.

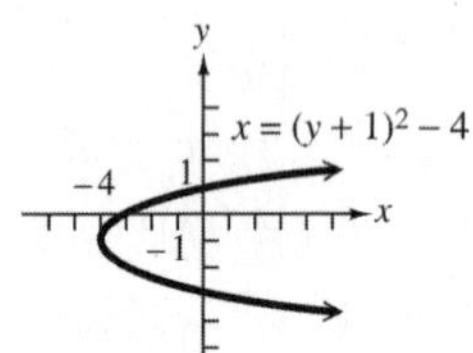

9. $x = -y^2 + 2y + 5$
$x = -(y^2 - 2y) + 5$

Complete the square on $y^2 - 2y$.

$$\left[\tfrac{1}{2}(-2)\right]^2 = (-1)^2 = 1$$

Add and subtract 1.

$$\begin{aligned} x &= -(y^2 - 2y + 1 - 1) + 5 \\ &= -(y^2 - 2y + 1) - (-1) + 5 \\ x &= -(y - 1)^2 + 6 \end{aligned}$$

The vertex is $(6, 1)$ and the axis is $y = 1$. Since $a = -1 < 0$, the graph opens to the left, and has the same shape as $y = x^2$. The domain is $(-\infty, 6]$, and the range is $(-\infty, \infty)$.

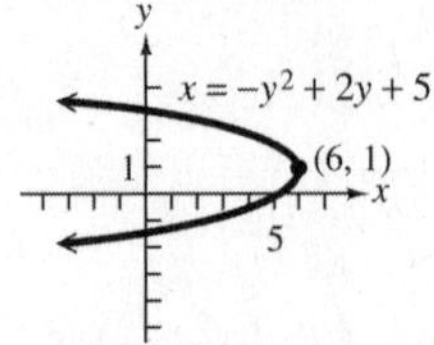

10. **(a)** $x = 2y^2 - 6y + 5$
$x = 2(y^2 - 3y) + 5$

Complete the square on $y^2 - 3y$.

$$\left[\tfrac{1}{2}(-3)\right]^2 = (-\tfrac{3}{2})^2 = \tfrac{9}{4}$$

Add and subtract $\frac{9}{4}$.

$$\begin{aligned} x &= 2(y^2 - 3y + \tfrac{9}{4} - \tfrac{9}{4}) + 5 \\ &= 2(y^2 - 3y + \tfrac{9}{4}) - 2(\tfrac{9}{4}) + 5 \\ &= 2(y^2 - 3y + \tfrac{9}{4}) - \tfrac{9}{2} + 5 \\ x &= 2(y - \tfrac{3}{2})^2 + \tfrac{1}{2} \end{aligned}$$

The vertex is $(\frac{1}{2}, \frac{3}{2})$. Since $a = 2 > 0$, the graph opens to the right. Since y can be any real number, the range is $(-\infty, \infty)$. The domain is $[\frac{1}{2}, \infty)$.

(b) From Margin Exercise 9, the vertex is $(6, 1)$ and the axis is $y = 1$. Since $a = -1 < 0$, the graph opens to the left. The domain is $(-\infty, 6]$, and the range is $(-\infty, \infty)$.

11. **(a)** **A.** $y = -x^2 + 20x + 80$ is of the form $y = ax^2 + bx + c$, so it has a *vertical* parabola as its graph.

B. $x = 2y^2 + 6y + 5$ is of the form $x = ay^2 + by + c$, so it has a *horizontal* parabola as its graph.

C. $x + 1 = (y + 2)^2$

$$\begin{aligned} x + 1 &= y^2 + 4y + 4 \\ x &= y^2 + 4y + 3 \end{aligned}$$

As in **B**, this is a *horizontal* parabola.

D. $f(x) = (x - 4)^2$

$$y = x^2 - 8x + 16$$

As in **A**, this is a *vertical* parabola.

(b) Only vertical parabolas are functions, so equations **A** and **D** represent functions.

10.6 Section Exercises

1. If there is an x^2-term in the equation, the axis is vertical. If there is a y^2-term, the axis is horizontal.

3. Use the discriminant, $b^2 - 4ac$, of the corresponding quadratic equation. If it is positive, there are two x-intercepts. If it is zero, there is one x-intercept (at the vertex), and if it is negative, there is no x-intercept.

5. As in Example 2, we'll complete the square to find the vertex.

$$\begin{aligned} y &= 2x^2 + 4x + 5 \\ &= 2(x^2 + 2x) + 5 \\ &= 2(x^2 + 2x + 1 - 1) + 5 \\ &= 2(x^2 + 2x + 1) + 2(-1) + 5 \\ &= 2(x + 1)^2 - 2 + 5 \\ y &= 2(x + 1)^2 + 3 \end{aligned}$$

The vertex is $(-1, 3)$.
Because $a = 2 > 1$, the graph opens up and is narrower than the graph of $y = x^2$.

For $y = 2x^2 + 4x + 5$, $a = 2$, $b = 4$, and $c = 5$. The discriminant is

$$\begin{aligned} b^2 - 4ac &= 4^2 - 4(2)(5) \\ &= 16 - 40 = -24. \end{aligned}$$

The discriminant is negative, so the parabola has no x-intercepts.

7. $y = f(x) = -x^2 + 5x + 3$

Use the vertex formula with $a = -1$ and $b = 5$.

The x-coordinate of the vertex is

$$\frac{-b}{2a} = \frac{-5}{2(-1)} = \frac{5}{2}.$$

The y-coordinate of the vertex is

$$\begin{aligned} f\left(\frac{-b}{2a}\right) &= f\left(\frac{5}{2}\right) \\ &= -\left(\frac{5}{2}\right)^2 + 5\left(\frac{5}{2}\right) + 3 \\ &= -\frac{25}{4} + \frac{25}{2} + 3 \\ &= \frac{-25 + 50 + 12}{4} = \frac{37}{4}. \end{aligned}$$

The vertex is

$$\left(\frac{-b}{2a}, f\left(\frac{-b}{2a}\right)\right) = \left(\frac{5}{2}, \frac{37}{4}\right).$$

Because $a = -1$, the parabola opens down and has the same shape as the graph of $y = x^2$.

$$\begin{aligned} b^2 - 4ac &= 5^2 - 4(-1)(3) \\ &= 25 + 12 = 37 \end{aligned}$$

The discriminant is positive, so the parabola has two x-intercepts.

9. Complete the square on the y-terms to find the vertex.

$$\begin{aligned} x &= \tfrac{1}{3}y^2 + 6y + 24 \\ &= \tfrac{1}{3}(y^2 + 18y) + 24 \\ &= \tfrac{1}{3}(y^2 + 18y + 81 - 81) + 24 \\ &= \tfrac{1}{3}(y^2 + 18y + 81) + \tfrac{1}{3}(-81) + 24 \\ &= \tfrac{1}{3}(y + 9)^2 - 27 + 24 \\ x &= \tfrac{1}{3}(y + 9)^2 - 3 \end{aligned}$$

The vertex is $(-3, -9)$.
The graph is a horizontal parabola. The graph opens to the right since $a = 0.\overline{3} > 0$ and is wider than the graph of $y = x^2$ since $|a| = |0.\overline{3}| < 1$.

11. $y = f(x) = x^2 + 4x + 3$

Step 1
Because $a = 1$, the graph opens up and has the same shape as the graph of $f(x) = x^2$.

Step 2
Here $a = 1$, $b = 4$, and $c = 3$. The vertex formula gives the x-coordinate as

$$\frac{-b}{2a} = \frac{-4}{2(1)} = -2,$$

and the y-coordinate as

$$f(-2) = (-2)^2 + 4(-2) + 3 = -1.$$

The vertex is at $(-2, -1)$. Since the graph opens up, the axis goes through the x-coordinate of the vertex—its equation is $x = -2$.

Step 3
Evaluate $f(0)$ to find the y-intercept.

$$f(0) = 0^2 + 4(0) + 3 = 3$$

The y-intercept is $(0, 3)$.

Solve $f(x) = 0$ to find any x-intercepts.

$$0 = x^2 + 4x + 3$$
$$0 = (x + 3)(x + 1)$$

$$x + 3 = 0 \quad \text{or} \quad x + 1 = 0$$
$$x = -3 \quad \text{or} \quad x = -1$$

The x-intercepts are $(-3, 0)$ and $(-1, 0)$.

Step 4
By symmetry, another point on the graph is $(-4, 3)$. Since x can be any real number, the domain is $(-\infty, \infty)$. Since $(-2, -1)$ is the lowest point of the parabola, the range is $[-1, \infty)$.

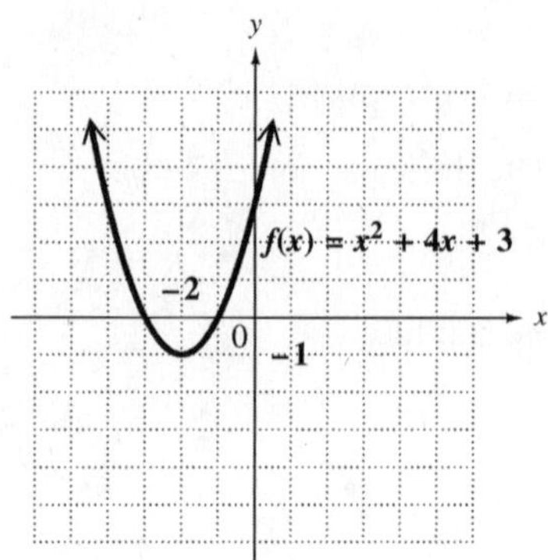

13. $y = f(x) = -2x^2 + 4x - 5$

Step 1
Since $a = -2$, the graph opens down and is narrower than the graph of $y = x^2$.

Step 2
Use the vertex formula to find the vertex.

$$x = \frac{-b}{2a} = \frac{-4}{2(-2)} = 1$$

$$f(1) = -2(1)^2 + 4(1) - 5 = -3$$

The vertex is at $(1, -3)$. Since the graph opens down, the axis goes through the x-coordinate of the vertex—its equation is $x = 1$.

Step 3
If $x = 0$, $y = -5$, so the y-intercept is $(0, -5)$.
To find the x-intercepts, let $y = 0$.

$$0 = -2x^2 + 4x - 5$$

Here $a = -2$, $b = 4$, and $c = -5$.

$$x = \frac{-4 \pm \sqrt{16 - 40}}{2(-2)}$$

The discriminant is negative, so there are no x-intercepts.

Step 4
By symmetry, $(2, -5)$ is also on the graph.

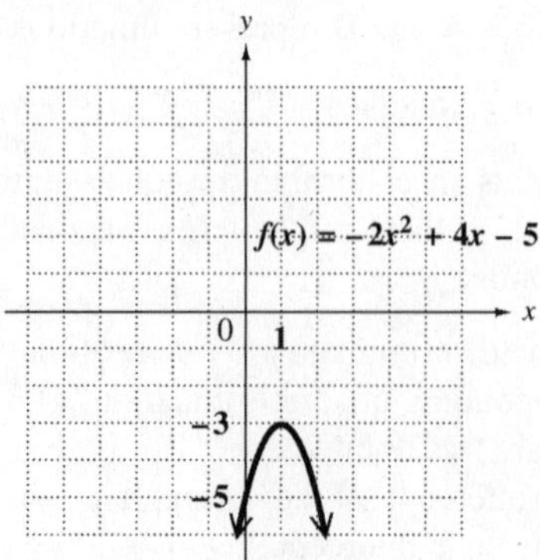

From the graph, we see that the domain is $(-\infty, \infty)$ and the range is $(-\infty, -3]$.

15. $x = -\frac{1}{5}y^2 + 2y - 4$

The roles of x and y are reversed, so this is a horizontal parabola.

Step 1
Since $a = -\frac{1}{5} < 0$, the graph opens to the left and is wider than the graph of $y = x^2$.

Step 2
The y-coordinate of the vertex is

$$\frac{-b}{2a} = \frac{-2}{2(-\frac{1}{5})} = \frac{-2}{-\frac{2}{5}} = 5.$$

The x-coordinate of the vertex is

$$-\frac{1}{5}(5)^2 + 2(5) - 4 = -5 + 10 - 4 = 1.$$

Thus, the vertex is $(1, 5)$. Since the graph opens left, the axis goes through the y-coordinate of the vertex—its equation is $y = 5$.

Step 3
To find the x-intercept, let $y = 0$.
If $y = 0$, $x = -4$, so the x-intercept is $(-4, 0)$.
To find the y-intercepts, let $x = 0$.

$$0 = -\frac{1}{5}y^2 + 2y - 4$$
$$0 = y^2 - 10y + 20 \quad \textit{Multiply by } -5.$$
$$y = \frac{10 \pm \sqrt{100 - 80}}{2} = \frac{10 \pm \sqrt{20}}{2}$$
$$= \frac{10 \pm 2\sqrt{5}}{2} = 5 \pm \sqrt{5}$$

The y-intercepts are approximately $(0, 7.2)$ and $(0, 2.8)$.

Step 4
For an additional point on the graph, let $y = 7$ (two units above the axis) to get $x = \frac{1}{5}$. So the point $(\frac{1}{5}, 7)$ is on the graph. By symmetry, the point $(\frac{1}{5}, 3)$ (two units below the axis) is on the graph.

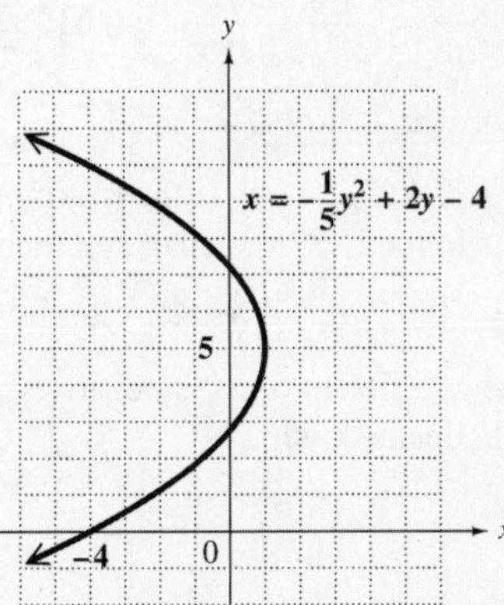

From the graph, we see that the domain is $(-\infty, 1]$ and the range is $(-\infty, \infty)$.

17. $x = 3y^2 + 12y + 5$

The roles of x and y are reversed, so this is a horizontal parabola.

Step 1
Since $a = 3 > 0$, the graph opens to the right and is narrower than the graph of $y = x^2$.

Step 2
Use the formula to find the y-value of the vertex.

$$\frac{-b}{2a} = \frac{-12}{2(3)} = -2$$

If $y = -2$, $x = -7$, so the vertex is $(-7, -2)$. Since the graph opens right, the axis goes through the y-coordinate of the vertex—its equation is $y = -2$.

Step 3
If $y = 0$, $x = 5$, so the x-intercept is $(5, 0)$.
To find the y-intercepts, let $x = 0$.

$$0 = 3y^2 + 12y + 5$$

Here $a = 3$, $b = 12$, and $c = 5$.

$$y = \frac{-12 \pm \sqrt{144 - 60}}{6} = \frac{-12 \pm \sqrt{84}}{6}$$
$$= \frac{-12 \pm 2\sqrt{21}}{6} = -2 \pm \frac{1}{3}\sqrt{21}$$

The y-intercepts are approximately $(0, -0.5)$ and $(0, -3.5)$.

Step 4
By symmetry, $(5, -4)$ is also on the graph.

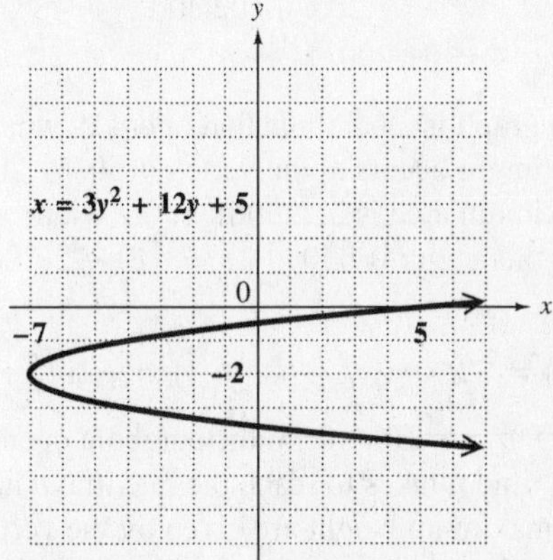

From the graph, we see that the domain is $[-7, \infty)$ and the range is $(-\infty, \infty)$.

19. The graph of $y = 2x^2 + 4x - 3$ is a vertical parabola opening up, so choice **F** is correct. **(F)**

21. The graph of $y = -\frac{1}{2}x^2 - x + 1$ is a vertical parabola opening down, so choices **A** and **C** are possibilities. The graph in **C** is wider than the graph in **A**, so it must correspond to $a = -\frac{1}{2}$ while the graph in **A** must correspond to $a = -1$. **(C)**

23. The graph of $x = -y^2 - 2y + 4$ is a horizontal parabola opening to the left, so choice **D** is correct. **(D)**

25. Let $x =$ one number, $60 - x =$ the other number, and $P =$ the product.

$$\begin{aligned} P &= x(60 - x) \\ &= 60x - x^2 \quad \text{or} \quad -x^2 + 60x \end{aligned}$$

This parabola opens down so the maximum occurs at the vertex.

Here $a = -1$, $b = 60$, and $c = 0$. The x-coordinate of the vertex is

$$\frac{-b}{2a} = \frac{-60}{2(-1)} = 30.$$

$x = 30$ when the product is a maximum.

Since $x = 30$, $60 - x = 30$, and the two numbers are 30 and 30.

27. Let $x =$ the width of the lot. Then $640 - 2x =$ the length of the lot.

Area A is length times width.

$$\begin{aligned} A &= x(640 - 2x) \\ A(x) &= 640x - 2x^2 = -2x^2 + 640x \end{aligned}$$

Use the vertex formula.

$$x = \frac{-b}{2a} = \frac{-640}{2(-2)} = 160$$

$$\begin{aligned} A(160) &= -2(160)^2 + 640(160) \\ &= -51{,}200 + 102{,}400 \\ &= 51{,}200 \end{aligned}$$

The graph is a parabola that opens down, so the maximum occurs at the vertex $(160, 51{,}200)$. The maximum area is 51,200 ft^2 if the width x is 160 ft and the length is $640 - 2x = 640 - 2(160) = 320$ ft.

29. $h(t) = 32t - 16t^2$ or $h(t) = -16t^2 + 32t$

Here, $a = -16 < 0$, so the parabola opens down. The time it takes to reach the maximum height and the maximum height are given by the vertex of the parabola. Use the vertex formula to find that

$$t = \frac{-b}{2a} = \frac{-32}{2(-16)} = \frac{-32}{-32} = 1,$$

and $h(1) = -16(1)^2 + 32(1) = -16 + 32 = 16$.

The vertex is $(1, 16)$, so the maximum height is 16 feet which occurs when the time is 1 second. The object hits the ground when $h = 0$.

$$\begin{aligned} 0 &= -16t^2 + 32t \\ 0 &= -16t(t - 2) \end{aligned}$$

$$\begin{aligned} -16t &= 0 \quad \text{or} \quad t - 2 = 0 \\ t &= 0 \quad \text{or} \quad\quad t = 2 \end{aligned}$$

It takes 2 seconds for the object to hit the ground.

31. The number of people on the plane is $100 - x$ since x is the number of unsold seats. The price per seat is $200 + 4x$.

(a) The total revenue received for the flight is found by multiplying the number of seats by the price per seat. Thus, the revenue is

$$\begin{aligned} R(x) &= (100 - x)(200 + 4x) \\ &= 20{,}000 + 400x - 200x - 4x^2 \\ \text{or} \quad R(x) &= 20{,}000 + 200x - 4x^2. \end{aligned}$$

(b) Find the vertex.
Here $a = -4$, $b = 200$, and $c = 20{,}000$.

$$x = \frac{-b}{2a} = \frac{-200}{2(-4)} = 25$$

The number of unsold seats that will produce the maximum revenue is 25.

(c) $$\begin{aligned} R(25) &= 20{,}000 + 200(25) - 4(25)^2 \\ &= 20{,}000 + 5000 - 2500 \\ &= 22{,}500 \end{aligned}$$

The maximum revenue is \$22,500.

33. $f(x) = -0.0334x^2 + 0.2351x + 12.79$

(a) Since the graph opens down, the y-value of the vertex is a maximum.

(b) The x-value of the vertex is given by

$$x = \frac{-b}{2a} = \frac{-0.2351}{2(-0.0334)} \approx 3.519 \approx 3.5$$

The year was $1990 + 3 = 1993$.

$f(3.5) \approx 13.2\%$, which is the maximum percent of births in the U.S. to teenage mothers.

35. $f(x) = -20.57x^2 + 758.9x - 3140$

(a) The coefficient of x^2 is negative because the parabola opens down.

(b) Use the vertex formula.

$$x = \frac{-b}{2a} = \frac{-758.9}{2(-20.57)} \approx 18.45$$
$$f(18.45) \approx 3860$$

The vertex is approximately $(18.45, 3860)$.

(c) 18 corresponds to 2018, so in 2018 Social Security assets will reach their maximum value of \$3860 billion.

10.7 Polynomial and Rational Inequalities

10.7 Margin Exercises

1. **(a)** $x^2 + 6x + 8 > 0$

The x-intercepts are $(-2, 0)$ and $(-4, 0)$.

Notice from the graph that x-values less than -4 or greater than -2 result in y-values *greater than* 0. Therefore, the solution set is $(-\infty, -4) \cup (-2, \infty)$.

(b) $x^2 + 6x + 8 < 0$

Notice from the graph that x-values between -4 and -2 result in y-values *less than* 0. Therefore, the solution set is $(-4, -2)$.

2. $f(x) = x^2 + 3x - 4$

$$x = \frac{-b}{2a} = \frac{-3}{2} \text{ and } f\left(\frac{-3}{2}\right) = -\frac{25}{4}.$$

The vertex is $(-\frac{3}{2}, -\frac{25}{4})$.

The y-intercept is $(0, -4)$ and the x-intercepts are $(-4, 0)$ and $(1, 0)$, obtained by solving the equation

$$f(x) = x^2 + 3x - 4 = (x + 4)(x - 1) = 0.$$

The parabola opens up.

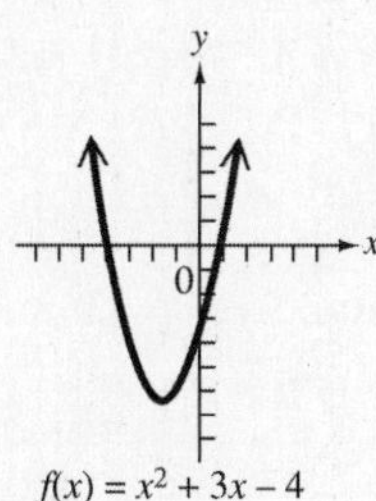

$f(x) = x^2 + 3x - 4$

(a) $x^2 + 3x - 4 \geq 0$

Notice from the graph that x-values less than or equal to -4 or greater than or equal to 1 result in y-values *greater than or equal to* 0. Therefore, the solution set is $(-\infty, -4] \cup [1, \infty)$.

(b) $x^2 + 3x - 4 \leq 0$

Notice from the graph that x-values between -4 and 1, including -4 and 1, result in y-values *less than or equal to* 0. Therefore, the solution set is $[-4, -1]$.

3. $x^2 - x - 12 > 0$

Let $x = 5$ in the inequality.

$$5^2 - 5 - 12 \overset{?}{>} 0$$
$$25 - 5 - 12 \overset{?}{>} 0$$
$$8 > 0 \quad \textit{True}$$

Yes, 5 satisfies $x^2 - x - 12 > 0$.

4. **(a)** $x^2 + x - 6 > 0$

Use factoring to solve the quadratic equation

$$x^2 + x - 6 = 0.$$
$$(x + 3)(x - 2) = 0$$
$$x + 3 = 0 \quad \text{or} \quad x - 2 = 0$$
$$x = -3 \quad \text{or} \quad x = 2$$

Locate the numbers -3 and 2 that divide the number line into three intervals A, B, and C.

Choose a number from each interval to substitute in the inequality

$$x^2 + x - 6 > 0.$$

Interval A: Let $x = -4$.

$$(-4)^2 + (-4) - 6 \overset{?}{>} 0$$
$$16 - 4 - 6 \overset{?}{>} 0$$
$$6 > 0 \quad \textit{True}$$

Interval B: Let $x = 0$.

$$0^2 + 0 - 6 \overset{?}{>} 0$$
$$-6 > 0 \quad \textit{False}$$

Interval C: Let $x = 3$.

$$3^2 + 3 - 6 \overset{?}{>} 0$$
$$9 + 3 - 6 \overset{?}{>} 0$$
$$6 > 0 \quad \textit{True}$$

The numbers in Intervals A and C are solutions. The numbers -3 and 2 are not included because of $>$.

Solution set: $(-\infty, -3) \cup (2, \infty)$

(b) $3m^2 - 13m - 10 \leq 0$

Solve the quadratic equation

$3m^2 - 13m - 10 = 0.$

$(3m + 2)(m - 5) = 0$

$$3m + 2 = 0 \quad \text{or} \quad m - 5 = 0$$
$$m = -\tfrac{2}{3} \quad \text{or} \quad m = 5$$

Locate these numbers on a number line.

A B C

$-\frac{2}{3}$ 5

Choose a number from each interval to substitute in the inequality

$$3m^2 - 13m - 10 \leq 0.$$

Interval A: Let $m = -1$.

$$3(-1)^2 - 13(-1) - 10 \overset{?}{\leq} 0$$
$$3 + 13 - 10 \overset{?}{\leq} 0$$
$$6 \leq 0 \qquad \textit{False}$$

Interval B: Let $m = 0$.

$$3(0)^2 - 13(0) - 10 \overset{?}{\leq} 0$$
$$-10 \leq 0 \qquad \textit{True}$$

Interval C: Let $m = 6$.

$$3(6)^2 - 13(6) - 10 \overset{?}{\leq} 0$$
$$108 - 78 - 10 \overset{?}{\leq} 0$$
$$20 \leq 0 \qquad \textit{False}$$

The numbers in interval B, including $-\frac{2}{3}$ and 5 because of $\leq$, are solutions.

Solution set: $[-\frac{2}{3}, 5]$

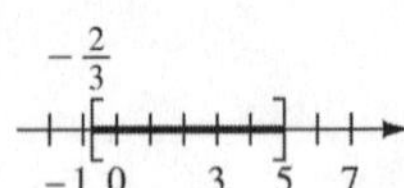

5. **(a)** $(3x - 2)^2 > -2$

Since $(3x - 2)^2$ is either 0 or positive, $(3x - 2)^2$ will always be greater than -2. Therefore, the solution set is $(-\infty, \infty)$.

(b) $(3x - 2)^2 < -2$

Since $(3x - 2)^2$ is never negative, $(3x - 2)^2$ will never be less than or equal to a negative number. Therefore, the solution set is $\emptyset$.

6. **(a)** $(x - 3)(x + 2)(x + 1) > 0$

Solve the equation

$(x - 3)(x + 2)(x + 1) = 0.$

$$x - 3 = 0 \quad \text{or} \quad x + 2 = 0 \quad \text{or} \quad x + 1 = 0$$
$$x = 3 \quad \text{or} \quad x = -2 \quad \text{or} \quad x = -1$$

Locate these numbers and the intervals A, B, C, and D on a number line.

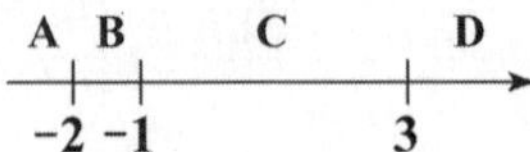

Test a number from each interval in the inequality

$$(x - 3)(x + 2)(x + 1) > 0.$$

Interval A: Let $x = -3$.

$$(-3 - 3)(-3 + 2)(-3 + 1) \overset{?}{>} 0$$
$$-6(-1)(-2) \overset{?}{>} 0$$
$$-12 > 0 \qquad \textit{False}$$

Interval B: Let $x = -\frac{3}{2}$.

$$(-\tfrac{3}{2} - 3)(-\tfrac{3}{2} + 2)(-\tfrac{3}{2} + 1) \overset{?}{>} 0$$
$$-\tfrac{9}{2}(\tfrac{1}{2})(-\tfrac{1}{2}) \overset{?}{>} 0$$
$$\tfrac{9}{8} > 0 \qquad \textit{True}$$

Interval C: Let $x = 0$.

$$(0 - 3)(0 + 2)(0 + 1) \overset{?}{>} 0$$
$$-3(2)(1) \overset{?}{>} 0$$
$$-6 > 0 \qquad \textit{False}$$

Interval D: Let $x = 4$.

$$(4 - 3)(4 + 2)(4 + 1) \overset{?}{>} 0$$
$$1(6)(5) \overset{?}{>} 0$$
$$30 > 0 \qquad \textit{True}$$

The numbers in Intervals B and D, not including $-2, -1,$ or 3 because of $>$, are solutions.

Solution set: $(-2, -1) \cup (3, \infty)$

-2 -1 0 1 2 3 4

(b) $(x - 5)(x + 1)(x - 3) \leq 0$

Solve the equation

$(x - 5)(x + 1)(x - 3) = 0.$

$$x - 5 = 0 \quad \text{or} \quad x + 1 = 0 \quad \text{or} \quad x - 3 = 0$$
$$x = 5 \quad \text{or} \quad x = -1 \quad \text{or} \quad x = 3$$

Locate these numbers on a number line.

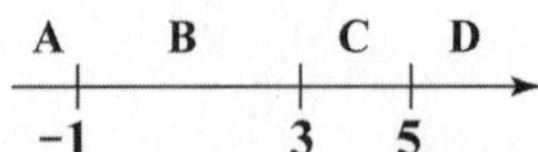

Test a number from each interval in the inequality

$$(x - 5)(x + 1)(x - 3) \leq 0.$$

Interval A: Let $x = -2$.

$$(-2-5)(-2+1)(-2-3) \stackrel{?}{\le} 0$$
$$-7(-1)(-5) \stackrel{?}{\le} 0$$
$$-35 \le 0 \quad \textit{True}$$

Interval B: Let $x = 0$.

$$(0-5)(0+1)(0-3) \stackrel{?}{\le} 0$$
$$-5(1)(-3) \stackrel{?}{\le} 0$$
$$15 \le 0 \quad \textit{False}$$

Interval C: Let $x = 4$.

$$(4-5)(4+1)(4-3) \stackrel{?}{\le} 0$$
$$-1(5)(1) \stackrel{?}{\le} 0$$
$$-5 \le 0 \quad \textit{True}$$

Interval D: Let $x = 6$.

$$(6-5)(6+1)(6-3) \stackrel{?}{\le} 0$$
$$1(7)(3) \stackrel{?}{\le} 0$$
$$21 \le 0 \quad \textit{False}$$

The numbers in Intervals A and C, including -1, 3, and 5 because of $\le$, are solutions.

Solution set: $(-\infty, -1] \cup [3, 5]$

-1 0 1 3 5

7. (a)

$$\frac{2}{x-4} < 3$$

Write the inequality so that 0 is on one side.

$$\frac{2}{x-4} - 3 < 0$$
$$\frac{2}{x-4} - \frac{3(x-4)}{x-4} < 0$$
$$\frac{2-3x+12}{x-4} < 0$$
$$\frac{-3x+14}{x-4} < 0$$

The number $\frac{14}{3}$ makes the numerator 0, and 4 makes the denominator 0. These two numbers determine three intervals.

A B C

4 $\frac{14}{3}$

Test a number from each interval in the inequality

$$\frac{2}{x-4} < 3.$$

Interval A: Let $x = 0$.

$$\frac{2}{0-4} \stackrel{?}{<} 3$$
$$-\tfrac{1}{2} < 3 \quad \textit{True}$$

Interval B: Let $x = \frac{13}{3}$.

$$\frac{2}{\frac{13}{3}-4} \stackrel{?}{<} 3$$
$$6 < 3 \quad \textit{False}$$

Interval C: Let $x = 5$.

$$\frac{2}{5-4} \stackrel{?}{<} 3$$
$$2 < 3 \quad \textit{True}$$

The solution set includes numbers in Intervals A and C, excluding endpoints.

Solution set: $(-\infty, 4) \cup (\frac{14}{3}, \infty)$

$\frac{14}{3}$

0 1 2 3 4 5 6

(b)

$$\frac{5}{x+1} > 4$$

Write the inequality so that 0 is on one side.

$$\frac{5}{x+1} - 4 > 0$$
$$\frac{5}{x+1} - \frac{4(x+1)}{x+1} > 0$$
$$\frac{5-4x-4}{x+1} > 0$$
$$\frac{-4x+1}{x+1} > 0$$

The number $\frac{1}{4}$ makes the numerator 0, and -1 makes the denominator 0. These two numbers determine three intervals.

A B C

-1 $\frac{1}{4}$

Test a number from each interval in the inequality

$$\frac{5}{x+1} > 4.$$

continued

Interval A: Let $x = -2$.

$$\frac{5}{-2+1} \stackrel{?}{>} 4$$

$$-5 > 4 \quad \textit{False}$$

Interval B: Let $x = 0$.

$$\frac{5}{0+1} \stackrel{?}{>} 4$$

$$5 > 4 \quad \textit{True}$$

Interval C: Let $x = 1$.

$$\frac{5}{1+1} \stackrel{?}{>} 4$$

$$\frac{5}{2} > 4 \quad \textit{False}$$

The numbers in Interval B, not including -1 or $\frac{1}{4}$, are solutions.

Solution set: $(-1, \frac{1}{4})$

$\frac{1}{4}$

−2 −1 0 1 2

8.

$$\frac{x+2}{x-1} \leq 5$$

Write the inequality so that 0 is on one side.

$$\frac{x+2}{x-1} - 5 \leq 0$$

$$\frac{x+2}{x-1} - \frac{5(x-1)}{x-1} \leq 0$$

$$\frac{x+2-5x+5}{x-1} \leq 0$$

$$\frac{-4x+7}{x-1} \leq 0$$

The number $\frac{7}{4}$ makes the numerator 0, and 1 makes the denominator 0. These two numbers determine three intervals.

A B C

1 $\frac{7}{4}$

Test a number from each interval in the inequality

$$\frac{x+2}{x-1} \leq 5.$$

Interval A: Let $x = 0$.

$$\frac{0+2}{0-1} \stackrel{?}{\leq} 5$$

$$-2 \leq 5 \quad \textit{True}$$

Interval B: Let $x = \frac{3}{2}$.

$$\frac{\frac{3}{2}+2}{\frac{3}{2}-1} \stackrel{?}{\leq} 5$$

$$7 \leq 5 \quad \textit{False}$$

Interval C: Let $x = 2$.

$$\frac{2+2}{2-1} \stackrel{?}{\leq} 5$$

$$4 \leq 5 \quad \textit{True}$$

The numbers in Intervals A and C are solutions. 1 is not in the solution set (since it makes the denominator 0), but $\frac{7}{4}$ is.

Solution set: $(-\infty, 1) \cup [\frac{7}{4}, \infty)$

0 1 $\frac{7}{4}$ 3

10.7 Section Exercises

1. **(a)** The x-intercepts determine the solutions of the equation $x^2 - 4x + 3 = 0$. From the graph, the solution set is $\{1, 3\}$.

(b) The x-values of the points on the graph that are *above* the x-axis form the solution set of the inequality $x^2 - 4x + 3 > 0$. From the graph, the solution set is $(-\infty, 1) \cup (3, \infty)$.

(c) The x-values of the points on the graph that are *below* the x-axis form the solution set of the inequality $x^2 - 4x + 3 < 0$. From the graph, the solution set is $(1, 3)$.

3. **(a)** The x-intercepts determine the solutions of the equation $-2x^2 - x + 15 = 0$. From the graph, the solution set is $\{-3, \frac{5}{2}\}$.

(b) The x-values of the points on the graph that are *above* the x-axis form the solution set of the inequality $-2x^2 - x + 15 > 0$. From the graph, the solution set for $-2x^2 - x + 15 \geq 0$ is $[-3, \frac{5}{2}]$.

(c) The x-values of the points on the graph that are *below* the x-axis form the solution set of the inequality $-2x^2 - x + 15 < 0$. From the graph, the solution set for $-2x^2 - x + 15 \leq 0$ is $(-\infty, -3] \cup [\frac{5}{2}, \infty)$.

5. Include the endpoints if the symbol is $\leq$ or $\geq$. Exclude the endpoints if the symbol is $<$ or $>$.

7. $(x+1)(x-5) > 0$

Solve the equation $(x+1)(x-5) = 0$.

$$x + 1 = 0 \quad \text{or} \quad x - 5 = 0$$

$$x = -1 \quad \text{or} \quad x = 5$$

The numbers -1 and 5 divide a number line into three intervals: A, B, and C.

A B C

−1 5

Test a number from each interval in the original inequality.

Interval A: Let $x = -2$.

$$(x+1)(x-5) > 0$$
$$(-2+1)(-2-5) \overset{?}{>} 0$$
$$-1(-7) \overset{?}{>} 0$$
$$7 > 0 \quad \textit{True}$$

Interval B: Let $x = 0$.

$$(0+1)(0-5) \overset{?}{>} 0$$
$$-5 > 0 \quad \textit{False}$$

Interval C: Let $x = 6$.

$$(6+1)(6-5) \overset{?}{>} 0$$
$$7 > 0 \quad \textit{True}$$

The solution set includes the numbers in Intervals A and C, excluding -1 and 5 because of $>$.
Solution set: $(-\infty, -1) \cup (5, \infty)$

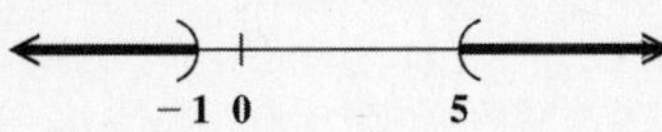

9. $(r+4)(r-6) < 0$
Solve the equation
$(r+4)(r-6) = 0$.

$$r+4=0 \quad \text{or} \quad r-6=0$$
$$r=-4 \quad \text{or} \quad r=6$$

These numbers divide a number line into three intervals: A, B, and C.

A B C

−4 6

Test a number from each interval in the original inequality.

Interval A: Let $r = -5$.

$$(r+4)(r-6) < 0$$
$$(-5+4)(-5-6) \overset{?}{<} 0$$
$$-1(-11) \overset{?}{<} 0$$
$$11 < 0 \quad \textit{False}$$

Interval B: Let $r = 0$.

$$4(-6) \overset{?}{<} 0$$
$$-24 < 0 \quad \textit{True}$$

Interval C: Let $r = 7$.

$$(7+4)(7-6) \overset{?}{<} 0$$
$$11(1) \overset{?}{<} 0$$
$$11 < 0 \quad \textit{False}$$

The solution set includes Interval B, excluding -4 and 6 because of $<$.
Solution set: $(-4, 6)$

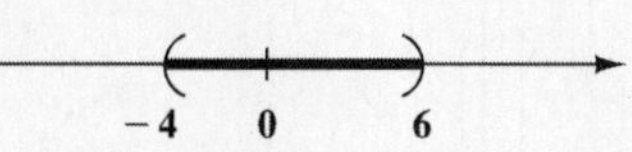

11. $x^2 - 4x + 3 \geq 0$
Solve the equation
$x^2 - 4x + 3 = 0$.
$(x-1)(x-3) = 0$

$$x-1=0 \quad \text{or} \quad x-3=0$$
$$x=1 \quad \text{or} \quad x=3$$

A B C

1 3

Test a number from each interval in the original inequality.

Interval A: Let $x = 0$.

$$3 \geq 0 \quad \textit{True}$$

Interval B: Let $x = 2$.

$$2^2 - 4(2) + 3 \overset{?}{\geq} 0$$
$$-1 \geq 0 \quad \textit{False}$$

Interval C: Let $x = 4$.

$$4^2 - 4(4) + 3 \overset{?}{\geq} 0$$
$$3 \geq 0 \quad \textit{True}$$

The solution set includes the numbers in Intervals A and C, including 1 and 3 because of $\geq$.
Solution set: $(-\infty, 1] \cup [3, \infty)$

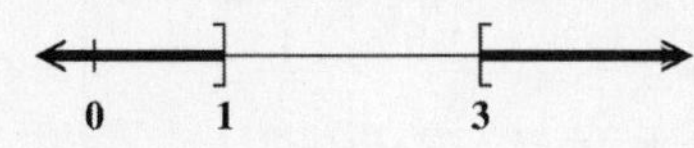

13. $10t^2 + 9t \geq 9$
$10t^2 + 9t - 9 \geq 0$
Solve the equation
$10t^2 + 9t - 9 = 0$.
$(2t+3)(5t-3) = 0$

$$2t+3=0 \quad \text{or} \quad 5t-3=0$$
$$t=-\tfrac{3}{2} \quad \text{or} \quad t=\tfrac{3}{5}$$

A B C

$-\frac{3}{2}$ $\frac{3}{5}$

Test a number from each interval in the original inequality.

continued

Interval A: Let $t = -2$.

$$10(-2)^2 + 9(-2) \overset{?}{\geq} 9$$
$$40 - 18 \overset{?}{\geq} 9$$
$$22 \geq 9 \quad \textit{True}$$

Interval B: Let $t = 0$.

$$0 \geq 9 \quad \textit{False}$$

Interval C: Let $t = 1$.

$$10(1)^2 + 9(1) \overset{?}{\geq} 9$$
$$10 + 9 \overset{?}{\geq} 9$$
$$19 \geq 9 \quad \textit{True}$$

The solution set includes the numbers in Intervals A and C, including $-\frac{3}{2}$ and $\frac{3}{5}$ because of $\geq$.

Solution set: $(-\infty, -\frac{3}{2}] \cup [\frac{3}{5}, \infty)$

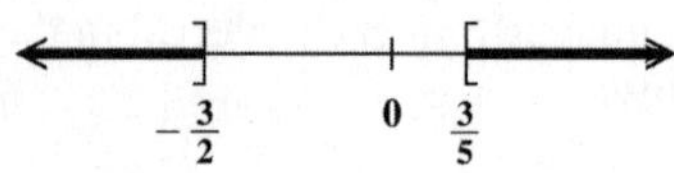

15. $9p^2 + 3p < 2$

Solve the equation

$$9p^2 + 3p = 2.$$
$$9p^2 + 3p - 2 = 0$$
$$(3p - 1)(3p + 2) = 0$$

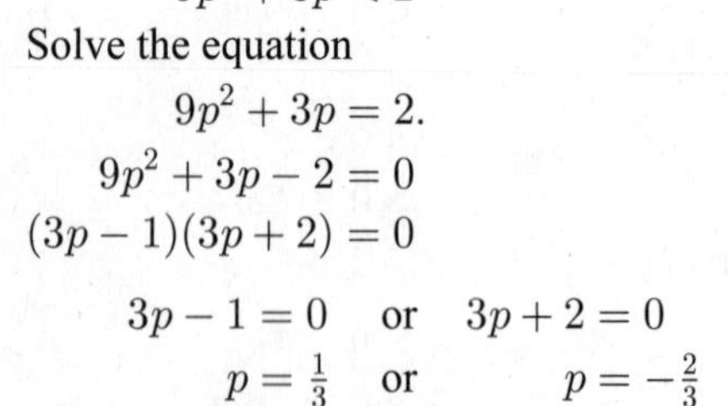

$$3p - 1 = 0 \quad \text{or} \quad 3p + 2 = 0$$
$$p = \tfrac{1}{3} \quad \text{or} \quad p = -\tfrac{2}{3}$$

These numbers divide a number line into three intervals: A, B, and C.

Test a number from each interval in the original inequality.

Interval A: Let $p = -1$.

$$9p^2 + 3p < 2$$
$$9(-1)^2 + 3(-1) \overset{?}{<} 2$$
$$9 - 3 \overset{?}{<} 2$$
$$6 < 2 \quad \textit{False}$$

Interval B: Let $p = 0$.

$$0 < 2 \quad \textit{True}$$

Interval C: Let $p = 1$.

$$9(1)^2 + 3(1) \overset{?}{<} 2$$
$$9 + 3 \overset{?}{<} 2$$
$$12 < 2 \quad \textit{False}$$

The solution set includes Interval B, excluding $-\frac{2}{3}$ and $\frac{1}{3}$ because of $<$.

Solution set: $(-\frac{2}{3}, \frac{1}{3})$

17. $6x^2 + x \geq 1$

$$6x^2 + x - 1 \geq 0$$

Solve the equation

$$6x^2 + x - 1 = 0.$$
$$(2x + 1)(3x - 1) = 0$$
$$2x + 1 = 0 \quad \text{or} \quad 3x - 1 = 0$$
$$x = -\tfrac{1}{2} \quad \text{or} \quad x = \tfrac{1}{3}$$

Test a number from each interval in the inequality.

$$6x^2 + x \geq 1.$$

Interval A: Let $x = -1$.

$$6(-1)^2 + (-1) \overset{?}{\geq} 1$$
$$5 \geq 1 \quad \textit{True}$$

Interval B: Let $x = 0$.

$$0 \geq 1 \quad \textit{False}$$

Interval C: Let $x = 1$.

$$6(1)^2 + 1 \overset{?}{\geq} 1$$
$$7 \geq 1 \quad \textit{True}$$

The solution set includes the numbers in Intervals A and C, including $-\frac{1}{2}$ and $\frac{1}{3}$ because of $\geq$.

Solution set: $(-\infty, -\frac{1}{2}] \cup [\frac{1}{3}, \infty)$

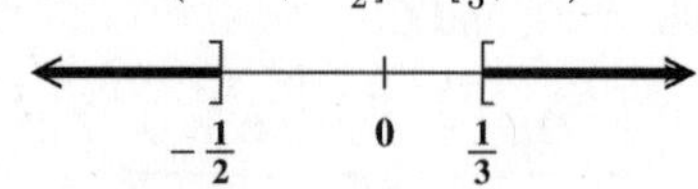

19. $x^2 - 6x + 6 \geq 0$

Solve the equation

$x^2 - 6x + 6 = 0.$

Since $x^2 - 6x + 6$ does not factor, let $a = 1$, $b = -6$, and $c = 6$ in the quadratic formula.

$$x = \frac{-(-6) \pm \sqrt{(-6)^2 - 4(1)(6)}}{2(1)}$$
$$= \frac{6 \pm \sqrt{12}}{2} = \frac{6 \pm 2\sqrt{3}}{2}$$
$$= \frac{2(3 \pm \sqrt{3})}{2} = 3 \pm \sqrt{3}$$
$$x = 3 + \sqrt{3} \quad \text{or} \quad x = 3 - \sqrt{3}$$
$$x \approx 4.7 \quad \text{or} \quad x \approx 1.3$$

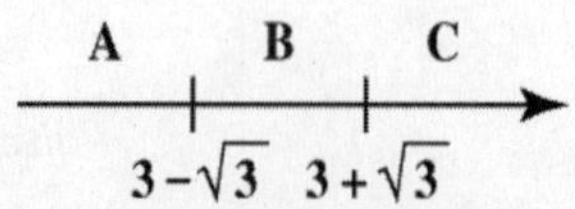

Test a number from each interval in the inequality

$$x^2 - 6x + 6 \geq 0.$$

Interval A: Let $x = 0$.
$6 \geq 0$ *True*

Interval B: Let $x = 3$.
$3^2 - 6(3) + 6 \overset{?}{\geq} 0$
$-3 \geq 0$ *False*

Interval C: Let $x = 5$.
$5^2 - 6(5) + 6 \overset{?}{\geq} 0$
$1 \geq 0$ *True*

The solution set includes the numbers in Intervals A and C, including $3 - \sqrt{3}$ and $3 + \sqrt{3}$ because of $\geq$.

Solution set: $(-\infty, 3 - \sqrt{3}] \cup [3 + \sqrt{3}, \infty)$

0 $3 - \sqrt{3}$ $3 + \sqrt{3}$

21. $(4 - 3x)^2 \geq -2$

Since $(4 - 3x)^2$ is either 0 or positive, $(4 - 3x)^2$ will always be greater than -2. Therefore, the solution set is $(-\infty, \infty)$.

23. $(3x + 5)^2 \leq -4$

Since $(3x + 5)^2$ is never negative, $(3x + 5)^2$ will never be less than or equal to a negative number. Therefore, the solution set is $\emptyset$.

25. $(p - 1)(p - 2)(p - 4) < 0$
The numbers 1, 2, and 4 are solutions of the cubic equation

$$(p - 1)(p - 2)(p - 4) = 0.$$

These numbers divide a number line into four intervals.

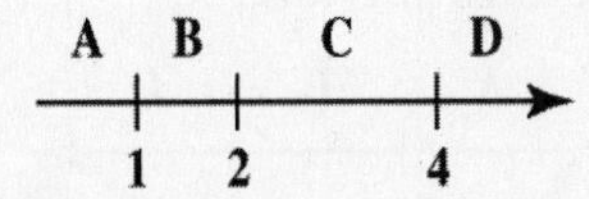

A B C D
1 2 4

Test a number from each interval in the inequality

$$(p - 1)(p - 2)(p - 4) < 0.$$

Interval A: Let $p = 0$.
$-1(-2)(-4) \overset{?}{<} 0$
$-8 < 0$ *True*

Interval B: Let $p = 1.5$.
$(1.5 - 1)(1.5 - 2)(1.5 - 4) \overset{?}{<} 0$
$0.5(-0.5)(-2.5) \overset{?}{<} 0$
$0.625 < 0$ *False*

Interval C: Let $p = 3$.
$(3 - 1)(3 - 2)(3 - 4) \overset{?}{<} 0$
$2(1)(-1) \overset{?}{<} 0$
$-2 < 0$ *True*

Interval D: Let $p = 5$.
$(5 - 1)(5 - 2)(5 - 4) \overset{?}{<} 0$
$4(3)(1) \overset{?}{<} 0$
$12 < 0$ *False*

The numbers in Intervals A and C, not including 1, 2, or 4, are solutions.

Solution set: $(-\infty, 1) \cup (2, 4)$

0 1 2 4

27. $(x - 4)(2x + 3)(3x - 1) \geq 0$
The numbers 4, $-\frac{3}{2}$, and $\frac{1}{3}$ are solutions of the cubic equation

$$(x - 4)(2x + 3)(3x - 1) = 0.$$

These numbers divide a number line into 4 intervals.

A B C D
$-\frac{3}{2}$ $\frac{1}{3}$ 4

Test a number from each interval in the original inequality.

Interval A: Let $x = -2$.
$-6(-1)(-7) \overset{?}{\geq} 0$
$-42 \geq 0$ *False*

Interval B: Let $x = 0$.
$-4(3)(-1) \overset{?}{\geq} 0$
$12 \geq 0$ *True*

Interval C: Let $x = 1$.
$-3(5)(2) \overset{?}{\geq} 0$
$-30 \geq 0$ *False*

Interval D: Let $x = 5$.
$1(13)(14) \overset{?}{\geq} 0$
$182 \geq 0$ *True*

The solution set includes numbers in Intervals B and D, including the endpoints.

Solution set: $[-\frac{3}{2}, \frac{1}{3}] \cup [4, \infty)$

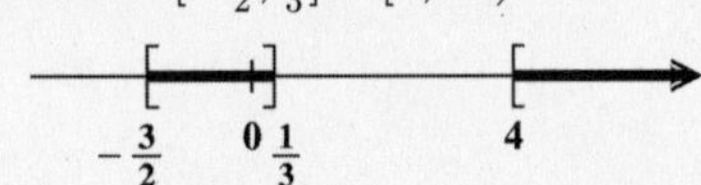

29. $\dfrac{x-1}{x-4} > 0$

The number 1 makes the numerator 0, and 4 makes the denominator 0. These two numbers determine three intervals.

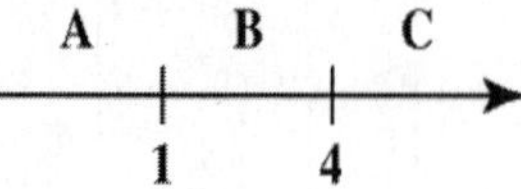

Test a number from each interval in the inequality

$$\frac{x-1}{x-4} > 0.$$

Interval A: Let $x = 0$.

$$\frac{0-1}{0-4} \overset{?}{>} 0$$
$$\frac{1}{4} > 0 \quad \textit{True}$$

Interval B: Let $x = 2$.

$$\frac{2-1}{2-4} \overset{?}{>} 0$$
$$\frac{1}{-2} > 0 \quad \textit{False}$$

Interval C: Let $x = 5$.

$$\frac{5-1}{5-4} \overset{?}{>} 0$$
$$4 > 0 \quad \textit{True}$$

The solution set includes numbers in Intervals A and C, excluding endpoints.

Solution set: $(-\infty, 1) \cup (4, \infty)$

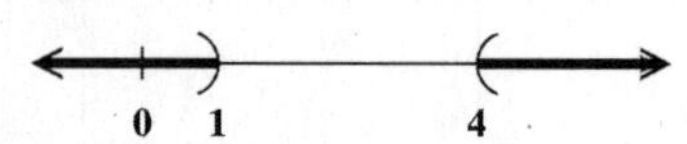

31. $\dfrac{2n+3}{n-5} \leq 0$

The number $-\frac{3}{2}$ makes the numerator 0, and 5 makes the denominator 0. These two numbers determine three intervals.

Test a number from each interval in the inequality

$$\frac{2n+3}{n-5} \leq 0.$$

Interval A: Let $n = -2$.

$$\frac{2(-2)+3}{(-2)-5} \overset{?}{\leq} 0$$
$$\frac{1}{7} \leq 0 \quad \textit{False}$$

Interval B: Let $n = 0$.

$$\frac{2(0)+3}{0-5} \overset{?}{\leq} 0$$
$$-\frac{3}{5} \leq 0 \quad \textit{True}$$

Interval C: Let $n = 6$.

$$\frac{2(6)+3}{6-5} \overset{?}{\leq} 0$$
$$15 \leq 0 \quad \textit{False}$$

The solution set includes the points in Interval B. The endpoint 5 is not included since it makes the left side undefined. The endpoint $-\frac{3}{2}$ is included because it makes the left side equal to 0.

Solution set: $[-\frac{3}{2}, 5)$

33. $\dfrac{8}{x-2} \geq 2$

Write the inequality so that 0 is on one side.

$$\frac{8}{x-2} - 2 \geq 0$$
$$\frac{8}{x-2} - \frac{2(x-2)}{x-2} \geq 0$$
$$\frac{8-2x+4}{x-2} \geq 0$$
$$\frac{-2x+12}{x-2} \geq 0$$

The number 6 makes the numerator 0, and 2 makes the denominator 0. These two numbers determine three intervals.

Test a number from each interval in the inequality

$$\frac{8}{x-2} \geq 2.$$

Interval A: Let $x = 0$.

$$\frac{8}{0-2} \overset{?}{\geq} 2$$
$$-4 \geq 2 \quad \textit{False}$$

Interval B: Let $x = 3$.

$$\frac{8}{3-2} \overset{?}{\geq} 2$$

$$8 \geq 2 \quad \textit{True}$$

Interval C: Let $x = 7$.

$$\frac{8}{7-2} \overset{?}{\geq} 2$$

$$\frac{8}{5} \geq 2 \quad \textit{False}$$

The solution set includes numbers in Interval B, including 6 but excluding 2, which makes the fraction undefined.

Solution set: $(2, 6]$

0 2 6

35.

$$\frac{3}{2t-1} < 2$$

Write the inequality so that 0 is on one side.

$$\frac{3}{2t-1} - 2 < 0$$

$$\frac{3}{2t-1} - \frac{2(2t-1)}{2t-1} < 0$$

$$\frac{3-4t+2}{2t-1} < 0$$

$$\frac{-4t+5}{2t-1} < 0$$

The number $\frac{5}{4}$ makes the numerator 0, and $\frac{1}{2}$ makes the denominator 0. These two numbers determine three intervals.

A B C

$\frac{1}{2}$ $\frac{5}{4}$

Test a number from each interval in the inequality

$$\frac{3}{2t-1} < 2.$$

Interval A: Let $t = 0$.

$$\frac{3}{2(0)-1} \overset{?}{<} 2$$

$$-3 < 2 \quad \textit{True}$$

Interval B: Let $t = 1$.

$$\frac{3}{2(1)-1} \overset{?}{<} 2$$

$$3 < 2 \quad \textit{False}$$

Interval C: Let $t = 2$.

$$\frac{3}{2(2)-1} \overset{?}{<} 2$$

$$1 < 2 \quad \textit{True}$$

The solution set includes numbers in Intervals A and C, excluding endpoints.

Solution set: $(-\infty, \frac{1}{2}) \cup (\frac{5}{4}, \infty)$

0 $\frac{1}{2}$ $\frac{5}{4}$

37.

$$\frac{w}{w+2} \geq 2$$

Write the inequality so that 0 is on one side.

$$\frac{w}{w+2} - 2 \geq 0$$

$$\frac{w}{w+2} - \frac{2(w+2)}{w+2} \geq 0$$

$$\frac{w-2w-4}{w+2} \geq 0$$

$$\frac{-w-4}{w+2} \geq 0$$

The number -4 makes the numerator 0, and -2 makes the denominator 0. These two numbers determine three intervals.

A B C

-4 -2

Test a number from each interval in the inequality

$$\frac{w}{w+2} \geq 2.$$

Interval A: Let $w = -5$.

$$\frac{-5}{-3} \overset{?}{\geq} 2$$

$$\frac{5}{3} \geq 2 \quad \textit{False}$$

Interval B: Let $w = -3$.

$$\frac{-3}{-1} \overset{?}{\geq} 2$$

$$3 \geq 2 \quad \textit{True}$$

Interval C: Let $w = 0$.

$$\frac{0}{2} \overset{?}{\geq} 2$$

$$0 \geq 2 \quad \textit{False}$$

The solution set includes numbers in Interval B, including -4 but excluding -2, which makes the fraction undefined.

Solution set: $[-4, -2)$

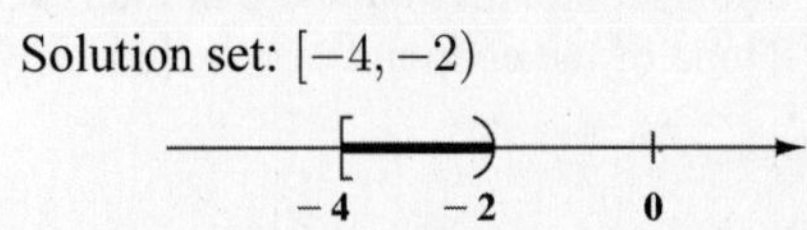

39. $$\frac{4k}{2k-1} < k$$

Write the inequality so that 0 is on one side.

$$\frac{4k}{2k-1} - k < 0$$

$$\frac{4k}{2k-1} - \frac{k(2k-1)}{2k-1} < 0$$

$$\frac{4k - 2k^2 + k}{2k-1} < 0$$

$$\frac{-2k^2 + 5k}{2k-1} < 0$$

$$\frac{k(-2k+5)}{2k-1} < 0$$

The numbers 0 and $\frac{5}{2}$ make the numerator 0, and $\frac{1}{2}$ makes the denominator 0. These three numbers determine four intervals.

A B C D

0 $\frac{1}{2}$ $\frac{5}{2}$

Test a number from each interval in the inequality

$$\frac{4k}{2k-1} < k.$$

Interval A: Let $k = -1$.

$$\frac{4(-1)}{2(-1)-1} \stackrel{?}{<} -1$$

$$\frac{4}{3} < -1 \quad \textit{False}$$

Interval B: Let $k = \frac{1}{4}$.

$$\frac{4(\frac{1}{4})}{2(\frac{1}{4})-1} \stackrel{?}{<} \frac{1}{4}$$

$$-2 < \tfrac{1}{4} \quad \textit{True}$$

Interval C: Let $k = 1$.

$$\frac{4(1)}{2(1)-1} \stackrel{?}{<} 1$$

$$4 < 1 \quad \textit{False}$$

Interval D: Let $k = 3$.

$$\frac{4(3)}{2(3)-1} \stackrel{?}{<} 3$$

$$\frac{12}{5} < 3 \quad \textit{True}$$

The solution set includes numbers in Intervals B and D. None of the endpoints are included.

Solution set: $(0, \frac{1}{2}) \cup (\frac{5}{2}, \infty)$

0 $\frac{1}{2}$ $\frac{5}{2}$

41. $$\frac{x-8}{x-4} \le 3$$

Write the inequality so that 0 is on one side.

$$\frac{x-8}{x-4} - 3 \le 0$$

$$\frac{x-8}{x-4} - \frac{3(x-4)}{x-4} \le 0$$

$$\frac{x - 8 - 3x + 12}{x-4} \le 0$$

$$\frac{-2x+4}{x-4} \le 0$$

The number 2 makes the numerator 0, and 4 makes the denominator 0. These two numbers determine three intervals.

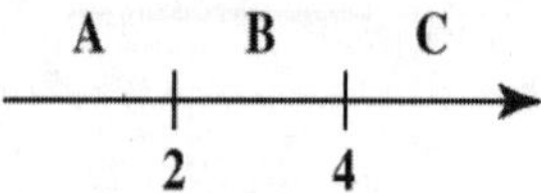

Test a number from each interval in the inequality

$$\frac{x-8}{x-4} \le 3.$$

Interval A: Let $x = 0$.

$$\frac{-8}{-4} \stackrel{?}{\le} 3$$

$$2 \le 3 \quad \textit{True}$$

Interval B: Let $x = 3$.

$$\frac{-5}{-1} \stackrel{?}{\le} 3$$

$$5 \le 3 \quad \textit{False}$$

Interval C: Let $x = 5$.

$$\frac{-3}{1} \le 3 \quad \textit{True}$$

The solution set includes numbers in Intervals A and C, including 2 but excluding 4, which makes the fraction undefined.

Solution set: $(-\infty, 2] \cup (4, \infty)$

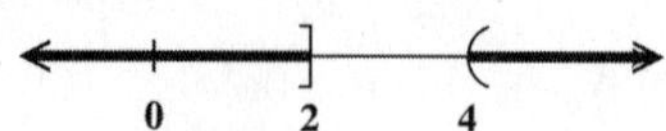

43.
$$s(t) = 624$$
$$-16t^2 + 256t = 624$$
$$t^2 - 16t = -39 \quad \textit{Divide by } -16.$$
$$t^2 - 16t + 39 = 0$$
$$(t-3)(t-13) = 0$$

$$t - 3 = 0 \quad \text{or} \quad t - 13 = 0$$
$$t = 3 \quad \text{or} \quad t = 13$$

The rock will be 624 feet above the ground after 3 seconds and after 13 seconds.

44. See Exercise 43 for the solution to the equation $s(t) = 624$.

For this problem, our common sense tells us that $s(t) > 624$ between 3 seconds and 13 seconds.

45.
$$s(t) = 0$$
$$-16t^2 + 256t = 0$$
$$t^2 - 16t = 0 \quad \textit{Divide by } -16.$$
$$t(t-16) = 0$$
$$t = 0 \quad \text{or} \quad t - 16 = 0$$
$$t = 16$$

The rock is at ground level at 0 seconds (the time when it is initially projected) and at 16 seconds (the time when it hits the ground).

46. Based on the answers for Exercises 43 and 45, we deduce that $s(t) < 624$ between 0 and 3 seconds and between 13 and 16 seconds.

Chapter 10 Review Exercises

1. $t^2 = 121$

$$t = \sqrt{121} \quad \text{or} \quad t = -\sqrt{121}$$
$$t = 11 \quad \text{or} \quad t = -11$$

Solution set: $\{-11, 11\}$

2. $p^2 = 3$

$$p = \sqrt{3} \quad \text{or} \quad p = -\sqrt{3}$$

Solution set: $\{-\sqrt{3}, \sqrt{3}\}$

3. $(2x+5)^2 = 100$

$$2x + 5 = \sqrt{100} \quad \text{or} \quad 2x + 5 = -\sqrt{100}$$
$$2x + 5 = 10 \quad \text{or} \quad 2x + 5 = -10$$
$$2x = 5 \quad \text{or} \quad 2x = -15$$
$$x = \tfrac{5}{2} \quad \text{or} \quad x = -\tfrac{15}{2}$$

Solution set: $\{-\frac{15}{2}, \frac{5}{2}\}$

4. $(3k-2)^2 = -25$

$$3k - 2 = \sqrt{-25} \quad \text{or} \quad 3k - 2 = -\sqrt{-25}$$
$$3k - 2 = 5i \quad \text{or} \quad 3k - 2 = -5i$$
$$3k = 2 + 5i \quad \text{or} \quad 3k = 2 - 5i$$
$$k = \frac{2+5i}{3} \quad \text{or} \quad k = \frac{2-5i}{3}$$

Solution set: $\{\frac{2}{3} + \frac{5}{3}i, \frac{2}{3} - \frac{5}{3}i\}$

5.
$$x^2 + 4x = 15$$

Complete the square.

$$\left[\tfrac{1}{2}(4)\right]^2 = 2^2 = 4$$

Add 4 to each side.

$$x^2 + 4x + 4 = 15 + 4$$
$$(x+2)^2 = 19$$
$$x + 2 = \sqrt{19} \quad \text{or} \quad x + 2 = -\sqrt{19}$$
$$x = -2 + \sqrt{19} \quad \text{or} \quad x = -2 - \sqrt{19}$$

Solution set: $\{-2 + \sqrt{19}, -2 - \sqrt{19}\}$

6.
$$2m^2 - 3m = -1$$
$$m^2 - \tfrac{3}{2}m = -\tfrac{1}{2} \quad \textit{Divide by 2.}$$

Complete the square.

$$\left[\tfrac{1}{2}(-\tfrac{3}{2})\right]^2 = (-\tfrac{3}{4})^2 = \tfrac{9}{16}$$

Add $\frac{9}{16}$ to each side.

$$m^2 - \tfrac{3}{2}m + \tfrac{9}{16} = -\tfrac{1}{2} + \tfrac{9}{16}$$
$$(m - \tfrac{3}{4})^2 = -\tfrac{8}{16} + \tfrac{9}{16}$$
$$(m - \tfrac{3}{4})^2 = \tfrac{1}{16}$$
$$m - \tfrac{3}{4} = \sqrt{\tfrac{1}{16}} \quad \text{or} \quad m - \tfrac{3}{4} = -\sqrt{\tfrac{1}{16}}$$
$$m - \tfrac{3}{4} = \tfrac{1}{4} \quad \text{or} \quad m - \tfrac{3}{4} = -\tfrac{1}{4}$$
$$m = \tfrac{3}{4} + \tfrac{1}{4} \quad \text{or} \quad m = \tfrac{3}{4} - \tfrac{1}{4}$$
$$m = 1 \quad \text{or} \quad m = \tfrac{1}{2}$$

Solution set: $\{\frac{1}{2}, 1\}$

7. By the square root property, the first step should be

$$x = \sqrt{12} \quad \text{or} \quad x = -\sqrt{12}.$$

The solution set is $\{-2\sqrt{3}, 2\sqrt{3}\}$.

8.
$$4.9t^2 = d$$
$$4.9t^2 = 165 \quad \textit{Let } d = 165.$$
$$t^2 = \frac{165}{4.9}$$
$$t = \sqrt{\frac{165}{4.9}} \quad t \geq 0$$
$$t \approx 5.8 \text{ seconds}$$

It would take about 5.8 seconds for the wallet to fall 165 meters.

9. $2x^2 + x - 21 = 0$

Here $a = 2$, $b = 1$, and $c = -21$.

$$x = \frac{-b \pm \sqrt{b^2 - 4ac}}{2a}$$
$$x = \frac{-1 \pm \sqrt{1^2 - 4(2)(-21)}}{2(2)}$$
$$= \frac{-1 \pm \sqrt{1 + 168}}{4}$$
$$= \frac{-1 \pm \sqrt{169}}{4} = \frac{-1 \pm 13}{4}$$
$$x = \frac{-1 + 13}{4} = \frac{12}{4} = 3 \text{ or}$$
$$x = \frac{-1 - 13}{4} = -\frac{14}{4} = -\frac{7}{2}$$

Solution set: $\{-\frac{7}{2}, 3\}$

10. $k^2 + 5k = 7$

$k^2 + 5k - 7 = 0$

Here $a = 1$, $b = 5$, and $c = -7$.

$$k = \frac{-b \pm \sqrt{b^2 - 4ac}}{2a}$$
$$k = \frac{-5 \pm \sqrt{5^2 - 4(1)(-7)}}{2(1)}$$
$$= \frac{-5 \pm \sqrt{25 + 28}}{2}$$
$$= \frac{-5 \pm \sqrt{53}}{2}$$

Solution set: $\left\{\frac{-5 + \sqrt{53}}{2}, \frac{-5 - \sqrt{53}}{2}\right\}$

11. $(t + 3)(t - 4) = -2$

$t^2 - t - 12 = -2$

$t^2 - t - 10 = 0$

Here $a = 1$, $b = -1$, and $c = -10$.

$$t = \frac{-b \pm \sqrt{b^2 - 4ac}}{2a}$$
$$t = \frac{-(-1) \pm \sqrt{(-1)^2 - 4(1)(-10)}}{2(1)}$$
$$= \frac{1 \pm \sqrt{1 + 40}}{2} = \frac{1 \pm \sqrt{41}}{2}$$

Solution set: $\left\{\frac{1 + \sqrt{41}}{2}, \frac{1 - \sqrt{41}}{2}\right\}$

12. $2x^2 + 3x + 4 = 0$

Here $a = 2$, $b = 3$, and $c = 4$.

$$x = \frac{-b \pm \sqrt{b^2 - 4ac}}{2a}$$
$$x = \frac{-3 \pm \sqrt{3^2 - 4(2)(4)}}{2(2)}$$
$$= \frac{-3 \pm \sqrt{9 - 32}}{4}$$
$$= \frac{-3 \pm \sqrt{-23}}{4}$$
$$= \frac{-3 \pm i\sqrt{23}}{4}$$

Solution set: $\left\{-\frac{3}{4} + \frac{\sqrt{23}}{4}i, -\frac{3}{4} - \frac{\sqrt{23}}{4}i\right\}$

13. $3p^2 = 2(2p - 1)$

$3p^2 = 4p - 2$

$3p^2 - 4p + 2 = 0$

Here $a = 3$, $b = -4$, and $c = 2$.

$$p = \frac{-b \pm \sqrt{b^2 - 4ac}}{2a}$$
$$p = \frac{-(-4) \pm \sqrt{(-4)^2 - 4(3)(2)}}{2(3)}$$
$$= \frac{4 \pm \sqrt{16 - 24}}{6} = \frac{4 \pm \sqrt{-8}}{6}$$
$$= \frac{4 \pm 2i\sqrt{2}}{6} = \frac{2(2 \pm i\sqrt{2})}{6}$$
$$= \frac{2 \pm i\sqrt{2}}{3}$$

Solution set: $\left\{\frac{2}{3} + \frac{\sqrt{2}}{3}i, \frac{2}{3} - \frac{\sqrt{2}}{3}i\right\}$

14. $m(2m - 7) = 3m^2 + 3$

$2m^2 - 7m = 3m^2 + 3$

$0 = m^2 + 7m + 3$

Here $a = 1$, $b = 7$, and $c = 3$.

$$m = \frac{-b \pm \sqrt{b^2 - 4ac}}{2a}$$
$$m = \frac{-7 \pm \sqrt{7^2 - 4(1)(3)}}{2(1)}$$
$$= \frac{-7 \pm \sqrt{49 - 12}}{2}$$
$$= \frac{-7 \pm \sqrt{37}}{2}$$

Solution set: $\left\{\frac{-7 + \sqrt{37}}{2}, \frac{-7 - \sqrt{37}}{2}\right\}$

15. $x^2 + 5x + 2 = 0$
Here $a = 1$, $b = 5$, and $c = 2$.

$$b^2 - 4ac = 5^2 - 4(1)(2) = 25 - 8 = 17$$

Since the discriminant is positive, but not a perfect square, there are two distinct irrational number solutions. The answer is **C**.

16. $4t^2 = 3 - 4t$
$4t^2 + 4t - 3 = 0$
Here $a = 4$, $b = 4$, and $c = -3$.

$$b^2 - 4ac = 4^2 - 4(4)(-3) = 16 + 48 = 64, \text{ or } 8^2$$

Since the discriminant is positive, and a perfect square, there are two distinct rational number solutions. The answer is **A**.

17. $4x^2 = 6x - 8$
$4x^2 - 6x + 8 = 0$
Here $a = 4$, $b = -6$, and $c = 8$.

$$b^2 - 4ac = (-6)^2 - 4(4)(8) = 36 - 128 = -92$$

Since the discriminant is negative, there are two distinct nonreal complex number solutions. The answer is **D**.

18. $9z^2 + 30z + 25 = 0$
Here $a = 9$, $b = 30$, and $c = 25$.

$$b^2 - 4ac = 30^2 - 4(9)(25) = 900 - 900 = 0$$

Since the discriminant is zero, there is exactly one rational number solution. The answer is **B**.

19. $\frac{15}{x} = 2x - 1$

$x\left(\frac{15}{x}\right) = x(2x - 1)$ *Multiply by the LCD, x.*

$15 = 2x^2 - x$
$0 = 2x^2 - x - 15$
$0 = (2x + 5)(x - 3)$

$2x + 5 = 0$ or $x - 3 = 0$
$x = -\frac{5}{2}$ or $x = 3$

Check $x = -\frac{5}{2}$: $-6 = -5 - 1$ *True*
Check $x = 3$: $5 = 6 - 1$ *True*

Solution set: $\{-\frac{5}{2}, 3\}$

20. $\frac{1}{n} + \frac{2}{n+1} = 2$

$n(n+1)\left(\frac{1}{n} + \frac{2}{n+1}\right) = n(n+1) \cdot 2$
Multiply by the LCD, n(n+1).

$(n+1) + 2n = 2n^2 + 2n$
$0 = 2n^2 - n - 1$
$0 = (2n + 1)(n - 1)$

$2n + 1 = 0$ or $n - 1 = 0$
$n = -\frac{1}{2}$ or $n = 1$

Check $n = -\frac{1}{2}$: $-2 + 4 = 2$ *True*
Check $n = 1$: $1 + 1 = 2$ *True*

Solution set: $\{-\frac{1}{2}, 1\}$

21. $-2r = \sqrt{\frac{48 - 20r}{2}}$

$(-2r)^2 = \left(\sqrt{\frac{48 - 20r}{2}}\right)^2$ *Square.*

$4r^2 = \frac{48 - 20r}{2}$
$4r^2 = 24 - 10r$
$4r^2 + 10r - 24 = 0$
$2r^2 + 5r - 12 = 0$
$(r + 4)(2r - 3) = 0$

$r + 4 = 0$ or $2r - 3 = 0$
$r = -4$ or $r = \frac{3}{2}$

Check $r = -4$: $8 = \sqrt{64}$ *True*
Check $r = \frac{3}{2}$: $-3 = \sqrt{9}$ *False*

Solution set: $\{-4\}$

22. $8(3x + 5)^2 + 2(3x + 5) - 1 = 0$
Let $u = 3x + 5$. The equation becomes

$8u^2 + 2u - 1 = 0.$
$(2u + 1)(4u - 1) = 0$

$2u + 1 = 0$ or $4u - 1 = 0$
$u = -\frac{1}{2}$ or $u = \frac{1}{4}$

To find x, substitute $3x + 5$ for u.

$3x + 5 = -\frac{1}{2}$ or $3x + 5 = \frac{1}{4}$
$3x = -\frac{11}{2}$ or $3x = -\frac{19}{4}$
$x = -\frac{11}{6}$ or $x = -\frac{19}{12}$

Check $x = -\frac{11}{6}$: $2 - 1 - 1 = 0$ *True*
Check $x = -\frac{19}{12}$: $0.5 + 0.5 - 1 = 0$ *True*

Solution set: $\{-\frac{11}{6}, -\frac{19}{12}\}$

23. $2x^{2/3} - x^{1/3} - 28 = 0$

Let $u = x^{1/3}$, so $u^2 = (x^{1/3})^2 = x^{2/3}$.

The equation becomes

$$2u^2 - u - 28 = 0.$$
$$(2u+7)(u-4) = 0$$

$$2u + 7 = 0 \quad \text{or} \quad u - 4 = 0$$
$$u = -\tfrac{7}{2} \quad \text{or} \quad u = 4$$

To find x, substitute $x^{1/3}$ for u.

$$x^{1/3} = -\tfrac{7}{2} \quad \text{or} \quad x^{1/3} = 4$$
$$(x^{1/3})^3 = (-\tfrac{7}{2})^3 \quad \text{or} \quad (x^{1/3})^3 = 4^3$$
$$x = -\tfrac{343}{8} \quad \text{or} \quad x = 64$$

Check $x = -\frac{343}{8}$: $24.5 + 3.5 - 28 = 0$ *True*

Check $x = 64$: $32 - 4 - 28 = 0$ *True*

Solution set: $\{-\frac{343}{8}, 64\}$

24. $p^4 - 5p^2 + 4 = 0$

Let $x = p^2$, so $x^2 = p^4$.

$$x^2 - 5x + 4 = 0$$
$$(x-1)(x-4) = 0$$

$$x - 1 = 0 \quad \text{or} \quad x - 4 = 0$$
$$x = 1 \quad \text{or} \quad x = 4$$

To find p, substitute p^2 for x.

$$p^2 = 1 \quad \text{or} \quad p^2 = 4$$
$$p = \pm\sqrt{1} \quad \text{or} \quad p = \pm\sqrt{4}$$
$$p = \pm 1 \quad \text{or} \quad p = \pm 2$$

Check $p = \pm 1$: $1 - 5 + 4 = 0$ *True*

Check $p = \pm 2$: $16 - 20 + 4 = 0$ *True*

Solution set: $\{-2, -1, 1, 2\}$

25. Let $x =$ Matthew's speed on the trip to pick up Jack.

Make a chart. Use $d = rt$, or $t = d/r$.

	Distance	Rate	Time
To Jack	8	x	$\frac{8}{x}$
To the Mall	11	$x+15$	$\frac{11}{x+15}$

Time to pick up Jack + time to mall = 24 min (or 0.4 hr).

$$\frac{8}{x} + \frac{11}{x+15} = 0.4$$

Multiply each side by the LCD, $x(x+15)$.

$$x(x+15)\left(\frac{8}{x} + \frac{11}{x+15}\right) = x(x+15)(0.4)$$
$$8(x+15) + 11x = 0.4x(x+15)$$
$$8x + 120 + 11x = 0.4x^2 + 6x$$
$$0 = 0.4x^2 - 13x - 120$$

Multiply by 5 to clear the decimal.

$$0 = 2x^2 - 65x - 600$$
$$0 = (x-40)(2x+15)$$

$$x - 40 = 0 \quad \text{or} \quad 2x + 15 = 0$$
$$x = 40 \quad \text{or} \quad x = -\tfrac{15}{2}$$

Speed cannot be negative, so $-\frac{15}{2}$ is not a solution. Matthew's speed on the trip to pick up Jack was 40 mph.

26. Let $x =$ the amount of time for the old machine alone and

$x - 1 =$ the amount of time for the new machine alone.

Make a chart.

Machine	Rate	Time Together	Fractional Part of the Job Done
Old	$\frac{1}{x}$	2	$\frac{2}{x}$
New	$\frac{1}{x-1}$	2	$\frac{2}{x-1}$

Part done by old machine + Part done by new machine = 1 whole job.

$$\frac{2}{x} + \frac{2}{x-1} = 1$$

Multiply by the LCD, $x(x-1)$.

$$x(x-1)\left(\frac{2}{x} + \frac{2}{x-1}\right) = x(x-1) \cdot 1$$
$$2(x-1) + 2x = x^2 - x$$
$$2x - 2 + 2x = x^2 - x$$
$$0 = x^2 - 5x + 2$$

Use the quadratic formula.

$$x = \frac{-b \pm \sqrt{b^2 - 4ac}}{2a}$$
$$x = \frac{-(-5) \pm \sqrt{(-5)^2 - 4(1)(2)}}{2(1)}$$
$$= \frac{5 \pm \sqrt{25 - 8}}{2} = \frac{5 \pm \sqrt{17}}{2}$$
$$x = \frac{5 + \sqrt{17}}{2} \quad \text{or} \quad x = \frac{5 - \sqrt{17}}{2}$$
$$x \approx 4.6 \quad \text{or} \quad x \approx 0.4$$

Reject 0.4 as the time for the old machine, because that would yield a negative time for the new machine. Thus, the old machine takes about 4.6 hours.

27. Solve $k = \dfrac{rF}{wv^2}$ for v.

Multiply both sides by v^2, then divide by k.

$$v^2 = \frac{rF}{kw}$$
$$v = \pm\sqrt{\frac{rF}{kw}} = \frac{\pm\sqrt{rF}}{\sqrt{kw}}$$
$$= \frac{\pm\sqrt{rF}}{\sqrt{kw}} \bullet \frac{\sqrt{kw}}{\sqrt{kw}}$$
$$v = \frac{\pm\sqrt{rFkw}}{kw}$$

28. Solve $mt^2 = 3mt + 6$ for t.

$mt^2 - 3mt - 6 = 0$

Use the quadratic formula with $a = m$, $b = -3m$, and $c = -6$.

$$t = \frac{-b \pm \sqrt{b^2 - 4ac}}{2a}$$
$$t = \frac{3m \pm \sqrt{(-3m)^2 - 4(m)(-6)}}{2m}$$
$$= \frac{3m \pm \sqrt{9m^2 + 24m}}{2m}$$

29. Let $x =$ the length of the longer leg;
$\frac{3}{4}x =$ the length of the shorter leg;
$2x - 9 =$ the length of the hypotenuse.

Use the Pythagorean formula.

$$c^2 = a^2 + b^2$$
$$(2x-9)^2 = x^2 + (\tfrac{3}{4}x)^2$$
$$4x^2 - 36x + 81 = x^2 + \tfrac{9}{16}x^2$$
$$16(4x^2 - 36x + 81) = 16(x^2 + \tfrac{9}{16}x^2)$$
$$64x^2 - 576x + 1296 = 16x^2 + 9x^2$$
$$39x^2 - 576x + 1296 = 0$$
$$13x^2 - 192x + 432 = 0 \qquad \textit{Divide by 3.}$$
$$(13x - 36)(x - 12) = 0$$

$$13x - 36 = 0 \quad \text{or} \quad x - 12 = 0$$
$$x = \tfrac{36}{13} \quad \text{or} \quad x = 12$$

Reject $x = \frac{36}{13}$ since $2(\frac{36}{13}) - 9$ is negative.
Since $x = 12$, $\frac{3}{4}x = \frac{3}{4}(12) = 9$, and
$2x - 9 = 2(12) - 9 = 15$.

The lengths of the three sides are 9 feet, 12 feet, and 15 feet.

30. Let $x =$ the amount removed from one dimension.

The area of the square is 256 cm^2, so the length of one side is $\sqrt{256}$ or 16 cm. The dimensions of the new rectangle are $16 + x$ and $16 - x$ cm. The area of the new rectangle is 16 cm^2 less than the area of the square.

$$(16 + x)(16 - x) = 256 - 16$$
$$256 - x^2 = 240$$
$$-x^2 = -16$$
$$x^2 = 16$$
$$x = \pm\sqrt{16} = \pm 4$$

Length cannot be negative, so reject -4. If $x = 4$, then $16 + x = 20$, and $16 - x = 12$. The dimensions are 20 cm by 12 cm.

31. Let $x =$ the width of the border.

Area of mat = length • width

$$352 = (2x + 20)(2x + 14)$$
$$352 = 4x^2 + 68x + 280$$
$$0 = 4x^2 + 68x - 72$$
$$0 = x^2 + 17x - 18$$
$$0 = (x + 18)(x - 1)$$

$$x + 18 = 0 \quad \text{or} \quad x - 1 = 0$$
$$x = -18 \quad \text{or} \quad x = 1$$

Reject the negative answer for length. The border is 1 inch wide.

32. $f(t) = -16t^2 + 45t + 400$

$$200 = -16t^2 + 45t + 400 \qquad \textit{Let } f(t) = 200.$$
$$0 = -16t^2 + 45t + 200$$
$$0 = 16t^2 - 45t - 200 \qquad \textit{Divide by } -1.$$

Here $a = 16$, $b = -45$, and $c = -200$.

$$t = \frac{-b \pm \sqrt{b^2 - 4ac}}{2a}$$
$$t = \frac{-(-45) \pm \sqrt{(-45)^2 - 4(16)(-200)}}{2(16)}$$
$$= \frac{45 \pm \sqrt{2025 + 12{,}800}}{32}$$
$$= \frac{45 \pm \sqrt{14{,}825}}{32}$$
$$t = \frac{45 + \sqrt{14{,}825}}{32} \quad \text{or} \quad t = \frac{45 - \sqrt{14{,}825}}{32}$$
$$t \approx 5.2 \quad \text{or} \quad t \approx -2.4$$

Reject the negative solution since time cannot be negative. The ball will reach a height of 200 ft above the ground after about 5.2 seconds.

33. $f(x) = -(x-1)^2$
Write in $y = a(x-h)^2 + k$ form as $y = -1(x-1)^2 + 0$. The vertex (h,k) is $(1,0)$.

34. $y = (x-3)^2 + 7$
The equation is in the form $y = a(x-h)^2 + k$, so the vertex (h,k) is $(3,7)$.

35. $y = f(x) = -3x^2 + 4x - 2$

Use the vertex formula with $a = -3$ and $b = 4$.

The x-coordinate of the vertex is

$$\frac{-b}{2a} = \frac{-4}{2(-3)} = \frac{2}{3}.$$

The y-coordinate of the vertex is

$$\begin{aligned} f\left(\frac{-b}{2a}\right) &= f\left(\frac{2}{3}\right) \\ &= -3(\tfrac{2}{3})^2 + 4(\tfrac{2}{3}) - 2 \\ &= -\tfrac{4}{3} + \tfrac{8}{3} - 2 \\ &= -\tfrac{2}{3}. \end{aligned}$$

The vertex is $(\frac{2}{3}, -\frac{2}{3})$.

36. $x = (y-3)^2 - 4$
$= (y-3)^2 + (-4)$

Since the roles of x and y are reversed, this is a horizontal parabola. The equation is in the form

$$x = a(y-k)^2 + h,$$

so the vertex (h,k) is $(-4,3)$.

37. $y = 2(x-2)^2 - 3$

The graph opens up since $a = 2 > 0$.

The vertex is $(2,-3)$ and the axis is $x = 2$. If $x = 0$, we get the y-intercept $(0,5)$, and by symmetry, the point $(4,5)$ is also on the graph. The domain is $(-\infty, \infty)$. The smallest y-value is -3, so the range is $[-3, \infty)$.

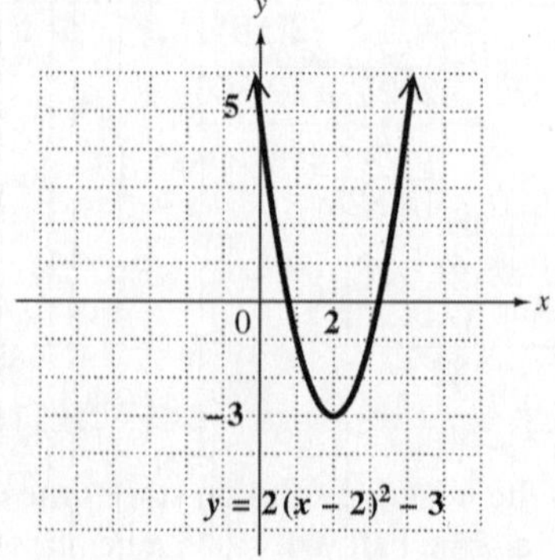

38. $f(x) = -2x^2 + 8x - 5$
Complete the square to find the vertex.

$$\begin{aligned} f(x) &= -2(x^2 - 4x) - 5 \\ &= -2(x^2 - 4x + 4 - 4) - 5 \\ &= -2(x^2 - 4x + 4) - 2(-4) - 5 \\ &= -2(x-2)^2 + 8 - 5 \\ &= -2(x-2)^2 + 3 \end{aligned}$$

The equation is in the form $y = a(x-h)^2 + k$, so the vertex (h,k) is $(2,3)$ and the axis is $x = 2$. Here, $a = -2 < 0$, so the parabola opens down.

Also, $|a| = |-2| = 2 > 1$, so the graph is narrower than the graph of $y = x^2$. The points $(0,-5)$, $(1,1)$, and $(3,1)$ are on the graph.

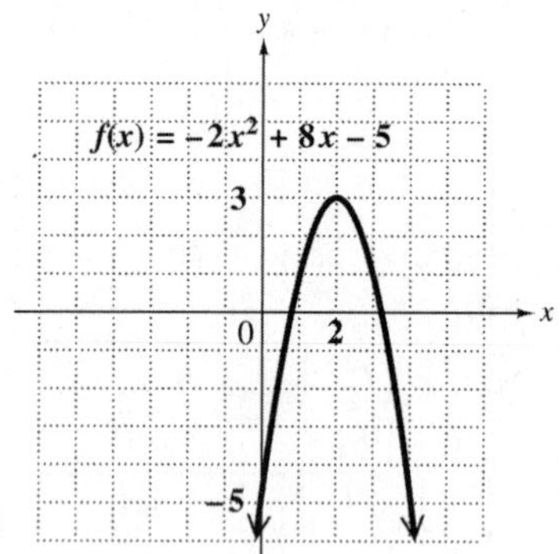

The domain is $(-\infty, \infty)$. The largest y-value is 3, so the range is $(-\infty, 3]$.

39. $x = 2(y+3)^2 - 4$
$= 2[y - (-3)]^2 + (-4)$

Since the roles of x and y are reversed, this is a horizontal parabola. The equation is in the form

$$x = a(y-k)^2 + h,$$

so the vertex (h,k) is $(-4,-3)$ and the axis is $y = -3$. Here, $a = 2 > 0$, so the parabola opens to the right and is narrower than the graph of $y = x^2$.

Two other points on the graph are $(4,-1)$ and $(4,-5)$.

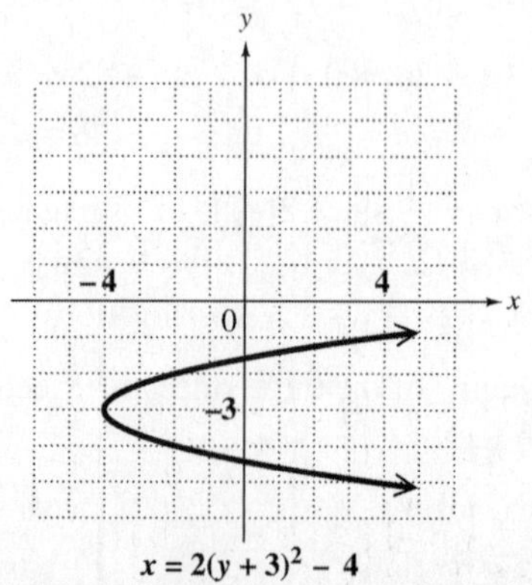

The smallest x-value is -4, so the domain is $[-4, \infty)$. The range is $(-\infty, \infty)$.

40. $x = -\frac{1}{2}y^2 + 6y - 14$

Since the roles of x and y are reversed, this is a horizontal parabola. Complete the square to find the vertex.

$$\begin{aligned} x &= -\tfrac{1}{2}y^2 + 6y - 14 \\ &= -\tfrac{1}{2}(y^2 - 12y) - 14 \\ &= -\tfrac{1}{2}(y^2 - 12y + 36 - 36) - 14 \\ &= -\tfrac{1}{2}(y^2 - 12y + 36) - \tfrac{1}{2}(-36) - 14 \\ &= -\tfrac{1}{2}(y - 6)^2 + 18 - 14 \\ x &= -\tfrac{1}{2}(y - 6)^2 + 4 \end{aligned}$$

The equation is in the form $x = a(y - k)^2 + h$, so the vertex (h, k) is $(4, 6)$ and the axis is $y = 6$. Here, $a = -\frac{1}{2} < 0$, so the parabola opens to the left.
Also, $|a| = \left|-\frac{1}{2}\right| = \frac{1}{2} < 1$, so the graph is wider than the graph of $y = x^2$. The points $(-14, 0)$, $(2, 4)$, and $(2, 8)$ are on the graph.

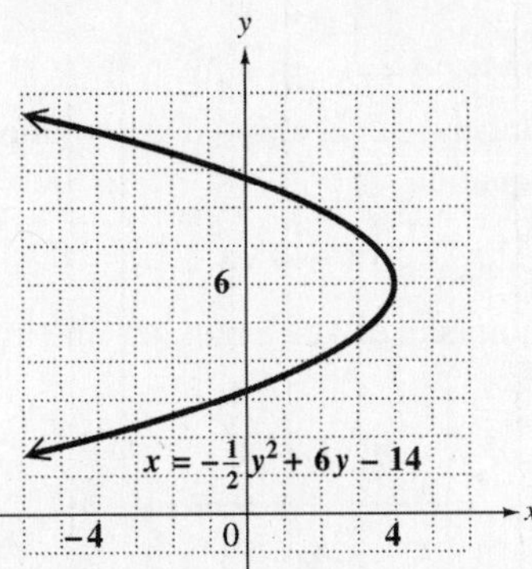

The largest x-value is 4, so the domain is $(-\infty, 4]$. The range is $(-\infty, \infty)$.

41. **(a)** Use $ax^2 + bx + c = y$ with $(0, 2.9)$, $(10, 24.3)$, and $(20, 56.5)$.

$$\begin{aligned} c &= 2.9 && (1) \\ 100a + 10b + c &= 24.3 && (2) \\ 400a + 20b + c &= 56.5 && (3) \end{aligned}$$

(b) Rewrite equations (2) and (3) with $c = 2.9$.

$$\begin{aligned} 100a + 10b + 2.9 &= 24.3 \\ 400a + 20b + 2.9 &= 56.5 \end{aligned}$$

$$\begin{aligned} 100a + 10b &= 21.4 && (4) \\ 400a + 20b &= 53.6 && (5) \end{aligned}$$

Now eliminate b.

$$\begin{aligned} -200a - 20b &= -42.8 && -2 \times (4) \\ 400a + 20b &= 53.6 && (5) \\ \hline 200a &= 10.8 \\ a &= \tfrac{10.8}{200} = 0.054 \end{aligned}$$

Use (4) to find b.

$$\begin{aligned} 100a + 10b &= 21.4 \quad (4) \\ 100(0.054) + 10b &= 21.4 \\ 5.4 + 10b &= 21.4 \\ 10b &= 16 \\ b &= \tfrac{16}{10} = 1.6 \end{aligned}$$

Thus, we get the quadratic function

$$f(x) = 0.054x^2 + 1.6x + 2.9.$$

(c) Use $f(x) = 0.054x^2 + 1.6x + 2.9$ with $x = 21$ for the year 2006.

$$\begin{aligned} f(21) &= 0.054(21)^2 + 1.6(21) + 2.9 \\ &= 60.314 \approx \$60.3 \text{ billion} \end{aligned}$$

The result using the model is close to the table value of \$61.4 billion, but slightly low.

42. $s(t) = -16t^2 + 160t$
The equation represents a parabola. Since $a = -16 < 0$, the parabola opens down. The time and maximum height occur at the vertex (h, k) of the parabola, given by

$$\left(\frac{-b}{2a}, s\left(\frac{-b}{2a}\right)\right).$$

Using the standard form of the equation, $a = -16$ and $b = 160$, so

$$h = \frac{-b}{2a} = \frac{-160}{2(-16)} = 5,$$

$$\begin{aligned} \text{and } k = s(5) &= -16(5)^2 + 160(5) \\ &= -400 + 800 = 400. \end{aligned}$$

The vertex is $(5, 400)$. The time at which the maximum height is reached is 5 seconds. The maximum height is 400 feet.

43. Let L = the length of the rectangle and W = the width.

The perimeter of the rectangle is 200 m, so

$$\begin{aligned} 2L + 2W &= 200 \\ 2W &= 200 - 2L \\ W &= 100 - L. \end{aligned}$$

Since the area is length times width, substitute $100 - L$ for W.

continued

$$A = LW$$
$$= L(100 - L)$$
$$= 100L - L^2 \text{ or } -L^2 + 100L$$

The graph of this equation is a parabola. Since $a = -1 < 0$, the parabola turns down, so the maximum occurs at the vertex. Use the vertex formula.

$$L = \frac{-b}{2a} = \frac{-100}{2(-1)} = 50$$

So $L = 50$ meters and
$W = 100 - L = 100 - 50 = 50$ meters.

44. $(x - 4)(2x + 3) > 0$
Solve the equation
$(x - 4)(2x + 3) = 0.$

$$x - 4 = 0 \quad \text{or} \quad 2x + 3 = 0$$
$$x = 4 \quad \text{or} \quad x = -\tfrac{3}{2}$$

The numbers $-\frac{3}{2}$ and 4 divide a number line into three intervals.

A B C
$-\frac{3}{2}$ 4

Test a number from each interval in the inequality

$$(x - 4)(2x + 3) > 0.$$

Interval A: Let $x = -2$.
$$-6(-1) \stackrel{?}{>} 0$$
$$6 > 0 \quad \textit{True}$$
Interval B: Let $x = 0$.
$$-4(3) \stackrel{?}{>} 0$$
$$-12 > 0 \quad \textit{False}$$
Interval C: Let $x = 5$.
$$1(13) \stackrel{?}{>} 0$$
$$13 > 0 \quad \textit{True}$$

The solution set includes numbers in Intervals A and C, excluding endpoints.

Solution set: $(-\infty, -\frac{3}{2}) \cup (4, \infty)$

$-\frac{3}{2}$ 0 4

45. $x^2 + x \le 12$
Solve the equation
$$x^2 + x = 12.$$
$$x^2 + x - 12 = 0$$
$$(x + 4)(x - 3) = 0$$

$$x + 4 = 0 \quad \text{or} \quad x - 3 = 0$$
$$x = -4 \quad \text{or} \quad x = 3$$

The numbers -4 and 3 divide a number line into three intervals.

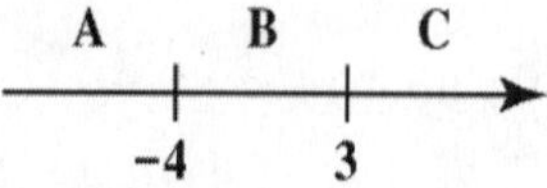

Test a number from each interval in the inequality

$$x^2 + x \le 12.$$

Interval A: Let $x = -5$.
$$25 - 5 \stackrel{?}{\le} 12$$
$$20 \le 12 \quad \textit{False}$$
Interval B: Let $x = 0$.
$$0 \le 12 \quad \textit{True}$$
Interval C: Let $x = 4$.
$$16 + 4 \stackrel{?}{\le} 12$$
$$20 \le 12 \quad \textit{False}$$

The numbers in Interval B, including -4 and 3, are solutions.

Solution set: $[-4, 3]$

-4 0 3

46. $(x + 2)(x - 3)(x + 5) \le 0$
The numbers -2, 3, and -5 are solutions of the cubic equation

$$(x + 2)(x - 3)(x + 5) = 0.$$

These numbers divide a number line into four intervals.

A B C D
-5 -2 3

Test a number from each interval in the inequality

$$(x + 2)(x - 3)(x + 5) \le 0.$$

Interval A: Let $x = -6$.
$$(-4)(-9)(-1) \le 0 \quad \textit{True}$$
Interval B: Let $x = -3$.
$$(-1)(-6)(2) \le 0 \quad \textit{False}$$
Interval C: Let $x = 0$.
$$(2)(-3)(5) \le 0 \quad \textit{True}$$
Interval D: Let $x = 4$.
$$(6)(1)(9) \le 0 \quad \textit{False}$$

The numbers in Intervals A and C, including -5, -2, and 3, are solutions.

Solution set: $(-\infty, -5] \cup [-2, 3]$

-5 -2 0 3

47. $(4m+3)^2 \le -4$

Since $(4m+3)^2$ is never negative, $(4m+3)^2$ will never be less than or equal to a negative number. Therefore, the solution set is $\emptyset$.

48. $$\frac{6}{2z-1} < 2$$

Write the inequality so that 0 is on one side.

$$\frac{6}{2z-1} - 2 < 0$$
$$\frac{6}{2z-1} - \frac{2(2z-1)}{2z-1} < 0$$
$$\frac{6-4z+2}{2z-1} < 0$$
$$\frac{-4z+8}{2z-1} < 0$$

The number 2 makes the numerator 0, and $\frac{1}{2}$ makes the denominator 0. These two numbers determine three intervals.

Test a number from each interval in the inequality

$$\frac{6}{2z-1} < 2.$$

Interval A: Let $z = 0$.
$-6 < 2$ *True*

Interval B: Let $z = 1$.
$6 < 2$ *False*

Interval C: Let $z = 3$.
$\frac{6}{5} < 2$ *True*

The solution set includes numbers in Intervals A and C, excluding endpoints.

Solution set: $(-\infty, \frac{1}{2}) \cup (2, \infty)$

49. $$\frac{3t+4}{t-2} \le 1$$

Write the inequality so that 0 is on one side.

$$\frac{3t+4}{t-2} - 1 \le 0$$
$$\frac{3t+4}{t-2} - \frac{1(t-2)}{t-2} \le 0$$
$$\frac{3t+4-t+2}{t-2} \le 0$$
$$\frac{2t+6}{t-2} \le 0$$

The number -3 makes the numerator 0, and 2 makes the denominator 0. These two numbers determine three intervals.

Test a number from each interval in the inequality

$$\frac{3t+4}{t-2} \le 1.$$

Interval A: Let $t = -4$.
$\frac{-8}{-6} \overset{?}{\le} 1$
$\frac{4}{3} \le 1$ *False*

Interval B: Let $t = 0$.
$\frac{4}{-2} \overset{?}{\le} 1$
$-2 \le 1$ *True*

Interval C: Let $t = 3$.
$\frac{13}{1} \overset{?}{\le} 1$
$13 \le 1$ *False*

The numbers in Interval B, including -3 but not 2, are solutions.

Solution set: $[-3, 2)$

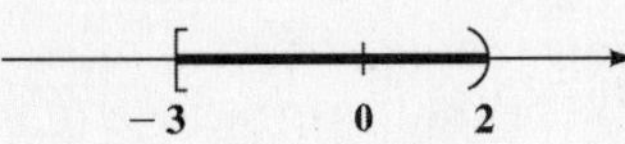

50. **[10.4]** Solve $V = r^2 + R^2 h$ for R.

$$V - r^2 = R^2 h$$
$$R^2 h = V - r^2$$
$$R^2 = \frac{V - r^2}{h}$$
$$R = \pm\sqrt{\frac{V-r^2}{h}} = \frac{\pm\sqrt{V-r^2}}{\sqrt{h}}$$
$$= \frac{\pm\sqrt{V-r^2}}{\sqrt{h}} \cdot \frac{\sqrt{h}}{\sqrt{h}}$$
$$= \frac{\pm\sqrt{Vh - r^2 h}}{h}$$

51. **[10.2]** $3t^2 - 6t = -4$

$3t^2 - 6t + 4 = 0$

Use the quadratic formula.

$$t = \frac{-b \pm \sqrt{b^2 - 4ac}}{2a}$$

$$t = \frac{-(-6) \pm \sqrt{(-6)^2 - 4(3)(4)}}{2(3)}$$

$$= \frac{6 \pm \sqrt{-12}}{6} = \frac{6 \pm 2i\sqrt{3}}{6}$$

$$= \frac{2(3 \pm i\sqrt{3})}{6} = \frac{3 \pm i\sqrt{3}}{3}$$

$$= \frac{3}{3} \pm \frac{i\sqrt{3}}{3} = 1 \pm \frac{\sqrt{3}}{3}i$$

Solution set: $\left\{1 + \frac{\sqrt{3}}{3}i, 1 - \frac{\sqrt{3}}{3}i\right\}$

52. **[10.3]** $(x^2 - 2x)^2 = 11(x^2 - 2x) - 24$

Let $u = x^2 - 2x$. The equation becomes

$$u^2 = 11u - 24.$$

$u^2 - 11u + 24 = 0$

$(u - 8)(u - 3) = 0$

$u - 8 = 0$ or $u - 3 = 0$

$u = 8$ or $u = 3$

To find x, substitute $x^2 - 2x$ for u.

$x^2 - 2x = 8$ or $x^2 - 2x = 3$

$x^2 - 2x - 8 = 0$ $\quad$ $x^2 - 2x - 3 = 0$

$(x - 4)(x + 2) = 0$ $\quad$ $(x - 3)(x + 1) = 0$

$x - 4 = 0$ or $x + 2 = 0$ $\quad$ $x - 3 = 0$ or $x + 1 = 0$

$x = 4$ or $x = -2$ or $x = 3$ or $x = -1$

The proposed solutions all check.

Solution set: $\{-2, -1, 3, 4\}$

53. **[10.7]** $(r - 1)(2r + 3)(r + 6) < 0$

Solve the equation

$(r - 1)(2r + 3)(r + 6) = 0.$

$r - 1 = 0$ or $2r + 3 = 0$ or $r + 6 = 0$

$r = 1$ or $r = -\frac{3}{2}$ or $r = -6$

The numbers -6, $-\frac{3}{2}$, and 1 divide a number line into four intervals.

A B C D

-6 $\quad -\frac{3}{2}$ $\quad 1$

Test a number from each interval in the inequality

$$(r - 1)(2r + 3)(r + 6) < 0.$$

Interval A: Let $r = -7$.

$-8(-11)(-1) \overset{?}{<} 0$

$-88 < 0$ *True*

Interval B: Let $r = -2$.

$-3(-1)(4) \overset{?}{<} 0$

$12 < 0$ *False*

Interval C: Let $r = 0$.

$-1(3)(6) \overset{?}{<} 0$

$-18 < 0$ *True*

Interval D: Let $r = 2$.

$1(7)(8) \overset{?}{<} 0$

$56 < 0$ *False*

The numbers in Intervals A and C, not including -6, $-\frac{3}{2}$, or 1, are solutions.

Solution set: $(-\infty, -6) \cup (-\frac{3}{2}, 1)$

54. **[10.1]** $(3k + 11)^2 = 7$

$3k + 11 = \sqrt{7}$ or $3k + 11 = -\sqrt{7}$

$3k = -11 + \sqrt{7}$ or $3k = -11 - \sqrt{7}$

$k = \frac{-11 + \sqrt{7}}{3}$ or $k = \frac{-11 - \sqrt{7}}{3}$

Solution set: $\left\{\frac{-11 + \sqrt{7}}{3}, \frac{-11 - \sqrt{7}}{3}\right\}$

55. **[10.4]** Solve $S = \frac{Id^2}{k}$ for d.

Multiply both sides by k, then divide by I.

$$\frac{Sk}{I} = d^2$$

$$d = \pm\sqrt{\frac{Sk}{I}} = \frac{\pm\sqrt{Sk}}{\sqrt{I}}$$

$$= \frac{\pm\sqrt{Sk}}{\sqrt{I}} \cdot \frac{\sqrt{I}}{\sqrt{I}}$$

$$d = \frac{\pm\sqrt{SkI}}{I}$$

56. **[10.3]** $2x - \sqrt{x} = 6$

$2x - \sqrt{x} - 6 = 0$

Let $u = \sqrt{x}$, so $u^2 = x$.

$2u^2 - u - 6 = 0$

$(2u + 3)(u - 2) = 0$

$2u + 3 = 0$ or $u - 2 = 0$

$u = -\frac{3}{2}$ or $u = 2$

$\sqrt{x} = -\frac{3}{2}$ or $\sqrt{x} = 2$

Since $\sqrt{x} \geq 0$, we must have $\sqrt{x} = 2$, or $x = 4$.

Check $x = 4$: $8 - 2 = 6$ *True*

Solution set: $\{4\}$

57. [10.3] $6+\frac{15}{s^2}=-\frac{19}{s}$

Multiply by the LCD, s^2.

$$s^2\left(6+\frac{15}{s^2}\right)=s^2\left(-\frac{19}{s}\right)$$
$$6s^2+15=-19s$$
$$6s^2+19s+15=0$$
$$(3s+5)(2s+3)=0$$
$$3s+5=0 \quad \text{or} \quad 2s+3=0$$
$$s=-\tfrac{5}{3} \quad \text{or} \quad s=-\tfrac{3}{2}$$

Check $s=-\frac{5}{3}$: $6+\frac{27}{5}=\frac{57}{5}$ *True*

Check $s=-\frac{3}{2}$: $6+\frac{20}{3}=\frac{38}{3}$ *True*

Solution set: $\{-\frac{5}{3},-\frac{3}{2}\}$

58. [10.7] $\frac{-2}{x+5}\le -5$

Write the inequality so that 0 is on one side.

$$\frac{-2}{x+5}+5\le 0$$
$$\frac{-2}{x+5}+\frac{5(x+5)}{x+5}\le 0$$
$$\frac{-2+5x+25}{x+5}\le 0$$
$$\frac{5x+23}{x+5}\le 0$$

The number $-\frac{23}{5}$ makes the numerator 0, and -5 makes the denominator 0. These two numbers determine three intervals.

A B C; -5, $-\frac{23}{5}$

Test a number from each interval in the inequality

$$\frac{-2}{x+5}\le -5.$$

Interval A: Let $x=-6$.

$$\frac{-2}{-1}\overset{?}{\le}-5$$
$$2\le -5 \quad \textit{False}$$

Interval B: Let $x=-\frac{24}{5}$.

$$-\frac{2}{\frac{1}{5}}\overset{?}{\le}-5$$
$$-10\le -5 \quad \textit{True}$$

Interval C: Let $x=0$.

$$\frac{-2}{5}\le -5 \quad \textit{False}$$

The numbers in Interval B, including $-\frac{23}{5}$ but not -5, are solutions.

Solution set: $(-5,-\frac{23}{5}]$

Chapter 10 Test

1. $t^2=54$

$$t=\sqrt{54} \quad \text{or} \quad t=-\sqrt{54}$$
$$t=3\sqrt{6} \quad \text{or} \quad t=-3\sqrt{6}$$

Solution set: $\{3\sqrt{6},-3\sqrt{6}\}$

2. $(7x+3)^2=25$

$$7x+3=\sqrt{25} \quad \text{or} \quad 7x+3=-\sqrt{25}$$
$$7x+3=5 \quad \text{or} \quad 7x+3=-5$$
$$7x=2 \quad \text{or} \quad 7x=-8$$
$$x=\tfrac{2}{7} \quad \text{or} \quad x=-\tfrac{8}{7}$$

Solution set: $\{-\frac{8}{7},\frac{2}{7}\}$

3. $x^2+2x=1$

Complete the square on x^2+2x.

$$[\tfrac{1}{2}(2)]^2=1^2=1$$

Add 1 to each side.

$$x^2+2x+1=1+1$$
$$(x+1)^2=2$$
$$x+1=\sqrt{2} \quad \text{or} \quad x+1=-\sqrt{2}$$
$$x=-1+\sqrt{2} \quad \text{or} \quad x=-1-\sqrt{2}$$

Solution set: $\{-1+\sqrt{2},-1-\sqrt{2}\}$

4. $2x^2-3x-1=0$

Here $a=2$, $b=-3$, and $c=-1$.

$$x=\frac{-b\pm\sqrt{b^2-4ac}}{2a}$$
$$x=\frac{-(-3)\pm\sqrt{(-3)^2-4(2)(-1)}}{2(2)}$$
$$=\frac{3\pm\sqrt{17}}{4}$$

Solution set: $\left\{\frac{3+\sqrt{17}}{4},\frac{3-\sqrt{17}}{4}\right\}$

5. $3t^2-4t=-5$

$3t^2-4t+5=0$

Here $a=3$, $b=-4$, and $c=5$.

$$t=\frac{-b\pm\sqrt{b^2-4ac}}{2a}$$
$$t=\frac{-(-4)\pm\sqrt{(-4)^2-4(3)(5)}}{2(3)}$$
$$=\frac{4\pm\sqrt{-44}}{6}=\frac{4\pm 2i\sqrt{11}}{6}$$
$$=\frac{2(2\pm i\sqrt{11})}{6}=\frac{2\pm i\sqrt{11}}{3}$$

Solution set: $\left\{\frac{2}{3}+\frac{\sqrt{11}}{3}i,\frac{2}{3}-\frac{\sqrt{11}}{3}i\right\}$

6. $$3x = \sqrt{\frac{9x+2}{2}}$$

Square both sides.

$$\begin{aligned} 9x^2 &= \frac{9x+2}{2} \\ 18x^2 &= 9x+2 \\ 18x^2 - 9x - 2 &= 0 \end{aligned}$$

As directed, use the quadratic formula with $a = 18$, $b = -9$, and $c = -2$.

$$x = \frac{-b \pm \sqrt{b^2 - 4ac}}{2a}$$

$$x = \frac{-(-9) \pm \sqrt{(-9)^2 - 4(18)(-2)}}{2(18)}$$

$$= \frac{9 \pm \sqrt{225}}{36} = \frac{9 \pm 15}{36}$$

$$x = \frac{9+15}{36} = \frac{24}{36} = \frac{2}{3} \text{ or}$$

$$x = \frac{9-15}{36} = \frac{-6}{36} = -\frac{1}{6}$$

Check $x = \frac{2}{3}$: $\quad 2 = \sqrt{4}$ *True*

Check $x = -\frac{1}{6}$: $\quad -\frac{1}{2} = \sqrt{\frac{1}{4}}$ *False*

Solution set: $\{\frac{2}{3}\}$

7. If k is a negative number, then $4k$ is also negative, so the equation $x^2 = 4k$ will have two nonreal complex solutions. The answer is **A**.

8. $2x^2 - 8x - 3 = 0$

$$\begin{aligned} b^2 - 4ac &= (-8)^2 - 4(2)(-3) \\ &= 64 + 24 = 88 \end{aligned}$$

The discriminant, 88, is positive but not a perfect square, so there will be two distinct irrational number solutions.

9. $$3 - \frac{16}{x} - \frac{12}{x^2} = 0$$

$$x^2\left(3 - \frac{16}{x} - \frac{12}{x^2}\right) = x^2 \cdot 0 \quad \textit{Multiply by the LCD, } x^2.$$

$$3x^2 - 16x - 12 = 0$$

$$(3x+2)(x-6) = 0 \quad \textit{Factor.}$$

$$3x + 2 = 0 \quad \text{or} \quad x - 6 = 0$$

$$x = -\tfrac{2}{3} \quad \text{or} \quad x = 6$$

Check $x = -\frac{2}{3}$: $\quad 3 + 24 - 27 = 0$ *True*

Check $x = 6$: $\quad 3 - \frac{8}{3} - \frac{1}{3} = 0$ *True*

Solution set: $\{-\frac{2}{3}, 6\}$

10. $4x^2 + 7x - 3 = 0$

Use the quadratic formula with $a = 4$, $b = 7$, and $c = -3$.

$$x = \frac{-b \pm \sqrt{b^2 - 4ac}}{2a}$$

$$x = \frac{-7 \pm \sqrt{7^2 - 4(4)(-3)}}{2(4)}$$

$$= \frac{-7 \pm \sqrt{97}}{8}$$

Solution set: $\left\{\frac{-7+\sqrt{97}}{8}, \frac{-7-\sqrt{97}}{8}\right\}$

11. $$9x^4 + 4 = 37x^2$$

$$9x^4 - 37x^2 + 4 = 0$$

Let $u = x^2$, so $u^2 = (x^2)^2 = x^4$.

$$9u^2 - 37u + 4 = 0$$

$$(9u - 1)(u - 4) = 0$$

$$9u - 1 = 0 \quad \text{or} \quad u - 4 = 0$$

$$u = \tfrac{1}{9} \quad \text{or} \quad u = 4$$

To find x, substitute x^2 for u.

$$x^2 = \tfrac{1}{9} \quad \text{or} \quad x^2 = 4$$

$$x = \pm\sqrt{\tfrac{1}{9}} \quad \text{or} \quad x = \pm\sqrt{4}$$

$$x = \pm\tfrac{1}{3} \quad \text{or} \quad x = \pm 2$$

Check $x = \pm\frac{1}{3}$: $\quad \frac{1}{9} + 4 = \frac{37}{9}$ *True*

Check $x = \pm 2$: $\quad 144 + 4 = 37(4)$ *True*

Solution set: $\{-2, -\frac{1}{3}, \frac{1}{3}, 2\}$

12. $12 = (2n+1)^2 + (2n+1)$

Let $u = 2n + 1$. The equation becomes

$$12 = u^2 + u.$$

$$0 = u^2 + u - 12$$

$$0 = (u+4)(u-3)$$

$$u + 4 = 0 \quad \text{or} \quad u - 3 = 0$$

$$u = -4 \quad \text{or} \quad u = 3$$

To find n, substitute $2n + 1$ for u.

$$2n + 1 = -4 \quad \text{or} \quad 2n + 1 = 3$$

$$2n = -5 \quad \text{or} \quad 2n = 2$$

$$n = -\tfrac{5}{2} \quad \text{or} \quad n = 1$$

Check $n = -\frac{5}{2}$: $\quad 12 = 16 - 4$ *True*

Check $n = 1$: $\quad 12 = 9 + 3$ *True*

Solution set: $\{-\frac{5}{2}, 1\}$

13. Solve $S = 4\pi r^2$ for r.

$$\frac{S}{4\pi} = r^2$$

$$r = \pm\sqrt{\frac{S}{4\pi}} = \frac{\pm\sqrt{S}}{2\sqrt{\pi}}$$

$$= \frac{\pm\sqrt{S}}{2\sqrt{\pi}} \bullet \frac{\sqrt{\pi}}{\sqrt{\pi}}$$

$$r = \frac{\pm\sqrt{\pi S}}{2\pi}$$

14. Let $x =$ Andrew's time alone.
Then $x - 2 =$ Kent's time alone.

Make a table.

	Rate	Time Together	Fractional Part of the Job Done
Andrew	$\frac{1}{x}$	5	$\frac{5}{x}$
Kent	$\frac{1}{x-2}$	5	$\frac{5}{x-2}$

Part done by Andrew plus part done by Kent equals 1 whole job.

$$\frac{5}{x} + \frac{5}{x-2} = 1$$

Multiply both sides by the LCD, $x(x-2)$.

$$x(x-2)\left(\frac{5}{x} + \frac{5}{x-2}\right) = x(x-2)\cdot 1$$

$$5x - 10 + 5x = x^2 - 2x$$

$$0 = x^2 - 12x + 10$$

Use the quadratic formula with $a = 1$, $b = -12$, and $c = 10$.

$$x = \frac{-b \pm \sqrt{b^2 - 4ac}}{2a}$$

$$x = \frac{-(-12) \pm \sqrt{(-12)^2 - 4(1)(10)}}{2(1)}$$

$$= \frac{12 \pm \sqrt{104}}{2} = \frac{12 \pm 2\sqrt{26}}{2}$$

$$= \frac{2(6 \pm \sqrt{26})}{2} = 6 \pm \sqrt{26}$$

$$x = 6 + \sqrt{26} \approx 11.1 \text{ or}$$

$$x = 6 - \sqrt{26} \approx 0.9$$

Reject 0.9 for Andrew's time, because that would yield a negative time for Kent. Thus, Andrew's time is about 11.1 hours and Kent's time is $x - 2 \approx 9.1$ hours.

15. Let $x =$ Bryn's rate.

Make a table. Use $d = rt$, or $t = d/r$.

	Distance	Rate	Time
Upstream	10	$x - 3$	$\frac{10}{x-3}$
Downstream	10	$x + 3$	$\frac{10}{x+3}$

Time upstream plus Time downstream equals 3.5 hr.

$$\frac{10}{x-3} + \frac{10}{x+3} = \frac{7}{2}$$

Multiply both sides by the LCD, $2(x+3)(x-3)$.

$$2(x+3)(x-3)\left(\frac{10}{x-3} + \frac{10}{x+3}\right)$$

$$= 2(x+3)(x-3)(\tfrac{7}{2})$$

$$20(x+3) + 20(x-3) = 7(x^2 - 9)$$

$$20x + 60 + 20x - 60 = 7x^2 - 63$$

$$0 = 7x^2 - 40x - 63$$

$$0 = (x-7)(7x+9)$$

$$x - 7 = 0 \quad \text{or} \quad 7x + 9 = 0$$

$$x = 7 \quad \text{or} \quad x = -\tfrac{9}{7}$$

Reject $-\frac{9}{7}$ since the rate can't be negative.
Bryn's rate was 7 mph.

16. Let $x =$ the width of the walk.
The area of the walk is equal to the area of the outer figure minus the area of the pool.

$$152 = (10 + 2x)(24 + 2x) - (24)(10)$$

$$152 = 240 + 68x + 4x^2 - 240$$

$$0 = 4x^2 + 68x - 152$$

$$0 = x^2 + 17x - 38$$

$$0 = (x + 19)(x - 2)$$

$$x + 19 = 0 \quad \text{or} \quad x - 2 = 0$$

$$x = -19 \quad \text{or} \quad x = 2$$

Reject -19 since width can't be negative.
The walk is 2 feet wide.

17. Let $x =$ the height of the tower. Then
$2x + 2 =$ the distance from the point to the top.

The distance from the base to the point is 30 m. These three segments form a right triangle, so the Pythagorean formula applies.

continued

$$a^2 + b^2 = c^2$$
$$x^2 + 30^2 = (2x + 2)^2$$
$$x^2 + 900 = 4x^2 + 8x + 4$$
$$0 = 3x^2 + 8x - 896$$
$$0 = (x - 16)(3x + 56)$$

$$x - 16 = 0 \quad \text{or} \quad 3x + 56 = 0$$
$$x = 16 \quad \text{or} \quad x = -\tfrac{56}{3}$$

Reject $-\frac{56}{3}$ since height can't be negative. The tower is 16 m high.

18. **(a)** $S(x) = -x^2 + 20x + 80$

Find the vertex of the graph of S.

$$\frac{-b}{2a} = \frac{-20}{2(-1)} = 10$$

The center should be open 10 hours to get a maximum of students to attend the algebra class.

(b) $S(x) = -x^2 + 20x + 80$

$$S(10) = -10^2 + 20(10) + 80$$
$$= 180$$

The maximum number of students is 180.

19. $f(x) = a(x - h)^2 + k$
Since $a < 0$, the parabola opens down. Since $h > 0$ and $k < 0$, the x-coordinate of the vertex is positive and the y-coordinate of the vertex is negative. Therefore, the vertex is in quadrant IV. The correct graph is **A**.

20. $f(x) = \frac{1}{2}x^2 - 2$
$f(x) = \frac{1}{2}(x - 0)^2 - 2$
The graph is a parabola in $f(x) = a(x - h)^2 + k$ form with vertex (h, k) at $(0, -2)$. The axis is $x = 0$. Since $a = \frac{1}{2} > 0$, the parabola opens up. Also, $|a| = \left|\frac{1}{2}\right| = \frac{1}{2} < 1$, so the graph of the parabola is wider than the graph of $f(x) = x^2$. The points $(2, 0)$ and $(-2, 0)$ are on the graph.

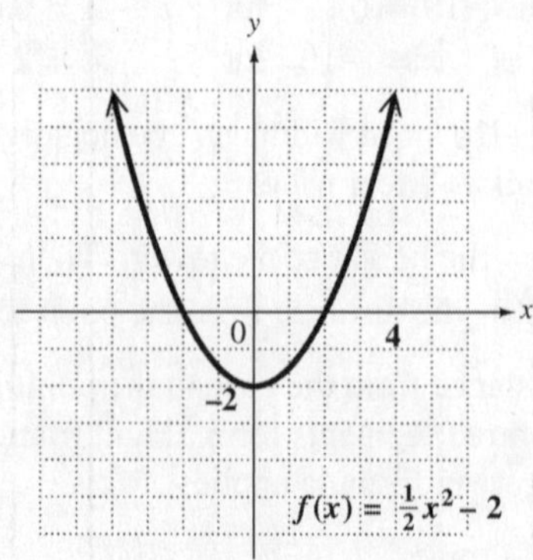

From the graph, we see that the x-values can be any real number, so the domain is $(-\infty, \infty)$. The y-values are greater than or equal to -2, so the range is $[-2, \infty)$.

21. $f(x) = -x^2 + 4x - 1$

The x-coordinate of the vertex is

$$x = \frac{-b}{2a} = \frac{-4}{2(-1)} = 2.$$

The y-coordinate of the vertex is

$$f(2) = -4 + 8 - 1 = 3.$$

The graph is a parabola with vertex (h, k) at $(2, 3)$ and axis $x = 2$. Since $a = -1 < 0$, the parabola opens down.
Also, $|a| = |-1| = 1$, so the graph has the same shape as the graph of $f(x) = x^2$. The points $(0, -1)$ and $(4, -1)$ are on the graph.

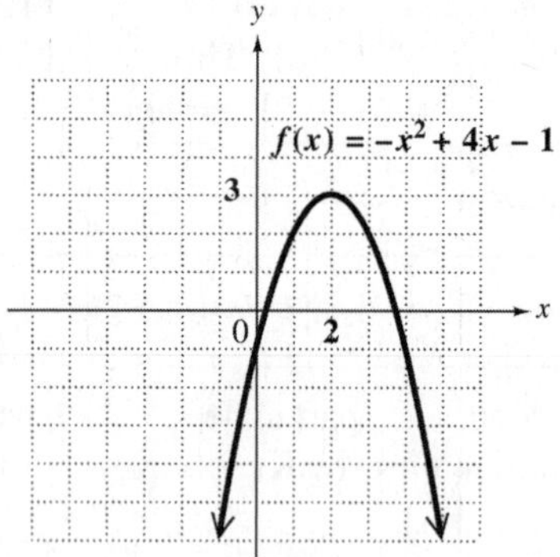

From the graph, we see that the x-values can be any real number, so the domain is $(-\infty, \infty)$. The y-values are less than or equal to 3, so the range is $(-\infty, 3]$.

22. $x = 2y^2 + 8y + 3$

The graph of this equation is a horizontal parabola. Complete the square to find the vertex.

$$x = 2(y^2 + 4y) + 3$$
$$= 2(y^2 + 4y + 4 - 4) + 3$$
$$= 2(y^2 + 4y + 4) + 2(-4) + 3$$
$$= 2(y + 2)^2 - 5$$

The equation is in the form $x = a(y - k)^2 + h$, so the vertex (h, k) is $(-5, -2)$ and the axis is $y = -2$. If $y = 0$, $x = 3$, so the x-intercept is $(3, 0)$. By symmetry, $(3, -4)$ is also on the graph.

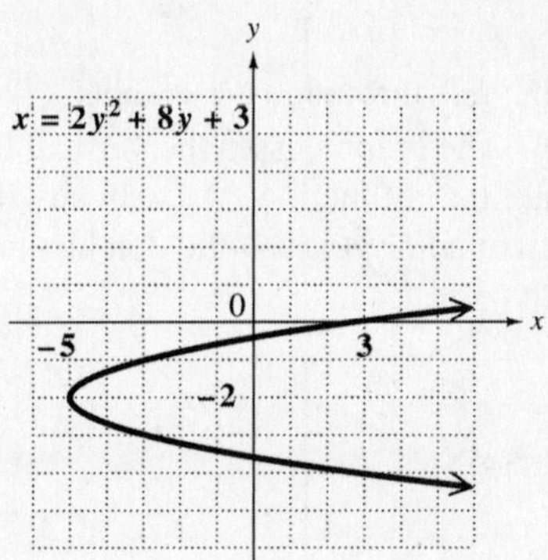

Since x is always greater than or equal to -5, the domain is $[-5, \infty)$; y can be any real number, so the range is $(-\infty, \infty)$.

23. Let x represent the length of the two equal sides, and let $280 - 2x$ represent the length of the remaining side (the side parallel to the highway). Then substitute x for L and $280 - 2x$ for W in the formula for the area of a rectangle, $A = LW$.

$$A = x(280 - 2x)$$
$$= 280x - 2x^2, \text{ or } -2x^2 + 280x$$

The maximum area will occur at the vertex.

$$x = \frac{-b}{2a} = \frac{-280}{2(-2)} = 70$$

When $x = 70$, $280 - 2x = 140$, and $A = 9800$. Thus, the dimensions of the lot with maximum area are 140 feet by 70 feet and the maximum area is 9800 square feet.

24.

$$2x^2 + 7x > 15$$
$$2x^2 + 7x - 15 > 0$$

Solve the equation

$$2x^2 + 7x - 15 = 0.$$
$$(2x - 3)(x + 5) = 0$$

$$2x - 3 = 0 \quad \text{or} \quad x + 5 = 0$$
$$x = \tfrac{3}{2} \quad \text{or} \quad x = -5$$

The numbers -5 and $\frac{3}{2}$ divide a number line into three intervals.

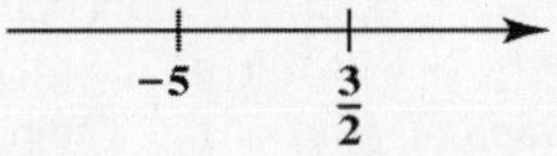

A B C — number line with marks at -5 and $\frac{3}{2}$

Test a number from each interval in the inequality

$$2x^2 + 7x > 15.$$

Interval A: Let $x = -6$.

$$72 - 42 \overset{?}{>} 15$$
$$30 > 15 \quad \textit{True}$$

Interval B: Let $x = 0$.

$$0 > 15 \quad \textit{False}$$

Interval C: Let $x = 2$.

$$8 + 14 \overset{?}{>} 15$$
$$22 > 15 \quad \textit{True}$$

The numbers in Intervals A and C, not including -5 and $\frac{3}{2}$, are solutions.

Solution set: $(-\infty, -5) \cup (\frac{3}{2}, \infty)$

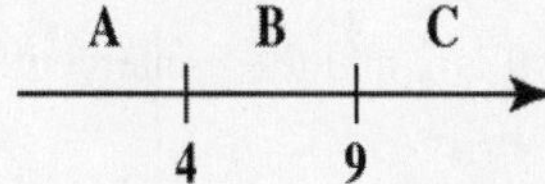

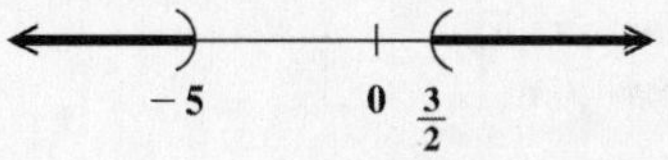

25.

$$\frac{5}{t-4} \le 1$$

Write the inequality so that 0 is on one side.

$$\frac{5}{t-4} - 1 \le 0$$
$$\frac{5}{t-4} - \frac{1(t-4)}{t-4} \le 0$$
$$\frac{5 - t + 4}{t-4} \le 0$$
$$\frac{-t + 9}{t-4} \le 0$$

The number 9 makes the numerator 0, and 4 makes the denominator 0. These two numbers determine three intervals.

A B C — number line with marks at 4 and 9

Test a number from each interval in the inequality

$$\frac{5}{t-4} \le 1.$$

Interval A: Let $t = 0$.

$$\tfrac{5}{-4} \le 1 \quad \textit{True}$$

Interval B: Let $t = 7$.

$$\tfrac{5}{3} \le 1 \quad \textit{False}$$

Interval C: Let $t = 10$.

$$\tfrac{5}{6} \le 1 \quad \textit{True}$$

The numbers in Intervals A and C, including 9 but not 4, are solutions.

Solution set: $(-\infty, 4) \cup [9, \infty)$

Number line graph with labels 0, 4, 9

Cumulative Review Exercises (Chapters 1–10)

1.
$$\begin{aligned} -2x+4 &= 5(x-4)+17 \\ -2x+4 &= 5x-20+17 \\ -2x+4 &= 5x-3 \\ 7 &= 7x \\ 1 &= x \end{aligned}$$

Solution set: $\{1\}$

2.
$$\begin{aligned} -2x+4 &\le -x+3 \\ -x &\le -1 \end{aligned}$$

Multiply by -1, and reverse the direction of the inequality.

$$x \ge 1$$

Solution set: $[1, \infty)$

3.
$$\begin{aligned} |3x-7| &\le 1 \\ -1 \le 3x-7 &\le 1 \\ 6 \le \ 3x \ &\le 8 \\ 2 \le \ x \ &\le \tfrac{8}{3} \end{aligned}$$

Solution set: $[2, \frac{8}{3}]$

4. $2x - 4y = 7$

Solve the equation for y to get the slope-intercept form, $y = mx + b$.

$$\begin{aligned} 2x - 4y &= 7 \\ -4y &= -2x + 7 \\ y &= \tfrac{1}{2}x - \tfrac{7}{4} \end{aligned}$$

The slope m is $\frac{1}{2}$ and the y-intercept, since $b = -\frac{7}{4}$, is $(0, -\frac{7}{4})$.

5. Through $(2, -1)$; perpendicular to $-3x + y = 5$

Find the slope of the given line.

$$\begin{aligned} -3x + y &= 5 \\ y &= 3x + 5 \end{aligned}$$

The slope is 3. The negative reciprocal of 3 is $-\frac{1}{3}$, so use the point-slope form with $m = -\frac{1}{3}$ and $(x_1, y_1) = (2, -1)$ to find the equation of the desired line.

$$\begin{aligned} y - y_1 &= m(x - x_1) \\ y - (-1) &= -\tfrac{1}{3}(x-2) \\ y + 1 &= -\tfrac{1}{3}(x-2) \\ 3y + 3 &= -1(x-2) \quad \textit{Multiply by 3.} \\ 3y + 3 &= -x + 2 \\ x + 3y &= -1 \end{aligned}$$

6. $4x - 5y = 15$

Draw the line through its intercepts, $(\frac{15}{4}, 0)$ and $(0, -3)$. The graph passes the vertical line test, so the relation is a function. As with any line that is not horizontal or vertical, the domain and range are both $(-\infty, \infty)$.

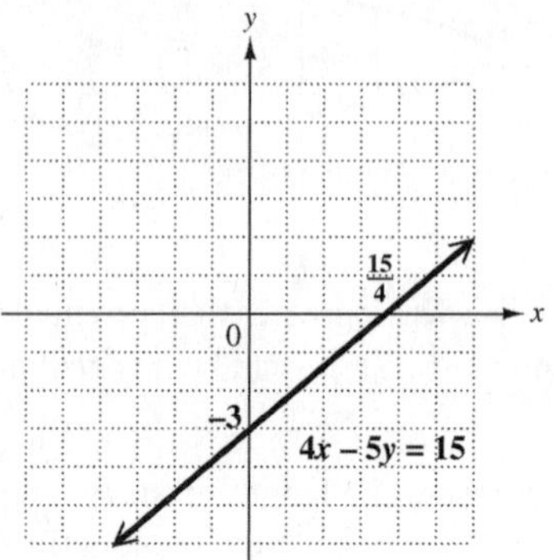

7. $4x - 5y < 15$

Draw a dashed line through the points $(\frac{15}{4}, 0)$ and $(0, -3)$. Check the origin:

$$\begin{aligned} 4(0) - 5(0) &\overset{?}{<} 15 \\ 0 &< 15 \quad \textit{True} \end{aligned}$$

Shade the region that contains the origin.

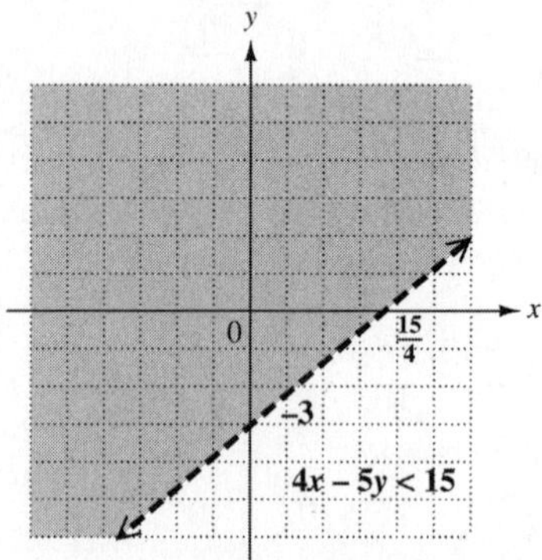

The relation is not a function since for any value of x, there is more than one value of y.

8. $y = -2(x-1)^2 + 3$

The equation is in $y = a(x-h)^2 + k$ form, so the graph is a parabola with vertex (h, k) at $(1, 3)$. Since $a = -2 < 0$, the parabola opens down. Also $|a| = |-2| = 2 > 1$, so the graph is narrower than the graph of $f(x) = x^2$. The points $(0, 1)$ and $(2, 1)$ are on the graph.

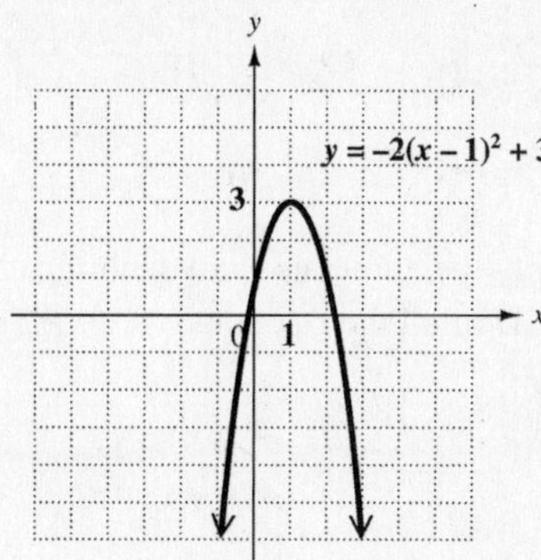

The relation is a function since it passes the vertical line test. The domain is $(-\infty, \infty)$. The largest value of y is 3, so the range is $(-\infty, 3]$.

9. $2x - 4y = 10$ (1)
$9x + 3y = 3$ (2)
Simplify the equations.

$$\begin{aligned} x - 2y &= 5 \quad (3) \quad \tfrac{1}{2} \times (1) \\ 3x + y &= 1 \quad (4) \quad \tfrac{1}{3} \times (2) \end{aligned}$$

To eliminate y, multiply (4) by 2 and add the result to (3).

$$\begin{aligned} x - 2y &= 5 \quad (3) \\ 6x + 2y &= 2 \quad 2 \times (4) \\ \hline 7x &= 7 \\ x &= 1 \end{aligned}$$

Substitute $x = 1$ into (4).

$$\begin{aligned} 3x + y &= 1 \quad (4) \\ 3(1) + y &= 1 \\ y &= -2 \end{aligned}$$

Solution set: $\{(1, -2)\}$

10. $x + y + 2z = 3$ (1)
$-x + y + z = -5$ (2)
$2x + 3y - z = -8$ (3)
Eliminate z by adding (2) and (3).

$$\begin{aligned} -x + y + z &= -5 \quad (2) \\ 2x + 3y - z &= -8 \quad (3) \\ \hline x + 4y &= -13 \quad (4) \end{aligned}$$

To get another equation without z, multiply equation (3) by 2 and add the result to equation (1).

$$\begin{aligned} x + y + 2z &= 3 \quad (1) \\ 4x + 6y - 2z &= -16 \quad 2 \times (3) \\ \hline 5x + 7y &= -13 \quad (5) \end{aligned}$$

To eliminate x, multiply (4) by -5 and add the result to (5).

$$\begin{aligned} -5x - 20y &= 65 \quad -5 \times (4) \\ 5x + 7y &= -13 \quad (5) \\ \hline -13y &= 52 \\ y &= -4 \end{aligned}$$

Use (4) to find x.

$$\begin{aligned} x + 4y &= -13 \quad (4) \\ x + 4(-4) &= -13 \\ x - 16 &= -13 \\ x &= 3 \end{aligned}$$

Use (2) to find z.

$$\begin{aligned} -x + y + z &= -5 \quad (2) \\ -3 - 4 + z &= -5 \\ -7 + z &= -5 \\ z &= 2 \end{aligned}$$

Solution set: $\{(3, -4, 2)\}$

11.
$$\begin{aligned} \left(\frac{x^{-3}y^2}{x^5y^{-2}}\right)^{-1} &= \left(x^{-3-5}y^{2-(-2)}\right)^{-1} \\ &= (x^{-8}y^4)^{-1} \\ &= x^8y^{-4} \\ &= \frac{x^8}{y^4} \end{aligned}$$

12.
$$\begin{aligned} \frac{(4x^{-2})^2(2y^3)}{8x^{-3}y^5} &= \frac{16x^{-4}(2y^3)}{8x^{-3}y^5} \\ &= \frac{4x^{-4}y^3}{x^{-3}y^5} \\ &= 4x^{-4-(-3)}y^{3-5} \\ &= 4x^{-1}y^{-2} \\ &= \frac{4}{xy^2} \end{aligned}$$

13.
$$\begin{aligned} (\tfrac{2}{3}t + 9)^2 &= (\tfrac{2}{3}t)^2 + 2(\tfrac{2}{3}t)(9) + 9^2 \\ &= \tfrac{4}{9}t^2 + 12t + 81 \end{aligned}$$

14. Divide $4x^3 + 2x^2 - x + 26$ by $x + 2$.

$$\begin{array}{r|l} & 4x^2 - 6x + 11 \\ \hline x + 2 & 4x^3 + 2x^2 - x + 26 \\ & \underline{4x^3 + 8x^2} \\ & \quad -6x^2 - x \\ & \quad \underline{-6x^2 - 12x} \\ & \qquad 11x + 26 \\ & \qquad \underline{11x + 22} \\ & \qquad\qquad 4 \end{array}$$

The answer is

$$4x^2 - 6x + 11 + \frac{4}{x+2}.$$

15.
$$\begin{aligned} 16x - x^3 &= x(16 - x^2) \\ &= x(4 + x)(4 - x) \end{aligned}$$

16. $24m^2 + 2m - 15$

The two integers whose product is $24(-15) = -360$ and whose sum is 2 are 20 and -18.

$$\begin{aligned} &24m^2 + 2m - 15 \\ &= 24m^2 + 20m - 18m - 15 \\ &= 4m(6m+5) - 3(6m+5) \\ &= (6m+5)(4m-3) \end{aligned}$$

17. $9x^2 - 30xy + 25y^2$
Use the perfect square formula,

$$a^2 - 2ab + b^2 = (a-b)^2,$$

with $a = 3x$ and $b = 5y$.

$$\begin{aligned} &9x^2 - 30xy + 25y^2 \\ &= [(3x)^2 - 2(3x)(5y) + (5y)^2] \\ &= (3x - 5y)^2 \end{aligned}$$

18. $\dfrac{5t+2}{-6} \div \dfrac{15t+6}{5}$

Multiply by the reciprocal.

$$\begin{aligned} &= \frac{5t+2}{-6} \cdot \frac{5}{15t+6} \\ &= \frac{5t+2}{-6} \cdot \frac{5}{3(5t+2)} \\ &= \frac{5}{-18} \text{ or } -\frac{5}{18} \end{aligned}$$

19. $\dfrac{3}{2-k} - \dfrac{5}{k} + \dfrac{6}{k^2 - 2k}$

$$\begin{aligned} &= \frac{3}{2-k} - \frac{5}{k} + \frac{6}{k(k-2)} \\ &= \frac{-3}{k-2} - \frac{5}{k} + \frac{6}{k(k-2)} \end{aligned}$$

The LCD is $k(k-2)$.

$$\begin{aligned} &= \frac{-3k}{(k-2)k} - \frac{5(k-2)}{k(k-2)} + \frac{6}{k(k-2)} \\ &= \frac{-3k - 5(k-2) + 6}{k(k-2)} \\ &= \frac{-3k - 5k + 10 + 6}{k(k-2)} \\ &= \frac{-8k + 16}{k(k-2)} \\ &= \frac{-8(k-2)}{k(k-2)} = -\frac{8}{k} \end{aligned}$$

20. $\dfrac{\frac{r}{s} - \frac{s}{r}}{\frac{r}{s} + 1}$

Multiply the numerator and denominator by the LCD of all the fractions, rs.

$$\begin{aligned} &= \frac{\left(\frac{r}{s} - \frac{s}{r}\right)rs}{\left(\frac{r}{s} + 1\right)rs} = \frac{r^2 - s^2}{r^2 + rs} \\ &= \frac{(r-s)(r+s)}{r(r+s)} = \frac{r-s}{r} \end{aligned}$$

21. $\sqrt[3]{\dfrac{27}{16}} = \dfrac{\sqrt[3]{27}}{\sqrt[3]{16}} = \dfrac{\sqrt[3]{3^3}}{\sqrt[3]{8 \cdot 2}} = \dfrac{3}{2\sqrt[3]{2}}$

$$\begin{aligned} &= \frac{3 \cdot \sqrt[3]{4}}{2\sqrt[3]{2} \cdot \sqrt[3]{4}} = \frac{3\sqrt[3]{4}}{2\sqrt[3]{8}} \\ &= \frac{3\sqrt[3]{4}}{2 \cdot 2} = \frac{3\sqrt[3]{4}}{4} \end{aligned}$$

22. $\dfrac{2}{\sqrt{7} - \sqrt{5}} = \dfrac{2(\sqrt{7} + \sqrt{5})}{(\sqrt{7} - \sqrt{5})(\sqrt{7} + \sqrt{5})}$

$$\begin{aligned} &= \frac{2(\sqrt{7} + \sqrt{5})}{7 - 5} \\ &= \frac{2(\sqrt{7} + \sqrt{5})}{2} = \sqrt{7} + \sqrt{5} \end{aligned}$$

23.

$$\begin{aligned} 2x &= \sqrt{\frac{5x+2}{3}} \\ (2x)^2 &= \left(\sqrt{\frac{5x+2}{3}}\right)^2 \quad \textit{Square.} \\ 4x^2 &= \frac{5x+2}{3} \\ 12x^2 &= 5x + 2 \\ 12x^2 - 5x - 2 &= 0 \\ (3x-2)(4x+1) &= 0 \end{aligned}$$

$$\begin{aligned} 3x - 2 = 0 \quad &\text{or} \quad 4x + 1 = 0 \\ x = \tfrac{2}{3} \quad &\text{or} \quad x = -\tfrac{1}{4} \end{aligned}$$

Check $x = -\frac{1}{4}$: $\quad -\frac{1}{2} = \frac{1}{2}$ *False*

Check $x = \frac{2}{3}$: $\quad \frac{4}{3} = \sqrt{\frac{16}{9}}$ *True*

Solution set: $\{\frac{2}{3}\}$

24. $2x^2 - 4x - 3 = 0$

Use $a = 2$, $b = -4$, and $c = -3$ in the quadratic formula.

$$x = \frac{-b \pm \sqrt{b^2 - 4ac}}{2a}$$

$$x = \frac{-(-4) \pm \sqrt{(-4)^2 - 4(2)(-3)}}{2(2)}$$

$$= \frac{4 \pm \sqrt{16 + 24}}{4}$$

$$= \frac{4 \pm \sqrt{40}}{4} = \frac{4 \pm 2\sqrt{10}}{4}$$

$$= \frac{2(2 \pm \sqrt{10})}{4} = \frac{2 \pm \sqrt{10}}{2}$$

Solution set: $\left\{\frac{2+\sqrt{10}}{2}, \frac{2-\sqrt{10}}{2}\right\}$

25. $z^2 - 2z = 15$

To complete the square on $z^2 - 2z$, take half the coefficient of z and square it.

$$\left[\tfrac{1}{2}(-2)\right]^2 = (-1)^2 = 1$$

Add 1 to each side.

$$z^2 - 2z + 1 = 15 + 1$$
$$(z - 1)^2 = 16$$

$$z - 1 = 4 \quad \text{or} \quad z - 1 = -4$$
$$z = 5 \quad \text{or} \quad z = -3$$

Solution set: $\{-3, 5\}$

26. $\frac{3}{x-3} - \frac{2}{x-2} = \frac{3}{x^2 - 5x + 6}$

$$\frac{3}{x-3} - \frac{2}{x-2} = \frac{3}{(x-3)(x-2)}$$

Multiply each side by the LCD, $(x-3)(x-2)$. $(x \neq 2, 3)$

$$(x-3)(x-2)\left(\frac{3}{x-3} - \frac{2}{x-2}\right)$$
$$= (x-3)(x-2)\left(\frac{3}{(x-3)(x-2)}\right)$$

$$3(x-2) - 2(x-3) = 3$$
$$3x - 6 - 2x + 6 = 3$$
$$x = 3$$

But $x \neq 3$ since it makes a denominator 0.

Solution set: $\emptyset$

27. $p^4 - 10p^2 + 9 = 0$

Let $u = p^2$, so $u^2 = p^4$.

$$u^2 - 10u + 9 = 0$$
$$(u - 1)(u - 9) = 0$$

$$u - 1 = 0 \quad \text{or} \quad u - 9 = 0$$
$$u = 1 \qquad\qquad u = 9$$
$$p^2 = 1 \quad \text{or} \quad p^2 = 9$$
$$p = \pm 1 \qquad\qquad p = \pm 3$$

Solution set: $\{-3, -1, 1, 3\}$

28. Let $x =$ the distance traveled by the southbound car and $2x - 38 =$ the distance traveled by the eastbound car.

Since the cars are traveling at right angles with one another, the Pythagorean formula can be applied.

$$a^2 + b^2 = c^2$$
$$x^2 + (2x - 38)^2 = 95^2$$
$$x^2 + 4x^2 - 152x + 1444 = 9025$$
$$5x^2 - 152x - 7581 = 0$$

Use $a = 5$, $b = -152$, and $c = -7581$ in the quadratic formula.

$$x = \frac{-(-152) \pm \sqrt{(-152)^2 - 4(5)(-7581)}}{2(5)}$$

$$= \frac{152 \pm \sqrt{174{,}724}}{10} = \frac{152 \pm 418}{10}$$

Thus, $x = \frac{152 + 418}{10} = 57$ (the other value is negative). The southbound car traveled 57 miles, and the eastbound car traveled $2x - 38 = 2(57) - 38 = 76$ miles.

CHAPTER 11 INVERSE, EXPONENTIAL, AND LOGARITHMIC FUNCTIONS

11.1 Inverse Functions

11.1 Margin Exercises

1. **(a)** $\{(1,2),(2,4),(3,3),(4,5)\}$

Each x-value in the relation corresponds to only one y-value, so the relation is indeed a function. Furthermore, each y-value corresponds to only one x-value, so this is a one-to-one function.

The inverse function is found by interchanging the x- and y-values in each ordered pair.

The inverse is

$$\{(2,1),(4,2),(3,3),(5,4)\}.$$

(b) $\{(0,3),(-1,2),(1,3)\}$

Since the y-value 3 corresponds to x-values 0 and 1, the function is not one-to-one. Therefore, it does not have an inverse.

(c) Since each time corresponds to only one distance, the function is one-to-one. The inverse of the function defined in the table is found by interchanging the columns.

Time	Distance
4:22	1.5 K
9:18	3 K
16:00	5 K
33:40	10 K

2. **(a)** Since every horizontal line will intersect the graph at most once, the function is one-to-one.

(b) Since a horizontal line could intersect the graph in two points, the function is not one-to-one.

3. **(a)** $f(x) = 3x - 4$

The graph of f is a line. By the horizontal line test, it is a one-to-one function. To find the inverse, first replace $f(x)$ with y.

Step 1 Interchange x and y.

$$y = 3x - 4$$
$$x = 3y - 4$$

Step 2 Solve for y.

$$x + 4 = 3y$$
$$\frac{x+4}{3} = y$$

Step 3 Replace y with $f^{-1}(x)$.

$$f^{-1}(x) = \frac{x+4}{3}, \quad \text{or} \quad \frac{1}{3}x + \frac{4}{3}$$

(b) $f(x) = x^3 + 1$

Refer to Section 6.3 to see from its graph that a cubing function like this is a one-to-one function. To find the inverse, first replace $f(x)$ with y.

Step 1 Interchange x and y.

$$y = x^3 + 1$$
$$x = y^3 + 1$$

Step 2 Solve for y.

$$x - 1 = y^3$$

Take the cube root on each side.

$$\sqrt[3]{x-1} = y$$

Step 3 Replace y with $f^{-1}(x)$.

$$f^{-1}(x) = \sqrt[3]{x-1}$$

(c) $f(x) = (x-3)^2$

This equation has a vertical parabola as its graph, so some horizontal lines will intersect the graph at two points. For example, both $x = 1$ and $x = 5$ correspond to $y = 4$. Thus, the function is not a one-to-one function and does not have an inverse.

4. **(a)** The given graph goes through $(-5,1)$ and $(1,-1)$. So the inverse will go through $(1,-5)$ and $(-1,1)$. Use these points and symmetry about $y = x$ to complete the graph of the inverse.

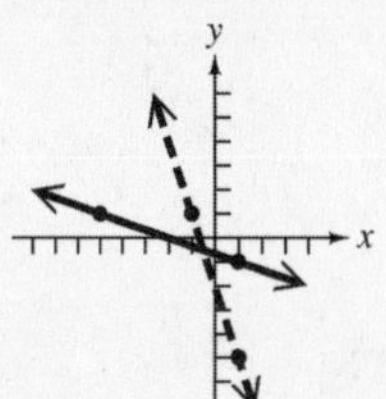

(b) Points on the given graph are $(-3, 0)$, $(-2, 2)$, and $(0, 3)$. So, the inverse will contain $(0, -3)$, $(2, -2)$, and $(3, 0)$. Use these points and symmetry about $y = x$ to complete the graph of the inverse.

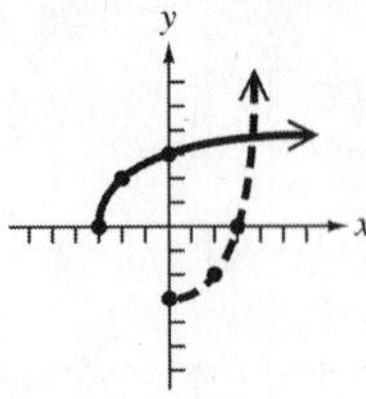

(c) The given graph goes through $(-1, \frac{1}{2})$, $(0, 1)$, $(1, 2)$, and $(2, 4)$. So, the inverse will go through $(\frac{1}{2}, -1)$, $(1, 0)$, $(2, 1)$, and $(4, 2)$. Use these points and symmetry about $y = x$ to complete the graph of the inverse.

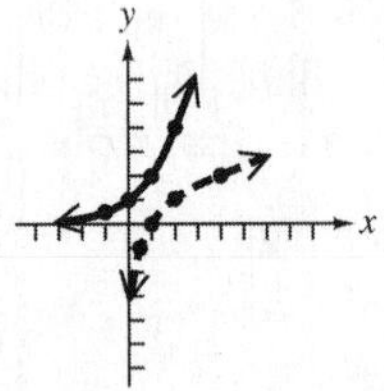

11.1 Section Exercises

1. It is not one-to-one. France and the United States are paired with the same trans fat percentage, 11.

3. Consider the ordered pair (x, y) where x is a student in a math class and y is that student's mother. If two siblings were attending that class, the function would not be one-to-one since there would exist two ordered pairs that have different x-values but the same y-value.

5. If a function is made up of ordered pairs in such a way that the same y-value appears in a correspondence with two different x-values, then the function is not one-to-one. Choice **B**

7. All of the graphs pass the vertical line test, so they all represent functions. The graph in choice **A** is the only one that passes the horizontal line test, so it is the one-to-one function.

9. $\{(3, 6), (2, 10), (5, 12)\}$
This function is one-to-one since each y-value corresponds to only one x-value. To find the inverse, interchange x and y in each ordered pair. The inverse is
$$\{(6, 3), (10, 2), (12, 5)\}.$$

11. $\{(-1, 3), (2, 7), (4, 3), (5, 8)\}$ is not a one-to-one function. The ordered pairs $(-1, 3)$ and $(4, 3)$ have the same y-values for two different x-values.

13. The graph of $f(x) = 2x + 4$ is a nonvertical, nonhorizontal line. By the horizontal line test, $f(x)$ is a one-to-one function. To find the inverse, first replace $f(x)$ with y.
$$y = 2x + 4$$
Interchange x and y.
$$x = 2y + 4$$
Solve for y.
$$2y = x - 4$$
$$y = \frac{x - 4}{2}$$
Replace y with $f^{-1}(x)$.
$$f^{-1}(x) = \frac{x - 4}{2}, \text{ or } f^{-1}(x) = \frac{1}{2}x - 2$$

15. Write $g(x) = \sqrt{x - 3}$ as $y = \sqrt{x - 3}$. Since $x \geq 3$, $y \geq 0$. The graph of g is half of a horizontal parabola that opens to the right. The graph passes the horizontal line test, so g is one-to-one. To find the inverse, interchange x and y to get
$$x = \sqrt{y - 3}.$$
Note that now $y \geq 3$, so $x \geq 0$.
Solve for y by squaring both sides.
$$x^2 = y - 3$$
$$x^2 + 3 = y$$
Replace y with $g^{-1}(x)$.
$$g^{-1}(x) = x^2 + 3, x \geq 0$$

17. $f(x) = 3x^2 + 2$ is not a one-to-one function because two x-values, such as 1 and -1, both have the same y-value, in this case 5. Also, the graph of this function is a vertical parabola which does not pass the horizontal line test.

19. The graph of $f(x) = x^3 - 4$ is the graph of $g(x) = x^3$ shifted down 4 units. (Recall that $g(x) = x^3$ is the elongated S-shaped curve.) The graph of f passes the horizontal line test, so f is one-to-one.

Replace $f(x)$ with y.

$$y = x^3 - 4$$

Interchange x and y.

$$x = y^3 - 4$$

Solve for y.

$$x + 4 = y^3$$

Take the cube root of each side.

$$\sqrt[3]{x+4} = y$$

Replace y with $f^{-1}(x)$.

$$f^{-1}(x) = \sqrt[3]{x+4}$$

In Exercises 21–24, $f(x) = 2^x$ is a one-to-one function.

21. **(a)** To find $f(3)$, substitute 3 for x.

$f(x) = 2^x$, so $f(3) = 2^3 = 8$.

(b) Since f is one-to-one and $f(3) = 8$, it follows that $f^{-1}(8) = 3$.

23. **(a)** To find $f(0)$, substitute 0 for x.

$f(x) = 2^x$, so $f(0) = 2^0 = 1$.

(b) Since f is one-to-one and $f(0) = 1$, it follows that $f^{-1}(1) = 0$.

25. **(a)** The function is one-to-one since any horizontal line intersects the graph at most once.

(b) The points $(-1, 5)$ and $(2, -1)$ are points of f, so $(5, -1)$ and $(-1, 2)$ are points of f^{-1}. Plot these points, then draw a dashed line (symmetric to the original graph about the line $y = x$) through them to obtain the graph of the inverse function.

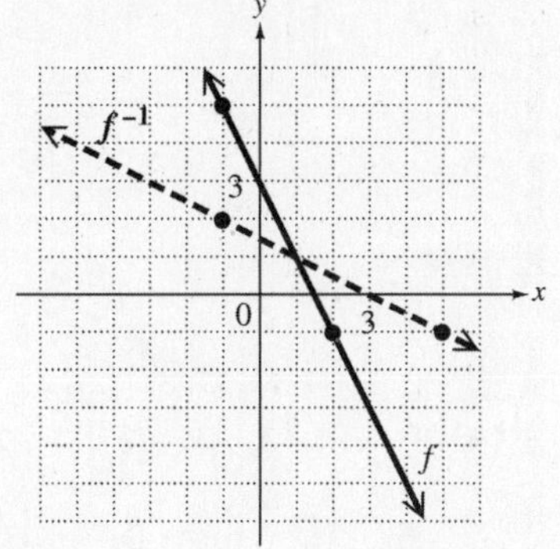

27. **(a)** The function is not one-to-one since there are horizontal lines that intersect the graph more than once. For example, the line $y = 1$ intersects the graph twice.

29. **(a)** The function is one-to-one since any horizontal line intersects the graph at most once.

(b) The points $(-4, 2)$, $(-1, 1)$, $(1, -1)$, and $(4, -2)$ are points of f, so $(2, -4)$, $(1, -1)$, $(-1, 1)$, and $(-2, 4)$ are points of f^{-1}. Plot these points, then draw a smooth dashed curve (symmetric to the original graph about the line $y = x$) through them to obtain the graph of the inverse function.

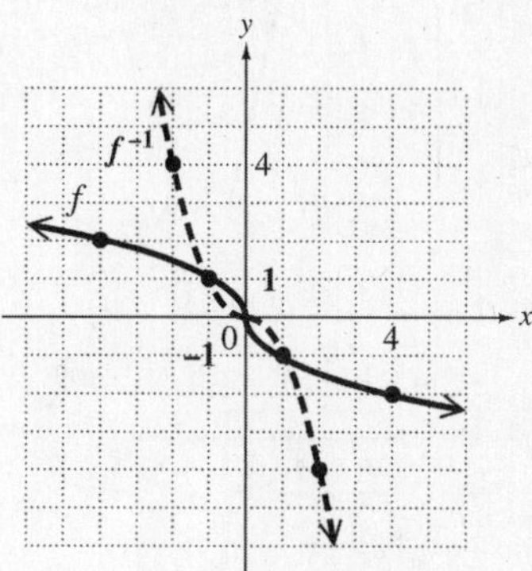

31. $f(x) = 2x - 1$ or $y = 2x - 1$

The graph of f is a line through $(0, -1)$ and $(2, 3)$. Plot these points and draw a solid line through them. The graph of f^{-1} is a line through $(-1, 0)$ and $(3, 2)$. Plot these points, then draw a dashed line (symmetric to the graph of f about the line $y = x$) through them to obtain the graph of f^{-1}.

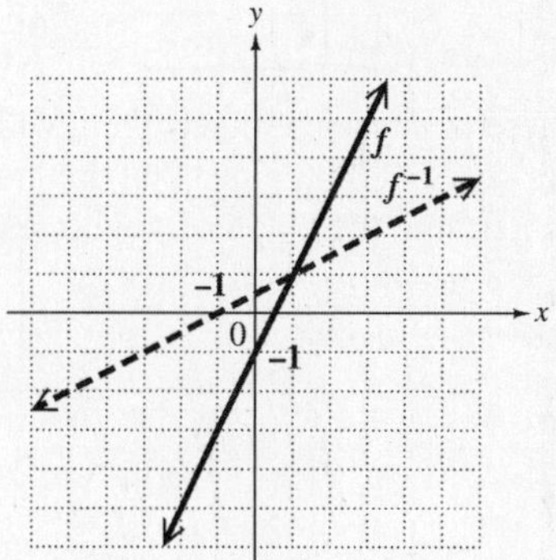

33. $g(x) = -4x$ or $y = -4x$

The graph of g is a line through $(0, 0)$ and $(1, -4)$. Plot these points and draw a solid line through them. The graph of g^{-1} is a line through $(0, 0)$ and $(-4, 1)$. Plot these points, then draw a dashed line (symmetric to the graph of g about the line $y = x$) through them to obtain the graph of g^{-1}.

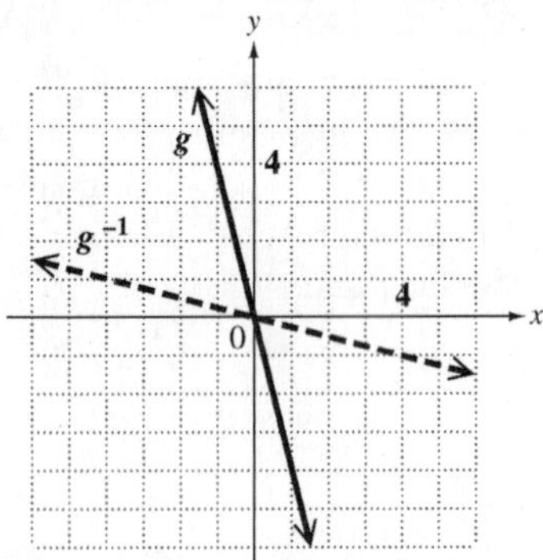

35. $f(x) = y = \sqrt{x},\ x \geq 0$

Complete the table of values.

x	$f(x) = y$
0	0
1	1
4	2

Plot these points and connect them with a solid smooth curve.
Since $f(x)$ is one-to-one, make a table of values for $f^{-1}(x)$ by interchanging x and y.

x	$f^{-1}(x)$
0	0
1	1
2	4

Plot these points and connect them with a dashed smooth curve. Use the fact that the graph of f^{-1} is symmetric to the graph of f with respect to the line $y = x$.

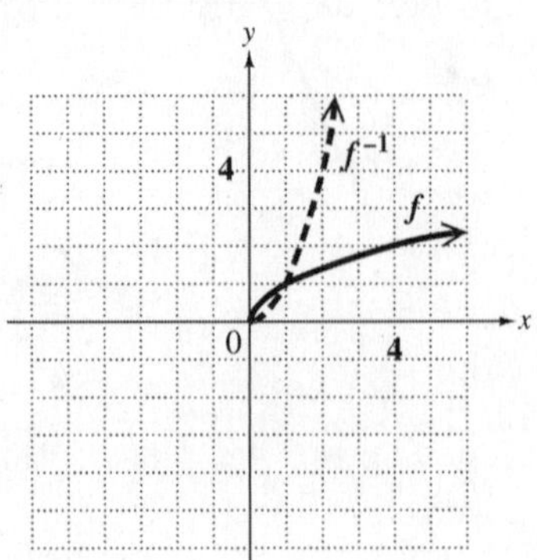

37. $f(x) = y = x^3 - 2$

Complete the table of values.

x	x^3	$f(x) = y$
-1	-1	-3
0	0	-2
1	1	-1
2	8	6

Plot these points and connect them with a solid smooth curve.
Since $f(x)$ is one-to-one, make a table of values for $f^{-1}(x)$ by interchanging x and y.

x	$f^{-1}(x)$
-3	-1
-2	0
-1	1
6	2

Plot these points and connect them with a dashed smooth curve. Use the fact that the graph of f^{-1} is symmetric to the graph of f with respect to the line $y = x$.

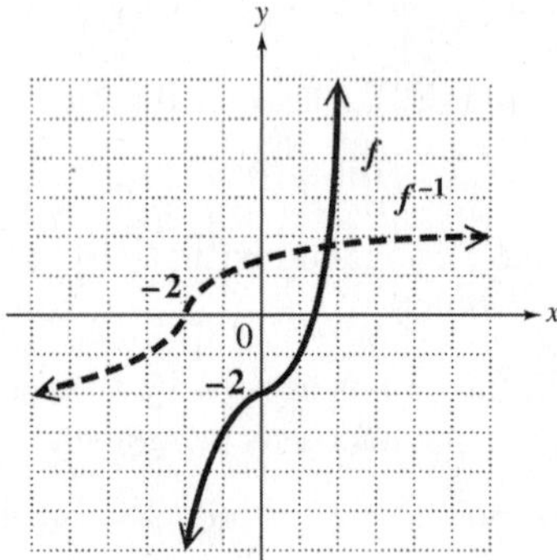

11.2 Exponential Functions

11.2 Margin Exercises

In Exercises 1 and 2, choose values of x and find the corresponding values of y. Then plot the points, and draw a smooth curve through them.

1. **(a)** $f(x) = 10^x$

x	-2	-1	0	1	2
$f(x) = 10^x$	0.01	0.1	1	10	100

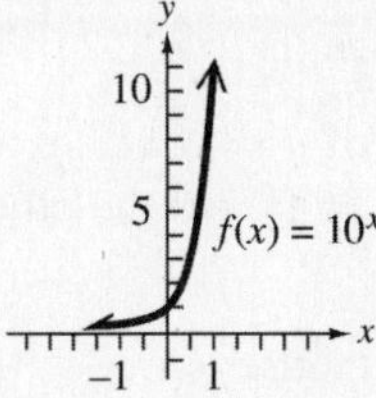

(b) $g(x) = (\frac{1}{10})^x$

x	-2	-1	0	1	2
$g(x) = (\frac{1}{10})^x$	100	10	1	$\frac{1}{10}$	$\frac{1}{100}$

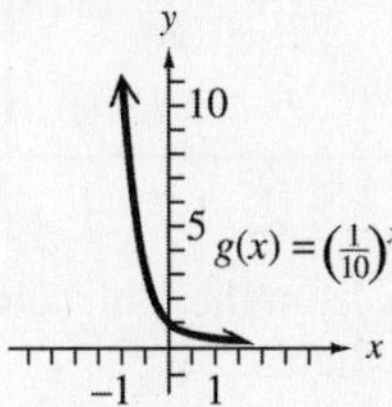

2. $y = 2^{4x-3}$

Make a table of values. It will help to find values for $4x - 3$ before you find y.

x	$4x-3$	$y = 2^{4x-3}$
0	-3	$\frac{1}{8}$
$\frac{3}{4}$	0	1
1	1	2
-1	-7	$\frac{1}{128}$

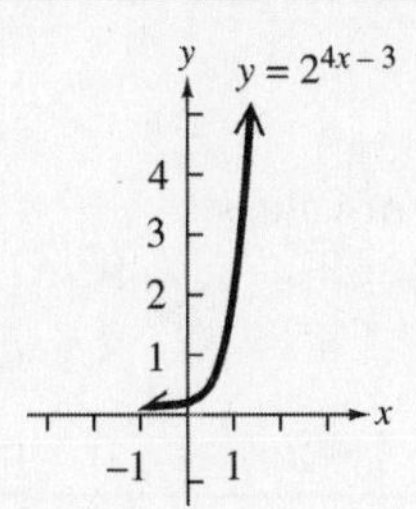

3. **(a)** $25^x = 125$

Step 1 Write each side with the base 5.

$$(5^2)^x = 5^3$$

Step 2 Simplify exponents.

$$5^{2x} = 5^3$$

Step 3 Set the exponents equal.

$$2x = 3$$

Step 4 Solve.

$$x = \frac{3}{2}$$

Check $x = \frac{3}{2}$ in the original equation.

$$25^{3/2} \stackrel{?}{=} 125$$
$$(25^{1/2})^3 \stackrel{?}{=} 125$$
$$5^3 \stackrel{?}{=} 125$$
$$125 = 125 \quad \textit{True}$$

Solution set: $\{\frac{3}{2}\}$

(b) $4^x = 32$

Step 1 Write each side with the base 2.

$$(2^2)^x = 2^5$$

Step 2 Simplify exponents.

$$2^{2x} = 2^5$$

Step 3 Set the exponents equal.

$$2x = 5$$

Step 4 Solve.

$$x = \frac{5}{2}$$

Check $x = \frac{5}{2}$: $4^{5/2} = 2^5 = 32$

Solution set: $\{\frac{5}{2}\}$

(c) $81^p = 27$

Step 1 Write each side with the base 3.

$$(3^4)^p = 3^3$$

Step 2 Simplify exponents.

$$3^{4p} = 3^3$$

Step 3 Set the exponents equal.

$$4p = 3$$

Step 4 Solve.

$$p = \frac{3}{4}$$

Check $p = \frac{3}{4}$: $81^{3/4} = 3^3 = 27$

Solution set: $\{\frac{3}{4}\}$

4. **(a)**
$$25^{x-2} = 125^x$$
$$(5^2)^{x-2} = (5^3)^x \quad \textit{Same base}$$
$$5^{2(x-2)} = 5^{3x} \quad \textit{Simplify exponents.}$$
$$2(x-2) = 3x \quad \textit{Set exponents equal.}$$
$$2x - 4 = 3x$$
$$-4 = x$$

Check $x = -4$:

$$25^{-4-2} \stackrel{?}{=} 125^{-4}$$
$$25^{-6} \stackrel{?}{=} 125^{-4}$$
$$4.096 \times 10^{-9} = 4.096 \times 10^{-9} \quad \textit{True}$$

Solution set: $\{-4\}$

(b)
$$4^x = \frac{1}{32}$$
$$(2^2)^x = \frac{1}{2^5}$$
$$2^{2x} = 2^{-5} \quad \textit{Same base}$$
$$2x = -5 \quad \textit{Set exponents equal.}$$
$$x = -\tfrac{5}{2}$$

Check $x = -\frac{5}{2}$: $4^{-5/2} = \frac{1}{4^{5/2}} = \frac{1}{2^5} = \frac{1}{32}$

Solution set: $\{-\frac{5}{2}\}$

(c)
$$(\tfrac{3}{4})^x = \tfrac{16}{9}$$
$$(\tfrac{3}{4})^x = (\tfrac{4}{3})^2$$
$$(\tfrac{3}{4})^x = (\tfrac{3}{4})^{-2}$$
$$x = -2$$

Check $x = -2$: $(\frac{3}{4})^{-2} = (\frac{4}{3})^2 = \frac{16}{9}$

Solution set: $\{-2\}$

5. $x = 1925 - 1750 = 175$

$$f(x) = 266(1.001)^x$$
$$f(175) = 266(1.001)^{175}$$
$$\approx 317$$

The carbon dioxide concentration in 1925 was about 317 parts per million.

6.
$$f(x) = 1038(1.000134)^{-x}$$
$$f(8000) = 1038(1.000134)^{-8000}$$
$$\approx 355$$

The pressure is approximately 355 millibars.

11.2 Section Exercises

1. Since the graph of $F(x) = a^x$ always contains the point $(0, 1)$, the correct response is **C**.

3. **A.** The y-intercept of the graph of $f(x) = 10^x$ is $(0, 1)$, not $(0, 10)$. *False*

B. For any $a > 1$, the graph of $f(x) = a^x$ *rises* from left to right, not *falls*. *False*

C. If $f(x) = 5^x$, then $f(\frac{1}{2}) = 5^{1/2} = \sqrt{5}$. Thus the point $(\frac{1}{2}, \sqrt{5})$ lies on the graph of $f(x) = 5^x$ and the statement is *true*.

D. The graph of $y = 10^x$ rises at a faster rate than the graph of $y = 4^x$, not the other way around. *False*

The correct response is **C**.

5. $f(x) = 3^x$
Make a table of values.
$f(-2) = 3^{-2} = \frac{1}{3^2} = \frac{1}{9}$,
$f(-1) = 3^{-1} = \frac{1}{3^1} = \frac{1}{3}$, and so on.

x	-2	-1	0	1	2
$f(x)$	$\frac{1}{9}$	$\frac{1}{3}$	1	3	9

Plot the points from the table and draw a smooth curve through them.

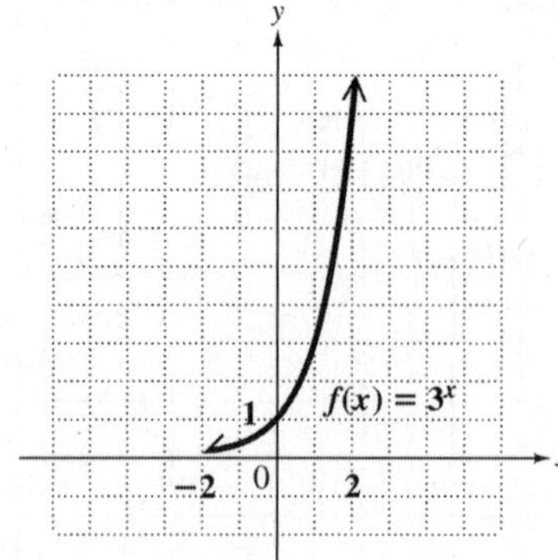

7. $g(x) = (\frac{1}{3})^x$
Make a table of values.
$g(-2) = (\frac{1}{3})^{-2} = (\frac{3}{1})^2 = 9$,
$g(-1) = (\frac{1}{3})^{-1} = (\frac{3}{1})^1 = 3$, and so on.

x	-2	-1	0	1	2
$g(x)$	9	3	1	$\frac{1}{3}$	$\frac{1}{9}$

Plot the points from the table and draw a smooth curve through them.

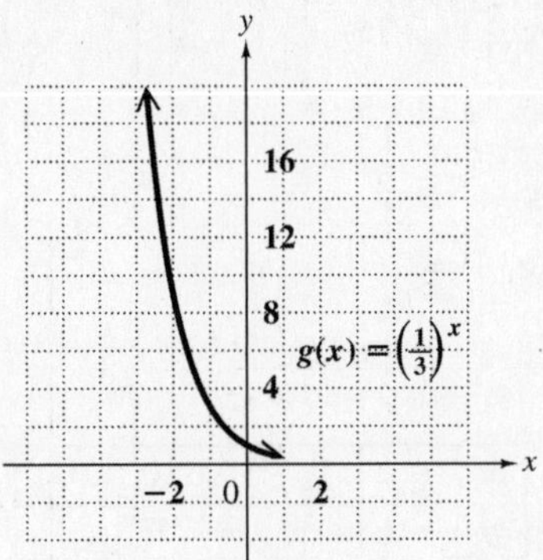

9. $y = 2^{2x-2}$

Make a table of values. It will help to find values for $2x - 2$ before you find y.

x	-2	-1	0	1	2	3
$2x-2$	-6	-4	-2	0	2	4
y	$\frac{1}{64}$	$\frac{1}{16}$	$\frac{1}{4}$	1	4	16

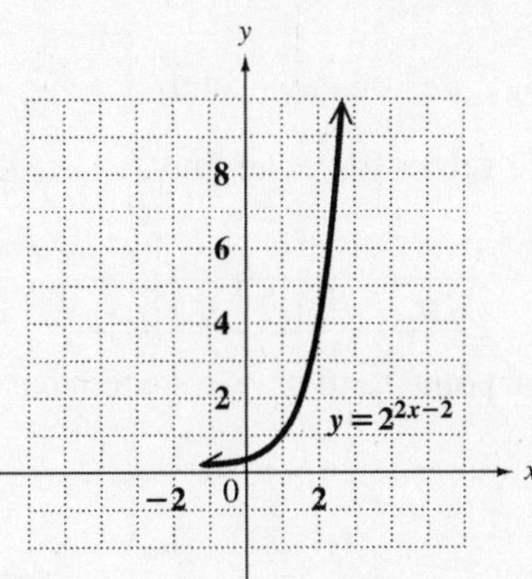

11. $6^x = 36$

$6^x = 6^2$ *Write each side as a power of 6.*

For $a > 0$ and $a \neq 1$, if $a^x = a^y$, then $x = y$. Setting the exponents equal to each other, we see that $x = 2$.

Check $x = 2$: $6^2 = 36$ *True*

Solution set: $\{2\}$

13. $100^x = 1000$

$(10^2)^x = 10^3$ *Write as powers of 10.*

$10^{2x} = 10^3$

For $a > 0$ and $a \neq 1$, if $a^x = a^y$, then $x = y$. Set the exponents equal to each other.

$$2x = 3$$
$$x = \tfrac{3}{2}$$

Check $x = \frac{3}{2}$: $100^{3/2} = 1000$ *True*

Solution set: $\{\frac{3}{2}\}$

15. $16^{2x+1} = 64^{x+3}$

$(4^2)^{2x+1} = (4^3)^{x+3}$ *Write as powers of 4.*

$4^{4x+2} = 4^{3x+9}$

$4x + 2 = 3x + 9$ *Set the exponents equal.*

$x = 7$

Check $x = 7$: $16^{15} = 64^{10}$ *True*

Solution set: $\{7\}$

17. $5^x = \dfrac{1}{125}$

$5^x = (\frac{1}{5})^3$

$5^x = 5^{-3}$ *Write as powers of 5.*

$x = -3$ *Set the exponents equal.*

Check $x = -3$: $5^{-3} = \frac{1}{125}$ *True*

Solution set: $\{-3\}$

19. $5^x = 0.2$

$5^x = \frac{2}{10} = \frac{1}{5}$

$5^x = 5^{-1}$ *Write as powers of 5.*

$x = -1$ *Set the exponents equal.*

Check $x = -1$: $5^{-1} = 0.2$ *True*

Solution set: $\{-1\}$

21. $(\frac{3}{2})^x = \frac{8}{27}$

$(\frac{3}{2})^x = (\frac{2}{3})^3$

$(\frac{3}{2})^x = (\frac{3}{2})^{-3}$ *Write as powers of $\frac{3}{2}$.*

$x = -3$ *Set the exponents equal.*

Check $x = -3$: $(\frac{3}{2})^{-3} = \frac{8}{27}$ *True*

Solution set: $\{-3\}$

23. **(a)** For an exponential function defined by $f(x) = a^x$, if $a > 1$, the graph rises from left to right. (See Example 1, $f(x) = 2^x$, in your text.) If $0 < a < 1$, the graph falls from left to right. (See Example 2, $g(x) = (\frac{1}{2})^x = 2^{-x}$, in your text.)

(b) An exponential function defined by $f(x) = a^x$ is one-to-one and has an inverse, since each value of $f(x)$ corresponds to one and only one value of x.

25. **(a)** The increase for the exponential-type curve in the year 2010 is about 1.0°C.

(b) The increase for the linear graph in the year 2010 is about 0.4°C.

27. **(a)** The increase for the exponential-type curve in the year 2040 is about 3.0°C.

(b) The increase for the linear graph in the year 2040 is about 0.7°C.

11.3 Logarithmic Functions

11.3 Margin Exercises

1.

Exponential Form	Logarithmic Form
$2^5 = 32$	$\log_2 32 = 5$
$100^{1/2} = 10$	$\log_{100} 10 = \frac{1}{2}$
$8^{2/3} = 4$	$\log_8 4 = \frac{2}{3}$
$6^{-4} = \frac{1}{1296}$	$\log_6 \frac{1}{1296} = -4$

2. **(a)** $\log_3 27 = x$

$3^x = 27$ *Write in exponential form.*

$3^x = 3^3$ *Write with the same base.*

$x = 3$ *Set the exponents equal.*

Check $x = 3$:

$\log_3 27 = 3$ is *true* since $3^3 = 27$.

Solution set: $\{3\}$

(b) $\log_5 p = 2$

$5^2 = p$ *Write in exponential form.*

$25 = p$

Check $p = 25$:

$\log_5 25 = 2$ is *true* since $5^2 = 25$.

Solution set: $\{25\}$

(c) $\log_m \dfrac{1}{16} = -4$

$m^{-4} = \dfrac{1}{16}$ *Write in exponential form.*

$\dfrac{1}{m^4} = \dfrac{1}{2^4}$

$m^4 = 2^4$

$m = 2$ *Take principal fourth root.*

Check $m = 2$:

$\log_2 \frac{1}{16} = -4$ is *true* since $2^{-4} = \frac{1}{16}$.

Solution set: $\{2\}$

(d) $\log_x 12 = 3$

$x^3 = 12$ *Write in exponential form.*

$x = \sqrt[3]{12}$

Check $x = \sqrt[3]{12}$:

$\log_{\sqrt[3]{12}} 12 = 3$ is *true* since $(\sqrt[3]{12})^3 = 12$.

Solution set: $\{\sqrt[3]{12}\}$

3. Use the properties of logarithms,

$\log_b b = 1$ and $\log_b 1 = 0$,

for $b > 0, b \neq 1$.

(a) $\log_{2/5} \frac{2}{5} = 1$

(b) $\log_\pi \pi = 1$

(c) $\log_{0.4} 1 = 0$

(d) $\log_6 1 = 0$

4. $y = \log_{10} x$ is equivalent to $x = 10^y$. Choose values for y and find x.

x	$\frac{1}{100}$	$\frac{1}{10}$	1	10	100
y	-2	-1	0	1	2

Plot the points, and draw a smooth curve through them.

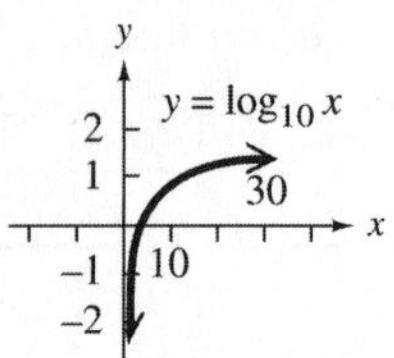

5. $y = \log_{1/10} x$ is equivalent to $x = (\frac{1}{10})^y$.

Choose values for y and find x.

x	10	1	$\frac{1}{10}$	$\frac{1}{100}$
y	-1	0	1	2

Plot the points, and draw a smooth curve through them.

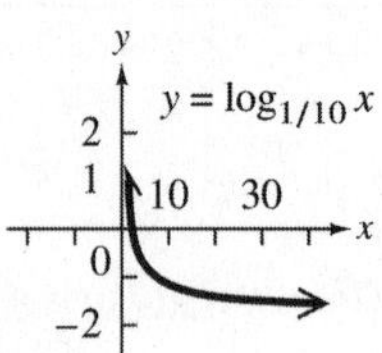

6. $P(t) = 80 \log_{10} (t + 10)$

(a) $P(0) = 80 \log_{10} (0 + 10)$ *Let t = 0.*

$= 80 \log_{10} 10$

$= 80(1)$ *$\log_{10} 10 = 1$*

$= 80$

The number of mites at the beginning of the study is 80.

(b) $P(90) = 80 \log_{10} (90 + 10)$ *Let t = 90.*

$= 80 \log_{10} 100$

$= 80(2)$ *$\log_{10} 100 = 2$*

$= 160$

The population after 90 days is 160 mites.

11.3 Section Exercises

1. **(a)** $\log_4 16$ is equal to 2, because 2 is the exponent to which 4 must be raised to obtain 16. **(C)**

(b) $\log_3 81$ is equal to 4, because 4 is the exponent to which 3 must be raised to obtain 81. **(F)**

(c) $\log_3(\frac{1}{3})$ is equal to -1, because -1 is the exponent to which 3 must be raised to obtain $\frac{1}{3}$. **(B)**

(d) $\log_{10} 0.01$ is equal to -2, because -2 is the exponent to which 10 must be raised to obtain 0.01. **(A)**

(e) $\log_5 \sqrt{5}$ is equal to $\frac{1}{2}$, because $\frac{1}{2}$ is the exponent to which 5 must be raised to obtain $\sqrt{5}$. **(E)**

(f) $\log_{13} 1$ is equal to 0, because 0 is the exponent to which 13 must be raised to obtain 1. **(D)**

3. The base is 4, the exponent (logarithm) is 5, and the number is 1024, so $4^5 = 1024$ becomes $\log_4 1024 = 5$ in logarithmic form.

5. $\frac{1}{2}$ is the base and -3 is the exponent, so $(\frac{1}{2})^{-3} = 8$ becomes $\log_{1/2} 8 = -3$ in logarithmic form.

7. The base is 10, the exponent (logarithm) is -3, and the number is 0.001, so $10^{-3} = 0.001$ becomes $\log_{10} 0.001 = -3$ in logarithmic form.

9. $\sqrt[4]{625} = 625^{1/4} = 5$

The base is 625, the exponent (logarithm) is $\frac{1}{4}$, and the number is 5, so $\sqrt[4]{625} = 5$ becomes $\log_{625} 5 = \frac{1}{4}$ in logarithmic form.

11. In $\log_4 64 = 3$, 4 is the base and 3 is the logarithm (exponent), so $\log_4 64 = 3$ becomes $4^3 = 64$ in exponential form.

13. In $\log_{10} \frac{1}{10{,}000} = -4$, the base is 10, the logarithm (exponent) is -4, and the number is $\frac{1}{10{,}000}$, so $\log_{10} \frac{1}{10{,}000} = -4$ becomes $10^{-4} = \frac{1}{10{,}000}$ in exponential form.

15. In $\log_6 1 = 0$, 6 is the base and 0 is the logarithm (exponent), so $\log_6 1 = 0$ becomes $6^0 = 1$ in exponential form.

17. In $\log_9 3 = \frac{1}{2}$, the base is 9, the logarithm (exponent) is $\frac{1}{2}$, and the number is 3, so $\log_9 3 = \frac{1}{2}$ becomes $9^{1/2} = 3$ in exponential form.

19. To evaluate $\log_9 3$, one has to ask "9 raised to what power gives you a result of 3?" We know that the square root (a radical) of 9 is 3. Therefore, the teacher's hint was to see what root of 9 equals 3. The answer is the reciprocal of the root index, $\frac{1}{2}$.

21.
$$
\begin{aligned}
x &= \log_{27} 3 \\
27^x &= 3 && \textit{Write in exponential form.} \\
(3^3)^x &= 3 && \textit{Write as powers of 3.} \\
3^{3x} &= 3^1 \\
3x &= 1 && \textit{Set the exponents equal.} \\
x &= \tfrac{1}{3}
\end{aligned}
$$

Check $x = \frac{1}{3}$: $\frac{1}{3} = \log_{27} 3$ since $27^{1/3} = 3$.
Solution set: $\{\frac{1}{3}\}$

23.
$$
\begin{aligned}
\log_x 9 &= \tfrac{1}{2} \\
x^{1/2} &= 9 && \textit{Write in exponential form.} \\
(x^{1/2})^2 &= 9^2 && \textit{Square.} \\
x^1 &= 81 \\
x &= 81
\end{aligned}
$$

$x = 81$ is an acceptable base since it is a positive number (not equal to 1).

Check $x = 81$: $\log_{81} 9 = \frac{1}{2}$ since $81^{1/2} = 9$.
Solution set: $\{81\}$

25.
$$
\begin{aligned}
\log_x 125 &= -3 \\
x^{-3} &= 125 && \textit{Write in exponential form.} \\
\frac{1}{x^3} &= 125 \\
1 &= 125(x^3) \\
\tfrac{1}{125} &= x^3 \\
\sqrt[3]{\tfrac{1}{125}} &= \sqrt[3]{x^3} && \textit{Take cube roots.} \\
\tfrac{1}{5} &= x
\end{aligned}
$$

$x = \frac{1}{5}$ is an acceptable base since it is a positive number (not equal to 1).

Check $x = \frac{1}{5}$:

$\log_{1/5} 125 = -3$ since $(\frac{1}{5})^{-3} = 5^3 = 125$.

Solution set: $\{\frac{1}{5}\}$

27.
$$
\begin{aligned}
\log_{12} x &= 0 \\
12^0 &= x && \textit{Write in exponential form.} \\
1 &= x
\end{aligned}
$$

The argument (the input of the logarithm) must be a positive number, so $x = 1$ is acceptable.

Check $x = 1$: $\log_{12} 1 = 0$ since $12^0 = 1$.
Solution set: $\{1\}$

29.
$$
\begin{aligned}
\log_x x &= 1 \\
x^1 &= x && \textit{Write in exponential form.}
\end{aligned}
$$

This equation is true for all the numbers x that are allowed as the base of a logarithm; that is, all positive numbers x, $x \neq 1$.

Solution set: $\{x \mid x > 0,\ x \neq 1\}$

31. $\log_x \dfrac{1}{25} = -2$

$x^{-2} = \dfrac{1}{25}$ *Write in exponential form.*

$\dfrac{1}{x^2} = \dfrac{1}{25}$

$x^2 = 25$ *Denominators must be equal.*

$x = \pm 5$

Reject $x = -5$ since the base of a logarithm must be positive and not equal to 1.

Check $x = 5$:

$\log_5 \frac{1}{25} = -2$ since $5^{-2} = (\frac{1}{5})^2 = \frac{1}{25}$.

Solution set: $\{5\}$

33. $\log_8 32 = x$

$8^x = 32$ *Exponential form*

$(2^3)^x = 2^5$ *Write as powers of 2.*

$2^{3x} = 2^5$

$3x = 5$ *Equate exponents.*

$x = \frac{5}{3}$

Check $x = \frac{5}{3}$:

$\log_8 32 = \frac{5}{3}$ since $8^{5/3} = (2^3)^{5/3} = 2^5 = 32$.

Solution set: $\{\frac{5}{3}\}$

35. $\log_\pi \pi^4 = x$

$\pi^x = \pi^4$ *Exponential form*

$x = 4$ *Equate exponents.*

Check $x = 4$: $\log_\pi \pi^4 = 4$ since $\pi^4 = \pi^4$.
Solution set: $\{4\}$

37. $\log_6 \sqrt{216} = x$

$\log_6 216^{1/2} = x$ *Equivalent form*

$6^x = 216^{1/2}$ *Exponential form*

$6^x = (6^3)^{1/2}$ *Same base*

$6^x = 6^{3/2}$

$x = \frac{3}{2}$ *Equate exponents.*

Check $x = \frac{3}{2}$:

$\log_6 \sqrt{216} = \frac{3}{2}$ since $6^{3/2} = \sqrt{6^3} = \sqrt{216}$.

Solution set: $\{\frac{3}{2}\}$

39. $y = \log_3 x$

$3^y = x$ *Write in exponential form.*

Refer to Section 11.2, Exercise 5, for the graph of $f(x) = 3^x$. Since $y = \log_3 x$ (or $3^y = x$) is the inverse of $f(x) = y = 3^x$, its graph is symmetric about the line $y = x$ to the graph of $f(x) = 3^x$.

The graph can be plotted by reversing the ordered pairs in the table of values belonging to $f(x) = 3^x$.

x	$\frac{1}{9}$	$\frac{1}{3}$	1	3	9
y	-2	-1	0	1	2

Plot the points, and draw a smooth curve through them.

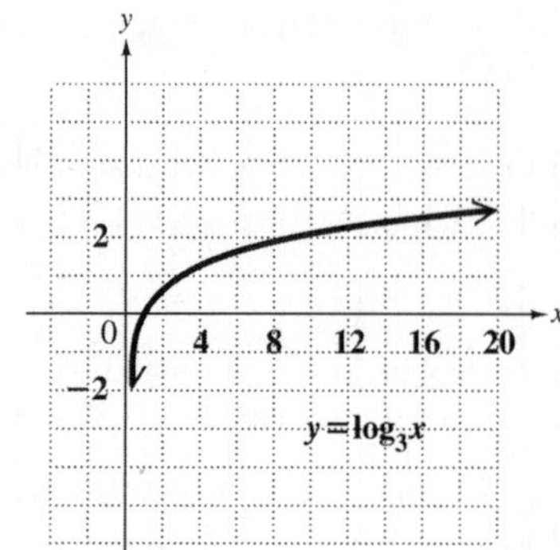

41. $y = \log_{1/3} x$

$(\frac{1}{3})^y = x$ *Write in exponential form.*

Refer to Section 11.2, Exercise 7, for the graph of $g(x) = (\frac{1}{3})^x$. Since $y = \log_{1/3} x$ (or $(\frac{1}{3})^y = x$)is the inverse of $y = (\frac{1}{3})^x$, its graph is symmetric about the line $y = x$ to the graph of $y = (\frac{1}{3})^x$. The graph can be plotted by reversing the ordered pairs in the table of values belonging to $g(x) = (\frac{1}{3})^x$.

x	9	3	1	$\frac{1}{3}$	$\frac{1}{9}$
y	-2	-1	0	1	2

Plot the points, and draw a smooth curve through them.

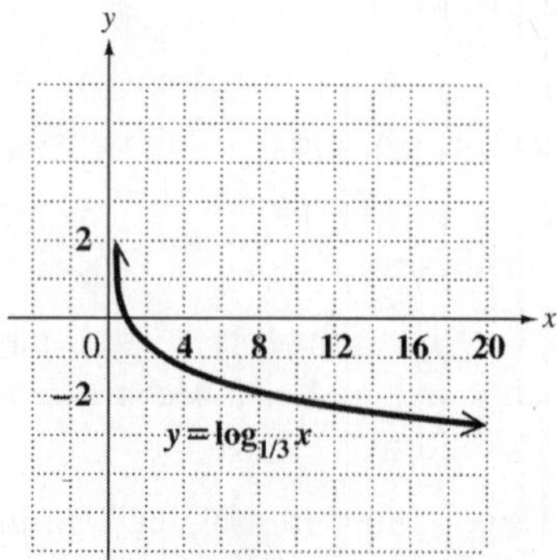

43. The following characteristics reinforce the concept that $F(x) = a^x$ and $G(x) = \log_a x$ are inverse functions.

(1) The graph of F will always contain $(0, 1)$; the graph of G will always contain $(1, 0)$.

(2) When $a > 1$, the graph of F will rise from left to right from the second quadrant to the first. When $a > 1$, the graph of G will rise from left to right from the fourth quadrant to the first, thus making G a mirror image of F with respect to the line $y = x$. Also, when $0 < a < 1$, the graph of F will fall from left to right from the second quadrant to the first, while the graph of G will fall from left to right from the first quadrant to the fourth, also making G a mirror image of F with respect to the line $y = x$.

(3) The graph of F approaches the x-axis but never touches it. The graph of G approaches the y-axis but never touches it.

(4) The domain of F is $(-\infty, \infty)$, and the range is $(0, \infty)$. The domain of G is $(0, \infty)$, and the range is $(-\infty, \infty)$.

45. The values of t are on the horizontal axis, and the values of $f(t)$ are on the vertical axis. Read the value of $f(t)$ from the graph for the given value of t. At $t = 0$, $f(0) = 8$.

47. To find $f(60)$, find 60 on the t-axis, then go up to the graph and across to the $f(t)$ axis to read the value of $f(60)$. At $t = 60$, $f(60) = 24$.

49. Since every real number power of 1 equals 1, if $y = \log_1 x$, then $x = 1^y$ and so $x = 1$ for every y. This contradicts the definition of a function.

51. The graph of $f(x) = 3^x$ (see Exercise 5 in Section 11.2) rises at a faster rate from left to right than the graph of $g(x) = \log_3 x$ (see Exercise 39 in this section).

53. $f(x) = 3800 + 585 \log_2 x$

(a) $x = 1982 - 1980 = 2$

$$\begin{aligned} f(2) &= 3800 + 585 \log_2 2 \\ &= 3800 + 585(1) \\ &= 4385 \end{aligned}$$

The model gives an approximate withdrawal of 4385 billion ft^3 of natural gas from crude oil wells in the United States for 1982.

(b) $x = 1988 - 1980 = 8$

$$\begin{aligned} f(8) &= 3800 + 585 \log_2 8 \\ &= 3800 + 585(3) \\ &= 5555 \end{aligned}$$

The model gives an approximate withdrawal of 5555 billion ft^3 of natural gas from crude oil wells in the United States for 1988.

(c) $x = 1996 - 1980 = 16$

$$\begin{aligned} f(16) &= 3800 + 585 \log_2 16 \\ &= 3800 + 585(4) \\ &= 6140 \end{aligned}$$

The model gives an approximate withdrawal of 6140 billion ft^3 of natural gas from crude oil wells in the United States for 1996.

55. $R = \log_{10} \dfrac{x}{x_0}$

Change to exponential form.

$$10^R = \frac{x}{x_0}, \text{ so } x = x_0\, 10^R.$$

Let $R = 6.7$ for the Northridge earthquake, which had intensity x_1.

$$x_1 = x_0 10^{6.7}$$

Let $R = 7.3$ for the Landers earthquake, which had intensity x_2.

$$x_2 = x_0 10^{7.3}$$

The ratio of x_2 to x_1 is

$$\frac{x_2}{x_1} = \frac{x_0 10^{7.3}}{x_0 10^{6.7}} = 10^{0.6} \approx 3.98.$$

The Landers earthquake was about 4 times more powerful than the Northridge earthquake.

11.4 Properties of Logarithms

11.4 Margin Exercises

1. Use the product rule for logarithms.

$$\log_b xy = \log_b x + \log_b y$$

(a) $\log_6 (5 \cdot 8) = \log_6 5 + \log_6 8$

(b) $\log_4 3 + \log_4 7 = \log_4 (3 \cdot 7) = \log_4 21$

(c) $\log_8 8k = \log_8 8 + \log_8 k = 1 + \log_8 k \quad (k > 0)$

(d) $\log_5 m^2 = \log_5 (m \cdot m) = \log_5 m + \log_5 m = 2\log_5 m \quad (m > 0)$

2. Use the quotient rule for logarithms.

$$\log_b \frac{x}{y} = \log_b x - \log_b y$$

(a) $\log_7 \frac{9}{4} = \log_7 9 - \log_7 4$

(b) $\log_3 p - \log_3 q = \log_3 \dfrac{p}{q} \quad (p > 0,\ q > 0)$

(c) $\log_4 \frac{3}{16} = \log_4 3 - \log_4 16 = \log_4 3 - 2$

3. Use the power rule for logarithms.

$$\log_b x^r = r \log_b x$$

(a) $\log_3 5^2 = 2\log_3 5$

(b) $\log_a x^4 = 4\log_a x \quad (x > 0)$

(c) $\log_b \sqrt{8} = \log_b 8^{1/2}$
$= \frac{1}{2}\log_b 8 \quad (b > 0)$

(d) $\log_2 \sqrt[3]{2} = \log_2 2^{1/3}$
$= \frac{1}{3}\log_2 2$
$= \frac{1}{3}(1) = \frac{1}{3}$

4. Use the Special Properties,

$$b^{\log_b x} = x \text{ or } \log_b b^x = x.$$

(a) By the second property,

$$\log_{10} 10^3 = 3.$$

(b) Using the second property,

$$\log_2 8 = \log_2 2^3 = 3.$$

(c) By the first property,

$$5^{\log_5 3} = 3.$$

5. Use the Properties of Logarithms.

(a) $\log_6 36m^5$
$= \log_6 36 + \log_6 m^5$ *Product rule*
$= 2 + 5\log_6 m$ *$\log_6 6^2 = 2$; Power rule*

(b) $\log_2 \sqrt{9z}$
$= \log_2 (9z)^{1/2}$
$= \frac{1}{2}\log_2 (9z)$ *Power rule*
$= \frac{1}{2}(\log_2 9 + \log_2 z)$ *Product rule*
$= \frac{1}{2}\log_2 3^2 + \frac{1}{2}\log_2 z$
$= \frac{1}{2}(2\log_2 3) + \frac{1}{2}\log_2 z$ *Power rule*
$= \log_2 3 + \frac{1}{2}\log_2 z$

(c) $\log_q \dfrac{8r^2}{m-1} \quad (m > 1, q \neq 1)$
$= \log_q (8r^2) - \log_q (m-1)$ *Quotient rule*
$= \log_q 8 + \log_q r^2 - \log_q (m-1)$ *Product rule*
$= \log_q 8 + 2\log_q r - \log_q (m-1)$ *Power rule*

(d) $2\log_a x + 3\log_a y \quad (a \neq 1)$
$= \log_a x^2 + \log_a y^3$ *Power rule*
$= \log_a x^2y^3$ *Product rule*

(e) $\log_4 (3x + y)$ cannot be written as a sum of logarithms, since

$$\log_b (M + N) \neq \log_b M + \log_b N.$$

There is no property of logarithms to rewrite the logarithm of a sum.

6. **(a)** $\log_6 36 - \log_6 6 = \log_6 30$

Evaluate both sides.

$\text{LS} = \log_6 36 - \log_6 6 = \log_6 \frac{36}{6}$
$= \log_6 6 = 1$

$\text{RS} = \log_6 30$

The statement is *false* because $\log_6 30 \neq 1$.

(b) $\log_4 (\log_2 16) = \dfrac{\log_6 6}{\log_6 30}$

Evaluate both sides.

$\text{LS} = \log_4 (\log_2 16) = \log_4 (\log_2 2^4)$
$= \log_4 4 = 1$

$\text{RS} = \dfrac{\log_6 6}{\log_6 36} = \dfrac{1}{\log_6 6^2} = \dfrac{1}{2}$

The statement is *false* because $1 \neq \frac{1}{2}$.

11.4 Section Exercises

1. $\log_b x + \log_b y = \log_b (x + y)$ is *false*;
$\log_b x + \log_b y = \log_b xy$

3. $\log_b b^x = x$ is *true*.

5. $\log_7 \frac{4}{5} = \log_7 4 - \log_7 5$ *Quotient rule*

7. $\log_2 8^{1/4} = \frac{1}{4}\log_2 8$ *Power rule*
$= \frac{1}{4}\log_2 2^3$ *Match bases.*
$= \frac{1}{4}(3)$ *$\log_b b^x = x$*
$= \frac{3}{4}$

9. $\log_4 \dfrac{3\sqrt{x}}{y}$
$= \log_4 (3\sqrt{x}) - \log_4 y$ *Quotient rule*
$= \log_4 3 + \log_4 x^{1/2} - \log_4 y$ *Product rule*
$= \log_4 3 + \frac{1}{2}\log_4 x - \log_4 y$ *Power rule*

11. $\log_3 \dfrac{\sqrt[3]{4}}{x^2y}$
$= \log_3 \dfrac{4^{1/3}}{x^2y}$
$= \log_3 4^{1/3} - \log_3 (x^2y)$ *Quotient rule*
$= \log_3 4^{1/3} - (\log_3 x^2 + \log_3 y)$ *Product rule*
$= \log_3 4^{1/3} - \log_3 x^2 - \log_3 y$
$= \frac{1}{3}\log_3 4 - 2\log_3 x - \log_3 y$ *Power rule*

13. $\log_3 \sqrt{\frac{xy}{5}}$

$= \log_3 \left(\frac{xy}{5}\right)^{1/2}$

$= \frac{1}{2}\log_3 \left(\frac{xy}{5}\right)$ *Power rule*

$= \frac{1}{2}[\log_3 (xy) - \log_3 5]$ *Quotient rule*

$= \frac{1}{2}(\log_3 x + \log_3 y - \log_3 5)$ *Product rule*

$= \frac{1}{2}\log_3 x + \frac{1}{2}\log_3 y - \frac{1}{2}\log_3 5$

15. $\log_2 \frac{\sqrt[3]{x} \cdot \sqrt[5]{y}}{r^2}$

$= \log_2 \frac{x^{1/3}y^{1/5}}{r^2}$

$= \log_2 (x^{1/3}y^{1/5}) - \log_2 r^2$ *Quotient rule*

$= \log_2 x^{1/3} + \log_2 y^{1/5} - \log_2 r^2$ *Product rule*

$= \frac{1}{3}\log_2 x + \frac{1}{5}\log_2 y - 2\log_2 r$ *Power rule*

17. The distributive property tells us that the *product* $a(x + y)$ equals the sum $ax + ay$. In the notation $\log_a (x + y)$, the parentheses do not indicate multiplication. They indicate that $x + y$ is the result of raising a to some power.

19. By the product rule for logarithms,

$$\log_b x + \log_b y = \log_b xy.$$

21. $3\log_a m - \log_a n$

$= \log_a m^3 - \log_a n$ *Power rule*

$= \log_a \frac{m^3}{n}$ *Quotient rule*

23. $(\log_a r - \log_a s) + 3\log_a t$

$= \log_a \frac{r}{s} + \log_a t^3$ *Quotient and power rules*

$= \log_a \frac{rt^3}{s}$ *Product rule*

25. $3\log_a 5 - 4\log_a 3$

$= \log_a 5^3 - \log_a 3^4$ *Power rule*

$= \log_a \frac{5^3}{3^4}$ *Quotient rule*

$= \log_a \frac{125}{81}$

27. $\log_{10} (x + 3) + \log_{10} (x - 3)$

$= \log_{10} (x + 3)(x - 3)$ *Product rule*

$= \log_{10} (x^2 - 9)$

29. $3\log_p x + \frac{1}{2}\log_p y - \frac{3}{2}\log_p z - 3\log_p a$

$= \log_p x^3 + \log_p y^{1/2} - \log_p z^{3/2} - \log_p a^3$

Power rule

Group the terms into sums.

$= (\log_p x^3 + \log_p y^{1/2}) - (\log_p z^{3/2} + \log_p a^3)$

$= \log_p x^3y^{1/2} - \log_p z^{3/2}a^3$ *Product rule*

$= \log_p \frac{x^3y^{1/2}}{z^{3/2}a^3}$ *Quotient rule*

31. $\text{LS} = \log_2 (8 + 32) = \log_2 40$

$\text{RS} = \log_2 8 + \log_2 32 = \log_2 (8 \cdot 32)$

$= \log_2 256$

LS $\neq$ RS, so the statement is *false.*

33. $\log_3 7 + \log_3 7^{-1} = \log_3 7 + (-1)\log_3 7$

$= 0$

The statement is *true.*

35. $\log_6 60 - \log_6 10 = \log_6 \frac{60}{10}$

$= \log_6 6 = 1$

The statement is *true.*

37. $\frac{\log_{10} 7}{\log_{10} 14} \stackrel{?}{=} \frac{1}{2}$

$2\log_{10} 7 \stackrel{?}{=} 1\log_{10} (7 \cdot 2)$

Cross products are equal

$2\log_{10} 7 \stackrel{?}{=} \log_{10} 7 + \log_{10} 2$

$\log_{10} 7 \stackrel{?}{=} \log_{10} 2$ *Subtract $\log_{10} 7$.*

The statement is *false.*

39. $\log_3 81 = \log_3 3^4 = 4$

40. $\log_3 81$ is the exponent to which 3 must be raised to obtain 81.

41. Using the result from Exercise 39,

$$3^{\log_3 81} = 3^4 = 81.$$

42. $\log_2 19$ is the exponent to which 2 must be raised to obtain 19.

43. Keeping in mind the result from Exercise 41,

$$2^{\log_2 19} = 19.$$

44. To find $k^{\log_k m}$, first assume $\log_k m = y$. This means, changing to an exponential equation, $k^y = m$. Therefore,

$$k^{\log_k m} = k^y = m.$$

11.5 Common and Natural Logarithms

11.5 Margin Exercises

1. Evaluate each logarithm using a calculator.

(a) $\log 41{,}600 \approx 4.6191$

(b) $\log 43.5 \approx 1.6385$

(c) $\log 0.442 \approx -0.3546$

2.
$$\begin{aligned} \text{pH} &= -\log[\text{H}_3\text{O}^+] \\ &= -\log(1.2 \times 10^{-3}) \\ &= -(\log 1.2 + \log 10^{-3}) \\ &= -(\log 1.2 - 3\log 10) \\ &\approx -[0.0792 - 3(1)] \\ &= -0.0792 + 3 \\ &= 2.9208 \approx 2.9 \end{aligned}$$

Since the pH is less than 3.0, the wetland is a bog.

3. **(a)**
$$\begin{aligned} \text{pH} &= -\log[\text{H}_3\text{O}^+] \\ 4.6 &= -\log[\text{H}_3\text{O}^+] && \textit{Let pH = 4.6.} \\ -4.6 &= \log[\text{H}_3\text{O}^+] \\ -4.6 &= \log_{10}[\text{H}_3\text{O}^+] \\ 10^{-4.6} &= [\text{H}_3\text{O}^+] && \textit{Exponential form} \\ [\text{H}_3\text{O}^+] &\approx 2.5 \times 10^{-5} \end{aligned}$$

(b)
$$\begin{aligned} \text{pH} &= -\log[\text{H}_3\text{O}^+] \\ 7.5 &= -\log[\text{H}_3\text{O}^+] && \textit{Let pH = 7.5.} \\ -7.5 &= \log[\text{H}_3\text{O}^+] \\ -7.5 &= \log_{10}[\text{H}_3\text{O}^+] \\ 10^{-7.5} &= [\text{H}_3\text{O}^+] && \textit{Exponential form} \\ [\text{H}_3\text{O}^+] &\approx 3.2 \times 10^{-8} \end{aligned}$$

4.
$$\begin{aligned} D &= 10\log\left(\frac{I}{I_0}\right) \\ &= 10\log\left(\frac{115I_0}{I_0}\right) \\ &= 10\log 115 \\ &\approx 10(2.06) \approx 21 \end{aligned}$$

The level is about 21 dB.

5. Evaluate each logarithm using a calculator.

(a) $\ln 0.01 \approx -4.6052$

(b) $\ln 27 \approx 3.2958$

(c) $\ln 529 \approx 6.2710$

6.
$$\begin{aligned} f(x) &= 51{,}600 - 7457\ln x \\ f(700) &= 51{,}600 - 7457\ln 700 \\ &\approx 2748.6 \approx 2700 \end{aligned}$$

The altitude at 700 millibars of pressure is approximately 2700 m.

11.5 Section Exercises

1. Since $\log x = \log_{10} x$, the base is 10. The correct response is **C**.

3. $10^0 = 1$ and $10^1 = 10$, so $\log 1 = 0$ and $\log 10 = 1$. Thus, the value of $\log 5.6$ must lie between 0 and 1. The correct response is **C**.

5. $\log 10^{19.2} = \log_{10} 10^{19.2} = 19.2$ by the special property, $\log_b b^x = x$.

7. To four decimal places,

$$\log 328.4 \approx 2.5164.$$

9. $\log 0.0326 \approx -1.4868$

11. $\log(4.76 \times 10^9) \approx 9.6776$
On a TI-83, enter

[LOG] 4.76 [2nd] [EE] 9).

13. $\ln 7.84 \approx 2.0592$

15. $\ln 0.0556 \approx -2.8896$

17. $\ln 10 \approx 2.3026$

19.
$$\begin{aligned} \text{pH} &= -\log[\text{H}_3\text{O}^+] \\ &= -\log(2.5 \times 10^{-5}) \approx 4.6 \end{aligned}$$

Since the pH is between 3.0 and 6.0, the wetland is classified as a *poor fen*.

21.
$$\begin{aligned} \text{pH} &= -\log[\text{H}_3\text{O}^+] \\ &= -\log(2.5 \times 10^{-7}) \approx 6.6 \end{aligned}$$

Since the pH is between 6.0 and 7.5, the wetland is classified as a *rich fen*.

23. Tuna has a hydronium ion concentration of 1.3×10^{-6}.

$$\begin{aligned} \text{pH} &= -\log[\text{H}_3\text{O}^+] \\ &= -\log(1.3 \times 10^{-6}) \approx 5.9 \end{aligned}$$

25. Human gastric contents have a pH of 2.0.

$$\begin{aligned} \text{pH} &= -\log[\text{H}_3\text{O}^+] \\ 2.0 &= -\log[\text{H}_3\text{O}^+] \\ \log_{10}[\text{H}_3\text{O}^+] &= -2.0 \\ [\text{H}_3\text{O}^+] &= 10^{-2} = 1.0 \times 10^{-2} \end{aligned}$$

27. Bananas have a pH of 4.6.

$$\begin{aligned} \text{pH} &= -\log[\text{H}_3\text{O}^+] \\ 4.6 &= -\log[\text{H}_3\text{O}^+] \\ \log_{10}[\text{H}_3\text{O}^+] &= -4.6 \\ [\text{H}_3\text{O}^+] &= 10^{-4.6} \approx 2.5 \times 10^{-5} \end{aligned}$$

29. $t = t(r) = \dfrac{\ln 2}{\ln(1+r)}$

(a) 2% (or 0.02);

$$t(0.02) = \frac{\ln 2}{\ln(1+0.02)} \approx 35.0$$

The doubling time for 2% is about 35.0 years.

(b) 5% (or 0.05);

$$t(0.05) = \frac{\ln 2}{\ln(1+0.05)} \approx 14.2$$

The doubling time for 5% is about 14.2 years.

31. $p(h) = 86.3 \ln h - 680$

(a) $p(5000) = 86.3 \ln 5000 - 680 \approx 55$

The percent of moisture at 5000 feet that falls as snow rather than rain is 55%.

(b) $p(7500) = 86.3 \ln 7500 - 680 \approx 90$

The percent of moisture at 7500 feet that falls as snow rather than rain is 90%.

33. $t = -2.57 \ln\left(\dfrac{87-L}{63}\right)$

(a) $t = -2.57 \ln\left(\dfrac{87-80}{63}\right) \approx 5.6$

The age of a female blue whale that measures 80 feet is 5.6 years.

(b) t must be positive since it represents years. Because t is the product of a negative number, -2.57, and a natural logarithm, the natural logarithm must be negative to make it positive. For the natural logarithm to be negative (and still defined), its argument (the input of the logarithm) must be between 0 and 1, that is,

$$\begin{aligned} 0 &< \frac{87-L}{63} < 1 \\ 0(63) &< 87-L < 1(63) \\ -87 &< -L < -24 \\ 87 &> L > 24 \end{aligned}$$

Thus, $t > 0$ and $\dfrac{87-L}{63}$ is positive and in the domain of the function only if $24 < L < 87$.

11.6 Exponential and Logarithmic Equations; Further Applications

11.6 Margin Exercises

1. (a)

$$\begin{aligned} 2^x &= 9 \\ \log 2^x &= \log 9 && \textit{Property 3} \\ x \log 2 &= \log 9 && \textit{Power rule} \\ x &= \frac{\log 9}{\log 2} && \textit{Divide.} \\ x &\approx 3.170 \end{aligned}$$

Check $x = 3.170$: $2^{3.170} \approx 9$

Solution set: $\{3.170\}$

(b)

$$\begin{aligned} 10^x &= 4 \\ \log 10^x &= \log 4 && \textit{Property 3} \\ x \log 10 &= \log 4 && \textit{Power rule} \\ x &\approx 0.602 && \textit{log 10 = 1} \end{aligned}$$

Check $x = 0.602$: $10^{0.602} \approx 4$

Solution set: $\{0.602\}$

2. $e^{-0.01t} = 0.38$

Take base e logarithms on both sides.

$$\begin{aligned} \ln e^{-0.01t} &= \ln 0.38 \\ -0.01t \ln e &= \ln 0.38 && \textit{Power rule} \\ -0.01t &= \ln 0.38 && \textit{ln e = 1} \\ t &= \frac{\ln 0.38}{-0.01} \approx 96.8 \end{aligned}$$

Check $t = 96.8$: $e^{-0.01(96.8)} \approx 0.38$

Solution set: $\{96.8\}$

3.

$$\begin{aligned} \log_3 (x+1)^5 &= 3 \\ (x+1)^5 &= 3^3 && \textit{Exponential form} \\ (x+1)^5 &= 27 \\ x+1 &= \sqrt[5]{27} && \textit{Fifth root} \\ x &= -1+\sqrt[5]{27} && \textit{Subtract 1.} \end{aligned}$$

Check $x = -1+\sqrt[5]{27}$:

$$\begin{aligned} &\log_3 \left(-1+\sqrt[5]{27}+1\right)^5 \\ &= \log_3 \left(\sqrt[5]{27}\right)^5 \\ &= \log_3 27 = \log_3 3^3 = 3 \end{aligned}$$

Solution set: $\{-1+\sqrt[5]{27}\}$

4.

$$\begin{aligned} \log_8 (2x+5) + \log_8 3 &= \log_8 33 \\ \log_8 [3(2x+5)] &= \log_8 33 && \textit{Product rule} \\ 3(2x+5) &= 33 && \textit{Property 4} \\ 2x+5 &= 11 && \textit{Divide by 3.} \\ 2x &= 6 \\ x &= 3 \end{aligned}$$

Check $x = 3$: $\log_8 11 + \log_8 3 = \log_8 33$ *True*

Solution set: $\{3\}$

5. $\log_3 2x - \log_3 (3x+15) = -2$

$\log_3 \dfrac{2x}{3x+15} = -2$ *Quotient rule*

$3^{-2} = \dfrac{2x}{3x+15}$ *Exponential form*

$\dfrac{1}{9} = \dfrac{2x}{3x+15}$

$3x + 15 = 18x$ *Multiply by 9(3x + 15).*

$15 = 15x$

$1 = x$

Check $x = 1$:

$\log_3 2 - \log_3 18 = \log_3 \frac{2}{18}$

$= \log_3 \frac{1}{9} = \log_3 3^{-2} = -2$

Solution set: $\{1\}$

6. Use $A = P\left(1 + \dfrac{r}{n}\right)^{nt}$ with $P = 2000$, $r = 5\% = 0.05$, $n = 1$ (compounded annually), and $t = 10$.

$$A = 2000\left(1 + \frac{0.05}{1}\right)^{1(10)}$$

$$= 2000(1.05)^{10} \approx 3257.79$$

The value is about \$3257.79.

7. $A = P\left(1 + \dfrac{r}{n}\right)^{nt}$

$2(500) = 500\left(1 + \dfrac{0.04}{2}\right)^{2\cdot t}$ *Semiannually (n = 2)*

$2(500) = 500(1.02)^{2t}$

$2 = (1.02)^{2t}$ *Divide by 500.*

$\log 2 = \log (1.02)^{2t}$ *Property 3*

$\log 2 = 2t \log 1.02$ *Power rule*

$2t = \dfrac{\log 2}{\log 1.02}$ *Divide by log 1.02.*

$t = \dfrac{\log 2}{2 \log 1.02}$ *Divide by 2.*

$t \approx 17.50$

It will take about 17.50 yr for \$500 in an account paying 4% interest compounded semiannually to double.

8. **(a)** $A = Pe^{rt}$

$A = 2500e^{(0.04)3}$

$A \approx 2818.74$

The \$2500 will grow to \$2818.74.

(b) $A = Pe^{rt}$

$2(2500) = 2500e^{0.04t}$

$2 = e^{0.04t}$ *Divide by 2500.*

$\ln 2 = 0.04t$ *Property 3 with ln; $\ln e^k = k$*

$t = \dfrac{\ln 2}{0.04}$ *Divide by 0.04.*

$t \approx 17.33$

It would take about 17.33 yr for the initial investment to double.

9. $y = y_0 e^{-0.0239t}$

(a) $y = 12e^{-0.0239(35)}$ *Let t = 35.*

$y \approx 5.20$

After 35 yr, there will be about 5.20 g.

(b) $\frac{1}{2}(12) = 12e^{-0.0239t}$

$\frac{1}{2} = e^{-0.0239t}$

$\ln \frac{1}{2} = -0.0239t$ *Property 3 with ln*

$t = \dfrac{\ln \frac{1}{2}}{-0.0239} \approx 29$

The half-life is about 29 years.

10. **(a)** $\log_3 17 = \dfrac{\log 17}{\log 3} \approx 2.5789$

(b) $\log_3 17 = \dfrac{\ln 17}{\ln 3} \approx 2.5789$

11.6 Section Exercises

1. $5^x = 125$

$\log 5^x = \log 125$

2. $x\log 5 = \log 125$

3. $\dfrac{x\log 5}{\log 5} = \dfrac{\log 125}{\log 5}$

$x = \dfrac{\log 125}{\log 5}$

4. $\dfrac{\log 125}{\log 5} = 3$ (from calculator)

Solution set: $\{3\}$

5. $7^x = 5$

Take the logarithm of each side.

$\log 7^x = \log 5$

Use the power rule for logarithms.

$x\log 7 = \log 5$

$x = \dfrac{\log 5}{\log 7} \approx 0.827$

Solution set: $\{0.827\}$

7.
$$\begin{aligned} 9^{-x+2} &= 13 \\ \log 9^{-x+2} &= \log 13 \\ (-x+2)\log 9 &= \log 13\,(*) \\ -x\log 9 + 2\log 9 &= \log 13 \\ -x\log 9 &= \log 13 - 2\log 9 \\ x\log 9 &= 2\log 9 - \log 13 \\ x &= \frac{2\log 9 - \log 13}{\log 9} \\ &\approx 0.833 \end{aligned}$$

$(*)$ Alternative solution steps:
$$\begin{aligned} (-x+2)\log 9 &= \log 13 \\ -x+2 &= \frac{\log 13}{\log 9} \\ 2 - \frac{\log 13}{\log 9} &= x \end{aligned}$$

Solution set: $\{0.833\}$

9.
$$\begin{aligned} 3^{2x} &= 14 \\ \log 3^{2x} &= \log 14 \\ 2x\log 3 &= \log 14 \\ x &= \frac{\log 14}{2\log 3} \approx 1.201 \end{aligned}$$

Solution set: $\{1.201\}$

11.
$$\begin{aligned} 2^{y+3} &= 5^y \\ \log 2^{y+3} &= \log 5^y \\ (y+3)\log 2 &= y\log 5 \\ y\log 2 + 3\log 2 &= y\log 5 \end{aligned}$$

Get the y-terms on one side.
$$\begin{aligned} y\log 2 - y\log 5 &= -3\log 2 \\ y(\log 2 - \log 5) &= -3\log 2 \quad \textit{Factor out } y. \\ y &= \frac{-3\log 2}{\log 2 - \log 5} \\ &\approx 2.269 \end{aligned}$$

Solution set: $\{2.269\}$

13.
$$\begin{aligned} 2^{x+3} &= 3^{x-4} \\ \log 2^{x+3} &= \log 3^{x-4} \\ (x+3)\log 2 &= (x-4)\log 3 \\ x\log 2 + 3\log 2 &= x\log 3 - 4\log 3 \end{aligned}$$
Distributive property

Get the x-terms on one side.
$$x\log 2 - x\log 3 = -3\log 2 - 4\log 3$$
Factor out x.
$$\begin{aligned} x(\log 2 - \log 3) &= -3\log 2 - 4\log 3 \\ x &= \frac{-3\log 2 - 4\log 3}{\log 2 - \log 3} \\ &\approx 15.967 \end{aligned}$$

Solution set: $\{15.967\}$

15.
$$\begin{aligned} e^{0.012x} &= 23 \\ \ln e^{0.012x} &= \ln 23 \\ 0.012x\,(\ln e) &= \ln 23 \\ 0.012x &= \ln 23 \quad \textit{ln e = 1} \\ x &= \frac{\ln 23}{0.012} \approx 261.291 \end{aligned}$$

Solution set: $\{261.291\}$

17.
$$\begin{aligned} e^{-0.205x} &= 9 \\ \ln e^{-0.205x} &= \ln 9 \\ -0.205x\,(\ln e) &= \ln 9 \\ -0.205x &= \ln 9 \quad \textit{ln e = 1} \\ x &= \frac{\ln 9}{-0.205} \approx -10.718 \end{aligned}$$

Solution set: $\{-10.718\}$

19.
$$\begin{aligned} \ln e^{3x} &= 9 \\ 3x &= 9 \quad \textit{Property 3 with ln} \\ x &= 3 \end{aligned}$$

Solution set: $\{3\}$

21.
$$\begin{aligned} \ln e^{0.45x} &= \sqrt{7} \\ 0.45x &= \sqrt{7} \\ x &= \frac{\sqrt{7}}{0.45} \approx 5.879 \end{aligned}$$

Solution set: $\{5.879\}$

23. Let's try Exercise 14.
$$\begin{aligned} e^{0.006x} &= 30 \\ \log e^{0.006x} &= \log 30 \\ 0.006x\,(\log e) &= \log 30 \\ x &= \frac{\log 30}{0.006\log e} \approx 566.866 \end{aligned}$$

The natural logarithm is easier because $\ln e = 1$, whereas $\log e$ needs to be calculated.

25.
$$\begin{aligned} \log_3(6x+5) &= 2 \\ 6x+5 &= 3^2 \quad \textit{Exponential form} \\ 6x+5 &= 9 \\ 6x &= 4 \\ x &= \tfrac{4}{6} = \tfrac{2}{3} \end{aligned}$$

$x = \frac{2}{3}$ is acceptable since $6x + 5 = 9 > 0$; that is, it yields a logarithm of a positive number in the original equation.

Check $x = \frac{2}{3}$: $\log_3 9 = \log_3 3^2 = 2$

Solution set: $\{\frac{2}{3}\}$

27. $\log_2 (2x - 1) = 5$

$$2x - 1 = 2^5 \quad \textit{Exponential form}$$
$$2x - 1 = 32$$
$$2x = 33$$
$$x = \tfrac{33}{2}$$

$x = \frac{33}{2}$ is acceptable since $2x - 1 = 32 > 0$.

Check $x = \frac{33}{2}$: $\log_2 32 = \log_2 2^5 = 5$
Solution set: $\{\frac{33}{2}\}$

29. $\log_7 (x + 1)^3 = 2$

$$(x + 1)^3 = 7^2 \quad \textit{Exponential form}$$
$$x + 1 = \sqrt[3]{49} \quad \textit{Cube root}$$
$$x = -1 + \sqrt[3]{49}$$

$x = -1 + \sqrt[3]{49}$ is acceptable since $(x + 1)^3 = 49 > 0$.

Check $x = -1 + \sqrt[3]{49}$: $\log_7 49 = \log_7 7^2 = 2$
Solution set: $\{-1 + \sqrt[3]{49}\}$

31. 2 cannot be a solution because $\log (2 - 3) = \log (-1)$, and -1 is not in the domain of $\log x$.

33. $\log (6x + 1) = \log 3$

$$6x + 1 = 3 \quad \textit{Property 4}$$
$$6x = 2$$
$$x = \tfrac{2}{6} = \tfrac{1}{3}$$

$x = \frac{1}{3}$ is acceptable since $6x + 1 = 3 > 0$.

Check $x = \frac{1}{3}$: $\log (2 + 1) = \log 3$ *True*
Solution set: $\{\frac{1}{3}\}$

35. $\log_5 (3t + 2) - \log_5 t = \log_5 4$

$$\log_5 \frac{3t + 2}{t} = \log_5 4$$
$$\frac{3t + 2}{t} = 4$$
$$3t + 2 = 4t$$
$$2 = t$$

$t = 2$ is acceptable since $3t + 2 = 8 > 0$ and $t = 2 > 0$.

Check $t = 2$: $\log_5 8 - \log_5 2 = \log_5 \frac{8}{2} = \log_5 4$
Solution set: $\{2\}$

37. $\log 4x - \log (x - 3) = \log 2$

$$\log \frac{4x}{x - 3} = \log 2$$
$$\frac{4x}{x - 3} = 2$$
$$4x = 2(x - 3)$$
$$4x = 2x - 6$$
$$2x = -6$$
$$x = -3$$

Reject $x = -3$, because $4x = -12$, which yields an equation in which the logarithm of a negative number must be found.

Solution set: $\emptyset$

39. $\log_2 x + \log_2 (x - 7) = 3$

$$\log_2 [x(x - 7)] = 3$$
$$x(x - 7) = 2^3 \quad \textit{Exponential form}$$
$$x^2 - 7x = 8$$
$$x^2 - 7x - 8 = 0$$
$$(x - 8)(x + 1) = 0$$
$$x - 8 = 0 \quad \text{or} \quad x + 1 = 0$$
$$x = 8 \quad \text{or} \quad x = -1$$

Reject $x = -1$, because it yields an equation in which the logarithm of a negative number must be found.

Check $x = 8$: $\log_2 8 + \log_2 1 = \log_2 2^3 + 0 = 3$
Solution set: $\{8\}$

41. $\log 5x - \log (2x - 1) = \log 4$

$$\log \frac{5x}{2x - 1} = \log 4$$
$$\frac{5x}{2x - 1} = 4$$
$$5x = 8x - 4$$
$$4 = 3x$$
$$\tfrac{4}{3} = x$$

$x = \frac{4}{3}$ is acceptable since $5x = \frac{20}{3} > 0$ and $2x - 1 = \frac{5}{3} > 0$.

Check $x = \frac{4}{3}$: $\log \frac{20}{3} - \log \frac{5}{3} = \log \frac{20/3}{5/3} = \log 4$
Solution set: $\{\frac{4}{3}\}$

43. $\log_2 x + \log_2 (x - 6) = 4$

$$\log_2 [x(x - 6)] = 4$$
$$x(x - 6) = 2^4 \quad \textit{Exponential form}$$
$$x^2 - 6x = 16$$
$$x^2 - 6x - 16 = 0$$
$$(x - 8)(x + 2) = 0$$
$$x - 8 = 0 \quad \text{or} \quad x + 2 = 0$$
$$x = 8 \quad \text{or} \quad x = -2$$

Reject $x = -2$, because it yields an equation in which the logarithm of a negative number must be found.

Check $x = 8$:

$$\log_2 8 + \log_2 2 = \log_2 16 = \log_2 2^4 = 4$$

Solution set: $\{8\}$

45. **(a)** Use the formula $A = P\left(1 + \frac{r}{n}\right)^{nt}$ with $P = 2000$, $r = 0.04$, $n = 4$, and $t = 6$.

$$\begin{aligned} A &= 2000\left(1 + \frac{0.04}{4}\right)^{4\cdot 6} \\ &= 2000(1.01)^{24} \approx 2539.47 \end{aligned}$$

The account will contain \$2539.47.

(b)
$$\begin{aligned} 3000 &= 2000\left(1 + \frac{0.04}{4}\right)^{4t} \\ \tfrac{3000}{2000} &= (1.01)^{4t} \\ \log\left(\tfrac{3}{2}\right) &= \log(1.01)^{4t} \\ \log\left(\tfrac{3}{2}\right) &= 4t\log(1.01) \\ t &= \frac{\log\left(\frac{3}{2}\right)}{4\log(1.01)} \approx 10.2 \end{aligned}$$

It will take about 10.2 years for the account to grow to \$3000.

47. Use the formula $A = Pe^{rt}$ with $P = 4000$, $r = 0.035$, and $t = 6$.

$$\begin{aligned} A &= 4000e^{(0.035)(6)} \\ &= 4000e^{0.21} \approx 4934.71 \end{aligned}$$

There will be \$4934.71 in the account.

49. Here $r = 4.5\%$ (0.045), and if the principal doubles, then $A = 2P$.

$$\begin{aligned} A &= Pe^{rt} \\ 2P &= Pe^{0.045t} \\ 2 &= e^{0.045t} && \textit{Divide by P.} \\ \ln 2 &= \ln e^{0.045t} && \textit{Property 4 with ln} \\ \ln 2 &= 0.045t \\ t &= \frac{\ln 2}{0.045} \approx 15.4 \end{aligned}$$

The money will double in about 15.4 yr.

51.
$$\begin{aligned} A(t) &= 400e^{-0.032t} \\ A(25) &= 400e^{-0.032(25)} && \textit{Let } t = 25. \\ &= 400e^{-0.8} \approx 179.73 \end{aligned}$$

About 180 g of lead will be left in the sample after 25 yr.

53. $\log_6 13 = \frac{\log 13}{\log 6} \approx 1.4315$

55. $\log_{\sqrt{2}}\pi = \frac{\log \pi}{\log\sqrt{2}} \approx 3.3030$

57. $\log_{21} 0.7496 = \frac{\log 0.7496}{\log 21} \approx -0.0947$

59. $\log_{1/2} 5 = \frac{\ln 5}{\ln \frac{1}{2}} \approx -2.3219$

61. $\log_{0.3} 12 = \frac{\ln 12}{\ln 0.3} \approx -2.0639$

63. There are 60 of one species and 40 of another, so

$$p_1 = \tfrac{60}{100} = 0.6 \quad\text{and}\quad p_2 = \tfrac{40}{100} = 0.4.$$

Thus, the index of diversity is

$$\begin{aligned} &-(p_1 \ln p_1 + p_2 \ln p_2) \\ &= -(0.6\ln 0.6 + 0.4\ln 0.4) \\ &\approx 0.673. \end{aligned}$$

Chapter 11 Review Exercises

1. Since a horizontal line intersects the graph in two points, the function is not one-to-one.

2. Since every horizontal line intersects the graph in no more than one point, the function is one-to-one.

3. This function is not one-to-one because two sodas in the list have 41 mg of caffeine.

4. The function $f(x) = -3x + 7$ is a linear function. By the horizontal line test, it is a one-to-one function. To find the inverse, first replace $f(x)$ with y.

$$y = -3x + 7$$

Interchange x and y.

$$x = -3y + 7$$

Solve for y.

$$\begin{aligned} x - 7 &= -3y \\ \frac{x-7}{-3} &= y \quad\text{or}\quad \frac{7-x}{3} = y \end{aligned}$$

Replace y with $f^{-1}(x)$.

$$f^{-1}(x) = \frac{x-7}{-3}, \quad\text{or}\quad f^{-1}(x) = -\frac{1}{3}x + \frac{7}{3}$$

5. $f(x) = \sqrt[3]{6x - 4}$
The graph of f is similar to the graph in Example 3(b) of Section 9.1. The graph passes the horizontal line test, so the function is one-to-one. To find the inverse, first replace $f(x)$ with y.

$$y = \sqrt[3]{6x - 4}$$

Interchange x and y.

$$x = \sqrt[3]{6y - 4}$$

Solve for y. Cube both sides first.

$$\begin{aligned} x^3 &= 6y - 4 \\ x^3 + 4 &= 6y \\ \frac{x^3 + 4}{6} &= y \end{aligned}$$

Replace y with $f^{-1}(x)$.

$$\frac{x^3 + 4}{6} = f^{-1}(x)$$

6. $f(x) = -x^2 + 3$
This is an equation of a vertical parabola which opens down. Since a horizontal line will intersect the graph in two points, the function is not one-to-one.

7. The graph is a linear function through $(0, 1)$ and $(3, 0)$. The graph of $f^{-1}(x)$ will include the points $(1, 0)$ and $(0, 3)$, found by interchanging x and y. Plot these points, and draw a straight line through them.

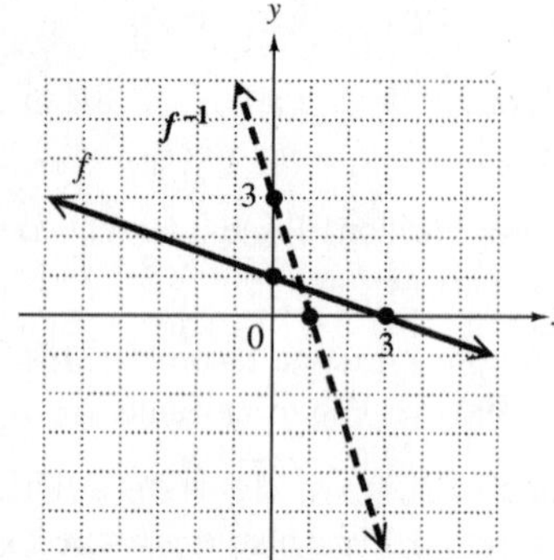

8. The graph is a curve through $(1, 2)$, $(0, 1)$, and $(-1, \frac{1}{2})$. Interchange x and y to get $(2, 1)$, $(1, 0)$, and $(\frac{1}{2}, -1)$, which are on the graph of $f^{-1}(x)$. Plot these points, and draw a smooth curve through them. Use the fact that the graph of f^{-1} is symmetric to the graph of f with respect to the line $y = x$.

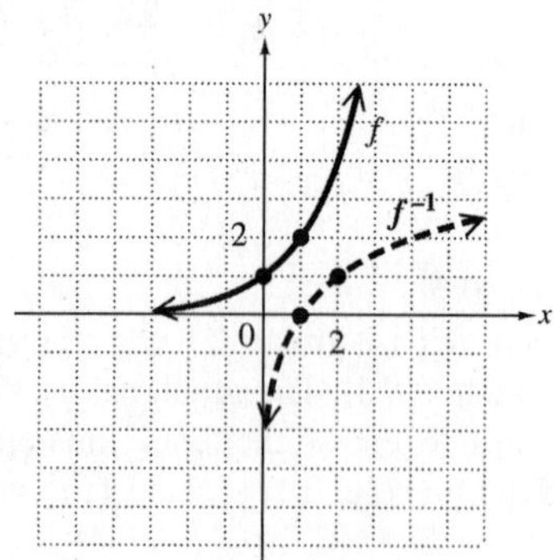

9. $f(x) = 4^x$
Make a table of values.

x	-2	-1	0	1	2
$f(x)$	$\frac{1}{16}$	$\frac{1}{4}$	1	4	16

Plot the points from the table and draw a smooth curve through them.

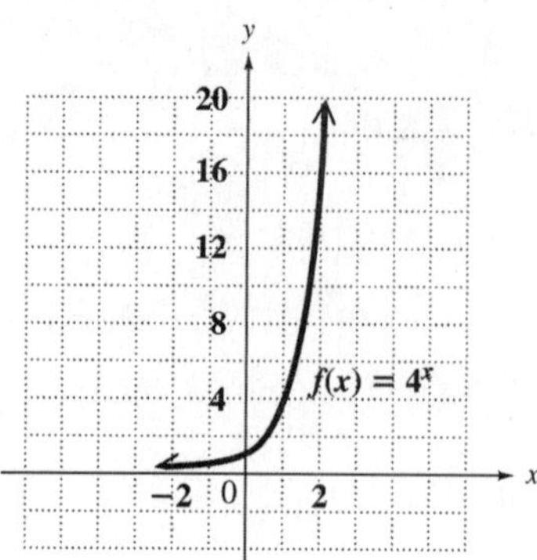

10. $f(x) = (\frac{1}{4})^x$
Make a table of values.

x	-2	-1	0	1	2
$f(x)$	16	4	1	$\frac{1}{4}$	$\frac{1}{16}$

Plot the points from the table and draw a smooth curve through them.

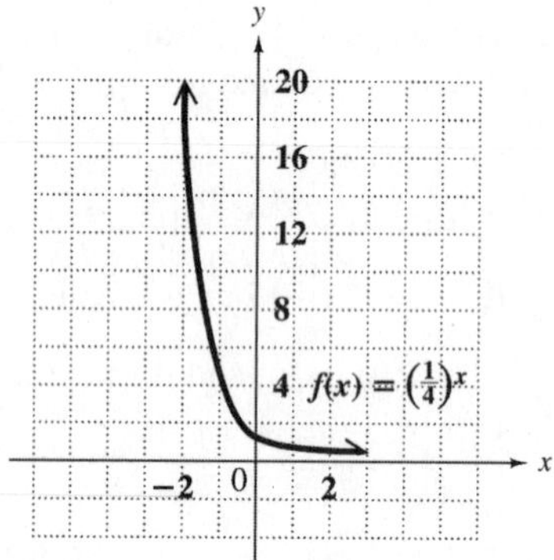

11. $f(x) = 4^{x+1}$
Make a table of values.

x	-3	-2	-1	0	1
$f(x)$	$\frac{1}{16}$	$\frac{1}{4}$	1	4	16

Plot the points from the table and draw a smooth curve through them.

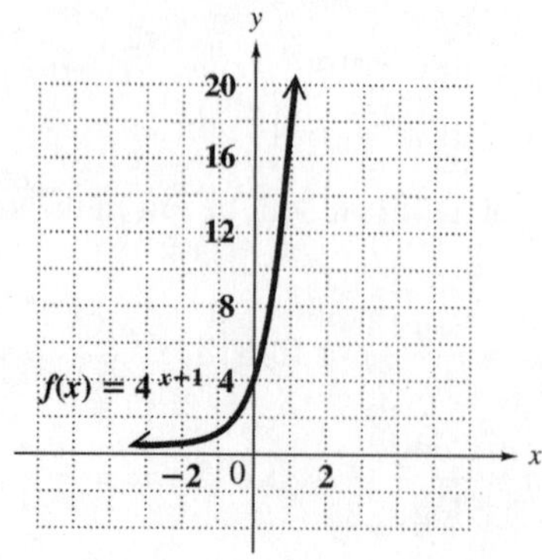

12. $4^{3x} = 8^{x+4}$

$(2^2)^{3x} = (2^3)^{(x+4)}$ *Write as powers of 2.*

$2^{6x} = 2^{(3x+12)}$

$6x = 3x + 12$ *Equate exponents.*

$3x = 12$

$x = 4$

Check $x = 4$: $4^{12} = 8^8$ *True*

Solution set: $\{4\}$

13. $(\frac{1}{27})^{x-1} = 9^{2x}$

$[(\frac{1}{3})^3]^{x-1} = (3^2)^{2x}$

$(3^{-3})^{x-1} = (3^2)^{2x}$ *Write as powers of 3.*

$3^{(-3x+3)} = 3^{4x}$

$-3x + 3 = 4x$ *Equate exponents.*

$3 = 7x$

$\frac{3}{7} = x$

Check $x = \frac{3}{7}$: $(\frac{1}{27})^{-4/7} = 9^{6/7}$ *True*

Use a calculator or show that both sides are equal to $3^{12/7}$ for the check.

Solution set: $\{\frac{3}{7}\}$

14. $5^x = 1$

$5^x = 5^0$ *Write as powers of 5.*

$x = 0$ *Equate exponents.*

Solution set: $\{0\}$

15. $f(x) = 35.6(2)^{0.0306x}$

(a) 2015: $x = 2015 - 2000 = 15$

$$f(15) = 35.6(2)^{0.0306(15)} \approx 48.9$$

The U.S. Hispanic population estimate for 2015 is 48.9 million.

(b) 2030: $x = 2030 - 2000 = 30$

$$f(30) = 35.6(2)^{0.0306(30)} \approx 67.3$$

The U.S. Hispanic population estimate for 2030 is 67.3 million.

16. **(a)** The base is 5, the logarithm (exponent) is 4, and the number is 625, so $\log_5 625 = 4$ becomes $5^4 = 625$ in exponential form.

(b) The base is 5, the exponent (logarithm) is -2, and the number is 0.04, so $5^{-2} = 0.04$ becomes $\log_5 0.04 = -2$ in logarithmic form.

17. **(a)** $\log_b a$ is the exponent to which b must be raised to obtain a.

(b) From part (a),

$$b^{\log_b a} = a.$$

18. $g(x) = \log_4 x$

$y = \log_4 x$ *Replace $g(x)$ with y.*

$4^y = x$ *Write in exponential form.*

Make a table of values. Since $x = 4^y$ is the inverse of $f(x) = y = 4^x$ in Exercise 9, simply reverse the ordered pairs in the table of values belonging to $f(x) = 4^x$.

x	$\frac{1}{16}$	$\frac{1}{4}$	1	4	16
y	-2	-1	0	1	2

Plot the points from the table and draw a smooth curve through them.

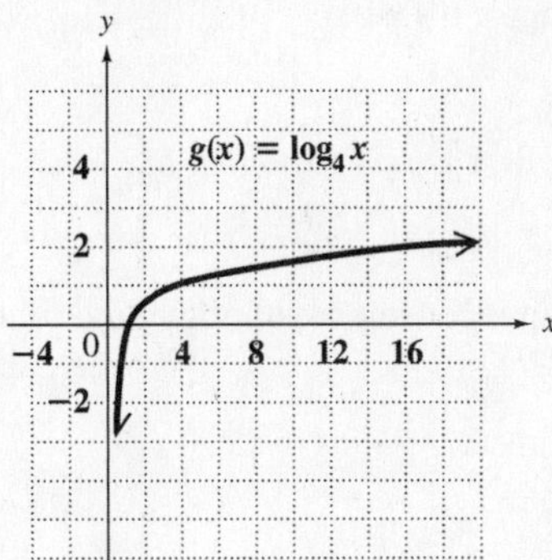

19. $g(x) = \log_{1/4} x$

$y = \log_{1/4} x$ *Replace $g(x)$ with y.*

$(\frac{1}{4})^y = x$ *Write in exponential form.*

Make a table of values. Since $x = (\frac{1}{4})^y$ is the inverse of $f(x) = y = (\frac{1}{4})^x$ in Exercise 10, simply reverse the ordered pairs in the table of values belonging to $f(x) = (\frac{1}{4})^x$.

x	16	4	1	$\frac{1}{4}$	$\frac{1}{16}$
y	-2	-1	0	1	2

Plot the points from the table and draw a smooth curve through them.

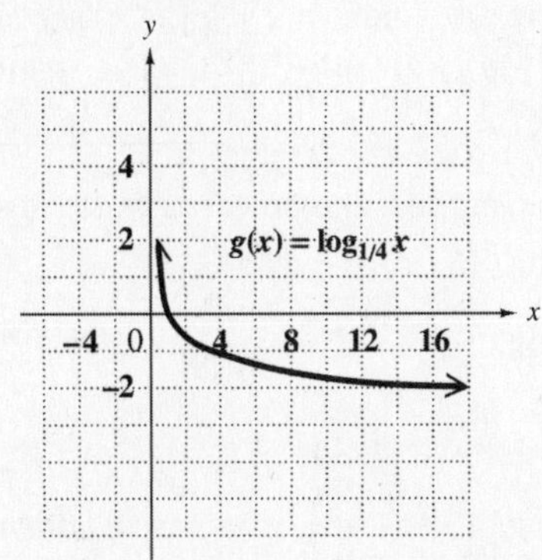

20. $\log_8 64 = x$

$8^x = 64$ *Exponential form*

$8^x = 8^2$ *Write as powers of 8.*

$x = 2$ *Equate exponents.*

Solution set: $\{2\}$

21. $\log_7 \left(\frac{1}{49}\right) = x$

$$7^x = \frac{1}{49}$$
$$7^x = \frac{1}{7^2}$$
$$7^x = 7^{-2}$$
$$x = -2$$

Solution set: $\{-2\}$

22. $\log_4 x = \frac{3}{2}$

$x = 4^{3/2}$ *Exponential form*

$x = \left(\sqrt{4}\right)^3 = 2^3 = 8$

The argument (the input of the logarithm) must be a positive number, so $x = 8$ is acceptable.

Solution set: $\{8\}$

23. $\log_b b^2 = 2$

$b^2 = b^2$ *Exponential form*

This is an identity. Thus, b can be any real number, $b > 0$ and $b \neq 1$.

Solution set: $\{b \mid b > 0,\ b \neq 1\}$

24. $\log_4 3x^2$

$= \log_4 3 + \log_4 x^2$ *Product rule*

$= \log_4 3 + 2\log_4 x$ *Power rule*

25. $\log_2 \dfrac{p^2 r}{\sqrt{z}}$

$= \log_2 \dfrac{p^2 r}{z^{1/2}}$

$= \log_2 (p^2 r) - \log_2 z^{1/2}$ *Quotient rule*

$= \log_2 p^2 + \log_2 r - \log_2 z^{1/2}$ *Product rule*

$= 2\log_2 p + \log_2 r - \frac{1}{2}\log_2 z$ *Power rule*

26. $\log_b 3 + \log_b x - 2\log_b y$

Use the product and power rules for logarithms.

$= \log_b (3 \cdot x) - \log_b y^2$

$= \log_b \dfrac{3x}{y^2}$ *Quotient rule*

27. $\log_3 (x+7) - \log_3 (4x+6)$

$= \log_3 \left(\dfrac{x+7}{4x+6}\right)$ *Quotient rule*

28. $\log 28.9 \approx 1.4609$

29. $\log 0.257 \approx -0.5901$

30. $\ln 28.9 \approx 3.3638$

31. $\ln 0.257 \approx -1.3587$

32. Milk has a hydronium ion concentration of 4.0×10^{-7}.

$$\text{pH} = -\log\,[\text{H}_3\text{O}^+]$$
$$\text{pH} = -\log\,(4.0 \times 10^{-7}) \approx 6.4$$

33. Crackers have a hydronium ion concentration of 3.8×10^{-9}.

$$\text{pH} = -\log\,[\text{H}_3\text{O}^+]$$
$$\text{pH} = -\log\,(3.8 \times 10^{-9}) \approx 8.4$$

34. Orange juice has a pH of 4.6.

$$\text{pH} = -\log\,[\text{H}_3\text{O}^+]$$
$$4.6 = -\log\,[\text{H}_3\text{O}^+]$$
$$\log_{10}\,[\text{H}_3\text{O}^+] = -4.6$$
$$[\text{H}_3\text{O}^+] = 10^{-4.6} \approx 2.5 \times 10^{-5}$$

35. $t = t(r) = \dfrac{\ln 2}{\ln(1+r)}$

(a) $4\% = 0.04;\ t(0.04) = \dfrac{\ln 2}{\ln(1+0.04)} \approx 18$

At 4%, it would take about 18 years.

(b) $6\% = 0.06;\ t(0.06) = \dfrac{\ln 2}{\ln(1+0.06)} \approx 12$

At 6%, it would take about 12 years.

(c) $10\% = 0.10;\ t(0.10) = \dfrac{\ln 2}{\ln(1+0.10)} \approx 7$

At 10%, it would take about 7 years.

(d) $12\% = 0.12;\ t(0.12) = \dfrac{\ln 2}{\ln(1+0.12)} \approx 6$

At 12%, it would take about 6 years.

(e) Each comparison shows approximately the same number. For example, in part (a) the doubling time is 18 yr (rounded) and $\frac{72}{4} = 18$.

Thus, the formula $t = \dfrac{72}{100r}$ (called the *rule of 72*) is an excellent approximation of the doubling time formula.

36. $T = \dfrac{1}{k}\ln\dfrac{C_2}{C_1}$

$T = \dfrac{1}{\frac{1}{3}}\ln\dfrac{5}{2} = 3\ln 2.5 \approx 2$ (rounded down)

It should be administered every 2 hours.

37.
$$\begin{aligned} 3^x &= 9.42 \\ \log 3^x &= \log 9.42 \\ x \log 3 &= \log 9.42 \\ x &= \frac{\log 9.42}{\log 3} \approx 2.042 \end{aligned}$$

Check $x = 2.042$: $3^{2.042} \approx 9.425$
Solution set: $\{2.042\}$

38.
$$\begin{aligned} 2^{x-1} &= 15 \\ \log 2^{x-1} &= \log 15 \\ (x-1) \log 2 &= \log 15 \\ x - 1 &= \frac{\log 15}{\log 2} \\ x &= \frac{\log 15}{\log 2} + 1 \approx 4.907 \end{aligned}$$

Check $x = 4.907$: $2^{4.907} \approx 15.0$
Solution set: $\{4.907\}$

39. $e^{0.06x} = 3$

Take base e logarithms on both sides.
$$\begin{aligned} \ln e^{0.06x} &= \ln 3 \\ 0.06x \ln e &= \ln 3 \\ 0.06x &= \ln 3 && \textit{ln e = 1} \\ x &= \frac{\ln 3}{0.06} \approx 18.310 \end{aligned}$$

Check $x = 18.310$: $e^{1.0986} \approx 3.0$
Solution set: $\{18.310\}$

40.
$$\begin{aligned} \log_3 (9x+8) &= 2 \\ 9x + 8 &= 3^2 && \textit{Exponential form} \\ 9x + 8 &= 9 \\ 9x &= 1 \\ x &= \tfrac{1}{9} \end{aligned}$$

$x = \frac{1}{9}$ is acceptable since $9x + 8 = 9 > 0$.

Check $x = \frac{1}{9}$: $\log_3 9 = \log_3 3^2 = 2$
Solution set: $\{\frac{1}{9}\}$

41.
$$\begin{aligned} \log_5 (x+6)^3 &= 2 \\ (x+6)^3 &= 5^2 && \textit{Exponential form} \\ (x+6)^3 &= 25 \\ x + 6 &= \sqrt[3]{25} && \textit{Take cube roots.} \\ x &= \sqrt[3]{25} - 6 \end{aligned}$$

$x = -6 + \sqrt[3]{25}$ is acceptable since $(x+6)^3 = 25 > 0$.

Check $y = -6 + \sqrt[3]{25}$: $\log_5 25 = \log_5 5^2 = 2$
Solution set: $\{-6 + \sqrt[3]{25}\}$

42.
$$\begin{aligned} \log_3 (p+2) - \log_3 p &= \log_3 2 \\ \log_3 \frac{p+2}{p} &= \log_3 2 && \textit{Quotient rule} \\ \frac{p+2}{p} &= 2 && \textit{Property 4} \\ p + 2 &= 2p \\ 2 &= p \end{aligned}$$

$p = 2$ is acceptable since $p + 2 = 4 > 0$ and $p = 2 > 0$.

Check $p = 2$: $\log_3 4 - \log_3 2 = \log_3 \frac{4}{2} = \log_3 2$
Solution set: $\{2\}$

43.
$$\begin{aligned} \log (2x+3) - \log x &= 1 \\ \log_{10} (2x+3) - \log_{10} x &= 1 \\ \log_{10} \frac{2x+3}{x} &= 1 && \textit{Quotient rule} \\ 10^1 &= \frac{2x+3}{x} && \textit{Exponen. form} \\ 10x &= 2x + 3 \\ 8x &= 3 \\ x &= \tfrac{3}{8} \end{aligned}$$

$x = \frac{3}{8}$ is acceptable since $2x + 3 = \frac{15}{4} > 0$ and $x = \frac{3}{8} > 0$.

Check $x = \frac{3}{8}$:

LS $= \log (\frac{3}{4} + 3) = \log \frac{15}{4}$

RS $= \log 10 + \log \frac{3}{8} = \log \frac{30}{8} = \log \frac{15}{4}$

Solution set: $\{\frac{3}{8}\}$

44.
$$\begin{aligned} \log_4 x + \log_4 (8-x) &= 2 \\ \log_4 [x(8-x)] &= 2 && \textit{Product rule} \\ x(8-x) &= 4^2 && \textit{Exponential form} \\ 8x - x^2 &= 16 \\ x^2 - 8x + 16 &= 0 \\ (x-4)(x-4) &= 0 \\ x - 4 &= 0 \\ x &= 4 \end{aligned}$$

$x = 4$ is acceptable since $x = 4 > 0$ and $8 - x = 4 > 0$.

Check $x = 4$: $\log_4 4 + \log_4 4 = 1 + 1 = 2$
Solution set: $\{4\}$

45. $\log_2 x + \log_2 (x+15) = 4$

$$\log_2 [x(x+15)] = 4 \quad \textit{Product rule}$$
$$x(x+15) = 2^4 \quad \textit{Exponential form}$$
$$x^2 + 15x = 16$$
$$x^2 + 15x - 16 = 0$$
$$(x+16)(x-1) = 0$$
$$x + 16 = 0 \quad \text{or} \quad x - 1 = 0$$
$$x = -16 \quad \text{or} \quad x = 1$$

Reject $x = -16$, because it yields an equation in which the logarithm of a negative number must be found.

Check $x = 1$: $\log_2 1 + \log_2 16 = 0 + \log_2 16$
$= \log_2 16 = \log_2 2^4 = 4$

Solution set: $\{1\}$

46. Use $A = P\left(1 + \frac{r}{n}\right)^{nt}$ with $P = 6500$, $r = 3\% = 0.03$, $n = 365$ (compounded daily), and $t = 3$.

$$A = 6500\left(1 + \frac{0.03}{365}\right)^{365(3)} \approx 7112.11$$

In 3 yr, the value would be about \$7112.11.

47. Use $A = P\left(1 + \frac{r}{n}\right)^{nt}$.

Plan A:
Let $P = 1000$, $r = 0.04$, $n = 4$, and $t = 3$.

$$A = 1000\left(1 + \frac{0.04}{4}\right)^{4 \cdot 3} \approx 1126.83$$

Plan B:
Let $P = 1000$, $r = 0.039$, $n = 12$, and $t = 3$.

$$A = 1000\left(1 + \frac{0.039}{12}\right)^{12 \cdot 3} \approx 1123.91$$

Plan A is the better plan by \$2.92.

48. $S = C(1-r)^n$

Let $C = 30{,}000$, $r = 0.15$, and $n = 12$.

$$S = 30{,}000(1 - 0.15)^{12}$$
$$= 30{,}000(0.85)^{12} \approx 4267$$

The scrap value is about \$4267.

49. $S = C(1-r)^n$

Let $S = \frac{1}{2}C$ and $n = 6$.

$$S = C(1-r)^n$$
$$\tfrac{1}{2}C = C(1-r)^6$$
$$0.5 = (1-r)^6 \quad (*)$$
$$\ln 0.5 = \ln (1-r)^6$$
$$\ln 0.5 = 6 \ln (1-r)$$
$$\ln (1-r) = \frac{\ln 0.5}{6}$$
$$\ln (1-r) \approx -0.1155$$
$$1 - r = e^{-0.1155}$$
$$1 - r \approx 0.89$$
$$r = 0.11$$

The rate is approximately 11%.

$(*)$ Alternative solution steps without logarithms:

$$0.5 = (1-r)^6$$
$$\sqrt[6]{0.5} = 1 - r$$
$$r = 1 - \sqrt[6]{0.5} \approx .11$$

Note that $1 - r$ must be positive, so $\pm \sqrt[6]{0.5}$ is not needed.

In Exercises 50–52, use the change-of-base rule,

$$\boldsymbol{\log_a x = \frac{\log_b x}{\log_b a}}.$$

50. $\log_{16} 13 = \dfrac{\log 13}{\log 16} \approx 0.9251$

51. $\log_4 12 = \dfrac{\log 12}{\log 4} \approx 1.7925$

52. $\log_{\sqrt{6}} \sqrt{13} = \dfrac{\ln \sqrt{13}}{\ln \sqrt{6}} \approx 1.4315$

53. **[11.6]** $\log_3 (x+9) = 4$

$$x + 9 = 3^4 \quad \textit{Exponential form}$$
$$x + 9 = 81$$
$$x = 72$$

$x = 72$ is acceptable since $x + 9 = 81 > 0$.

Check $x = 72$: $\log_3 81 = \log_3 3^4 = 4$
Solution set: $\{72\}$

54. **[11.3]** $\log_2 32 = x$

$$2^x = 32 \quad \textit{Exponential form}$$
$$2^x = 2^5 \quad \textit{Same base}$$
$$x = 5 \quad \textit{Set the exponents equal.}$$

Check $x = 5$: $\log_2 32 = \log_2 2^5 = 5$
Solution set: $\{5\}$

55. **[11.3]** $\log_x \frac{1}{81} = 2$

$$x^2 = \frac{1}{81} \quad \textit{Exponential form}$$
$$x = \pm\sqrt{\frac{1}{81}} \quad \textit{Square root property}$$
$$x = \frac{1}{9} \quad \textit{Reject } -\tfrac{1}{9}.$$

$x = \frac{1}{9}$ is an acceptable base since it is a positive number (not equal to 1).

Check $x = \frac{1}{9}$: $\log_{1/9} \frac{1}{81} = 2$ since $(\frac{1}{9})^2 = \frac{1}{81}$.
Solution set: $\{\frac{1}{9}\}$

56. **[11.2]** $27^x = 81$

$$(3^3)^x = 3^4 \quad \textit{Write as powers of 3.}$$
$$3^{3x} = 3^4$$
$$3x = 4 \quad \textit{Equate exponents.}$$
$$x = \frac{4}{3}$$

Check $x = \frac{4}{3}$: $27^{4/3} = 3^4 = 81$
Solution set: $\{\frac{4}{3}\}$

57. **[11.2]** $2^{2x-3} = 8$

$$2^{2x-3} = 2^3 \quad \textit{Write as powers of 2.}$$
$$2x - 3 = 3 \quad \textit{Equate exponents.}$$
$$2x = 6$$
$$x = 3$$

Check $x = 3$: $2^3 = 8$ *True*
Solution set: $\{3\}$

58. **[11.6]**
$\log_3 (x+1) - \log_3 x = 2$

$$\log_3 \frac{x+1}{x} = 2 \quad \textit{Quotient rule}$$
$$\frac{x+1}{x} = 3^2 \quad \textit{Exponential form}$$
$$9x = x + 1$$
$$8x = 1$$
$$x = \frac{1}{8}$$

$x = \frac{1}{8}$ is acceptable since $x + 1 = \frac{9}{8} > 0$ and $x = \frac{1}{8} > 0$.

Check $x = \frac{1}{8}$:

$$\log_3 \tfrac{9}{8} - \log_3 \tfrac{1}{8} = \log_3 9 = \log_3 3^2 = 2$$

Solution set: $\{\frac{1}{8}\}$

59. **[11.6]** $\log (3x - 1) = \log 10$

$$3x - 1 = 10$$
$$3x = 11$$
$$x = \frac{11}{3}$$

$x = \frac{11}{3}$ is acceptable since $3x - 1 = 10 > 0$.

Check $x = \frac{11}{3}$: $\log (11 - 1) = \log 10$ *True*
Solution set: $\{\frac{11}{3}\}$

60. **[11.6]** $S = C(1 - r)^n$

Let $S = 10{,}000$, $C = 30{,}000$, and $r = 15\% = 0.15$.

$$10{,}000 = 30{,}000(1 - 0.15)^n$$
$$\frac{1}{3} = (0.85)^n$$
$$\log \frac{1}{3} = \log (0.85)^n$$
$$\log \frac{1}{3} = n \log (0.85)$$
$$n = \frac{\log \frac{1}{3}}{\log (0.85)} \approx 6.8$$

The useful life is 6.8 years.

Chapter 11 Test

1. **(a)** $f(x) = x^2 + 9$

This function is not one-to-one. The graph of $f(x)$ is a vertical parabola. A horizontal line will intersect the graph more than once.

(b) This function is one-to-one. A horizontal line will not intersect the graph in more than one point.

2. $f(x) = \sqrt[3]{x + 7}$

Replace $f(x)$ with y.

$$y = \sqrt[3]{x + 7}$$

Interchange x and y.

$$x = \sqrt[3]{y + 7}$$

Solve for y.

$$x^3 = y + 7$$
$$x^3 - 7 = y$$

Replace y with $f^{-1}(x)$.

$$f^{-1}(x) = x^3 - 7$$

3. By the horizontal line test, $f(x)$ is a one-to-one function and has an inverse. Choose some points on the graph of $f(x)$, such as $(4, 0)$, $(3, -1)$, and $(0, -2)$. To graph the inverse, interchange the x- and y-values to get $(0, 4)$, $(-1, 3)$, and $(-2, 0)$. Plot these points and draw a smooth curve through them.

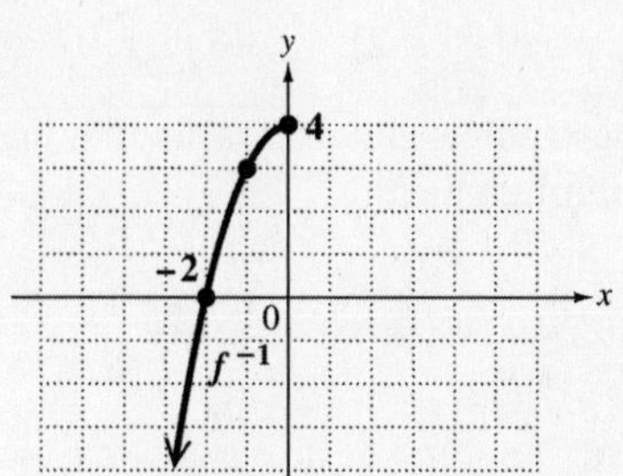

4. $y = f(x) = 6^x$

Make a table of values.

x	-2	-1	0	1
$f(x)$	$\frac{1}{36}$	$\frac{1}{6}$	1	6

Plot these points and draw a smooth exponential curve through them.

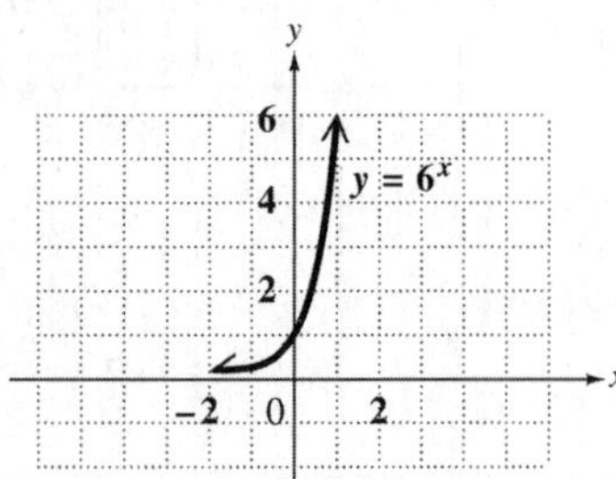

5. $y = g(x) = \log_6 x$

Make a table of values.

Powers of 6	6^{-2}	6^{-1}	6^0	6^1
x	$\frac{1}{36}$	$\frac{1}{6}$	1	6
$g(x)$	-2	-1	0	1

Plot these points and draw a smooth logarithmic curve through them.

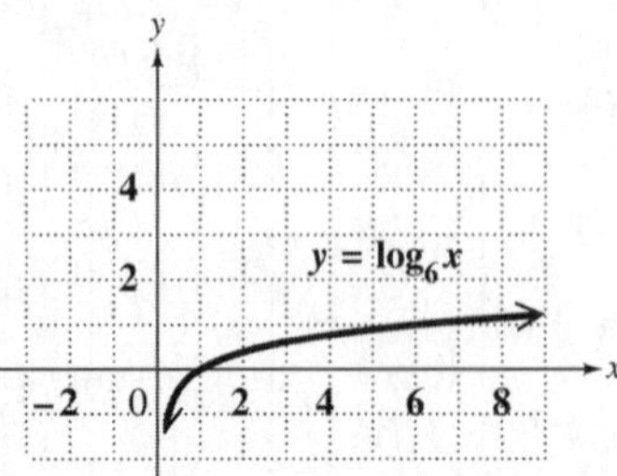

6. $y = 6^x$ and $y = \log_6 x$ are inverse functions. To use the graph from Exercise 4 to obtain the graph of the function in Exercise 5, interchange the x- and y-coordinates of the ordered pairs

$$(-2, \tfrac{1}{36}), (-1, \tfrac{1}{6}), (0, 1), \text{ and } (1, 6)$$

to get $(\tfrac{1}{36}, -2), (\tfrac{1}{6}, -1), (1, 0), \text{ and } (6, 1).$

Plot these points and draw a smooth logarithmic curve through them.

7.
$$5^x = \frac{1}{625}$$
$$5^x = (\tfrac{1}{5})^4$$
$$5^x = (5^{-1})^4 \quad \textit{Write as powers of 5.}$$
$$5^x = 5^{-4}$$
$$x = -4 \quad \textit{Equate exponents.}$$

Check $x = -4$: $5^{-4} = (\frac{1}{5})^4 = \frac{1}{625}$ *True*

Solution set: $\{-4\}$

8.
$$2^{3x-7} = 8^{2x+2}$$
$$2^{3x-7} = 2^{3(2x+2)} \quad \textit{Write as powers of 5.}$$
$$3x - 7 = 3(2x+2) \quad \textit{Equate exponents.}$$
$$3x - 7 = 6x + 6$$
$$-13 = 3x$$
$$-\tfrac{13}{3} = x$$

Check $x = -\frac{13}{3}$:

$$\text{LS} = 2^{-13-7} = 2^{-20}$$
$$\text{RS} = 8^{-20/3} = (2^3)^{-20/3} = 2^{-20}$$

Solution set: $\{-\frac{13}{3}\}$

9. $f(x) = 1013e^{-0.0001341x}$

(a) $f(2000) = 1013e^{-0.0001341(2000)}$
≈ 775

The atmospheric pressure at 2000 m is about 775 millibars.

(b) $f(10{,}000) = 1013e^{-0.0001341(10{,}000)}$
≈ 265

The atmospheric pressure at 10,000 m is about 265 millibars.

10. The base is 4, the exponent (logarithm) is -2, and the number is 0.0625, so

$$4^{-2} = 0.0625 \quad \text{becomes} \quad \log_4 0.0625 = -2$$

in logarithmic form.

11. The base is 7, the logarithm (exponent) is 2, and the number is 49, so

$$\log_7 49 = 2 \quad \text{becomes} \quad 7^2 = 49$$

in exponential form.

12.
$$\log_{1/2} x = -5$$
$$x = (\tfrac{1}{2})^{-5} \quad \textit{Exponential form}$$
$$x = (\tfrac{2}{1})^5 = 32$$

$x = 32$ is acceptable since it is positive.

Check $x = 32$:

$$\log_{1/2} 32 = -5 \text{ since } (\tfrac{1}{2})^{-5} = 2^5 = 32$$

Solution set: $\{32\}$

13.
$$x = \log_9 3$$
$$9^x = 3 \quad \textit{Exponential form}$$
$$(3^2)^x = 3 \quad \textit{Write as powers of 3.}$$
$$3^{2x} = 3^1$$
$$2x = 1 \quad \textit{Equate exponents.}$$
$$x = \tfrac{1}{2}$$

Check $x = \frac{1}{2}$:

$$\tfrac{1}{2} = \log_9 3 \text{ since } 9^{1/2} = \sqrt{9} = 3$$

Solution set: $\{\frac{1}{2}\}$

14. $\log_x 16 = 4$

$x^4 = 16$ *Exponential form*

$x^2 = \pm 4$ *Square root property*

Reject -4 since $x^2 \geq 0$.

$x^2 = 4$

$x = \pm 2$ *Square root property*

Reject -2 since the base cannot be negative.

Check $x = 2$: $\log_2 16 = 4$ since $2^4 = 16$
Solution set: $\{2\}$

15. $\log_3 x^2 y$

$= \log_3 x^2 + \log_3 y$ *Product rule*

$= 2\log_3 x + \log_3 y$ *Power rule*

16. $\frac{1}{4}\log_b r + 2\log_b s - \frac{2}{3}\log_b t$
Use the power rule for logarithms.
$= \log_b r^{1/4} + \log_b s^2 - \log_b t^{2/3}$
Use the product rule for logarithms.
$= \log_b (r^{1/4} s^2) - \log_b t^{2/3}$
Use the quotient rule for logarithms.
$= \log_b \dfrac{r^{1/4} s^2}{t^{2/3}}$

17. **(a)** $\log 21.3 \approx 1.3284$

(b) $\ln 0.43 \approx -0.8440$

(c) $\log_6 45 = \dfrac{\log 45}{\log 6} \approx 2.1245$

18. $3^x = 78$

$\ln 3^x = \ln 78$

$x \ln 3 = \ln 78$ *Power rule*

$x = \dfrac{\ln 78}{\ln 3} \approx 3.9656$

Check $x = 3.9656$: $3^{3.9656} \approx 78.0$
Solution set: $\{3.9656\}$

19. $\log_8 (x+5) + \log_8 (x-2) = \log_8 8$
$\log_8 (x+5) + \log_8 (x-2) = 1$
Use the product rule for logarithms.

$\log_8 [(x+5)(x-2)] = 1$

$(x+5)(x-2) = 8^1$ *Exp. form*

$x^2 + 3x - 10 = 8$

$x^2 + 3x - 18 = 0$

$(x+6)(x-3) = 0$

$x + 6 = 0$ or $x - 3 = 0$

$x = -6$ or $x = 3$

Reject $x = -6$, because $x + 5 = -1$, which yields an equation in which the logarithm of a negative number must be found.

Check $x = 3$: $\log_8 8 + \log_8 1$
$= \log_8 (8 \cdot 1) = \log_8 8$
Solution set: $\{3\}$

20. **(a)** $A = P\left(1 + \dfrac{r}{n}\right)^{nt}$

$A = 10{,}000\left(1 + \dfrac{0.045}{4}\right)^{4 \cdot 5} \approx 12{,}507.51$

\$10,000 invested at 4.5% annual interest, compounded quarterly, will increase to \$12,507.51 in 5 years.

(b) $2(10{,}000) = 10{,}000\left(1 + \dfrac{0.045}{4}\right)^{4 \cdot t}$

$2 = (1.01125)^{4t}$

$\log 2 = \log (1.01125)^{4t}$

$\log 2 = 4t \log (1.01125)$

$t = \dfrac{\log 2}{4 \log (1.01125)}$

$t \approx 15.5$

It will take about 15.5 years for the initial principal to double.

Cumulative Review Exercises (Chapters 1–11)

For Exercises 1–3,

$S = \{-\frac{9}{4}, -2, -\sqrt{2}, 0, 0.6, \sqrt{11}, \sqrt{-8}, 6, \frac{30}{3}\}$.

1. The integers are -2, 0, 6, and $\frac{30}{3}$ (or 10).

2. The rational numbers are $-\frac{9}{4}$, -2, 0, 0.6, 6, and $\frac{30}{3}$ (or 10). Each can be expressed as a quotient of two integers.

3. The irrational numbers are $-\sqrt{2}$ and $\sqrt{11}$.

4. $7 - (3 + 4x) + 2x = -5(x-1) - 3$

$7 - 3 - 4x + 2x = -5x + 5 - 3$

$4 - 2x = -5x + 2$

$3x = -2$

$x = -\frac{2}{3}$

Solution set: $\{-\frac{2}{3}\}$

5. $2x + 2 \leq 5x - 1$

$-3x \leq -3$

Divide by -3; reverse the inequality.

$x \geq 1$

Solution set: $[1, \infty)$

6. $|2x-5|=9$

$$\begin{aligned} 2x-5&=9 \quad\text{or}\quad 2x-5=-9\\ 2x&=14 \quad\text{or}\quad 2x=-4\\ x&=7 \quad\text{or}\quad x=-2 \end{aligned}$$

Solution set: $\{-2,7\}$

7. $|4x+2|>10$

$$\begin{aligned} 4x+2&>10 \quad\text{or}\quad 4x+2<-10\\ 4x&>8 \quad\text{or}\quad 4x<-12\\ x&>2 \quad\text{or}\quad x<-3 \end{aligned}$$

Solution set: $(-\infty,-3)\cup(2,\infty)$

8. **(a)** Yes, this graph is the graph of a function because it passes the vertical line test.

(b) $(x_1,y_1)=(2003, 41{,}218)$ and $(x_2,y_2)=(2006, 50{,}980)$.

$$\begin{aligned} m=\frac{y_2-y_1}{x_2-x_1}&=\frac{50{,}980-41{,}218}{2006-2003}\\ &=\frac{9762}{3}=3254 \end{aligned}$$

The slope of the line in the graph is 3254 and can be interpreted as follows: The number of travelers increased by an average of 3254 thousand per year during the period 2003-2006.

9.
$$\begin{aligned} 5x-3y&=14 \quad (1)\\ 2x+5y&=18 \quad (2) \end{aligned}$$

Multiply equation (1) by 5 and equation (2) by 3. Then add the results.

$$\begin{array}{rcrl} 25x-15y&=&70 & 5\times(1)\\ 6x+15y&=&54 & 3\times(2)\\ \hline 31x&=&124 &\\ x&=&4 & \end{array}$$

Substitute 4 for x in equation (1) to find y.

$$\begin{aligned} 5x-3y&=14 \quad (1)\\ 5(4)-3y&=14\\ 20-3y&=14\\ -3y&=-6\\ y&=2 \end{aligned}$$

Solution set: $\{(4,2)\}$

10.
$$\begin{aligned} x+2y+3z&=11 \quad (1)\\ 3x-y+z&=8 \quad (2)\\ 2x+2y-3z&=-12 \quad (3) \end{aligned}$$

To eliminate z, add equations (1) and (3).

$$\begin{array}{rcrl} x+2y+3z&=&11 & (1)\\ 2x+2y-3z&=&-12 & (3)\\ \hline 3x+4y&=&-1 & (4) \end{array}$$

To eliminate z again, multiply equation (2) by 3 and add the result to equation (3).

$$\begin{array}{rcrl} 9x-3y+3z&=&24 & 3\times(2)\\ 2x+2y-3z&=&-12 & (3)\\ \hline 11x-y&=&12 & (5) \end{array}$$

Multiply equation (5) by 4 and add the result to equation (4).

$$\begin{array}{rcrl} 44x-4y&=&48 & 4\times(5)\\ 3x+4y&=&-1 & (4)\\ \hline 47x&=&47 &\\ x&=&1 & \end{array}$$

Substitute 1 for x in equation (5) to find y.

$$\begin{aligned} 11x-y&=12 \quad (5)\\ 11(1)-y&=12\\ 11-y&=12\\ -y&=1\\ y&=-1 \end{aligned}$$

Substitute 1 for x and -1 for y in equation (2) to find z.

$$\begin{aligned} 3x-y+z&=8 \quad (2)\\ 3(1)-(-1)+z&=8\\ 3+1+z&=8\\ 4+z&=8\\ z&=4 \end{aligned}$$

Solution set: $\{(1,-1,4)\}$

11. $(2p+3)(3p-1)=6p^2-2p+9p-3$
$$=6p^2+7p-3$$

12. $(4k-3)^2=(4k)^2-2(4k)(3)+3^2$
$$=16k^2-24k+9$$

13. $(3m^3+2m^2-5m)-(8m^3+2m-4)$
$$\begin{aligned} &=3m^3+2m^2-5m-8m^3-2m+4\\ &=3m^3-8m^3+2m^2-5m-2m+4\\ &=-5m^3+2m^2-7m+4 \end{aligned}$$

14. Divide $6t^4+17t^3-4t^2+9t+4$ by $3t+1$.

$$\begin{array}{r|l} & 2t^3+5t^2-3t+4\\ \hline 3t+1 & 6t^4+17t^3-4t^2+9t+4\\ & 6t^4+2t^3\\ \hline & 15t^3-4t^2\\ & 15t^3+5t^2\\ \hline & -9t^2+9t\\ & -9t^2-3t\\ \hline & 12t+4\\ & 12t+4\\ \hline & 0 \end{array}$$

The quotient is

$$2t^3+5t^2-3t+4.$$

15. $5z^3 - 19z^2 - 4z = z(5z^2 - 19z - 4)$
$= z(5z+1)(z-4)$

16. $16a^2 - 25b^4$

Use the difference of squares formula,

$$x^2 - y^2 = (x+y)(x-y),$$

where $x = 4a$ and $y = 5b^2$.

$16a^2 - 25b^4 = (4a + 5b^2)(4a - 5b^2)$

17. $8c^3 + d^3$

Use the sum of cubes formula,

$$x^3 + y^3 = (x+y)(x^2 - xy + y^2),$$

where $x = 2c$ and $y = d$.

$8c^3 + d^3 = (2c + d)(4c^2 - 2cd + d^2)$

18. $\dfrac{(5p^3)^4(-3p^7)}{2p^2(4p^4)} = \dfrac{(5^4p^{12})(-3p^7)}{8p^6}$
$= \dfrac{(625)(-3)p^{19}}{8p^6}$
$= -\dfrac{1875p^{13}}{8}$

19. $\dfrac{x^2-9}{x^2+7x+12} \div \dfrac{x-3}{x+5}$

Multiply by the reciprocal.

$= \dfrac{x^2-9}{x^2+7x+12} \cdot \dfrac{x+5}{x-3}$

$= \dfrac{(x+3)(x-3)}{(x+3)(x+4)} \cdot \dfrac{(x+5)}{(x-3)}$ *Factor.*

$= \dfrac{x+5}{x+4}$

20. $\dfrac{2}{k+3} - \dfrac{5}{k-2}$

The LCD is $(k+3)(k-2)$.

$= \dfrac{2(k-2)}{(k+3)(k-2)} - \dfrac{5(k+3)}{(k-2)(k+3)}$

$= \dfrac{2k - 4 - 5k - 15}{(k+3)(k-2)}$

$= \dfrac{-3k - 19}{(k+3)(k-2)}$

21. $\sqrt{288} = \sqrt{144 \cdot 2} = \sqrt{144}\sqrt{2} = 12\sqrt{2}$

22. $\dfrac{-8^{4/3}}{8^2} = -8^{4/3-2} = -8^{4/3-6/3}$

$= -8^{-2/3} = -\dfrac{1}{8^{2/3}}$

$= -\dfrac{1}{\left(\sqrt[3]{8}\right)^2} = -\dfrac{1}{2^2} = -\dfrac{1}{4}$

23. $2\sqrt{32} - 5\sqrt{98} = 2\sqrt{16 \cdot 2} - 5\sqrt{49 \cdot 2}$
$= 2 \cdot 4\sqrt{2} - 5 \cdot 7\sqrt{2}$
$= 8\sqrt{2} - 35\sqrt{2}$
$= -27\sqrt{2}$

24. $\sqrt{2x+1} - \sqrt{x} = 1$
$\sqrt{2x+1} = 1 + \sqrt{x}$
$\left(\sqrt{2x+1}\right)^2 = \left(1 + \sqrt{x}\right)^2$
$2x + 1 = 1 + 2\sqrt{x} + x$
$x = 2\sqrt{x}$
$(x)^2 = \left(2\sqrt{x}\right)^2$
$x^2 = 4x$
$x^2 - 4x = 0$
$x(x-4) = 0$

$x = 0$ or $x = 4$

Check $x = 0$: $\sqrt{1} - \sqrt{0} = 1$ *True*

Check $x = 4$: $\sqrt{9} - \sqrt{4} = 1$ *True*

Solution set: $\{0, 4\}$

25. $(5+4i)(5-4i) = 5^2 - (4i)^2$
$= 25 - 16i^2$
$= 25 - 16(-1)$
$= 25 + 16 = 41$

26. $3x^2 = x + 1$
$3x^2 - x - 1 = 0$

Use $a = 3$, $b = -1$, and $c = -1$ in the quadratic formula.

$x = \dfrac{-b \pm \sqrt{b^2 - 4ac}}{2a}$

$x = \dfrac{-(-1) \pm \sqrt{(-1)^2 - 4(3)(-1)}}{2(3)}$

$= \dfrac{1 \pm \sqrt{1+12}}{6} = \dfrac{1 \pm \sqrt{13}}{6}$

Solution set: $\left\{\dfrac{1+\sqrt{13}}{6}, \dfrac{1-\sqrt{13}}{6}\right\}$

27. $x^2 + 2x - 8 > 0$
Solve the equation

$$x^2 + 2x - 8 = 0.$$
$$(x+4)(x-2) = 0$$

$$x + 4 = 0 \quad \text{or} \quad x - 2 = 0$$
$$x = -4 \quad \text{or} \quad x = 2$$

The numbers -4 and 2 divide a number line into three intervals.

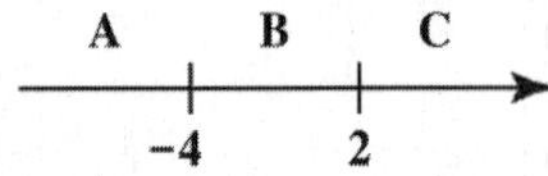

Test a number from each interval in the inequality

$$x^2 + 2x - 8 > 0.$$

Interval A: Let $x = -5$.
$$25 - 10 - 8 \overset{?}{>} 0$$
$$7 > 0 \quad \textit{True}$$
Interval B: Let $x = 0$.
$$-8 > 0 \quad \textit{False}$$
Interval C: Let $x = 3$.
$$9 + 6 - 8 \overset{?}{>} 0$$
$$7 > 0 \quad \textit{True}$$

The numbers in Intervals A and C, not including -4 or 2 because of $>$, are solutions.

Solution set: $(-\infty, -4) \cup (2, \infty)$

28. $x^4 - 5x^2 + 4 = 0$
Let $u = x^2$, so $u^2 = (x^2)^2 = x^4$.
$$u^2 - 5u + 4 = 0$$
$$(u-1)(u-4) = 0$$

$$u - 1 = 0 \quad \text{or} \quad u - 4 = 0$$
$$u = 1 \quad \text{or} \quad u = 4$$

To find x, substitute x^2 for u.

$$x^2 = 1 \quad \text{or} \quad x^2 = 4$$
$$x = \pm 1 \quad \text{or} \quad x = \pm 2$$

Solution set: $\{-2, -1, 1, 2\}$

29.
$$5^{x+3} = \left(\tfrac{1}{25}\right)^{3x+2}$$
$$5^{x+3} = [(\tfrac{1}{5})^2]^{3x+2}$$
$$5^{x+3} = (5^{-2})^{(3x+2)} \quad \textit{Write as powers of 5.}$$
$$5^{x+3} = 5^{-2(3x+2)}$$
$$x + 3 = -2(3x+2)$$
$$x + 3 = -6x - 4$$
$$7x = -7$$
$$x = -1$$

Check $x = -1$: $5^2 = (\frac{1}{25})^{-1}$ *True*
Solution set: $\{-1\}$

30.
$$\log_5 x + \log_5 (x+4) = 1$$
$$\log_5 [x(x+4)] = 1 \quad \textit{Product rule}$$
$$x(x+4) = 5^1 \quad \textit{Exponential form}$$
$$x^2 + 4x = 5$$
$$x^2 + 4x - 5 =$$
$$(x+5)(x-1) = 0$$

$$x + 5 = 0 \quad \text{or} \quad x - 1 = 0$$
$$x = -5 \quad \text{or} \quad x = 1$$

Check both proposed solutions. The number 1 makes the equation true. The number -5 must be rejected because it yields an equation in which the logarithm of a negative number must be found.

Check $x = 1$: $\log_5 1 + \log_5 5 = 0 + 1 = 1$
Solution set: $\{1\}$

31.
$$\log_5 125 = 3 \quad \textit{Logarithmic form}$$
$$5^3 = 125 \quad \textit{Exponential form}$$

32.
$$\log \frac{x^3\sqrt{y}}{z}$$
$$= \log \frac{x^3 y^{1/2}}{z}$$
$$= \log (x^3 y^{1/2}) - \log z \quad \textit{Quotient rule}$$
$$= \log x^3 + \log y^{1/2} - \log z \quad \textit{Product rule}$$
$$= 3 \log x + \tfrac{1}{2} \log y - \log z \quad \textit{Power rule}$$

33. $y = f(x) = \frac{1}{3}(x-1)^2 + 2$ is in $f(x) = a(x-h)^2 + k$ form. The graph is a vertical parabola with vertex (h, k) at $(1, 2)$. Since $a = \frac{1}{3} > 0$, the graph opens up. Also, $|a| = \left|\frac{1}{3}\right| = \frac{1}{3} < 1$, so the graph is wider than the graph of $f(x) = x^2$. The points $(0, 2\frac{1}{3})$, $(-2, 5)$, and $(4, 5)$ are also on the graph.

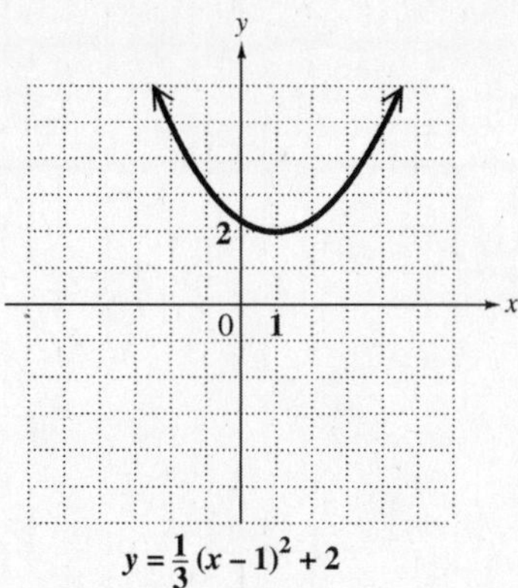

34. $f(x) = 2^x$
Make a table of values.

x	-2	-1	0	1	2
$f(x)$	$\frac{1}{4}$	$\frac{1}{2}$	1	2	4

Plot the ordered pairs from the table, and draw a smooth exponential curve through the points.

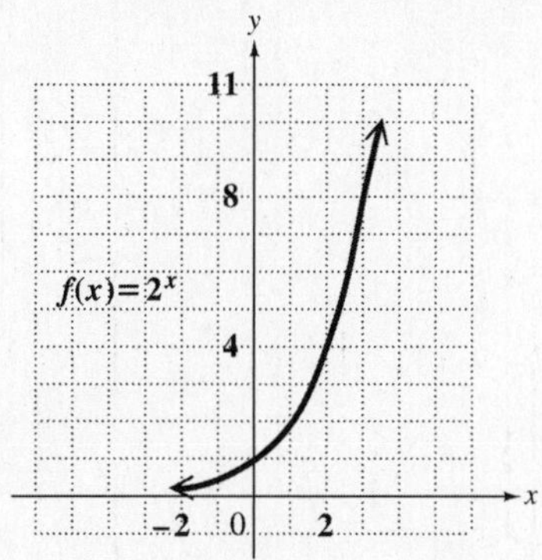

35. $f(x) = \log_3 x$
Make a table of values.

Powers of 3	3^{-2}	3^{-1}	3^0	3^1	3^2
x	$\frac{1}{9}$	$\frac{1}{3}$	1	3	9
y	-2	-1	0	1	2

Plot the ordered pairs and draw a smooth logarithmic curve through the points.

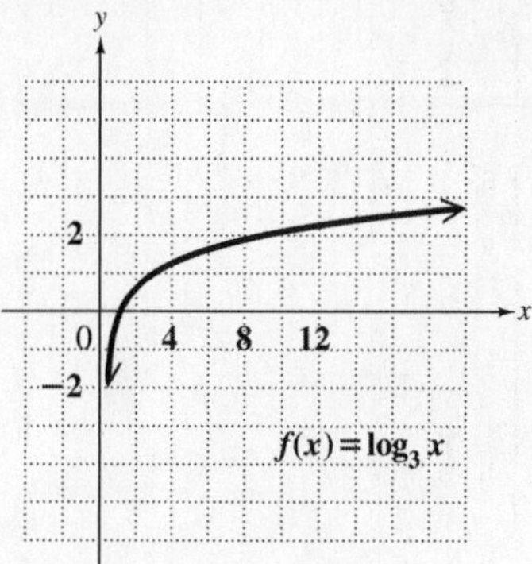

CHAPTER 12 NONLINEAR FUNCTIONS, CONIC SECTIONS, AND NONLINEAR SYSTEMS

12.1 Additional Graphs of Functions; Composition

12.1 Margin Exercises

1. $f(x) = \sqrt{x+4}$

The graph of this function is found by shifting the graph of the square root function, $f(x) = \sqrt{x}$, 4 units to the left.

Since x must be at least -4 for f to be defined, the domain of this function is $[-4, \infty)$. Since y is nonnegative, the range of this function is $[0, \infty)$.

x	y
-4	0
-3	1
0	2

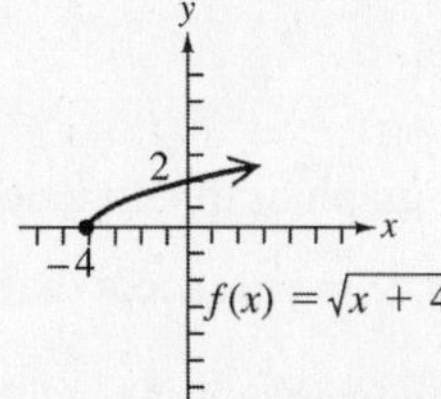

2. $f(x) = \dfrac{1}{x} - 2$

This is the graph of the reciprocal function, $f(x) = \frac{1}{x}$, shifted 2 units down. Since $x \neq 0$ (or a denominator of 0 results), the line $x = 0$ is a vertical asymptote and the domain of this function is $(-\infty, 0) \cup (0, \infty)$.

Since $\frac{1}{x} \neq 0$, the line $y = -2$ is a horizontal asymptote, and the range of this function is $(-\infty, -2) \cup (-2, \infty)$.

x	$\frac{1}{x}$	y
-2	$-\frac{1}{2}$	$-\frac{5}{2}$
-1	-1	-3
$-\frac{1}{2}$	-2	-4
$\frac{1}{2}$	2	0
1	1	-1
2	$\frac{1}{2}$	$-\frac{3}{2}$

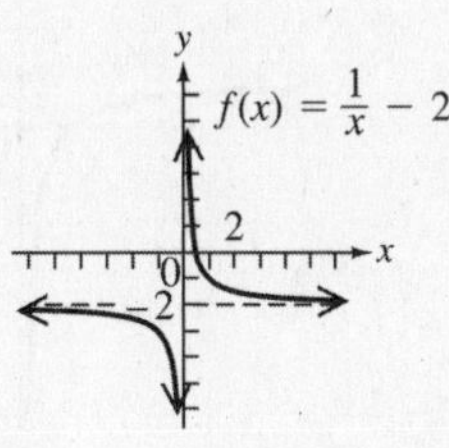

3. $f(x) = |x+2| + 1$

This is the graph of the absolute value function, $f(x) = |x|$, shifted 2 units to the left (since $x + 2 = 0$ if $x = -2$) and 1 unit upward (because of the $+1$).

The domain of this function is $(-\infty, \infty)$ and its range is $[1, \infty)$.

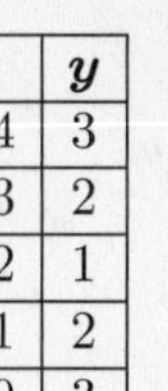

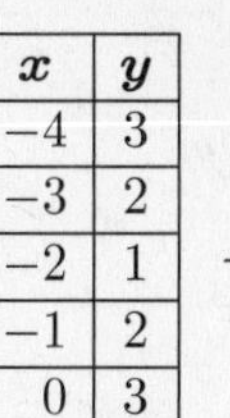

x	y
-4	3
-3	2
-2	1
-1	2
0	3

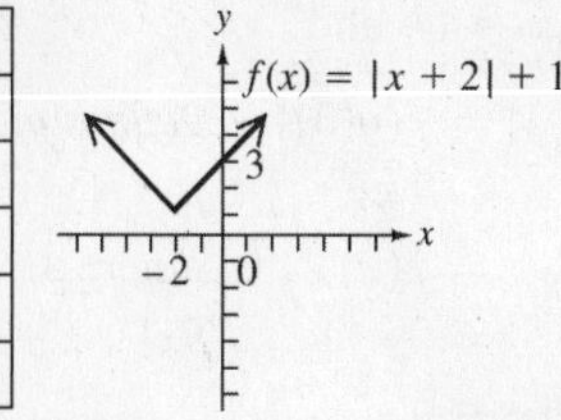

4. In each case, the answer is the largest integer that is less than or equal to the number in the greatest integer symbol.

(a) $[\![18]\!] = 18 \quad (18 \leq 18)$

(b) $[\![8.7]\!] = 8 \quad (8 \leq 8.7)$

(c) $[\![-5]\!] = -5 \quad (-5 \leq -5)$

(d) $[\![-6.9]\!] = -7 \quad (-7 \leq -6.9)$

(e) $[\![\frac{1}{2}]\!] = 0 \quad (0 \leq \frac{1}{2})$

5. $f(x) = [\![x+1]\!]$

The graph of f is the same as the graph of $y = [\![x]\!]$ shifted one unit to the left. Shifting the graph of $y = [\![x]\!]$ (right, left, up, down) does not change the domain and range. The domain of f is $(-\infty, \infty)$ and the range of f is the set of integers, that is, $\{\ldots, -2, -1, 0, 1, 2, \ldots\}$.

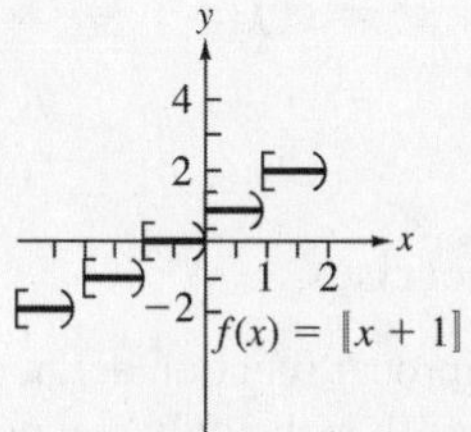

6. This function is similar to the greatest integer function, but in this case, we use the integer that is *greater than* or equal to the number. For example, for a $\frac{1}{2}$ oz letter, you would be charged for 1 oz.

Interval	Ounces Charged for	Cost
$(0, 1]$	1	\$0.80
$(1, 2]$	2	\$1.60
$(2, 3]$	3	\$2.40
$(3, 4]$	4	\$3.20

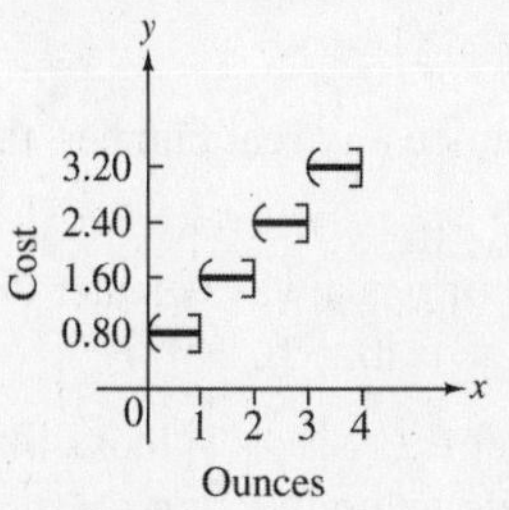

7. $f(x) = x - 4$ and $g(x) = x^2$

$$\begin{aligned}(f \circ g)(3) &= f(g(3)) && \textit{Definition} \\ &= f(3^2) && g(x) = x^2 \\ &= f(9) && \textit{Square 3.} \\ &= 9 - 4 && f(x) = x - 4 \\ &= 5\end{aligned}$$

8. $f(x) = 3x + 6, g(x) = x^3$

(a) $$\begin{aligned}(f \circ g)(2) &= f(g(2)) \\ &= f(2^3) && g(x) = x^3 \\ &= f(8) \\ &= 3(8) + 6 && f(x) = 3x + 6 \\ &= 30\end{aligned}$$

(b) $$\begin{aligned}(g \circ f)(2) &= g(f(2)) \\ &= g(3(2) + 6) && f(x) = 3x + 6 \\ &= g(12) \\ &= 12^3 && g(x) = x^3 \\ &= 1728\end{aligned}$$

(c) $$\begin{aligned}(f \circ g)(x) &= f(g(x)) \\ &= 3(g(x)) + 6 && f(x) = 3x + 6 \\ &= 3(x^3) + 6 && g(x) = x^3 \\ &= 3x^3 + 6\end{aligned}$$

(d) $$\begin{aligned}(g \circ f)(x) &= g(f(x)) \\ &= [f(x)]^3 && g(x) = x^3 \\ &= (3x + 6)^3 && f(x) = 3x + 6\end{aligned}$$

12.1 Section Exercises

1. For the reciprocal function defined by $f(x) = 1/x$, $\underline{0}$ is the only real number not in the domain since division by 0 is undefined.

3. The lowest point on the graph of $f(x) = |x|$ has coordinates $(\underline{0}, \underline{0})$.

5. $f(x) = |x - 2| + 2$

The graph of this function has its "vertex" at $(2, 2)$, so the correct graph is **B**.

7. $f(x) = |x - 2| - 2$

The graph of this function has its "vertex" at $(2, -2)$, so the correct graph is **A**.

9. $f(x) = |x + 1|$

Since x can be any real number, the domain is $(-\infty, \infty)$.

The value of y is always greater than or equal to 0, so the range is $[0, \infty)$.

The graph of $y = |x + 1|$ looks like the graph of the absolute value function $y = |x|$, but the graph is translated 1 unit to the left. The x-value of its "vertex" is obtained by setting $x + 1 = 0$ and solving for x:

$$\begin{aligned}x + 1 &= 0 \\ x &= -1.\end{aligned}$$

Since the corresponding y-value is 0, the "vertex" is $(-1, 0)$. The axis of symmetry is the vertical line $x = -1$. Some additional points are $(-3, 2)$, $(-2, 1)$, $(0, 1)$, and $(1, 2)$.

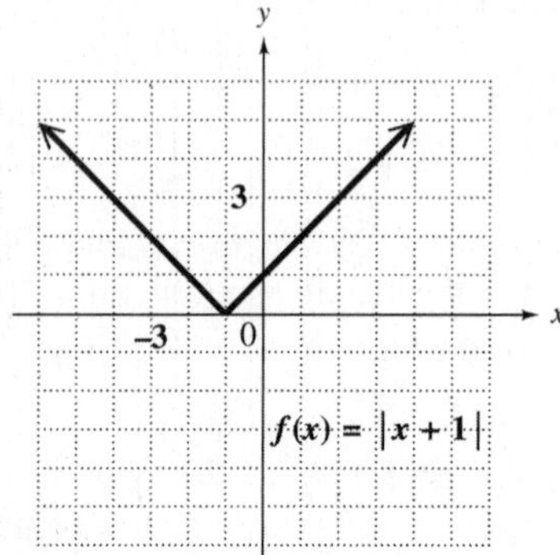

11. $f(x) = \dfrac{1}{x} + 1$

The graph of this function is similar to the graph of $g(x) = \dfrac{1}{x}$, except that each point is translated 1 unit upward. Just as with $g(x) = \dfrac{1}{x}$, $x = 0$ is the vertical asymptote, but this graph has $y = 1$ as its horizontal asymptote.

The domain is all real numbers except 0, that is, $(-\infty, 0) \cup (0, \infty)$.

The range is all real numbers except 1, that is, $(-\infty, 1) \cup (1, \infty)$.

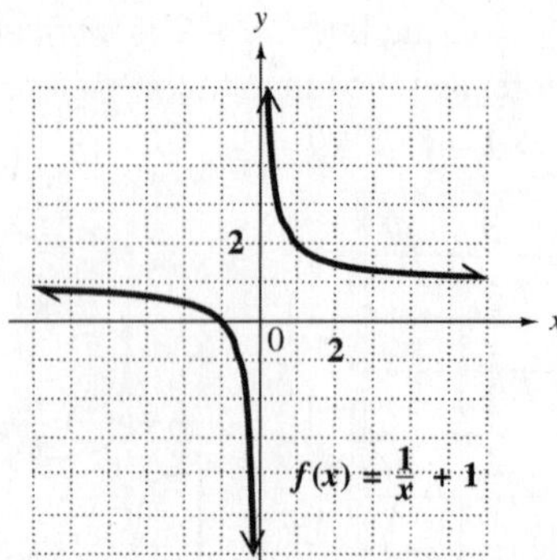

13. $f(x) = \sqrt{x - 2}$

The graph is found by shifting the graph of $g(x) = \sqrt{x}$ two units to the right. The following table of ordered pairs gives some specific points the graph passes through.

x	2	3	6
y	0	1	2

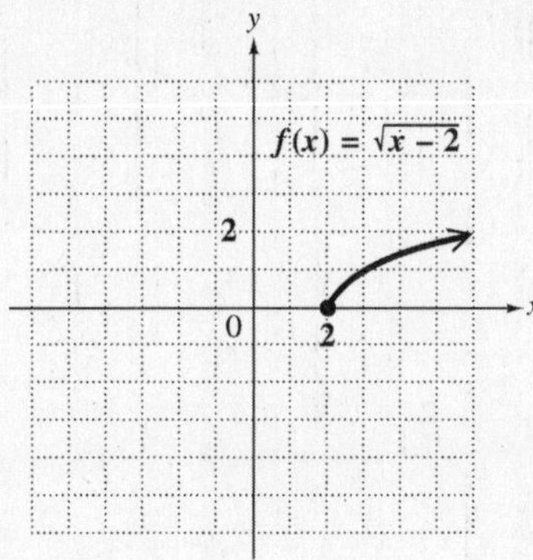

The domain of the function is $[2, \infty)$ and its range is $[0, \infty)$.

15. $f(x) = \dfrac{1}{x-2}$

This is the graph of the reciprocal function, $g(x) = \dfrac{1}{x}$, shifted 2 units to the right. Since $x \neq 2$ (if $x = 2$, the denominator equals 0), the line $x = 2$ is a vertical asymptote. Since $\dfrac{1}{x-2} \neq 0$, the line $y = 0$ is a horizontal asymptote.

x	$\frac{5}{2}$	3	4	5
y	2	1	$\frac{1}{2}$	$\frac{1}{3}$

x	-1	0	1	$\frac{3}{2}$
y	$-\frac{1}{3}$	$-\frac{1}{2}$	-1	-2

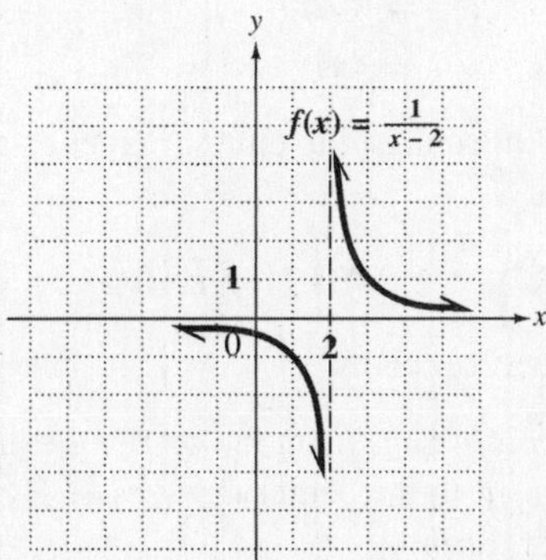

The domain of the function is $(-\infty, 2) \cup (2, \infty)$ and its range is $(-\infty, 0) \cup (0, \infty)$.

17. $f(x) = \sqrt{x+3} - 3$

The graph is found by shifting the graph of $g(x) = \sqrt{x}$ three units to the left and three units down. The following table of ordered pairs gives some specific points the graph passes through.

x	-3	-2	1
y	-3	-2	-1

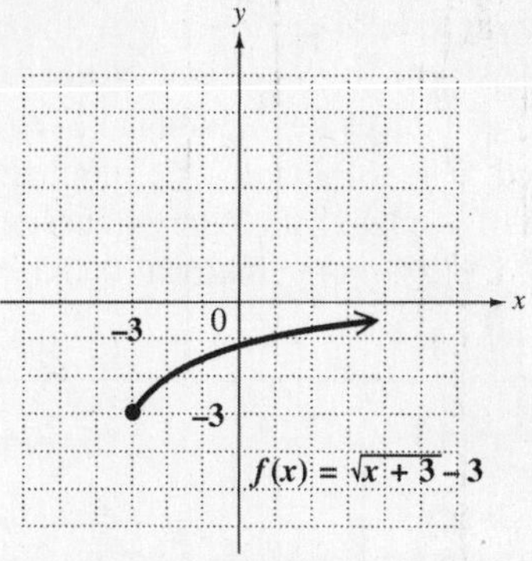

The domain of the function is $[-3, \infty)$ and its range is $[-3, \infty)$.

19. $f(x) = [\![-x]\!]$

In general, if $f(x) = [\![-x]\!]$, then

$$\begin{array}{llllll}
\text{for} & -3 & < & x & \leq & -2, \quad f(x) = 2, \\
\text{for} & -2 & < & x & \leq & -1, \quad f(x) = 1, \\
\text{for} & -1 & < & x & \leq & 0, \quad f(x) = 0, \\
\text{for} & 0 & < & x & \leq & 1, \quad f(x) = -1, \\
\text{for} & 1 & < & x & \leq & 2, \quad f(x) = -2,
\end{array}$$

and so on.

This is the graph of the greatest integer function, $g(x) = [\![x]\!]$, "flipped" about the y-axis.

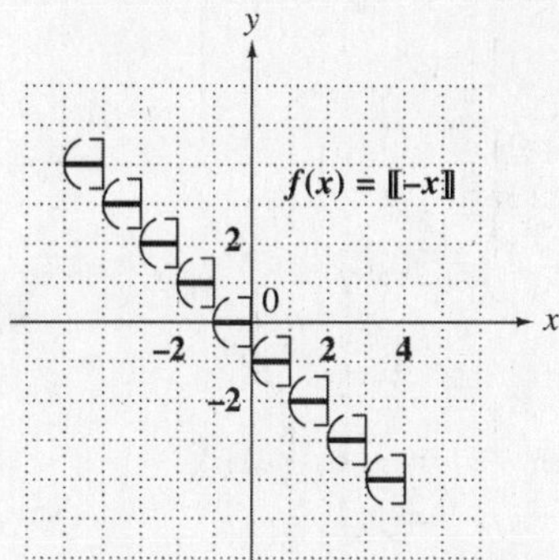

21. $f(x) = [\![x - 3]\!]$

This is the graph of the greatest integer function, $g(x) = [\![x]\!]$, shifted 3 units to the right.

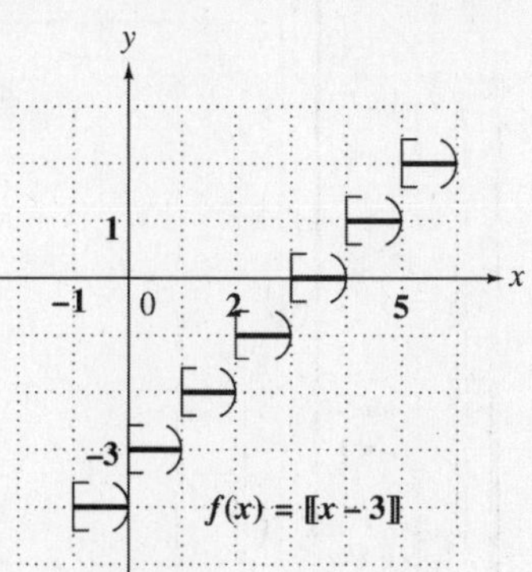

23. For any portion of the first ounce, the cost will be one 42¢ stamp. If the weight exceeds one ounce (up to two ounces), an additional 17¢ stamp is required. The following table summarizes the weight of a letter, x, and the number of stamps required, $p(x)$, on the interval $(0, 5]$.

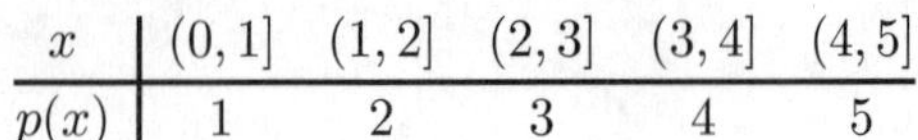

x	$(0,1]$	$(1,2]$	$(2,3]$	$(3,4]$	$(4,5]$
$p(x)$	1	2	3	4	5

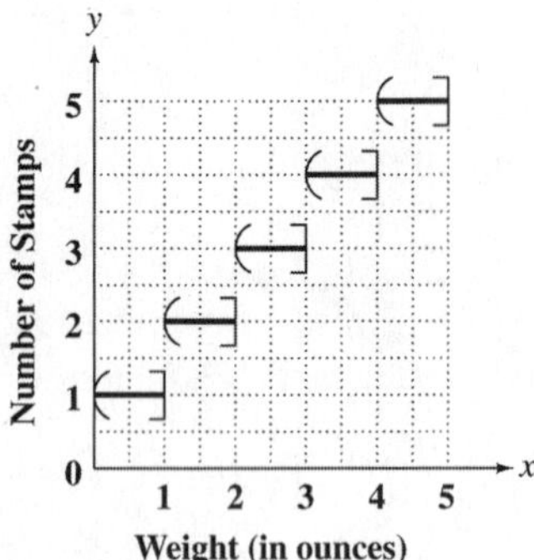

For Exercises 25–40,

$f(x) = x^2 + 4$, $g(x) = 2x + 3$, and $h(x) = x + 5$.

25. $$\begin{aligned}(h \circ g)(4) &= h(g(4)) \\ &= h(2 \cdot 4 + 3) \\ &= h(11) \\ &= 11 + 5 \\ &= 16\end{aligned}$$

27. $$\begin{aligned}(g \circ f)(6) &= g(f(6)) \\ &= g(6^2 + 4) \\ &= g(40) \\ &= 2 \cdot 40 + 3 \\ &= 83\end{aligned}$$

29. $$\begin{aligned}(f \circ h)(-2) &= f(h(-2)) \\ &= f(-2 + 5) \\ &= f(3) \\ &= 3^2 + 4 \\ &= 13\end{aligned}$$

31. $$\begin{aligned}(f \circ g)(x) &= f(g(x)) \\ &= f(2x + 3) \\ &= (2x + 3)^2 + 4 \\ &= 4x^2 + 12x + 9 + 4 \\ &= 4x^2 + 12x + 13\end{aligned}$$

33. $$\begin{aligned}(f \circ h)(x) &= f(h(x)) \\ &= f(x + 5) \\ &= (x + 5)^2 + 4 \\ &= x^2 + 10x + 25 + 4 \\ &= x^2 + 10x + 29\end{aligned}$$

35. $$\begin{aligned}(h \circ g)(x) &= h(g(x)) \\ &= h(2x + 3) \\ &= 2x + 3 + 5 \\ &= 2x + 8\end{aligned}$$

37. $$\begin{aligned}(f \circ h)(\tfrac{1}{2}) &= f(h(\tfrac{1}{2})) \\ &= f(\tfrac{1}{2} + 5) \\ &= f(\tfrac{11}{2}) \\ &= (\tfrac{11}{2})^2 + 4 \\ &= \tfrac{121}{4} + \tfrac{16}{4} = \tfrac{137}{4}\end{aligned}$$

39. $$\begin{aligned}(f \circ g)(-\tfrac{1}{2}) &= f(g(-\tfrac{1}{2})) \\ &= f(2(-\tfrac{1}{2}) + 3) \\ &= f(2) \\ &= 2^2 + 4 \\ &= 4 + 4 = 8\end{aligned}$$

41. $f(x) = 12x$, $g(x) = 5280x$

$$\begin{aligned}(f \circ g)(x) &= f(g(x)) \\ &= f(5280x) \\ &= 12(5280x) \\ &= 63{,}360x\end{aligned}$$

$(f \circ g)(x)$ computes the number of inches in x mi.

43. $r(t) = 2t$, $A(r) = \pi r^2$

$$\begin{aligned}(A \circ r)(t) &= A[r(t)] \\ &= A(2t) \\ &= \pi(2t)^2 \\ &= 4\pi t^2\end{aligned}$$

This is the area of the circular layer as a function of time.

12.2 The Circle and the Ellipse

12.2 Margin Exercises

1. If the point (x, y) is on the circle, the distance from (x, y) to the center $(0, 0)$ is 4. By the distance formula,

$$\sqrt{(x_2 - x_1)^2 + (y_2 - y_1)^2} = d \quad \textit{Distance formula}$$

$$\sqrt{(x - 0)^2 + (y - 0)^2} = 4$$

$$\sqrt{x^2 + y^2} = 4$$

$$x^2 + y^2 = 16. \quad \textit{Square both sides.}$$

An equation of this circle is $x^2 + y^2 = 16$.

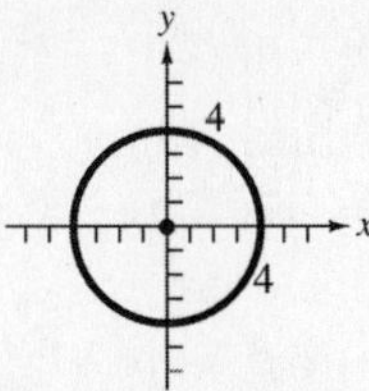

2. **(a)** Center at $(3, -2)$; radius 4

$$\sqrt{(x_2 - x_1)^2 + (y_2 - y_1)^2} = d$$
$$\sqrt{(x - 3)^2 + [y - (-2)]^2} = 4$$
$$\sqrt{(x - 3)^2 + (y + 2)^2} = 4$$
$$(x - 3)^2 + (y + 2)^2 = 16 \quad \textit{Square.}$$

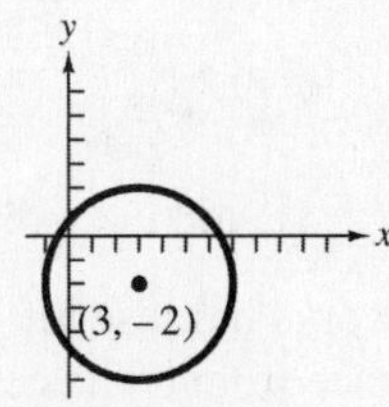

(b) $(x - 5)^2 + (y + 2)^2 = 9$

The equation of the circle is an equation of the form

$$(x - h)^2 + (y - k)^2 = r^2$$
$$(x - 5)^2 + [y - (-2)]^2 = 3^2.$$

The center (h, k) is $(5, -2)$. The radius r is 3.

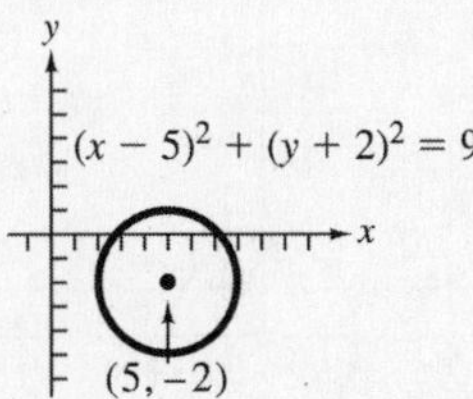

3. To find the center and radius, add the appropriate constants to complete the squares on x and y.

$$x^2 + y^2 - 6x + 8y - 11 = 0$$
$$(x^2 - 6x) + (y^2 + 8y) = 11$$
$$(x^2 - 6x + 9) + (y^2 + 8y + 16) = 11 + 9 + 16$$
$$(x - 3)^2 + (y + 4)^2 = 36$$
$$(x - 3)^2 + [y - (-4)]^2 = 6^2$$

The circle has center at $(3, -4)$ and radius 6.

4. **(a)** $\dfrac{x^2}{4} + \dfrac{y^2}{25} = 1$

This ellipse has an equation of the form

$$\frac{x^2}{a^2} + \frac{y^2}{b^2} = 1.$$

Here, $a^2 = 4$, so $a = 2$ and the x-intercepts are $(2, 0)$ and $(-2, 0)$. Similarly, $b^2 = 25$, so $b = 5$ and the y-intercepts are $(0, 5)$ and $(0, -5)$. Plot the intercepts, and draw the ellipse through them.

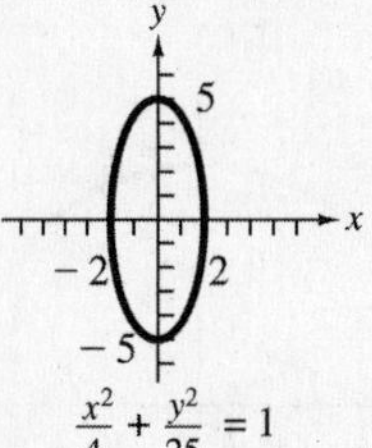

(b) $\dfrac{x^2}{64} + \dfrac{y^2}{49} = 1$ or $\dfrac{x^2}{8^2} + \dfrac{y^2}{7^2} = 1$

The x-intercepts for this ellipse are $(8, 0)$ and $(-8, 0)$. The y-intercepts are $(0, 7)$ and $(0, -7)$. Plot the intercepts, and draw the ellipse through them.

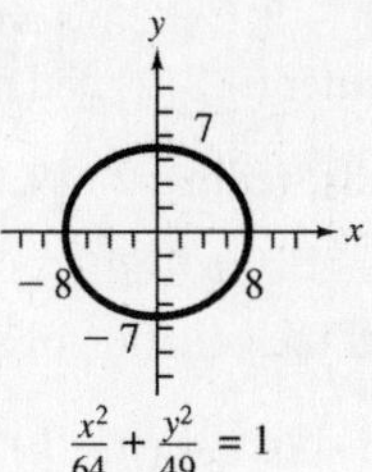

5. $$\frac{(x + 4)^2}{16} + \frac{(y - 1)^2}{36} = 1$$
$$\frac{[x - (-4)]^2}{4^2} + \frac{(y - 1)^2}{6^2} = 1$$

The center is $(-4, 1)$.

The ellipse passes through the four points:

$$(-4 - 4, 1) = (-8, 1)$$
$$(-4 + 4, 1) = (0, 1)$$
$$(-4, 1 + 6) = (-4, 7)$$
$$(-4, 1 - 6) = (-4, -5)$$

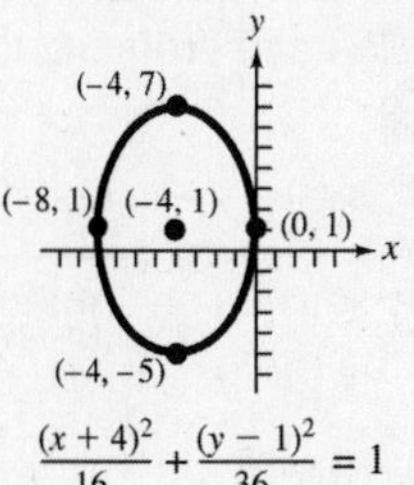

12.2 Section Exercises

1. **(a)** $x^2 + y^2 = 25$ can be written in the center-radius form as

$$(x - 0)^2 + (y - 0)^2 = 5^2.$$

The center is the point $(0, 0)$.

(b) The radius is 5.

(c) The x-intercepts are $(5, 0)$ and $(-5, 0)$.
The y-intercepts are $(0, 5)$ and $(0, -5)$.

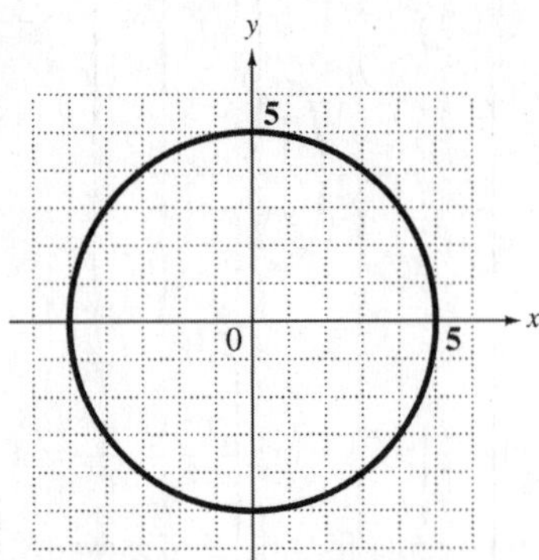

3. $(x-3)^2 + (y-2)^2 = 25$ is an equation of a circle with center $(3, 2)$ and radius 5, choice **B**.

5. $(x+3)^2 + (y-2)^2 = 25$ is an equation of a circle with center $(-3, 2)$ and radius 5, choice **D**.

7. Center: $(-4, 3)$; radius: 2

Substitute $h = -4$, $k = 3$, and $r = 2$ in the center-radius form of the equation of a circle.

$$(x-h)^2 + (y-k)^2 = r^2$$
$$[x-(-4)]^2 + (y-3)^2 = 2^2$$
$$(x+4)^2 + (y-3)^2 = 4$$

9. Center: $(-8, -5)$; radius: $\sqrt{5}$

Substitute $h = -8$, $k = -5$, and $r = \sqrt{5}$ in the center-radius form of the equation of a circle.

$$(x-h)^2 + (y-k)^2 = r^2$$
$$[x-(-8)]^2 + [y-(-5)]^2 = (\sqrt{5})^2$$
$$(x+8)^2 + (y+5)^2 = 5$$

11. $x^2 + y^2 + 4x + 6y + 9 = 0$

Rewrite the equation keeping only the variable terms on the left and grouping the x-terms and y-terms.

$$x^2 + 4x + y^2 + 6y = -9$$

Complete the squares on x and y and add the same constants to the right.

$$(x^2 + 4x + \underline{4}) + (y^2 + 6y + \underline{9}) = -9 + \underline{4} + \underline{9}$$
$$(x+2)^2 + (y+3)^2 = 4$$

From the form $(x-h)^2 + (y-k)^2 = r^2$,
we have $h = -2$, $k = -3$, and $r = 2$.
The center is $(-2, -3)$, and the radius r is 2.

13. $x^2 + y^2 + 10x - 14y - 7 = 0$

$$(x^2 + 10x \quad\quad) + (y^2 - 14y \quad\quad) = 7$$
$$(x^2 + 10x + \underline{25}) + (y^2 - 14y + \underline{49})$$
$$= 7 + \underline{25} + \underline{49}$$
$$(x+5)^2 + (y-7)^2 = 81$$

The center is $(-5, 7)$, and the radius is $\sqrt{81} = 9$.

15.
$$3x^2 + 3y^2 - 12x - 24y + 12 = 0$$
$$3(x^2 - 4x \quad\quad) + 3(y^2 - 8y \quad\quad) = -12$$

Divide by 3.

$$(x^2 - 4x \quad\quad) + (y^2 - 8y \quad\quad) = -4$$
$$(x^2 - 4x + \underline{4}) + (y^2 - 8y + \underline{16})$$
$$= -4 + \underline{4} + \underline{16}$$
$$(x-2)^2 + (y-4)^2 = 16$$

The center is $(2, 4)$, and the radius is $\sqrt{16} = 4$.

17. This method works because the pencil is always the same distance from the fastened end. The fastened end works as the center, and the length of the string from the fastened end to the pencil is the radius.

19.
$$x^2 + y^2 = 4$$
$$(x-0)^2 + (y-0)^2 = 2^2$$

The center of the circle is $(0, 0)$ and $r = 2$.

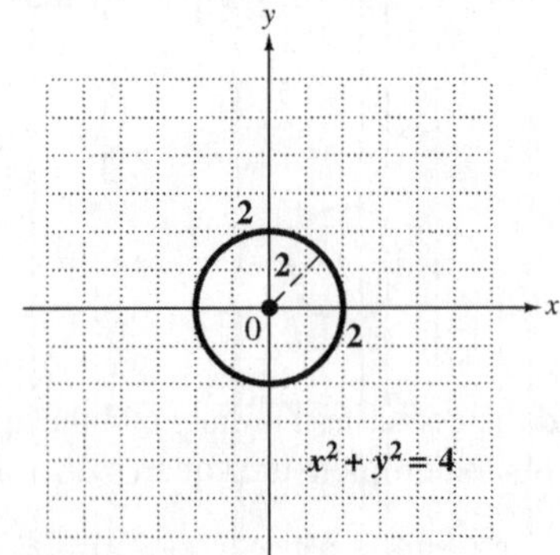

21.
$$3x^2 = 48 - 3y^2$$
$$3x^2 + 3y^2 = 48$$
$$x^2 + y^2 = 16 = 4^2 \quad \textit{Divide by 3.}$$

This is an equation of a circle with center $(0, 0)$ and radius 4.

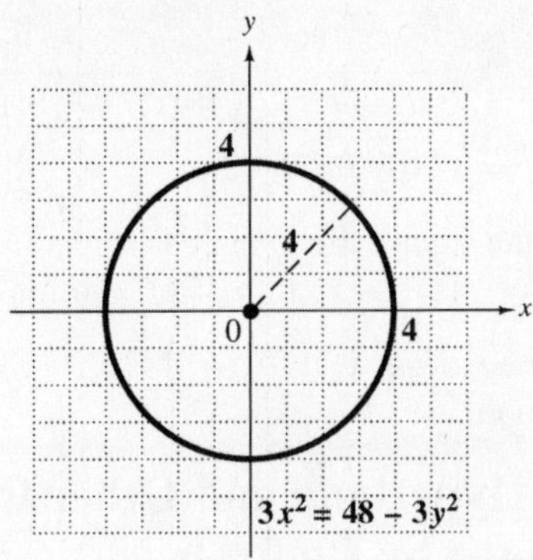

23. $(x - 1)^2 + (y + 3)^2 = 16$

Here, $h = 1$, $k = -3$, and $r = \sqrt{16} = 4$. The graph is a circle with center $(1, -3)$ and radius 4.

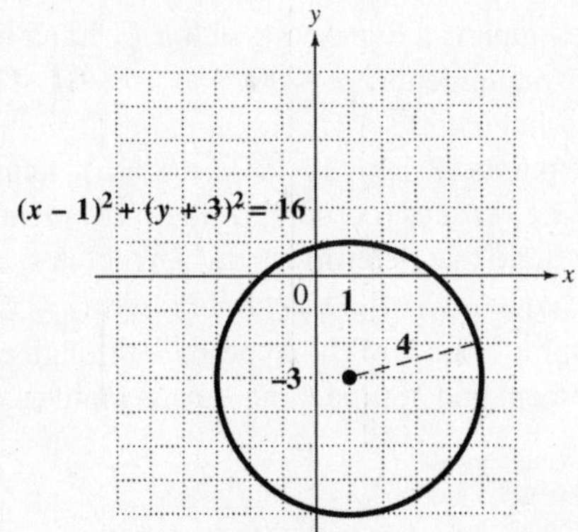

25.
$$\begin{aligned} x^2 + y^2 + 8x + 2y - 8 &= 0 \\ (x^2 + 8x \qquad) + (y^2 + 2y \qquad) &= 8 \\ (x^2 + 8x + \underline{16}) + (y^2 + 2y + \underline{1}) & \\ &= 8 + \underline{16} + \underline{1} \\ (x + 4)^2 + (y + 1)^2 &= 25 \end{aligned}$$

Here, $h = -4$, $k = -1$, and $r = \sqrt{25} = 5$. The graph is a circle with center $(-4, -1)$ and radius 5.

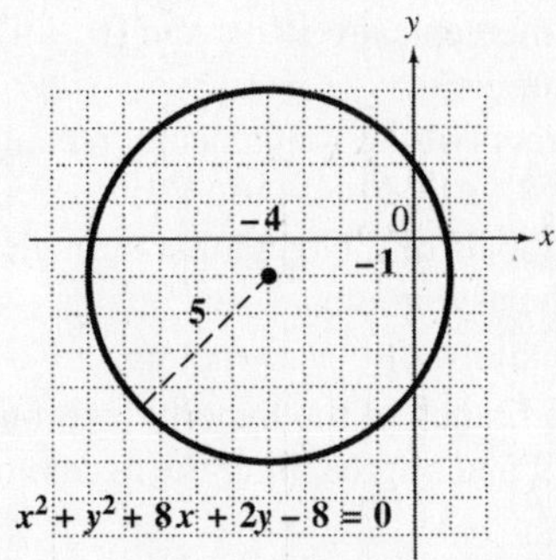

27. As the pencil traces the graph along the taut string, notice that the distances from the pencil to the two fastened ends continue to change, but the sum of the two distances is always the same, namely, the length of the string between the two fastened ends. This is exactly the definition of an ellipse—the set of all points in a plane the sum of whose distances from two fixed points (the fastened ends) is constant.

29. The equation $\dfrac{x^2}{9} + \dfrac{y^2}{25} = 1$ is in the form $\dfrac{x^2}{a^2} + \dfrac{y^2}{b^2} = 1$. The graph is an ellipse with $a^2 = 9$ and $b^2 = 25$, so $a = 3$ and $b = 5$. The x-intercepts are $(3, 0)$ and $(-3, 0)$. The y-intercepts are $(0, 5)$ and $(0, -5)$. Plot the intercepts, and draw the ellipse through them.

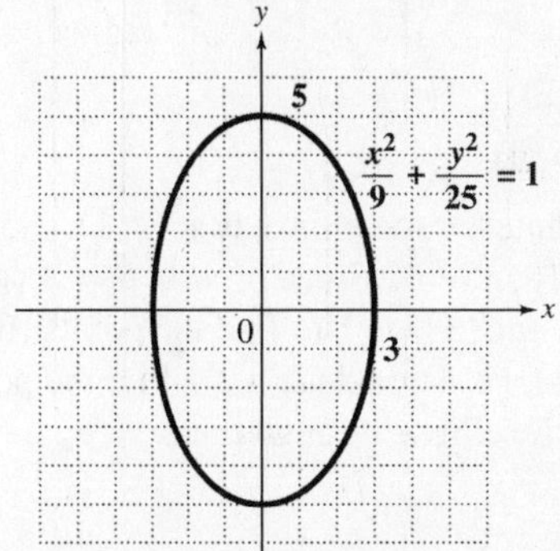

31. $\dfrac{x^2}{36} + \dfrac{y^2}{16} = 1$ is in the form $\dfrac{x^2}{a^2} + \dfrac{y^2}{b^2} = 1$. The graph is an ellipse with $a^2 = 36$ and $b^2 = 16$, so $a = 6$ and $b = 4$. The x-intercepts are $(6, 0)$ and $(-6, 0)$. The y-intercepts are $(0, 4)$ and $(0, -4)$. Plot the intercepts, and draw the ellipse through them.

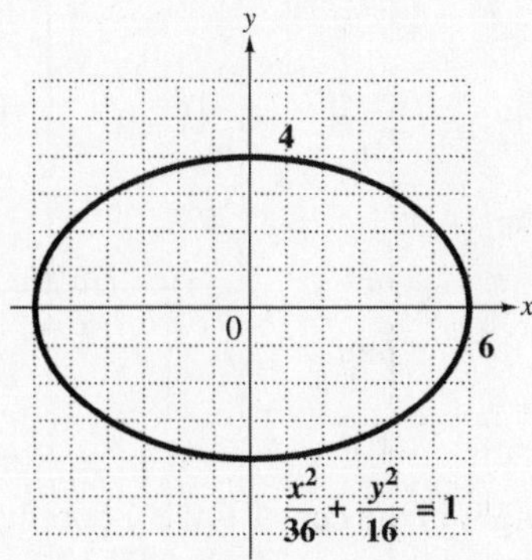

33. $\frac{x^2}{49}+\frac{y^2}{25}=1$ is in the form $\frac{x^2}{a^2}+\frac{y^2}{b^2}=1$. The graph is an ellipse with $a^2=49$ and $b^2=25$, so $a=7$ and $b=5$. The x-intercepts are $(7,0)$ and $(-7,0)$. The y-intercepts are $(0,5)$ and $(0,-5)$. Plot the intercepts, and draw the ellipse through them.

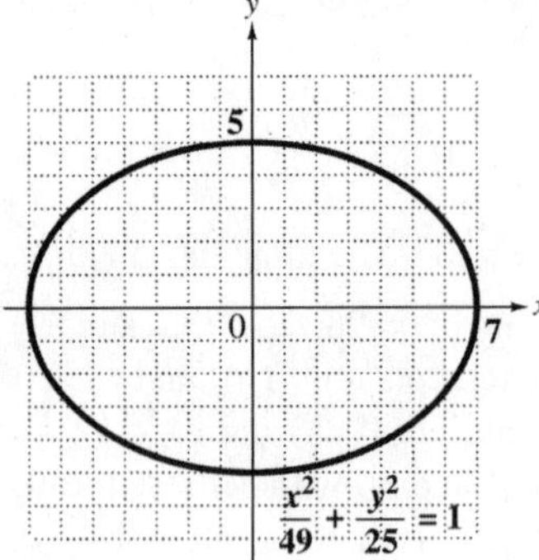

35. $\frac{(x-2)^2}{16}+\frac{(y-1)^2}{9}=1$

The center of the ellipse is at $(2,1)$. Since $a^2=16$, $a=4$. Since $b^2=9$, $b=3$. Add ± 4 to 2, and add ± 3 to 1 to find the points $(6,1)$, $(-2,1)$, $(2,4)$, and $(2,-2)$. Plot the points, and draw the ellipse through them.

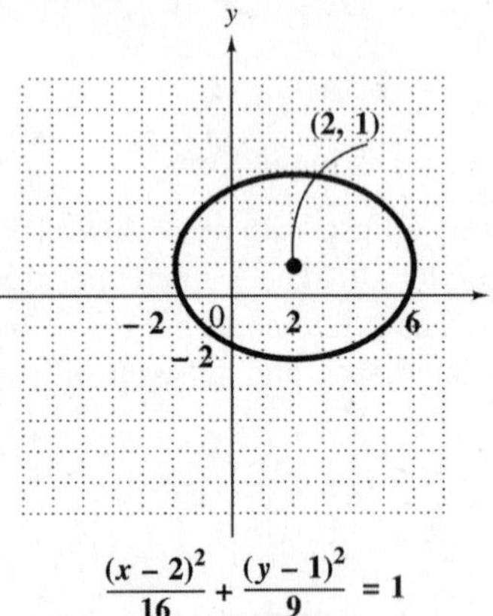

37. **(a)** $100x^2+324y^2=32{,}400$

$$\frac{x^2}{324}+\frac{y^2}{100}=1 \qquad \textit{Divide by 32,400.}$$

$$\frac{x^2}{18^2}+\frac{y^2}{10^2}=1$$

The height in the center is the y-coordinate of the positive y-intercept. The height is 10 meters.

(b) The width of the ellipse is the distance between the x-intercepts, $(-18,0)$ and $(18,0)$. The width across the bottom of the arch is $18+18=36$ meters.

39. $\frac{x^2}{141.7^2}+\frac{y^2}{141.1^2}=1$

(a) $c^2=a^2-b^2$, so

$$c=\sqrt{a^2-b^2}=\sqrt{141.7^2-141.1^2}$$
$$=\sqrt{169.68}\approx 13.0$$

From the figure, the *apogee* is $a+c=141.7+13.0=154.7$ million miles.

(b) The *perigee* is $a-c=141.7-13.0=128.7$ million miles.

12.3 The Hyperbola and Other Functions Defined by Radicals

12.3 Margin Exercises

1. $\frac{x^2}{4}-\frac{y^2}{25}=1$ or $\frac{x^2}{2^2}-\frac{y^2}{5^2}=1$

The graph is a hyperbola with $a=2$ and $b=5$. The x-intercepts are $(2,0)$ and $(-2,0)$. There are no y-intercepts.
The points $(2,5)$, $(2,-5)$, $(-2,-5)$, and $(-2,5)$ are the vertices (corners) of the fundamental rectangle that determines the asymptotes.
The equations of the asymptotes are $y=\pm\frac{5}{2}x$.
Graph a branch of the hyperbola through each intercept and approaching the asymptotes.

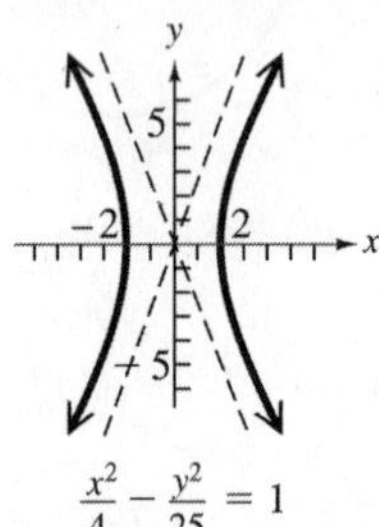

2. $\frac{y^2}{81}-\frac{x^2}{64}=1$ or $\frac{y^2}{9^2}-\frac{x^2}{8^2}=1$

The graph is a hyperbola with $a=8$ and $b=9$. The y-intercepts are $(0,9)$ and $(0,-9)$. There are no x-intercepts.
The corners of the fundamental rectangle are $(8,9)$, $(8,-9)$, $(-8,-9)$, and $(-8,9)$. Extend the diagonals of the rectangle through these points to get the asymptotes.
The equations of the asymptotes are $y=\pm\frac{9}{8}x$.
Graph a branch of the hyperbola through each intercept and approaching the asymptotes.

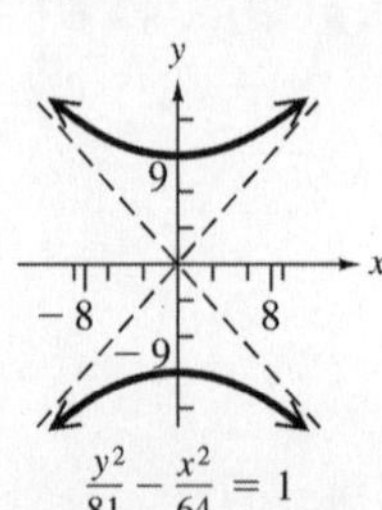

3. (a)

$$3x^2 = 27 - 4y^2$$
$$3x^2 + 4y^2 = 27$$
$$\frac{x^2}{9} + \frac{4y^2}{27} = 1 \quad \textit{Divide by 27; reduce.}$$

Since the x^2- and y^2-terms have different positive coefficients, the graph of the equation is an ellipse.

(b)

$$6x^2 = 100 + 2y^2$$
$$6x^2 - 2y^2 = 100$$
$$\frac{x^2}{\frac{50}{3}} - \frac{y^2}{50} = 1 \quad \textit{Divide by 100; reduce.}$$

Because of the minus sign and since both variables are squared, the graph of the equation is a hyperbola.

(c) $3x^2 = 27 - 4y$

$$4y = -3x^2 + 27$$

This is an equation of a vertical parabola since only one variable, x, is squared.

(d)

$$3x^2 = 27 - 3y^2$$
$$3x^2 + 3y^2 = 27$$
$$x^2 + y^2 = 9$$

This is an equation of the form $(x-h)^2 + (y-k)^2 = r^2$, where $(h,k) = (0,0)$ and $r = 3$, so the graph of the equation is a circle.

4.

$$f(x) = \sqrt{36 - x^2}$$
$$y = \sqrt{36 - x^2} \quad f(x) = y$$
$$y^2 = 36 - x^2 \quad \textit{Square both sides.}$$
$$x^2 + y^2 = 36 \quad \textit{Add } x^2.$$

This equation is the graph of a circle with center $(0,0)$ and radius 6. Since $f(x)$ represents a principal square root in the original equation, $f(x)$ is nonnegative and its graph is the upper half of the circle.

The domain of the function is $[-6, 6]$ and its range is $[0, 6]$.

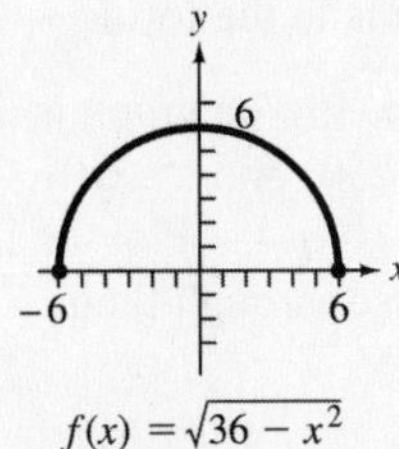

$f(x) = \sqrt{36 - x^2}$

5.

$$\frac{y}{3} = -\sqrt{1 - \frac{x^2}{4}}$$
$$\frac{y^2}{9} = 1 - \frac{x^2}{4} \quad \textit{Square both sides.}$$
$$\frac{x^2}{4} + \frac{y^2}{9} = 1 \quad \textit{Add } \frac{x^2}{4}.$$

This is the equation of an ellipse with intercepts $(2,0)$, $(-2,0)$, $(0,3)$, and $(0,-3)$. Since $\frac{y}{3}$ equals a negative square root in the original equation, y must be nonpositive, restricting the graph to the lower half of the ellipse.

The domain of the function is $[-2, 2]$ and its range is $[-3, 0]$.

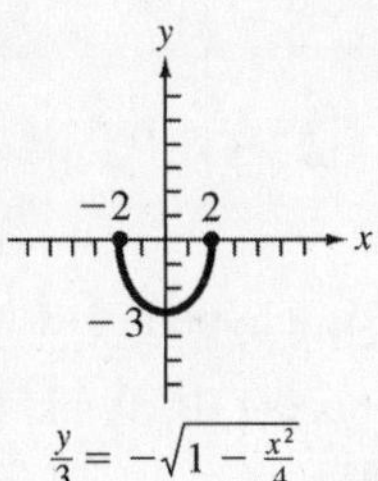

$\frac{y}{3} = -\sqrt{1 - \frac{x^2}{4}}$

12.3 Section Exercises

1. $\dfrac{x^2}{25} + \dfrac{y^2}{9} = 1$

This is the standard form for the equation of an ellipse with x-intercepts $(5,0)$ and $(-5,0)$ and y-intercepts $(0,3)$ and $(0,-3)$. This is graph **C**.

3. $\dfrac{x^2}{9} - \dfrac{y^2}{25} = 1$

This is the standard form for the equation of a hyperbola that opens left and right. Its x-intercepts are $(3,0)$ and $(-3,0)$. This is graph **D**.

5. If the equation of a hyperbola is in standard form (that is, equal to one), the hyperbola opens to the left and right if the x^2-term is positive. It opens up and down if the y^2-term is positive.

7. The equation $\dfrac{x^2}{16} - \dfrac{y^2}{9} = 1$ is in the form $\dfrac{x^2}{a^2} - \dfrac{y^2}{b^2} = 1$. The graph is a hyperbola with $a = 4$ and $b = 3$. The x-intercepts are $(4,0)$ and $(-4,0)$. There are no y-intercepts.
The vertices of the fundamental rectangle are $(4,3)$, $(4,-3)$, $(-4,-3)$, and $(-4,3)$. Extend the diagonals of the rectangle through these points to get the asymptotes.
The equations of the asymptotes are $y = \pm\frac{3}{4}x$. Graph a branch of the hyperbola through each intercept and approaching the asymptotes.

continued

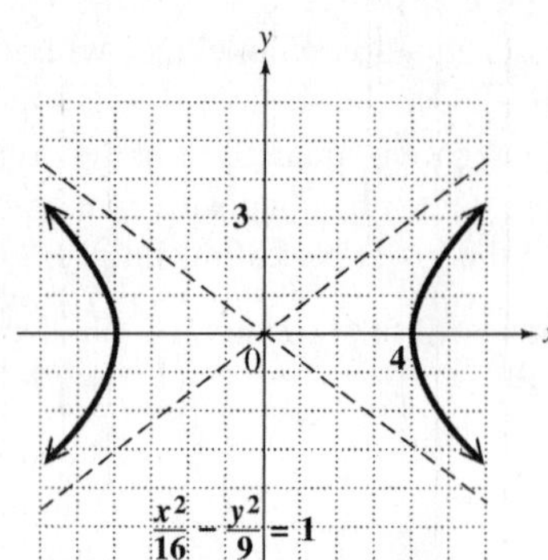

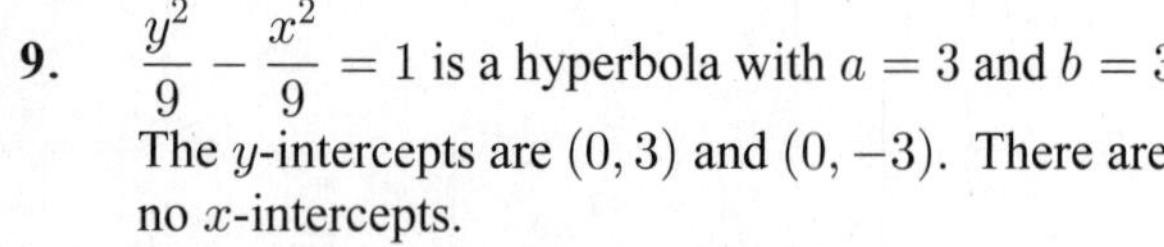

9. $\frac{y^2}{9} - \frac{x^2}{9} = 1$ is a hyperbola with $a = 3$ and $b = 3$. The y-intercepts are $(0, 3)$ and $(0, -3)$. There are no x-intercepts.
One asymptote passes through $(3, 3)$ and $(-3, -3)$. The other asymptote passes through $(-3, 3)$ and $(3, -3)$.
The equations of the asymptotes are $y = \pm\frac{3}{3}x = \pm x$. Draw the asymptotes and sketch the hyperbola through the intercepts and approaching the asymptotes.

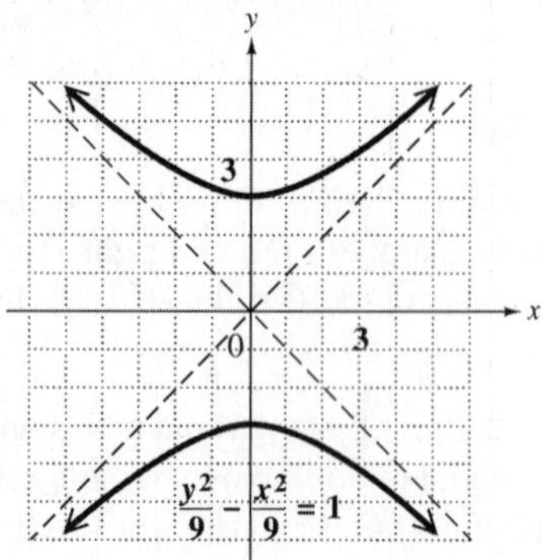

11. $\frac{x^2}{25} - \frac{y^2}{36} = 1$ is a hyperbola with $a = 5$ and $b = 6$. The x-intercepts are $(5, 0)$ and $(-5, 0)$. There are no y-intercepts.
To sketch the graph, draw the diagonals of the fundamental rectangle with vertices $(5, 6)$, $(5, -6)$, $(-5, -6)$ and $(-5, 6)$.
The equations of the asymptotes are $y = \pm\frac{6}{5}x$. Graph a branch of the hyperbola through each intercept and approaching the asymptotes.

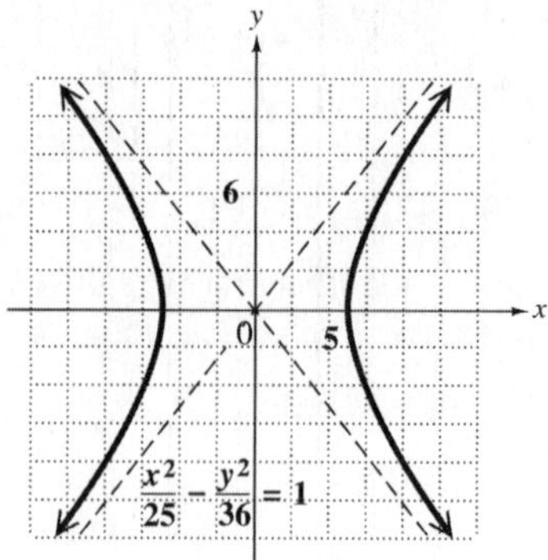

13. $x^2 - y^2 = 16$

$\frac{x^2}{16} - \frac{y^2}{16} = 1$ *Divide by 16.*

This equation is in the form $\frac{x^2}{a^2} - \frac{y^2}{b^2} = 1$ with $a = 4$ and $b = 4$. The graph is a hyperbola with x-intercepts $(4, 0)$ and $(-4, 0)$ and no y-intercepts.
One asymptote passes through $(4, 4)$ and $(-4, -4)$. The other asymptote passes through $(-4, 4)$ and $(4, -4)$.
The equations of the asymptotes are $y = \pm\frac{4}{4}x = \pm x$. Sketch the graph through the intercepts and approaching the asymptotes.

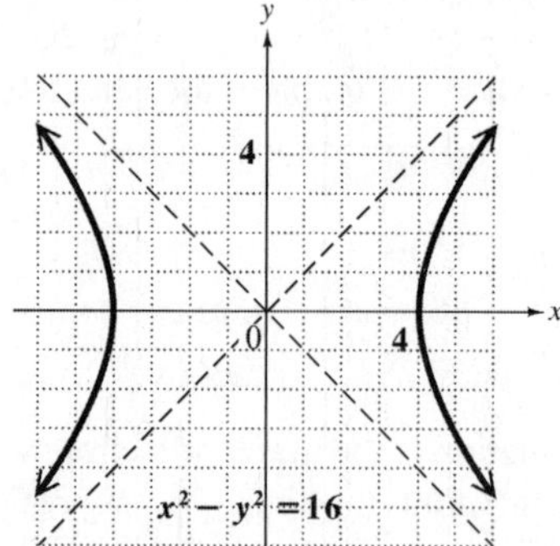

15. $4x^2 + y^2 = 16$

$\frac{x^2}{4} + \frac{y^2}{16} = 1$ *Divide by 16.*

This equation is in the form $\frac{x^2}{a^2} + \frac{y^2}{b^2} = 1$ with $a = 2$ and $b = 4$. The graph is an ellipse. The x-intercepts $(2, 0)$ and $(-2, 0)$. The y-intercepts are $(0, 4)$ and $(0, -4)$. Plot the intercepts and draw the ellipse through them.

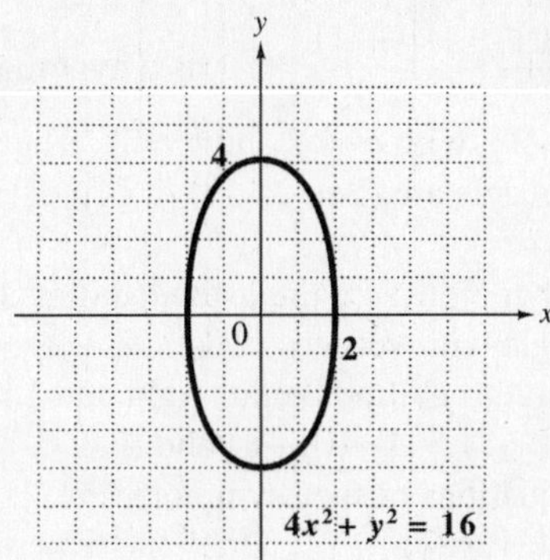

17.
$$y^2 = 36 - x^2$$
$$x^2 + y^2 = 36$$
$$(x-0)^2 + (y-0)^2 = 36$$

The graph is a circle with center at $(0, 0)$ and radius $\sqrt{36} = 6$.

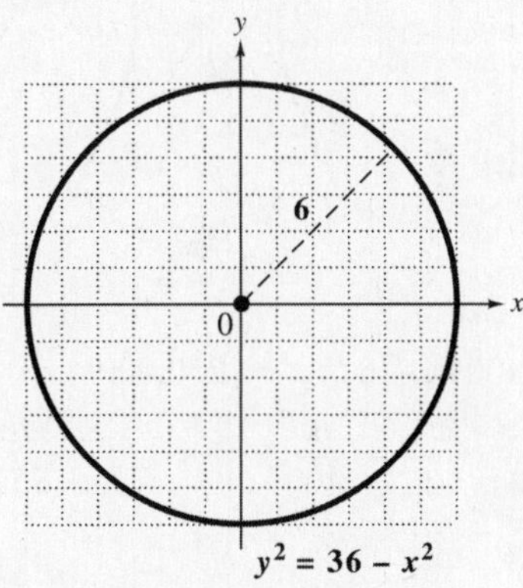

19.
$$9x^2 = 144 + 16y^2$$
$$9x^2 - 16y^2 = 144$$
$$\frac{x^2}{16} - \frac{y^2}{9} = 1 \qquad \textit{Divide by 144.}$$

The equation is a hyperbola in the form $\frac{x^2}{a^2} - \frac{y^2}{b^2} = 1$ with $a = 4$ and $b = 3$. The x-intercepts are $(4, 0)$ and $(-4, 0)$. There are no y-intercepts.
To sketch the graph, draw the diagonals of the fundamental rectangle with vertices $(4, 3)$, $(4, -3)$, $(-4, -3)$ and $(-4, 3)$.
The equations of the asymptotes are $y = \pm\frac{3}{4}x$.
Graph a branch of the hyperbola through each intercept approaching the asymptotes.

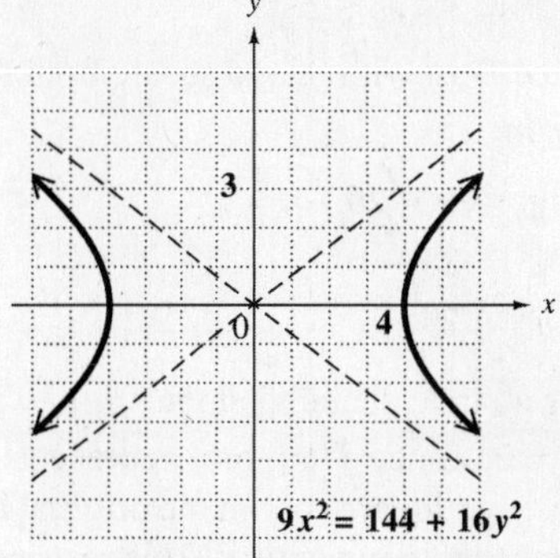

21.
$$x^2 + 9y^2 = 9$$
$$\frac{x^2}{9} + \frac{y^2}{1} = 1 \quad \textit{Divide by 9.}$$

The graph is an ellipse with x-intercepts $(3, 0)$ and $(-3, 0)$ and y-intercepts $(0, 1)$ and $(0, -1)$.

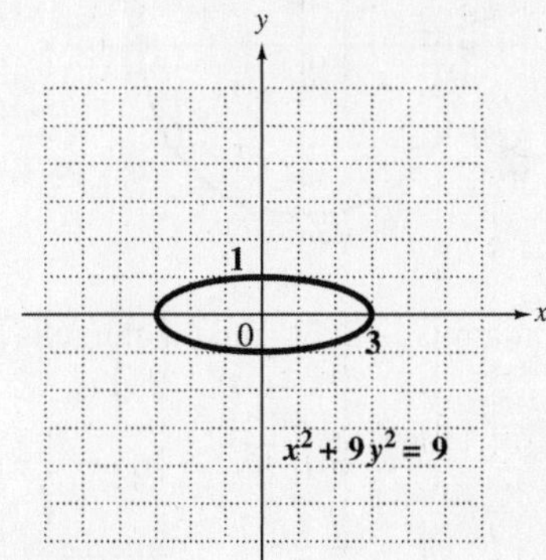

23. $f(x) = \sqrt{9 - x^2}$

Replace $f(x)$ with y and square both sides to get the equation

$$y^2 = 9 - x^2 \quad \text{or} \quad x^2 + y^2 = 9.$$

This is the graph of a circle with center $(0, 0)$ and radius 3. Since $f(x)$, or y, represents a principal square root in the original equation, $f(x)$ must be nonnegative. This restricts the graph to the upper half of the circle.

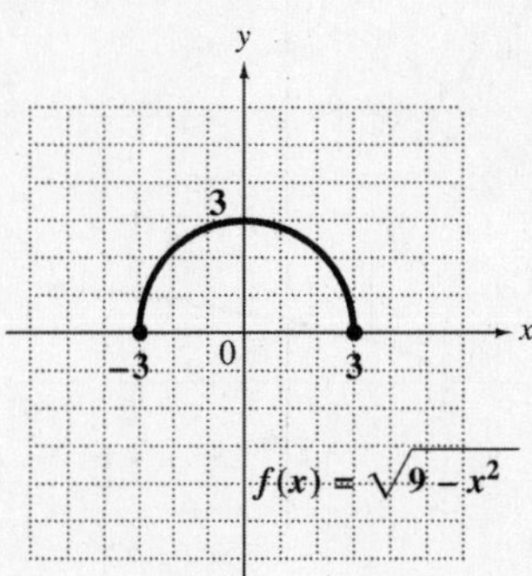

The domain is $[-3, 3]$, and the range is $[0, 3]$.

25. $f(x) = -\sqrt{25 - x^2}$

Replace $f(x)$ with y, and square both sides of the equation.

$$y = -\sqrt{25 - x^2}$$
$$y^2 = 25 - x^2$$
$$x^2 + y^2 = 25$$

This is a circle centered at the origin with radius $\sqrt{25} = 5$. Since $f(x)$, or y, represents a nonpositive square root in the original equation, $f(x)$ must be nonpositive. This restricts the graph to the bottom half of the circle.

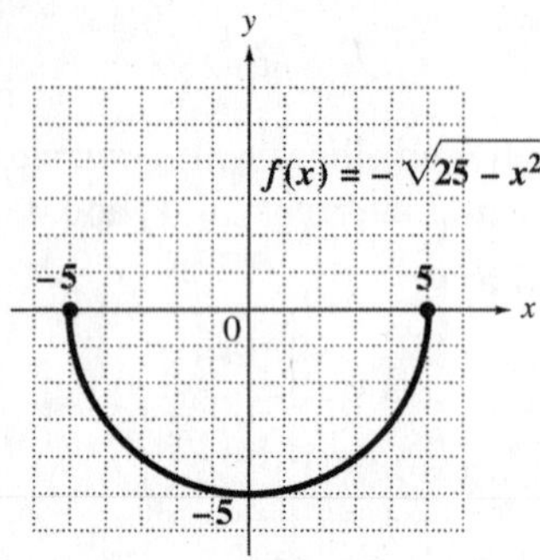

The domain is $[-5, 5]$, and the range is $[-5, 0]$.

27.
$$y = \sqrt{\frac{x+4}{2}}$$
$$y^2 = \frac{x+4}{2} \qquad \textit{Square both sides.}$$
$$2y^2 = x + 4$$
$$2y^2 - 4 = x$$
$$2(y-0)^2 - 4 = x$$

This is a parabola that opens to the right with vertex $(-4, 0)$. However, y is nonnegative in the original equation, so only the top half of the parabola is included in the graph.

x	-2	0	4
y	1	$\sqrt{2}$	2

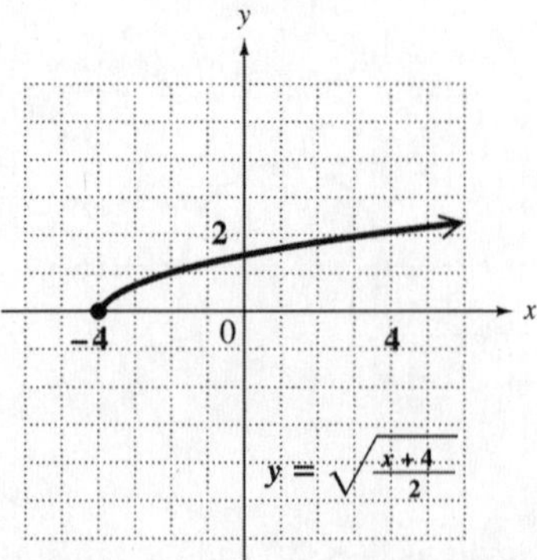

The domain is $[-4, \infty)$, and the range is $[0, \infty)$.

29. $\dfrac{(x-2)^2}{4} - \dfrac{(y+1)^2}{9} = 1$ is a hyperbola centered at $(2, -1)$, with $a = 2$ and $b = 3$. The x-intercepts are $(2 \pm 2, -1)$ or $(4, -1)$ and $(0, -1)$.
The asymptotes are the extended diagonals of the rectangle with vertices $(2, 3)$, $(2, -3)$, $(-2, -3)$ and $(-2, 3)$ shifted 2 units right and 1 unit down, or $(4, 2)$, $(4, -4)$, $(0, -4)$ and $(0, 2)$.
The equations of the asymptotes are $y + 1 = \pm\frac{3}{2}(x - 2)$. Draw the hyperbola.

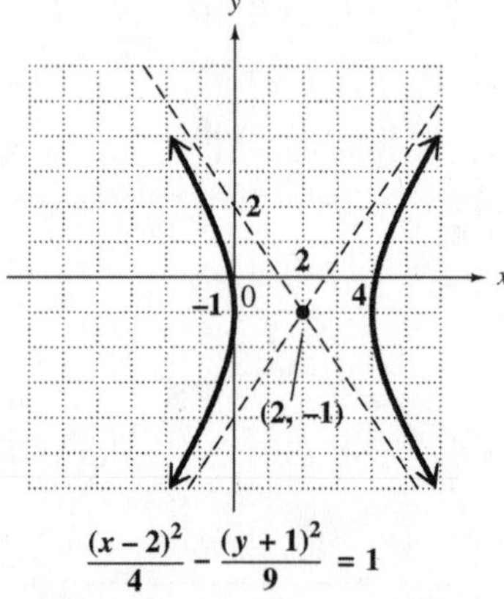

31. **(a)** $400x^2 - 625y^2 = 250{,}000$

$$\frac{x^2}{625} - \frac{y^2}{400} = 1 \qquad \textit{Divide by 250,000.}$$
$$\frac{x^2}{25^2} - \frac{y^2}{20^2} = 1$$

The x-intercepts are $(25, 0)$ and $(-25, 0)$. The distance between the buildings is the distance between the x-intercepts. The buildings are $25 + 25 = 50$ meters apart at their closest point.

(b) At $x = 50$, $y = \dfrac{d}{2}$, so $d = 2y$.

$$400(50)^2 - 625y^2 = 250{,}000$$
$$1{,}000{,}000 - 625y^2 = 250{,}000$$
$$-625y^2 = -750{,}000$$
$$y^2 = 1200$$
$$y = \sqrt{1200}$$

The distance d is $2\sqrt{1200} \approx 69.3$ meters.

12.4 Nonlinear Systems of Equations

12.4 Margin Exercises

1. **(a)** $x^2 + y^2 = 10$ (1)
$x = y + 2$ (2)

Substitute $y + 2$ for x in equation (1).

$$\begin{aligned} x^2 + y^2 &= 10 \quad (1) \\ (y+2)^2 + y^2 &= 10 \\ y^2 + 4y + 4 + y^2 &= 10 \\ 2y^2 + 4y - 6 &= 0 \\ y^2 + 2y - 3 &= 0 \\ (y+3)(y-1) &= 0 \end{aligned}$$

$$y + 3 = 0 \quad \text{or} \quad y - 1 = 0$$
$$y = -3 \quad \text{or} \quad y = 1$$

From (2) with $y = -3$, $x = -3 + 2 = -1$.
From (2) with $y = 1$, $x = 1 + 2 = 3$.

Check $(-1, -3)$: $1 + 9 = 10$, $-1 = -3 + 2$
Check $(3, 1)$: $9 + 1 = 10$, $3 = 1 + 2$

Solution set: $\{(3, 1), (-1, -3)\}$

(b) $x^2 - 2y^2 = 8$ (1)
$y + x = 6$ (2)

Solve equation (2) for y.

$$y = 6 - x \quad (3)$$

Substitute $6 - x$ for y in equation (1).

$$\begin{aligned} x^2 - 2y^2 &= 8 \quad (1) \\ x^2 - 2(6-x)^2 &= 8 \\ x^2 - 2(x^2 - 12x + 36) &= 8 \\ x^2 - 2x^2 + 24x - 72 &= 8 \\ -x^2 + 24x - 80 &= 0 \\ x^2 - 24x + 80 &= 0 \\ (x-4)(x-20) &= 0 \end{aligned}$$

$$x - 4 = 0 \quad \text{or} \quad x - 20 = 0$$
$$x = 4 \quad \text{or} \quad x = 20$$

From (3) with $x = 4$, $y = 6 - 4 = 2$.
From (3) with $x = 20$, $y = 6 - 20 = -14$.

Check $(4, 2)$: $16 - 8 = 8$, $2 + 4 = 6$
Check $(20, -14)$: $400 - 392 = 8$, $-14 + 20 = 6$

Solution set: $\{(4, 2), (20, -14)\}$

2. **(a)** $xy = 8$ (1)
$x + y = 6$ (2)

Solve equation (2) for x.

$$x = 6 - y \quad (3)$$

Substitute $6 - y$ for x in equation (1).

$$\begin{aligned} xy &= 8 \quad (1) \\ (6-y)y &= 8 \\ 6y - y^2 &= 8 \\ -y^2 + 6y - 8 &= 0 \\ y^2 - 6y + 8 &= 0 \\ (y-2)(y-4) &= 0 \end{aligned}$$

$$y - 2 = 0 \quad \text{or} \quad y - 4 = 0$$
$$y = 2 \quad \text{or} \quad y = 4$$

From (3) with $y = 2$, $x = 6 - 2 = 4$.
From (3) with $y = 4$, $x = 6 - 4 = 2$.

Check $(4, 2)$: $4(2) = 8$, $4 + 2 = 6$
Check $(2, 4)$: $2(4) = 8$, $2 + 4 = 6$

Solution set: $\{(4, 2), (2, 4)\}$

(b) $xy + 10 = 0$ (1)
$4x + 9y = -2$ (2)

Solve equation (1) for y.

$$y = -\frac{10}{x} \quad (3)$$

Substitute $-\frac{10}{x}$ for y in equation (2).

$$\begin{aligned} 4x + 9y &= -2 \quad (2) \\ 4x + 9\left(-\frac{10}{x}\right) &= -2 \\ 4x - \frac{90}{x} &= -2 \\ 4x^2 - 90 &= -2x \quad \textit{Multiply by x.} \\ 4x^2 + 2x - 90 &= 0 \\ 2x^2 + x - 45 &= 0 \\ (2x-9)(x+5) &= 0 \end{aligned}$$

$$2x - 9 = 0 \quad \text{or} \quad x + 5 = 0$$
$$x = \tfrac{9}{2} \quad \text{or} \quad x = -5$$

From (3) with $x = \frac{9}{2}$, $y = -\frac{10}{\frac{9}{2}} = -\frac{20}{9}$.

From (3) with $x = -5$, $y = -\frac{10}{-5} = 2$.

Check $(\frac{9}{2}, -\frac{20}{9})$: $-10 + 10 = 0$, $18 - 20 = -2$
Check $(-5, 2)$: $-10 + 10 = 0$, $-20 + 18 = -2$

Solution set: $\{(-5, 2), (\frac{9}{2}, -\frac{20}{9})\}$

3. **(a)**
$$\begin{array}{rcll} x^2 + y^2 &=& 41 & (1) \\ x^2 - y^2 &=& 9 & (2) \\ \hline 2x^2 &=& 50 & \textit{Add} \\ x^2 &=& 25 & \end{array}$$

$$x = 5 \quad \text{or} \quad x = -5$$

Substitute each value of x in equation (1) to find y. First substitute 5 for x.

$$\begin{aligned} x^2 + y^2 &= 41 \qquad (1) \\ 5^2 + y^2 &= 41 \\ 25 + y^2 &= 41 \\ y^2 &= 16 \end{aligned}$$

$$y = 4 \quad \text{or} \quad y = -4$$

Now substitute -5 for x in equation (1).

$$\begin{aligned} x^2 + y^2 &= 41 \qquad (1) \\ (-5)^2 + y^2 &= 41 \\ 25 + y^2 &= 41 \\ y^2 &= 16 \end{aligned}$$

$$y = 4 \quad \text{or} \quad y = -4$$

Check: If $x = \pm 5$, $x^2 = 25$, and if $y = \pm 4$, $y^2 = 16$. Thus, in any case, we get $25 + 16 = 41$ in (1) and $25 - 16 = 9$ in (2). All four ordered pairs check.

Solution set: $\{(5, 4), (5, -4), (-5, 4), (-5, -4)\}$

(b)
$$\begin{aligned} x^2 + 3y^2 &= 40 \qquad (1) \\ 4x^2 - y^2 &= 4 \qquad (2) \end{aligned}$$

Multiply equation (2) by 3 and add the result to equation (1).

$$\begin{array}{rcll} x^2 + 3y^2 &=& 40 & (1) \\ 12x^2 - 3y^2 &=& 12 & 3 \times (2) \\ \hline 13x^2 &=& 52 & \\ x^2 &=& 4 & \end{array}$$

$$x = 2 \quad \text{or} \quad x = -2$$

Substitute 2 for x in equation (1) to find y.

$$\begin{aligned} x^2 + 3y^2 &= 40 \quad (1) \\ 2^2 + 3y^2 &= 40 \\ 4 + 3y^2 &= 40 \\ 3y^2 &= 36 \\ y^2 &= 12 \end{aligned}$$

$$y = 2\sqrt{3} \quad \text{or} \quad y = -2\sqrt{3}$$

Since x is squared in equation (1), the values of y, $2\sqrt{3}$ or $-2\sqrt{3}$, are the same for $x = 2$ and $x = -2$. So we get four solutions as in part (a).

Solution set:
$\{(2, 2\sqrt{3}), (2, -2\sqrt{3}), (-2, 2\sqrt{3}), (-2, -2\sqrt{3})\}$

4. **(a)**
$$\begin{aligned} x^2 + xy + y^2 &= 3 \qquad (1) \\ x^2 + y^2 &= 5 \qquad (2) \end{aligned}$$

Multiply equation (2) by -1 and add the result to equation (1).

$$\begin{array}{rcll} x^2 + xy + y^2 &=& 3 & (1) \\ -x^2 \qquad - y^2 &=& -5 & -1 \times (2) \\ \hline xy &=& -2 & (3) \end{array}$$

Solve equation (3) for y to get

$$y = -\frac{2}{x}. \quad (4)$$

Substitute $-\frac{2}{x}$ for y in equation (2).

$$\begin{aligned} x^2 + y^2 &= 5 \qquad (2) \\ x^2 + \left(-\frac{2}{x}\right)^2 &= 5 \\ x^2 + \frac{4}{x^2} &= 5 \\ x^4 + 4 &= 5x^2 \qquad \textit{Multiply by } x^2. \\ x^4 - 5x^2 + 4 &= 0 \\ (x^2 - 4)(x^2 - 1) &= 0 \end{aligned}$$

$$\begin{array}{lcl} x^2 - 4 = 0 & \text{or} & x^2 - 1 = 0 \\ x^2 = 4 & \text{or} & x^2 = 1 \\ x = \pm 2 & \text{or} & x = \pm 1 \end{array}$$

Substitute these values for x in equation (4) to find the values of y.

If $x = 1$, then $y = -2$. If $x = -1$, then $y = 2$.
If $x = 2$, then $y = -1$. If $x = -2$, then $y = 1$.

Solution set: $\{(1, -2), (-1, 2), (2, -1), (-2, 1)\}$

(b)
$$\begin{aligned} x^2 + 7xy - 2y^2 &= -8 \qquad (1) \\ -2x^2 \qquad + 4y^2 &= 16 \qquad (2) \end{aligned}$$

Multiply equation (1) by 2 and add the result to equation (2).

$$\begin{array}{rcll} 2x^2 + 14xy - 4y^2 &=& -16 & 2 \times (1) \\ -2x^2 \qquad + 4y^2 &=& 16 & (2) \\ \hline 14xy &=& 0 & \\ xy &=& 0 & \end{array}$$

If $xy = 0$, then either $x = 0$ or $y = 0$.

If $x = 0$, then substitute 0 for x in equation (1).

$$\begin{aligned} x^2 + 7xy - 2y^2 &= -8 \qquad (1) \\ 0 + 0 - 2y^2 &= -8 \\ y^2 &= 4 \end{aligned}$$

$$y = 2 \quad \text{or} \quad y = -2$$

If $y = 0$, then substitute 0 for y in equation (1).

$$x^2 + 7xy - 2y^2 = -8 \quad (1)$$
$$x^2 + 0 + 0 = -8$$

$$x = \sqrt{-8} \quad \text{or} \quad x = -\sqrt{-8}$$
$$x = 2i\sqrt{2} \quad \text{or} \quad x = -2i\sqrt{2}$$

Solution set:
$\{(0,2), (0,-2), (2i\sqrt{2},0), (-2i\sqrt{2},0)\}$

12.4 Section Exercises

1. The line intersects the ellipse in exactly one point, so there is *one* point in the solution set of the system.

3. The line does not intersect the hyperbola, so there are no points in the solution set of the system.

5. A line and a circle; no points

Draw any circle, and then draw a line that does not cross the circle.

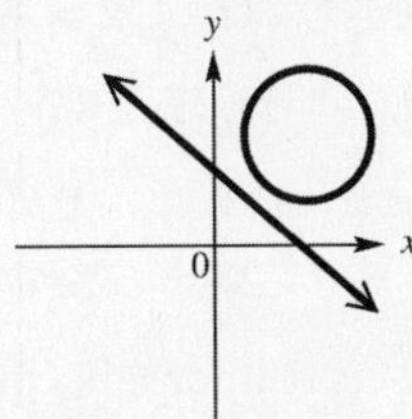

7. A line and an ellipse; two points

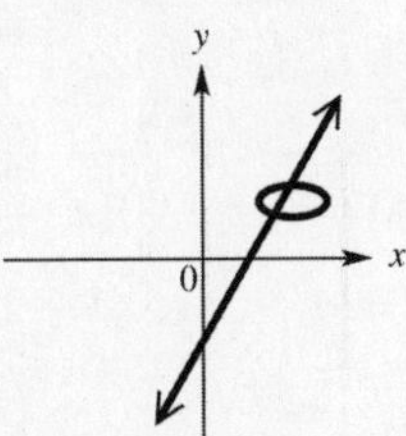

9. A circle and an ellipse; four points

Draw any ellipse, and then draw a circle with the same center whose radius is just large enough so that there are four points of intersection. (If the radius of the circle is too large or too small, there may be fewer points of intersection.)

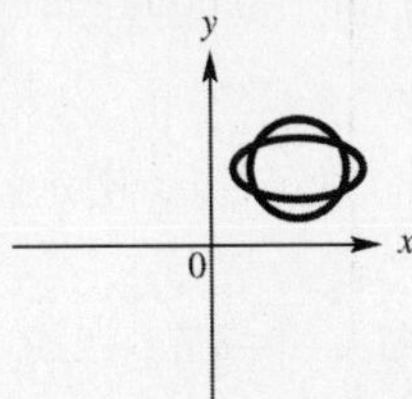

11. $y = 4x^2 - x \quad (1)$
$y = x \quad (2)$

Substitute x for y in equation (1).

$$y = 4x^2 - x \quad (1)$$
$$x = 4x^2 - x$$
$$0 = 4x^2 - 2x$$
$$0 = 2x(2x - 1)$$

$$2x = 0 \quad \text{or} \quad 2x - 1 = 0$$
$$x = 0 \quad \text{or} \quad x = \tfrac{1}{2}$$

Use equation (2) to find y for each x-value.

If $x = 0$, then $y = 0$.
If $x = \frac{1}{2}$, then $y = \frac{1}{2}$.

Solution set: $\{(0,0), (\frac{1}{2}, \frac{1}{2})\}$

13. $y = x^2 + 6x + 9 \quad (1)$
$x + y = 3 \quad (2)$

Substitute $x^2 + 6x + 9$ for y in equation (2).

$$x + y = 3 \quad (2)$$
$$x + (x^2 + 6x + 9) = 3$$
$$x^2 + 7x + 9 = 3$$
$$x^2 + 7x + 6 = 0$$
$$(x + 6)(x + 1) = 0$$

$$x + 6 = 0 \quad \text{or} \quad x + 1 = 0$$
$$x = -6 \quad \text{or} \quad x = -1$$

Substitute these values for x in equation (2) and solve for y.

If $x = -6$, then
$$x + y = 3 \quad (2)$$
$$-6 + y = 3$$
$$y = 9.$$

If $x = -1$, then
$$x + y = 3 \quad (2)$$
$$-1 + y = 3$$
$$y = 4.$$

Solution set: $\{(-6,9), (-1,4)\}$

15. $x^2 + y^2 = 2$ (1)
$2x + y = 1$ (2)

Solve equation (2) for y.

$$y = 1 - 2x \quad (3)$$

Substitute $1 - 2x$ for y in equation (1).

$$\begin{aligned} x^2 + y^2 &= 2 \quad (1) \\ x^2 + (1-2x)^2 &= 2 \\ x^2 + 1 - 4x + 4x^2 &= 2 \\ 5x^2 - 4x - 1 &= 0 \\ (5x+1)(x-1) &= 0 \end{aligned}$$

$5x + 1 = 0$ or $x - 1 = 0$
$x = -\frac{1}{5}$ or $x = 1$

Use equation (3) to find y for each x-value.

If $x = -\frac{1}{5}$, then

$$y = 1 - 2(-\tfrac{1}{5}) = 1 + \tfrac{2}{5} = \tfrac{7}{5}.$$

If $x = 1$, then

$$y = 1 - 2(1) = -1.$$

Solution set: $\{(-\frac{1}{5}, \frac{7}{5}), (1, -1)\}$

17. $xy = 4$ (1)
$3x + 2y = -10$ (2)

Solve equation (1) for y to get

$$y = \frac{4}{x}. \quad (3)$$

Substitute $\frac{4}{x}$ for y in equation (2) to find x.

$$\begin{aligned} 3x + 2y &= -10 \quad (2) \\ 3x + 2\left(\frac{4}{x}\right) &= -10 \end{aligned}$$

Multiply by the LCD, x.

$$\begin{aligned} 3x^2 + 8 &= -10x \\ 3x^2 + 10x + 8 &= 0 \\ (3x+4)(x+2) &= 0 \end{aligned}$$

$3x + 4 = 0$ or $x + 2 = 0$
$x = -\frac{4}{3}$ or $x = -2$

Substitute these values for x in equation (3).
If $x = -\frac{4}{3}$, then $y = \frac{4}{-\frac{4}{3}} = -3$.
If $x = -2$, then $y = \frac{4}{-2} = -2$.

Solution set: $\{(-2, -2), (-\frac{4}{3}, -3)\}$

19. $xy = -3$ (1)
$x + y = -2$ (2)

Solve equation (2) for y.

$$y = -x - 2 \quad (3)$$

Substitute $-x - 2$ for y in equation (1).

$$\begin{aligned} xy &= -3 \quad (1) \\ x(-x-2) &= -3 \\ -x^2 - 2x &= -3 \\ -x^2 - 2x + 3 &= 0 \\ x^2 + 2x - 3 &= 0 \quad \textit{Multiply by } -1. \\ (x+3)(x-1) &= 0 \end{aligned}$$

$x + 3 = 0$ or $x - 1 = 0$
$x = -3$ or $x = 1$

Use equation (3) to find y for each x-value.

If $x = -3$, then $y = -(-3) - 2 = 1$.

If $x = 1$, then $y = -(1) - 2 = -3$.

Solution set: $\{(-3, 1), (1, -3)\}$

21. $y = 3x^2 + 6x$ (1)
$y = x^2 - x - 6$ (2)

Substitute $x^2 - x - 6$ for y in equation (1) to find x.

$$\begin{aligned} y &= 3x^2 + 6x \quad (1) \\ x^2 - x - 6 &= 3x^2 + 6x \\ 0 &= 2x^2 + 7x + 6 \\ 0 &= (2x+3)(x+2) \end{aligned}$$

$2x + 3 = 0$ or $x + 2 = 0$
$x = -\frac{3}{2}$ or $x = -2$

Substitute $-\frac{3}{2}$ for x in equation (1) to find y.

$$\begin{aligned} y &= 3x^2 + 6x \quad (1) \\ y &= 3(-\tfrac{3}{2})^2 + 6(-\tfrac{3}{2}) \\ &= 3(\tfrac{9}{4}) + 6(-\tfrac{6}{4}) \\ &= \tfrac{27}{4} - \tfrac{36}{4} = -\tfrac{9}{4} \end{aligned}$$

Substitute -2 for x in equation (1) to find y.

$$\begin{aligned} y &= 3x^2 + 6x \quad (1) \\ y &= 3(-2)^2 + 6(-2) \\ &= 12 - 12 = 0 \end{aligned}$$

Solution set: $\{(-\frac{3}{2}, -\frac{9}{4}), (-2, 0)\}$

23. $2x^2 - y^2 = 6 \quad (1)$
$y = x^2 - 3 \quad (2)$

Solve equation (2) for x^2.

$$x^2 = y + 3 \quad (3)$$

Substitute $y+3$ for x^2 in equation (1).

$$\begin{aligned} 2x^2 - y^2 &= 6 \quad (1) \\ 2(y+3) - y^2 &= 6 \\ 2y + 6 - y^2 &= 6 \\ 0 &= y^2 - 2y \\ 0 &= y(y-2) \end{aligned}$$

$$y = 0 \quad \text{or} \quad y = 2$$

From (3) with $y = 0$, $x^2 = 3$, so $x = \pm\sqrt{3}$.

From (3) with $y = 2$, $x^2 = 5$, so $x = \pm\sqrt{5}$.

Solution set:
$\{(-\sqrt{3}, 0), (\sqrt{3}, 0), (-\sqrt{5}, 2), (\sqrt{5}, 2)\}$

25. $3x^2 + 2y^2 = 12 \quad (1)$
$x^2 + 2y^2 = 4 \quad (2)$

Multiply equation (2) by -1 and add the result to equation (1).

$$\begin{array}{rcrl} 3x^2 + 2y^2 &=& 12 & (1) \\ -x^2 - 2y^2 &=& -4 & -1 \times (2) \\ \hline 2x^2 &=& 8 & \\ x^2 &=& 4 & \\ x &=& \pm 2 & \end{array}$$

Substitute ± 2 for x in equation (2) to find y.

$$\begin{aligned} x^2 + 2y^2 &= 4 \quad (2) \\ (\pm 2)^2 + 2y^2 &= 4 \\ 4 + 2y^2 &= 4 \\ 2y^2 &= 0 \\ y^2 &= 0 \\ y &= 0 \end{aligned}$$

Solution set: $\{(-2, 0), (2, 0)\}$

27. $xy = 6 \quad (1)$
$3x^2 - y^2 = 12 \quad (2)$

Solve equation (1) for y to get

$$y = \frac{6}{x}. \quad (3)$$

Substitute $\frac{6}{x}$ for y in equation (2).

$$\begin{aligned} 3x^2 - \left(\frac{6}{x}\right)^2 &= 12 \\ 3x^2 - \frac{36}{x^2} &= 12 \\ 3x^4 - 36 &= 12x^2 \\ 3x^4 - 12x^2 - 36 &= 0 \\ x^4 - 4x^2 - 12 &= 0 \\ (x^2 - 6)(x^2 + 2) &= 0 \end{aligned}$$

$$\begin{array}{ccc} x^2 - 6 = 0 & \text{or} & x^2 + 2 = 0 \\ x^2 = 6 & \text{or} & x^2 = -2 \\ x = \pm\sqrt{6} & \text{or} & x = \pm i\sqrt{2} \end{array}$$

Use equation (3) to find y.

If $x = i\sqrt{2}$,

$$y = \frac{6}{i\sqrt{2}} = \frac{6 \cdot i\sqrt{2}}{i\sqrt{2} \cdot i\sqrt{2}} = \frac{6i\sqrt{2}}{2i^2} = -3i\sqrt{2}.$$

If $x = -i\sqrt{2}$, $y = 3i\sqrt{2}$.

If $x = \sqrt{6}$, $y = \frac{6}{\sqrt{6}} = \frac{6 \cdot \sqrt{6}}{\sqrt{6} \cdot \sqrt{6}} = \frac{6\sqrt{6}}{6} = \sqrt{6}$.

If $x = -\sqrt{6}$, $y = -\sqrt{6}$.

Solution set: $\{(i\sqrt{2}, -3i\sqrt{2}), (-i\sqrt{2}, 3i\sqrt{2}),$
$(-\sqrt{6}, -\sqrt{6}), (\sqrt{6}, \sqrt{6})\}$

29. $2x^2 + 2y^2 = 8 \quad (1)$
$3x^2 + 4y^2 = 24 \quad (2)$

Multiply equation (1) by -2 and add the result to equation (2).

$$\begin{array}{rcrl} -4x^2 - 4y^2 &=& -16 & -2 \times (1) \\ 3x^2 + 4y^2 &=& 24 & (2) \\ \hline -x^2 &=& 8 & \\ x^2 &=& -8 & \\ x &=& \pm\sqrt{-8} = \pm 2i\sqrt{2} & \end{array}$$

Substitute $\pm 2i\sqrt{2}$ for x in equation (1).

continued

$$\begin{aligned} 2x^2 &= 8 - 2y^2 \qquad (1) \\ 2(\pm 2i\sqrt{2})^2 &= 8 - 2y^2 \\ 2(-8) &= 8 - 2y^2 \\ -16 &= 8 - 2y^2 \\ 2y^2 &= 24 \\ y^2 &= 12 \\ y &= \pm\sqrt{12} = \pm 2\sqrt{3} \end{aligned}$$

Since $2i\sqrt{2}$ can be paired with either $2\sqrt{3}$ or $-2\sqrt{3}$ and $-2i\sqrt{2}$ can be paired with either $2\sqrt{3}$ or $-2\sqrt{3}$, there are four possible solutions.

Solution set:
$\{(-2i\sqrt{2}, -2\sqrt{3}), (-2i\sqrt{2}, 2\sqrt{3}), (2i\sqrt{2}, -2\sqrt{3}), (2i\sqrt{2}, 2\sqrt{3})\}$

31. $x^2 + xy + y^2 = 15 \quad (1)$
$x^2 + y^2 = 10 \quad (2)$

Multiply equation (2) by -1 and add the result to equation (1).

$$\begin{aligned} x^2 + xy + y^2 &= 15 \qquad (1) \\ -x^2 \qquad - y^2 &= -10 \qquad -1 \times (2) \\ \hline xy &= 5 \\ y &= \frac{5}{x} \qquad (3) \end{aligned}$$

Substitute $\frac{5}{x}$ for y in equation (2).

$$\begin{aligned} x^2 + y^2 &= 10 \qquad (2) \\ x^2 + \left(\frac{5}{x}\right)^2 &= 10 \\ x^2 + \frac{25}{x^2} &= 10 \\ x^4 + 25 &= 10x^2 \qquad \textit{Multiply by } x^2. \\ x^4 - 10x^2 + 25 &= 0 \end{aligned}$$

Let $z = x^2$, so $z^2 = x^4$.

$$\begin{aligned} z^2 - 10z + 25 &= 0 \\ (z-5)^2 &= 0 \\ z - 5 &= 0 \\ z &= 5 \end{aligned}$$

Since $z = x^2$, $x^2 = 5$, so $x = \pm\sqrt{5}$.

Using equation (3) we get the following.

If $x = -\sqrt{5}$, then

$$y = \frac{5}{-\sqrt{5}} = \frac{5 \cdot \sqrt{5}}{-\sqrt{5} \cdot \sqrt{5}} = \frac{5\sqrt{5}}{-5} = -\sqrt{5}.$$

Similarly, if $x = \sqrt{5}$, then $y = \sqrt{5}$.

Solution set: $\{(-\sqrt{5}, -\sqrt{5}), (\sqrt{5}, \sqrt{5})\}$

33. Let W = the width, and L = the length.

The formula for the area of a rectangle is $LW = A$, so

$$LW = 84. \quad (1)$$

The perimeter of a rectangle is given by $2L + 2W = P$, so

$$2L + 2W = 38. \quad (2)$$

Solve equation (2) for L to get

$$L = 19 - W. \quad (3)$$

Substitute $19 - W$ for L in equation (1).

$$\begin{aligned} LW &= 84 \qquad (1) \\ (19 - W)W &= 84 \\ 19W - W^2 &= 84 \\ -W^2 + 19W - 84 &= 0 \\ W^2 - 19W + 84 &= 0 \qquad \textit{Multiply by } -1. \\ (W-7)(W-12) &= 0 \end{aligned}$$

$$W - 7 = 0 \quad \text{or} \quad W - 12 = 0$$
$$W = 7 \quad \text{or} \quad W = 12$$

Using equation (3), with $W = 7$,

$$L = 19 - 7 = 12.$$

If $W = 12$, then $L = 7$, which are the same two numbers. Length must be greater than width, so the length is 12 feet and the width is 7 feet.

12.5 Second-Degree Inequalities and Systems of Inequalities

12.5 Margin Exercises

1. $y \geq (x+1)^2 - 5$

The boundary, $y = (x+1)^2 - 5$ is a parabola opening upward with its vertex at $(-1, -5)$. Use the point $(0, 0)$ as a test point.

$$\begin{aligned} 0 &\overset{?}{\geq} (0+1)^2 - 5 \\ 0 &\overset{?}{\geq} 1 - 5 \\ 0 &\geq -4 \qquad \textit{True} \end{aligned}$$

Shade the region that contains $(0, 0)$. This is the region inside the parabola. The parabola is drawn as a solid curve since the points of the parabola satisfy the inequality.

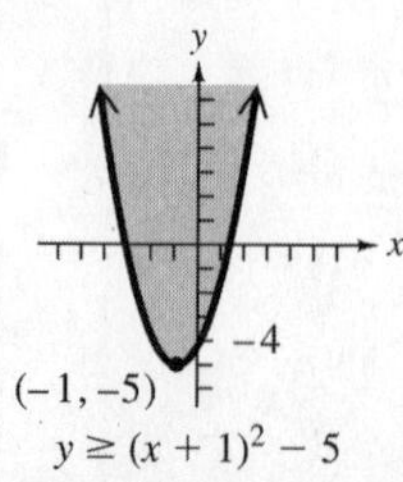

2. $x^2 + 4y^2 > 36$

$$\frac{x^2}{36} + \frac{y^2}{9} > 1 \quad \textit{Divide by 36.}$$

The boundary is the ellipse

$$\frac{x^2}{36} + \frac{y^2}{9} = 1,$$

which has intercepts $(6, 0)$, $(-6, 0)$, $(0, 3)$, and $(0, -3)$. Use the point $(0, 0)$ as a test point.

$$0^2 + 4(0)^2 \overset{?}{>} 36$$
$$0 > 36 \quad \textit{False}$$

Shade the region that does not contain $(0, 0)$. This is the region outside the ellipse. The ellipse is drawn as a dashed curve since the points of the ellipse do not satisfy the inequality.

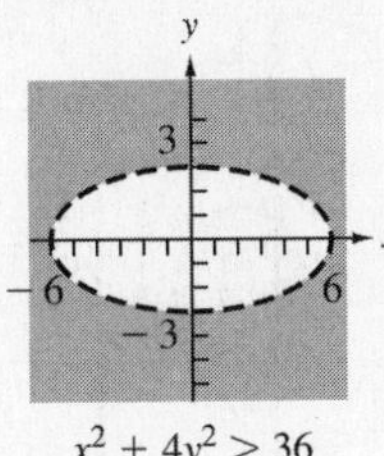

$x^2 + 4y^2 > 36$

3. $x^2 + y^2 \le 25$
$x + y \le 3$

The boundary $x^2 + y^2 = 25$ is a circle with center $(0, 0)$ and radius 5. Draw as a solid curve. Test $(0, 0)$.

$$0^2 + 0^2 \overset{?}{\le} 25$$
$$0 \le 25 \quad \textit{True}$$

Shade the region inside the circle.

The boundary $x + y = 3$ is a line with x-intercept $(3, 0)$ and y-intercept $(0, 3)$. Draw as a solid line. Test $(0, 0)$.

$$0 + 0 \overset{?}{\le} 3$$
$$0 \le 3 \quad \textit{True}$$

Shade the region on the side of the line containing $(0, 0)$.

The graph of the solution set of the system includes all points common to the two graphs, that is, the intersection of the two shaded regions.

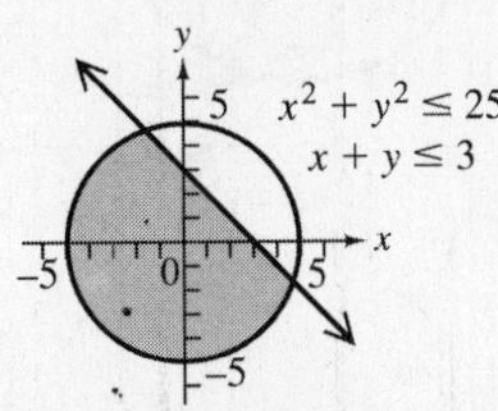

4. $3x - 4y \ge 12$
$x + 3y \ge 6$
$y \le 2$

The boundary $3x - 4y = 12$ is a solid line with intercepts $(4, 0)$ and $(0, -3)$. Test $(0, 0)$.

$$3(0) - 4(0) \overset{?}{\ge} 12$$
$$0 \ge 12 \quad \textit{False}$$

Shade the side of the line that does not contain $(0, 0)$.

The boundary $x + 3y = 6$ is a solid line with intercepts $(6, 0)$ and $(0, 2)$. Test $(0, 0)$.

$$0 + 3(0) \overset{?}{\ge} 6$$
$$0 \ge 6 \quad \textit{False}$$

Shade the side of the line that does not contain $(0, 0)$.

The boundary $y = 2$ is a solid horizontal line through $(0, 2)$. Shade the region below the line $y \le 2$.

The intersection of the three shaded regions is the graph of the solution set of the system.

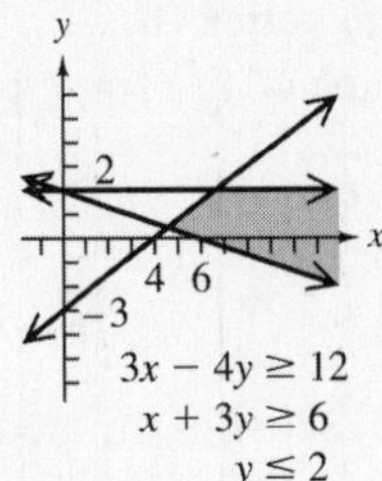

5. $y \ge x^2 + 1$

$$\frac{x^2}{9} + \frac{y^2}{4} \ge 1$$
$$y \le 5$$

The boundary $y = x^2 + 1$ is a parabola opening upward with vertex at $(0, 1)$. It is drawn as a solid curve because of the $\ge$ symbol. Test $(0, 0)$.

$$0 \overset{?}{\ge} 0^2 + 1$$
$$0 \ge 1 \quad \textit{False}$$

Shade the region that does not contain $(0, 0)$. This is the region inside the parabola.

The boundary $\frac{x^2}{9} + \frac{y^2}{4} = 1$ is an ellipse with intercepts $(3, 0)$, $(-3, 0)$, $(0, 2)$, and $(0, -2)$. It is drawn as a solid curve because of the $\ge$ symbol. Test $(0, 0)$.

continued

$$\frac{0^2}{9} + \frac{0^2}{4} \overset{?}{\geq} 1$$
$$0 \geq 1 \quad \textit{False}$$

Shade the region that does not contain $(0, 0)$. This is the region outside the ellipse.

The boundary $y = 5$ is a solid horizontal line through $(0, 5)$. Shade the region below the line since $y \leq 5$.

The graph of the solution set of the system includes all points common to the three graphs, that is, the intersection of the three shaded regions.

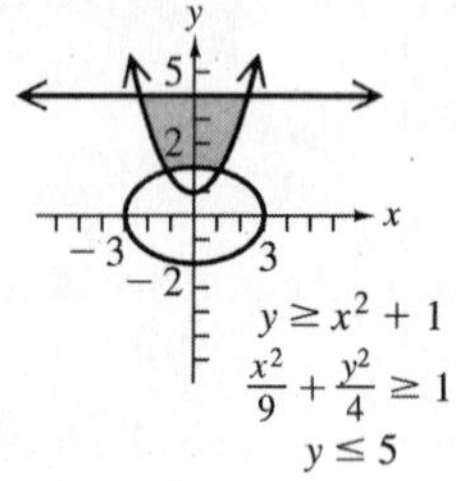

12.5 Section Exercises

1. $y > x^2 - 1$

The boundary, $y = x^2 - 1$, is a dashed parabola opening up with vertex $(0, -1)$. Its shape is the same as the graph of $f(x) = x^2$. Use $(0, 0)$ as a test point.

$$0 \overset{?}{>} 0^2 - 1$$
$$0 > -1 \quad \textit{True}$$

The point $(0, 0)$ satisfies the inequality so the graph includes it. Shade the region that includes $(0, 0)$.

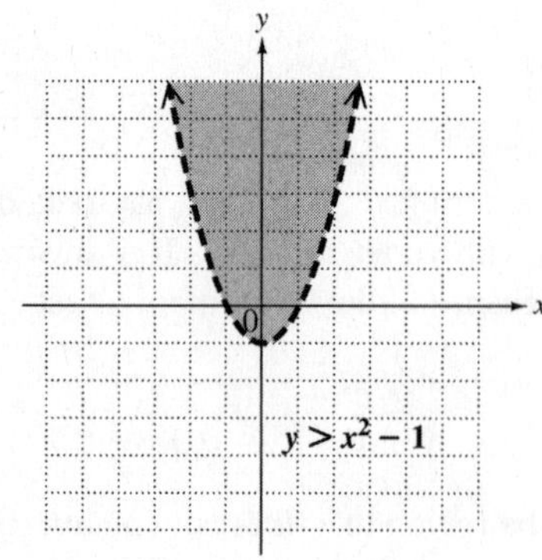

3. $y^2 \leq 4 - 2x^2$

The boundary is

$$y^2 = 4 - 2x^2$$
$$2x^2 + y^2 = 4$$
$$\frac{x^2}{2} + \frac{y^2}{4} = 1.$$

The boundary is a solid ellipse with x-intercepts $(\sqrt{2}, 0)$ and $(-\sqrt{2}, 0)$ and y-intercepts $(0, 2)$ and $(0, -2)$. Test $(0, 0)$.

$$0^2 \overset{?}{\leq} 4 - 2(0)^2$$
$$0 \leq 4 \quad \textit{True}$$

Shade the region that includes $(0, 0)$.

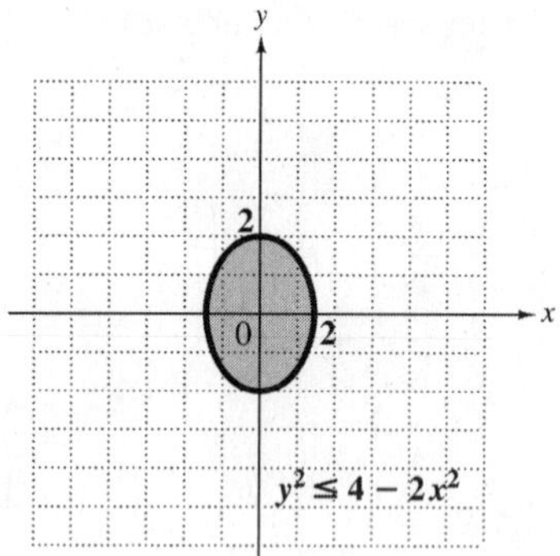

5. $x^2 \leq 16 - y^2$

The boundary is

$$x^2 = 16 - y^2$$
$$x^2 + y^2 = 16.$$

The boundary is a solid circle with center $(0, 0)$ and radius 4. Test $(0, 0)$.

$$0^2 \overset{?}{\leq} 16 - 0^2$$
$$0 \leq 16 \quad \textit{True}$$

Shade the region that includes $(0, 0)$.

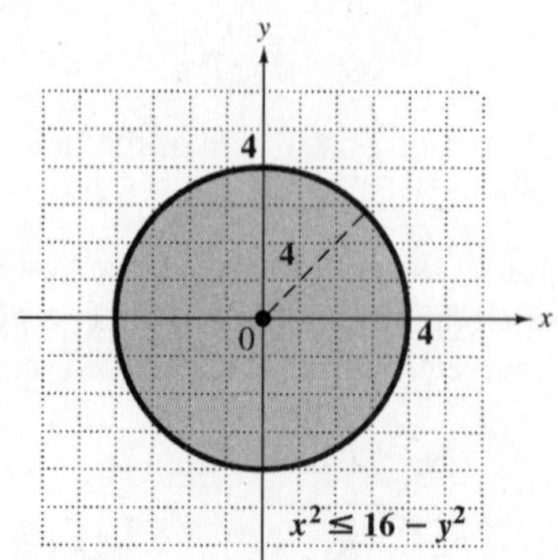

7. $x^2 \le 16 + 4y^2$

The boundary is

$$x^2 = 16 + 4y^2$$
$$x^2 - 4y^2 = 16$$
$$\frac{x^2}{16} - \frac{y^2}{4} = 1.$$

The boundary is a solid hyperbola with x-intercepts $(4, 0)$ and $(-4, 0)$. The asymptotes go through $(4, 2)$ and $(-4, -2)$ and through $(4, -2)$ and $(-4, 2)$. Test $(0, 0)$.

$$0^2 \stackrel{?}{\le} 16 + 4(0)^2$$
$$0 \le 16 \qquad \textit{True}$$

Shade the region between the branches of the hyperbola, which includes $(0, 0)$.

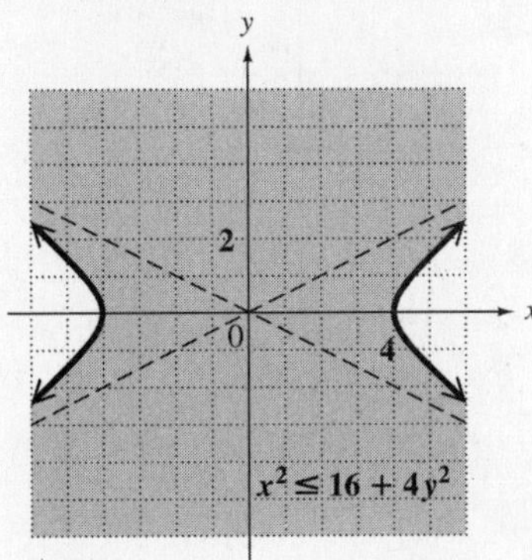

9. $9x^2 < 16y^2 - 144$

The boundary is

$$9x^2 = 16y^2 - 144$$
$$144 = 16y^2 - 9x^2$$
$$1 = \frac{y^2}{9} - \frac{x^2}{16}.$$

The boundary is a dashed hyperbola with y-intercepts $(0, 3)$ and $(0, -3)$. The asymptotes go through $(4, 3)$ and $(-4, -3)$ and through $(4, -3)$ and $(-4, 3)$. Test $(0, 0)$.

$$9(0)^2 \stackrel{?}{<} 16(0)^2 - 144$$
$$0 < -144 \qquad \textit{False}$$

Shade the region that does *not* include $(0, 0)$.

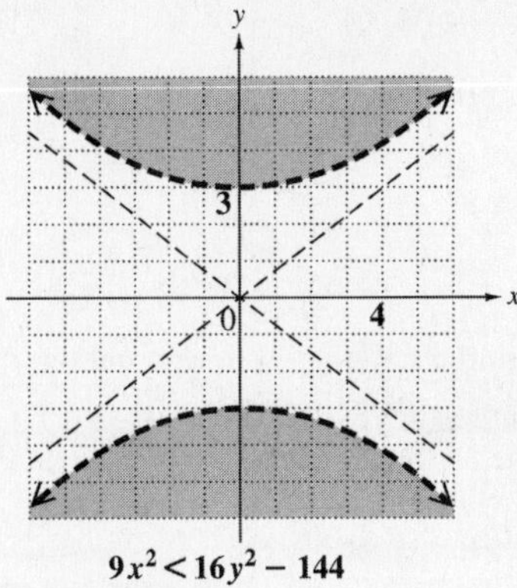

11. $4y^2 \le 36 - 9x^2$

The boundary is

$$4y^2 = 36 - 9x^2$$
$$9x^2 + 4y^2 = 36$$
$$\frac{x^2}{4} + \frac{y^2}{9} = 1.$$

The boundary is a solid ellipse with x-intercepts $(2, 0)$ and $(-2, 0)$ and y-intercepts $(0, 3)$ and $(0, -3)$. Test $(0, 0)$.

$$4(0)^2 \stackrel{?}{\le} 36 - 9(0)^2$$
$$0 \le 36 \qquad \textit{True}$$

Shade the region that includes $(0, 0)$.

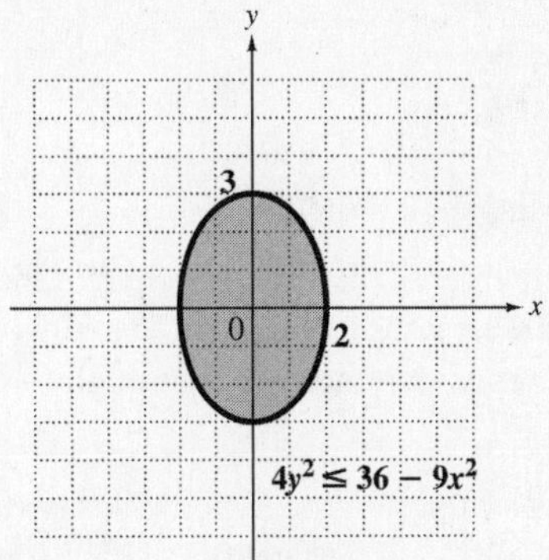

13. $x \geq y^2 - 8y + 14$

The boundary is

$$x = y^2 - 8y + 14$$
$$x = y^2 - 8y + 16 - 16 + 14$$
$$x = (y - 4)^2 - 2.$$

The boundary is a solid horizontal parabola with vertex $(-2, 4)$ opening to the right. Its shape is the same as the graph of $x = y^2$. Test $(0, 0)$.

$$0 \stackrel{?}{\geq} 0^2 - 8(0) + 14$$
$$0 \geq 14 \qquad \textit{False}$$

Shade the region that does *not* include $(0, 0)$.

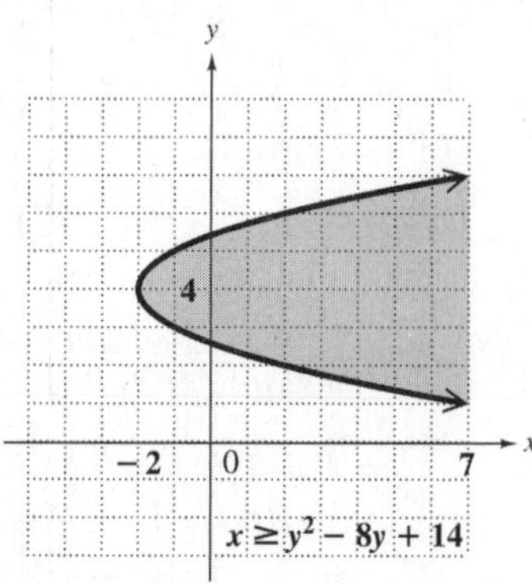

15.

$$25x^2 \leq 9y^2 + 225$$
$$25x^2 - 9y^2 \leq 225$$
$$\frac{x^2}{9} - \frac{y^2}{25} \leq 1$$

Graph the solid hyperbola $\frac{x^2}{9} - \frac{y^2}{25} = 1$ through the x-intercepts $(3, 0)$ and $(-3, 0)$. One asymptote passes through $(3, 5)$ and $(-3, -5)$. The other asymptote passes through $(-3, 5)$ and $(3, -5)$.

Test a point not on the hyperbola, say $(0, 0)$, in $25x^2 \leq 9y^2 + 225$ to get $0 \leq 225$, a true statement.

Shade the part of the graph that contains $(0, 0)$. This is the region between the two branches of the hyperbola.

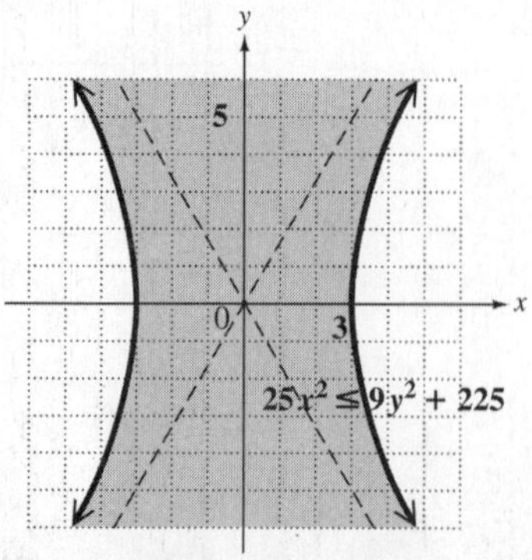

17. $3x - y > -6 \quad (1)$
$4x + 3y > 12 \quad (2)$

The graph of (1) has dashed boundary line $3x - y = -6$ through $(-2, 0)$ and $(0, 6)$. Since $3(0) - 0 > -6$ is true, the graph contains the region on the side of the line that includes the test point $(0, 0)$.
The graph of (2) has dashed boundary line $4x + 3y = 12$ through $(3, 0)$ and $(0, 4)$. Since $4(0) + 3(0) > 12$ is false, the graph contains the region that does not include $(0, 0)$. The graph of the system is the intersection of these graphs.

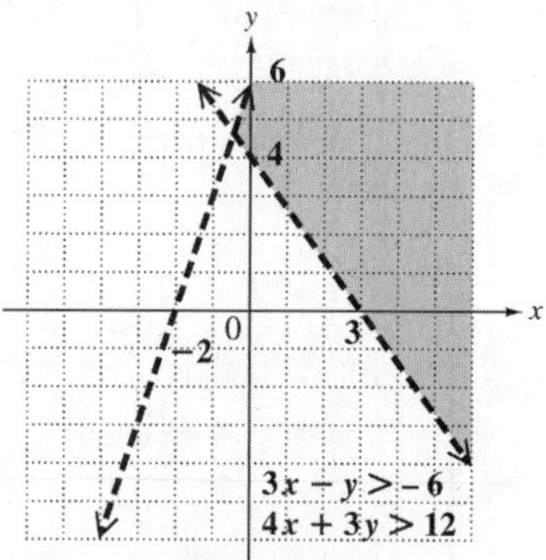

19. $4x - 3y \leq 0 \quad (1)$
$x + y \leq 5 \quad (2)$

The graph of (1) has solid boundary line $4x - 3y = 0$ through $(0, 0)$ and $(3, 4)$.
Since $4(-1) - 3(0) \leq 0$ is true, the graph contains the region that includes $(-1, 0)$.
The graph of (2) has solid boundary line $x + y = 5$ through $(5, 0)$ and $(0, 5)$.

Since $0 + 0 \leq 5$ is true, the graph contains the region that includes $(0, 0)$.

The graph of the system is the intersection of these graphs.

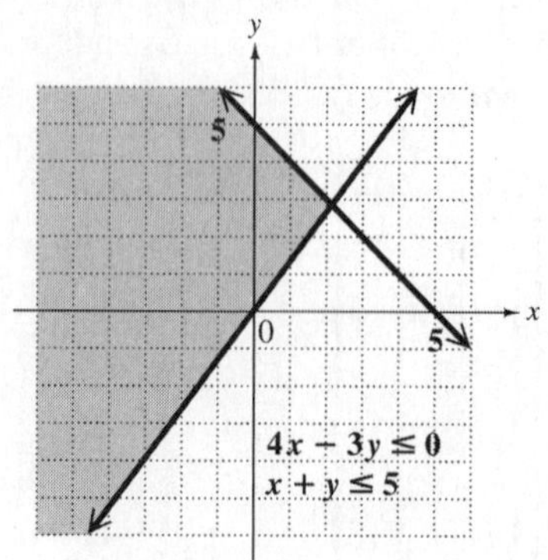

21. $x^2 - y^2 \geq 9 \quad (1)$

$\dfrac{x^2}{16} + \dfrac{y^2}{9} \leq 1 \quad (2)$

The graph of (1) has the solid hyperbola $\dfrac{x^2}{9} - \dfrac{y^2}{9} = 1$ with x-intercepts $(3, 0)$ and $(-3, 0)$ as its boundary. The asymptotes go through $(3, 3)$ and $(-3, -3)$ and through $(3, -3)$ and $(-3, 3)$. Since $0^2 - 0^2 \geq 9$ is false, the graph includes points to the right of the right branch of the hyperbola and to the left of the left branch of the hyperbola.

The graph of (2) has the solid ellipse $\dfrac{x^2}{16} + \dfrac{y^2}{9} = 1$ with x-intercepts $(4, 0)$ and $(-4, 0)$ and y-intercepts $(0, 3)$ and $(0, -3)$ as its boundary. Since $\dfrac{0^2}{16} + \dfrac{0^2}{9} \leq 1$ is true, the graph includes points inside the ellipse.
The graph of the system is the intersection of these graphs.

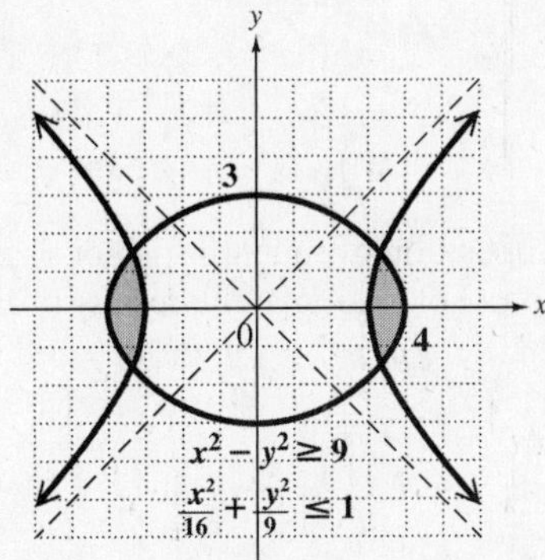

23.

$y < x^2$	**OR**	$y < x^2$
$y > -2$		$y > -2$
$x + y < 3$		$y < -x + 3$
$3x - 2y > -6$		$y < \frac{3}{2}x + 3$

The graph of the first inequality is the region below the dashed parabola with vertex $(0, 0)$ opening up. The graph of the second inequality is the region above the dashed horizontal line $y = -2$. The graph of the third inequality is the region below the dashed line with intercepts $(3, 0)$ and $(0, 3)$. The graph of the fourth inequality is the region below the dashed line with intercepts $(-2, 0)$ and $(0, 3)$.
The graph of the system is the intersection of the four graphs.

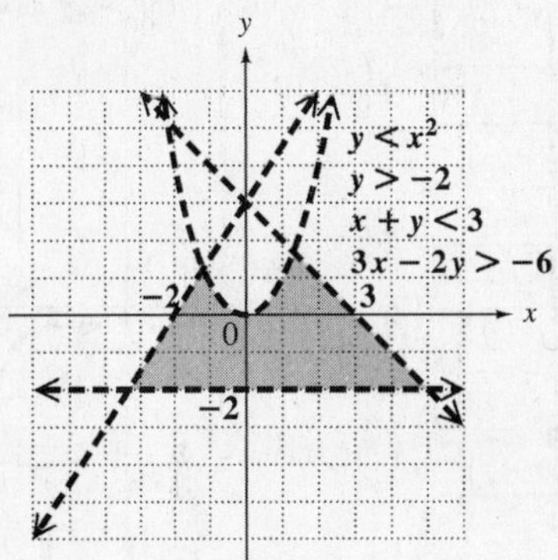

Chapter 12 Review Exercises

1. $f(x) = |x + 4|$

This is the graph of the absolute value function $g(x) = |x|$ shifted 4 units to the left (since $x + 4 = 0$ if $x = -4$).

Since x can be any real number, the domain is $(-\infty, \infty)$.

The value of y is always greater than or equal to 0, so the range is $[0, \infty)$.

x	y
-6	2
-5	1
-4	0
-3	1
-2	2

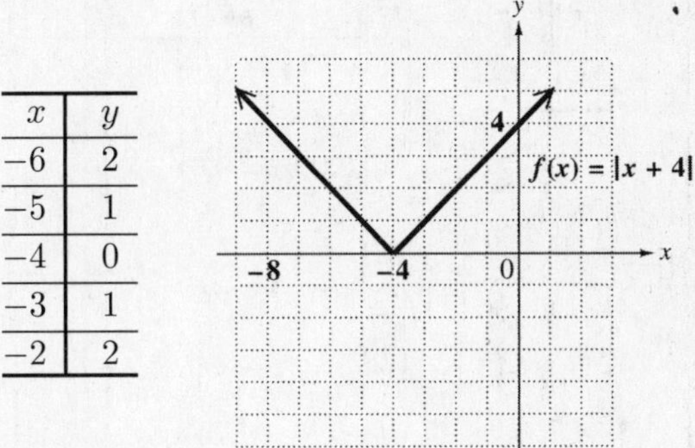

2. $f(x) = \dfrac{1}{x - 4}$

This is the graph of the reciprocal function, $g(x) = \dfrac{1}{x}$, shifted 4 units to the right. Since $x \neq 4$ (if $x = 4$, the denominator equals 0), the line $x = 4$ is a vertical asymptote. Since $\dfrac{1}{x - 4} \neq 0$, the line $y = 0$ is a horizontal asymptote.

The domain is all real numbers except 4, that is, $(-\infty, 4) \cup (4, \infty)$.

The range is all real numbers except 0, that is, $(-\infty, 0) \cup (0, \infty)$.

continued

x	y
2	$-\frac{1}{2}$
3	-1
$3\frac{1}{2}$	-2
$4\frac{1}{2}$	2
5	1
6	$\frac{1}{2}$

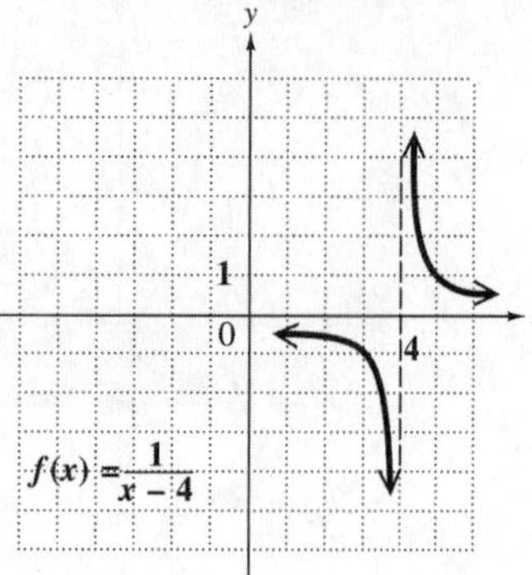

3. $f(x) = \sqrt{x} + 3$

This is the graph of the square root function, $g(x) = \sqrt{x}$, shifted 3 units upward.

The domain of the function is $[0, \infty)$ and its range is $[3, \infty)$.

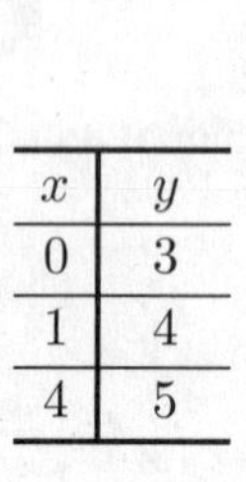

x	y
0	3
1	4
4	5

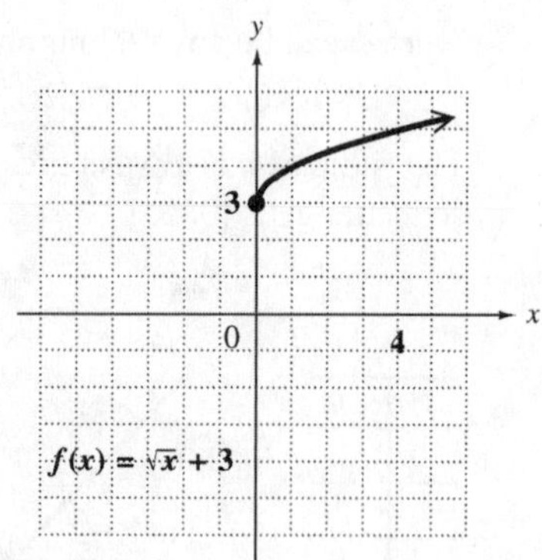

4. $[\![12]\!] = 12 \quad (12 \le 12)$

5. $[\![2\frac{1}{4}]\!] = 2 \quad (2 \le 2\frac{1}{4})$

6. $[\![-4.75]\!] = -5 \quad (-5 \le -4.75)$

In Exercises 7–9, $f(x) = 3x^2 + 2x - 1$ and $g(x) = 5x + 7$.

7. **(a)** $$\begin{aligned}(g \circ f)(3) &= g(f(3)) \\ &= g(3 \cdot 3^2 + 2 \cdot 3 - 1) \\ &= g(32) \\ &= 5 \cdot 32 + 7 \\ &= 167\end{aligned}$$

(b) $$\begin{aligned}(f \circ g)(3) &= f(g(3)) \\ &= f(5 \cdot 3 + 7) \\ &= f(22) \\ &= 3 \cdot 22^2 + 2 \cdot 22 - 1 \\ &= 1495\end{aligned}$$

8. **(a)** $$\begin{aligned}(f \circ g)(-2) &= f(g(-2)) \\ &= f(5(-2) + 7) \\ &= f(-3) \\ &= 3(-3)^2 + 2(-3) - 1 \\ &= 20\end{aligned}$$

(b) $$\begin{aligned}(g \circ f)(-2) &= g(f(-2)) \\ &= g(3(-2)^2 + 2(-2) - 1) \\ &= g(7) \\ &= 5 \cdot 7 + 7 \\ &= 42\end{aligned}$$

9. **(a)** $$\begin{aligned}&(f \circ g)(x) \\ &= f(g(x)) \\ &= f(5x + 7) \\ &= 3(5x + 7)^2 + 2(5x + 7) - 1 \\ &= 3(25x^2 + 70x + 49) + 10x + 14 - 1 \\ &= 75x^2 + 210x + 147 + 10x + 13 \\ &= 75x^2 + 220x + 160\end{aligned}$$

(b) $$\begin{aligned}(g \circ f)(x) &= g(f(x)) \\ &= g(3x^2 + 2x - 1) \\ &= 5(3x^2 + 2x - 1) + 7 \\ &= 15x^2 + 10x - 5 + 7 \\ &= 15x^2 + 10x + 2\end{aligned}$$

10. No, composition of functions is not a commutative operation. For example, the results of Exercise 9 show that $(f \circ g)(x) \ne (g \circ f)(x)$ in this case.

11. Center $(-2, 4)$, $r = 3$
Here $h = -2$, $k = 4$, and $r = 3$, so an equation of the circle is

$$\begin{aligned}(x - h)^2 + (y - k)^2 &= r^2 \\ [x - (-2)]^2 + (y - 4)^2 &= 3^2 \\ (x + 2)^2 + (y - 4)^2 &= 9.\end{aligned}$$

12. Center $(-1, -3)$, $r = 5$
Here $h = -1$, $k = -3$, and $r = 5$, so an equation of the circle is

$$\begin{aligned}(x - h)^2 + (y - k)^2 &= r^2 \\ [x - (-1)]^2 + [y - (-3)]^2 &= 5^2 \\ (x + 1)^2 + (y + 3)^2 &= 25.\end{aligned}$$

13. Center $(4, 2)$, $r = 6$
Here $h = 4$, $k = 2$, and $r = 6$, so an equation of the circle is

$$\begin{aligned}(x - h)^2 + (y - k)^2 &= r^2 \\ (x - 4)^2 + (y - 2)^2 &= 6^2 \\ (x - 4)^2 + (y - 2)^2 &= 36.\end{aligned}$$

14. $x^2 + y^2 + 6x - 4y - 3 = 0$

Write the equation in center-radius form,

$$(x - h)^2 + (y - k)^2 = r^2,$$

by completing the squares on x and y.

$$(x^2 + 6x \qquad) + (y^2 - 4y \qquad) = 3$$
$$(x^2 + 6x + \underline{9}) + (y^2 - 4y + \underline{4})$$
$$= 3 + \underline{9} + \underline{4}$$
$$(x + 3)^2 + (y - 2)^2 = 16$$
$$[x - (-3)]^2 + (y - 2)^2 = 16$$

The circle has center (h, k) at $(-3, 2)$ and radius $\sqrt{16} = 4$.

15. $x^2 + y^2 - 8x - 2y + 13 = 0$

Write the equation in center-radius form,

$$(x - h)^2 + (y - k)^2 = r^2,$$

by completing the squares on x and y.

$$(x^2 - 8x \qquad) + (y^2 - 2y \qquad) = -13$$
$$(x^2 - 8x + \underline{16}) + (y^2 - 2y + \underline{1})$$
$$= -13 + \underline{16} + \underline{1}$$
$$(x - 4)^2 + (y - 1)^2 = 4$$

The circle has center (h, k) at $(4, 1)$ and radius $\sqrt{4} = 2$.

16. $2x^2 + 2y^2 + 4x + 20y = -34$

$x^2 + y^2 + 2x + 10y = -17$

Write the equation in center-radius form,

$$(x - h)^2 + (y - k)^2 = r^2,$$

by completing the squares on x and y.

$$(x^2 + 2x \qquad) + (y^2 + 10y \qquad) = -17$$
$$(x^2 + 2x + \underline{1}) + (y^2 + 10y + \underline{25})$$
$$= -17 + \underline{1} + \underline{25}$$
$$(x + 1)^2 + (y + 5)^2 = 9$$
$$[x - (-1)]^2 + [y - (-5)]^2 = 9$$

The circle has center (h, k) at $(-1, -5)$ and radius $\sqrt{9} = 3$.

17. $4x^2 + 4y^2 - 24x + 16y = 48$

$x^2 + y^2 - 6x + 4y = 12$

Write the equation in center-radius form,

$$(x - h)^2 + (y - k)^2 = r^2,$$

by completing the squares on x and y.

$$(x^2 - 6x \qquad) + (y^2 + 4y \qquad) = 12$$
$$(x^2 - 6x + \underline{9}) + (y^2 + 4y + \underline{4})$$
$$= 12 + \underline{9} + \underline{4}$$
$$(x - 3)^2 + (y + 2)^2 = 25$$
$$(x - 3)^2 + [y - (-2)]^2 = 25$$

The circle has center (h, k) at $(3, -2)$ and radius $\sqrt{25} = 5$.

18. $x^2 + y^2 = 16$

$(x - 0)^2 + (y - 0)^2 = 4^2$

This is a circle with center $(0, 0)$ and radius 4.

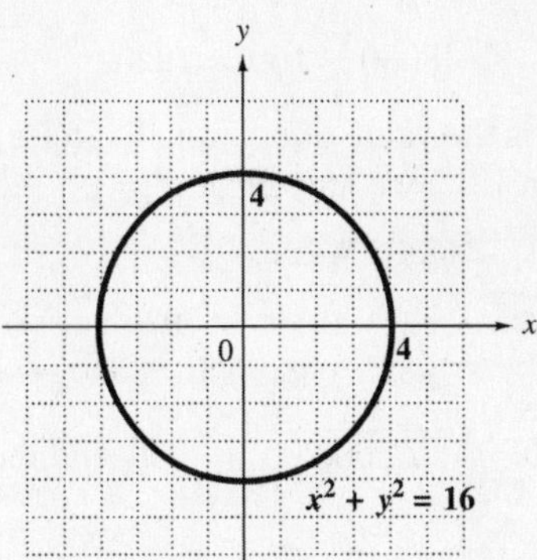

19. $\frac{x^2}{16} + \frac{y^2}{9} = 1$ is in $\frac{x^2}{a^2} + \frac{y^2}{b^2} = 1$ form with $a = 4$ and $b = 3$. The graph is an ellipse with x-intercepts $(4, 0)$ and $(-4, 0)$ and y-intercepts $(0, 3)$ and $(0, -3)$. Plot the intercepts, and draw the ellipse through them.

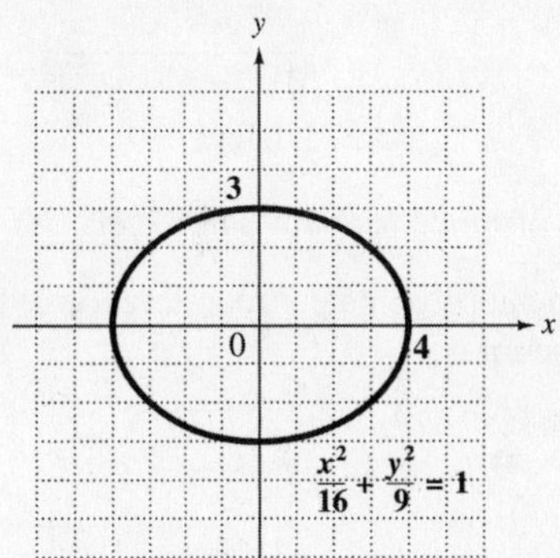

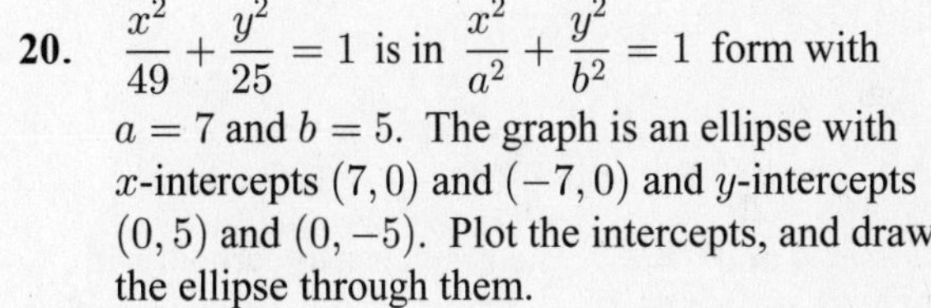

20. $\frac{x^2}{49} + \frac{y^2}{25} = 1$ is in $\frac{x^2}{a^2} + \frac{y^2}{b^2} = 1$ form with $a = 7$ and $b = 5$. The graph is an ellipse with x-intercepts $(7, 0)$ and $(-7, 0)$ and y-intercepts $(0, 5)$ and $(0, -5)$. Plot the intercepts, and draw the ellipse through them.

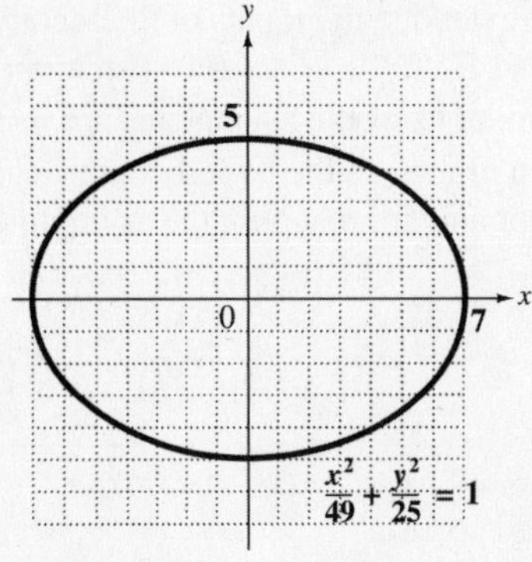

21. The total distance on the horizontal axis is $160 + 16{,}000 = 16{,}160$ km. This represents $2a$, so $a = \frac{1}{2}(16{,}160) = 8080$. The distance from Earth to the center of the ellipse is

$$8080 - 160 = 7920,$$

which is the value of c. From Exercise 39 in Section 12.2, we know that $c^2 = a^2 - b^2$, so

$$\begin{aligned} b^2 &= a^2 - c^2. \\ b^2 &= 8080^2 - 7920^2 \\ &= 2{,}560{,}000 \end{aligned}$$

Thus, $b = \sqrt{2{,}560{,}000} = 1600$, and the equation is

$$\frac{x^2}{8080^2} + \frac{y^2}{1600^2} = 1$$

$$\text{or} \quad \frac{x^2}{65{,}286{,}400} + \frac{y^2}{2{,}560{,}000} = 1.$$

22. **(a)** The distance between the foci is $2c$, where c can be found using the relationship

$$\begin{aligned} c &= \sqrt{a^2 - b^2}. \\ &= \sqrt{310^2 - (513/2)^2} \\ &\approx 174.1 \text{ feet} \end{aligned}$$

So the distance is about 348.2 feet.

(b) The approximate circumference of the Roman Colosseum is

$$\begin{aligned} C &\approx 2\pi\sqrt{\frac{a^2 + b^2}{2}}. \\ &= 2\pi\sqrt{\frac{310^2 + (513/2)^2}{2}} \\ &\approx 1787.6 \text{ feet} \end{aligned}$$

23. $\frac{x^2}{16} - \frac{y^2}{25} = 1$ is in $\frac{x^2}{a^2} - \frac{y^2}{b^2} = 1$ form with $a = 4$ and $b = 5$. The graph is a hyperbola with x-intercepts $(4, 0)$ and $(-4, 0)$ and asymptotes that are the extended diagonals of the rectangle with vertices $(4, 5)$, $(4, -5)$, $(-4, -5)$, and $(-4, 5)$. The equations of the asymptotes are $y = \pm\frac{5}{4}x$. Graph a branch of the hyperbola through each intercept and approaching the asymptotes.

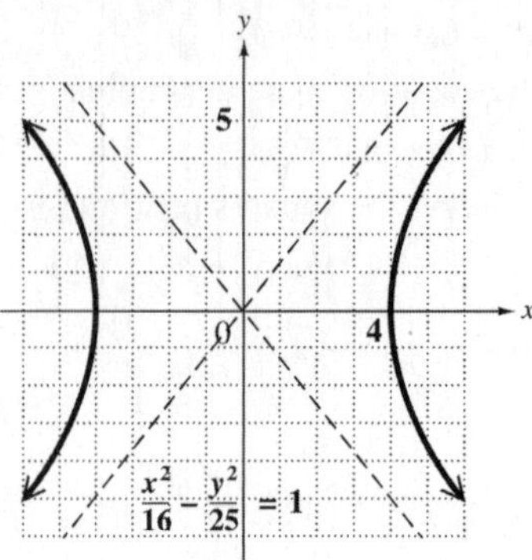

24. $\frac{y^2}{25} - \frac{x^2}{4} = 1$ is in $\frac{y^2}{b^2} - \frac{x^2}{a^2} = 1$ form with $a = 2$ and $b = 5$. The graph is a hyperbola with y-intercepts $(0, 5)$ and $(0, -5)$ and asymptotes that are the extended diagonals of the rectangle with vertices $(2, 5)$, $(2, -5)$, $(-2, -5)$, and $(-2, 5)$. The equations of the asymptotes are $y = \pm\frac{5}{2}x$. Graph a branch of the hyperbola through each intercept and approaching the asymptotes.

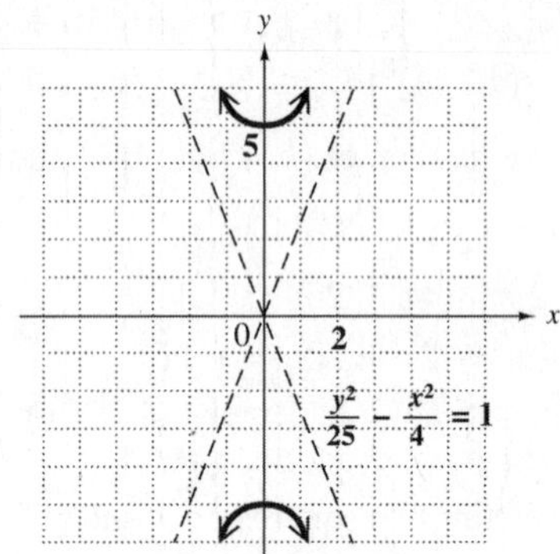

25. $f(x) = -\sqrt{16 - x^2}$

Replace $f(x)$ with y.

$$y = -\sqrt{16 - x^2}$$

Square both sides.

$$\begin{aligned} y^2 &= 16 - x^2 \\ x^2 + y^2 &= 16 \end{aligned}$$

This equation is the graph of a circle with center $(0, 0)$ and radius 4. Since $f(x)$ represents a nonpositive square root, $f(x)$ is nonpositive and its graph is the lower half of the circle.

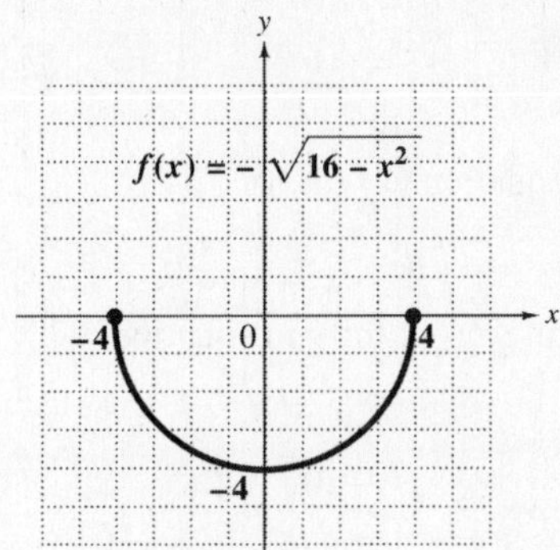

26.
$$x^2 + y^2 = 64$$
$$(x - 0)^2 + (y - 0)^2 = 8^2$$

The last equation is in $(x - h)^2 + (y - k)^2 = r^2$ form. The graph is a *circle*.

27. $y = 2x^2 - 3$
$$y = 2(x - 0)^2 - 3$$

The last equation is in $y = a(x - h)^2 + k$ form. The graph is a *parabola*.

28.
$$y^2 = 2x^2 - 8$$
$$2x^2 - y^2 = 8$$
$$\frac{x^2}{4} - \frac{y^2}{8} = 1 \quad \textit{Divide by 8.}$$

The last equation is in $\frac{x^2}{a^2} - \frac{y^2}{b^2} = 1$ form, so the graph is a *hyperbola*.

29.
$$y^2 = 8 - 2x^2$$
$$2x^2 + y^2 = 8$$
$$\frac{x^2}{4} + \frac{y^2}{8} = 1 \quad \textit{Divide by 8.}$$

The last equation is in $\frac{x^2}{a^2} + \frac{y^2}{b^2} = 1$ form, so the graph is an *ellipse*.

30. $x = y^2 + 4$
$$x = (y - 0)^2 + 4$$

The last equation is in $x = a(y - k)^2 + h$ form, so the graph is a *parabola*.

31. $x^2 - y^2 = 64$
$$\frac{x^2}{64} - \frac{y^2}{64} = 1 \quad \textit{Divide by 64.}$$

The last equation is in $\frac{x^2}{a^2} - \frac{y^2}{b^2} = 1$ form, so the graph is a *hyperbola*.

32. A hyperbola is defined as the set of all points in a plane such that the absolute value of the *difference* of the distances from two fixed points (called *foci*) is constant. The hyperbola shown in the text will have an equation of the form

$$\frac{x^2}{a^2} - \frac{y^2}{b^2} = 1. \quad (1)$$

The constant difference for this hyperbola is

$$|d_1 - d_2| = |80 - 30| = 50.$$

Let Q be the point on the right branch of the hyperbola that is on $\overline{MS}$. Let $x = \overline{MQ}$ and $y = \overline{QS}$. Since $\overline{MS} = 100$ and $\overline{MQ} - \overline{QS}$ must be 50, we have the following system.

$$\begin{aligned} x + y &= 100 \quad (2) \\ x - y &= 50 \\ \hline 2x &= 150 \quad \textit{Add.} \\ x &= 75 \end{aligned}$$

From (2), we see that $y = 25$. The distance from the center of the hyperbola, C, to S, is 50, so $a = \overline{CQ} = 50 - 25 = 25$.
To find b^2, we'll find a point on the hyperbola, substitute for a, x, and y, and then solve for b^2. Let P have coordinates (d, e). If we draw a perpendicular line from P to $\overline{MS}$ (intersecting at point R), we see two right triangles, PRM and PRS. Using the Pythagorean formula, we get the following system of equations.

$$\begin{aligned} (50 + d)^2 + e^2 &= 80^2 \\ (50 - d)^2 + e^2 &= 30^2 \quad (3) \\ \hline (2500 + 100d + d^2) - (2500 - 100d + d^2)& \\ &= 6400 - 900 \quad \textit{Subtract.} \\ 200d &= 5500 \\ d &= 27.5 \end{aligned}$$

From (3) with $d = 27.5$, we get $e^2 = 30^2 - 22.5^2$, so $e = \sqrt{393.75}$. Now substitute 25 for a, 27.5 for x, and $\sqrt{393.75}$ for y in (1) to solve for b^2.

$$\frac{27.5^2}{25^2} - \frac{393.75}{b^2} = 1$$
$$\frac{756.25}{625} - \frac{625}{625} = \frac{393.75}{b^2}$$
$$\frac{131.25}{625} = \frac{393.75}{b^2}$$
$$\frac{625}{131.25} = \frac{b^2}{393.75}$$
$$b^2 = \frac{625}{131.25}(393.75)$$
$$b^2 = 1875$$

Thus, equation (1) becomes

$$\frac{x^2}{625} - \frac{y^2}{1875} = 1.$$

33.
$$2y = 3x - x^2 \quad (1)$$
$$x + 2y = -12 \quad (2)$$

Substitute $3x - x^2$ for $2y$ in equation (2).

$$\begin{aligned} x + 2y &= -12 \quad (2) \\ x + (3x - x^2) &= -12 \\ -x^2 + 4x + 12 &= 0 \\ x^2 - 4x - 12 &= 0 \\ (x-6)(x+2) &= 0 \end{aligned}$$

$$x - 6 = 0 \quad \text{or} \quad x + 2 = 0$$
$$x = 6 \quad \text{or} \quad x = -2$$

Substitute these values for x in equation (2) to find y.

If $x = 6$, then

$$\begin{aligned} x + 2y &= -12 \quad (2) \\ 6 + 2y &= -12 \\ 2y &= -18 \\ y &= -9. \end{aligned}$$

If $x = -2$ then

$$\begin{aligned} x + 2y &= -12 \quad (2) \\ -2 + 2y &= -12 \\ 2y &= -10 \\ y &= -5 \end{aligned}$$

Solution set: $\{(6, -9), (-2, -5)\}$

34.
$$y + 1 = x^2 + 2x \quad (1)$$
$$y + 2x = 4 \quad (2)$$

Solve equation (2) for y.

$$y = 4 - 2x \quad (3)$$

Substitute $4 - 2x$ for y in (1).

$$\begin{aligned} (4 - 2x) + 1 &= x^2 + 2x \\ 0 &= x^2 + 4x - 5 \\ 0 &= (x+5)(x-1) \end{aligned}$$

$$x + 5 = 0 \quad \text{or} \quad x - 1 = 0$$
$$x = -5 \quad \text{or} \quad x = 1$$

Substitute these values for x in equation (3) to find y.
If $x = -5$, then $y = 4 - 2(-5) = 14$.
If $x = 1$, then $y = 4 - 2(1) = 2$.

Solution set: $\{(1, 2), (-5, 14)\}$

35.
$$x^2 + 3y^2 = 28 \quad (1)$$
$$y - x = -2 \quad (2)$$

Solve equation (2) for y.

$$y = x - 2 \quad (3)$$

Substitute $x - 2$ for y in equation (1).

$$\begin{aligned} x^2 + 3y^2 &= 28 \quad (1) \\ x^2 + 3(x-2)^2 &= 28 \\ x^2 + 3(x^2 - 4x + 4) - 28 &= 0 \\ x^2 + 3x^2 - 12x + 12 - 28 &= 0 \\ 4x^2 - 12x - 16 &= 0 \\ 4(x^2 - 3x - 4) &= 0 \\ 4(x-4)(x+1) &= 0 \end{aligned}$$

$$x - 4 = 0 \quad \text{or} \quad x + 1 = 0$$
$$x = 4 \quad \text{or} \quad x = -1$$

Substitute these values for x in equation (3) to find y.
If $x = 4$, then $y = 4 - 2 = 2$.
If $x = -1$, then $y = -1 - 2 = -3$.

Solution set: $\{(4, 2), (-1, -3)\}$

36.
$$xy = 8 \quad (1)$$
$$x - 2y = 6 \quad (2)$$

Solve equation (2) for x.

$$x = 2y + 6 \quad (3)$$

Substitute $2y + 6$ for x in equation (1) to find y.

$$\begin{aligned} xy &= 8 \quad (1) \\ (2y + 6)y &= 8 \\ 2y^2 + 6y - 8 &= 0 \\ 2(y^2 + 3y - 4) &= 0 \\ 2(y+4)(y-1) &= 0 \end{aligned}$$

$$y + 4 = 0 \quad \text{or} \quad y - 1 = 0$$
$$y = -4 \quad \text{or} \quad y = 1$$

Substitute these values for y in equation (3) to find x.
If $y = -4$, then $x = 2(-4) + 6 = -2$.
If $y = 1$, then $x = 2(1) + 6 = 8$.

Solution set: $\{(-2, -4), (8, 1)\}$

37. $x^2 + y^2 = 6 \quad (1)$
$x^2 - 2y^2 = -6 \quad (2)$

Multiply equation (2) by -1 and add the result to equation (1).

$$\begin{aligned} x^2 + y^2 &= 6 \quad (1) \\ -x^2 + 2y^2 &= 6 \qquad -1 \times (2) \\ \hline 3y^2 &= 12 \\ y^2 &= 4 \end{aligned}$$

$$y = 2 \quad \text{or} \quad y = -2$$

Substitute these values for y in equation (1) to find x.
If $y = \pm 2$, then

$$\begin{aligned} x^2 + y^2 &= 6 \qquad (1) \\ x^2 + (\pm 2)^2 &= 6 \\ x^2 + 4 &= 6 \\ x^2 &= 2. \end{aligned}$$

$$x = \sqrt{2} \quad \text{or} \quad x = -\sqrt{2}$$

Since each value of x can be paired with each value of y, there are four points of intersection.

Solution set:
$\{(\sqrt{2}, 2), (-\sqrt{2}, 2), (\sqrt{2}, -2), (-\sqrt{2}, -2)\}$

38. $3x^2 - 2y^2 = 12 \quad (1)$
$x^2 + 4y^2 = 18 \quad (2)$

Multiply equation (1) by 2 and add the result to equation (2).

$$\begin{aligned} 6x^2 - 4y^2 &= 24 \qquad 2 \times (1) \\ x^2 + 4y^2 &= 18 \quad (2) \\ \hline 7x^2 &= 42 \\ x^2 &= 6 \end{aligned}$$

$$x = \sqrt{6} \quad \text{or} \quad x = -\sqrt{6}$$

Substitute these values for x in equation (2) to find y.
If $x = \pm\sqrt{6}$, then

$$\begin{aligned} x^2 + 4y^2 &= 18. \qquad (2) \\ (\pm\sqrt{6})^2 + 4y^2 &= 18 \\ 6 + 4y^2 &= 18 \\ 4y^2 &= 12 \\ y^2 &= 3 \end{aligned}$$

$$y = \sqrt{3} \quad \text{or} \quad y = -\sqrt{3}$$

Since each value of x can be paired with each value of y, there are four points of intersection.

Solution set: $\{(\sqrt{6}, \sqrt{3}), (\sqrt{6}, -\sqrt{3}),$
$(-\sqrt{6}, \sqrt{3}), (-\sqrt{6}, -\sqrt{3})\}$

39. A circle and a line can intersect in zero, one, or two points, so zero, one, or two solutions are possible.

40. A parabola and a hyperbola can intersect in zero, one, two, three, or four points, so zero, one, two, three, or four solutions are possible.

41.
$$\begin{aligned} 9x^2 &\ge 16y^2 + 144 \\ 9x^2 - 16y^2 &\ge 144 \\ \frac{x^2}{16} - \frac{y^2}{9} &\ge 1 \end{aligned}$$

The boundary, $\frac{x^2}{16} - \frac{y^2}{9} = 1$, is a hyperbola with x-intercepts $(4, 0)$ and $(-4, 0)$. The asymptotes are the extended diagonals of the rectangle with vertices $(4, 3)$, $(4, -3)$, $(-4, -3)$, and $(-4, 3)$. Test $(0, 0)$.

$$\begin{aligned} 9(0)^2 &\overset{?}{\ge} 16(0)^2 + 144 \\ 0 &\ge 144 \qquad \textit{False} \end{aligned}$$

Shade the sides of the hyperbola that do not contain $(0, 0)$. These are the regions inside the branches of the hyperbola. The hyperbola is drawn as a solid curve since the points of the hyperbola satisfy the inequality.

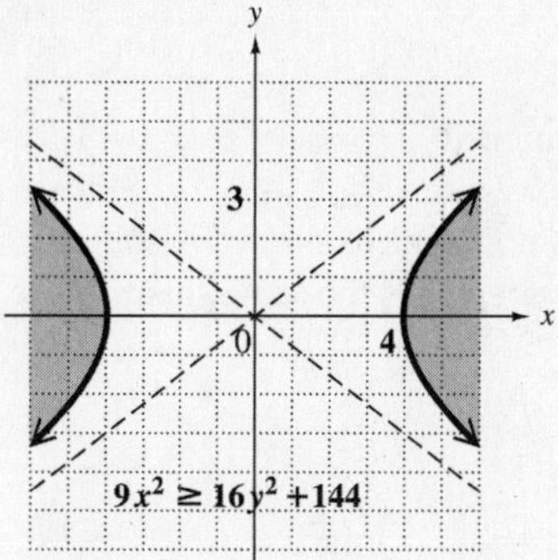

42. $4x^2 + y^2 \geq 16$

$$\frac{x^2}{4} + \frac{y^2}{16} \geq 1$$

The boundary, $\frac{x^2}{4} + \frac{y^2}{16} = 1$, is an ellipse with intercepts $(2,0)$, $(-2,0)$, $(0,4)$, and $(0,-4)$. Test $(0,0)$.

$$4(0)^2 + 0^2 \overset{?}{\geq} 16$$
$$0 \geq 16 \quad \textit{False}$$

Shade the side of ellipse that does not contain $(0,0)$. This is the region outside the ellipse. The ellipse is drawn as a solid curve since the points of the ellipse satisfy the inequality.

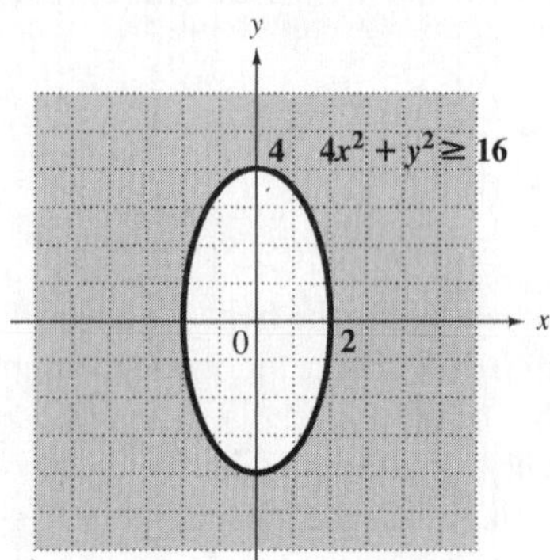

43. $y < -(x+2)^2 + 1$

The boundary, $y = -(x+2)^2 + 1$, is a vertical parabola with vertex $(-2,1)$. Since $a = -1 < 0$, the parabola opens down. Also, $|a| = |-1| = 1$, so the graph has the same shape as the graph of $y = x^2$. Test $(0,0)$.

$$0 \overset{?}{<} -(0+2)^2 + 1$$
$$0 \overset{?}{<} -(4) + 1$$
$$0 < -3 \qquad \textit{False}$$

Shade the side of the parabola that does not contain $(0,0)$. This is the region inside the parabola. The parabola is drawn as a dashed curve since the points of the parabola do not satisfy the inequality.

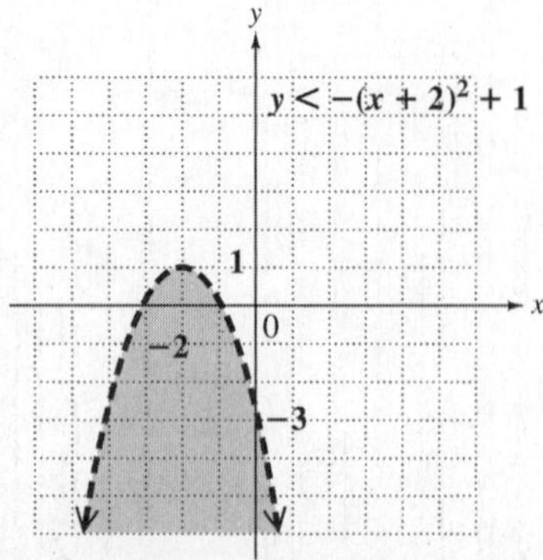

44. $2x + 5y \leq 10$
$3x - y \leq 6$

The boundary, $2x + 5y = 10$ is a solid line with intercepts $(5,0)$ and $(0,2)$. Test $(0,0)$.

$$2(0) + 5(0) \overset{?}{\leq} 10$$
$$0 \leq 10 \quad \textit{True}$$

Shade the side of the line that contains $(0,0)$. The boundary, $3x - y = 6$, is a solid line with intercepts $(2,0)$ and $(0,-6)$. Test $(0,0)$.

$$3(0) - 0 \overset{?}{\leq} 6$$
$$0 \leq 6 \quad \textit{True}$$

Shade the side of the line that contains $(0,0)$. The graph of the system is the intersection of the two shaded regions.

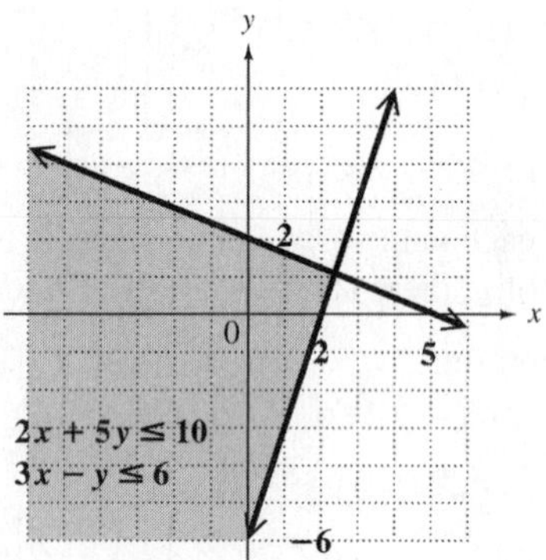

45. $|x| \leq 2$
$|y| > 1$
$4x^2 + 9y^2 \leq 36$

The equation of the boundary, $|x| = 2$, can be written as

$$x = -2 \quad \text{or} \quad x = 2.$$

The graph is these two solid vertical lines. Since $|0| \leq 2$ is true, the region between the lines, containing $(0,0)$, is shaded.
The boundary, $|y| = 1$, consists of the two dashed horizontal lines $y = 1$ and $y = -1$. Since $|0| > 1$ is false, the regions above and below the lines, not containing $(0,0)$, are shaded.
The boundary given by

$$4x^2 + 9y^2 = 36$$
$$\text{or} \quad \frac{x^2}{9} + \frac{y^2}{4} = 1$$

is graphed as a solid ellipse with intercepts $(3,0)$, $(-3,0)$, $(0,2)$, and $(0,-2)$. Test $(0,0)$.

$$4(0)^2 + 9(0)^2 \overset{?}{\leq} 36$$
$$0 \leq 36 \quad \textit{True}$$

The region inside the ellipse, containing $(0, 0)$, is shaded.
The graph of the system consists of the regions that include the common points of the three shaded regions.

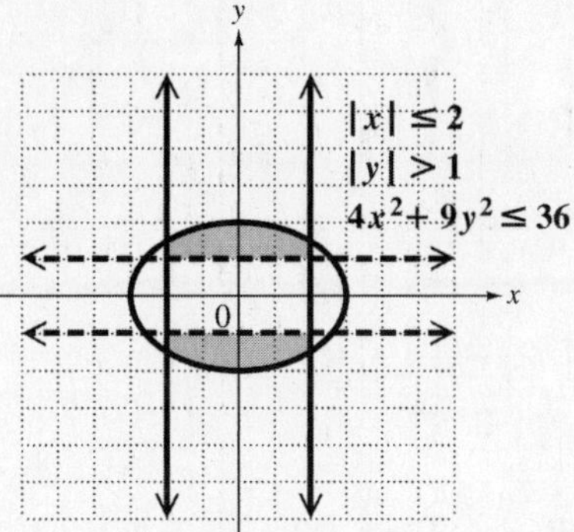

46. $$9x^2 \le 4y^2 + 36$$
$$x^2 + y^2 \le 16$$

The equation of the first boundary is

$$9x^2 = 4y^2 + 36$$
$$9x^2 - 4y^2 = 36$$
$$\frac{x^2}{4} - \frac{y^2}{9} = 1.$$

The graph is a solid hyperbola with x-intercepts $(2, 0)$ and $(-2, 0)$. The asymptotes are the extended diagonals of the rectangle with vertices $(2, 3)$, $(2, -3)$, $(-2, -3)$, and $(-2, 3)$. Test $(0, 0)$.

$$9(0)^2 \stackrel{?}{\le} 4(0)^2 + 36$$
$$0 \le 36 \qquad \textit{True}$$

Shade the region between the branches of the hyperbola that contains $(0, 0)$.
The equation of the second boundary is $x^2 + y^2 = 16$. This is a solid circle with center $(0, 0)$ and radius 4. Test $(0, 0)$.

$$0^2 + 0^2 \stackrel{?}{\le} 16$$
$$0 \le 16 \quad \textit{True}$$

Shade the region inside the circle.
The graph of the system is the intersection of the shaded regions which is between the two branches of the hyperbola and inside the circle.

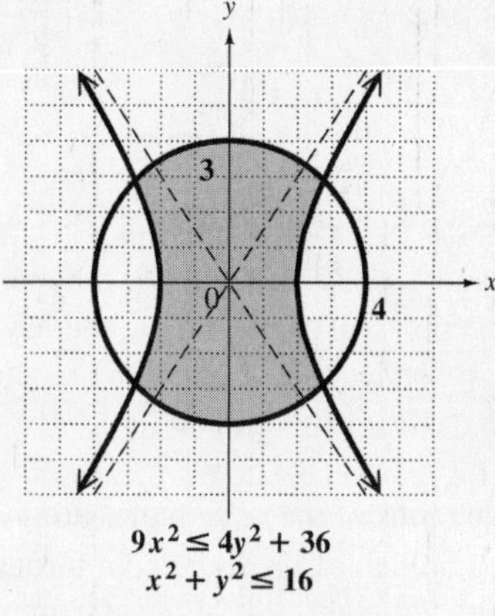

47. **[12.2]** $x^2 + y^2 = 25$ is in the form $(x - h)^2 + (y - k)^2 = r^2$. The graph is a circle with center at $(0, 0)$ and radius 5.

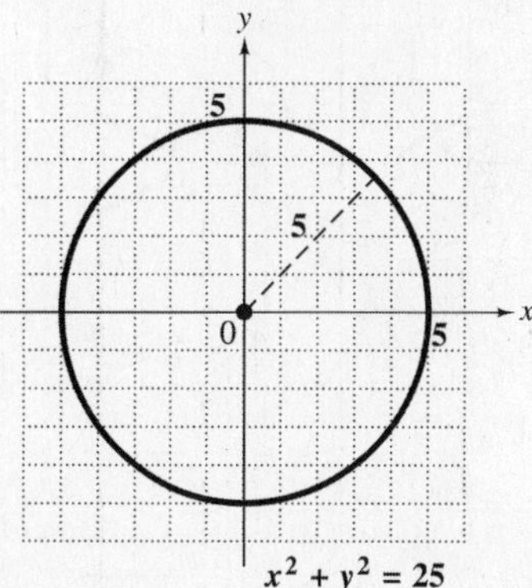

48. **[12.2]** $x^2 + 9y^2 = 9$ in $\frac{x^2}{a^2} + \frac{y^2}{b^2} = 1$ form is $\frac{x^2}{9} + \frac{y^2}{1} = 1$ with $a = 3$ and $b = 1$. The graph is an ellipse with x-intercepts $(3, 0)$ and $(-3, 0)$ and y-intercepts $(0, 1)$ and $(0, -1)$. Plot the intercepts, and draw the ellipse through them.

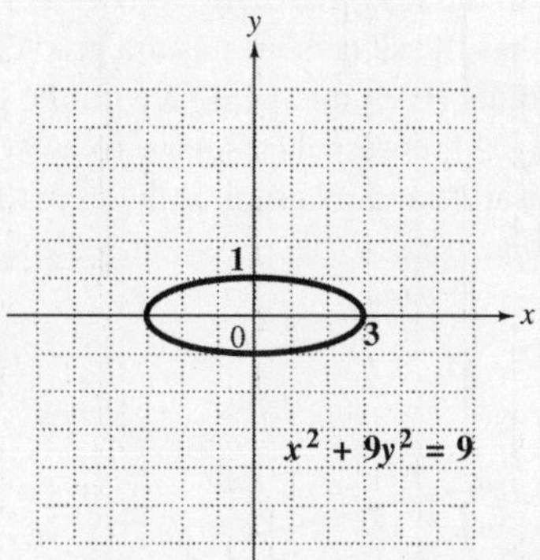

49. **[12.3]** $x^2 - 9y^2 = 9$

$$\frac{x^2}{9} - \frac{y^2}{1} = 1$$

The equation is in the $\frac{x^2}{a^2} - \frac{y^2}{b^2} = 1$ form with $a = 3$ and $b = 1$. The graph is a hyperbola with x-intercepts $(3, 0)$ and $(-3, 0)$ and asymptotes that are the extended diagonals of the rectangle with vertices $(3, 1)$, $(3, -1)$, and $(-3, -1)$, and $(-3, 1)$.
The equations of the asymptotes are $y = \pm\frac{1}{3}x$.
Graph a branch of the hyperbola through each intercept and approaching the asymptotes.

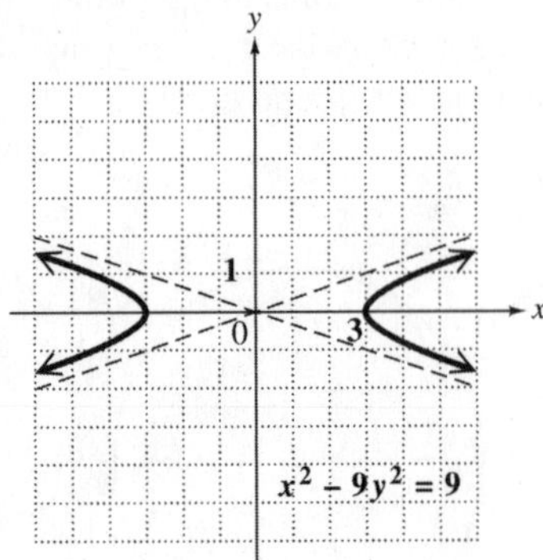

50. **[12.3]** $f(x) = \sqrt{4 - x}$
Replace $f(x)$ with y.

$$y = \sqrt{4 - x}$$

Square both sides.

$$y^2 = 4 - x$$
$$x = -y^2 + 4$$
$$x = -1(y - 0)^2 + 4$$

This equation is the graph of a horizontal parabola with vertex $(4, 0)$. Since $a = -1 < 0$, the graph opens to the left. Also, $|a| = |-1| = 1$, so the graph has the same shape as the graph of $y = x^2$. The points $(0, 2)$ and $(3, 1)$ are on the graph. Since $f(x)$ represents a square root, $f(x)$ is nonnegative and its graph is the upper half of the parabola.

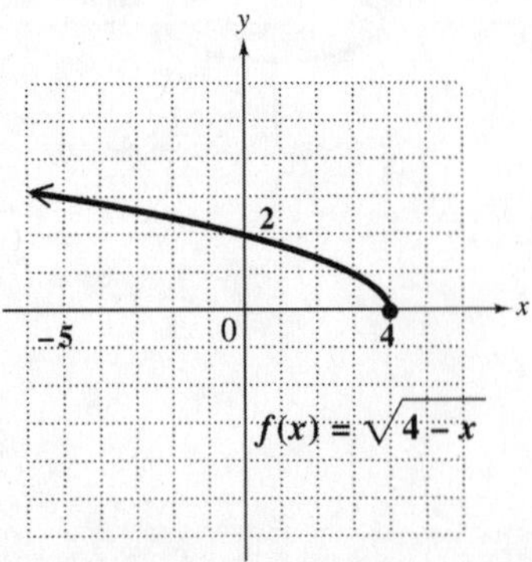

51. **[12.1]** $f(x) = [\![x]\!] - 1$

This is the graph of the greatest integer function, $g(x) = [\![x]\!]$, shifted down 1 unit.

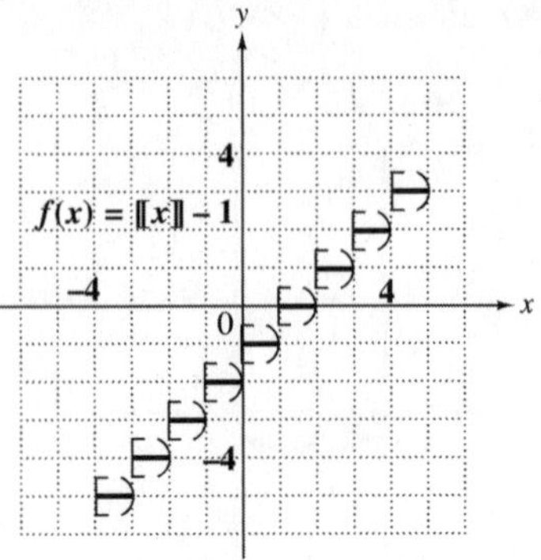

52. **[12.5]** $4y > 3x - 12$
$x^2 < 16 - y^2$

The boundary $4y = 3x - 12$ is a dashed line with intercepts $(0, -3)$ and $(4, 0)$. Test $(0, 0)$.

$$4(0) \overset{?}{>} 3(0) - 12$$
$$0 > -12 \qquad \textit{True}$$

Shade the side of the line that contains $(0, 0)$.
The boundary $x^2 = 16 - y^2$, or $x^2 + y^2 = 16$, is a dashed circle with center at $(0, 0)$ and radius 4.
Test $(0, 0)$.

$$0^2 \overset{?}{<} 16 - 0^2$$
$$0 < 16 \qquad \textit{True}$$

Shade the region inside the circle.
The graph of the system is the intersection of the two shaded regions.

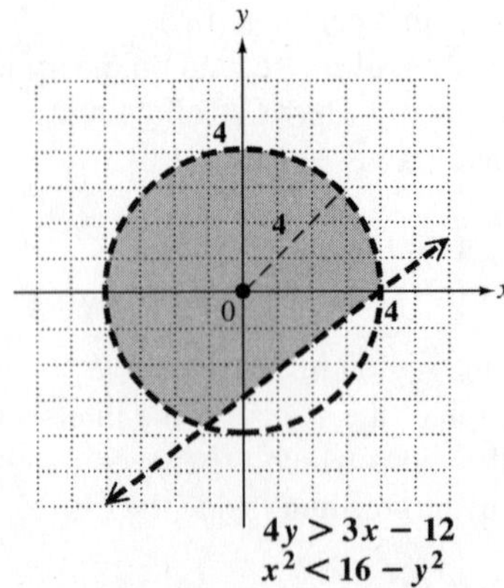

53. **[12.2]** There are cases where one x-value will yield two y-values. In a function, every x yields one and only one y.

Chapter 12 Test

1. $f(x) = \sqrt{x - 2}$ is the graph of $g(x) = \sqrt{x}$ shifted 2 units right. The graph is **C**.

2. $f(x) = \sqrt{x+2}$ is the graph of $g(x) = \sqrt{x}$ shifted 2 units left. The graph is **A**.

3. $f(x) = \sqrt{x} + 2$ is the graph of $g(x) = \sqrt{x}$ shifted 2 units up. The graph is **D**.

4. $f(x) = \sqrt{x} - 2$ is the graph of $g(x) = \sqrt{x}$ shifted 2 units down. The graph is **B**.

5. $f(x) = |x - 3| + 4$

This is the graph of the absolute value function, $g(x) = |x|$, shifted 3 units to the right (since $x - 3 = 0$ if $x = 3$) and 4 units upward (because of the $+4$). Its vertex is at $(3, 4)$.

x	0	1	2	3	4	5	6
y	7	6	5	4	5	6	7

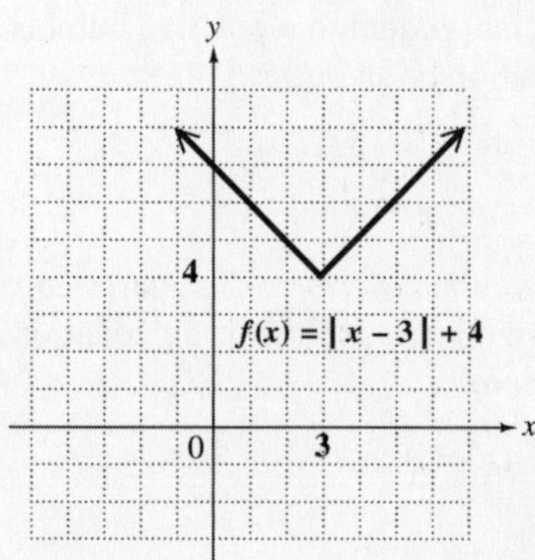

Since x can be any real number, the domain is $(-\infty, \infty)$.

The value of y is always greater than or equal to 4, so the range is $[4, \infty)$.

6. $f(x) = 3x + 5, g(x) = x^2 + 2$

(a)
$$\begin{aligned}(f \circ g)(-2) &= f(g(-2)) \\ &= f((-2)^2 + 2) \\ &= f(6) \\ &= 3 \cdot 6 + 5 \\ &= 23\end{aligned}$$

(b)
$$\begin{aligned}(f \circ g)(x) &= f(g(x)) \\ &= f(x^2 + 2) \\ &= 3(x^2 + 2) + 5 \\ &= 3x^2 + 6 + 5 \\ &= 3x^2 + 11\end{aligned}$$

(c)
$$\begin{aligned}(g \circ f)(x) &= g(f(x)) \\ &= g(3x + 5) \\ &= (3x + 5)^2 + 2 \\ &= 9x^2 + 30x + 25 + 2 \\ &= 9x^2 + 30x + 27\end{aligned}$$

7.
$$\begin{aligned}(x-2)^2 + (y+3)^2 &= 16 \\ (x-2)^2 + [y-(-3)]^2 &= 4^2\end{aligned}$$

The graph is a circle with center $(2, -3)$ and radius 4.

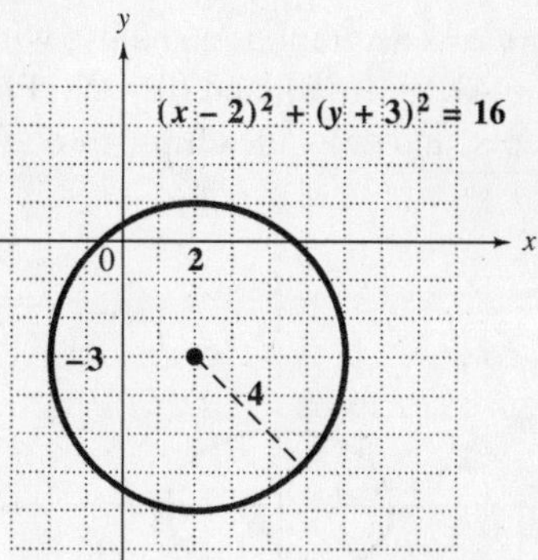

8. $x^2 + y^2 + 8x - 2y = 8$

To find the center and radius, complete the squares on x and y.

$$\begin{aligned}(x^2 + 8x \quad\quad) + (y^2 - 2y \quad\quad) &= 8 \\ (x^2 + 8x + \underline{16}) + (y^2 - 2y + \underline{1}) &= 8 + \underline{16} + \underline{1} \\ (x+4)^2 + (y-1)^2 &= 25\end{aligned}$$

The graph is a circle with center $(-4, 1)$ and radius $\sqrt{25} = 5$.

9. $f(x) = \sqrt{9 - x^2}$

Replace $f(x)$ with y.

$$y = \sqrt{9 - x^2}$$

Square each side.

$$\begin{aligned}y^2 &= 9 - x^2 \\ x^2 + y^2 &= 9\end{aligned}$$

The graph of $x^2 + y^2 = 9$ is a circle of radius $\sqrt{9} = 3$ centered at the origin. Since $f(x)$ is nonnegative, only the top half of the circle is graphed.

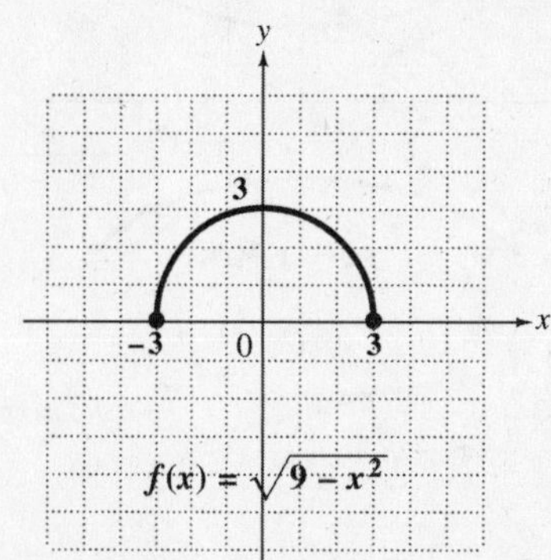

10. $4x^2 + 9y^2 = 36$

$$\frac{x^2}{9} + \frac{y^2}{4} = 1 \quad \textit{Divide by 36.}$$

The equation is in $\frac{x^2}{a^2} + \frac{y^2}{b^2} = 1$ form with $a = 3$ and $b = 2$. The graph is an ellipse with intercepts $(3, 0)$, $(-3, 0)$, $(0, 2)$, and $(0, -2)$. Plot these intercepts, and draw the ellipse through them.

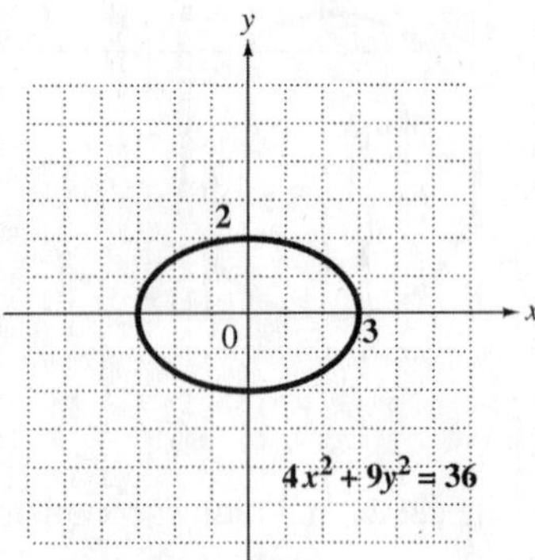

11. $16y^2 - 4x^2 = 64$

$$\frac{y^2}{4} - \frac{x^2}{16} = 1 \quad \textit{Divide by 64.}$$

The equation is in $\frac{y^2}{b^2} - \frac{x^2}{a^2} = 1$ form with $a = 4$ and $b = 2$. The graph is a hyperbola with y-intercepts $(0, 2)$ and $(0, -2)$ and asymptotes that are the extended diagonals of the rectangle with vertices $(4, 2)$, $(4, -2)$, $(-4, -2)$, and $(-4, 2)$.

The equations of the asymptotes are $y = \pm\frac{2}{4}x = \pm\frac{1}{2}x$. Draw a branch of the hyperbola through each intercept and approaching the asymptotes.

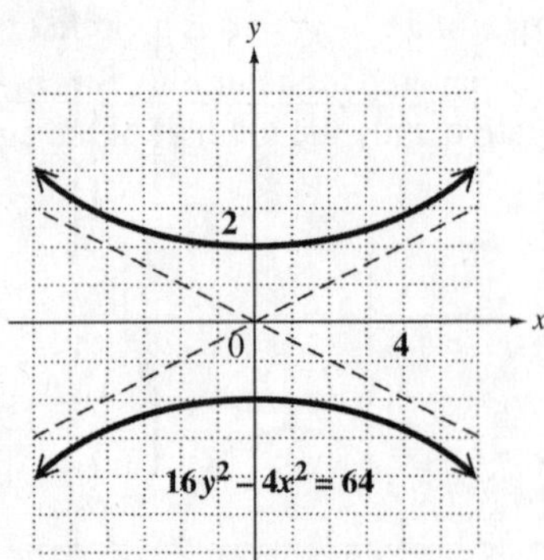

12.

$$\frac{y}{2} = -\sqrt{1 - \frac{x^2}{9}}$$

$$\frac{y^2}{4} = 1 - \frac{x^2}{9} \quad \textit{Square both sides.}$$

$$\frac{x^2}{9} + \frac{y^2}{4} = 1$$

This is an ellipse with x-intercepts $(3, 0)$ and $(-3, 0)$ and y-intercepts $(0, 2)$ and $(0, -2)$. Since y represents a negative square root in the original equation, y must be nonpositive. This restricts the graph to the lower half of the ellipse.

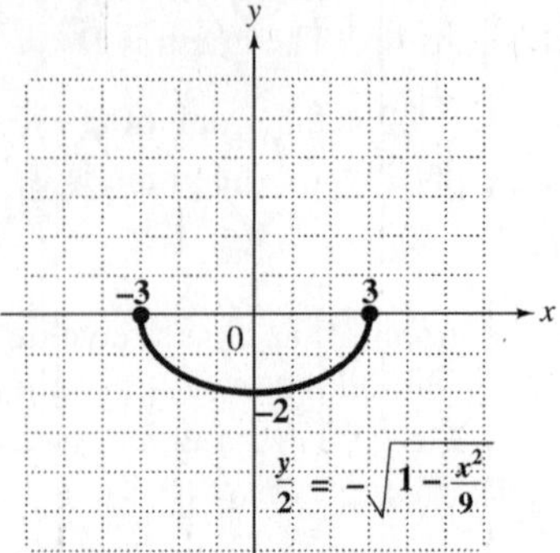

13. $6x^2 + 4y^2 = 12$

We have the *sum* of squares with different coefficients equal to a positive number, so this is an equation of an *ellipse*.

14.

$$16x^2 = 144 + 9y^2$$
$$16x^2 - 9y^2 = 144$$

We have the *difference* of squares equal to a positive number, so this is an equation of a *hyperbola*.

15.

$$4y^2 + 4x = 9$$
$$4x = -4y^2 + 9$$

We have an x-term and a y^2-term, so this is an equation of a horizontal *parabola*.

16.

$$2x - y = 9 \quad (1)$$
$$xy = 5 \quad (2)$$

Solve equation (1) for y.

$$y = 2x - 9 \quad (3)$$

Substitute $2x - 9$ for y in equation (2).

$$xy = 5 \quad (2)$$
$$x(2x - 9) = 5$$
$$2x^2 - 9x = 5$$
$$2x^2 - 9x - 5 = 0$$
$$(2x + 1)(x - 5) = 0$$

$$2x + 1 = 0 \quad \text{or} \quad x - 5 = 0$$
$$x = -\tfrac{1}{2} \quad \text{or} \quad x = 5$$

Substitute these values for x in equation (3) to find y.
If $x = -\frac{1}{2}$, then $y = 2(-\frac{1}{2}) - 9 = -10$.
If $x = 5$, then $y = 2(5) - 9 = 1$.

Solution set: $\{(-\frac{1}{2}, -10), (5, 1)\}$

17. $x - 4 = 3y \quad (1)$
$x^2 + y^2 = 8 \quad (2)$

Solve equation (1) for x.

$$x = 3y + 4 \quad (3)$$

Substitute $3y + 4$ for x in equation (2).

$$\begin{aligned} x^2 + y^2 &= 8 \quad (2) \\ (3y+4)^2 + y^2 &= 8 \\ 9y^2 + 24y + 16 + y^2 &= 8 \\ 10y^2 + 24y + 8 &= 0 \\ 2(5y^2 + 12y + 4) &= 0 \\ 2(5y+2)(y+2) &= 0 \end{aligned}$$

$$5y + 2 = 0 \quad \text{or} \quad y + 2 = 0$$
$$y = -\tfrac{2}{5} \quad \text{or} \quad y = -2$$

Substitute these values for y in equation (3) to find x.

If $y = -\frac{2}{5}$, then

$$x = 3(-\tfrac{2}{5}) + 4 = -\tfrac{6}{5} + 4 = \tfrac{14}{5}.$$

If $y = -2$, then $x = 3(-2) + 4 = -2$.

Solution set: $\{(-2, -2), (\frac{14}{5}, -\frac{2}{5})\}$

18. $x^2 + y^2 = 25 \quad (1)$
$x^2 - 2y^2 = 16 \quad (2)$

Multiply equation (1) by 2 and add the result to equation (2).

$$\begin{aligned} 2x^2 + 2y^2 &= 50 \quad 2 \times (1) \\ x^2 - 2y^2 &= 16 \quad (2) \\ \hline 3x^2 &= 66 \\ x^2 &= 22 \end{aligned}$$

$$x = \sqrt{22} \quad \text{or} \quad x = -\sqrt{22}$$

Substitute 22 for x^2 in equation (1).

$$\begin{aligned} x^2 + y^2 &= 25 \quad (1) \\ 22 + y^2 &= 25 \\ y^2 &= 3 \end{aligned}$$

$$y = \sqrt{3} \quad \text{or} \quad y = -\sqrt{3}$$

Since $\sqrt{22}$ can be paired with either $\sqrt{3}$ or $-\sqrt{3}$ and $-\sqrt{22}$ can be paired with either $\sqrt{3}$ or $-\sqrt{3}$, there are four possible solutions.

Solution set: $\{(\sqrt{22}, \sqrt{3}), (\sqrt{22}, -\sqrt{3}), (-\sqrt{22}, \sqrt{3}), (-\sqrt{22}, -\sqrt{3})\}$

19. $y < x^2 - 2$

The boundary, $y = x^2 - 2$, is a parabola in $y = a(x-h)^2 + k$ form with vertex (h, k) at $(0, -2)$. Since $a = 1 > 0$, the parabola opens up. It also has the same shape as $y = x^2$. The points $(2, 2)$ and $(-2, 2)$ are on the graph. Test $(0, 0)$.

$$0 \overset{?}{\le} (0)^2 - 2$$
$$0 \le -2 \quad \textit{False}$$

Shade the side of the parabola that does not contain $(0, 0)$. This is the region outside the parabola. The parabola is drawn as a dashed curve since the points of the parabola do not satisfy the inequality.

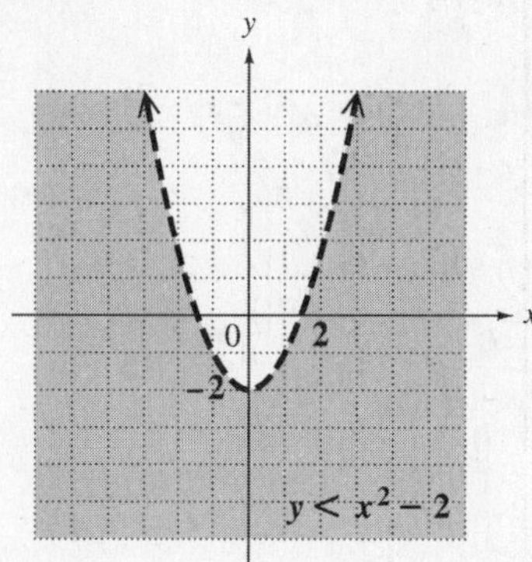

20. $x^2 + 25y^2 \le 25$
$x^2 + y^2 \le 9$

The first boundary, $\frac{x^2}{25} + \frac{y^2}{1} = 1$, is a solid ellipse with intercepts $(5, 0)$, $(-5, 0)$, $(0, 1)$, and $(0, -1)$. Test $(0, 0)$.

$$0^2 + 25 \cdot 0^2 \overset{?}{\le} 25$$
$$0 \le 25 \quad \textit{True}$$

Shade the region inside the ellipse.
The second boundary, $x^2 + y^2 = 9$, is a solid circle with center $(0, 0)$ and radius 3. Test $(0, 0)$.

$$0^2 + 0^2 \overset{?}{\le} 9$$
$$0 \le 9 \quad \textit{True}$$

Shade the region inside the circle. The solution of the system is the intersection of the two shaded regions.

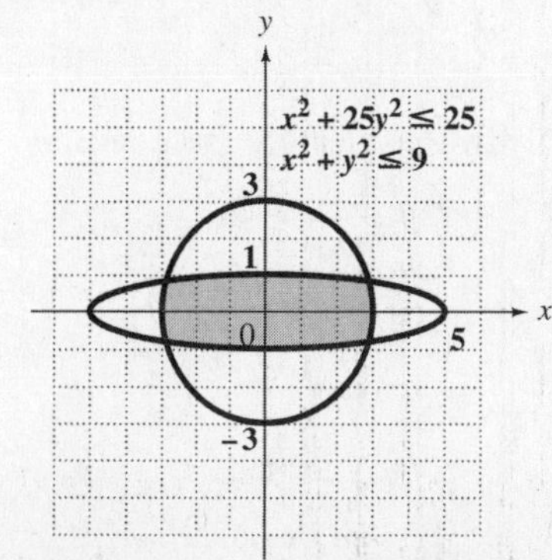

Cumulative Review Exercises (Chapters 1–12)

1.
$$\begin{aligned}4-(2x+3)+x&=5x-3\\4-2x-3+x&=5x-3\\-x+1&=5x-3\\-6x&=-4\\x&=\tfrac{2}{3}\end{aligned}$$

Solution set: $\{\frac{2}{3}\}$

2.
$$\begin{aligned}-4k+7&\geq 6k+1\\-10k&\geq -6\end{aligned}$$

Divide by -10; reverse the direction of the inequality.

$$\begin{aligned}k&\leq \frac{-6}{-10}\\k&\leq \frac{3}{5}\end{aligned}$$

Solution set: $(-\infty, \frac{3}{5}]$

3.
$$\begin{aligned}|5m|-6&=14\\|5m|&=20\end{aligned}$$

$$\begin{aligned}5m=20 \quad &\text{or} \quad 5m=-20\\m=4 \quad &\text{or} \quad m=-4\end{aligned}$$

Solution set: $\{-4, 4\}$

4. Let $(x_1, y_1) = (2, 5)$ and $(x_2, y_2) = (-4, 1)$.

$$m = \frac{y_2 - y_1}{x_2 - x_1} = \frac{1-5}{-4-2} = \frac{-4}{-6} = \frac{2}{3}$$

5. Through $(-3, -2)$; perpendicular to $2x - 3y = 7$

Write $2x - 3y = 7$ in slope-intercept form.

$$\begin{aligned}-3y&=-2x+7\\y&=\tfrac{2}{3}x-\tfrac{7}{3}\end{aligned}$$

The slope is $\frac{2}{3}$. Perpendicular lines have slopes that are negative reciprocals of each other, so a line perpendicular to the given line will have slope $-\frac{3}{2}$. Let $m = -\frac{3}{2}$ and $(x_1, y_1) = (-3, -2)$ in the point-slope form.

$$\begin{aligned}y-y_1&=m(x-x_1)\\y-(-2)&=-\tfrac{3}{2}[x-(-3)]\\y+2&=-\tfrac{3}{2}(x+3)\end{aligned}$$

Multiply by 2 to clear the fraction.

$$\begin{aligned}2y+4&=-3(x+3)\\2y+4&=-3x-9\\3x+2y&=-13\end{aligned}$$

6.
$$\begin{aligned}3x - y &= 12 \quad (1)\\2x + 3y &= -3 \quad (2)\end{aligned}$$

Multiply equation (1) by 3 and add the result to equation (2).

$$\begin{array}{rcll}9x - 3y &=& 36 & 3\times(1)\\2x + 3y &=& -3 & (2)\\\hline 11x &=& 33 &\\x &=& 3 &\end{array}$$

Substitute 3 for x in equation (1) to find y.

$$\begin{aligned}3x-y&=12 \quad (1)\\3(3)-y&=12\\9-y&=12\\-y&=3\\y&=-3\end{aligned}$$

Solution set: $\{(3, -3)\}$

7.
$$\begin{aligned}x + y - 2z &= 9 \quad (1)\\2x + y + z &= 7 \quad (2)\\3x - y - z &= 13 \quad (3)\end{aligned}$$

Add equation (2) and equation (3).

$$\begin{array}{rcll}2x + y + z &=& 7 & (2)\\3x - y - z &=& 13 & (3)\\\hline 5x &=& 20 &\\x &=& 4 &\end{array}$$

Multiply equation (1) by -1 and add the result to equation (2).

$$\begin{array}{rcll}-x - y + 2z &=& -9 & -1\times(1)\\2x + y + z &=& 7 & (2)\\\hline x + 3z &=& -2 & (4)\end{array}$$

Substitute 4 for x in equation (4) to find z.

$$\begin{aligned}x+3z&=-2 \quad (4)\\4+3z&=-2\\3z&=-6\\z&=-2\end{aligned}$$

Substitute 4 for x and -2 for z in equation (2) to find y.

$$\begin{aligned}2x+y+z&=7 \quad (2)\\2(4)+y-2&=7\\y+6&=7\\y&=1\end{aligned}$$

Solution set: $\{(4, 1, -2)\}$

8. $xy = -5 \quad (1)$
$2x + y = 3 \quad (2)$

Solve equation (2) for y.

$$y = -2x + 3 \quad (3)$$

Substitute $-2x + 3$ for y in equation (1).

$$\begin{aligned} xy &= -5 \quad (1) \\ x(-2x+3) &= -5 \\ -2x^2 + 3x &= -5 \\ -2x^2 + 3x + 5 &= 0 \\ 2x^2 - 3x - 5 &= 0 \\ (2x-5)(x+1) &= 0 \end{aligned}$$

$$2x - 5 = 0 \quad \text{or} \quad x + 1 = 0$$
$$x = \tfrac{5}{2} \quad \text{or} \quad x = -1$$

Substitute these values for x in equation (3) to find y.
If $x = \frac{5}{2}$, then $y = -2(\frac{5}{2}) + 3 = -2.$

If $x = -1$, then $y = -2(-1) + 3 = 5.$

Solution set: $\{(-1, 5), (\frac{5}{2}, -2)\}$

9. $(5y-3)^2 = (5y)^2 - 2(5y)3 + 3^2$
$= 25y^2 - 30y + 9$

10. $(2r+7)(6r-1)$
$= 12r^2 - 2r + 42r - 7$ *FOIL*
$= 12r^2 + 40r - 7$

11. $\dfrac{8x^4 - 4x^3 + 2x^2 + 13x + 8}{2x+1}$

$$\begin{array}{r|l} & 4x^3 - 4x^2 + 3x + 5 \\ \hline 2x+1 & 8x^4 - 4x^3 + 2x^2 + 13x + 8 \\ & \underline{8x^4 + 4x^3} \\ & -8x^3 + 2x^2 \\ & \underline{-8x^3 - 4x^2} \\ & 6x^2 + 13x \\ & \underline{6x^2 + 3x} \\ & 10x + 8 \\ & \underline{10x + 5} \\ & 3 \end{array}$$

The answer is

$$4x^3 - 4x^2 + 3x + 5 + \frac{3}{2x+1}.$$

12. $12x^2 - 7x - 10 = (4x-5)(3x+2)$

13. $z^4 - 1$
$= (z^2+1)(z^2-1)$ *Diff. of squares*
$= (z^2+1)(z+1)(z-1)$ *Diff. of squares*

14. $a^3 - 27b^3$
$= a^3 - (3b)^3$
$= (a-3b)(a^2 + 3ab + 9b^2)$ *Diff. of cubes*

15. $\dfrac{y^2-4}{y^2-y-6} \div \dfrac{y^2-2y}{y-1}$

$= \dfrac{y^2-4}{y^2-y-6} \cdot \dfrac{y-1}{y^2-2y}$ *Multiply by the reciprocal.*

$= \dfrac{(y+2)(y-2)}{(y-3)(y+2)} \cdot \dfrac{(y-1)}{y(y-2)}$ *Factor.*

$= \dfrac{y-1}{y(y-3)}$ *Simplify.*

16. $\dfrac{5}{c+5} - \dfrac{2}{c+3}$

The LCD is $(c+5)(c+3)$.

$= \dfrac{5(c+3)}{(c+5)(c+3)} - \dfrac{2(c+5)}{(c+3)(c+5)}$

$= \dfrac{5c + 15 - 2c - 10}{(c+5)(c+3)}$

$= \dfrac{3c+5}{(c+5)(c+3)}$

17. $\dfrac{p}{p^2+p} + \dfrac{1}{p^2+p} = \dfrac{p+1}{p^2+p}$

$= \dfrac{p+1}{p(p+1)} = \dfrac{1}{p}$

18. Let $x =$ the time to do the job working together.

Make a chart.

Worker	Rate	Time Together	Part of Job Done
Kareem	$\frac{1}{3}$	x	$\frac{x}{3}$
Jamal	$\frac{1}{2}$	x	$\frac{x}{2}$

Part done by Kareem plus part done by Jamal equals 1 whole job.

$$\frac{x}{3} + \frac{x}{2} = 1$$

Multiply by the LCD, 6.

$$\begin{aligned} 6\left(\frac{x}{3} + \frac{x}{2}\right) &= 6 \cdot 1 \\ 2x + 3x &= 6 \\ 5x &= 6 \\ x &= \tfrac{6}{5}, \text{ or } 1\tfrac{1}{5} \end{aligned}$$

It takes $\frac{6}{5}$, or $1\frac{1}{5}$, hours to do the job together.

19. $\dfrac{(2a)^{-2}a^4}{a^{-3}} = \dfrac{2^{-2}a^{-2}a^4}{a^{-3}}$

$= \dfrac{2^{-2}a^2}{a^{-3}} = \dfrac{a^2a^3}{2^2} = \dfrac{a^5}{4}$

20. $4\sqrt[3]{16} - 2\sqrt[3]{54} = 4\sqrt[3]{8 \cdot 2} - 2\sqrt[3]{27 \cdot 2}$
$= 4 \cdot 2\sqrt[3]{2} - 2 \cdot 3\sqrt[3]{2}$
$= 8\sqrt[3]{2} - 6\sqrt[3]{2} = 2\sqrt[3]{2}$

21. $\frac{3\sqrt{5x}}{\sqrt{2x}} = \frac{3\sqrt{5x} \cdot \sqrt{2x}}{\sqrt{2x} \cdot \sqrt{2x}}$
$= \frac{3\sqrt{10x^2}}{2x} = \frac{3x\sqrt{10}}{2x} = \frac{3\sqrt{10}}{2}$

22. $\frac{5+3i}{2-i}$

Multiply the numerator and denominator by the conjugate of the denominator, $2+i$.

$= \frac{(5+3i)(2+i)}{(2-i)(2+i)}$
$= \frac{10+5i+6i+3i^2}{4-i^2}$
$= \frac{10+11i+3(-1)}{4-(-1)}$
$= \frac{7+11i}{5} = \frac{7}{5} + \frac{11}{5}i$

23. $2\sqrt{k} = \sqrt{5k+3}$
$4k = 5k + 3$ *Square both sides.*
$-k = 3$ *Subtract 5k.*
$k = -3$ *Multiply by −1.*

Since k must be nonnegative so that $\sqrt{k}$ is a real number, -3 cannot be a solution. The solution set is $\emptyset$.

24. $10q^2 + 13q = 3$
$10q^2 + 13q - 3 = 0$
$(5q-1)(2q+3) = 0$

$5q - 1 = 0$ or $2q + 3 = 0$
$q = \frac{1}{5}$ or $q = -\frac{3}{2}$

Solution set: $\{\frac{1}{5}, -\frac{3}{2}\}$

25. $3k^2 - 3k - 2 = 0$

Use the quadratic formula with $a = 3$, $b = -3$, and $c = -2$.

$k = \frac{-b \pm \sqrt{b^2 - 4ac}}{2a}$
$k = \frac{-(-3) \pm \sqrt{(-3)^2 - 4(3)(-2)}}{2(3)}$
$= \frac{3 \pm \sqrt{9+24}}{6} = \frac{3 \pm \sqrt{33}}{6}$

Solution set: $\left\{\frac{3+\sqrt{33}}{6}, \frac{3-\sqrt{33}}{6}\right\}$

26. $2(x^2-3)^2 - 5(x^2-3) = 12$
Let $u = (x^2 - 3)$.
$2u^2 - 5u = 12$
$2u^2 - 5u - 12 = 0$
$(2u+3)(u-4) = 0$

$2u + 3 = 0$ or $u - 4 = 0$
$u = -\frac{3}{2}$ or $u = 4$

Substitute $x^2 - 3$ for u to find x.

If $u = -\frac{3}{2}$, then

$x^2 - 3 = -\frac{3}{2}$
$x^2 = \frac{3}{2}$
$x = \pm\sqrt{\frac{3}{2}} = \pm\frac{\sqrt{3}}{\sqrt{2}} \cdot \frac{\sqrt{2}}{\sqrt{2}} = \pm\frac{\sqrt{6}}{2}.$

If $u = 4$, then

$x^2 - 3 = 4$
$x^2 = 7$
$x = \pm\sqrt{7}.$

Solution set: $\left\{-\frac{\sqrt{6}}{2}, \frac{\sqrt{6}}{2}, -\sqrt{7}, \sqrt{7}\right\}$

27. $\log(x+2) + \log(x-1) = 1$
$\log_{10}[(x+2)(x-1)] = 1$
$(x+2)(x-1) = 10^1$
$x^2 + x - 2 = 10$
$x^2 + x - 12 = 0$
$(x+4)(x-3) = 0$

$x + 4 = 0$ or $x - 3 = 0$
$x = -4$ or $x = 3$

The original equation is undefined when $x = -4$.

Check $x = 3$: $\log 5 + \log 2 = \log 10 = 1$
Solution set: $\{3\}$

28. Solve $F = \frac{kwv^2}{r}$ for v.

$Fr = kwv^2$ *Multiply by r.*
$v^2 = \frac{Fr}{kw}$ *Divide by kw.*

Take the square root of each side and then rationalize the denominator.

$v = \pm\sqrt{\frac{Fr}{kw}} = \frac{\pm\sqrt{Fr}}{\sqrt{kw}} \cdot \frac{\sqrt{kw}}{\sqrt{kw}} = \frac{\pm\sqrt{Frkw}}{kw}$

29. $f(x) = y = x^3 + 4$

Interchange x and y and solve for y.

$$\begin{aligned} x &= y^3 + 4 \\ y^3 &= x - 4 \\ y &= \sqrt[3]{x - 4} \\ f^{-1}(x) &= \sqrt[3]{x - 4} \end{aligned}$$

30. **(a)** $a^{\log_a x} = x$, so $3^{\log_3 4} = 4$.

(b) $a^{\log_a x} = x$, so $e^{\ln x} = e^{\log_e 7} = 7$.

31. $2\log(3x + 7) - \log 4$

$= \log(3x + 7)^2 - \log 4$ *Power rule*

$= \log \dfrac{(3x + 7)^2}{4}$ *Quotient rule*

32. $f(x) = x^2 + 2x - 4$, $g(x) = 3x + 2$

(a) $$\begin{aligned} (g \circ f)(1) &= g(f(1)) \\ &= g(1^2 + 2 \cdot 1 - 4) \\ &= g(-1) \\ &= 3(-1) + 2 \\ &= -1 \end{aligned}$$

(b) $$\begin{aligned} (f \circ g)(x) &= f(g(x)) \\ &= f(3x + 2) \\ &= (3x + 2)^2 + 2(3x + 2) - 4 \\ &= 9x^2 + 12x + 4 + 6x + 4 - 4 \\ &= 9x^2 + 18x + 4 \end{aligned}$$

33. $f(x) = -3x + 5$

The equation is in slope-intercept form, so the y-intercept is $(0, 5)$ and $m = -3$ or $\frac{-3}{1}$.

Plot $(0, 5)$. From $(0, 5)$, move down 3 units and right 1 unit. Draw the line through these two points.

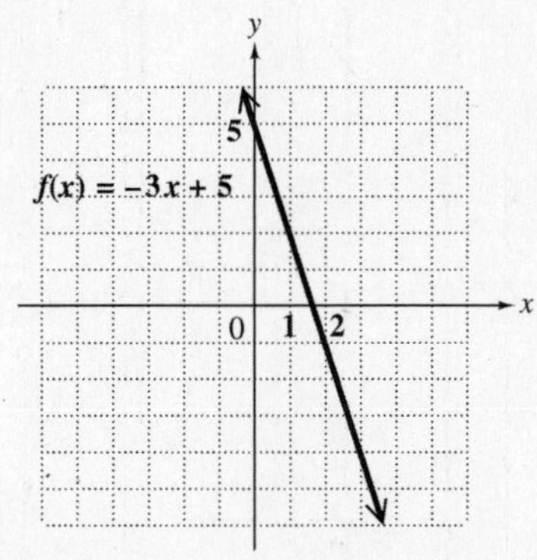

34. $f(x) = -2(x - 1)^2 + 3$

The graph is a parabola that has been shifted 1 unit to the right and 3 units upward from $(0, 0)$, so its vertex is at $(1, 3)$. Since $a = -2 < 0$, the parabola opens down. Also $|a| = |-2| = 2 > 1$, so the graph is narrower than the graph of $f(x) = x^2$. The points $(0, 1)$ and $(2, 1)$ are on the graph.

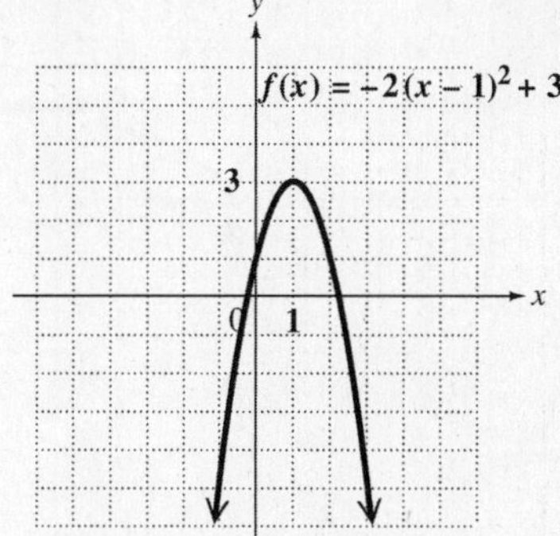

35. $\dfrac{x^2}{25} + \dfrac{y^2}{16} \leq 1$

The boundary, $\dfrac{x^2}{25} + \dfrac{y^2}{16} = 1$, is an ellipse in $\dfrac{x^2}{a^2} + \dfrac{y^2}{b^2} = 1$ form with intercepts $(5, 0)$, $(-5, 0)$, $(0, 4)$, and $(0, -4)$. Test $(0, 0)$.

$$\frac{0^2}{25} + \frac{0^2}{16} \overset{?}{\leq} 1$$

$$0 \leq 1 \quad \textit{True}$$

Shade the region inside the ellipse. The ellipse is drawn as a solid curve since the points of the ellipse satisfy the inequality.

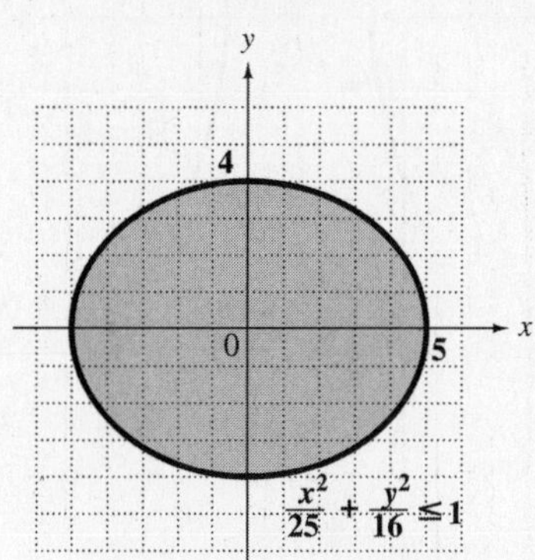

36. $f(x) = \sqrt{x-2}$

This is the graph of $g(x) = \sqrt{x}$ shifted 2 units right.

x	2	3	6
$f(x)$	0	1	2

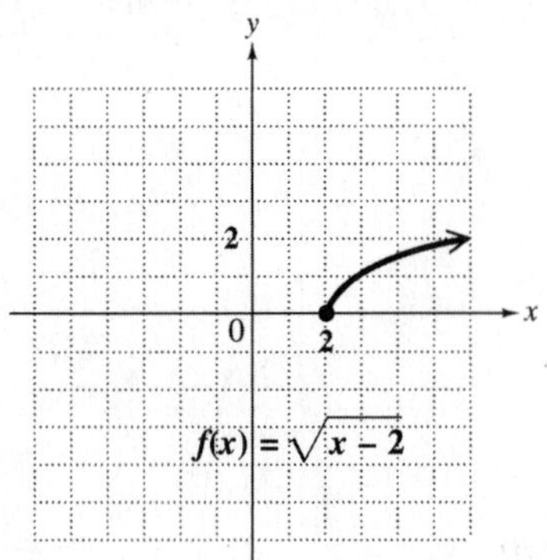

37. $\frac{x^2}{4} - \frac{y^2}{16} = 1$ is in $\frac{x^2}{a^2} - \frac{y^2}{b^2} = 1$ form.

The graph is a hyperbola with x-intercepts $(2, 0)$ and $(-2, 0)$ and asymptotes that are the extended diagonals of the rectangle with vertices $(2, 4)$, $(2, -4)$, $(-2, -4)$, and $(-2, 4)$.
The equations of the asymptotes are $y = \pm \frac{4}{2}x = \pm 2x$. Draw a branch of the hyperbola through each intercept approaching the asymptotes.

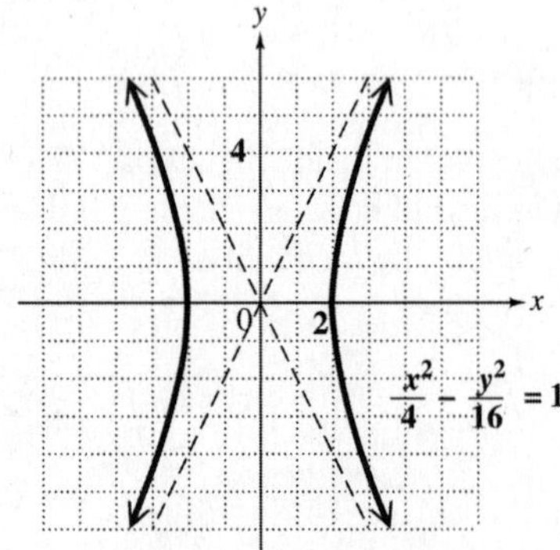

38. $f(x) = 3^x$

The graph of f is an increasing exponential.

x	-1	0	1	2
$f(x)$	$\frac{1}{3}$	1	3	9

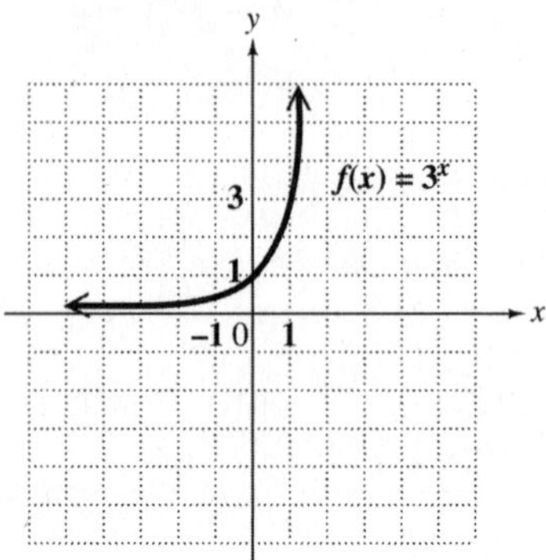

APPENDIX A STRATEGIES FOR PROBLEM SOLVING

Appendix A Margin Exercises

1. Step 1 compares to Polya's first step, Steps 2 and 3 compare to his second step, Step 4 compares to his third step, and Step 6 compares to his fourth step.

2.

Chosen Terms	First chosen term	Fourth chosen term	Product	Second chosen term	Third chosen term	Product
4th–7th	3	13	39	5	8	40
6th–9th	8	34	272	13	21	273
8th–11th	21	89	1869	34	55	1870

The product of the two middle terms is always 1 more than the product of the first and last term.

3. **(a)** Suppose she started with x dollars. She bought a book for \$10, so now she has $x - 10$ dollars. She spent half her remaining money on a train ticket, so she has $\frac{1}{2}(x-10) = \frac{1}{2}x - 5$ dollars remaining. She bought lunch for \$4, so she has $(\frac{1}{2}x - 5) - 4 = \frac{1}{2}x - 9$ dollars remaining. She spent half her remaining money at a bazaar, so she has $\frac{1}{2}(\frac{1}{2}x - 9) = \frac{1}{4}x - \frac{9}{2}$ dollars remaining. This must equal the \$20 she left the bazaar with, so solve the equation

$$\frac{1}{4}x - \frac{9}{2} = 20.$$

$$\frac{1}{4}x = \frac{49}{2} \quad \textit{Add } \tfrac{9}{2}.$$

$$x = 98 \quad \textit{Multiply by 4.}$$

She started with \$98.

(b) The given subtraction problem:

$$\begin{array}{r} 7a2 \\ -\,48b \\ \hline c73 \end{array}$$

Change to an addition problem:

$$\begin{array}{r} 48b \\ +\,c73 \\ \hline 7a2 \end{array}$$

Now $b + 3 = 2$, so $b = 9$ and we carry 1 to the tens column. Thus, $1 + 8 + 7 = 16$, so $a = 6$ and we carry 1 to the hundreds column. Thus, $1 + 4 + c = 7$, so $c = 2$. Hence, $a + b + c = 6 + 9 + 2 = 17$, which is choice **D**.

4. **(a)** By inspection, $44^2 = 1936$, $45^2 = 2025$, and $46^2 = 2116$. Only 2025 makes sense for this problem, and $2025 - 76 = 1949$, which must be the year he was born.

(b) Use trial and error for this problem. Here is one possible solution.

	3	5	
7	1	8	2
	4	6	

5. **(a)** Let x represent the positive number. "If I square it" gives x^2. "Double the result" gives $2x^2$. "Take half of that result" gives $\frac{1}{2}(2x^2) = x^2$ (the last two steps canceled each other). "Then add 12" gives $x^2 + 12$. "I get 21" implies that $x^2 + 12 = 21$. Solve this equation.

$$x^2 + 12 = 21$$

$$x^2 = 9 \quad \textit{Subtract 12.}$$

$$x = 3 \quad \textit{Positive square root}$$

The number is 3.

(b) Condition (2) leads us to a year in the 1960s. If x is the ones digit, then condition (1) gives $1 + 9 + 6 + x = 23$, so $x = 7$, and the year is 1967.

6. **(a)** Examine the units digit in powers of 7.

$7^1 =$	**7**	$7^5 =$	16,80**7**
$7^2 =$	4**9**	$7^6 =$	117,64**9**
$7^3 =$	34**3**	$7^7 =$	823,54**3**
$7^4 =$	240**1**	$7^8 =$	5,764,80**1**

We see that the units digit cycles through 7, 9, 3, and 1. Since $491 = 4 \cdot 122 + 3$ (122 full cycles), we know that 7^{491} and 7^3 have the same units digit. Thus, the units digit in 7^{491} is 3.

(b) Since $\frac{1}{11} = 0.0909\ldots$, we see that every odd-numbered digit in the decimal representation of $\frac{1}{11}$ is 0. Since 103 is odd, the 103rd digit in the decimal representation of $\frac{1}{11}$ is 0.

7. **(a)** Since there are 3 rows, and there could only be a maximum of 2 crosses per row, we know that there are a maximum of 6 crosses in the 9 squares. In the first row, put a cross in the second and third squares. In the second row, put a cross in the first square (since the first column already has one blank square), and then put a cross in the third square (if we put a cross in the second square, we wouldn't be able to put a cross in the first square of the third row—why?). In the third row, put a cross in the first and second squares. This gives us the maximum number of crosses, 6.

	X	X
X		X
X	X	

(b) The sum of the numbers 1 to 12 on the face of a clock is 78, and 78 divided by 3 is 26. Start with the number 12 and investigate what numbers can be combined with 12 to add up to 26. Moving left, $12 + 11 = 23$. We can't include 10 with 12 and 11 because the sum is greater than 26, so we'll move right and include 1 and 2 for a sum of 26. Draw a line that separates 11, 12, 1, and 2 from the rest of the numbers.

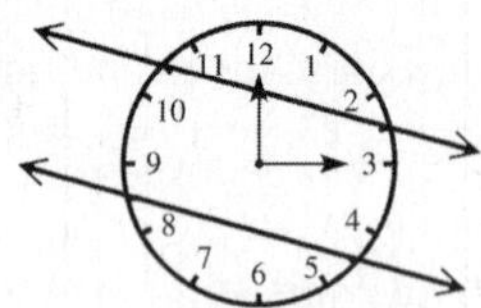

Now we can group the 10 and 9 together (but not with the 8—why?) along with the 3 and the 4 for a sum of 26. Draw another line and check that the remaining numbers add up to 26.

8. **(a)** Neither is correct, since $3^3 = 27$. However, "three cubed *is* twenty-seven" would be correct.

(b) If you take 7 bowling pins from 10 bowling pins, you have 7 bowling pins.

Appendix A Exercises

1. You could choose a sock from the box labeled *red and green socks*. Since it is mislabeled, it contains only red socks or only green socks, determined by the sock you choose. If the sock is green, relabel this box *green socks*. Since the other two boxes were mislabeled, switch the remaining label to the other box and place the label that says *red and green socks* on the unlabeled box. No other choice guarantees a correct relabeling, since you can remove only one sock.

3. There are 4 choices to place your next O.

(1) Row 2, column 2: If you place an O here, then your friend will place an X in row 3, column 3, which forces you to get three Os in a row, and you lose.

(2) Row 2, column 3: If you place an O here, then your friend will place an X in row 3, column 1, which forces you to get three Os in a row, and you lose.

(3) Row 3, column 1: If you place an O here, then your friend can't use row 2, column 2 (Why?), so they must use row 2, column 3 or row 3, column 3.

(A) If they place an X in row 2, column 3, then you would use row 3, column 3 and they are forced to place an X in row 2, column 2, which gives them 3 Xs in a row (2nd column).

(B) If they place an X in row 3, column 3, then you would use row 2, column 3 and they are forced to place an X in row 2, column 2, which gives them 3 Xs in a row (2nd column).

Thus, you win in all cases by placing an O in the *bottom-left square.*

(4) Row 3, column 3: If you place an O here, then your friend would place an X in row 3, column 1, you would place an O in row 2, column 2, your friend would place an X in row 2, column 3, and there would not be a winner.

5. Fill the big bucket. Pour into the small bucket. This leaves 4 gallons in the larger bucket. Empty the small bucket. Pour from the big bucket to fill up the small bucket. This leaves 1 gallon in the big bucket. Empty the small bucket. Pour 1 gallon from the big bucket to the small bucket. Fill up the big bucket. Pour into the small bucket. This leaves 5 gallons in the big bucket. Pour out the small bucket. This leaves exactly 5 gallons in the big bucket to take home. The above sequence is indicated by the following table.

Big	7	4	4	1	1	0	7	5	5
Small	0	3	0	3	0	1	1	3	0

7. One strategy is to organize a table such as the one which follows. Let $x =$ Chris's current age.

	Current Age	Past Age	Elapsed No. of Years
Pat	24	x	$24 - x$
Chris	x	$x - (24 - x)$	

Since Chris's past age can be represented as

$$x - (24 - x) = -24 + 2x = 2x - 24,$$

and Pat's current age, 24, is twice that of Chris's past age, we have

$$\begin{aligned} 24 &= 2(2x - 24) \\ 24 &= 4x - 48 \\ 72 &= 4x \\ 18 &= x. \end{aligned}$$

Thus, Chris's current age is 18 years.

9. Similar to Example 5 in the text, we might examine the units place and tens place for repetitive powers of 7 in order to explore possible patterns.

$$\begin{array}{llll} 7^1 = & \mathbf{07} & 7^5 = & 16{,}8\mathbf{07} \\ 7^2 = & \mathbf{49} & 7^6 = & 117{,}6\mathbf{49} \\ 7^3 = & 3\mathbf{43} & 7^7 = & 823{,}5\mathbf{43} \\ 7^4 = & 24\mathbf{01} & 7^8 = & 5{,}764{,}8\mathbf{01} \end{array}$$

Since the final two digits cycle over four values, we might consider dividing the successive exponents by 4 and examining their remainders. (Note: We are using inductive reasoning when we assume that this pattern will continue and will apply when the exponent is 1997.) Dividing the exponent 1997 by 4, we get a remainder of 1. This is the same remainder we get when dividing the exponent 1 (on 7^1) and 5 (on 7^5). Thus, we expect that the last two digits for 7^{1997} would be 07 as well.

11. At the end of 1st day, the frog has a net progression of 1 foot; day 2: 2 feet; day 3: 3 feet; ...; day 16: 16 feet (it crawls up 4 feet from 15 to 19 feet and then falls back 3 feet to 16 feet); on the 17th day it crawls up 4 feet from the 16-foot level, which takes it to the top.

13. Set 4 opposite 12, then 8 is halfway around from 4 to 12. Opposite 8 must be 16 to allow for three equally spaced numbers (children) between each of these values. Note that children 1, 2, and 3 stand between 16 and 4. So there are 16 children in the circle.

15. Add the given diagonal elements together to get 15. Each row, column, and other diagonal must also add to 15. This yields the following perfect square.

6	1	8
7	5	3
2	9	4

17. 25 pitches (The visiting team's pitcher retires 24 consecutive batters through the first eight innings, using only one pitch per batter. His team does not score either. Going into the bottom of the ninth inning tied 0–0, the first batter for the home team hits his first pitch for a homerun. The pitcher threw 25 pitches and loses the game by a score of 1–0.)

19. For three weighings, first balance four against four. Of the lighter four, balance two against the other two. Finally, of the lighter two, balance them one against the other.

To find the bad coin in two weighings, divide the eight coins into groups of 3, 3, 2. Weigh the groups of three against each other on the scale. If the groups weigh the same, the fake is in the two left out and can be found in one additional weighing. If the two groups of three do not weigh the same, pick the lighter group. Choose any two of the coins and weigh them. If one of these is lighter, it is the fake; if they weigh the same, then the third coin is the fake.

21. This may be worked algebraically or in reverse as Example 2 in the text. Multiplying 2 by 10 gives 20. Subtract 8 to give 12 and then square to get 144. Add 52 to get 196. This represents a number times itself. The number is 14, from the fact that $14 \times 14 = 196$ (or the square root of $196 = 14$). The quotient must be 21 since $21 - \frac{1}{3} \times 21 = 14$. Multiplying 21 by 7, we get 147, which represents 3 times the original number plus $\frac{3}{4}$ of that same product. The original number must be 28 since $3 \times 28 = 84$ and $\frac{3}{4}$ of 84 is 63, and $84 + 63 = 147$.

23. A solution, found by trial and error, is shown here.

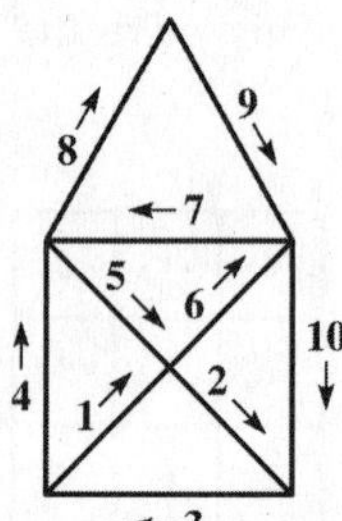

25. Jessica is married to James or Dan. Since Jessica is married to the oldest person in the group, she is not married to James, who is younger than Cathy. So Jessica is married to Dan, and Cathy is married to James. Since Jessica is married to the oldest person, we know that Dan is 36. Since James is older than Jessica but younger than Cathy, we conclude that Cathy is 31, Jame is 30, and Jessica is 29.

27. This is a problem with a "catch." Someone reading this problem might go ahead and calculate the volume of a cube 6 feet on each side, to get the answer 216 cubic feet. However, common sense tells us that since holes are by definition empty, there is no dirt in a hole.

29. One solution is
$1 + 2 + 3 + 4 + 5 + 6 + 7 + 8 \times 9 = 100.$

31. The first digit in the answer cannot be 0, 2, 3, or 5, since these digits have already been used. It cannot be more than 3, since one of the factors is a number in the 30's, making it impossible to get a product over 45,000. Thus, the first digit of the answer must be 1. To find the first digit in the 3-digit factor, use estimation. Dividing a number between 15,000 and 16,000 by a number between 30 and 40 could give a result with a first digit of 3, 4, or 5. Since 3 and 5 have already been used, this first digit must be 4. Thus, the 3-digit factor is 402. We now have the following.

$$\begin{array}{r} \underline{4}\ 0\ 2 \\ \times \qquad 3 \\ \hline \underline{1}\ 5, \qquad \end{array}$$

To find the units digit of the 2-digit factor, use trial and error with the digits that have not yet been used: 6, 7, 8, and 9.

$36 \times 402 = 14{,}472$ (Too small and reuses 2 and 4)
$37 \times 402 = 14{,}874$ (Too small and reuses 4)
$38 \times 402 = 15{,}276$ (Reuses 2)
$39 \times 402 = 15{,}678$ (Correct)

The correct problem is as follows.

$$\begin{array}{r} \underline{4}\ 0\ 2 \\ \times \quad 3\ \underline{9} \\ \hline \underline{1}\ 5,\underline{6}\ \underline{7}\ \underline{8} \end{array}$$

Notice that a combination of strategies was used to solve this problem.

33. To count the triangles, it helps to draw sketches of the figure several times. There are 5 triangles formed by two sides of the pentagon and a diagonal. There are 4 triangles formed with each side of the pentagon as a base, so there are $4 \times 5 = 20$ triangles formed in this way. Each point of the star forms a small triangle, so there are 5 of these. Finally, there are 5 triangles formed with a diagonal as a base. In each, the other two sides are inside the pentagon. (None of these triangles has a side common to the pentagon.) Thus, the total number of triangles in the figure is $5 + 20 + 5 + 5 = 35$.

35. It will still take $7\frac{1}{2}$ minutes to boil 5 eggs since you can boil them all at the same time.

APPENDIX B REVIEW OF FRACTIONS

Appendix B Margin Exercises

1. **(a)** Since 12 can be divided by 2, it has more than two different factors, so it is a *composite* number. Note: There are always two factors of any number—the number itself and 1.

   (b) 13 has exactly two different factors, 1 and 13, so it is a *prime* number.

   (c) Since 27 can be divided by 3, it has more than two different factors, so it is *composite.*

   (d) 59 has exactly two different factors, 1 and 59, so it is *prime.*

   (e) 1806 can be divided by 2, so it is *composite.*

2. **(a)** To write 70 in prime factored form, first divide by the smallest prime, 2, to get

$$70 = 2 \cdot 35.$$

Since 35 can be factored as $5 \cdot 7$, we have

$$70 = 2 \cdot 5 \cdot 7.$$

(b)

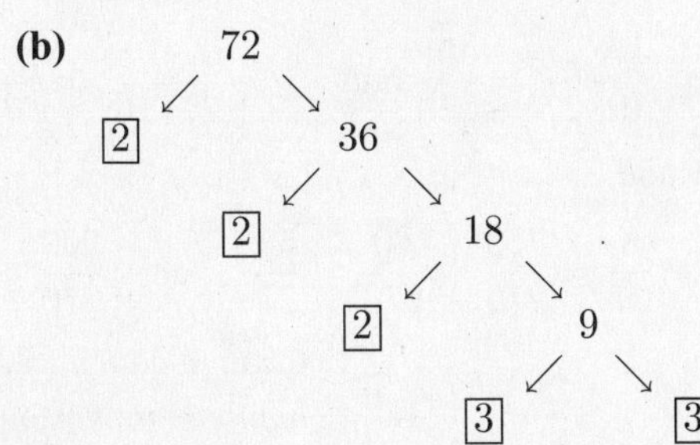

$72 = 2 \cdot 2 \cdot 2 \cdot 3 \cdot 3$

(c) 693 → 3, 231; 231 → 3, 77; 77 → 7, 11

$693 = 3 \cdot 3 \cdot 7 \cdot 11$

(d) Since 97 is a prime number, its prime factored form is just 97.

3. **(a)** $\frac{8}{14} = \frac{4 \cdot 2}{7 \cdot 2} = \frac{4}{7} \cdot \frac{2}{2} = \frac{4}{7} \cdot 1 = \frac{4}{7}$

   (b) $\frac{35}{42} = \frac{5 \cdot 7}{6 \cdot 7} = \frac{5}{6} \cdot \frac{7}{7} = \frac{5}{6} \cdot 1 = \frac{5}{6}$

   (c) $\frac{72}{120} = \frac{3 \cdot 24}{5 \cdot 24} = \frac{3}{5} \cdot \frac{24}{24} = \frac{3}{5} \cdot 1 = \frac{3}{5}$

4. $\frac{92}{5}$ Divide 92 by 5.

$$\begin{array}{r} 18 \\ 5\overline{)92} \\ \underline{5} \\ 42 \\ \underline{40} \\ 2 \end{array}$$

Thus, $\frac{92}{5} = 18\frac{2}{5}$.

5. To write $11\frac{2}{3}$ as an improper fraction, the numerator is

$$3 \cdot 11 + 2 = 33 + 2 = 35.$$

The denominator is 3. Thus, $11\frac{2}{3} = \frac{35}{3}$.

6. **(a)**
$$\begin{aligned} \frac{5}{8} \cdot \frac{2}{10} &= \frac{5 \cdot 2}{8 \cdot 10} && \textit{Multiply numerators. Multiply denominators.} \\ &= \frac{5 \cdot 2}{2 \cdot 4 \cdot 2 \cdot 5} && \textit{Factor.} \\ &= \frac{1}{2 \cdot 4} = \frac{1}{8} && \textit{Write in lowest terms.} \end{aligned}$$

   (b)
$$\begin{aligned} \frac{1}{10} \cdot \frac{12}{5} &= \frac{1 \cdot 12}{10 \cdot 5} && \textit{Multiply numerators. Multiply denominators.} \\ &= \frac{1 \cdot 2 \cdot 6}{2 \cdot 5 \cdot 5} && \textit{Factor.} \\ &= \frac{6}{5 \cdot 5} = \frac{6}{25} && \textit{Write in lowest terms.} \end{aligned}$$

   (c)
$$\begin{aligned} \frac{7}{9} \cdot \frac{12}{14} &= \frac{7 \cdot 12}{9 \cdot 14} && \textit{Multiply numerators. Multiply denominators.} \\ &= \frac{7 \cdot 2 \cdot 2 \cdot 3}{3 \cdot 3 \cdot 2 \cdot 7} && \textit{Factor.} \\ &= \frac{2}{3} && \textit{Write in lowest terms.} \end{aligned}$$

   (d)
$$\begin{aligned} 3\frac{1}{3} \cdot 1\frac{3}{4} &= \frac{10}{3} \cdot \frac{7}{4} && \textit{Write as improper fractions.} \\ &= \frac{10 \cdot 7}{3 \cdot 4} && \textit{Multiply numerators. Multiply denominators.} \\ &= \frac{2 \cdot 5 \cdot 7}{3 \cdot 2 \cdot 2} && \textit{Factor.} \\ &= \frac{35}{6}, \text{ or } 5\frac{5}{6} && \textit{Write as a mixed number.} \end{aligned}$$

7. **(a)** $\frac{3}{10} \div \frac{2}{7} = \frac{3}{10} \cdot \frac{7}{2}$ *Multiply by the reciprocal of the second fraction.*

$= \frac{21}{20}$, or $1\frac{1}{20}$

(b) $\frac{3}{4} \div \frac{7}{16} = \frac{3}{4} \cdot \frac{16}{7}$ *Multiply by the reciprocal of the second fraction.*

$= \frac{3 \cdot 4 \cdot 4}{7 \cdot 4}$

$= \frac{12}{7}$, or $1\frac{5}{7}$

(c) $\frac{4}{3} \div 6 = \frac{4}{3} \div \frac{6}{1}$

$= \frac{4}{3} \cdot \frac{1}{6}$ *Multiply by the reciprocal of the second fraction.*

$= \frac{2 \cdot 2}{3 \cdot 2 \cdot 3}$

$= \frac{2}{3 \cdot 3} = \frac{2}{9}$

(d) $3\frac{1}{4} \div 1\frac{2}{5} = \frac{13}{4} \div \frac{7}{5}$ *Change both mixed numbers to improper fractions.*

$= \frac{13}{4} \cdot \frac{5}{7}$ *Multiply by the reciprocal of the second fraction.*

$= \frac{65}{28}$, or $2\frac{9}{28}$

8. **(a)** $\frac{3}{5} + \frac{4}{5} = \frac{3+4}{5}$ Add numerators; denominator does not change.

$= \frac{7}{5}$, or $1\frac{2}{5}$

(b) $\frac{5}{14} + \frac{3}{14} = \frac{5+3}{14}$ Add numerators; denominator does not change.

$= \frac{8}{14}$

$= \frac{2 \cdot 4}{2 \cdot 7}$ Factor.

$= \frac{4}{7}$

9. **(a)** $\frac{7}{30} + \frac{2}{45}$

Since $30 = 2 \cdot 3 \cdot 5$ and $45 = 3 \cdot 3 \cdot 5$, the least common denominator must have one factor of 2 (from 30), two factors of 3 (from 45), and one factor of 5 (from either 30 or 45), so it is $2 \cdot 3 \cdot 3 \cdot 5 = 90$.

Write each fraction with a denominator of 90.

$\frac{7}{30} = \frac{7 \cdot 3}{30 \cdot 3} = \frac{21}{90}$ and $\frac{2}{45} = \frac{2 \cdot 2}{45 \cdot 2} = \frac{4}{90}$

Now add.

$\frac{7}{30} + \frac{2}{45} = \frac{21}{90} + \frac{4}{90} = \frac{21+4}{90} = \frac{25}{90}$

$\frac{25}{90}$ can be simplified.

$\frac{25}{90} = \frac{5 \cdot 5}{5 \cdot 18} = \frac{5}{18}$

(b) $\frac{17}{10} + \frac{8}{27}$

Since $10 = 2 \cdot 5$ and $27 = 3 \cdot 3 \cdot 3$, the least common denominator is $2 \cdot 5 \cdot 3 \cdot 3 \cdot 3 = 270$.

Write each fraction with a denominator of 270.

$\frac{17}{10} = \frac{17 \cdot 27}{10 \cdot 27} = \frac{459}{270}$ and $\frac{8}{27} = \frac{8 \cdot 10}{27 \cdot 10} = \frac{80}{270}$

Now add.

$\frac{17}{10} + \frac{8}{27} = \frac{459}{270} + \frac{80}{270} = \frac{539}{270}$, or $1\frac{269}{270}$

(c) $2\frac{1}{8} + 1\frac{2}{3} = \frac{17}{8} + \frac{5}{3}$ *Change both mixed numbers to improper fractions.*

The least common denominator is 24, so write each fraction with a denominator of 24.

$\frac{17}{8} = \frac{17 \cdot 3}{8 \cdot 3} = \frac{51}{24}$ and $\frac{5}{3} = \frac{5 \cdot 8}{3 \cdot 8} = \frac{40}{24}$

Now add.

$\frac{17}{8} + \frac{5}{3} = \frac{51}{24} + \frac{40}{24} = \frac{51+40}{24}$

$= \frac{91}{24}$, or $3\frac{19}{24}$

(d) $132\frac{4}{5} + 28\frac{3}{4}$

We will use a vertical method. LCD = 20

$$\begin{array}{rcl} 132\frac{4}{5} = 132\frac{16}{20} & & \text{Add the whole numbers and the fractions separately.} \\ +\,28\frac{3}{4} = 28\frac{15}{20} & & \\ \hline 160\frac{31}{20} & = 160 + \left(1 + \frac{11}{20}\right) = 161\frac{11}{20} & \end{array}$$

10. **(a)** $\frac{9}{11} - \frac{3}{11} = \frac{9-3}{11}$ Subtract numerators; denominator does not change.

$= \frac{6}{11}$

(b) $\frac{13}{15} - \frac{5}{6}$

Since $15 = 3 \cdot 5$ and $6 = 2 \cdot 3$, the least common denominator is $3 \cdot 5 \cdot 2 = 30$. Write each fraction with a denominator of 30.

$$\frac{13}{15} = \frac{13 \cdot 2}{15 \cdot 2} = \frac{26}{30} \text{ and } \frac{5}{6} = \frac{5 \cdot 5}{6 \cdot 5} = \frac{25}{30}$$

Now subtract.

$$\frac{13}{15} - \frac{5}{6} = \frac{26}{30} - \frac{25}{30} = \frac{1}{30}$$

(c) $2\frac{3}{8} - 1\frac{1}{2} = \frac{19}{8} - \frac{3}{2}$ *Change each mixed number into an improper fraction.*

The least common denominator is 8. Write each fraction with a denominator of 8. $\frac{19}{8}$ remains unchanged, and

$$\frac{3}{2} = \frac{3 \cdot 4}{2 \cdot 4} = \frac{12}{8}.$$

Now subtract.

$$\frac{19}{8} - \frac{3}{2} = \frac{19}{8} - \frac{12}{8} = \frac{19-12}{8} = \frac{7}{8}$$

(d) $50\frac{1}{4} - 32\frac{2}{3} = \frac{201}{4} - \frac{98}{3}$ *Write as improper fractions.*

The least common denominator is 12.

$$\frac{201}{4} = \frac{201 \cdot 3}{4 \cdot 3} = \frac{603}{12} \text{ and } \frac{98}{3} = \frac{98 \cdot 4}{3 \cdot 4} = \frac{392}{12}$$

Now subtract.

$$\frac{603}{12} - \frac{392}{12} = \frac{211}{12}, \text{ or } 17\frac{7}{12}$$

Appendix B Section Exercises

1. True; the number above the fraction bar is called the numerator and the number below the fraction bar is called the denominator.

3. False; the fraction $\frac{17}{51}$ can be reduced to $\frac{1}{3}$ since $\frac{17}{51} = \frac{17 \cdot 1}{17 \cdot 3} = \frac{1}{3}$.

5. False; *product* refers to multiplication, so the product of 8 and 2 is 16. The *sum* of 8 and 2 is 10.

7. Since 19 has only itself and 1 as factors, it is a prime number.

9. The number 52 is composite since it has factors other than 1 and itself (2, for example).

11. 2468 can be divided by 2. It has more than two different factors, so it is composite.

13. As stated in the text, the number 1 is neither prime nor composite, by agreement.

15. $30 = 2 \cdot 15$
$= 2 \cdot 3 \cdot 5$

17. $252 = 2 \cdot 126$
$= 2 \cdot 2 \cdot 63$
$= 2 \cdot 2 \cdot 3 \cdot 21$
$= 2 \cdot 2 \cdot 3 \cdot 3 \cdot 7$

19. $124 = 2 \cdot 62$
$= 2 \cdot 2 \cdot 31$

21. Since 29 has only itself and 1 as factors, it is a prime number. Its prime factored form is just 29.

23. $\frac{8}{16} = \frac{1 \cdot 8}{2 \cdot 8} = \frac{1}{2}$

25. $\frac{15}{18} = \frac{3 \cdot 5}{3 \cdot 6} = \frac{5}{6}$

27. $\frac{15}{75} = \frac{1 \cdot 15}{5 \cdot 15} = \frac{1}{5}$

29. $\frac{144}{120} = \frac{6 \cdot 24}{5 \cdot 24} = \frac{6}{5}$

31. $\frac{12}{7}$ Divide 12 by 7.

$$\begin{array}{r} 1 \\ 7\overline{)12} \\ \underline{7} \\ 5 \end{array}$$

Thus, $\frac{12}{7} = 1\frac{5}{7}$.

33. $\frac{77}{12}$ Divide 77 by 12.

$$\begin{array}{r} 6 \\ 12\overline{)77} \\ \underline{72} \\ 5 \end{array}$$

Thus, $\frac{77}{12} = 6\frac{5}{12}$.

35. $\frac{83}{11}$ Divide 83 by 11.

$$\begin{array}{r} 7 \\ 11\overline{)83} \\ \underline{77} \\ 6 \end{array}$$

Thus, $\frac{83}{11} = 7\frac{6}{11}$.

37. $2\frac{3}{5}$ The numerator is

$$5 \cdot 2 + 3 = 10 + 3 = 13.$$

The denominator is 5. Thus, $2\frac{3}{5} = \frac{13}{5}$.

39. $10\frac{3}{8}$ The numerator is

$$8 \cdot 10 + 3 = 80 + 3 = 83.$$

The denominator is 8. Thus, $10\frac{3}{8} = \frac{83}{8}$.

41. $10\frac{4}{5}$ The numerator is

$$5 \cdot 10 + 4 = 50 + 4 = 54.$$

The denominator is 5. Thus, $10\frac{4}{5} = \frac{54}{5}$.

43. A common denominator for $\frac{p}{q}$ and $\frac{r}{s}$ must be a multiple of both denominators, q and s. Such a number is $q \cdot s$. Therefore, **A** is correct.

45. $\frac{4}{5} \cdot \frac{6}{7} = \frac{4 \cdot 6}{5 \cdot 7} = \frac{24}{35}$

47. $\frac{1}{10} \cdot \frac{12}{5} = \frac{1 \cdot 12}{10 \cdot 5} = \frac{1 \cdot 2 \cdot 6}{2 \cdot 5 \cdot 5} = \frac{6}{25}$

49.
$$\begin{aligned} \frac{15}{4} \cdot \frac{8}{25} &= \frac{15 \cdot 8}{4 \cdot 25} \\ &= \frac{3 \cdot 5 \cdot 4 \cdot 2}{4 \cdot 5 \cdot 5} \\ &= \frac{3 \cdot 2}{5} \\ &= \frac{6}{5}, \text{ or } 1\frac{1}{5} \end{aligned}$$

51. $2\frac{2}{3} \cdot 5\frac{4}{5}$

Change both mixed numbers to improper fractions.

$$2\frac{2}{3} = \frac{3 \cdot 2 + 2}{3} = \frac{8}{3}$$
$$5\frac{4}{5} = \frac{5 \cdot 5 + 4}{5} = \frac{29}{5}$$

$$\begin{aligned} 2\frac{2}{3} \cdot 5\frac{4}{5} &= \frac{8}{3} \cdot \frac{29}{5} \\ &= \frac{8 \cdot 29}{3 \cdot 5} \\ &= \frac{232}{15}, \text{ or } 15\frac{7}{15} \end{aligned}$$

53. $\frac{5}{4} \div \frac{3}{8} = \frac{5}{4} \cdot \frac{8}{3}$ *Multiply by the reciprocal of the second fraction.*

$$\begin{aligned} &= \frac{5 \cdot 8}{4 \cdot 3} \\ &= \frac{5 \cdot 4 \cdot 2}{4 \cdot 3} \\ &= \frac{5 \cdot 2}{3} \\ &= \frac{10}{3}, \text{ or } 3\frac{1}{3} \end{aligned}$$

55. $\frac{32}{5} \div \frac{8}{15} = \frac{32}{5} \cdot \frac{15}{8}$ *Multiply by the reciprocal of the second fraction.*

$$\begin{aligned} &= \frac{32 \cdot 15}{5 \cdot 8} \\ &= \frac{8 \cdot 4 \cdot 3 \cdot 5}{1 \cdot 5 \cdot 8} \\ &= \frac{4 \cdot 3}{1} = 12 \end{aligned}$$

57. $\frac{3}{4} \div 12 = \frac{3}{4} \cdot \frac{1}{12}$ *Multiply by the reciprocal of 12.*

$$= \frac{3 \cdot 1}{4 \cdot 12}$$
$$= \frac{3 \cdot 1}{4 \cdot 3 \cdot 4}$$
$$= \frac{1}{4 \cdot 4} = \frac{1}{16}$$

59. $2\frac{5}{8} \div 1\frac{15}{32}$

Change both mixed numbers to improper fractions.

$$2\frac{5}{8} = \frac{8 \cdot 2 + 5}{8} = \frac{21}{8}$$
$$1\frac{15}{32} = \frac{32 \cdot 1 + 15}{32} = \frac{47}{32}$$

$$2\frac{5}{8} \div 1\frac{15}{32} = \frac{21}{8} \div \frac{47}{32}$$
$$= \frac{21}{8} \cdot \frac{32}{47}$$
$$= \frac{21 \cdot 32}{8 \cdot 47}$$
$$= \frac{21 \cdot 8 \cdot 4}{8 \cdot 47}$$
$$= \frac{21 \cdot 4}{47}$$
$$= \frac{84}{47}, \text{ or } 1\frac{37}{47}$$

61. $\frac{7}{12} + \frac{1}{12} = \frac{7+1}{12}$

$$= \frac{8}{12}$$
$$= \frac{2 \cdot 4}{3 \cdot 4} = \frac{2}{3}$$

63. $\frac{5}{9} + \frac{1}{3}$

Since $9 = 3 \cdot 3$, and 3 is prime, the LCD (least common denominator) is $3 \cdot 3 = 9$.

$$\frac{1}{3} = \frac{1}{3} \cdot \frac{3}{3} = \frac{3}{9}$$

Now add the two fractions with the same denominator.

$$\frac{5}{9} + \frac{1}{3} = \frac{5}{9} + \frac{3}{9} = \frac{8}{9}$$

65. $3\frac{1}{8} + \frac{1}{4}$

Change the mixed number to an improper fraction.

$$3\frac{1}{8} = \frac{8 \cdot 3 + 1}{8} = \frac{25}{8}$$
$$3\frac{1}{8} + \frac{1}{4} = \frac{25}{8} + \frac{1}{4}$$

Since $8 = 2 \cdot 2 \cdot 2$ and $4 = 2 \cdot 2$, the LCD is $2 \cdot 2 \cdot 2$ or 8.

$$= \frac{25}{8} + \frac{1 \cdot 2}{4 \cdot 2}$$
$$= \frac{25}{8} + \frac{2}{8}$$
$$= \frac{27}{8}, \text{ or } 3\frac{3}{8}$$

67. $\frac{7}{12} - \frac{1}{9}$

Since $12 = 2 \cdot 2 \cdot 3$ and $9 = 3 \cdot 3$, the LCD is $2 \cdot 2 \cdot 3 \cdot 3 = 36$.

$$\frac{7}{12} = \frac{7}{12} \cdot \frac{3}{3} = \frac{21}{36} \quad \text{and} \quad \frac{1}{9} \cdot \frac{4}{4} = \frac{4}{36}$$

Now subtract fractions with the same denominator.

$$\frac{7}{12} - \frac{1}{9} = \frac{21}{36} - \frac{4}{36} = \frac{17}{36}$$

69. $6\frac{1}{4} - 5\frac{1}{3}$

Change both mixed numbers to improper fractions.

$$6\frac{1}{4} = \frac{4 \cdot 6 + 1}{4} = \frac{25}{4}$$
$$5\frac{1}{3} = \frac{3 \cdot 5 + 1}{3} = \frac{16}{3}$$
$$6\frac{1}{4} - 5\frac{1}{3} = \frac{25}{4} - \frac{16}{3}$$

Since $4 = 2 \cdot 2$, and 3 is prime, the LCD is $2 \cdot 2 \cdot 3 = 12$.

$$= \frac{25 \cdot 3}{4 \cdot 3} - \frac{16 \cdot 4}{3 \cdot 4}$$
$$= \frac{75}{12} - \frac{64}{12}$$
$$= \frac{11}{12}$$

71. $\frac{5}{3}+\frac{1}{6}-\frac{1}{2}$

Since 2 and 3 are prime, and $6 = 2 \cdot 3$, the LCD is $2 \cdot 3 = 6$. Write $\frac{5}{3} \cdot \frac{2}{2} = \frac{10}{6}$ and $\frac{1}{2} = \frac{1}{2} \cdot \frac{3}{3} = \frac{3}{6}$.

Now add and subtract, then write the answer in lowest terms.

$$\begin{aligned}\frac{5}{3}+\frac{1}{6}-\frac{1}{2} &= \frac{10}{6}+\frac{1}{6}-\frac{3}{6}\\ &= \frac{8}{6} = \frac{4 \cdot 2}{3 \cdot 2} = \frac{4}{3}, \text{ or } 1\frac{1}{3}\end{aligned}$$

73. Multiply the number of cups of water per serving by the number of servings.

$$\begin{aligned}\frac{3}{4} \cdot 8 &= \frac{3}{4} \cdot \frac{8}{1}\\ &= \frac{3 \cdot 8}{4 \cdot 1}\\ &= \frac{3 \cdot 2 \cdot 4}{4 \cdot 1}\\ &= \frac{3 \cdot 2}{1} = 6 \text{ cups}\end{aligned}$$

For 8 microwave servings, 6 cups of water will be needed.

75. The perimeter is the sum of the measures of the 5 sides.

$$\begin{aligned}&196 + 98\frac{3}{4} + 146\frac{1}{2} + 100\frac{7}{8} + 76\frac{5}{8}\\ &= 196 + 98\frac{6}{8} + 146\frac{4}{8} + 100\frac{7}{8} + 76\frac{5}{8}\\ &= 196 + 98 + 146 + 100 + 76 + \frac{6+4+7+5}{8}\\ &= 616 + \frac{22}{8} \quad \left(\frac{22}{8} = 2\frac{6}{8} = 2\frac{3}{4}\right)\\ &= 618\frac{3}{4} \text{ feet}\end{aligned}$$

The perimeter is $618\frac{3}{4}$ feet.

77. The difference between the two measures is found by subtracting, using 16 as the LCD.

$$\begin{aligned}\frac{3}{4} - \frac{3}{16} &= \frac{3 \cdot 4}{4 \cdot 4} - \frac{3}{16}\\ &= \frac{12}{16} - \frac{3}{16}\\ &= \frac{12-3}{16} = \frac{9}{16}\end{aligned}$$

The difference is $\frac{9}{16}$ inch.

79. Subtract $\frac{3}{8}$ from $\frac{11}{16}$ using 16 as the LCD.

$$\begin{aligned}\frac{11}{16} - \frac{3}{8} &= \frac{11}{16} - \frac{3 \cdot 2}{8 \cdot 2}\\ &= \frac{11}{16} - \frac{6}{16}\\ &= \frac{5}{16}\end{aligned}$$

Thus, $\frac{3}{8}$ inch is $\frac{5}{16}$ inch smaller than $\frac{11}{16}$ inch.

81. Multiply the amount of fabric it takes to make one costume by the number of costumes.

$$\begin{aligned}2\frac{3}{8} \cdot 7 &= \frac{19}{8} \cdot \frac{7}{1}\\ &= \frac{19 \cdot 7}{8 \cdot 1}\\ &= \frac{133}{8}, \text{ or } 16\frac{5}{8} \text{ yd}\end{aligned}$$

For 7 costumes, $16\frac{5}{8}$ yards of fabric would be needed.

83. The sum of the fractions representing the U.S. foreign-born population from Latin America, Asia, or Europe is

$$\begin{aligned}\frac{27}{50}+\frac{1}{4}+\frac{7}{50} &= \frac{27 \cdot 2}{50 \cdot 2}+\frac{1 \cdot 25}{4 \cdot 25}+\frac{7 \cdot 2}{50 \cdot 2}\\ &= \frac{54+25+14}{100}\\ &= \frac{93}{100}.\end{aligned}$$

So the fraction representing the U.S. foreign-born population from other regions is

$$\begin{aligned}1 - \frac{93}{100} &= \frac{100}{100} - \frac{93}{100}\\ &= \frac{7}{100}.\end{aligned}$$

85. Multiply the fraction representing the U.S. foreign-born population from Europe, $\frac{7}{50}$, by the total number of foreign-born people in the U.S., approximately 34 million.

$$\frac{7}{50} \cdot 34 = \frac{7}{50} \cdot \frac{34}{1} = \frac{7 \cdot 2 \cdot 17}{2 \cdot 25} = \frac{119}{25}, \text{ or } 4\frac{19}{25}$$

There were approximately $4\frac{19}{25}$ million, or 4,760,000, foreign-born people in the U.S. in 2004 who were born in Europe.

APPENDIX C SYNTHETIC DIVISION

Appendix C Margin Exercises

1. **(a)** $\dfrac{3z^2+10z-8}{z+4}$

$$\begin{array}{r|rrr} -4 & 3 & 10 & -8 \\ & & -12 & 8 \\ \hline & 3 & -2 & 0 \end{array} \leftarrow \textit{Remainder}$$

Read the quotient from the bottom row.

Answer: $3z-2$

(b) $(2x^2+3x-5) \div (x+1)$

$$\begin{array}{r|rrr} -1 & 2 & 3 & -5 \\ & & -2 & -1 \\ \hline & 2 & 1 & -6 \end{array} \leftarrow \textit{Remainder}$$

Answer: $2x+1+\dfrac{-6}{x+1}$

2. **(a)** $\dfrac{3a^3-2a+21}{a+2}$

Insert a 0 for the missing a^2-term.

$$\begin{array}{r|rrrr} -2 & 3 & 0 & -2 & 21 \\ & & -6 & 12 & -20 \\ \hline & 3 & -6 & 10 & 1 \end{array} \leftarrow \textit{Remainder}$$

Answer: $3a^2-6a+10+\dfrac{1}{a+2}$

(b) $(-4x^4+3x^3+18x+2) \div (x-2)$

$$\begin{array}{r|rrrrr} 2 & -4 & 3 & 0 & 18 & 2 \\ & & -8 & -10 & -20 & -4 \\ \hline & -4 & -5 & -10 & -2 & -2 \end{array} \leftarrow \textit{Remainder}$$

Answer: $-4x^3-5x^2-10x-2+\dfrac{-2}{x-2}$

3. Let $P(x)=x^3-5x^2+7x-3$.

To find $P(a)$, divide the polynomial by $x-a$ using synthetic division. $P(a)$ will be the remainder.

(a) Find $P(1)$.

Divide $P(x)$ by $x-1$.

$$\begin{array}{r|rrrr} 1 & 1 & -5 & 7 & -3 \\ & & 1 & -4 & 3 \\ \hline & 1 & -4 & 3 & 0 \end{array} \leftarrow \textit{Remainder}$$

The remainder is 0, so by the remainder theorem, $P(1)=0$.

(b) Find $P(-2)$.

Divide $P(x)$ by $x-(-2)=x+2$.

$$\begin{array}{r|rrrr} -2 & 1 & -5 & 7 & -3 \\ & & -2 & 14 & -42 \\ \hline & 1 & -7 & 21 & -45 \end{array} \leftarrow \textit{Remainder}$$

By the remainder theorem, $P(-2)=-45$.

4. To check whether 2 is a solution to $P(x)=0$, divide the polynomial by $x-2$. If the remainder is 0, then 2 is a solution. Otherwise, it is not.

(a) $3x^3-11x^2+17x-14=0$

$$\begin{array}{r|rrrr} 2 & 3 & -11 & 17 & -14 \\ & & 6 & -10 & 14 \\ \hline & 3 & -5 & 7 & 0 \end{array} \leftarrow \textit{Remainder}$$

Since the remainder is 0, 2 is a solution of the given equation.

(b) $4x^5-7x^4-11x^2+2x+6=0$

Insert a 0 for the missing x^3-term.

$$\begin{array}{r|rrrrrr} 2 & 4 & -7 & 0 & -11 & 2 & 6 \\ & & 8 & 2 & 4 & -14 & -24 \\ \hline & 4 & 1 & 2 & -7 & -12 & -18 \end{array} \leftarrow \textit{Remainder}$$

Since the remainder is not 0, 2 is not a solution of the given equation.

Appendix C Section Exercises

1. $\dfrac{x^2+3x-6}{x-2}$

The correct set-up is **C**. $2\,\overline{)\,1 \quad 3 \quad -6}$

3. $\dfrac{x^2-6x+5}{x-1}$

$$\begin{array}{r|rrr} 1 & 1 & -6 & 5 \\ & & 1 & -5 \\ \hline & 1 & -5 & 0 \end{array}$$

$\leftarrow$ *Coefficients of numerator*

Write the answer from the bottom row.

$\downarrow \; \downarrow$

$x-5$

Answer: $x-5$

5. $\dfrac{4m^2+19m-5}{m+5}$

$m+5=m-(-5)$, so use -5.

$$\begin{array}{r|rrr} -5 & 4 & 19 & -5 \\ & & -20 & 5 \\ \hline & 4 & -1 & 0 \end{array}$$

Answer: $4m-1$

7. $\dfrac{2a^2+8a+13}{a+2}$

$a+2=a-(-2)$, so use -2.

$$\begin{array}{r|rrr} -2 & 2 & 8 & 13 \\ & & -4 & -8 \\ \hline & 2 & 4 & 5 \end{array} \leftarrow \textit{Remainder}$$

Answer: $2a+4+\dfrac{5}{a+2}$

9. $(p^2-3p+5)\div(p+1)$

$$\begin{array}{r|rrr} -1 & 1 & -3 & 5 \\ & & -1 & 4 \\ \hline & 1 & -4 & 9 \end{array}$$

Answer: $p-4+\dfrac{9}{p+1}$

11. $\dfrac{4a^3-3a^2+2a-3}{a-1}$

$$\begin{array}{r|rrrr} 1 & 4 & -3 & 2 & -3 \\ & & 4 & 1 & 3 \\ \hline & 4 & 1 & 3 & 0 \end{array}$$

Answer: $4a^2+a+3$

13. $(x^5-2x^3+3x^2-4x-2)\div(x-2)$

Insert 0 for the missing x^4-term.

$$\begin{array}{r|rrrrrr} 2 & 1 & 0 & -2 & 3 & -4 & -2 \\ & & 2 & 4 & 4 & 14 & 20 \\ \hline & 1 & 2 & 2 & 7 & 10 & 18 \end{array} \leftarrow \textit{Remainder}$$

Answer: $x^4+2x^3+2x^2+7x+10+\dfrac{18}{x-2}$

15. $(-4r^6-3r^5-3r^4+5r^3-6r^2+3r)\div(r-1)$

$$\begin{array}{r|rrrrrrr} 1 & -4 & -3 & -3 & 5 & -6 & 3 & 0 \\ & & -4 & -7 & -10 & -5 & -11 & -8 \\ \hline & -4 & -7 & -10 & -5 & -11 & -8 & -8 \end{array} \leftarrow \textit{Remainder}$$

Answer:

$-4r^5-7r^4-10r^3-5r^2-11r-8+\dfrac{-8}{r-1}$

17. $(-3y^5+2y^4-5y^3-6y^2-1)\div(y+2)$

Insert 0 for the missing y-term.

$$\begin{array}{r|rrrrrr} -2 & -3 & 2 & -5 & -6 & 0 & -1 \\ & & 6 & -16 & 42 & -72 & 144 \\ \hline & -3 & 8 & -21 & 36 & -72 & 143 \end{array} \leftarrow \textit{Remainder}$$

Answer:

$-3y^4+8y^3-21y^2+36y-72+\dfrac{143}{y+2}$

19. $\dfrac{y^3+1}{y-1}=\dfrac{y^3+0y^2+0y+1}{y-1}$

$$\begin{array}{r|rrrr} 1 & 1 & 0 & 0 & 1 \\ & & 1 & 1 & 1 \\ \hline & 1 & 1 & 1 & 2 \end{array} \leftarrow \textit{Remainder}$$

Answer: $y^2+y+1+\dfrac{2}{y-1}$

21. $P(x) = 2x^3 - 4x^2 + 5x - 3$; $k = 2$

To find $P(2)$, divide the polynomial by $x - 2$. $P(2)$ will be the remainder.

$$\begin{array}{r|rrrr} 2 & 2 & -4 & 5 & -3 \\ & & 4 & 0 & 10 \\ \hline & 2 & 0 & 5 & 7 \end{array} \leftarrow \textit{Remainder}$$

By the remainder theorem, $P(2) = 7$.

23. $P(r) = -r^3 - 5r^2 - 4r - 2$; $k = -4$

Divide by $r + 4$. The remainder is equal to $P(-4)$.

$$\begin{array}{r|rrrr} -4 & -1 & -5 & -4 & -2 \\ & & 4 & 4 & 0 \\ \hline & -1 & -1 & 0 & -2 \end{array} \leftarrow \textit{Remainder}$$

By the remainder theorem, $P(-4) = -2$.

25. $P(x) = 2x^3 - 4x^2 + 5x - 33$; $k = 3$

Divide by $x - 3$. The remainder is equal to $P(3)$.

$$\begin{array}{r|rrrr} 3 & 2 & -4 & 5 & -33 \\ & & 6 & 6 & 33 \\ \hline & 2 & 2 & 11 & 0 \end{array} \leftarrow \textit{Remainder}$$

By the remainder theorem, $P(3) = 0$.

27. Is $x = -2$ a solution of

$$x^3 - 2x^2 - 3x + 10 = 0?$$

To decide whether -2 is a solution to the given equation, divide the polynomial by $x + 2$.

$$\begin{array}{r|rrrr} -2 & 1 & -2 & -3 & 10 \\ & & -2 & 8 & -10 \\ \hline & 1 & -4 & 5 & 0 \end{array} \leftarrow \textit{Remainder}$$

Since the remainder is 0, -2 is a solution of the equation.

29. Is $m = -2$ a solution of

$$m^4 + 2m^3 - 3m^2 + 8m - 8 = 0?$$

To decide whether -2 is a solution to the given equation, divide the polynomial by $m + 2$.

$$\begin{array}{r|rrrrr} -2 & 1 & 2 & -3 & 8 & -8 \\ & & -2 & 0 & 6 & -28 \\ \hline & 1 & 0 & -3 & 14 & -36 \end{array} \leftarrow \textit{Remainder}$$

Since the remainder is not 0, -2 is not a solution of the equation.

31. Is $x = -2$ a solution of

$$3x^3 + 2x^2 - 2x + 11 = 0?$$

$$\begin{array}{r|rrrr} -2 & 3 & 2 & -2 & 11 \\ & & -6 & 8 & -12 \\ \hline & 3 & -4 & 6 & -1 \end{array} \leftarrow \textit{Remainder}$$

Since the remainder is not 0, -2 is not a solution of the equation.

33. Since the variables are not present, a missing term will not be noticed in synthetic division, so the quotient will be wrong if placeholders are not inserted.